Hazardous Materials
Management
Desk Reference

Hazardous Materials Management Desk Reference

Doye B. Cox, PE, CHMM
Editor-in-Chief

Adriane P. Borgias, MSEM, CHMM
Technical Editor

McGraw-Hill Inc.

New York • San Francisco • Washington, D.C.• Auckland • Bogotá
Caracas • Lisbon • London • Madrid • Mexico City • Milan
Montreal • New Delhi • San Juan • Singapore
Sydney • Tokyo • Toronto

McGraw-Hill

A Division of The McGraw-Hill Companies

3 4 5 6 7 8 9 0 KGP/KGP 0 4 3 2 1 0

ISBN 0-07-135173-6

The sponsoring editor for this book was Bob Esposito, the editing supervisor was David E. Fogarty, and the production supervisor was Sherri Souffrance. It was designed and typeset by fusion environment&energy, LLC.

Printed and bound by Quebecor/Kingsport.

 This book was printed on recycled, acid-free paper
containing a minimum of 50% recycled de-inked fiber.

McGraw-Hill books are available at special quantity discounts to use as premiums and sales promotions, or for use in corporate training programs. For more information, please write to the Director of Special Sales, McGraw-Hill, Inc. 11 West 19th Street, New York, NY 10011. Or contact your local bookstore.

Table of Contents

Part IV   Land and Natural Resources

Part VI   Air Quality

Chapter 29.   The Clean Air Act   371
Adriane P. Borgias, MSEM, CHMM

Chapter 30.   Prevention of Accidental Releases of Chemicals: Risk Management Programs   397
Ken B. Baier, CSP, CEA
Alan R. Hohl, CSP, CEA, CHMM
John S. Kirar, CSP, CEA, CHMM

Chapter 31.   A Step-by-Step Guide to Risk Management Planning   417
Dale M. Petroff, NRRPT, CHMM

Part VII   **Water Quality**

Chapter 32.   Clean Water Act   **441**
John M. Higgins, PhD, PE, CHMM

Chapter 33.   Oil Pollution Act   **449**
Mark L. Bricker, PE, CHMM

Chapter 34.   Ground Water   **461**
Michael R. Matthews, PE, PG, CHMM

Contributors

This edition of the *Hazardous Materials Mangement Desk Reference* has been prepared in collaboration with a large number and variety of professionals in this field. The support and vision of the contributors as well as the members of the Academy of Certified Hazardous Materials Managers are acknowledged here. Special thanks goes to Hugh T. (Tom) Carson for his efforts in the early development of this book. Specific contributors to this volume are noted below and at the end of each chapter.

Doye B. Cox, PE, CHMM, *Editor-in Chief.* Mr. Cox has been certified since 1985 and has served the Academy in several capacities at the national level, including President. He is also Past President of the Tennessee Section of the American Society of Civil Engineers. He has over 24 years experience in hazardous materials management and emergency response experience. His specialties include remedial and environmental construction project management and chemistry of hazardous materials. Mr. Cox is the Manager of the Chattanooga, Tennessee business unit for Koester Environmental Services, Inc.

Associate Editors

Charles M. Bessey, CHMM
Hugh T. (Tom) Carson, CHMM
Alan A. Eckmyre, CHMM
Harry S. Kemp, CHMM
Charley Kubler, CHMM

Adriane P. Borgias, MSEM, CHMM, *Technical Editor.* Ms. Borgias is a Past President of the Academy. She has over 20 years of experience in providing environmental and hazardous materials management services to the energy industry. Her specialities include regulatory collaboration, auditing, environmental management systems, and environmental permitting. She is a contributor to *Women in Chemistry and Physics: A Biobibliographic Sourcebook,* and is highlighted in *Northwest Women in Science: Women Making a Difference.* Ms. Borgias is the owner of *fusion environment&energy, LLC* in Spokane, Washington.

Todd A. Kuhn, PhD, CET, RPIH,
    CHCM, CHSP, CHMM
Donald W. Ott, CHMM
Cindy L. Savage, CHMM
Richard A. Senn, CHMM

Authors and Contributors

Ken B.Baier, CSP, CEA
Charles M. Bessey, CHMM
Dorothy Bloomer, CHMM
Adriane P. Borgias, MSEM, CHMM
Martha Boss, CIH, CSP
Mark L. Bricker, PE, CHMM
Harry A. Bryson, CHMM
Brett A. Burdick, PG, CHMM
Eldon L. Burkett, PE, CSP, CHMM
W. Scott Butterfield, MS, CHMM
Guy S. Camomilli, CSP, CHSP, CHMM
Mignon J. Clarke, MS, CHMM
Gregory C. DeCamp, MS, CHMM
Valentino P. De Rocili, PhD, CHMM
Daryl W. Dierwechter, REA, CHMM
Alan A. Eckmyre, CHMM
Michael Eyer, CHMM
Chris Gunther, CHMM
Gazi A. George, PhD, CHMM
David Green, MS
Richard E. Hagen, PhD, CHMM
Dawn Han, MS, CIH
Marilyn L. Hau, MS, RN-C, COHN-S,
      EMT-P, OHST, ASP
John M. Higgins, PhD, PE, CHMM
Thomas Hillmer, CHMM
Alan R. Hohl, CSP, CEA, CHMM

Patricia A. Kandziora, CHMM
John S. Kirar, CSP, CEA, CHMM
Harry S. Kemp, CHMM
K. Leigh Leonard, CHMM
Keith Liner, MCE, CHMM
Robert Lipscomb, CHMM
Louis Martino, CHMM
Michael R. Matthews, PE, PG, CHMM
Chris McKeeman, CSP, CHMM
John E. Milner, JD
George D. Mosho, CHMM
Margaret V. Naugle, EdD, CHMM
James L. Oliver
Dale M. Petroff, NRRPT, CHMM
Frank Pfeiffer, RS, CHMM
Ed Pinero, CPG
Stephen D. Riner, CHMM
Robert Roy, PhD, DABT
Steve Rowley, CHMM
Richard A. Senn, CHMM
Robert Skoglund, PhD, DABT, CIH
Keith Trombley, CHMM
Tony Uliano, Jr., MS, CIH, CHMM
Charles A. Waggoner, PhD, CHMM
Eugene R. Wasson, JD
Steven G. Weems, MPH, CIH, CHMM
Michael H. Ziskin, CHCM, CHMM

Consultants

Alabama Chapter of CHMMs
Timothy Anderson
Gregory N. Baker, CIH, CSP, CHMM
Terry L. Baker, MS, CHMM, CHSP
Kit Baldwin, CHMM
E. Rush Barnett
Irene Boone, CHMM
Adriane P. Borgias, MSEM, CHMM
Brandan A. Borgias, PhD
Bruce H. Bradford, PhD, PE
David B. Chandler, PhD, DABT,
      DABAT, CIH, CHMM
Beverly J. Curtis, CHMM, MS
Alison Dean, CHMM
Elwyn Dolecek, Ph.D

Beth S. Fifield, PE
Timothy J. Foelker, CHMM
Stephen G. Frantz, PE, CHP
Kevin B. George
E. Paine Gilly, Jr., PG
William Henle, CHMM
Dee Kaiser
Michael A. Kay, ScD, CHMM
Todd A. Kuhn, PhD, CET, RPIH,
      CHCM,CHSP,CHMM
Aaron A. Leritz, CHMM
Heidi L. Rosenberg, CHMM
Vince Runde, CHMM
Richard A. Senn, CHMM
Lynn St. Georges, CHMM
Daniel L. Todd, QEP, CHMM

Preface

The Academy of Certified Hazardous Materials Managers (ACHMM), established in 1985, is a non-profit membership organization dedicated to fostering professional development through continuing education, peer group interaction, the exchange of ideas and information relating to hazardous materials management. With 60 Chapters and over 6,000 members, the Academy is represented in 38 states plus the District of Columbia and Guam. The Academy's headquarters are located in Rockville, Maryland, just outside of Washington, DC.

The purposes of the Academy are to

- Educate and instruct Academy members in hazardous materials management, environmental health and safety

- Provide a means for hazardous materials managers to meet and communicate with each other

- Provide a meeting ground for members involved in different types of hazardous management work—academia, consulting, government, industry, and transportation

- Promote the CHMM certification as a standard of excellence in the hazardous materials management field

Eligibility in the Academy is achieved through education, experience, and the successful completion of an examination administered by the Institute of Hazardous Materials Management. The Academy would like to acknowledge the Institute of Hazardous Materials Management—the founding body for this organization and the administrator of the certification. It is the Institute's vision and leadership that has made the Academy the premier organization for hazardous materials management professionals. For more information about becoming a Certified Hazardous Materials Manager, please visit the Institute's Homepage at <http://www.ihmm.org> or call the Institute at (301) 984–8969. For more information about the Academy, please visit the Academy's Homepage at <http://www.achmm.org> or call toll-free (800) 437–0137.

Introduction

The *Hazardous Materials Management Desk Reference* is designed to be a comprehensive overview of the field of hazardous materials management. Its primary intent is to meet the management needs of over 8000 existing and prospective Certified Hazardous Materials Managers. The topics included in this book are also of interest to professionals in the safety, engineering, industrial hygiene, and environmental fields.

The authors and editors of the *Hazardous Materials Management Desk Reference* have been selected because of their knowledge and practical expertise in their fields. Most of the authors are Certified Hazardous Materials Managers (CHMMs), having met the certification's rigorous experience, peer review, and technical requirements. Each chapter of the book has been peer-reviewed for technical content.

Included in the desk reference are chapters on the science, laws and regulations, and management principles needed to properly manage hazardous materials. The chapters provide a general overview and guide to a comprehensive list of environmental and hazardous materials topics. The intent is to act as a reference or "jumping-off point" for more in-depth study. The comprehensive range of topics covered makes the book unique among currently available titles.

The topics covered by the desk reference are highly regulatory in focus and in constant change. Where appropriate, Internet references have been provided so that the reader can quickly link to reliable sources of current information. For those readers new to the field of hazardous materials management, a list of acronyms and a glossary are provided at the end of the book. Glossary items are noted in the text in **bold italic**.

Chapters in the reference are grouped by general topics. Part I, Regulatory and Management Perspectives, gives the reader a "big picture" understanding of the profession. Included in this section are chapters that overview the development of Federal laws and regulations, liability and compliance, environmental management systems auditing and risk analysis.

In Part II, Safety Principles, the reader will find topics on the science and regulation of safety, including chapters on industrial toxicology, OSHA requirements, personal protective equipment, laboratory safety and bloodborne pathogens. Community and worker right-to-know requirements are found in Part III,

The Right-to-Know. Part III overviews the fundamental requirements associated with release reporting, hazard communication, and Material Safety Data Sheets.

For information on land-related issues, the reader is urged to study Part IV, Land and Natural Resources. This part reviews the seminal National Environmental Policy Act, property assessments, and thoroughly researched state-by-state summaries of brownfields requirements.

In Part V, Management of Hazardous Materials and Substances, many of the topics that are of the most interest to hazardous materials managers are covered. These include chapters on the transportation of hazardous materials, underground storage tanks, pesticides, the Toxic Substances Control Act, incident response, as well as overviews on the hazardous substances asbestos and lead.

An overview of the Clean Air Act is provided in Part VI, Air Quality, as well as the requirements and implementation of the Risk Management Programs. Of particular interest is the step-by-step guide for preparation of a the risk management plan. Part VII, Water Quality overviews the Clean Water Act and the related Oil Pollution Act. The chapter on ground water supplements these overviews and provides the reader with a fundamental understanding of the movement of pollutants in ground water systems.

After reviewing the chapter on the Resource Conservation and Recovery Act (RCRA) in Part VIII, Management of Hazardous Wastes, the reader can learn more about the fundamentals of hazardous waste management from the complementary chapters on waste minimization/pollution prevention, treatment technologies, and RCRA corrective action. An additional chapter on mixed waste management highlights the requirements and challenges the disposal of mixed radioactive and hazardous waste pose to the hazardous materials manager.

Part IX, Chemical Perspectives (last, but not least important), is a basic chemistry foundation hazardous hazardous materials manager and is followed by an overview of the concepts of environmental sampling.

While the *Hazardous Materials Management Desk Reference* is comprehensive in scope, it is by no means complete. Complete coverage of these important topics would take far more than one volume to achieve and, given the rapidly changing regulatory environment, may not even be feasible. The reader, therefore, is urged to "take the leap"; to learn more by exploring the references provided in the bibliographies and Internet resource sections at the end of each chapter.

Adriane P. Borgias, MSEM, CHMM
Technical Editor

Hazardous Materials
Management
Desk Reference

Part I

Regulatory and Management Perspectives

CHAPTER 1

Overview of Federal Laws and Regulations

Stephen D. Riner, CHMM

Introduction

In order to do the job well, every hazardous materials manager must have some knowledge of the scope of Federal laws that exist to protect the environment and human health and safety. However, many professionals are unaware of how to find and read these laws, or of the mechanisms for implementing and administering these laws. This section will provide a framework for understanding the legal foundation for the enactment and administration of environmental laws. We will focus mainly on the Federal legislative and regulatory process, but most states also have similar processes.

The information discussed in this chapter applies to all forms of government regulation that the hazardous materials manager may be involved with: environment, worker health and safety, and hazardous materials transportation. However, to avoid constantly repeating this long phrase in the following discussion, we will use the term *environmental* as a shorthand terminology for all of these areas of regulatory law unless the context indicates otherwise.

Constitutional Framework for Enactment and Enforcement of Environmental Laws

As we were taught in high school civics, there are three branches of the Federal government: *executive*, *legislative*, and *judicial*. From the standpoint of environmental laws, this division of responsibilities can be summarized as follows.

Legislative Branch. The *legislative branch* (Congress) enacts laws that create regulatory agencies and gives them the authority to carry out their responsibilities, such as protecting the environment. These laws provide specific authority to regulatory agencies, and agencies cannot do anything that they have not been specifically given authority by Congress to do. Congress also enacts laws that govern how regulatory agencies perform their administrative functions such as rulemaking and enforcement.

Executive Branch. Regulatory agencies are located in the *executive branch*. The President heads the Executive Branch and names administrators of these agencies. The agencies may be independent like the United States Environmental Protection Agency (EPA), or cabinet departments like Labor and Transportation. These agencies administer the laws for which Congress has given

them responsibility, and develop regulations and program guidance. These agencies also administer the laws through enforcement actions, which often require involvement of the courts. Another agency within the executive branch, the Attorney General, may be involved in bringing action in the courts to enforce laws. Although Congress creates regulatory agencies and cabinet-level departments, the President and his subordinates have authority to organize the Executive Branch and executive departments and agencies, including agencies that assume responsibility for programs given by Congress to other agencies or departments. Neither the EPA nor the Occupational Safety and Health Administration (OSHA) were created by acts of Congress, but rather by Executive Reorganization Orders issued by the President (EPA) and Secretary of the Department of Labor (OSHA).

Judicial Branch. The *judicial branch* (the court system) is where environmental laws are interpreted and enforced. Regulated industries have recourse to the courts when governmental agencies take action to enforce the laws. In addition to trying civil and criminal cases for violations of the laws, courts review laws for constitutionality and review the actions of the regulatory agencies to make sure the actions conform to the law. Figure 1 below shows this flow of authority among the branches of government.

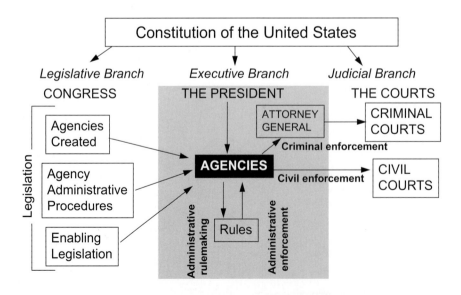

Figure 1. Flow of Authority to and from Regulatory Agencies

Legislation

The Federal government can take no action on any issue without a law that allows the government to act. We will not touch on the politically charged process of making laws, but will instead concentrate on building and understanding of laws.

The organization and structure of laws and the means of finding the most up-to-date version can be confusing. First, there are individual **acts** that are enacted by Congress, such as the Clean Air Act Amendments of 1990, the Superfund Amendments and Reauthorization Act of 1986, or the Toxic Substances Control Act of 1976. These names, or **short titles**, refer only to those portions of the law that are enacted by Congress at that time. These laws may be original acts that create an entirely new regulatory area, or (more often) acts that amend existing laws. In addition to the short title, each such act has a *Public Law* (PL) citation. For instance, the law commonly called the Clean Air Act Amendments of 1990 is PL 101–549, the 549th act passed by the 101st Congress.

Most major environmental laws have been revised many times. For example, the current Federal Water Pollution Control Act, as amended, has had a number of major and minor amendments, resulting in a law that now bears little resemblance to the original 1948 act. Laws that are amended must then be revised and republished to reflect all permanent amendments. This represents the **body of standing law**, or **statutes**.

Statutes are compiled in two ways. One way is to separately publish major laws, such as the Clean Air Act, as amended. (The use of the suffix *as amended* indicates that this is a compilation of the original Clean Air Act and all subsequent amendments, rather than just one amending act.) The second way is in the **United States Code** (USC), a complete compilation of Federal statutes with all amendments up to the date of publication. The USC is organized into *Titles* that place statutes of similar subject matter in the same volume. The section numbers within a law differ somewhat between these two methods of compilation, so be careful when looking up a reference. For example, Section 102 of the Clean Air Act, as amended (CAA §102) is found in Title 42, Section 7402 of the USC (42 USC §7402). To determine the current legal requirements in a particular subject area, consult one of these compilations

(after ensuring that all amendments to date have been included), rather than the individual acts of Congress.

Executive Branch Actions: Regulations, Permits, and Enforcement

Regulations. Laws enacted by Congress to establish regulatory programs can be very detailed, but they are usually general in their technical scope. The procedural matters covered by these laws (such as permit or enforcement programs) must be detailed more precisely than the law. So, agencies enact *regulations* that outline specific procedures for the administration and enforcement of environmental laws. The terms **rules** and **regulations** are synonymous, and both terms are used interchangeably in the law. For consistency, we will use the term *regulation* to refer to the final product (since it is compiled in the **Code of Federal Regulations**), but we will use **rulemaking** to refer to the process by which regulations are developed. Regulations cover such matters as the procedures for issuing permits, emission or effluent standards, and requirements for delegating authority from the Federal agencies to State environmental programs.

The statutory definition of a hazardous waste (42 USC §6903[5]) is a good example of the general language found in laws:

> The term *hazardous waste* means a solid waste, or combination of solid wastes, which because of its quantity, concentration, or physical, chemical, or infectious characteristics may—
>
> (A) cause, or significantly contribute to an increase in mortality or an increase in serious irreversible, or incapacitating reversible, illness; or
>
> (B) pose a substantial present or potential hazard to human health or the environment when improperly treated, stored, transported, or disposed of, or otherwise managed.

This statutory definition is inadequate to classify all of the potential industrial wastes without additional guidance. Congress did not intend the above definition to be the sole means of identifying hazardous wastes. RCRA also provides (42 USC §6921[a]):

[T]he Administrator shall, after notice and opportunity for public hearing, and after consultation with appropriate Federal and State agencies, develop and promulgate criteria for identifying the characteristics of hazardous waste, and for listing hazardous waste, which should be subject to the provisions of this subchapter, taking into account toxicity, persistence, and degradability in nature, potential for accumulation in tissue, and other related factors such as flammability, corrosiveness, and other hazardous characteristics. Such criteria shall be revised from time to time as may be appropriate.

EPA has developed regulations that list specific industrial waste streams, and also describe the measurable physical characteristics for nonlisted wastes. These regulations allow any generator of waste to determine whether that waste is hazardous or not.

The following examples show the difference in scope between laws and regulations:

- The Clean Air Act directs EPA to develop ambient air quality standards, but for the most part does not dictate what pollutants should be included in these standards. The air quality standards are developed and issued as regulations. Amendments to environmental laws can be used by Congress to correct any shortcomings in the regulatory process. In the Clean Air Act Amendments of 1990, Congress specifically listed 189 hazardous air pollutants that it required to be regulated. Congress took this step because of EPA's inability to enact standards for more than a few hazardous air pollutants.

- The Federal Water Pollution Control (Clean Water) Act requires facilities that discharge their wastewater into surface waters to obtain a National Pollutant Discharge Elimination System (NPDES) permit. EPA, through regulation, has set specific requirements for obtaining a permit. The EPA regulations also specifically limit the effluents for various types of industrial wastewater discharges.

Although regulations are developed by nonelected government employees rather than by elected legislators, *regulations have the force of law.* Violation of a regulation is the same as violation of a statute, and in some instances can be prosecuted criminally.

Permits. Many environmental regulatory programs require permits for specific activities. Here, *environmental* means environmental in the narrow, more specific sense. Permits are not required for actions relating to occupational safety and health nor hazardous materials transportation. Although there are variances, there are registrations and licenses under these regulations that may be considered analogous to a permit. **Permits** are issued by Federal agencies or by State agencies that have been delegated authority under Federal law. The permits allow a permittee to conduct an activity that has an impact on the environment. For example, environmental permits are issued for the operation of hazardous waste treatment, storage or disposal facilities, the emission of air pollutants, or discharges of wastewater. A regulation describes the procedures for application, public notice, and agency decision that govern the permitting process. Any violation of terms of a permit is usually considered violation of the regulation(s) that govern the actions allowed by the permit.

Enforcement. Regulatory agencies not only develop and administer regulations and permits, they also *enforce* them. There are two kinds of enforcement: *administrative* and *judicial*. **Administrative enforcement** covers all enforcement actions taken by the agency, while **judicial enforcement** occurs once the agency takes a case to the courthouse for lawsuit or criminal prosecution. Administrative enforcement generally follows this process:

- An inspection or administrative review resulting in discovery of a violation.

- Official notice to the regulated party in a form such as a Notice of Violation or citation.

- The scope of the enforcement action is determined by the agency. This action can take one of the following paths:

  - for minor violations, the regulated party may be directed to correct the problems with no penalties assessed.

  - for more serious violations, the agency may assess an administrative penalty (if permitted by law). This is a unilateral agency action that the regulated party may appeal either to an administrative review board or to the courts.

  - as an alternative, the agency may try to negotiate an administrative document called a *consent order* (or some similar name). This order represents a settlement of the issues between the agency and the regulated party under which the agency agrees not to sue in exchange for payment of negotiated penalties and correction of actions to achieve compliance.

If the violation is so serious that criminal prosecution is appropriate, or if the agency and the regulated party are not able to negotiate a settlement, the agency may seek prosecution in criminal court or sue the regulated party in civil court. At this point, the process becomes a matter of judicial enforcement.

Most disputes between agencies and the regulated community do not end up in court. Major agencies usually have administrative review boards that arbitrate and rule on disputes such as administrative penalties and issuance or denial of permits. Disagreements that go through these processes usually never reach the Judicial Branch unless one party appeals the board's decision.

Most states have administrative processes similar to the Federal processes described above.

Involvement of the Courts

The Judicial Branch does not become involved in a case until it is brought to them by regulatory agencies, regulated parties, or citizens. The courts become involved in environmental matters in several ways:

- Lawsuits involving agencies that are unable to negotiate consent agreements are heard in civil court. Courts are typically asked by the regulatory agency to assess monetary penalties and compel the violator to take corrective action.

- If a statute provides for criminal prosecution of knowing and willful violations, a regulatory agency may take a company and/or individuals to criminal court. The criminal complaint may result in monetary penalties and may compel corrective action like a civil lawsuit. In order to convict in criminal court, the government must meet the beyond-a-reasonable-doubt burden of proof, which is a higher level of proof than that required in civil court.

- Regulated parties may sue a governmental agency if they believe that a regulation does

not meet the requirements of the law. The underlying statute may also be challenged by a lawsuit that questions its constitutionality or conformance with other laws. Lawsuits may also be filed by regulated parties for agency actions such as denial of a permit.

- Some environmental statutes provide for citizen lawsuits. These laws allow members of the public, who may not otherwise have legal standing to sue, to file lawsuits against regulated parties for failure to comply with the law.

Summary of the Roles of Each Branch of the Federal Government in Regulation

Congress. Congress enacts legislation to create regulatory agencies and give them the authority to regulate in particular subject areas.

Regulatory Agencies. Regulatory agencies develop regulations that describe in detail the requirements of these regulatory programs, and enforce these regulations by issuing permits and taking action against violations.

Courts. Courts determine whether or not a regulated entity is civilly or criminally liable for violating laws and regulations. The courts also rule on the constitutionality of laws and the conformance of agency actions to laws and regulations.

Development of Regulations

It is very important for the hazardous materials manager to understand how regulations are developed. Many of the technical details of Federal and State regulations are determined during the development process. There are numerous opportunities during the rulemaking process to comment on and have input into the final regulations. Anyone tracking the rulemaking process must be familiar with the *Federal Register*, which is published daily. All official notices regarding a proposed regulation are published in the *Federal Register*.

Laws from several sources govern rulemaking. First, there are Federal administrative procedure laws, originally enacted as the Administrative Procedures Act (but now recodified with other administrative statues at 5 USC §§500–596). Second, the underlying statute that authorizes the agency to undertake the rulemaking may have specific procedural requirements for regulations enacted under that law (for example, Section 307[d] of the Clean Air Act, as amended, 42 USC §7607[d], which describes EPA's procedures for issuing regulations under this act). Finally, an agency may have administrative regulations that further lay out the details of rulemaking processes for that agency (*e.g.*, 49 CFR 106 and 29 CFR 1911, which respectively describe the rulemaking processes for the Department of Transportation, Research and Special Programs Administration and the Department of Labor, Occupational Safety and Health Administration). There are other laws also that specify requirements for rulemaking, especially concerning regulations that have a significant economic impact. These laws include the Small Business Regulatory Enforcement Fairness Act (SBREFA), the Unfunded Mandates Reform Act, and the Regulatory Flexibility Act.

A typical rulemaking process is described below. There may be variations in these steps depending on specific statutory requirements, the urgency of rulemaking, and statutory or judicial deadlines.

1) Congress enacts a law permitting or requiring an administrative agency to develop regulations. The law may require particular rulemaking within a given time, or an agency may determine that rulemaking is necessary in order to fulfill its statutory responsibilities. A court may determine that an agency is not fulfilling its statutory responsibilities, and direct the agency to develop regulations.

2) An agency may establish a rulemaking schedule to put interested parties on notice that a particular rulemaking process will reach the proposal stage soon. Some agencies periodically publish a list of regulations under development in the *Federal Register*.

3) The agency establishes a *docket*, or official administrative record, for a particular regulation. This is necessary to document that all administrative procedures, as well as any procedures specified in the enabling legislation, are followed. All comments that are submitted in response to a proposed regulation are kept with this docket.

4) The agency internally develops the concept for how the regulation will work. Agencies may develop several alternatives that accomplish the same end, and may work with an advisory committee of interested parties on these ideas. An Advance Notice of Proposed Rulemaking (ANPR) may be published in the *Federal Register* in order to solicit comments on one or more regulatory concepts.

5) Once a regulation's concept is defined, the agency develops the proposed regulation and publishes it in the *Federal Register*. An important part of each proposal is the **preamble**. The preamble describes the rationale for the regulation, how it complies with and/or fulfills the requirements of the underlying statutes. This section also summarizes the input received from interested parties to date and how this feedback has impacted development of the regulation. *The preamble contains the agency's interpretation of how the regulation will work*, and is a potential resource for any future questions of interpretation. The notice contains a deadline for comment.

6) If necessary, there may be a public hearing. Such proceedings can be quasi-judicial, with sworn witnesses, cross-examination and testimony kept by a court reporter. The hearing examiner issues findings on the issues of fact for the agency's consideration in the rulemaking process. An agency may choose to hold an informal hearing, intended strictly to gather information and discuss the agency's proposal.

7) After receiving comments, the agency may then withdraw the regulation for additional work or comment, make minor revisions to the regulation in response to comments, or publish the regulation in final form. If more than minor revisions are made, the agency must propose the regulation again and solicit further comment. The agency must also publish a preamble to the final regulation that discusses all comments received and the agency's response to these comments.

Amendments to existing regulations (including revocation) must go through the same process as a new regulation. Minor amendments may have short preambles, and technical amendments (corrections) may require little if any supporting data.

Once a final rule is published in the *Federal Register* (the entire rule or the parts proposed for revision are republished, including any changes that were made in response to comments), it is effective as of the date indicated in the notice. If the regulated community does not accept the final rule, there are options for legal review. In legal challenges to a regulation, litigants can argue that the agency did not follow administrative procedure requirements and/or specific requirements in either the enabling legislation or the agency's administrative rules. Litigants can also argue that the regulation is based on erroneous science or economic analysis. Litigants have a better case to challenge a regulation in court if they can show that their objections were raised during the rulemaking process and not properly considered by the agency, rather than raising their arguments for the first time in litigation. Any law that sets specific rulemaking requirements for the agency (*e.g.*, SBREFA) can be used to challenge a rulemaking process.

The rulemaking process can be relatively short (months) for noncontroversial proposals or may take years where there is significant controversy, lack of consistent data, or scientific uncertainty. Legal challenges to rules can delay these processes even longer. For example, the United States Department of Transportation (DOT) recently completed a ten-year effort to amend its regulations to extend DOT jurisdiction to most intrastate shipments of hazardous materials.

Congress can overturn or render regulations ineffective. Because the underlying statute gives an agency the authority to enact regulations, regulations become ineffective if Congress amends the underlying statute and changes the scope of the agency's authority. Also, under 1994 legislation, Congress can disapprove regulations by resolution within 60 days of enactment. The President can veto the resolution of disapproval but the veto is subject to override by a 2/3 vote of Congress.

Courts can overturn regulations if an agency fails to follow the rulemaking process. Examples of situations where agencies have failed to follow their administrative procedures include:

1) OSHA's permissible exposure levels (PELs) are limits on occupational exposure to toxic chemicals, listed at 29 CFR 1910.1000. The Occupational Safety and Health (OSH) Act of 1970 directed OSHA to adopt existing industry consensus standards for worker health and safety. OSHA adopted as its PELs existing consensus standards from two organizations that had developed chemical exposure limitations: the 1968 American Conference of Governmental Industrial Hygienists Threshold Limit Values (ACGIH TLVs), and American National Standards Institute (ANSI) standards. Because OSHA took action specifically directed by the enabling legislation, these PELs were not subject to legal challenge. By the mid-1980s it was clear that these standards were out of date. OSHA then sought to make a wholesale change to the standard by revising many of the standards in 29 CFR 1910.1000. These revisions were challenged on the grounds that OSHA is required by the OSH Act to provide detailed toxicity data for each compound for which it intended to revise the original standard; in this case, OSHA did not do so because of the sheer numbers of compounds involved. As a result, the revised standard was overturned, and the regulation reverted to the original pre-1970 criteria. OSHA's only recourse is to revise each compound's PEL individually, providing the necessary background (which is still subject to challenge in court), or to convince Congress to amend the OSH Act to allow OSHA to update the standards by adopting up-to-date ACGIH TLVs without having to provide the detailed supporting data.

2) EPA originally proposed hazardous waste regulations under the Resource Conservation and Recovery Act in 1978. The proposed regulation had the current *listed* and *characteristic* waste definitions, but some commenters were concerned that a listed waste could be rendered nonhazardous simply by mixing it with a nonhazardous waste. Because of this concern, EPA's final regulation, published in 1980, added the *mixture* and *derived-from* rules, whereby any waste mixed with or derived from a listed hazardous waste is also regulated as hazardous. In a subsequent challenge to this rule (Shell *vs.* EPA), which did not come to a decision for over 10 years, a United States District Court ruled that EPA went beyond what was permitted in its revisions to the proposed regulation without soliciting comments on the revised provisions. The addition of the mixture and derived-from rules was ruled to be so significantly different from the original proposal that opportunity for comment was required.

The outcome of this second case was so significant that EPA immediately reissued the offending provisions as *interim final* regulations, and sought assistance from Congress. Congress passed legislation allowing the provisions to stay in place for a period of time until EPA could finish its effort to redefine hazardous waste (the *hazardous waste identification rule*, or HWIR).

Keeping up with Current Regulations

Given that regulations change over time, how does a person find the current law? The **Code of Federal Regulations** (CFR) is the body of standing regulation, and is accurate as of the annual publication date of each title. There are 48 titles of the CFR, identified by number, that contain regulations pertaining to a particular *subject matter*, not a specific agency. For example, although the vast majority of EPA regulations are in Title 40 (Protection of the Environment), there are also EPA regulations in Titles 5 (Administrative Personnel), 41 (Public Contracts and Property Management) and 48 (Public Acquisitions Regulations System). Title 40 also contains regulations of the Council on Environmental Quality. The Titles of most interest to hazardous materials managers are summarized in Table 1.

Each CFR title is published once each year, and thus each volume contains only the amendments that were effective as of the publication date. Subsequent amendments and additions are not included until the next volume. Amendments published in the CFR that do not become effective until after the publication date have a note that indicates the date they will become effective; they also note any existing language that will cease to be effective at that time.

Even the most recently published CFR may not be accurate, so the regulated community needs to know how to locate the currently effective regulations, especially in subject areas where there are frequent changes. The tracking choices are:

1) Track the *Federal Register* daily or review its monthly index looking for proposed and final regulations.

2) Subscribe to the *List of CFR Sections Affected*, which is issued monthly by the Superintendent of Documents. This publication gives a cumulative listing of CFR sections that have been updated each year without having to subscribe to or review each issue of the *Federal Register*. However, once you find that a section of interest has been modified, you then have to find the *Federal Register* issue that contains the changes.

3) The Information Age has drastically improved access to current regulations. You can

Table 1. CFR Titles Most Often Used by Hazardous Materials Managers

Department/Agency	Title	Parts
Nuclear Regulatory Commission	10: Energy	0-171
Department of Labor (Occupational Safety and Health Administration)	29: Labor	1900-1926
Department of Labor (Mine Safety and Health Administration)	30: Mineral Resources	1-104
Coast Guard	33: Navigation and Navigable Waters	125-338
Environmental Protection Agency	40: Protection of the Environment	1-799
Department of Transportation (Research and Special Projects Administration) *Includes hazardous materials transportation and pipeline safety*	49: Transportation	106-195

subscribe to a CD-based or on-line service that updates Federal regulations as soon as amendments become effective. However, you need to be sure that these updates come out soon after the regulation's effective date; some services take several months to incorporate the changes into their databases. Also, be aware that *proposed* regulations are usually not included in the information these services provide, although they may be available through a separate subscription. These information services can cost hundreds to thousands of dollars per year.

4) The most significant development in providing ready access to up-to-date regulations to anyone with a personal computer and a modem is the Internet. The Government Printing Office operates a worldwide web site providing access to the current CFR. This site allows you to search for a particular subject or topic if you do not know the CFR citation. Current and recent past issues of the *Federal Register* are also accessible on-line. In addition, EPA and OSHA both have compilations of their regulations on their web sites, and many states have web sites providing access to their rules and statutes as well.

Those who need the text of current statutes can go to similar resources. The *United States Code* is also updated periodically, and some of the same commercial services that provide regulations also provide full text of statutes. There is at least one web site, operated by a law school that has a current compilation of the USC. Statutes change much less frequently than do regulations. Major statutory changes are generally publicized well enough that professionals in the field should be aware of their development and final adoption, and will know to look for the updated law.

Because of the rapid evolution of information technology, additional or more sophisticated sources of information may become available before this volume becomes obsolete. Hazardous materials managers need to remain aware of developments in this area, and seek the most up-to-date sources of information.

Regulatory laws and their associated standards form the basis of the regulatory programs that hazardous materials managers work under each day. All hazardous materials managers should understand these laws and the process of developing regulations. For those who find reading and application of laws a bit overwhelming, national and local organizations offer training courses geared toward technical professionals who are not lawyers.

History of Federal Environmental Legislation

First Federal Regulatory Laws

Regulation of business by the Federal government is mostly a late 20th century phenomenon, and most Federal environmental and related laws have been enacted after 1970. However, there are much earlier examples of Federal laws that are forerunners of today's regulatory statutes:

- The first law that had direct environmental impact was the Rivers and Harbors Act of 1899. The motivation for the law was to protect navigable waterways from deposition of debris that would impede navigation, and it is doubtful that the law's authors saw pathogenic microorganisms as hazards to navigation. However, around 1970 the law was resurrected to provide authority for an Army Corps of Engineers program to issue permits to wastewater dischargers.

- One of the earliest entities created by Congress as a regulatory agency was the Interstate Commerce Commission (ICC). The Transportation of Explosives Act of 1909 gave ICC authority to set manufacturing standards for shipping containers used to transport dangerous goods. The United States Department of Transportation (DOT) now has authority over many of the now-defunct ICC's regulatory areas, and some of the current manufacturing standards for compressed gas cylinders in DOT regulations date back to the ICC's original 1911 regulations.

- By the 1930s, Congress had enacted several laws directed at purveyors of snake oil patent medicines containing opium or other powerful drugs and carrying unsubstantiated claims of efficacy. While not significant from a strictly environmental standpoint, these laws represent a significant expansion of Federal agencies whose sole reason for existence is to regulate. The Federal Food Drug and Cosmetic Act of 1938 established the Food and Drug Administration.

Public Health–Directed Legislation

At this point we will digress briefly from this discussion of Federal laws, since many of the early efforts in protecting public health took place at the State level. In the early part of the 20th century, as major cities grew and provided both public water supplies and sewer systems, states began to enact legislation in response to the resulting public health concerns. Minnesota was a good example (and similar events no doubt occurred in other states): typhoid outbreaks resulted from the Minneapolis and St. Paul water intakes being downstream of the point where sewers discharged untreated sewage into the Mississippi River. To address the issue, the Board of Health (today's Minnesota Department of Health) was created in the 1920s, and by the 1930s had authority to regulate public water supplies. In 1947, the Water Pollution Control Commission was created within the Health Department to regulate discharges to surface waters; 20 years later, this commission became the separate Minnesota Pollution Control Agency. In many states, the environmental agency has evolved from the State health department, and in a few states is still part of the health department.

Signs of Environmental Deterioration

Except for the examples cited at the beginning of this chapter, there were no significant environmental laws enacted during the first half of the century. By the mid-1950s, early versions of the Clean Air Act and Clean Water Act were in place, but these laws imposed only planning and coordination requirements on the states. States with significant air quality problems (such as the dirty-air cities Pittsburgh and Los Angeles) enacted regulatory programs that had measurable impacts, but the environmental quality in other areas deteriorated with the economic and industrial boom after World War II. Specific environmental problems became apparent:

- Rachel Carson published her book *Silent Spring* in 1962. The book's impact was significant in increasing public awareness and concerns about environmental issues. While many disagree with her all-encompassing condemnation of pesticide use, evidence is strong that dichlorodiphenyltrichloroethane (DDT) residuals in certain bird species (*e.g.*, brown pelicans) resulted in thinning of eggshells, with a consequential decline in the bird populations. Following banning of DDT, bird populations rebounded.

- Gross water pollution from poorly controlled wastewater discharges in more heavily industrialized parts of the country became commonplace. Lake Erie was proclaimed to be dead (it was not, although it certainly was highly polluted), and chemical residuals floating on the Cuyahoga River in Cleveland, Ohio caught fire and burned for several days.

- Despite local efforts, air pollution was a continuing problem in major cities.

It was this evident deterioration, together with public awareness that increased by the late 1960s, that set the stage for the first significant Federal environmental legislation. Perhaps the first significant Federal act concerning the environment was the *National Environmental Policy Act* (NEPA) of 1969, which did not establish a regulatory program *per se*, but required analysis of the environmental impacts of significant Federal agency actions.

Creation of the Environmental Protection Agency; Medium-Specific Environmental Legislation

The first of the major regulatory laws was enacted in 1970, and that same year the United States Environmental Protection Agency (EPA) was created. Previously, diverse agencies such as the Corps of Engineers, Public Health Service, and Department of Health, Education and Welfare had responsibility for pieces of the environmental puzzle. Creation of the EPA put the responsibility in one place. Major laws were enacted during this era.

Clean Air Act, 1970. The intent of this act was to control gross air pollution problems by regulating the *dirty air* pollutants: ozone, nitrogen oxides, sulfur dioxide, and particulates. EPA was required to set ambient air quality standards that each state ultimately was to attain through pollution controls imposed on air pollutant sources under a State Implementation Plan (SIP).

Federal Water Pollution Control Act, 1972.

This law (with subsequent amendments, now called Clean Water Act), established a nationwide wastewater discharge permitting program, the National Pollutant Discharge Elimination System (NPDES) permit program. States were also required to set surface water quality standards under this act. The pollutants that were the primary focus were those that degrade water quality by depleting oxygen: Biochemical Oxygen

Resource Conservation and Recovery Act, 1976.

The Resource Conservation and Recovery Act (RCRA) extended protection to the last remaining environmental medium: the land. As the name implies, the law set forth an intent to promote conservation of resources through reduced reliance on landfilling. Both solid waste and hazardous waste are covered by this law, although by far the most attention is paid to the hazardous waste aspect. It was not until the late 1980s that EPA issued rules setting minimum technical standards for solid waste landfills, and most technical standards relating to solid waste disposal are developed by individual states.

Bits and Pieces

An illustration of how fragmented the United States environmental regulatory efforts were in 1970, is the list of the various programs that were consolidated to create the Environmental Protection Agency:

- **Department of Interior:** Federal Water Quality Administration

- **Department of HEW:** National Air Pollution Control Administration; Food and Drug Administration pesticide research group; Bureau of Solid Waste Management; Bureau of Water Hygiene; Bureau of Radiological Health

- **Department of Agriculture:** Agricultural Research Service pesticide activities

- **Atomic Energy Commission and Federal Radiation Council:** Radiation criteria and standards programs

---

Demand (BOD_5), and suspended solids. The act set *fishable and swimmable* waters as its near-term goal, and dictated that discharges would ultimately meet *Best Available Technology* rather than merely enough treatment to meet water quality standards.

Safe Drinking Water Act, 1974.

This law is directed at utilization of a resource rather than waste disposal or emission control. The primary focus is protection of public health by control of water pollutants that affect public health, but also addresses secondary parameters that affect the aesthetic value of drinking water. All *public* water supplies are covered by this law, which in effect means that only private residential wells are excluded from regulation.

Recognition of the Role of Toxic Substances

The laws described above had one thing in common: they were directed at the gross contaminants that made the air and water visibly dirty. During this same period, however, analytical technology improved to the point where very small amounts of contaminants thought to have toxic effects in small concentrations could be detected. Environmental advocacy organizations sued EPA to require it to regulate toxic pollutants in addition to conventional pollutants. The initial major Clean Water Act and Clean Air Act reauthorizations (both 1977) were directed at increasing control of toxic substances.

Toxic Substances Control Act (1976). This unique law regulates toxic chemicals when they are manufactured and used, rather than only when they enter the environment. Under this law, manufacturers and importers must file a pre-manufacture notification when they propose to manufacture or import a chemical not previously included on EPA's inventory of chemical substances. Other provisions of this law addressed the manufacture, use, and disposal of poly-chlorinated biphenyls (PCBs), effectively phasing out the use of PCBs.

Multimedia and Incident-Driven Legislation

By the late 1970s, it was apparent that control of individual media (air, water) was not sufficient to protect the environment. There was growing recognition of the interrelationship of environmental media, especially because many traditional end-of-pipe pollution control measures merely transferred pollutants from one medium to another. Another aspect to legislation following the late 1970s was to respond to high-profile environmental incidents with laws designed to prevent recurrences or a means of quick response by government.

Comprehensive Environmental Response, Compensation and Liability Act of 1980.[1] The Comprehensive Environmental Response, Compensation and Liability Act (CERCLA) is more commonly known by the unfortunate sobriquet *Superfund Act*. This law was enacted in large part in response to the contamination emanating from the Love Canal disposal site in Niagara Falls, New York, and the inability of State and Federal governments to provide quick relief to affected homeowners. CERCLA did not establish a regulatory program *per se*; rather, it established a fund which could be used for responses to hazardous substance release sites, and established liability for these releases.

---

[1] This law is often called *SARA Title III* because it was enacted as one title of the *Superfund Amendments and Reauthorization Act* (SARA). It is, however, a separate law, even though some of its subject matter overlaps with that of CERCLA.

The Emergency Planning and Community Right-to-Know Act (1986). The Emergency Planning and Community Right to Know Act (EPCRA), was enacted in response to the chemical release disaster that occurred in Bhopal, India in 1984. Under this law, states and local governments were required to establish emergency planning and response organizations, and users of hazardous and toxic chemicals were required to report their usage and (for some chemicals) emissions to these newly established State and local agencies.

Oil Pollution Act (1990). The Oil Pollution Act of 1990 was a direct result of the accident involving the oil tanker *Exxon Valdez*. It imposed new stringent standards on transportation and storage of oil.

Pollution Prevention Act (1990). The Pollution Prevention Act of 1990 is a multimedia law not brought about by any headline event. This law is more a statement of policy than another regulatory program, and did not directly impose any responsibilities upon the regulated community. It did formally establish reduction of contaminants at the source—rather than end-of-pipe treatment—as the official environmental control policy of the United States, and directed EPA to develop its regulations in the future and incentives to bring this about.

Major Reauthorizations of Existing Environmental Statutes

We have previously discussed the reauthorizations of the Clean Water and Clean Air Acts in 1977 and how these laws were revised to focus more attention on control of toxic substances. There are other significant reauthorizations of environmental laws that bear discussion.

The Hazardous and Solid Waste Amendments of 1984. The era of the early 1980s was the time when EPA was implementing the first RCRA regulations regulating the disposal of hazardous waste, which were published in 1980. However, this era was also part of the early Reagan years, and Reagan's first EPA administrator was openly hostile to environmental regulations. During this time the EPA budget was cut 23 percent, and many experienced staff left the agency. Opposition to these deregulatory efforts by the public and Congress resulted in a change in leadership at EPA. In response to the turmoil at EPA and perceived under-regulation of hazardous waste disposal, Congress took steps to ratchet down hazardous waste regulation by enacting the Hazardous and Solid Waste Amendments of 1984 (HSWA). This act reauthorized RCRA, severely limited land disposal of hazardous waste, and set stringent deadlines by which land disposal restrictions would become effective (*hammer provisions*) if EPA had not enacted rules. Its hammer provisions and level of detail were intended to minimize regulatory discretion on the part of EPA. Another significant aspect of this law was establishment of a program to regulate underground storage tanks holding petroleum products and hazardous substances.

Superfund Amendments and Reauthorization Act of 1986. Critics of the Superfund program expressed major concerns in the program's early years. One of these was that cleanups of the sites did not emphasize *permanent* solutions. Other concerns addressed included increasing participation of both states and the public in the remedy selection process, expanding health risk assessment capability in the Federal government, and providing a liability defense for *innocent landowners* who purchased property later found to be contaminated. This act was reauthorized as the Superfund Amendments and Reauthorization Act of 1986 (SARA) and also became a legislative vehicle for enactment of the Emergency Planning and Community Right to Know Act (EPCRA), discussed above.

Clean Air Act Amendments of 1990. This is the last major piece of environmental legislation that has passed Congress. The following are the major features of this law:

1) A nationwide permitting program was established for *major sources* of air pollution, administered by each state

---

Other Laws Establishing Significant Regulatory Programs

There have been several other significant laws enacted over the years, affecting the environment, safety, or public health. These laws do not fit neatly into the discussion of the progression of environmental legislation, but are important from the standpoint of environment, health, and safety professionals.

- **Federal Insecticide, Fungicide and Rodenticide Act (FIFRA), 1947.** Initially, the law was administered by the Department of Agriculture, and set labeling requirements for pesticides. Jurisdiction was transferred to EPA, and in 1972, a major amendment to this law established a program to control the manufacture, distribution and use of pesticides. EPA was given authority to register all pesticides, and to suspend, cancel, or restrict pesticides that could pose environmental hazards.

- **Occupational Safety and Health Act (OSHAct), 1970.** This law directed the Department of Labor to establish standards for physical, chemical, and biological hazards in the workplace. The Occupational Safety and Health Administration (OSHA) was set up within the Department of Labor to administer this program.

- **Hazardous Materials Transportation Act (HMTA), 1975.** The Department of Transportation regulates shipments of hazardous materials by highway, rail, vessel, and air. Previously, this regulation principally covered interstate shipments, and many intrastate shipments were unregulated (depending on the state). DOT has recently extended its authority to most intrastate hazardous materials shipments as well.

2) Replacement of the former National Emission Standards for Hazardous Air Pollutants (NESHAPS) provisions, under which only seven hazardous air pollutants were regulated, with a requirement to regulate hazardous air pollutants on a source-specific basis. A list of 189 hazardous air pollutants was included in the legislation

3) Increased efforts by states to bring non-attainment areas, especially those for ozone, into compliance with ambient air standards

Safe Drinking Water Act Amendments of 1996. Because this law was enacted after the "Republican Revolution" of 1994, it may be a harbinger of future reauthorizations, and is therefore discussed below.

An Era of Retrenchment and Reconsideration

The 1994 elections produced a major shift in control of Congress. A number of the new conservative members of Congress were openly hostile toward the existing regulatory structure. However, the new Congress's early attempts to make significant changes in environmental laws were mostly turned aside by a coalition of Democrats and moderate Republicans, and by adverse public opinion. A few laws that slightly weakened the ability of the Federal government and regulatory agencies to impose regulations were enacted, but stronger regulatory reforms that might have imposed strict cost-benefit analyses on regulations were defeated.

Reauthorization of several laws have been debated in the past and current Congresses, but the only significant piece of environmental legislation to pass since 1994 has been the Safe Drinking Water Act Amendments of 1996. The law did not gut the original act, but it did remove a requirement (liked by few, including EPA) that EPA regulate a large number of additional potential drinking water contaminants. Instead, EPA was directed to focus on those contaminants known to present significant hazards (including biological contaminants) and to enact regulations based on cost and feasibility as well as health impact.

Interrelationship of Federal Environmental Statutes

Despite the proliferation of laws, there has been an effort by Congress and regulatory agencies to minimize potential conflicts among the various statutes. These laws make frequent use of cross-referencing to draw in requirements from other laws. There are areas of potential inconsistency, which are more of an inconvenience than an impediment to compliance:

- Emergency planning requirements do not always track. For instance, an OSHA hazard communication program is not laid out the same way as a RCRA hazardous waste training program, even though the RCRA part would seem to be a logical subset of the OSHA part. To help regulated entities avoid the need to have several separate plans covering the same subject matter, an Integrated Contingency Plan concept has been developed outside the regulatory arena.

- Definitions may be inconsistent or duplicative between laws. Hazardous Substance has different meanings under OSHA and under CERCLA and the Clean Water Act.

A Look into the Future

One constant for the past 25 years has been a perpetual state of evolution in imposition and modification of environmental laws. The trend today is away from strict command-and-control regulation and toward performance-based requirements. Development of the ISO 14000 series of internationally recognized environmental management standards has laid the groundwork for responsible companies to develop voluntary internal programs that document their commitment to environmental protection.

Regulatory agencies are slowly making the changes in attitude and structure necessary to bring about these changes. However, a long period of time and incremental changes in environmental laws will be necessary to bring about this transition. Whatever changes come about, barring an unprecedented political movement and change in public attitude toward environmental protection, environmental regulation as we know it will not go away any time soon.

Stephen D. Riner is a Senior Environmental Scientist with UtiliCorp United, Inc. (UCU) in Rosemount, Minnesota, providing regulatory compliance assistance and employee training for UCU's natural gas and electrical distribution services in a three-state region. Mr. Riner has broad experience in environmental and safety management and auditing, including the areas of air quality, hazardous waste management, and hazardous materials transportation. He has a BS in Chemistry from Loyola University of Los Angeles and a MS in Chemistry from New Mexico State University. Mr. Riner has over 23 years of hazardous materials management experience, working in government, consulting, manufacturing, and utilities.

CHAPTER 2

Overview of the Law in an Environmental Context

Eugene R. Wasson, JD

This chapter provides a brief overview of the legal system that creates and implements what we know as *environmental law*. The discussion divides environmental law into two categories: ***common law*** and ***statutory law***. However, this division is strictly for ease of discussion. In real situations, it is imperative that the implications of both common law and statutory law be considered together.

Common Law

Common law originated in England and was adopted in the American colonies. It consists of various rules of law that have evolved over centuries from judicial decisions that relied on usages and customs of the people. It is often called *case law*, which is an easy way to distinguish it from statutory law that is enacted by legislative bodies. Common law is a work in progress, which is always evolving, case by case, to meet the needs of a changing society and new technology. New causes of action and rules are created as necessary to address new wrongs for which there is no existing common law or statutory cause of action.

For example, new causes of action have evolved in the past few decades for interference with business relations and for infliction of emotional distress.

Common law and statutory law directly affect each other and neither should be considered alone. For example, the common law tends to fill voids in the statutory law by providing remedies for wrongs that the statutory law does not address. It is simply impossible for a legislature to address every factual scenario that may arise. On the flip side, a legislative body may determine that a cause of action or rule of the common law is not consistent with current public policy and enact a statutory law that the court must then follow instead of the common law.

The common law causes of action applicable to environmental law are known as *torts*.[1] One definition of a tort is a legal wrong committed upon the person or property of another, independent of contract. Often the most costly environmental liability is not a government penalty for violation of a regulation, but is tort liability to an individual for harm to his person or property. Typically, actions filed by private individuals will allege numerous theories of liability, usually including negligence, strict liability, nuisance, and trespass, or some combination of these. The damages sought in a typical case may include diminution in property value, medical monitoring expenses, medical costs, lost earnings, reduced life expectancy, emotional distress, pain and suffering, and punitive damages. Such actions are often filed against companies when their activities adversely affect others. For example, the release of a harmful substance into the environment or onto the lands of another is likely to prompt surrounding landowners to file civil lawsuits. Lawsuits may be filed and potentially be successful even if the release of the harmful substance was not a violation of any regulatory requirement or permit condition. Private actions may be directed against excessive noise, light, odors, particulate fallout, vibrations, and physical invasion of another's property, just to name a few. In some cases, liability may be imposed even though the facility operator did everything humanly possible to operate his facility in a safe manner and was not aware of and did not intend any harm to the other party.

Although there are numerous tort causes of action, such as assault, defamation and invasion of privacy, the ones most likely to be applicable in an environmental matter are trespass, nuisance, negligence, and strict liability.[2]

Trespass

The tort of *trespass* is a physical invasion of another's rights, whether in his person, personal property, or land. We are concerned here with trespass to land. Trespass to land involves either (1) intentionally entering or causing something to enter onto the lands of another, with or without resultant harm or (2) recklessly, negligently, or as the result of an abnormally dangerous activity, entering or causing something to enter onto the lands of another with resultant harm. The second category, negligent entry onto the land of another, is the most likely source of liability in an environmental context. The entry onto the plaintiff's property may be either by a person or by some tangible matter, such as ash from a smokestack. Intangibles such as light, sound, vibrations, and odor do not provide grounds for a trespass claim, although such invasions may be actionable based on private nuisance as discussed below.

---

[1] Common law also provides ***contract-based remedies***. We do not address contract-based actions here, but it is important to note that contracts may have an important impact on a given environmental situation. For example, a lease agreement for a facility usually requires that the tenant comply with all applicable laws. Thus, failure to comply with environmental laws may be a basis for the landlord to require the tenant to clean up any releases of hazardous substances on the property. Similarly, a contract may require that a purchaser of a chemical product will ensure its lawful usage and disposal. A contract also might affect the factors such as control and knowledge that an environmental agency might consider in determining who has responsibility for complying with a particular environmental regulation.

[2] The common law of the various states is not identical, because each state has its own court system. Although all of the states started out with the common law adopted from England, except Louisiana which was a French territory and therefore has the Napoleonic Code as the foundation of its law, the courts of each state have developed their own common law case by case over the years so that each state now has its own legal peculiarities. For that reason, it is critical to consult an attorney familiar with the law of the state applicable to your situation.

For example, if a plant emits arsenic from its stack that is deposited on neighboring lands, the adjoining landowners may have a trespass action against the plant operator. The landowners might obtain judgments for damages such as diminution in the value of their land, harm to their cattle, personal injuries from exposure to the arsenic, *etc.*

As another example, the plant might have had an underground storage tank from which gasoline leaked into the groundwater and migrated onto a neighbor's land. The plant may have taken every precaution required by law and may not have even known that its tank was leaking, but it would still be liable for damage to the neighboring land. However, if the plant violated the law or knowingly allowed the tank to continue to leak, then the neighbor might have a case for punitive damages.

Nuisance

The tort of nuisance has traditionally been divided into public nuisance and private nuisance, although the line is often hard to draw in a particular case. *Public nuisance* is a use of one's property to intentionally cause or permit a condition to exist that injures or endangers the public health, welfare or safety. It is distinguishable from a private nuisance because it (1) affects all the residents of a particular area similarly, *e.g.*, a public stream becomes polluted and (2) does not require an interference with the right to use and enjoy one's land. Traditionally, a public nuisance was a crime subject to a fine, *e.g.*, operating a brothel. However, it may also provide grounds for a private cause of action if the plaintiff suffers a harm different than the public in general. For instance, if a lake is polluted so that commercial fishermen cannot practice their livelihood, then the fishermen may have a private cause of action for public nuisance because they have suffered a special damage over and above the harm to the general public, *i.e.*, the right to clean waters.

In contrast, a *private nuisance* requires an intentional act by the defendant which substantially and unreasonably interferes with the plaintiff's use and enjoyment of his land. The intent required is not that the defendant intend the resulting harm, but only that the defendant intend to commit the act when he knew or should have known that harm to others would result from his action. The interference must be of such a continuing and substantial nature that a reasonable person would expect compensation for exposure to such a condition. As a general rule, the interference constituting a private nuisance will be by an intangible means such as noise, light, vibrations, odor, gaseous emissions, or even fear or repulsion if reasonable. The interference is not actionable if the plaintiff is overly sensitive; rather, the standard is whether an ordinary person would be substantially affected under ordinary conditions.

A slaughterhouse is a prime example of a use of one's land that may constitute a private nuisance. The operation may be actionable if it emits odors that interfere with the adjoining landowners' use and enjoyment of their property. To be actionable, the defendant slaughterhouse must have been aware or should have been aware that the odors were interfering with the adjoining landowners' use of their property. The interference must have been such that a reasonable person could not have enjoyed the use of the property while such an interference was occurring. Finally, the plaintiff must have some legal interest in the land affected. If successful, the plaintiff might receive an injunction against the interference, damages for the diminution in the value of his property, or both.

A minority of courts have held that the plaintiff's *coming to a nuisance* provides the defendant a defense. Traditionally, this defense was applied where the defendant had an existing operation and the plaintiff subsequently bought land in the area affected by the defendant's operation. Obviously the thought was that the plaintiff should not have moved into the area if he did not like the effects of the defendant's operation. However, the majority of courts no longer recognizes this defense. One possible basis for the majority position is that the defense allows the defendant to diminish the value of the surrounding land without paying for it.

For harms for which money damages are not sufficient or are not readily determinable, a court may grant the plaintiff injunctive relief.[3] Injunctive relief is often sought to abate, or stop, a

---

[3] For example, injunctive relief may be appropriate where health or safety are threatened, because monetary damages cannot restore the plaintiff's health. However, flooding of the plaintiff's crop may be more suited to monetary damages than injunctive relief if there is a reasonable method for estimating the value

nuisance; however, it may also be available as relief against other torts such as trespass. An *injunction* is a court order directing the defendant to take some action or to refrain from an action. If the plaintiff needs immediate relief, he may seek a temporary restraining order. This is a form of injunctive relief that may be issued by the court without notice or with only short notice to the defendant upon the plaintiff demonstrating to the court (1) that he would suffer immediate and irreparable harm without the order and (2) that he has attempted to give the defendant notice of the hearing. A temporary restraining order, as evident by its name, lasts only a few days. For longer relief pending a full trial, the plaintiff must seek a preliminary injunction. To obtain a preliminary injunction, the plaintiff must show:

1) A substantial likelihood that he will prevail on the merits at trial.

2) A substantial threat that he will suffer irreparable injury if the injunction is not granted.

3) That the threatened injury to him outweighs the threatened harm the injunction may do to the defendant, and

4) That granting the preliminary injunction will not disserve the public interest. If the plaintiff obtains a preliminary injunction, then the defendant will be restrained until the matter reaches a full trial, at which the issue will be whether the plaintiff is entitled to have the defendant permanently enjoined.

Negligence

Negligence is another common environmental tort. The elements of a *negligence* claim include:

1) A duty owed by the defendant to the plaintiff

2) A breach of that duty by the defendant

3) The plaintiff was damaged by the breach, and

4) The breach was the proximate cause of damage

---

of the crop. Generally, the courts will not provide injunctive relief if monetary damages will make the plaintiff whole, *i.e.,* compensate the plaintiff for his injury.

The duty alleged in most environmental cases is the duty of the defendant to conduct his activities so that he does not harm other persons. The question of whether or not this duty has been breached is judged on an after-the-fact, objective standard; in other words, the jury exercises its hindsight to decide what a reasonable person would have done under similar circumstances. The issue of causation addresses whether or not the defendant's action or inaction did in fact cause the plaintiff's harm and whether that harm was a reasonably foreseeable result. The damages may include property damage or personal injuries, but must be supported by some evidence from the plaintiff.

In determining whether a defendant was *negligent*, which generally means that the defendant owed a duty and breached that duty—the first two factors in establishing a cause of action for negligence—it is not necessarily a defense that the defendant obeyed all applicable laws or followed the practice accepted in the particular industry. Compliance with law and industry standards is a minimum standard, and the jury may determine that more was required of the defendant under the particular facts. On the other hand, non-compliance with a law or an industry standard is very likely to establish that the defendant acted negligently. If the plaintiff can establish the remaining two elements, causation and damages, then the jury should return a judgment for the plaintiff.

In order for the violation of a law to establish the defendant's negligence, the plaintiff must show that the statute, regulation, or ordinance was enacted to (1) protect the class of individuals to which the plaintiff belongs and (2) prevent the kind of harm that occurred. If the plaintiff can prove the violation and that he was in the class of persons to be protected by the statute and the harm was the type intended to be prevented by the statute, then he still must prove causation and damages. The defendant's statutory violation is said to have been negligent *per se*, or, in itself.

As an example of a negligence action, a suit might arise from a company mishandling a hazardous waste so that it is accidentally released onto the ground or down a drain. If the waste migrates onto neighboring properties, the neighbors might be exposed to it through drinking water from shallow ground water wells, eating vegetables from their gardens, inhaling dust from roads, *etc.*

Lawsuits might result if, for example, a governmental health agency or the members of the neighborhood noticed a higher incidence of cancer or mental disorders in the area and zeroed in on the waste and traced it back to the defendant company. The resulting lawsuits would be based on the company's duty to conduct its operations so as not to harm others, the breach of that duty by carelessly handling a harmful waste, causation based on medical reports and the tracing of the waste back to the company, as well as damages based on medical evidence. Possible weak links in the plaintiff's case might be a lack of scientific evidence that the particular waste causes the types of harm reported by the plaintiffs, a lack of proof that the waste came from the defendant's plant, or the unforeseeability of harm due to an unlikely exposure pathway. However, a jury drawn from the community and blessed with 20/20 hindsight would likely return a judgment for the plaintiffs.

The plaintiff in the above example might establish duty and breach of duty through use of the doctrine of negligence *per se*, which was discussed above. For instance, if the company failed to report the release and the plaintiff's harm was within the scope of harms intended to be prevented by an applicable reporting statute, the plaintiff would not have to present evidence of the defendant's duty of care. Instead of debating with the jury whether a reasonable person would have notified everyone within 100 feet or 2 miles of the plant, the plaintiff need only show that the company violated its statutory duty to report the release.

In the above example, if the company had committed gross negligence which was likely to shock the conscience of the jury, such as throwing the waste into a nearby stream which it knew was used by other persons for drinking water or fishing, then punitive damages would likely be imposed. ***Punitive damages*** are damages imposed in addition to actual damages that compensate the plaintiff for his injury. The purpose of punitive damages is, as the name implies, to punish the defendant. Punitive damages may exceed the defendant's actual damages and may far exceed any civil or criminal penalty imposed by an enforcement agency. The United States Supreme Court has affirmed an award of punitive damages about 500 times greater than the actual damages in a case. Thus, a plaintiff who drank contaminated water might suffer only $1,000.00 in medical and other actual damages, but receive $500,000.00 in punitive damages. Furthermore, this 500 to 1 ratio is not a cap, just what has been allowed in a particular case.

Strict Liability

Strict liability for abnormally dangerous activities may cause a defendant to be liable for harm to the person, land, or personal property of another resulting from that activity, although the defendant has exercised the utmost care to prevent the harm. Generally, courts consider the following factors to determine whether a particular activity is abnormally dangerous:

1) The high risk of harm occurring from the activity

2) The likelihood that the resulting harm will be great

3) The ability to eliminate the risk by exercising ordinary care

4) The extent to which the activity is not common to the community

5) The appropriateness of the activity to the place where it was carried on, and

6) The activity's value to the community

The prime example of an ultrahazardous activity is the storing of dynamite in the middle of a city. Storage of dynamite in a populated area poses a significant threat to the population regardless of the care used in its handling. Therefore, any harm caused by an explosion of the dynamite would subject the owner to liability, regardless of how careful he was in storing the dynamite. Other examples of activities that have been subject to strict liability are gas wells, gasoline storage tanks, and pipelines operated in populated areas.

Torts in General

The above discussion of torts is not exhaustive, but is intended to place environmental managers on notice that there are considerations other than just what the Environmental Protection Agency (EPA) may require. There are numerous other existing tort causes of action, and new tort theories are limited only by the imagination of the plaintiff's attorney. Finally, as noted above, common law varies from state to state, as do the statutes that may alter the common law, so it is imperative to consult with an attorney familiar with the law of the particular state.

The Courts

Because common law is made by the courts, it is important to understand the structure of the courts. If there are conflicts in the cases that you find on an issue, you must know what weight to give those cases. Generally, State courts have control over out-of-state courts and Federal courts in your district, or circuit control over cases from other Federal jurisdictions. Whether State or Federal, higher courts have control over lower courts. As between State and Federal cases, Federal courts defer to State courts on matters of State law. On matters of Federal law, such as constitutional issues, the United States Supreme Court is the ultimate authority.

The State court systems usually consist of a hierarchical set of courts. At the bottom are trial courts at the city, county, or multicounty level. In most states, appeals from the trial courts are to one or more levels of intermediate appellate courts, and then to a court of final jurisdiction, usually called the supreme court. In states with lighter caseloads there may not be an intermediate appellate court, and appeals are instead taken directly to the supreme court.

The State courts are separate from the Federal courts. In contrast to the State courts, which have general jurisdiction, the Federal courts have limited jurisdiction. Basically, the Federal courts have jurisdiction over cases involving Federal questions or parties of diverse state citizenship. Federal question jurisdiction includes cases arising under Federal statutes, such as the Comprehensive Environmental Response, Compensation and Liability Act (CERCLA), the Resource Conservation and Recovery Act (RCRA), *etc.* Thus, Federal enforcement actions will be in Federal court. **Diversity jurisdiction** is intended to protect a defendant who is a citizen of one state from being subjected to a lawsuit in the plaintiff's state where the plaintiff might be shown favoritism. For instance, if the defendant injured the plaintiff while driving through the plaintiff's state, the plaintiff could file suit in his state's courts but the defendant could *remove* the case to the Federal district court due to the party's diverse citizenship. In addition to diversity of citizenship, diversity

jurisdiction requires that the amount in controversy be at least $75,000. If the Federal courts have jurisdiction in a case, they may also hear other claims over which they might not otherwise have had jurisdiction. For example, if a private plaintiff asserts both a CERCLA cost recovery claim and a negligence claim arising from the same facts, the Federal court will have Federal question jurisdiction over the CERCLA claim and may then hear the negligence claim as a supplemental matter.

The Federal court system is hierarchical like the state courts. In each state, there are one or more federal district courts that are the trial courts of the Federal system. Decisions of the district courts may be appealed to the Federal circuit court of which they are a part. There are twelve Federal circuit courts that are each comprised of several states. The circuit court for the District of Columbia is especially important in environmental law, because several of the environmental statutes give it jurisdiction over matters such as challenges to regulations. Of course, the final court of appeal is the Supreme Court.

Statutes

Sources and Limits of Statutory Law

Although common law plays an important role in *environmental law*, people generally think of statutes like the Clean Air Act and the Resource Conservation and Recovery Act when they hear the term *environmental law*. These are Federal statutes enacted by the United States Congress.[4] These statutes are addressed in detail in other chapters. Here, however, we will address only the enactment and implementation of such statutes.

Congress is the source of Federal statutes. Congress is limited to enacting legislation that

---

[4] In addition to the obvious environmental statutes such as RCRA, several provisions of the Federal criminal code may also be invoked in an environmental case. These statutes pertain to false statements to the Federal government, mail fraud, and conspiracy. Other laws may also be applicable even though the matter is primarily environmental.

falls within one of its enumerated powers under the United States Constitution.[5] For example, Congress often relies upon the **Commerce Clause of the Constitution** which authorizes it to regulate interstate and foreign commerce. The applicability of the Commerce Clause to some environmental statutes is readily apparent, such as the regulation of certain products under the Toxic Substances Control Act. However, the constitutional basis for other statutes is less apparent, such as the regulation of the quality of waters that do not cross State boundaries. Nonetheless the Supreme Court, which hears lawsuits challenging the validity of Federal legislation, has shown great deference in determining whether Congress has remained within its constitutional limitations.

Federal statutes are codified in the **United States Code**.[6] That is the starting point for the Federal system of environmental law. Generally, Congress will specify in a statute a vague requirement, *e.g.*, that there shall be no discharge of any pollutant

into the waters of the United States except pursuant to a permit that incorporates effluent limits. Congress will then delegate responsibility for implementing the statutory provision to a particular agency and will authorize the agency to take certain steps, such as implementing regulations to flesh out the statutory requirement and to enforce the statute and implementing regulations. However, Congress is prohibited by the constitutional doctrine of separation of powers from delegating unfettered and undefined power to an agency, and the agency is prohibited from exceeding the limits of its statutory delegation of authority. Thus, Congress cannot simply delegate to EPA the authority to establish whatever environmental regulations it deems necessary. Likewise, EPA could not establish regulations applicable to solid waste if the enabling statute only authorized regulation of hazardous waste.

The statutory delegation of authority is critical, because an agency has only the authority delegated to it by statute.[7] This limitation, however, is not as limiting as it may seem at first. An agency has discretion to work within the framework established by the enabling statute to construct the detailed program that Congress did not have the time or the expertise to supply.

If Congress acts beyond its constitutional limits or an administrative agency exceeds its statutory authorization, a party that is adversely affected has *standing* to bring a lawsuit in Federal court to challenge the law or regulation. As a practical matter, the Supreme Court has seldom held that Congress exceeded its constitutional limits. More frequently, courts have found that an administrative agency has exceeded its statutory authorization.

In reviewing challenges to agencies' actions, courts have typically divided the cases into three categories: interpretation of the statute, factual basis for the agency's action, and agency procedure. When the case involves interpretation of the authorizing statute, courts generally will follow the agency's interpretation of the statute it is charged with enforcing if that interpretation is reasonable but the court is not bound by that

---

[5] Unlike Congress, the states have broad authority to protect the health and safety of their citizens. However, when Congress enacts legislation it may preempt any state or local laws on the same subject. Preemption may occur when Congress specifically declares that it is preempting state and local law or if Congress has established a regulatory scheme that completely occupies or covers the particular field. Similarly, a state cannot enact laws that impermissibly burdens *interstate commerce* by discriminating against interstate commerce in favor of state businesses or that deny *equal protection* between different regulated entities without a reasonable basis for the distinction.

[6] When Congress passes a bill and it is sub-sequently approved by the President, either through signing it or simply not vetoing it, the bill becomes a statute. Each statute is then codified in the *United States Code*, which is the official cumulative set of all statutes organized by subject matter titles. For example, many environmental statutes are codified in Title 42–Health and Environment. The complete text of the new statute may not be included in the Code in a single undivided block. Instead, the Congress will designate in the statute how the various sections of the statute will affect the existing Code—a paragraph may be added here or there, or a word may be changed, or a whole new subchapter may be added. For example, the Resource Conservation and Recovery Act is a statute that was passed in 1976 and which was codified as amendments to Sections 6901 to 6992k of Chapter 82, entitled "Solid Waste Disposal," of Title 42 of the Code. That is why you may hear someone refer to Section 7002 of RCRA but the provision is codified at 42 USC §6972.

[7] Agencies also may receive authority in an area by **executive order** of the President. The President delegates authority that he may have as a direct result of his office or as a result of a Congressional delegation of authority to him.

interpretation. Instead, the court will focus primarily on the language of the statute. If the statute is not clear, then the court will consider other evidence of Congress's intent such as the legislative history of the statute. The court will also consider the purpose and public policy underlying the statute. If a party challenges the factual basis for an agency's action, such as the denial of a permit, then the court will usually apply an *arbitrary and capricious* standard which requires only that the agency had some evidence to support its action. In some cases, the court will apply the slightly more stringent standard of *substantial evidence*. Finally, if a party challenges an agency's action on the grounds that it did not follow procedural requirements, such as notice of the proposed action or the right to a hearing before the agency, the court will review the matter to determine if the agency complied with the procedural requirements of the particular statute, the Administrative Procedures Act,[8] the Constitution, and the agency's rules. In short, agency actions are generally upheld by the courts in the absence of a clear exceedance of statutory authority, complete lack of factual support, or a major failure to follow procedural requirements.

Enforcement

Each of the Federal environmental statutes provides for enforcement of its requirements through administrative orders, civil penalties and relief, as well as criminal sanctions. Generally, the statute will provide a maximum monetary penalty, which typically is applied cumulatively on a per-day-of-violation basis. While civil penalties are imposed on a strict liability basis, the criminal penalties may require either a knowing violation or even a negligent violation, depending on the statute. Although the statutes set forth the maximum penalties, various EPA and Department of Justice (DOJ) policies provide further guidance on the penalties which may actually be imposed in a given situation. It is important for an environmental manager to (1) be

aware of the civil and criminal penalties authorized under the Federal environmental statutes, (2) be familiar with how the statutory penalties may be influenced under EPA and DOJ enforcement policies, and (3) take steps to prevent violations and mitigate any possible penalties.

Statutory Penalty Amounts and Standards for Imposition. Although a full discussion of the penalty amounts and the standards for imposing penalties under each of the Federal environmental statutes is beyond the scope of this chapter, the standards of the RCRA and the CWA for *knowing* violations, which are discussed immediately below, are indicative of the standards under most of the statutes.

Resource Conservation and Recovery Act
Knowing violations of the requirements applicable to generation, transportation, treatment, storage and disposal of hazardous waste under the Resource Conservation and Recovery Acy (RCRA) are felony offenses subject to two to five years in prison and fines of up to $50,000 per day of violation. These include, but are not limited to, recordkeeping and manifesting violations.

A *knowing* violation of RCRA does not necessarily coincide with our common, everyday understanding of that term. For instance, in *United States (US) v. Hoflin*, 880 F.2d 1033, 1038–1040 (9th Cir. 1989), the defendant was convicted of disposing of hazardous waste without a permit even though he neither knew the material was a RCRA hazardous waste nor that the party actually doing the disposal lacked a RCRA permit. The court found *knowing* disposal of a hazardous waste where, even though the defendant did not know the paint was a RCRA hazardous waste, he did know the paint was not "an innocuous substance like water." *Accord, US v. Sellers*, 926 F.2d 410, 417 (5th Cir. 1991) (affirming conviction for unlawful disposal of RCRA hazardous waste where defendant knew paint solvent was "potentially dangerous to human beings and the environment", and would therefore be subject to some form of government regulation); *US v. Baytank (Houston), Inc.*, 934 F.2d 599, 613 (5th Cir. 1991) (*knowingly* means no more than that the defendant knows factually what he is doing—storing, what is being stored, and that what is being stored factually has the potential for harm to others or the environment—and it is not required that he know that there is a regulation which says what he is storing is hazardous waste under RCRA).

---

[8] For establishing new regulations, the federal Administrative Procedures Act generally requires publication in the *Federal Register*, an opportunity for the public to comment, a public hearing in some circumstances, response by the agency to the public's comments and final publication.

Knowing endangerment under RCRA is a more serious offense than a *knowing* violation of a RCRA requirement. Consequently, RCRA imposes harsher penalties for *knowing endangerment*. Specifically, 42 USC §6928(e) states:

> Any person who knowingly transports, treats, stores, disposes of, or exports any hazardous waste identified or listed [under Subtitle C of RCRA] or used oil not identified or listed as a hazardous waste . . . who knows at that time that he thereby places another person in eminent danger of death or serious bodily injury, shall, upon conviction, be subject to a fine of not more than $250,000.00 or imprisonment for not more than 15 years [felony], or both. A defendant that is an organization shall, upon conviction of violating this subsection, be subject to a fine of not more than $1,000,000.00.

For purposes of *knowing endangerment*, a defendant *knows* of the endangerment he creates if he is aware or believes that his conduct is substantially certain to cause danger of death or serious bodily injury. Circumstantial evidence may be used, including evidence that the defendant took affirmative steps to shield himself from relevant information. For example, if a manager tells his employee to "get rid of it but don't tell me how," it may be sufficient to establish *knowing endangerment* that the manager knows the hazardous waste could harm anyone exposed to it and that the employee is likely to dispose of the waste in an unsafe manner.

Clean Water Act. Of particular interest under the Clean Water Act (CWA) is the fact that criminal penalties may be imposed not only for *knowing* violations, but also for *negligent* violations. For *negligent* violations of the Clean Water Act or permits issued thereunder, the violator may be convicted of a misdemeanor with imprisonment of up to one year and/or a fine of between $2,500.00 and $25,000.00 per day of violation. For example, criminal misdemeanor charges based on negligence were prosecuted in the *Exxon Valdez* incident. The captain of the *Exxon Valdez* was convicted, but the conviction was overturned on procedural grounds.

The CWA also provides for penalties and imprisonment for *knowing* violations of the Act's provisions. Specifically, the Act states:

> Any person who commits a *knowing* violation may be convicted of a felony and shall be punished by a fine of not less than $5,000.00 nor more than $50,000.00 per day of violation, or by imprisonment for not more than 3 years, or by both.

In 1992, for instance, Chevron pleaded guilty to knowingly exceeding oil and gas discharge limits in its National Pollutant Discharge Elimination System (NPDES) permit at an offshore drilling platform and paid a criminal fine of $6.5 million. *US v. Chevron USA*, No. CV–88–7836 (D.C. Cal. May 18, 1992). Other charges included diluting samples, concealing test results, by-passing the wastewater treatment system, and dumping sandblast waste into the ocean.

To Whom Does the Penalty Apply? Environmental statutes may provide for monetary penalties and imprisonment for various categories of violations, but who has to pay the fine or go to jail? That may depend on the particular requirement that is violated. For example, the Clean Water Act requires that *any person in charge* of a facility shall report a discharge of oil or a hazardous substance as soon as he knows of such discharge. Under that provision, an employee in charge of the operation that causes the release and the corporation for which he works may both be responsible. Other requirements under the various environmental laws are applicable to the *owner* and/or *operator*. Under such an *owner or operator* standard, both an employee and his employer could be liable as operators. Still other statutory and regulatory requirements are applicable to *any person*. In that case, an employee and/or his employer might be responsible.

Generally, if a person is directly involved in a violation, then he is potentially subject to an enforcement action. For example, if the janitor dumps his mop bucket of trichloroethylene into the ditch behind the plant, then the janitor is personally responsible for his violation because of his direct involvement. Likewise, if a manager told the janitor to dispose of the waste in that manner, then the manager is also directly liable for the violation. The manager could also be liable if he knew the janitor had the waste and chose to remain ignorant of the janitor's method of disposal.

In addition to direct liability for violations, persons in positions of authority within a corporation may be held liable for indirect responsibility for violations. In its purest form, the ***Responsible Corporate Officer Doctrine*** allows the criminal conviction of a corporate officer based purely on the fact that he was in a position of responsibility and authority within the corporation which would have allowed him to prevent the violation. Under this theory, the corporate officer could be convicted even though he knew absolutely nothing about the violation; all that is necessary is that he had the authority to prevent the violation. The doctrine is not limited to corporate officers as the name implies, but may impose liability on any person with the authority to prevent a violation.

The Responsible Corporate Officer Doctrine was developed in two cases under the Federal Food, Drug and Cosmetic Act of 1938, *US v. Dotterweich*, 320 U.S. 277 (1943) and *US v. Park*, 421 US G58 (1975). In *Dotterweich,* the court set forth its rationale for the doctrine as follows:

> The prosecution to which Dotterweich was subjected is based on a now familiar type of legislation whereby penalties serve as effective means of regulation. Such legislation dispenses with the conventional requirement for criminal conduct—awareness of some wrongdoing. In the interest of the larger good it puts the burden of acting at hazard upon a person otherwise innocent but standing in responsible relation to a public danger.

Because both the Food and Drug Act and the environmental acts are intended to protect the public welfare, the Department of Justice has attempted in the past to apply the Responsible Corporate Officer Doctrine to environmental prosecutions. More recently, DOJ has taken the position that it will not prosecute based solely on the responsible corporate officer doctrine, but will require *some actual knowledge* of the violative acts. Despite DOJ's position, the Clean Water Act and the Clean Air Act expressly state that the *responsible corporate officers* may be persons liable under those Acts. However, because those acts did not define *responsible corporate officer*, it has been left to the courts to determine the application of this term in environmental prosecutions. Some courts have held that an officer cannot be convicted of a *knowing* violation based solely on knowledge imputed to him as a result of his position, but have allowed his position, authority, and responsibility

to serve as circumstantial evidence that he possessed knowledge of the violation.

Nonindividual entities, such as corporations, may also be liable for violations. Referring back to the example of the janitor, the company that employs him may also be liable for his violation. Generally, under the ***Doctrine of Respondeat Superior***, employers are responsible for the acts of their employees in the course of their employment. Therefore, because it was generally within the scope of the janitor's job to generate his mop waste and to dispose of that waste, his employer may also be liable. On the other hand, if the janitor was using some of his free time on the night shift to change the oil in his personal car and then dumped that oil into the ditch behind the plant, the employer would not be responsible because changing the oil in his personal car was not part of his job.

Administrative Civil Penalty Policies. Although the Federal environmental statutes set the maximum civil and criminal penalties for their violation, the penalties actually imposed are usually below those statutory maximums. The penalties are set in accordance with EPA penalty policies, the Department of Justice enforcement policy, and the United States Sentencing Guidelines (USSG).

On February 16, 1984, EPA issued two documents; one entitled "Policy on Civil Penalties" and the other entitled "A Framework for Statute-Specific Approaches to Penalty Assessments." These two documents jointly set forth EPA's general scheme for calculating civil penalties and authorized program-specific penalty policies. Over the years, penalty policies have been developed in programs such as the Toxic Substances Control Act (TSCA), RCRA, CWA, the Clean Air Act (CAA), the Emergency Planning and Community Right-to-Know Act (EPCRA), and CERCLA. In fact, some of these programs have multiple policies which are tailored to specific areas within the overall program; for instance, the TSCA program has a penalty policy specifically for polychlorinated biphenyls (PCBs). Because all of the penalty policies generally follow EPA's 1984 penalty framework, the RCRA penalty policy is discussed briefly below as an example.

The RCRA civil penalty policy calculates penalties by

1) Determining a gravity-based penalty for a particular violation from a penalty assessment matrix

2) Adding a *multiday* component, as appropriate, to account for a violation's duration

3) Adjusting the sum of the gravity-based and multiday components, up or down, for case specific circumstances, and

4) Adding to this amount the economic benefit gained through noncompliance

The sum of the gravity-based component plus the multiday component may be adjusted upward or downward based on good or bad faith, degree of willfulness and/or negligence, and other unique factors, including the risk and cost of litigation. The penalty may be adjusted upward for history of noncompliance and downward for an inability to pay the penalty or for projects to be undertaken by the violator. After the penalty has been adjusted, an amount will be added for the economic benefit the violator gained from noncompliance. The resulting amount is the figure EPA would take in a settlement of the complaint. If the complaint were to go to trial, no downward adjustments of the penalty amount would be taken into consideration.

Enforcement Agencies

Environmental Protection Agency. The EPA is the agency charged with enforcing most of the Federal environmental statutes. However, many of the Federal environmental statutes authorize states to establish permitting and enforcement programs to operate *in lieu of*, or to be *delegated*, the Federal program within their borders. Generally, to obtain *delegation* of a Federal program, a state must enact laws and establish regulations that are at least as stringent as their Federal counterparts. Upon receiving delegation of a program, the state is the primary enforcement authority for that program. However, the EPA retains the right to intervene if it believes the state has not been diligent in its enforcement of the Federal requirements. There is currently an issue as to whether the EPA has the right to take additional enforcement actions against a violator that has resolved its violation with a delegated state, or whether EPA is limited to revoking the state's delegation as to future enforcement.

Department of Justice. When the EPA decides to proceed beyond civil penalties and pursue criminal

penalties, it notifies the United States Attorney's office for the jurisdiction in which the violation occurred or the Environment and Natural Resources Division (ENRD) of the Department of Justice (DOJ) in Washington.

On July 1, 1991, DOJ issued its policy encouraging self-auditing, self-policing, and *voluntary disclosure* of environmental violations by the regulated community. The stated purpose of the DOJ document is "to describe the factors that the Department of Justice considers in deciding whether to bring a criminal prosecution for a violation of an environmental statute, so that such prosecutions do not create a disincentive to or undermine the goal of encouraging critical self-auditing, self-policing, and voluntary disclosure." DOJ set forth several nonexclusive factors which should be considered in determining whether and how to prosecute. These factors include voluntary disclosure, cooperation with investigators, preventative measures and compliance programs, the pervasiveness of noncompliance, internal disciplinary action, and subsequent compliance efforts.

A voluntary disclosure must be voluntary in the sense that it is not already required by law, regulation, or permit. The voluntary disclosure must also be timely and complete, such that the violator comes forward promptly as soon as possible after discovering the noncompliance and provides sufficient information to aid the government's investigation with information not already obtained by the government through another source.

The violator's full and prompt cooperation is another factor to be considered, including the violator's willingness to make all relevant information, including all internal and external investigations and the names of all potential witnesses, available to the government.

The existence and scope of any regularized, intensive, and comprehensive environmental compliance program maintained by the violator is also considered. The compliance program must have included sufficient measures to identify and prevent future noncompliance, and the program must have been adopted in good faith in a timely manner. The compliance program should also contain an effective internal disciplinary action program to make employees aware that unlawful conduct will not be condoned.

The DOJ memorandum includes an example of the application of the factors to a *good* defendant. The company conducted regular, comprehensive environmental compliance audits. It also had a compliance program, which included clear policies, employee training, and a hotline for suspected violations. When an audit uncovered employees disposing of hazardous waste by dumping in an unpermitted location, the company disclosed all pertinent information to the appropriate government agency after confirming the violations. The company undertook compliance planning with that agency and carried out satisfactory remediation measures. The company also undertook to correct any false information previously submitted to the government in relation to the violations, and the company disciplined the employees actually involved in the violations, including any supervisor who was lax in preventing or detecting the activity. Finally, the company reviewed its compliance program to determine how the violations slipped by and corrected the weaknesses found by that review. The company also disclosed the names of the responsible employees and cooperated in providing documentation necessary to investigate those employees. As a result of its actions, the DOJ memorandum states that the company would stand a good chance of being favorably considered for prosecutorial leniency, to the extent of not being criminally prosecuted at all.

State and Local Enforcement Agencies. All of the states have environmental programs that parallel all or a part of the Federal programs. They have their own statutes, regulations, penalty policies, and enforcement agencies. As noted above, many of the State programs operate in lieu of the Federal programs where the Federal agency determines that the State program is at least as stringent as the Federal program. The *at least as stringent as* language means that states may have additional requirements over and above those imposed by the Federal agency. This is especially common in environmentally active states such as California and New Jersey.

In addition, states and local governments may have laws that cover entirely different topics, such as ground water pollution, disclosure laws applicable to real property transfers, criminal litter laws, *etc.*

By and large, the discussion of the Federal system in this chapter is also applicable to State laws and enforcement. However, each state has its own quirks. Therefore, it is absolutely necessary to consult someone familiar with a particular state's laws and regulations if an issue arises in that state.

Citizen Suit Enforcement. Several of the Federal environmental statutes provide for enforcement by citizens acting as private attorneys general. (See, *e.g.,* RCRA §7002, 42 USC §6972, and CWA §505, 33 USC §1365.) These provisions allow any person who has an interest that is or may be adversely affected to file an enforcement action in Federal district court to compel compliance through court order or imposition of penalties and/ or to seek injunctive relief against imminent endangerments to human health or the environment. Prior to filing suit, a citizen plaintiff must notify the violator, the EPA, and any applicable State enforcement agency to allow them to correct the violation. Also, a citizen cannot file suit if an appropriate enforcement agency is already taking action. Citizens' suits are becoming more prevalent, due in large part to the possibility of obtaining their attorneys' fees and expert witnesses' fees if they substantially prevail in the case.

United States Sentencing Guidelines. If the EPA and the DOJ obtain a conviction of a violator, the judge must decide what sentence to impose. In the past, Federal district court judges had tremendous discretion in sentencing criminals convicted of Federal crimes. Furthermore, once sentenced, the actual length of time the violator spent in jail depended on parole guidelines. In 1984, Congress decided to standardize sentences for Federal criminals and pursuant to the Sentencing Reform Act of 1984 established the United States Sentencing Commission. The Commission has developed the *United States Sentencing Guidelines* with the objectives of increasing the uniformity of sentencing, removing the uncertainty of sentences caused by the parole system, and ensuring proportionality in sentencing criminal conduct of differing degrees of severity.

Persons acting either individually or as agents of a corporation may be held personally liable both civilly and criminally for their personal actions. The *Guidelines* provide a step-by-step process for determining a sentencing range for convicted individuals. Basically, the *Guidelines* operate on a point system by assigning an initial point value to the violation, and then adjusting that point value up or down based on several factors. The resulting point value and a number corresponding to the defendant's criminal history are used jointly to determine the range of sentences from a sentencing table.

For example, if a manager directs one of his employees to dispose of paint by pouring it out on the plant property, both the manager and the employee who did the actual dumping would be guilty of mishandling a hazardous waste under RCRA. The base offense level is adjusted based on five factors, including:

1) The nature of the victim of the crime

2) The defendant's role in the offense

3) Whether there was an attempted or actual obstruction of the administration of justice

4) Whether the defendant was convicted on multiple counts, and

5) Whether the defendant accepted responsibility for the crime

In considering the defendant's role in the offense, the base offense level may be increased if the defendant was an organizer, leader, manager, or supervisor in any criminal activity. If the defendant obstructs justice by committing perjury or producing a false or altered document or record during an official investigation or judicial proceeding, the base offense level may be increased. The base offense level may be decreased if the defendant clearly demonstrates acceptance of responsibility for his offense or if the defendant has timely notified the authorities of his intention to enter a plea of guilty prior to the government preparing for trial.

There are separate sentencing guidelines for crimes in general, not just environmental crimes, by organizations such as corporations. In 1993, draft organizational sentencing guidelines specific to environmental crimes were proposed, but they have not been adopted. Nonetheless, a few comments on the draft are worthwhile. Because an organization cannot be placed in jail, the draft provides for monetary penalties. Instead of operating on a points system as with individual violators, the draft organizational guidelines provide for a percentage of the maximum statutory fine. The base fine under the organizational guidelines is the greater of (1) the economic gain plus costs directly attributable to the offense or

(2) a percentage of the maximum statutory fine that could be imposed for the offenses.

Economic gain is defined as (1) the economic benefits that an offender realized by avoiding or delaying capital costs necessary to comply with the environmental statute, based upon the estimated costs of capital to the offender; (2) the continuing expenses (*e.g.*, labor, energy, leases, operation, and maintenance) the offender avoided or delayed by noncompliance; and other profits directly attributable to the offense. Costs include: (1) actual environmental harm including degradation of a natural resource and (2) harms incurred in remediation or other costs borne by others.

According to the draft guidelines, the base offense level could be increased if one or more members of the substantial authority personnel of the organization participated in, condoned, solicited, or concealed criminal conduct, or recklessly tolerated conditions or circumstances that created or perpetuated a significant risk that criminal behavior of the same general type or kind would occur or continue. If a corporate manager lacking the authority or responsibility to be classified as a member of the organization's substantial authority personnel, but having supervisory responsibility to detect, prevent, or abate the violation, engaged in the criminal conduct, the base fine may be increased by a different percentage. The draft guidelines define **substantial authority personnel** to include high-level personnel, individuals who exercise substantial supervisory authority (*e.g.*, a plant manager, a sales manager), and any other individuals who, although not a part of an organization's management, nevertheless exercise substantial discretion when acting within the scope of their authority (*e.g.*, an individual with authority in an organization to negotiate or set price levels or an individual authorized to negotiate or to approve significant contracts).

Conclusion

Environmental law is broader than the Federal environmental statutes and EPA's regulations. There may be existing court decisions that interpret or limit the pertinent statutes and regulations, or there may be grounds to challenge the validity of the statute or regulation or its enforcement in a particular case. Likewise, agencies such as EPA have policies that affect the enforcement of the statutes and regulations. In addition, private tort actions and State environmental laws may play a role in a particular situation. Thus, it is important to be familiar with the *big picture* of environmental law.

Eugene R. Wasson is a member of the law firm of Brunini, Grantham, Grower and Hewes, PLLC, in Jackson, Mississippi. Gene is a member of the firm's Environmental Group and devotes the majority of his practice to environmental administrative and litigation matters. After receiving a bachelor of science degree in biological engineering with honors from Mississippi State University, he graduated with honors from the University of Mississippi School of Law. Mr. Wasson currently is active in the American Bar Association and the Mississippi Bar Association.

CHAPTER 3

Federal Facility Compliance Act of 1992

Harry A. Bryson, CHMM

Federal Facility Compliance Act of 1992

Background/History

Regulatory Impact

FFCAct Description by Section

Title I. Federal Facility Compliance Act of 1992 • Title II. Metropolitan Washington Waste Management Study Act

Impacts of Key Sections of the FFCAct

§102 (Application of Certain Provisions to Federal Facilities) • §105 (Mixed Waste Inventory Reports and Plan) • §107 (Munitions) • §109 (Small Town Environmental Planning)

Bibliography

Laws • Hazardous Waste Management • Military Munitions • Mixed Waste

---

The Federal Facility Compliance Act (FFCAct) of 1992 (PL 102-386, October 6, 1992) amended the Solid Waste Disposal Act (SWDA) of 1965, as amended, with respect to a number of issues which affected primarily (but not only) Federal facilities and installations. The SWDA is alternatively referred to as the Resource Conservation and Recovery Act (RCRA), as amended, because RCRA was essentially a complete rewrite of the SWDA when it was enacted in 1976. RCRA will be used in this discussion instead of SWDA for clarity, since RCRA has become the most commonly used term in the regulatory arena. The acronym FFCAct will be used to avoid confusion with the acronym FFCA, which is generally used for Federal Facility Compliance Agreement by the United States (US) Department of Energy, the Federal agency most affected by the FFCAct.

The Federal Facility Compliance Act of 1992 consists of two titles:

* Title I–Federal Facility Compliance Act of 1992

* Title II–Metropolitan Washington Waste Management Study Act

Title I addressed a range of topics, and consisted of ten sections: §§101–110. By section, the topics are:

- §101: Short Title of Title I
- §102: Application of Certain General Provisions to Federal Facilities
- §103: Definition of Person
- §104: Facility Environmental Assessments
- §105: Mixed Waste Inventory Reports and Plan
- §106: Public Vessels
- §107: Munitions
- §108: Federally Owned Treatment Works
- §109: Small Town Environmental Planning
- §110: Chief Financial Officer Report

Title II was very narrowly focused on the federally owned Interstate Highway 95 (I-95) sanitary landfill near Lorton, Virginia. It consisted of Sections 201–204, as described below:

- §201: Short Title of Title II
- §202: Findings
- §203: Environmental Impact Statement
- §204: Definitions

Background/History

The primary purposes of Title I of the FFCAct were to waive Federal sovereign immunity from Federal and State RCRA enforcement actions and to force resolution of two hazardous waste issues. The waiving of sovereign immunity made it possible for regulatory agencies to carry out enforcement actions as they did for non-Federal entities. The two waste issues were treatment and disposal of radioactive mixed waste (MW) at Federal (pri-marily US Department of Energy) facilities and installations, and the determination of when military munitions became solid wastes subject to RCRA regulations. The MW issue itself affects non-Federal entities, and the munitions issue also affected nonmilitary munitions, since both wastes from Federal, other governmental, and non-governmental activities are ultimately under the same regulatory programs.

The other issues addressed by Title I of the FFCAct were clarifications regarding RCRA administra-

tive requirements, RCRA Subtitle C storage requirements for public vessels, allowable discharges into Federally owned wastewater treatment plants, and Federal review and evaluation of small town environmental compliance problems. Federal employee liability under RCRA was also clarified.

Title II of the Act addressed the singular issue of the I-95 Sanitary Landfill in Lorton, Virginia, near Washington, DC.

Regulatory Impact

The FFCAct for the most part has had little direct effect on the overall technical aspects of Federal regulations. Rather, by amending RCRA, it

1) Clarified a number of Federal, State, and local jurisdictional issues

2) Mandated greater US Environmental Protection Agency (EPA) oversight of Federal facilities, and enacted certain provisions which somewhat simplified EPA's execution of its Federal facility oversight mandate

3) Mandated a major effort by the US Department of Energy (DOE) to address radioactive mixed waste treatment, storage, and disposal, to include development of current status reports, Site Treatment Plans (STPs), and annual progress reports

4) Put other Federal agencies *on notice* regarding mixed waste management violations

5) Required EPA, in concert with the US Department of Defense (DoD), to specify when munitions (and more specifically, military munitions) were to be determined to be solid wastes, so that subsequent hazardous waste determinations could be made and management requirements identified

6) Clarified certain RCRA issues regarding hazardous waste storage on *public vessels*

7) Clarified certain issues regarding discharges to Federally Owned Treatment Works

The radioactive mixed waste and munitions issues did require certain actions. The MW issue, which has been an ongoing challenge since the RCRA regulations first went into effect in 1980, continues to be addressed within the RCRA and Atomic Energy Act (AEA) framework. The phase in of the RCRA Land Disposal Restrictions (LDRs, 40 CFR 268) has required cooperation between the EPA, US Nuclear Regulatory Commission (NRC), and DOE in addressing the MW issue. The FFCAct has forced an intensive effort by DOE to develop MW treatment technology so that stored MW could be treated to meet RCRA LDRs.

The munitions issue has generated additional and specific EPA rulemaking under RCRA. This rulemaking, known as the Military Munitions Rule (MMR), was promulgated as RCRA Subtitle C regulations on February 12, 1997 (62 FR 6622). The MMR included provisions not required by the FFCAct, but which were determined to be desirable in the course of rulemaking and within the authority of existing legislative authority. These additional actions are explained in the preamble to the final rule.

In addition, EPA's MMR has had a significant effect on the development of DoD's *Range Rule*, proposed 32 CFR 178. While the Range Rule does not cite authority under the FFCAct, the MMR, which was mandated by the FFCAct, was a key determinant on formulation of regulations for the operation, transfer, and disposal of former and current military bombing and artillery ranges.

FFCAct Description by Section

A brief description of each section of the FFCAct follows in numerical order. An expanded discussion of the significance (*i.e.*, technical, programmatic, and regulatory impact) of Sections 102, 103, 104, 105, 107, and 109 then follows.

Title I. Federal Facility Compliance Act of 1992

§101: Short Title. This section simply stated the short title of the Act.

§102: Application of Certain Provisions to Federal Facilities. This section amended RCRA Section 6001, and addressed several issues of Federal sovereign immunity; Federal, State, and local jurisdiction; Federal employee personal liability protection; and Federal government entity immunity from criminal sanction. Specifically, this section

1) Amended the Solid Waste Disposal Act (RCRA) to waive the sovereign immunity of the United States for purposes of enforcing Federal, State, interstate, and local requirements with respect to solid and hazardous waste management

2) Provided that such substantive and procedural requirements include all administrative orders and civil and administrative fines

3) For acts within the scope of their duties, absolved Federal government agents, employees and officers from personal liability under Federal, State, interstate or local solid and/or hazardous waste laws

4) Made Federal employees subject to criminal sanctions under such laws

5) Forbade Federal entities from being subject to criminal sanctions under such laws

6) Provided EPA authority for commencement of enforcement authority against a Federal government department, agency, or instrumentality

7) Required fines collected by the states from the Federal government for violations of hazardous and solid waste management requirements to be used only for projects to improve or protect the environment or to defray the costs of environmental protection or enforcement

8) Provided a 3-year lead time for waiver of sovereign immunity with respect to radioactive mixed waste storage time violations at DOE facilities (subject to certain conditions)

9) Provided that the above would be effective upon the date of enactment of the Act, with specified exceptions and delayed effective dates for the waiver of sovereign immunity for violations involving the storage of certain mixed waste

10) Defined ***mixed waste*** as waste that is a mixture of hazardous wastes as defined under RCRA, and source, special nuclear, or by-product (*i.e.*, radioactive) material subject to the Atomic Energy Act (AEA) of 1954

§103: Definition of Person. Section 103 amended RCRA Section 1004(15) to add departments, agencies, and instrumentalities of the US to the definition of *person* with respect to regulated entities.

§104: Facility Environmental Assessments. This section amended RCRA Section 3007(c) to require Federal agencies that own or operate hazardous waste facilities to reimburse EPA for costs of inspections. It also directed the EPA to conduct comprehensive ground water monitoring evaluations at such facilities unless an evaluation was conducted during the 12–month period preceding this Act's enactment date (October 6, 1992). In addition, states with authorized hazardous waste programs under RCRA were authorized to conduct inspections of Federal facilities for the purpose of enforcement.

§105: Mixed Waste Inventory Reports and Plan. Section 105 amended RCRA Section 1004 and added a new RCRA Section 3021. The amendment to RCRA Section 1004 required the Comptroller General to report, 18 months after enactment of the Act, on DOE's progress in complying with the new RCRA Section 3021(b), to include:

1) DOE's progress in submitting STPs to the states and EPA, and the status of the state and EPA review

2) DOE's progress on entering into Compliance Orders

3) An evaluation of the completeness and adequacy of each plan

4) Identification of any recurring problems in the plans

5) A description of MW treatment technologies and capacities developed by DOE since FFCAct enactment, and a list, by facility, of the wastes expected to be treated

6) DOE's progress in characterizing MW streams at each site

7) Identification and analysis of additional DOE actions to

    a) Complete submission of all plans

    b) Obtain Compliance Orders

    c) Develop MW treatment technologies and capacities

The new Section 3021 required the Secretary of the Department of Energy to submit to EPA and the states two reports: (1) a report containing a national inventory, on a state-by-state basis, of all mixed wastes at DOE facilities, regardless of time of generation and (2) a report containing a national

inventory of mixed waste treatments, capacities, and technologies at each site. The reports were to be made to the EPA Administrator and the governor of each state in which the DOE generates or stores mixed wastes.

This section also required DOE to prepare a plan for each DOE facility for developing treatment capacities and technologies to treat all the facility's mixed waste, regardless of when generated, to the standards promulgated pursuant to RCRA Section 3004(m), *i.e.*, EPA's Land Disposal Restrictions (LDRs). This plan has been termed the facility's Site Treatment Plan (STP). Exempted from such requirements were facilities which produced or stored no mixed waste, and any facility subject to a permit establishing a schedule for treatment of such wastes or any existing agreement or administrative or judicial order governing the treatment of such wastes to which the state is a party.

STPs were to be approved by: (1) State regulatory officials, in the case of facilities located in states with authority to prohibit land disposal of mixed waste, and to regulate hazardous components of such waste and (2) the EPA Administrator, in the case of states without such authority. A state is authorized to waive the requirement for STP submission if the state enters into an agreement with the Secretary that addresses compliance with respect to mixed waste and issues an order requiring compliance.

It should be noted that this section of the FFCAct applies only to US Department of Energy (DOE) facilities. However, the specific requirements for radioactive mixed waste management under RCRA and the AEA apply to all generators, transporters, storage, treatment, and disposal facilities.

§106: Public Vessels. This section added a new RCRA Section 3022 to address hazardous waste management on public vessels. This amendment stated that a public vessel was not subject to the storage, manifest, inspection, or recordkeeping requirements until the waste is transferred to a shore facility. Exceptions to this provision are: (1) if waste is stored on the vessel for more than 90 days after the vessel is placed in reserve or is otherwise no longer in service or (2) if the waste is transferred to another vessel within the territorial waters of the US and is stored on that vessel for more than

90 days after the date of transfer. As defined in the US Coast Guard's 33 CFR 151.1006, *Definitions,* **public vessel** means a vessel that:

1) Is owned, or demise chartered, and operated by the United States government or a government of a foreign country; and

2) Is not engaged in commercial service

Vessel means every description of watercraft or other artificial contrivance used, or capable of being used, as a means of transportation on water.

§107: Munitions. This section amended RCRA Section 3004 by adding a new Subsection 3004(y), which required EPA to develop regulations identifying when conventional and chemical military munitions became hazardous waste under RCRA, and providing for the safe transportation and storage of such wastes. EPA was to consult with DoD and the States in rule development. The regulations were to be proposed within 6 months of FFCAct enactment, and finalized within 24 months of enactment.

§108: Federally Owned Treatment Works. This section added a new RCRA Section 3023 to establish a prohibition against the introduction of a hazardous waste into *federally owned treatment works* (FOTWs) which operate under a permit issued under Section 402 of the Federal Water Pollution Control Act (Clean Water Act). This type of permit is normally issued under the National Pollutant Discharge Elimination System (NPDES). This provision applies to treatment works belonging to a Federal department, agency, or instrumentality and which is used primarily to treat domestic sewage. The provisions of Section 108 essentially further defined acceptable materials which can be discharged to the FOTW, and specifically prohibits discharge of hazardous waste to a FOTW. This section also defines an FOTW.

§109: Small Town Environmental Planning. This section did not amend RCRA. Essentially, this section of the Act was intended to address problems faced by small towns (defined as those with populations less than 2500) in complying with Federal environmental regulations. This section required EPA to establish the Small Town Environmental Planning Program and a Small Town Environmental Planning Task Force. The

program required EPA to publish and update annually a list of requirements under Federal environmental and public health statutes. A significant aspect of this program was a mandate for EPA to evaluate and report to the Congress on the feasibility of establishing a multimedia (*e.g.,* air, water, land) permitting program for small towns.

§110: Chief Financial Officer Report. This section did not amend RCRA. It addressed Federal agencies that generate mixed waste streams. It required the Chief Financial Officer of each *affected agency* to submit an annual report to Congress detailing mixed waste compliance activities and associated fines and penalties imposed for violations involving mixed wastes. While DOE seems to be the Federal agency most affected, Section 102 of the FFCAct references separately DOE and other "department, agencies, and instrumentalities of the executive branch of the Federal government" for violations of mixed waste storage regulations under RCRA Section 3004(j).

Title II. Metropolitan Washington Waste Management Study Act

Title II dealt with the I-95 Sanitary Landfill in Lorton, Virginia, and is not of general interest or significance. No Federal regulations were affected. A brief description of the four sections follows. No further discussion is provided.

§201: Short Title of Title II. This section designated the short title.

§202: Findings. This section was a statement by Congress that the landfill was on Federal land and as such was the responsibility of the Federal government.

§203: Environmental Impact Statement. Section 203 required an Environmental Impact Statement (EIS) before expansion of the landfill. It prohibited the expansion of the landfill on lands owned by the US Government unless (1) an Environmental Impact Statement regarding such expansion has been completed and approved by the EPA Administrator and (2) the costs incurred in completing such statement are paid from the landfill's enterprise fund or in accordance with another payment formula based on jurisdictional usage of the landfill. It allowed the landfill to be expanded for purposes of the ash monofill if it was used solely for the disposal of incinerator ash from the parties of the July, 1981 Memorandum of Understanding. It also allowed the use of the monofill for solid waste disposal for a maximum of 30 days whenever a resource recovery facility or an incinerator operated by or for such parties is unavailable because of an emergency shutdown. The landfill was prohibited from receiving or disposing of municipal or industrial waste other than incinerator ash, unless the environmental impact statement and cost-sharing requirements were met.

§204: Definitions. This section contains definitions of *expansion* (of the landfill), land ownership, and a previously existing Memorandum of Understanding (MOU).

Impacts of Key Sections of the FFCAct

The following is an assessment of the technical and regulatory impacts of key sections of the FFCAct: Sections 102, 104, 105, 107, and 109.

Impact of §102 (Application of Certain Provisions to Federal Facilities)

The impact of this amendment to RCRA was to rather dramatically alter the status of all Federal facilities with respect to RCRA compliance by mandating EPA oversight and by waiving Federal sovereign immunity to prosecution. It also provided a 3-year delay in the waiver with respect to MW storage time violations under RCRA Section 3004(j).

Impact of §104 (Facility Environmental Assessments)

This section focused sudden attention of Federal facilities with respect to RCRA. It specifically required a groundwater quality investigation at each Federal facility within 12 months of FFCAct enactment, unless an investigation had been performed within the previous 12-month period. The Act also required Federal facilities to reimburse EPA for inspection costs.

In the several years following enactment, activity at EPA Regional Offices was redirected to perform inspections of (primarily) DoD and DOE facilities, since such facilities typically were both large in areal extent, used large quantities of hazardous materials, and were substantial hazardous waste generators.

Impact of §105 (Mixed Waste Inventory Reports and Plan)

Section 105 has had a substantial and far-reaching effect on DOE mixed waste management activities. The Mixed Waste Inventory Reports and the Site Treatment Plans required a substantial effort to prepare and review by the affected parties, as discussed below. The MW treatment technology development is an ongoing activity which, in itself, constitutes a very large research and development program.

DOE possesses approximately 98% of the mixed waste in the United States. Of this, approximately 50% is stored at the Hanford site in Washington State. Approximately 72% of this mixed waste is High Level Waste (HLW), 20% low level waste (LLW), and 8% mixed transuranic (MTRU) waste. DOE's Waste Isolation Pilot Plant (WIPP) in New Mexico was designed to accept transuranic (TRU) and MTRU waste only. EPA published its certification of WIPP in the *Federal Register* on May 18, 1998 (63 FR 27354). Because the WIPP was certified as a *no migration* facility, RCRA LDRs do not apply. ***No migration*** means that waste constituents will not migrate from the facility; *i.e.*, there is no migration pathway.

Mixed Waste Inventory Reports. The waste inventory report was required to contain a detailed state-by-state inventory of mixed waste, regardless of the time of generation. The inventory had the following specific requirements:

1) A detailed description of each mixed waste

2) The amount of each type of mixed waste, differentiated by MW subject to and MW not subject to RCRA LDRs

3) An estimate of each MW type expected to be generated during the following five years

4) DOE waste minimization activities implemented

5) EPA hazardous waste code for each MW

6) An inventory of all noncharacterized MW

7) Basis of each hazardous waste code determination

8) Description of each waste source

9) LDR treatment technology specified for each MW

10) A statement regarding the affect of radionuclide content on the treatment technologies

Waste Treatment Capacities and Technologies Report. This report was required to include the following:

1) Estimate of treatment capacity for each inventoried waste based on existing treatment technology

2) A description of each treatment unit identified for the above

3) A description of existing treatment units not included, and the reasons they were not included

4) A description of proposed treatment units

5) Information needed to address treatment of waste for which no treatment technology existed

DOE Site Treatment Plans. The FFCAct amended RCRA Section 3021(b) to require DOE to prepare Site Treatment Plans for developing treatment capacities and technologies for mixed waste at each site where DOE stores or generates mixed waste. As used here, *mixed waste* is specifically a mixture of radioactive and RCRA hazardous wastes. Specifically, as defined in the FFCAct, MW is "waste that contains both hazardous waste and source, special nuclear, or by-product material subject to the Atomic Energy Act of 1954."

In 1995, DOE submitted 40 proposed STPs for sites in 20 states. In fiscal year 1997, the number of STPs required had dropped to 38; of these, 37 had been approved by the end of the fiscal year. The remaining STP has been approved in principle at the last site, Argonne National Laboratory–East

in Illinois. However, the State of Illinois placed a low priority of activity on this site, and is taking no further action to issue a Compliance Order, according to DOE.

The STPs were required to contain:

1) For MWs for which treatment technologies existed, a schedule for submitting all applicable permit applications, entering into contracts, initiating construction, *etc.*, and processing backlogged and currently generated MW

2) For MW with no existing treatment technology, a schedule for identifying and developing such technologies

3) An estimate of waste volumes that would exist based on various radionuclide separation technologies

An annual progress report for the STPs was also required from the DOE Secretary for three consecutive years after the Act passage (*i.e.*, October 1993, October 1994, and October 1995); these were submitted.

STPs were developed to enable DOE to meet RCRA LDR requirements. STPs address the treatment aspects of mixed waste management, including how, where, and when mixed waste is treated to meet LDRs.

After consultation with EPA and State regulators, DOE published its plan to submit the STPs in three stages (58 FR 17875, April 6, 1993). In Stage 1, conceptual STPs were submitted in October 1993, which described a wide range of possible treatment technologies for each mixed waste at each site. From these, Draft Site Treatment Plans (Stage 2) were submitted for review (59 FR 44979, August 31, 1994). DOE planned to submit the revised Proposed Plans (Stage 3) in February, 1995 (60 FR 10840, February 28, 1995), but issue was delayed until April 5, 1995 (60 FR 17346). A total of 37 proposed plans were prepared to address 40 sites in 20 states.

Radioactive Mixed Waste Regulatory History. Radioactive mixed wastes are problematic in that they can only be stored, treated, or disposed at facilities which are permitted for both hazardous and radioactive wastes. There are few such facilities. The preferred approach for managing existing wastes is to treat the wastes so that they are either hazardous, or radioactive. Pollution

Prevention (P2) initiatives are developed and employed to avoid generation of radioactive mixed wastes. Common P2 approaches include material substitution to reduce use of hazardous materials and/or reduce generation of hazardous wastes, and process modifications to reduce or eliminate wastes. Waste minimization techniques are employed to manage the wastes that are generated after P2 efforts have been employed.

MWs are wastes which are hazardous wastes under RCRA and radioactive wastes under the Atomic Energy Act (AEA). MWs are generated in particular by research facilities, medical facilities, laboratories, nuclear power plants, and nuclear weapons program activities.

There are three categories of MW: (1) high-level waste (HLW), (2) mixed transuranic waste (MTRU), and (3) low-level mixed waste (LLMW).

The regulatory history of MW begins with the passage of RCRA in 1976, with the systematic *cradle-to-grave* regulatory program for hazardous waste, as such wastes were defined at that time. Prior to that time, MWs were managed as radioactive wastes with varying degrees of success by various generators. However, RCRA specifically excluded source, special nuclear, and by-product material as defined in the Atomic Energy Act (AEA) of 1954. This effectively prevented these types of radioactive materials from being defined as solid waste under RCRA. Since hazardous wastes are a subset of solid wastes under RCRA, these wastes were not regulated under RCRA. They were, however, regulated by both the Nuclear Regulatory Commission (NRC) for commercial activities and DOE (Federal nuclear weapons related activities) under the AEA.

In 1981, NRC recognized the need for joint regulation of mixed waste under the AEA and RCRA while in the process of promulgating land disposal regulations for low-level radioactive waste (10 CFR 61). At the same time, DOE maintained that RCRA did not apply to DOE activities. This issue was resolved in Federal court in 1984 with the decision that RCRA did indeed apply to DOE activities.

Also in 1984, the Hazardous and Solid Waste Amendments (HSWA) to RCRA were enacted, which set the stage for more stringent land disposal regulations (*e.g.*, LDRs and treatment standards).

In 1986, the EPA published notice that it intended to exercise regulatory authority over the hazardous constituents of mixed waste. In 1986, EPA also promulgated the Land Disposal Restrictions for what were called the *California List* wastes.

In 1987, NRC and EPA published joint draft guidance on the definition and identification of LLMW. Also in 1987, the NRC and EPA published joint siting guidelines for disposal of LLMW and a conceptual design approach for commercial LLMW disposal facilities, and joint guidance on mixed waste storage (60 FR 40204, August 7, 1987).

From 1988 to 1992, consolidation of efforts continued between the three Federal agencies in regulation of MW. EPA had also banned land disposal of MW under its RCRA authority, if such MW exceeded LDRs.

In 1992, the FFCAct was enacted. Among other things, the definition of mixed waste was established for RCRA, and EPA was mandated to regulate mixed waste activities. At the same time, DOE was given the task of spearheading MW treatment technology.

From 1994 to the present, progress has been made in both the treatment of existing (stored) MW and in the reduction or elimination of generation of MW (pollution prevention). The EPA and NRC published joint guidance on the storage of LLMW.

In 1997, the EPA and NRC published joint guidance on testing requirements for MW (62 FR 62079, November 20, 1997).

In 1998, EPA certified DOE's WIPP (as codified in 40 CFR 194) for disposal of TRU and MTRU wastes. EPA also renewed its enforcement policy on MW storage violations under RCRA §3004(j), as progress continued to be made on treatment technologies.

EPA activity on a mixed waste disposal rule is progressing concurrently with EPA's Hazardous Waste Identification Rule (HWIR). MW is being addressed as a special hazardous waste identification issue in order to focus on the radioactive aspect of MW, with which relatively few hazardous generators have to deal. The schedule is to propose the rule by October 31, 1999, and finalize by April 30, 2001.

Note: Radioactive mixed wastes also exist which contain toxic substances (*e.g.*, polychlorinated biphenyls–PCBs) regulated under the Toxic Substances Control Act (TSCA) of 1976. Such waste is also subject to both AEA and TSCA regulations, and potentially RCRA LDRs. In cases where all three regulations apply, the most stringent requirements of each will normally apply.

Related DOE Initiatives. Concurrent with actions to achieve compliance with RCRA Subtitle C requirements, DOE has initiated two related actions: a Waste Management Programmatic Environmental Impact Statement (PEIS), and the Baseline Environmental Management Report (BEMR). The PEIS was begun to address MW treatment and disposal National Environmental Policy Act (NEPA) issues. The BEMR is a congressional budgetary requirement and addresses DOE (including FFCAct) budgetary issues.

Related EPA Initiatives. EPA has been working closely with NRC on the commercial MW disposal issue since the 1970s and with DOE since the early 1980's. The LDR requirement mandated by HSWA has added additional restrictions for land disposal of hazardous waste, and so mixed waste. Cur-

rently, EPA is pursuing a MW management rule (concurrently with the HWIR) which should simplify HW disposal requirements.

Impact of §107 (Munitions)

The primary regulatory action implemented pursuant to this section has been EPA's promulgation under RCRA Subtitle C at 40 CFR 260–266 and 270 of what became known as the Military Munitions Rule (MMR). This rule addresses both conventional and chemical military munitions, as required by the FFCAct. In addition, both military and nonmilitary conventional munitions are addressed by the MMR under other RCRA authority with respect to emergency response requirements.

Table 1 is a synopsis of RCRA regulatory changes pursuant to promulgation of the MMR.

EPA's Military Munitions Rule. The primary regulatory action implemented pursuant to this section has been EPA's promulgation under RCRA Subtitle C at 40 CFR 260–266 and 270 of what

Table 1. RCRA Regulatory Changes Pursuant to Promulgation of the MMR

Reference	RCRA Regulatory Change
40 CFR 260	Addition to 40 CFR 260.10 of definitions related to explosives
40 CFR 261	Addition of 40 CFR 261.2(a) (2) (iv): "A military munition identified as a solid waste in 40 CFR 26.202."
40 CFR 262	Exemption under 40 CFR 262.10 and 262.20 for emergency response to an explosives emergency
40 CFR 263	Exemption under 40 CFR 263.10 to transportation requirements for response to explosives emergencies
40 CFR 264	Exemptions under 40 CFR 264.1 and 264.70 for explosives emergency response
	New Subpart EE—Hazardous Waste Munitions and Explosives Storage as 40 CFR 264.1200-264.1202 (alternatives to general storage requirements)
40 CFR 265	Exemptions under 40 CFR 265.1 and 265.70 for explosives emergency response
	New Subpart EE—Hazardous Waste Munitions and Explosives Storage as 40 CFR 265.1200-265.1202 (alternatives to general storage requirements)
40 CFR 266	New Subpart M—Military Munitions as 40 CFR 266.200-266.206, which provide for identification of munitions as solid waste and, if hazardous under 40 CFR 261, applicable management standards as alternatives to 40 CFR 264 and 265 requirements
40 CFR 270	Conditional exemption for explosives emergency response under new 40 CFR 270.1(c)(3)(i)(D) and 270.1(c)(3)(iii)
	Permit modification for addition munitions waste acceptance under 40 CFR 270.42(h)

became known as the Military Munitions Rule. This rule is a subset of the rulemaking activity which is called the Hazardous Waste Identification Rule (HWIR). The MMR addresses both conventional and chemical military munitions, as required by the FFCAct. In addition, both military and nonmilitary munitions are addressed by the MMR under other RCRA authority with respect to emergency response requirements. Nuclear weapons and components were not subject to this action; they are regulated by DOE and DoD under (primarily) AEA authority.

As stated in the preamble to the Final Rule (62 FR 6622), the final rule established a process for identification of munitions as solid waste, and then further classification as either hazardous or non-hazardous solid waste. The rule also provides for safe transport and storage of hazardous waste munitions, and makes provisions for conditional exemptions from RCRA regulation for emergency response. The MMR allows for storage under either RCRA regulations or in accordance with the Department of Defense Explosive Safety Board (DDESB) requirements as specified in DoD 6055.9–STD–DoD Ammunition and Explosives Safety Standards.

Three categories of munitions (conventional and/ or chemical) are addressed by EPA's MMR:

1) Unused munitions

2) Munitions being used for their intended purpose

3) Used or fired munitions

Military munitions are now solid wastes when *being used for their intended purpose.* Conditional exemptions exist for RCRA manifesting and marking requirements for shipments from one DoD Treatment, Storage, and/or Disposal Facility (TSDF) to another, and for RCRA Subtitle C storage requirements for storage under Department of Defense Environmental Safety Board regulations.

The MMR also clarified that chemical munitions are not subject to RCRA Section 3004(j) storage prohibitions codified at 40 CFR 268.50. Such munitions are subject more specifically to a series of other management and disposal requirements specifically mandated by Congress beginning in 1985. A chemical munitions destruction program is in progress at all DoD storage sites. However, other RCRA Subtitle C requirements apply unless specifically superseded.

In promulgating the MMR, EPA included non-military munitions in most of the regulatory changes. Only 40 CFR 266, Subpart M is exclusively applicable to military munitions. It should also be noted that 40 CFR 266, Subpart M provides management standards tailored to DDESB standards, while 40 CFR 264/265, Subpart EE standards are applicable to military and non-military parties. The Subpart EE standards are also tailored alternative standards; other standards in 40 CFR 264 and 265 may be applied as appropriate, as specified in facility permits.

DoD's Range Rule. One reason the proposal and promulgation of EPA's MMR was delayed was to allow for close coordination between EPA and DoD, the entity which was primarily affected. The MMR has had a significant effect on the development of the Range Rule, since the MMR determined munitions waste management. DoD's long record of safe management of military munitions provided much useful input into EPA's rulemaking process. In fact, DoD's Explosives Safety Board (DDESB) had long-established standards in place, which served as both technical input and an alternative for EPA's rules, as discussed above. Terminology, definitions, and intent are consistent between the MMR and the Range Rule.

The Range Rule was proposed as 32 CFR 178, "Closed, Transferred, and Transferring Ranges Containing Military Munitions," on September 26, 1997 (62 FR 50796) under authority of the Defense Environmental Restoration Program, the DDESB, the Comprehensive Environmental Response and Compensation Act of 1980 (CERCLA) Section 104, as amended, and Executive Order 12580. The Range Rule was designed to be compatible with the response actions taken under CERCLA. It contains a five-part process which is not inconsistent with CERCLA, and is tailored to the special risks posed by military munitions and military ranges. DoD has five categories of ranges: Active, Inactive, Closed, Transferred, and Transferring. The Range Rule does not apply to Active and Inactive ranges.

While the Range Rule does not have authority under the FFCAct, the EPA's MMR, which was mandated by the FFCAct, was a key determinant on formulation of regulations for the operation, transfer, and disposal of former and current military bombing and artillery ranges.

Impact of §109 (Small Town Environmental Planning)

The Small Town Task Force (STTF) was created by EPA in 1992 as mandated by Congress. The purpose of the task force was to advise EPA on how to work better with small communities in order to improve compliance with environmental regulations. The STTF was presented with a challenging scope of work by Congress. The legislation identified five specific responsibilities. These were to

1) Identify regulations developed pursuant to Federal environmental laws, which pose significant compliance problems for small towns

2) Identify means to improve the working relationship between the Environmental Protection Agency and small towns

3) Review proposed regulations for the protection of the environmental and public health, and suggest revisions that could improve the ability of small towns to comply with such regulations

4) Identify means to promote regionalization of environmental treatment systems and infrastructure serving small towns, to improve the economic condition of such systems and infrastructure

5) Provide such other assistance to the Administrator as the Administrator deems appropriate.

The Task Force consisted of 14 members from towns with populations under 2,500. The STTF met six times between enactment of the legislation in October, 1992 and June, 1996. The STTF issued its draft report in August, 1994 and the final report in June, 1996. The report contained 39 recommendations, which include continued small town advisory work with EPA, establishment of a small town ombudsman, increased technical support, regulatory changes, increased flexibility, small town involvement in regulation development,

increasing financial resources for small towns, interagency coordination of environmental programs, requirements and processes and recommendations requiring Congressional action.

The Small Community Advisory Committee (SCAC) of the Local Government Advisory Committee (LGAC) took over the role of the STTF following submission of the STTF's Final Report in 1996. The LGAC operates under the Office of Congressional and Intergovernmental Relations (OCIR), which serves as EPA's principal liaison with the State and local government officials and the organizations which represent them. This office consolidates Section 109 requirements and those of other legislative statutes and Executive Orders with respect to Federal-State-local government interaction and cooperative efforts.

The Environmental Council of the States (ECOS), which was founded in December, 1993, has teamed up with US EPA's Office of State and Local Relations in a cooperative agreement aimed at improving how small towns and communities deal with environmental matters.

The ECOS/EPA Small Towns Project follows up on the work completed by the STTF. The primary goal of the project is to work towards the implementation of a few key recommendations presented by the STTF. In 1998, the major activities of the ECOS/EPA partnership were the National Environmental Performance Partnership System (NEPPS), under which a state and EPA negotiate a Performance Partnership Agreement (PPA) which lays out what each partner will do in the coming year. The PPA is a key element in the continuing process of devolving enforcement authority from the Federal to the State governments.

In essence, the original EPA Headquarters–STTF relationship has been merged into the overall and evolving relationship which involves EPA Headquarters, EPA Regions, states, local governments, and small communities. As the trend toward greater State responsibility for both Federal and State environmental protection and management programs continues, the unique problems faced by small communities are increasingly addressed at the State and local (city, county) level.

Bibliography

Laws

"Comprehensive Environmental Response, Compensation and Liability Act of 1980, as amended." PL 96–510, 94 *Statutes* 2767. *US Code.* Title 42, Sec. 9601–9675.

"Federal Facility Compliance Act of 1992." PL 102–386, 106 *Statutes* 1505. October 6, 1992.

"Resource Conservation and Recovery Act of 1976, as amended." PL 94–580. 90 *Statutes* 2795. *US Code.* Title 42, Sec. 6901–6992k.

Hazardous Waste Management

"EPA Administered Permit Programs: The Hazardous Waste Permit Program." *Code of Federal Regulations.* Title 40, Pt. 270.

"Hazardous Waste Management System. General." *Code of Federal Regulations.* Title 40, Pt. 260.

"Identification and Listing of Hazardous Wastes." *Code of Federal Regulations.* Title 40, Pt. 261.

"Interim Status Standards for Owners and Operators of Hazardous Waste Treatment, Storage and Disposal Facilities." *Code of Federal Regulations.* Title 40, Pt. 265.

"Procedures for Decisionmaking." *Code of Federal Regulations.* Title 40, Pt. 124.

"Standards Applicable to Generators of Hazardous Waste." *Code of Federal Regulations.* Title 40, Pt. 262.

"Standards Applicable to Transporters of Hazardous Waste." *Code of Federal Regulations.* Title 40, Pt. 263.

"Standards for Owners and Operators of Hazardous Waste Treatment, Storage, and Disposal Facilities." *Code of Federal Regulations.* Title 40, Pt. 264.

"Standards for the Management of Specific Hazardous Wastes and Specific Types of Hazardous Waste Management Facilities." *Code of Federal Regulations.* Title 40, Pt. 266.

Military Munitions

"Closed, Transferred, and Transferring Ranges Containing Military Munitions; Proposed Rule. (32 CFR 178.)" *Federal Register* 62 (26 September 1997): 50796–50843.

"Military Munitions Rule: Hazardous Waste Identification and Management; Explosives Emergencies, Manifest Exemption for Transport of Hazardous Waste on Right-of-Ways on Contiguous Properties, Final Rule. (40 CFR Parts 260, 261, 262, 263, 264, 265, 266, and 270.)" *Federal Register* 62 (12 February 1997): 6622–6657.

Mixed Waste

"Extension of the Policy on Enforcement of RCRA Section 3004(j) Storage Prohibitions at Facilities Generating Mixed Radioactive/Hazardous Waste." *Federal Register* 61 (26 April 1996): 18588–18592.

"Extension of the Policy on Enforcement of RCRA Section 3004(j) Storage Prohibition at Facilities Generating Mixed Radioactive/Hazardous Waste." *Federal Register* 63 (9 April 1998): 17414.

"Extension of the Policy on Enforcement of RCRA Section 3004(j) Storage Prohibitions at Facilities Generating Mixed Radioactive/Hazardous Waste." *Federal Register* 63 (6 November 1998): 59989–59992.

"Joint Nuclear Regulatory Commission/Environmental Protection Agency Guidance on the Storage of Mixed Radioactive and Hazardous Waste." *Federal Register* 60 (7 August 1995): 40204–40211.

"Joint Nuclear Regulatory Commission/Environmental Protection Agency Guidance on Testing Requirements for Mixed Radioactive and Hazardous Waste." *Federal Register* 62 (20 November 1997): 462079–62094.

"Office of Environmental Management Proposed Site Treatment Plan. Notice of Availability," *Federal Register* 60 (5 April 1995): 17346–17349.

Harry A. Bryson is a CHMM at the Master's level and a Senior Environmental Scientist with Earth Technologies, Inc. He is currently on assignment from Earth Technologies' Oak Ridge, Tennessee office to the US Army Kwajalein Atoll installation in the Republic of the Marshall Islands, where he provides support to the Army's Environmental Compliance Office in its oversight of the O & M Contractor's operations. He has 16 years experience in environmental regulatory compliance, and 26 years experience in military flight operations and emergency planning. He has Master's degrees in Environmental Engineering (University of Tennessee, Knoxville; 1984), and Biology (Butler University, Indianapolis; 1979) and Bachelor of Science degrees in Engineering Physics (University of Tennessee, Knoxville; 1981) and Life Sciences (USAF Academy, Colorado; 1971).

CHAPTER 4

Environmental Management Systems– ISO 14001

Harry S. Kemp, CHMM
Steve Rowley, CHMM
Ed Pinero, CPG

Introduction

Environmental management was the sole purview of the United States Environmental Protection Agency (EPA) throughout the 1970s and 1980s. This was a transition time when the emphasis was the use of legislation and regulation to correct past problems and practices by dictating *end of the pipeline* solutions. In the late 1980s this approach began to give way to legislative emphasis (without threat of the big stick) on the promotion of lessening the environmental impact of operations through waste minimization, implementation of best management practices, and pollution prevention.

Some forward-thinking companies, realizing that they could no longer leave management of their environmental affairs up to chance (it just wasn't good for business) made the decision to apply the proper resources to ensure the company would operate in compliance with the regulations. Environmental managers were placed in positions of authority, and procedures were developed that

45

created systems so companies consistently complied with the regulations. Over time the systems approach to compliance by larger companies has begun to take hold with smaller companies too.

Now there is a new development in the evolution of environmental management where rather than simply respond to regulation through compliance programs, companies are taking the initiative to identify all of the impacts on the environment from their operations, whether regulated or not, and are setting objectives and developing programs to consciously reduce these impacts. This effort goes beyond compliance and brings more power to industry in proactively charting the course of environmental management, rather than leaving it up to the government.

What Is an Environmental Management System?

Implementing a well-thought-out **Environmental Management System** (EMS) is probably the most important part of providing your organization with the structure to sustain continuous improvement in environmental performance while producing a positive impact on the bottom line. Every organization that has implemented policies, procedures, and/or work instructions to ensure some level of environmental management already has an EMS. The policies, procedures, and/or work instructions may be well documented, or they may not be documented at all, but nonetheless they exist and are a part of operations. These can be used to provide a basis for developing a formal EMS.

After the establishment of the progressive income tax during Woodrow Wilson's first term as President, business organizations were forced to upgrade and coordinate their existing accounting and reporting policies, procedures, and work instructions to ensure that proper record of financial activities would be created. Over the years, this has developed into what any businessperson today knows as an accounting and financial reporting system. Accounting and financial systems have become an everyday part

of business. They help to keep business owners and management out of trouble with the government, while generating information useful to management in making sound business decisions.

What is a system? *System* comes from the Greek word *sustema* which is defined as *a combined or organized whole from many parts.* In the accounting and financial reporting example above, at some point it becomes advantageous to bring all of the separate policies, procedure, and work instructions under one umbrella so that they are working together in concert to achieve a stated goal. In other words the accounting and financial reporting system is made up of multiple policies, procedures, rules, and work instructions, that when properly applied and coordinated by a system will achieve a desired result in a reliable manner. Exactly the same principle applies to developing an EMS.

What an Environmental Management System Is, and Is Not

An EMS is not a guarantee that your company will be in compliance. It does not guarantee that Greenpeace will never bother you. It is not a guarantee that *60 Minutes, Dateline, Prime Time Live,* or CNN will not visit you. It is not a guarantee that you will have optimal environmental performance. It does not guarantee you will have no regulatory visits. It provides no guarantee of increased market share or improved community image. It is not environmental Nirvana, Shangri La, Holy Grail, or Paradise.

What Is an Environmental Management System?
It is the organizational structure, planning activities, responsibilities, practices, procedures, processes, and resources for developing, implementing, maintaining, reviewing, and correcting/ improving an organization's approach to environmental management. An EMS is simply how an organization manages its environmental affairs and requirements. The system itself is the framework within which this occurs. Even though environmental management can vary from site to site, taking the system approach provides a role for all employees in ensuring that the company improves its environmental performance.

Benefits

We are all environmentalists, and we are all concerned about protecting the environment for future generations; but in business, changes in operations, which implementation of an EMS clearly is, have to make sense to the bottom line. Besides the personal satisfaction that employees and business owners will gain from knowing that they are doing the right thing, there are many benefits to the development and implementation of an EMS. These benefits can be broadly grouped into three basic categories:

1) Cost savings

2) Compliance

3) Marketing opportunity

Installing an EMS in your organization will usually be a sizable investment in time and money, but there are payoffs. For organizations in business for the long run it will be worth the effort.

Marketing. As consumers in the United States, and the world, have become more aware of the historical impact of manufacturing on our environment, and the importance of protecting our environment, they have begun to consider the *greenness* of products. In other words, they want to know the impact your product has on the health of the environment. This is beginning to affect purchasing decisions by end users and industry as well. As this trend continues, being able to demonstrate that your company has a robust environmental management system can make you more competitive through product acceptance. The word *robust* is used in the sense that the system will continue to operate well regardless of whether or not there is a change in management, and that the system is part of the company and not just there because one or two people make it work.

Our country's current environmental policy requires strict government enforcement to work. In the European communities, and Asia to some extent, enforcement is almost nonexistent. There are a lot of laws and regulations but industry is not afraid of the government policing them. The regulator is the general community itself. Companies are literally terrified that the public will think that they do not have a good environmental record. They fear that this might result in a boycott of their products and bad press.

Our society is moving in this direction with the strengthening of the *grassroots* environmental movements. A robust environmental management system will go a long way to promote good community relations which can then be used to market how friendly your company and its products are to the environment.

The mere fact that you have an environmental management system, particularly one as broadly advertised as ISO 14001, can be used as a marketing tool to promote your company and products.

It is anticipated that there will come a time when any company wanting to do business overseas, particularly in Europe and the Pacific Rim, will have to be ISO 14001–certified. Many of our current trading partners have been more attuned to the need for good environmental stewardship because of cultural differences, and the demands of scarce resources and limited space.

Developing nations in the Pacific Rim are very much aware of how we have damaged our environment, and we are now paying billions of dollars cleaning it up. The motivation of these nations to develop a culture of environmental stewardship is to protect their undamaged resources, and not make the mistakes that we did. This attitude of environmental stewardship will surely create a bias in the minds of the consumer in these societies. Certification to ISO 14001 should eliminate any barriers put up between our foreign trading partners due to concerns about environmental management.

Compliance. The ISO 14001 standard does not say anywhere in it that you must be in regulatory compliance. In fact, it is quite feasible that under any EMS your company will be out of compliance from time to time. Having an EMS really only means that you have committed to complying with applicable environmental regulatory requirements. It doesn't in and of itself ensure regulatory compliance. This type of commitment can, and should, help you to develop a relationship of trust between a company and the environmental regulatory agencies. On the surface we all know that perception is important to the successful conclusion of an inspection by a regulatory agency.

A well-documented environmental management system can create a positive perception in the minds of the regulators, which should lessen the impact of their oversight requirement. In fact several states are already pursuing the establishment of policies, procedures, and programs whereby organizations adopting qualifying EMSs can gain consideration. The Pennsylvania Department of Environmental Protection has been one of the leading states in this effort. They will consider granting site-specific regulatory relief in exchange for imposing standard requirements on the company. The state's minimum requirements are that the environmental management system has third-party registration, and that the system be robust.

Cost Savings. The history of environmental management is a study of reactivity. A new regulation is promulgated and everyone reacts to get into compliance. A company is inspected by the regulators and it reacts to the findings. A reporting deadline is getting close and the company reacts by scrambling to collect data and organize it into the required format. Living in a reactive mode is economically inefficient because it usually means damage control. The damage is already done, now you have to pay to recover.

Managing your environmental affairs by a proactive system like ISO 14001 means that your operation will comply with the regulatory requirements as a natural course of business. Having to research and pull together data at year-end in order to submit a required report is disruptive. The quality of the data is also questionable because it was not generated with the reporting need in mind. It was probably generated for the needs of purchasing, or operations. The environmental manager will usually have to massage the figures in order to complete the report. This is an inefficient and costly way to manage your environmental affairs.

Missing deadlines, or maybe not even knowing about them, can also be expensive. Having a well-thought-out EMS will provide you with timely information to use in managing your environmental affairs. There is money to be saved by managing your environmental affairs as a business function, and not as a necessary evil.

The area of greatest potential for a tangible effect on the bottom line is by putting environmental responsibility into the hands of all employees. With this increased responsibility comes increased awareness, and subsequently, performance improvements. Once empowered, the people who actually do the work can and often do come up with ways to save money by changing work processes in ways that benefit the environment.

The standard requires that prevention of pollution be part of the organization's operations. In today's regulatory environment, business has to pay for the privilege of polluting. It doesn't take a genius to understand that if you do not create pollution, then you will not have to pay for it. There are many potential opportunities to improve operations so that resources are not wasted, hard costs are not created due to pollution generation, and potential costs due to liability risks are minimized.

Integration with Other Systems

Integration of management systems allows an organization to take advantage of shared concepts.

ISO 14000 and ISO 9000. ISO 9000 was readily grasped by industry as a mechanism for managing product quality. Accordingly, the technical committee (TC) responsible for the development of the quality standard provided guidance to TC 207, which was tasked with the responsibility for the development of the environmental management standard. ISO 9000 and ISO 14000's greatest similarities lie in the fact that they are both based on the quality management cycle concept of *plan, do, check, and improve*. Additionally, the standards share several common elements, the most similar of which include

- The development of a policy
- Document control
- Corrective and preventive action
- EMS and quality audit
- Management review

There are other common elements which exist between the two standards but are not noted here; however, they can be referenced in Tables B.1 and B.2 of the standard.

As a result of these areas of overlap, there are certain efficiencies with regard to implementing and maintaining an organization's Quality Management System (QMS) and EMS. Some may choose to integrate the two systems by merging quality control documentation, records, training, and auditing into one system. A more common approach is to harmonize the two systems. As an example, the same procedures for document control, corrective action, and recordkeeping are used in their entirety or just used as the basis for the creation of EMS-specific procedures. In either case, documentation and records are maintained separately and auditing is conducted independently using a different set of auditors who have received specific EMS training. In all cases, the culture and skills that are developed through system implementation and maintenance are shared and utilized.

With regard to this discussion, ISO 9000 is currently under its normal revision cycle and is being revised to be more compatible with ISO 14001. ISO 9000:2000 is available in draft form and should be final in November, 2000. The purpose of this effort is to facilitate efficient implementation and maintenance of both systems while allowing third-party auditors to combine the audits for cost effectiveness.

ISO 14000 and Health and Safety. There is currently no existing ISO standard for the management of issues of occupational health and safety because efforts to prepare such a standard have not been successful. ISO 14001 deliberately excludes issues of occupational health and safety from the scope of the management system. However, there is a fine line between managing environmental issues and health and safety, and in some cases there is actual crossover. Furthermore, in most facilities, the environmental manager has also been tasked with managing health and safety.

Therefore, it comes as no surprise that many organizations who have implemented successful management systems, whether a Quality Management System (QMS) or EMS, have realized the benefits that such a system can provide and have created their own health and safety management systems or have incorporated health and safety issues into their existing EMS. In addition, organizations such as the American Industrial Hygiene Association, the British Standards Institute (BS 8800), and the Occupational Safety and Health Administration have created guidance for the creation of systems of occupational health and safety management. These management systems were prepared using the same concepts used for the development of the ISO 14001 standard.

ISO 14000 Background

ISO 14000 is a series of standards which were developed by the International Organization for Standardization (ISO) and includes nearly 30 standards covering topics such as product labeling, environmental life cycle assessment, environmental performance evaluation, systems auditing, and the elements of an Environmental Management System described in ISO 14001. ISO 14001 is the specification standard against which an EMS is prepared and audited.

ISO, which is composed of national standards bodies from well over 100 countries, was founded in 1947 for the purposes of developing product standards as a means of facilitating the international exchange of goods and services. The American National Standards Institute (ANSI) is the United States representative to ISO. The standards developed by ISO remained focused on product specifications until the focus was shifted toward management systems with the advent of the hugely popular ISO 9000 quality management and quality assurance standards.

The world had identified the need for the development of worldwide standards governing and defining environmental management dating back to the environmental movement of the early 1970s. The precursor to ISO 14001, and the basis upon which the standard was created, was the British Standard (BS) 7750, which was published in 1992. Also in 1992, the European Union developed the Eco-Management and Audit Scheme (EMAS). Direct similarities can be drawn between the requirements of ISO 14001 and EMAS; however, EMAS is more of a compliance tool than a management system tool.

While the British standard has been dropped for the International Standard, EMAS is a regulation that is still a requirement for industry. TC 207 was created to develop a management system standard and other guidance documents focused on the environment. The final text of the ISO 14001 international standard was published in September 1996. As with all ISO standards, these are voluntary standards; however, some countries and industries have adopted them as requirements of doing business.

Elements of the ISO 14001 Standard

In order to effectively manage an ISO 14001 EMS, it is important to have an understanding of the standard's requirements. A quick review of the standard shows that it is structured following the *Plan, Do, Check, Improve* philosophy of the Total Quality Management movement, as follows:

Plan
4.2   Policy
4.3   Planning

Do
4.4   Implementation and Operation

Check
4.5   Checking and Corrective Action

Improve
4.6   Management Review

Within these five elements are 17 subelements stating the various requirements.

4.2   Policy

4.3   Planning
    4.3.1   Environmental Aspects
    4.3.2   Legal and Other Requirements
    4.3.3   Objectives and Targets
    4.3.4   Environmental Management Programs

4.4   Implementation and Operation
    4.4.1   Structure and Responsibility
    4.4.2   Training Awareness and Competence
    4.4.3   Communications
    4.4.4   EMS Documentation
    4.4.5   Document Control
    4.4.6   Operation Control
    4.4.7   Emergency Planning and Response

4.5   Checking and Corrective Action
    4.5.1   Monitoring and Measurement
    4.5.2   Nonconformance, Corrective, and Preventive Action
    4.5.3   Records
    4.5.4   EMS Audit

4.6   Management Review

Within these 17 subelements are all of the requirements, or *shalls*, necessary to conform to ISO 14001. There is no substitute for reading the standard in terms of recognizing the requirements. As a matter of fact, no auditor should embark on an audit without having easily available the criteria to which they are doing the audit. However, below we briefly summarize the key points of the subelements. This summary is not intended to be a replacement for ISO 14001, and should not be used exclusively as such during an audit.

Detailed Description of the Elements

4.2 Policy. ISO 14001 requires that the organization have a policy statement to drive the EMS. These tend to be short (one page or less) documents, and simply affirm the commitments of the organization. There is no expectation that these commitments go beyond the requirements of the standard, but the organization may commit to anything it chooses to. It is important not to commit to anything the organization is not truly capable of, or willing to achieve, as the system will hold the organization accountable.

There is also no expectation for detailed descriptions of the organization's commitments. For example, the commitment to prevention of pollution can be stated simply by saying "We are committed to prevention of pollution". The policy must be clearly endorsed by top management and be available to the public and employees. Although the availability to the public can be rather passive (*i.e., it is here if they want it*), there is an expectation that communicating it to all employees will involve proactively developing their awareness. Section 4.2 of ISO 14001 lists the other requirements of the policy.

4.3.1 Environmental Aspects. This element requires a procedure that not only identifies the aspects and impacts, but also provides for determination of significance, and keeping the information up-to-date. ISO 14001 does not prescribe what aspects should be significant, or even how to determine significance. However, it is expected the organization will develop a consistent and verifiable process to do so.

4.3.2 Legal and Other Requirements. This is a requirement for a procedure that explains how the organization obtains information regarding its legal and other requirements, and makes that information known to key functions. This is not the assessment or compliance audit requirement, but rather a more up-front determination of requirements.

4.3.3 Objectives and Targets. There is no requirement for a procedure in this element, only that objectives and targets be documented. It does require that certain items be considered in developing the objectives, such as legal requirements and prevention of pollution. It is sometimes easiest to develop a procedure for this element to be able to verify that these considerations were made.

4.3.4 Environmental Management Programs. Environmental Management Programs (EMPs) are the detailed plans and programs explaining how the objectives and targets will be accomplished. These EMPs usually note responsible personnel, milestones and dates, and measurements of success. Noting monitoring and measurement parameters directly in the EMP facilitates conforming to 4.5.1 on Monitoring and Measurement discussed below.

4.4.1 Structure and Responsibility. ISO 14001 requires that the relevant management and accountability structure be defined in this element. This usually takes the form of an organizational chart. Also, the organization must denote the Management Representative who is responsible to oversee the EMS and report to management on its operation.

4.4.2 Training Awareness and Competence. The key point in this element is that personnel must receive applicable training regarding the EMS. Specific requirements are itemized in ISO 14001, and include general, companywide items such as knowing the policy, to more function-specific training on environmental aspects and emergency response. An organization usually responds to this element with a training matrix, cross-referencing to training materials and records.

4.4.3 Communications. Procedures are required for both internal and external communications. The process of internal communication between the various levels and functions of the organization must be developed. It is important to note that ISO 14001 also requires procedures for documenting and responding to external communication, and allows the organization to decide for itself the degree of openness and level of disclosure of information. This decision on the level of disclosure must be recorded.

4.4.4 EMS Documentation. This requirement is simply that the organization has documented the system in either electronic or paper form such that it addresses the elements of the standard and provides direction to related documentation. Not

all ISO 14001–required procedures need to be documented, as long as the system requirements can be verified.

4.4.5 Document Control. Procedures are required to control documents, such as system procedures and work instructions, and to ensure that current versions are distributed and obsolete versions are removed from the system.

4.4.6 Operational Control. This element is the one which connects the EMS with the organization as a whole. Here, the critical functions related to significant aspects and objectives and targets are identified and procedures and work instructions created to ensure proper execution of activities. Requirements for communicating applicable system requirements to contractors are also addressed.

4.4.7 Emergency Planning and Response. Although typically addressed through conventional emergency response plans, this element also requires that a process exist for identifying the potential emergencies, in addition to planning and mitigating them. A linkage to the aspects analysis, where impacts are assessed, is appropriate. Emergency incidents include those that may not be regulated, but may still cause significant impact as defined by the organization.

4.5.1 Monitoring and Measurement. Procedures are required describing how the organization will monitor and measure key parameters of operations. These parameters relate to the significant aspects, objectives, and targets and legal and regulatory compliance. In order to manage the system properly, measurements must be taken of its performance to provide data for action. Responses to this element usually cross-reference to many other specific procedures and work instructions describing measurement and equipment calibration. It is in this element that we find the requirement for what is commonly referred to as a compliance audit.

4.5.2 Nonconformance, Corrective, and Preventive Action. This element requires procedures for acting on nonconformances identified in the system, including corrective and preventive action. Nonconformances may be identified through audits, monitoring and measurement, and communications. The intent is to correct the system flaws. Typically, Corrective Action Report (CAR) forms are the norm, noting the nonconformance, the suggested fix, and closure of the action when completed. Note that this requirement does not imply in any way that the party identifying the nonconformance must be the one to suggest the fix. Instead, it is expected that the system provide for the information to be routed to the most appropriate party to address the concern.

4.5.3 Records. Records are expected to exist to serve as verification of the system that is in operation. For example, records include audit reports and training records. Unlike controlled documents, records are *once and done* documents, resulting from the execution of some process or procedure. Procedures in this element are required for the maintenance of records.

4.5.4 EMS Audits. ISO 14001 requires that the system provide for internal audits. This includes methodologies, schedules, and processes to conduct the audits. Interestingly, the EMS audit will, in essence, audit the audit process itself!

4.6 Management Review. This element requires that top management periodically review the EMS to ensure it is operating as planned. If not, resources must be provided for corrective action. For areas where there are no problems, the expectation is that with time, management will provide for improvement programs. Usually there is no detailed procedure for this element, although records of agendas, attendance, and agreed-upon action items are maintained as verification.

EMS Implementation Process

Whether you are implementing the ISO 14001 EMS standard to pursue certification and/or improve your market position or simply using the standard as guidance to improve your existing system of environmental management, it is important to understand what it will take to implement ISO 14001 and the benefits it may provide. The following tasks are arranged in a reasonably logical order, and detail one of any number of approaches for undertaking the implementation of ISO 14001.

Evaluation Stage

The first stage of this approach is referred to as the Evaluation Stage and is where *baseline conditions* are established (*i.e.*, assessing where your existing system is, relative to the require-

ments of the standard). The information obtained from the Evaluation Stage is used to estimate the level of effort needed to actually implement the EMS and prepare an implementation plan. Tasks one through five, explained below, detail the Evaluation Stage process.

1) Management Commitment. Obtaining top management commitment is a critical component in the development of an EMS. Without top management support, you won't have the resources and the overall support of the organization to effectively implement and maintain an EMS. It goes without saying that what gets rewarded gets done, and without top management support the efforts necessary to implement an EMS just won't happen. Accordingly, the first step in proceeding with the development of an EMS is to educate top management on the concept of EMS and the benefits that it can provide. One may not get the green light to proceed with implementation at this juncture; however, the intent of this effort, in most cases, is to develop enough top management interest that it is possible to proceed with the Evaluation Stage as shown in Figure 1.

The session to educate top management can introduce the following areas:

- Update on ISO 14000 standard series

- Introduction of ISO 14001

- What ISO 14001 is, and how it may impact an organization in general

- Review of the core elements of ISO 14001

- How to implement ISO 14001, the benefits it may provide, and how to determine what resources may be necessary

- Terminology, including *accreditation*, *certification*, *registration*, and how those relate to auditors and the EMS

- The legal and strategic reasons for the introduction of an EMS, such as ISO 14001

2) Identification of Project Leader. Once approval to proceed with the Evaluation Stage has been obtained, it will be necessary to identify a person to champion the effort to implement the EMS. This person will take the lead in coordinating the implementation of the EMS, overseeing development of EMS documentation, and obtaining employee buy-in into the program. In the vernacular of ISO 14000, this person is the *Management Representative*.

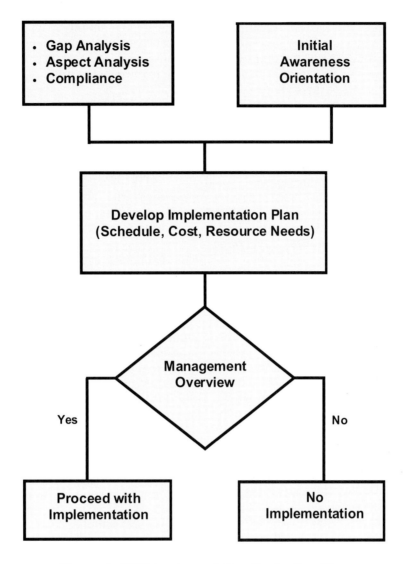

Figure 1. EMS Implementation Evaluation Stage

3) Gap Analysis. The goal of the *gap analysis* is to develop a clear understanding of the gap between current environmental management activities and documentation and ISO 14001 requirements. This is accomplished through a section-by-section review of current environmental management activities against the ISO 14001 criteria. This task is not a requirement of the standard but rather a planning tool to better determine an order-of-magnitude level of effort which will be needed to implement the EMS.

Specifically, the work under this task will include the following:

• Establish the criteria and protocols for conducting the gap analysis at the site. This will include preparing questionnaires and data evaluation methods.

• Briefly review and compare the following to the appropriate sections of ISO 14001:

– current management system procedures and structure

– the company's environmental policy (if already prepared)

– current environmental management records related to the site.

This is accomplished through interviews with appropriate personnel and review of documents to determine if there is an existing understanding of the specific requirement of the standard and whether corresponding procedures have been established and maintained which conform to the standard. If these exist, some minor adjustments will likely be needed to address the specific requirements of the standard. If significant modifications must be made to existing procedures, then corresponding training must be conducted. If there are no procedures in place that even begin to address the requirements of the standard, then the challenge to implementation will be clear.

Using this information, and understanding the resources the organization wishes to commit to the implementation process, it will then be possible to prepare an implementation plan.

4) Aspects Analysis. ISO 14001 requires that your facility or site establish and maintain a procedure to identify *environmental aspects* of activities, products and services ". . . in order to determine those which have or can have significant impacts on the environment" (ISO 14001, Section 4.3).

An environmental aspect is any element of the organization's activities, products, and services that can interact with the environment that it can control and over which it can be expected to have an influence. The impact is the change which occurs in the environment, positive or negative. Current as well as potential aspects and impacts are identified.

An organization must identify the environmental aspects that apply to its operations, products, and services in various aspect categories such as air emissions, water releases, waste management, ecosystem interaction, energy, and natural resources. The organization will then identify the related impacts on the environment. This is a very important step in the planning process, for it is on the basis of the environmental aspects that the organization will ultimately determine those which are significant using its own criteria; and from those significant aspects, establish objectives and targets, environmental management programs, and operational controls, and identify specific training needs.

The aspects analysis is typically conducted through the following:

•   Review of the site operations to prepare a list of categories for activities, products, and services. These categories may include manufacturing and production, chemical storage, wastewater treatment, and product transportation, among others.

•   Interviews with appropriate personnel, and a site reconnaissance to identify the facility's various operational units that interact with

the environment. Migration and transport pathways and environmental receptors will be noted, to document how the activity, product, or services of the unit are identified as aspects.

•   Review of documented corporate environmental policies and related documents (performance standards, procedures, and compliance management tools) to assist in determining the environmental character of the facility.

•   Preparation of a matrix, which presents the facility's aspects and associated impacts, sorted by activity, product, or service.

The significance of your environmental aspects can depend on a variety of site-specific conditions; *i.e.*, business type, geographic location, and sensitive receptors. Strictly speaking, ISO 14001 does not require use of enumerated criteria to determine significance. It is up to the company to apply appropriate criteria and determine significance in a way that makes sense to the business; the nature of the environmental aspects and impact risks; regulatory and legal liability factors; and the interests of the community and other stakeholders.

A *regulatory compliance review* can be an appropriate supplement to the initial aspect analysis to determine baseline conditions and the environmental complexity of the organization's operations. During implementation, aspect analysis and compliance auditing procedures are prepared as part of the EMS so that these functions can be repeated in the future. In other words, the standard requires developing procedures for those tasks. However, to effectively begin EMS implementation, the aspect analysis and compliance review should be conducted at an early stage; therefore, the procedures used in this early assessment can be incorporated later into EMS documentation.

5) Implementation Plan Development. Once the aforementioned steps are completed, an assessment of the level of effort needed to implement the EMS can be made and an implementation plan developed. An implementation plan is, again, not a requirement of the standard but rather a useful planning tool. An implementation plan may include such information as a timetable for implementation and an estimate of the internal and external resources that the organization will need to draw upon to effectively implement the EMS within the designated time frame. External resources may include consultants, software, and training courses, among others.

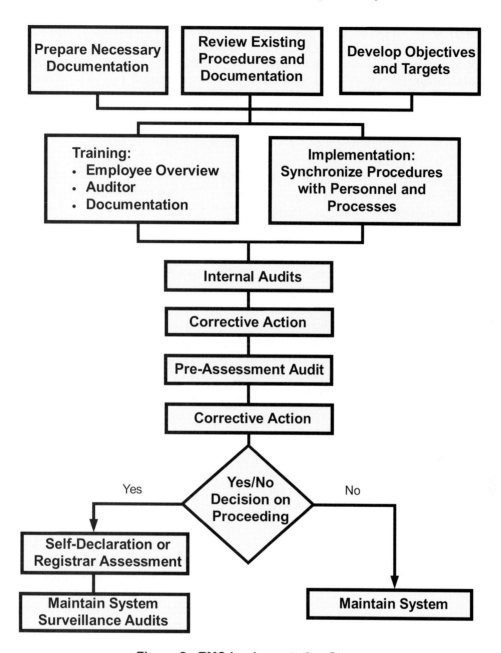

Figure 2. EMS Implementation Stage

Implementation Stage

Implementation is the most sensitive step in developing an effective EMS which has company buy-in. It is actually detrimental to have an external party completely develop the EMS for an organization. Once a decision is made to proceed with ISO 14001 and resources and personnel are allocated, the implementation plan for the EMS can commence. The implementation plan is based on four key factors: planning, implementation, checking, and continual improvement. Using the information collected in the initial assessment, appropriate procedures and documents are

prepared and auditing is conducted. The following tasks follow the sequence necessary to systematically develop the EMS (see Figure 2).

Environmental Policy. Using the information collected in the initial assessment, the organization must develop or modify its environmental policy so that it not only meets 4.2 requirements but also is conducive to EMS development.

Develop Objectives and Targets. With the baseline information and policy in hand, the company can establish specific performance targets and objectives and a schedule for achievement and review. *Objectives* consist of goals the organization sets for itself based on its environmental policy and significant aspects; *targets* define the detailed performance requirements arising from identified objectives. Objectives and targets should be developed where a significant environmental aspect is not already addressed under operational control.

Documentation Training. Communication of the documented EMS and its procedures is one of the key components needed in order to ensure employee buy-in and continuous improvement. Preparing the appropriate documents and procedures is therefore a critical step.

Procedures and Work Instructions. Current procedures, work instructions, and standard operating procedures associated with the EMS are reviewed to focus on identifying areas where adjustments may be needed in order to conform to the standard. Appropriate ISO 14001 documents are written and current documents are modified to address the standard's requirements.

Implementation. This is the actual deployment of the EMS into the organization. This step is predominantly performed by the company staff as part of normal operational responsibilities. This task is usually process-specific, involving small groups addressing particular items or aspects. Also included here is the introduction of new or revised work instructions. This task's goal is to ensure that the system, as documented, is in place throughout the organization. It is important during implementation to synchronize processes and personnel with existing procedures. Having the business units and all employees work together to develop the system leads to consistency throughout the company and a more effective EMS.

Overview Training. It is very important, and a key element of ISO 14001, to educate all employees, from the top to the shop floor, to ensure that everyone in the company is familiar with the EMS and its applications and each person's role in making the system functional and effective. A one-hour overview session (several may be needed to cover all employees) will introduce the ISO 14000 general concept and the company's position. This training is not operation-specific, but is to introduce the general employee population to the ISO 14000 project.

Internal Auditor Training. ISO 14001 requires that the EMS be periodically audited, independent of any third-party certification program. Although outside consultants can be used, organizations are highly encouraged to assemble their own EMS internal audit team to conduct the audits. This usually involves a two-day training course for your internal EMS audit team.

Internal Auditing. Normally, internal audits are conducted in conjunction with the auditor training to provide a hands-on way to use the techniques learned. In this way, the training and the initial EMS audits are coordinated for efficiency.

Corrective Action. In this task, nonconformances identified during the auditing process are corrected. The company will address the nonconformances identified through the internal audits and other information sources and take the appropriate action.

Preassessment Audit. A mock certification audit is conducted to identify any remaining nonconformances and prepare the site for the certification audit. Additional corrective action may occur after the preassessment audit. It is possible that the internal auditors will conduct the preassessment audit, although we recommend that the company consider using the selected registrar to provide an objective, fresh review of the system.

Registrar Assessment. This task involves retaining an accredited auditor to perform the third-party, certification audit. This effort requires that appropriate company staff be available. The auditors prefer not to interview external consultants to ensure that the organization is familiar and comfortable with their EMS system. It is strongly encouraged that the registrar be identified early in the implementation process to avoid disagreements in philosophy and surprises at the end. It is recommended that registrars be interviewed early.

Continuous Improvement. This will be the company's ongoing task and may involve researching other systems, benchmarking, developing strategies to achieve goals, establishing future corporate requirements, and linking them to environmental issues.

Once implemented, the EMS becomes standard operating procedure for the organization. Resources to support these functions must be incorporated into the environmental business function. Periodic internal audits and third-party surveillance audits are essential to the ongoing maintenance of the EMS.

Introduction to EMS Auditing Concepts

The Environmental Management System audit is based on the generic concept of auditing. Simply put, an *audit*, any audit, is the comparison of actual conditions to expected conditions, and a determination as to whether the actual conditions are in conformance or not in conformance with the expected conditions. This is the same philosophy used to perform financial, quality, regulatory compliance, and systems audits. It is prudent to first review what the common elements are, in order to better understand why audits are different.

There are several definitions of audit components that are common to any type of audit. ISO 14010 defines these terms for EMS audits, but they apply in other cases also. As a matter of fact, the ISO committee decided not to create auditing standards for other types of audits, such as compliance audits, although it was originally considered. The main reason for deleting the work items was because the concepts and processes defined in ISO 14011, originally intended for EMS audits, were generic enough to be applied *as is* to other types of audits.

An audit is fundamentally a comparison of audit *evidence* with audit *criteria* to determine findings. The evidence is the objective information collected through interviews, visual reconnaissance, and documentation review. The audit criteria are the expectations or *rules* of how conditions should be. It is the criteria that distinguish one audit from the next. For example, in compliance auditing, the criteria are the regulations. With an EMS audit, the criteria would be the description of the expected system elements. In this case, the EMS criteria would be that described in ISO 14001, the specification standard.

When evidence is compared to criteria, one can determine whether the audited entity does or does not conform. This determination is a *finding*. A finding can either be one of conformance, or non-conformance. Therefore, an audit will always produce findings, even if what is being audited is in full conformance with criteria.

Other key definitions to be aware of with auditing are: objectives, scope, auditee, and auditor. The *audit objective(s)* is simply why you are conducting an audit; usually the reason is to demonstrate conformance to stated criteria. The *audit scope* is what is being audited, and can be a company, a site, or unit within a site or company.

In the ISO 14000 realm, there is a clear distinction between the auditee and client. The *auditee* is the entity being audited. The *client* is the party commissioning the audit. For example, a client can be the customer, and the auditee a supplier to that customer. In ISO 14000 this distinction is important, because the client sets the scope, objectives, and plan for an audit, not the auditee, although it is expected the auditee will be involved and cooperate.

The *auditor* is the one actually collecting evidence and determining findings. The auditor can be comprised of several individuals on a team. There are requirements in ISO 14001 that state that those performing functions within the EMS, such as the auditors, be qualified in their tasks. This means the auditors must have received training in EMS auditing. However, there may be audit team members who do not have the training but are on the team because of some unique expertise, such as process, language, or regulatory knowledge.

The ISO 14000 Auditing Standards

The only standard in the ISO 14000 series that must be followed for third-party verification (*i.e.*, getting certified) is 14001. In that document, there is a requirement that the management system be periodically audited. Section 4.5.4 of ISO 14001 states that

the organization shall establish and maintain (a) program(s) and procedures for periodic EMS audits to be carried out in order to:

a) determine whether or not the EMS:

1) conforms to planned arrangements for environmental management including

the requirements of this international standard; and

2) has been properly implemented and maintained; and

b) provide information on the results of the audits to management

Because of this requirement, the ISO 14000 committee decided to prepare guidance standards for users describing techniques to help meet this requirement of ISO 14001. The resulting auditing standards were created as guidance documents. They do not need to be followed or used in order to obtain certification. There are three standards in the auditing series: ISO 14010, 14011, and 14012.

ISO 14010–General Auditing Principles. This standard is a general principles document that describes key definitions and general expectations of auditors. For example the terms defined above, such as auditee, are addressed in ISO 14010. ISO 14010 also addresses confidentiality and professionalism, and discusses the audit report which will be explained below.

ISO 14011–EMS Auditing Procedures. This standard is the auditing procedures standard that describes how to establish an audit program including planning, staffing, and reporting. Additional definitions are also addressed in ISO 14011 such as lead auditor and client. ISO 14011 defines the roles and responsibilities of the involved parties in an audit and also provides more information on reporting. These details are defined in the various sections below. The intent of ISO 14011 is to provide guidance to the user on addressing ISO 14001, Section 4.5.4, which states:

> The organization's audit program, including any schedule, shall be based on the environmental importance of the activity concerned and the results of previous audits. In order to be comprehensive, the audit procedures shall cover the audit scope, frequency and methodologies, as well as the responsibilities and requirements for conducting and reporting results.

ISO 14012–Auditor Criteria. This standard describes recommended EMS auditor qualifications in terms of education, training, and practical experience. In general, an EMS auditor should be familiar with management systems, regulatory and legal requirements, processes and operations involved, and environmental science issues related to the auditee. As a guidance standard, ISO 14012 can only recommend such qualifications, and the key is to ensure that the audit team is familiar with the EMS that they are responsible for and not all areas of environmental science or regulations. Secondly, it is understood that no single individual may have all of these qualifications, hence the concept of the audit team.

The auditing standards will provide complete detail on the mechanics of auditing. It is interesting to note that ISO 14001, 14010, 14011, and 14012 do not acknowledge the concept of certification and third-party auditors. The expectation is that the EMS will have its own internal auditors, usually employees of the site or company. The third-party auditor, commonly refered to as the registrar or certifier, is accredited by another organization. They are by definition completely objective, not having participated in the system development or implementation.

Accrediting bodies and registrars usually elevate ISO 14010, 14011, and 14012 to requirements for themselves. As a result, an auditor with an accredited registrar will usually meet or exceed the qualifications described in ISO 14012. Using this same terminology, first- and second-party auditors are usually the internal auditing staff, consultants, and customers or other interested parties, not holding accreditations to certify.

Compliance *vs.* System Auditing

We have discussed above what is fundamentally the same among all audit types, and what makes them different. Often, however, there is confusion between regulatory compliance auditing and EMS auditing. This is because there are many elements of regulatory compliance that overlap with the EMS. Recall that the criteria in a compliance audit is the applicable regulation, whereas the criteria in an EMS audit would be ISO 14001. But does not ISO 14001 address compliance? The answer is yes, but from a system standpoint, not performance. In other words, the standard requires that certain procedures exist regarding identification of legal and other requirements, that periodic compliance assessments be performed, that legal requirements be considered in setting objectives and targets, and that there be a commitment to compliance. However, actually being in compliance is a performance issue, and out of the purview of ISO 14001. Of course, a system that is constantly out of compliance or does not identify and initiate action to correct noncompliances will eventually fail due to system failure.

The subtle yet important point is that during an EMS audit, identified regulatory noncompliances are relevant only to the extent that they reflect a potential system problem. The finding therefore is not that the site is out of compliance with a given regulation, but that the noncompliance means some EMS element is not conformed to. For example, a regulatory noncompliance can be related to a problem with training, recordkeeping, or monitoring and measurement.

The EMS auditor is not to do a compliance audit as part of the EMS audit. If, as part of the statistical sampling to verify EMS element requirements, the auditor identifies a regulatory noncompliance, he or she treats it as any other evidence. This point has been difficult to accept, especially among industry in the United States because of our long history of regulatory enforcement. The EMS auditor needs to constantly remember that compliance auditing is being done separately as part of the EMS requirements itself (4.5.1, paragraph 3) and to stay focused on the criteria at hand—ISO 14001 and the site's EMS. There may be legal requirements regarding noncompliances encountered during the EMS audit, but these should be decided and addressed in the audit plan.

In summary, the goal of the compliance audit is to verify compliance with regulations, whereas the EMS audit's goal is to verify that the EMS conforms to planned arrangements, including ISO 14001.

EMS Auditing Interviews

The concept of verifying that the organization *does what it says it will do* is rooted in ISO 14001's use of the term *establish and maintain*. In regard to required procedures, ISO 14001 states "the organization shall establish and maintain a procedure to . . . ." This goes beyond preparing and documenting a procedure. It also means integrating the procedure into site operations, conducting related training, and periodically ensuring that the procedure works, is followed, and

is improved upon when there is a problem. The auditor in turn must verify this degree of implementation.

It is ironic that probably the most sensitive part of auditing is the most difficult to teach, and is more an acquired skill. Interviewing is essentially the technique of gathering information from another individual by asking a series of questions. This may sound easy, but there are varying styles of questions that will prompt different types of answers. For example, closed questions (*i.e.*, yes-no answers) will not yield details or explanations. It is not feasible to assess how well someone understands a concept by using closed questions. On the other hand there is a time for closed questions, usually when the auditor wishes to verify a point or time is short. Keep in mind also that the auditor can ask additional clarifying questions to elaborate on a point.

Other types of questions, such as antagonistic or leading, are not recommended. Keep in mind that silence, allowing the interviewee to think, is also a valid technique of obtaining information. In general, interviews should be characterized by structured, thoughtful questions, putting the auditee at ease, explaining what is required, listening to the response, and avoiding personal judgement.

Typical Interviewees. It is important to interview a representative sampling of people in any particular area of the organization in order to verify conformance. Some typical types of personnel to interview are

- Plant management

- Management representatives

- Department managers

- Document control and record departments

- Research and development

- Engineering

- Operations employees (plant, administration)

- Human resources and training

- Contractor management and purchasing

- Security

Typical Interview Questions. What types of questions are typically asked during ISO 14001

EMS audits? There are a few basic questions that are nearly always asked, at least to begin discussions. It should be noted that interviews are situation-specific, and many other clarifying questions may follow those listed below. However, either to create a checklist, or in lieu of a checklist, the following are good rules of thumb. You will note two sets of questions. The first set represents elements that all employees should be able to answer, and can be asked of anyone within the EMS. The second set consists of more specific questions, applicable to individuals involved with critical functions, as defined by ISO 14001.

First Set (all personnel).

- Are you familiar with the policy?

- Are you familiar with the EMS program?

- What do you do in case of a procedural nonconformance?

- What do you do in case of an emergency?

- What kind of training have you received?

- How do you communicate environmental concerns or ideas?

- What do you do if you receive environmental-related communication from external parties?

Second Set (critical function personnel).

- What are the significant aspects and impacts associated with your function?

- How do you know what to do? (Ask for procedures and operating criteria.)

- What specific training have you received?

- Are there any objectives and targets associated with your function?

- Are you responsible for any monitoring and measurement activities?

- What records do you keep?

- Any other specific questions prompted by answers to specific circumstances or the questions above.

Document Review. During an EMS audit, the auditor will be reviewing a wide variety of documentation. Documents will vary from high-level management policies and procedures to specific records. In general the documentation review is part of the overall evidence, gathering phase. More specifically, the auditor is looking for the following:

- Does your documented system respond to the standard?

- Do the procedures describe what's happening?

- Is the documentation controlled?

- Are all employees informed?

- Are the procedures followed by everyone all the time?

- Is there objective evidence that the procedures are being followed?

It is easy to quickly become overwhelmed by the sheer volume of documents that may exist. Once again, the auditor must remind him or herself that an audit is a statistical sampling in an instant of time of the EMS. There is no expectation that every document will be reviewed. Part of the art of auditing is knowing how to select a representative sampling. Although there is much latitude with sample size, one should definitely not continue auditing until one finds a nonconformance. Unless there is an indication of a problem within the pre-agreed-upon sample size, the audit is complete when that sampling is done, even if no non-conformances were noted. The nature and size of the documentation sample size is determined during the audit planning.

Listed below are typical documents reviewed when auditing against the various ISO 14001 elements. Obviously, the title and format will vary from site to site. However, the list shown in Figure 3 includes typical document types that will facilitate verifying conformance to the specific ISO 14001 element.

Closing

The ISO 14001 EMS Standard provides the framework within which an organization can build their EMS. The complexity of the EMS, the overall goals of the EMS, and the reasons for implementing an EMS will be very specific to each and every organization. The management system concept is not a passing fad, but a movement that is growing in popularity and has been embraced by numerous organizations as the means of achieving the primary goals of environmental

EMS Document Review List

Aspects Procedure

- Aspects list
- Significant determination information
- Significant aspects/impacts list

Legal and Other Requirements

- Listings of applicable legal and other requirements
- Appropriate instructions for compliance
- Permits, manifests, *etc.*

Objectives and Targets and Environmental Management Programs

- Minutes/notes of objectives and target development
- List of objectives and targets
- Related action plans

Structure and Responsibility

- Job descriptions
- Organizational charts

Training Awareness and Competence

- Training needs listings/matrix
- Manuals, course materials
- Sign-in sheets
- Test records, certificate copies, *etc.*

Communication

- Specific work instructions
- Records of communication and correspondence

Document Control

- Documents, procedures, and manuals

Operational Control

- Critical operations/aspects listing/matrix
- Specific work instructions
- Environmental issues and instructions within other work instructions
- Contractor policies, work orders, etc.
- Supplier requirements

Emergency Preparedness and Response

- Emergency plans and protocols
- Practice and drill results

Monitoring and Measurement

- Objectives and target action plans
- Specific procedures and work instructions
- Records of monitoring and measurement data collected, including calibration records

Nonconformance, Corrective and Preventive Action

- Corrective action reports
- Evidence of discussion and follow-up (meeting notes, *etc.*)

Records

- Records

EMS Audit

- Specific audit procedures, checklists, forms
- EMS audit notes and working documents
- EMS audit reports

Management Review

- Meeting agendas and attendance
- Meeting minutes and action items
- Evidence of follow-up actions, reports, *etc.*

Figure 3. EMS Document Review List

protection. The EMS concept is something new, as it integrates the past methods of environmental management, establishing the connection between objectives, employee empowerment, and environmental performance. It is the means for coordinating and prioritizing all environmental efforts under one comprehensive system contributing to the sustainable growth of the organization.

A truly robust EMS, having the goal of integration of environmental issues and functions into the business operation of an organization, will contribute to the adoption of **Strategic Environmental Management** (SEM) in the organization. SEM is the adaptation of traditional business and engineering theory and practice geared toward the maximization of profit, to the responsible management of environmental issues with the goal of minimizing costs and potential liability. It requires planning to a desired effect, as opposed to reacting to results of other actions. Both of these efforts are valid business functions which impact the bottom line. Why should one be strategically planned for and budgeted, and the other managed only with the expectation to achieve compliance and keep the boss out of jail? The effort put into each will determine whether the impact to the bottom line is a positive one or a negative one. Environmental management can have a positive impact.

SEM will provide the management of an organization with tools to evaluate product development and process management decisions and their impact on the environment. Costly mistakes can be avoided and cost-saving opportunities can be identified by using SEM components such as pollution prevention, activity-based environmental cost accounting, life-cycle cost assessment, measurement protocols and performance indicators, effective community involvement, and EMS. Using these tools is a key to improving your organization's environmental performance while reaping the tangible and intangible benefits of greater control, reduced costs, enhanced community relations, and potential liability mitigation.

Harry S. Kemp and *Steve Rowley* are environmental consultants working for EnSafe Inc. out of its Jackson, Tennessee and Lancaster, Pennsylvania branch offices, respectively. They have a combined experience in the environmental field of 32 years, and are now focusing their practices in the area of environmental management system (EMS) development and implementation. EnSafe Inc. is an environmental and management consulting firm headquartered in Memphis, Tennessee (www.ensafe.com).

Ed Pinero, who has been a practicing environmental consultant for over 10 years and a committee chair on the United States TAG for ISO 14000 for four years, recently left EnSafe to take an appointed position with the Pennsylvania Department of Environmental Protection as the Special Assistant to the Deputy Secretary in the Office of Pollution Prevention and Compliance Assistance, where he will be focusing on the development of State policy and processes regarding EMS.

CHAPTER 5

Environmental Compliance Audits

Robert Lipscomb, CHMM
Chris McKeeman, CSP, CHMM

Introduction

This chapter outlines a *strategy* for conducting a successful environmental compliance audit. By strategy, we mean a way of understanding and approaching the audit process rather than simply walking through a manufacturing plant with a clipboard and checklist in one hand and a printout of the current regulations in the other. The strategy is based on first understanding why management or the client would like the audit performed. With that information, you can determine what questions need to be answered or what problems need to be resolved so that the client can make informed decisions about any actions that need to be taken.

You must also understand that the very word *audit* conjures up many negative connotations. If you have any doubt about this, imagine yourself in an Internal Revenue Service audit. Sometimes employees at the facility may view you and the audit process as a threat to their career or livelihood. Therefore, you need to remain professional, courteous, and diplomatic at all times. By establishing good communications with your hosts, you can minimize misperceptions and alleviate undue anxiety.

Because certain processes, and the materials that are used in those processes, may be confidential (trade secrets), you must establish the appropriate lines of communication for collecting proprietary information and keep a record of the names of the primary contacts from whom you collect this information. Also, you need to verify the level of confidentiality that the company management requires for the list of names, phone numbers, and addresses of your sources of confidential information. If appropriate, fax copies of the list to company management. These copies will help document that appropriate lines of communication have been established. You should maintain a telephone log to document all formal contacts that are made during the audit process.

The Auditing Process

Some auditors come to a compliance audit with a cookie-cutter approach, a set of step-by-step procedures and fill-in-the-blank checklists. They go through this approach thinking that they have conducted an environmental compliance audit simply because they filled out the right forms with the appropriate information, compiled the information into a report, and presented it to management or their client. These auditors approach the process in this manner because they lack a fundamental understanding of the audit's overall purpose and their role in it.

However, a successful audit is the result of a pragmatic process that seeks to understand the audit scope and purpose as well as what regulations apply to the specific facility and the processes that are conducted within that facility. One key to a successful audit is to go through the process steps in the proper *sequence*. You will often be tempted to jump ahead in the process before you have completed the necessary preparatory work.

You should also understand that the process is *iterative* as well. That is, information you come across later in the discovery process may compel you to revisit earlier steps, either to integrate the newly discovered information into your understanding of the audit's purpose, to conduct additional research into applicable regulations, or for some other reason. The underlying principle is that auditing is a continuous process of discovery.

Steps in the Auditing Process

1) Determine the audit's purpose.

2) Determine which regulations apply.

3) Evaluate regulatory compliance.

4) Report information which fulfills the audit's purpose.

The Purpose of the Audit

Establish the *purpose* of the audit before proceeding with any other steps. Knowing ahead of time what you are trying to achieve and confirming it with everyone involved may seem to run counter to the intuitive concept of objectivity, but there is a difference in determining in advance the *purpose* and determining the *findings*. Poorly conceived and communicated reports—regardless of their technical accuracy—can mislead the reader as much as reports written to intentionally mislead.

To avoid miscommunication with the reader, you must know your audience. Find out early on who the report's primary audience will be. Knowing who will actually read the report (management, technical personnel, potential buyer, *etc.*) can help you understand the central purpose of the audit. For instance, attorneys and investors who are participating in a property transfer will read an audit report with a different goal in mind than, say, a plant manager who is assessing his facility's environmental management system. If the primary audience is an investor, the purpose of the report will usually be to identify major liabilities, whereas a plant manager may be interested in ensuring that the inspection logs have been properly completed.

Potential Readers

- Employees directly responsible for the facility's environmental management

- Facility managers outside the environmental management group

- Other facility employees

- Other managers in the company, *but* outside the facility

- Lending institutions and potential buyers and their staff of environmental specialists, accountants, and attorneys

- Government regulators

- Business competitors

- News media

- General public

Once you determine *who will read the report*, you need to determine *what the reader wants to know*. Usually the question will be: "Is the facility in compliance with environmental regulations?" However, environmental compliance audits can also lead to questions about broader issues such as those listed below.

Liability and Risk. What are the existing environmental liabilities associated with the facility? What probable future liabilities will the facility incur? What real environmental risks are being created by the facility?

Business Systems. Does the facility conform to the environmental management systems standard prepared by the International Organization for Standardization (ISO 14001)? Is the facility as good as its benchmark, and has the correct benchmark been chosen?

Internal Communications and Training. Are breakdowns in communication occurring? Is employee training adequate?

Internal Accountability, Supervision, and Staffing. Is the environmental management team properly staffed and supervised? How is accountability for facility compliance defined?

Resource. Do you have the proper staff and financial resources for adequate follow-up on environmental compliance issues?

Early on in the process, you need to let the client know what questions the report *will* and *will not* answer, in order to avoid misperceptions. Even if the report is not intended to immediately answer some of these larger questions or address specific needs, you can tailor the audit so that these questions or needs can be addressed more easily in the future. For instance, the report may not directly assess ISO 14001 conformity, but you can conduct and document the audit so that the information you gathered can be used for a future ISO audit.

The Regulations

The next step in the process is to conduct a preliminary review of State and local regulations in order to determine the scope of regulations that may apply to the facility. Many states regulate local environmental issues that are not addressed at the Federal level. Even when regulations duplicate one another, the State and local regulations may vary significantly in detail.

To determine which regulations apply, consider the industry or facility type and have an understanding of the industrial processes involved. This information is readily available from Environmental Protection Agency (EPA) industry sector publications and other government sources. You must also review the entirety of Federal, State, and local regulations—an onerous task by any reckoning. However, tools are available that make this goal a little easier to achieve. First, each section and subsection of the regulations are usually prefaced with criteria for applicability. Also, many guidance documents are available to help you with the process. The regulations are now available in electronic formats that allow you to conduct computerized searches. The EPA and many State/local agencies offer their regulations, including daily updates, via the Internet. For specific interpretations, however, contact the regulatory agency directly for guidance.

All of these tools and information sources are useful but they are no substitute for personal knowledge of the regulations. You must be well versed in regulatory details and complexities before you can effectively conduct a compliance audit.

The Facility

The next step is to determine *which regulations apply to this facility*. Use one or more of the following characteristics to determine the applicability of specific regulations.

Facility.   Type, location, and size.

Materials.   Raw materials, products, by-products, and waste streams.

Processes.   Types of processes and process capacity.

Equipment.   Types and sizes of equipment used.

To the greatest extent possible, determine which bodies of regulations apply *before you make the site visit*. Get a thorough description of the facility from facility personnel in order to make this determination.

One way to collect information from facility personnel is through a written questionnaire. The questionnaire should be designed to provide the auditor with an outline of the facility's environmental management system. In this manner, you can collect much of the facility information you need *prior* to the site visit. At the same time, request copies of the items listed below.

Reports.   Environmental reports, including previous environmental audit reports

Records.   Environmental permits and permit applications, environmental logs and inspection sheets, correspondence to and from regulatory agencies

Material Data.   Facility material balance, material/waste/emission inventories, and Material Safety Data Sheets.

Graphical Illustrations.   Process flow diagams and facility plot plans.

Usually, only part of this information can be collected before the visit. However, the more successful you are in obtaining and digesting this information before the visit, the more effective and efficient you can be during the visit. For this reason, stress to your facility contacts how beneficial it will be to have this information before the visit.

Publicly Available Information

Government environmental regulatory agencies are a second possible source of information that can also be obtained *prior to the site visit*. Depending on the governing agency, a great deal of information might be obtained through a personal visit, telephone call, or via the Internet. Third-party companies also provide regulatory information at a reasonable cost, but you should make sure the information provided is reliable. Regardless of the source you use, confirm ahead of time that you have the authority to make these information requests. You should document all direct communication with any agency.

Make the most of publicly available information if the company has authorized its use. Do not be surprised, though, if agency records are not as complete and current as you anticipated or if they are not as useful as you would like. For instance, much of what you may find in agency files will be copies of reports submitted by the facility you are auditing. If the agency experiences a high turnover of employees, its current personnel may have gaps in their knowledge of the facility.

There may not have been very much interaction between the agency and the facility. A small to medium-sized facility may have never been inspected by an agency. Even if an inspection has been conducted, it may have occurred a long time ago and been limited to checking the facility's compliance for one medium, such as hazardous waste excluding other media (*e.g.*, air, waste, water). A lack of notices of violations does not necessarily confirm absolute compliance. Conversely, a facility with a past history of violations and fines may now be a model environmental citizen. For these reasons, you must evaluate agency records and put them into their proper context. Remember, for all their limitations, the existing agency records define the facility's official compliance status history.

Limitations of Public Information

- Agency records may not be complete and current.
- Agency personnel may have gaps in their knowledge of the facility.
- The agency may not have inspected a small to medium-sized facility.
- Inspections may have been limited to checking compliance for one medium.
- Lack of notices does not confirm absolute compliance.
- A facility with past notices may now be in compliance.

The Site Visit

Preparation

While the site visit itself is only one part of the audit process, it may be the most important part. The site visit is usually a one-time opportunity; therefore you must be well prepared for the visit. By the time you arrive at the facility, you should be well acquainted with the nature of the facility, its regulatory history, and the applicable environmental regulations. Your preparedness will minimize disruption to the facility and the work schedules of your contacts. It will also increase your credibility with facility personnel, and help ensure an effective and efficient site visit. One word of caution: *Maintain your objectivity.* Avoid drawing premature conclusions based on preliminary information.

Initial Briefing

If possible, initiate the site visit by briefing your contacts within the facility. During this briefing, describe the audit process, the site visit process, and the scope of the final report in order to clarify your objectives and alleviate any anxiety. Be honest. Clearly explain that the purpose of the site visit is to obtain information, and that compliance will be assessed subsequent to the visit. Offer to participate in a debriefing session to summarize what information was and was not collected. Take

this opportunity to establish how you will conduct yourself during the site visit. Learn the parts of the safety program that apply to you. Learn to whom you can and cannot speak. Find out what latitude you have to make unescorted walks through the facility, take photographs, and copy documents.

During the Site Visit

You must remain aware that the site visit is just that—a *visit*—and you are a *visitor.* You may ask for information, but you have no right to demand it. As an auditor, you walk a fine line between being diligent in obtaining all relevant information and overstepping your authority in obtaining it. If you believe that you are not receiving enough cooperation to conduct a worthwhile audit, review the situation with your contacts at the management level and the site. As you talk with your contacts, don't let the issue become personal ("*I'm* not getting what *I want.*") Rather, try to emphasize mutual goals with statements such as "An acceptable audit cannot be performed without the requested information, and here are the reasons why."

When the Site Visit is Complete

Offer to conduct a debriefing meeting with the facility contact, to focus on the status of the data you collected. Identify any gaps in the data you collected and present them in writing at this time. Identify any original documents or copies that you are taking with you. *Avoid rendering preliminary assessments until **all** relevant information has been reviewed.* Your client may compel you to identify broad, red-flag issues. If so, use care and judgement when articulating these issues (*e.g.,* "Your facility *may* be a major source for air pollution, but I won't be able to verify this until I return to my office and review additional information.").

The Chemical Processes

General knowledge of the facility is important, but particularly significant will be your knowledge of the facility's chemical processes. You must define all the ways that chemicals are used and managed in the facility—from receiving to processing to shipping—and delineate the individual chemical processes (including pollution control systems). These processes should be documented with simplified process flow diagrams (PFDs). The next step is to calculate and construct a material flow diagram of the facility (including pollutant emissions and discharges) and its chemical storage capacities. Typically, this is the most difficult part of the audit—and the most important.

Use your best professional judgement to determine the appropriate levels of accuracy and precision for the material balance, in order to be reasonably confident in assessing which regulations are applicable to the facility. In essence, you must discern between *nice to know* and *need to know* information.

An auditor usually develops PFDs in consultation with production and maintenance employees from the facility. Try to obtain material flow information from these contacts as well. Even though you ask specific questions, be prepared for vague responses (*e.g.*, "We make 500 to 2,000 side panels a day"; "We refill the tank every couple of months"; "We use XYZ degreaser for the red panels"; "We don't use any chemicals here, we just paint things"; "I'm not sure what they do with the used oil.")

Even with such general responses, you should be able to obtain sufficient information to develop a conceptual PFD and an order-of-magnitude balance of materials by using information from other sources. (For example, the purchasing department is a good source of information for annual consumption of raw materials and can provide useful historical information.) Each area of the facility should have Material Safety Data Sheets (MSDSs) for materials it uses. Although MSDSs may be vague in defining some product compositions, they provide a good basis for identifying the presence of regulated chemicals.

You will often find one or more chemical compounds with which you may be unfamiliar, and you may be unable to determine whether the compound is regulated without referring to some other source. Carry a list of regulated compounds on site visits. Even better, carry a laptop computer with compound lists and regulations on CD-ROMs. If these resources are not available, you will have to wait until you have access to the information before making your final determination.

Compliance Assessment

At this point in the process, you should have identified the applicable regulations, and you are now ready to begin assessing the facility's compliance. Compliance is a matter of two basic questions:

- Are the physical facilities in compliance with the regulations?

- Are the processes and procedures used at the facility in compliance with the regulations?

Physical Facility Compliance

The first question can be further broken down into the following questions:

- Are all the necessary pieces of equipment, tools, and supplies present?

- Are the pieces built and stationed in conformance with the design regulations?

- Do the tools, equipment, and supplies meet the regulatory performance standards?

A facility walk-through can answer most of these questions. Avoid using the word *inspection* for the walk-through; in most states the term *inspection* carries significant legal implications for engineering services. The rest of the questions can be answered by reviewing equipment documents such as engineering drawings. In some cases (*e.g.*, secondary containment capacity) field measures may be necessary.

Facility Procedures Compliance

The list of generic questions in the accompanying table can be used to determine the facility's compliance with regulations for operating procedures. These are deceptively simple questions with potentially complex and complicated answers. No one has memorized all of the Federal regulations, much less the regulations for all of the states and municipalities. Copies of the regulations will probably not be available at the facility, but there are some practical alternatives. You should conduct a previsit assessment and screening of the facility and the potentially applicable regulations. Bring copies of the relevant sections of the regulations that apply to the site. Carry an audit handbook that has checklists and references, or bring a laptop with a CD-ROM or modem.

Compliance Assessment

- Have the necessary permits been obtained?

- Are training requirements met?

- Are operating standards met?

- Are inspection requirements met?

- Are monitoring requirements met?

- Are recordkeeping requirements met?

- Are certification requirements met?

- Are reporting requirements met?

The Report

Audience Considerations

You must keep the various audiences for your report in mind, since each report has more than one reader. Each reader has a different purpose for reading the report, so you must write the report to give the readers the information they need at a level of detail appropriate to their technical backgrounds. Although you may have established who your audiences are early on, confirm your audiences with the client before writing the report. Also confirm the expected distribution of the report and whether the report is to retain attorney-client privilege under the law. An attorney should supervise the structure of the document to ensure that legal privileges are maintained. Regardless of how limited the initial direct distribution is, be aware that the actual readership of the report may be much larger.

Differences in Technical Background. We have already mentioned that the report may have more than one audience and that each reader will differ in technical background. This difference makes it more of a challenge to communicate your audit findings accurately in terms each reader can understand. For example, environmental attorneys and environmental specialists will be familiar with the environmental technical language, while potential buyers or investors may not understand the significance of specific word choices such as hazardous waste *storage* and *accumulation*. Conversely, an investor or operations manager in the rubber industry will know what a *Banbury mixer* is, but the environmental specialist may not.

Differences in Sensitivity to Contents. Above all, you also need to be attuned to how sensitive readers will be to the contents of the report. Audit results may have significant health, financial, legal, or political implications for some readers. Although you have to write an understandable report that answers the basic questions the audit was conducted to address, keep in mind the fact that individual readers may be held responsible for any shortcomings found in the audit. Take care to avoid inflammatory language (*e.g.*, "The audit revealed numerous areas where management has been negligent. . .") or to present the results of the audit in ways that might overstate or understate potential liability.

Differences in Level of Detail Required. Readers also differ in the level of detail they need. Will the primary audience be interested in inspection log minutiae or big-ticket liability issues? Will most readers be familiar with the production processes or will they need explanations? How much supporting detail should be appended to the report and how much should be included in the body? Since you will have more than one audience, the report has to provide the level of detail each reader needs. A well-constructed executive summary from one to three pages long is the best bridge between the bottom-line conclusions one reader wants and the exhaustive details another reader needs.

Report Considerations

- Font size/type, line spacing, double sided, number of copies?

- To whom is the cover letter addressed? For whom is the report written?

- Color copies of photographs or originals?

- Executive summary, conclusions, recommendations, appendicies included?

- Draft executive summary faxed to receipients prior to final report?

- Detailed regulatory applicability analysis even for nonapplicable regulations?

Report Format and Distribution

You will need to reconfirm some of the small but important details about the format and distribution of the report. If possible and appropriate, submit a draft to all the stakeholders for review and comment. The draft provides a way to consolidate a large amount of information into a concise description of the situation. It provides a context for the discussion of many issues about which the participants may have not yet known. This step in the process can be invaluable in the preparation of a complete, accurate, and precise report.

The Audit File

Maintaining audit files may ultimately be the most important part of the audit process. At the time of this writing, several issues are developing regarding the use of audits in legal actions. The EPA is trying to create incentives for self policing which would reward facilities for using audits to find deficiencies and correct them in a timely manner. The EPA is also pushing the use of *any credible evidence* in enforcement actions. Layered atop these policies are the individual State policies that are evolving concurrently with EPA policies.

A discussion of the legal issues surrounding the possible uses of the audit report is beyond the scope of this chapter. It is enough to say that the report may ultimately be used as evidence in a civil or criminal action. The plaintiff in such legal actions may be the client, a lender, a new owner, the state, the Federal government, a neighbor to the facility, or an environmental activist group. Therefore, you must work closely with the client and the client's attorney to determine how to appropriately manage the file.

The audit may be performed under the *attorney work product doctrine*. If so, you must be in close communication with the attorney to ensure that the privilege is maintained. The attorney will be familiar with the protections that are afforded under the law in that state. The attorney can also assist you in properly labeling the report and the files.

Clearly define your responsibility for implementing custody of the file, and if applicable, access to the file. Determine whether you are or are not to retain the file or copies of the file. If you are to retain the file, how long are you expected to keep it? Knowing the answers to these questions will help you determine whether the file is to be maintained in hard copy, electronic format, or both. Also, find out who retains access to the file if there is a change in the ownership of the facility.

Summary

An audit is a structured process for the collection, analysis and reporting of information. It should not be confused with a site visit. An effective audit is the result of a skilled auditor thoroughly preparing for and implementing each stage of the process. To conduct a successful compliance audit, you must know the regulations, the facility, and your audience.

Robert L. Lipscomb is Senior Project Manager with Barge, Waggoner, Sumner and Cannon, Inc., in Nashville, Tennessee. Mr. Lipscomb has over 20 years of experience in the hazardous materials management field. His areas of expertise include managment of commercial and industrial hazardous materials, investigation and cleanup of contaminated property, and redevelopment of Brownfields sites. He is especially noted for managing the first in situ hazardous waste incinerator employed at a Superfund site. Mr. Lipscomb would like to acknowledge Grady, Dean, Lisette, Joyce, Linda and especially Alie Lipscomb.

Chris D. McKeeman is the Environmental Health and Safety Coordinator at Neste Resins Corporation. Mr. McKeeman has over 11 years experience in the environmental, health, and safety compliance/management field. His areas of expertise include managing and conducting comprehensive enviornmental and safety compliance audits; environmental (including ISO 14000), safety and transportation-related training; as well as environmental and safety permitting and program develpment. Mr. McKeeman is the founder and first president of the Tri-State Chapter of the Academy of Certified Hazardous Materials Managers in North Carolina. Mr. McKeeman offers a special thanks to Robert Lipscomb for the opportunity to participate in the writing of this chapter, and to Kim, Kenton, and Kyle McKeeman for their support.

CHAPTER 6

Environmental Health and Safety Risk Analysis

Tony Uliano, Jr., MS, CIH, CHMM

Historical Risk Assessment

Risk analysis, as defined by the Society for Risk Analysis, is the detailed examination including risk assessment, risk evaluation, and risk management alternatives, performed to understand the nature of unwanted, negative consequences to human life, health, property, or the environment; an analytical process of quantification of the probabilities and expected consequences for identified risks.

Risk analysis techniques have a wide variety of applications. Included among these are evaluation of Superfund sites, contaminated ecological systems, and property transfers involving potentially contaminated sites, currently called *brownfields.*

Risk analysis also helps to predict the risk for accidents at industrial facilities, to assess the hazards of newly introduced chemicals or pharmaceutical products, and even to evaluate such diverse risks as following certain dietary patterns, transportation safety, and smoking.

This chapter addresses the risks posed by hazardous substances to workers, the public and the environment. The most applicable assessments are in the area of occupational and environmental exposures.

Risk assessment, as it is understood today, is a relatively new field (20 years), though its origin dates back to unrefined estimations of toxicological hazards that were performed centuries ago. The immediate precursor of modern risk assessment originated in the United States Food and Drug Administration's *Margins of Safety* for food additives. The science of risk assessment evolved from qualitative opinion to more precise quantitative techniques of measurement and estimation of risk. This evolution paralleled the technological advancements that enabled us to detect toxins at much lower limits, and to gain a better understanding of the risks of exposure to hazards at levels much lower than were previously considered acceptable.

A number of epidemiological studies were performed in the 1950s and 1960s which revealed significantly high risks of cancer in workers who were exposed to substances such as asbestos, benzene, benzidine, 2-napthylamine, and radon gas derived from uranium mines.

In 1980, the Supreme Court ruled that regulatory decisions should contain quantitative risk analysis. This analysis is usually reserved for occupational carcinogens, as evidenced by the Occupational Safety and Health Administration's (OSHA's) more recent (since 1980s) "expanded health standards for formaldehyde, ethylene oxide, and others."

Two fundamental approaches to risk assessment are derived from toxicological/epidemiological methodologies and more recent integrated environmental assessment protocols.

Environmental assessments originated from the Environmental Protection Agency's (EPA) Comprehensive Environmental Response, Compensation, and Liability Act (CERCLA/Superfund) rules relating to releases of hazardous substances into the environment. Initial assessments were performed on petrochemical release sites, such as leaking underground storage tanks and oil spills.

EPA developed a hazard ranking system to determine whether a site contaminated with any hazardous substance would gain National Priority Listing (NPL Site) and be eligible for cleanup funds. Such assessments are employed in almost any situation where a potential hazard to the environment or public may exist. Violations of any of the environmental regulations might be applicable, such as the illegal dumping or storage of hazardous chemicals under the Resource Conservation and Recovery Act (RCRA). These regulations are intended to facilitate the cleanup of uncontrolled waste sites and to prohibit the indiscriminate dumping of untreated chemicals into landfills in order to protect the environment and the public. The Agency for Toxic Substances and Disease Registry was charged under CERCLA to perform health assessments of National Priorities List (NPL) Superfund sites.

Basic Elements of Risk Assessments

In 1983, the National Research Council developed a four-part risk assessment strategy that included:

- Hazard identification

- Dose-response assessment (also known as hazard assessment)

- Exposure assessment

- Risk characterization or risk analysis

Some form of this strategy is used today in most environmental risk analyses. Risk analysis is fraught with numerous uncertainties. It is as much an art as it is science, and the uncertainties make risk analysis problematic. If we are to protect the public and the environment, our risk analysis methodologies must be scientifically defensible and offer a level of protection that takes into account the degree of uncertainty in each situation.

Hazard Identification

Hazard identification is the first step in risk assessment. The goal is to determine the hazardous chemical's relative toxicity, concentration, extent of contamination, and toxicological endpoint(s) of concern.

Many studies have been published on the toxicity of chemical substances. The data that are most useful for analyzing public health risks are those that most closely approximate the risks posed to humans. Unfortunately, the human epidemiological studies that provide this information are limited and have inherent weaknesses. Improved confidence is achieved when the following guidelines are met:

- *Biological plausibility*—The investigated agent is known to cause the health effect.

- *Consistency*—Two or more studies yield comparable findings of excess risk, preferably using different test models.

- *Concordance*—The same end point (*e.g.*, type of tumor) arises in more than one species, strain, or sex, and concordance in the type of pathological change is found in both animals and humans.

Some of the techniques used to determine the toxicity of the numerous chemical compounds in use today are listed below.

Epidemiological Studies. Epidemiological studies are studies of the distribution and determinants of diseases and injuries in the human population. There are three basic types of epidemiological studies: cohort, case control, and cross-sectional or prevalence studies. A cohort, or incidence study, examines the incidence of disorders that accrue following an exposure, and generally is prospective in nature. Exposed and unexposed groups or cohorts are being compared.

A case control study starts with the presence of a disease and then attempts to trace it back to a suspected exposure (retrospective in nature). It compares the cases with disease to controls without disease.

A cross-sectional study simply measures the prevalence of disease in a group of individuals. It offers no information about causation, but may offer insights into a potential exposure. Cross-sectional studies can point out human data that involves accidental exposures resulting in disease. Many occupational disease-agent correlations were discovered in this way.

There are some drawbacks to epidemiological studies, however. The most notable problems include:

- Choosing an acceptable control group, one that is as similar to the study group as possible, with the exception of the exposure or disease being evaluated

- Numerous confounding factors, such as other exposures; gender, racial, and health differences; fluctuations in dose among subjects, and inconsistent sampling data

- Difficulty in obtaining information from a large population in the study, difficulty in obtaining old records, and extensive time to follow up disease occurrence in prospective studies

In Vivo **Bioassays.** *In vivo* bioassays are studies performed on live animals, usually rodents, though other animal classes have been utilized. These assays study the effects (*e.g.*, tumor formation, or other pathological changes) of high dose and exposure to chemical or other agents on the test animals. The doses must be sufficiently high to observe effects in a reasonable amount of time for the limited animal populations studied. Responses at the lower end of the dose-response curve must be interpolated, since these animals are dosed at the maximum tolerated dose (MTD). *In vivo* assays are among the most common research methods, and provide most of the current information on human toxicity potential. These studies provide data such as lethal dose (LD_{50}), lethal concentration (LC_{50}), and other exposure values that may be extrapolated to provide human exposure levels. However, extrapolation to human toxicity continues to be problematic and controversial.

In Vitro **Methods.** *In vitro* methods refer to laboratory scale studies performed in test tubes. These tests are performed on tissue or cells in order to observe changes in the structure or growth of cells.

One widely used test (the Ames test) looks for mutagenicity, or alterations, in the deoxyribonucleic acid (DNA) in Salmonella bacteria. This is a test that is used to screen chemicals that might pose a mutagenic risk. Studies on the teratogenic effect of various agents are performed on animal embryos. *In vitro* testing is yet another level removed from direct human studies, and extrapolation has intrinsic uncertainties, such as the discrepancies between cellular and human metabolic interactions. Other *in vitro* approaches are listed below.

Experimental animal-human comparisons examine existing human data and compare exposure levels and effects to studies performed on animals. This experimental methodology attempts to establish correlations between the effects observed in animals and those observed in humans.

Molecular and experimental embryology studies are performed in order to detect abnormal changes at the molecular and embryonic levels. Studies are currently being performed on endocrine disrupters, classes of chemical pollutants that adversely affect reproduction.

Biological effects of complex chemical or other mixtures examine the synergistic, additive, and antagonistic effects of exposure to a number of chemicals. This includes the correlation between chemical and physical agents, (*i.e.*, noise, temperature, humidity, vibration and stress), and between chemical and environmental interactions, (circadian rhythms, seasonal changes), *etc.*

Structure/activity relationships have long been studied in the field of toxicology, and involve the assessment of the toxicity of a chemical based upon its molecular structure. Physio-chemical properties such as molecular and atomic structure, molecular weight, and chemical class can be a predictor of toxicity. Adding chemical groups or altering the placement of some groups is known to either increase or decrease toxicity. The chemical structure provides hints about its ability to be introduced via critical human metabolic pathways.

Finally, the risk assessments for carcinogens and noncarcinogens follow from different assumptions. For noncarcinogens, there are threshold levels below which adverse effects are not likely to occur. Juxtaposed to this is the fact that little is known as to the threshold levels of the carcinogens that are capable of inducing tumors. The more conservative estimate states that there is no "safe" level of exposure for a carcinogen. The International Agency for Research on Cancer (IARC) publishes assessments on the carcinogenicity of over 600 chemicals. IARC has catalogued these agents by virtue of their likelihood to promote cancer in animals and/or humans. In the 1986 "Guidelines for Carcinogen Risk Assessment," a similar rating system was employed. The following table shows the rating numbers and interpretations.

Table 1. Weight-of-Evidence Classification for Carcinogens

Human Evidence	Animal Evidence				
	Sufficient	Limited	Inadequate	No Data	No Evidence
Sufficient	A	A	A	A	A
Limited	B1	B1	B1	B1	B1
Inadequate	B2	C	D	D	D
No Data	B2	C	D	D	E
No Evidence	B2	C	D	D	E

(51 FR 33992)

This chart is not the only scheme for the classification of carcinogens. Industrial Hygienists may be more familiar with the American Conference of Governmental Industrial Hygienists' classification:

A-1	Confirmed human carcinogen
A-2	Suspected human carcinogen
A-3	Confirmed animal carcinogen
A-4	Suspect animal carcinogen, no sufficient evidence for human carcinogenicity
A-5	Not a human carcinogen

Hazard or Toxicity Assessment. A *hazard or toxicity assessment* is also referred to as a *dose-response assessment* because it seeks to determine whether dose-response information is available for the chemical of interest. While the hazard identification provides a basis for establishing evidence of an agent's effects, a hazard assessment relies on toxicological and epidemiological data to ascertain specific threshold exposure levels. It examines and attempts to quantify the severity of a potential hazard. This assessment requires information to establish risks for both carcinogens and noncarcinogenic agents.

The documentation of risk differs between carcinogens and noncarcinogens, due to the different mechanisms that are understood to operate in carcinogenesis. In both cases, the availability of dose-response data is crucial.

For chemical carcinogens, slope factors define the potency of various agents, since we presume that no threshold dose has been established for carcinogens, and most animal studies rely on doses at the MTD levels. The linear portion of the low dose section of the dose-response curve is interpolated using a 95% confidence interval. This is referred to as the *slope factor*. Slope factors for various carcinogens are listed in EPA's Integrated Risk Information System (IRIS) database.

Varying models have been employed which impart a range of slope factors dependent upon the interpretation of the carcinogenesis process, from a conservative model (one molecule of exposure or less) to EPA's multistage and other models. (Carcinogenesis involves promotion, time factors, *etc.*)

Dose-response curves exist for many non-carcinogenic substances. These curves are estimated on the basis of human and animal studies, mostly from the latter. Dose-response may also be observed empirically through clinical evaluations of individuals who were accidentally exposed to known concentrations of chemicals.

Dose-response curves are only as valid as the organism or population studied. A dose-response curve for one chemical may vary amongst a population of a single species. This is due to individual variations in responses and confounding factors; for example, the response of individuals to low levels of chemical agents, such as sensitizers, or individuals' varying susceptibilities to other non-allergic response provoking chemicals, such as diesel exhaust or tobacco smoke. Little information is available for very low doses because any response is subclinical; or else the chemical may have no effect due to a lack of absorption or impedance by metabolic or other processes.

The intent of the dose-response curve is to establish *safe* or acceptable dose levels. These dose levels vary, depending on the hazardous chemical and its dose (concentration), the route of entry, duration of exposure, and other time factors.

There are a number of terms describing dose levels. The *no observed adverse effect level* (NOAEL) represents the highest dose at which no adverse effects have been detected.

Due to many uncertainties in the study methods, this dose level should contain built-in safety factors. Safety factors may range from 10–2000, depending upon the available information.

The safe dose level may be interpolated to calculate a *lifetime acceptable daily intake* (ADI), taking into account safety and other modifying factors. Interestingly, ADI levels initially were employed for food additives.

Another term, *reference dose* (RfD), takes the NOAEL and attaches uncertainty and modifying factors, given the differences in animal and human responses and the varying sensitivities of humans. If human data were available, adjustment factors might not be necessary:

$$RfD = NOAEL/(UF)(MF) \qquad (1)$$

Where:

RfD = the reference dose in mg agent/kg body weight-day

NOAEL = the no observed adverse effect level in mg agent/kg body weight-day

UF = the uncertainty factor

MF = the modifying factor

The dose is expressed in milligrams of agent/kilogram of body weight per day. RfDs vary according to the route of entry. An exposure dose, or *chronic daily intake* (CDI), may be calculated as follows:

$$CDI = \frac{(C_m)(I_m)(EF)(ED)}{(BW)(AT)} \qquad (2)$$

Where:

C_m = concentration of agent in affected media (*e.g.*, mg/l)

I_m = intake of affected media (*e.g.*, l/day)

EF = exposure frequency (days/yr)

ED = exposure duration (yr)

BW = body weight (average adult is 70 kg)

AT = average exposure time (days)

The human acceptable dose would equal the exposure dose divided by the product of the safety factors.

A ***virtually safe dose*** (VSD) is that dose in which the risk is very low (10^{-6} or one excess tumor, *etc.*, per one million persons). This is the value accepted as the *safe dose* for humans by regulatory policy.

Exposure Assessment

Once overlooked as a component of risk assessment, exposure assessment is now a critical element of an integrated risk evaluation. The exposure assessment examines all avenues of exposure and may utilize mathematical models and projections based upon different scenarios. Some exposures may be confined to the occupational setting and thus can be monitored fairly easily. In the case of environmental risk assessments, exposure may spread through various media and pathways, such as air, soil, water, or food, though the actual exposure will be through inhalation, ingestion, dermal contact, and/or absorption or injection (however rare). The important point is that the agent must be available to the receptor, and the actual intake or dose should be quantifiable or estimated using all available data.

Exposure assessments look at environmental transport mechanisms, such as ground water-surface water movement, geologic and soil conditions, sources of chemical leakage and leachate, fugitive dust emissions, off-gassing, and the presence of contamination in the daily intake of food and water products. These mechanisms affect the route of exposure being analyzed. The assessments may also include a detailed analysis of the human or ecological species lifestyle or life cycles.

Analyzing the average ingestion of toxins includes surveying consumptive products, rates, soil consumption (usually for children), body weight, age and sex. Inhalation studies can verify the number of hours at a site where inhalation exposure occurs (duration) as well as age, activity, and sex-adjusted ventilation rates.

Risk Characterization (Risk Analysis)

Risk analysis, the composite of the hazard assessment and exposure assessment, plus an uncertainty analysis forms the core of risk assessment. ***Risk analysis*** is a quantitative appraisal of the levels of exposure that may pose risks because they exceed the ambient levels found in an unexposed population. In the case of non-carcinogens, a ***hazard quotient*** is used. The hazard quotient (HQ) may be expressed as:

$$HQ = \frac{CDI}{MAD} \quad or \quad \frac{CDI}{RfD} \qquad (3)$$

Where:

CDI = chronic daily intake in mg/kg-day

MAD = maximum allowable daily intake in mg/kg-day

RfD = reference dose in mg/kg-day

Both the numerator and the denominator are expressed in mg/kg-day; therefore the HQ is dimensionless.

For carcinogens, risk is measured by extrapolating from a low dose to the level in which the risk (excess mortality, tumor growth, *etc.*) does not exceed 10^{-6}. In a formula, it is expressed as:

Risk = CDI x SF (4)

Where:

CDI = chronic daily intake in mg/kg-day

SF = a slope factor in (mg/kg-day)$^{-1}$

One of the numerous models of carcinogenesis is the classical linear model, which assumes no threshold for carcinogenesis. This mechanistic approach includes the one-hit model, in which a critical cell *hit* by a carcinogen will induce a tumor, and the multi-hit model, which assumes that several *hits* (which occur randomly and are time-dependent) are required to induce a tumor.

Other models include the more widely accepted linear multi-stage model (which assumes that cancer involves both initiation and promotion processes), the Armitage-Doll Multi-stage model, the Moolgavker version of the Knudsen model, the dose distribution model, and the probability model (which is also called the probit, logit, or Weibull model). The choice of model should be based upon sound risk management practices, although it may be influenced by economic or political factors.

Environmental Risk Assessments

The normal risk analysis occurs at complex sites where contamination may spread outside the boundaries of a single location. Environmental contamination may pose a hazard to both public health and ecological systems. One of the objectives of an environmental assessment is to provide a technically defensible strategy for decision making in a variety of remediation scenarios. This remediation is referred to as *risk-based corrective action* (RBCA).

The assessments have been structured by a number of regulatory and technical agencies and associations (see the American Society for Testing and Materials [ASTM], "Standard Guide for RBCA Applied at Petroleum Sites," and EPA's "Guidelines for Carcinogenic Risk Assessments"). They all have tiered levels of investigational steps from basic site assessment to the more specific and quantitative assessments outlined previously. The methods often utilize standard sampling techniques, identification of the chemicals of interest

(COIs), determination of receptor (animal and human) populations, critical habitat evaluation, exposure and transport pathways, risk estimates, and mathematical and computer modeling. Some methods employ a decision point system based upon all the variables, and most have built-in default assumptions for use when precise data is not available. These defaults might be assumptions for average daily intake (via inhalation, ingestion, or absorption) of contaminants found in water, soil, dust, gases, or vapors for a specific group of individuals.

Decisions about cleanup may be made after only one tier level of sampling; however, the assessed risk may be too conservative. Proceeding to the next tier provides more data for evaluating the true risk. In fact, it is often necessary to complete all of the tier levels of evaluation. The additional data may validate initial assumptions about the level of risk.

Risk Management and Communication

Risk management decisions may be as simple as *no action*, or the decision tree may require further investigation at the next tier level. The more complex the site, and the greater the difficulty in assigning risk, the more sophisticated the risk analysis should be. Often, **Monte Carlo** analyses are employed.

Monte Carlo analyses are valuable for evaluating the risk analyses for site conditions that involve many complex variables and a significant degree of uncertainty. This differs from the deterministic risk assessments that use numerical estimates of each variable in calculations. In a Monte Carlo analysis, the probability curves for various uncertainty parameters are inserted into the calculations of risk.

RBCA-type risk assessments attempt to minimize heroic efforts at cleanup and to manage sites according to the risks they represent.

Risk management decisions take into account assumptions about future site use, whether or not the area is a critical habitat for animal species, and many other factors. These factors are weighed in order to establish the minimum level of remediation that is required to satisfy both public and environmental health concerns while reducing the cost of cleanup.

Risk communication may be the most difficult step in this process. Risk communication requires both an understanding of the risks from the public's perspective and the ability to communicate difficult concepts in terms the general public can under-stand. The risk of not taking the concerns of the potentially exposed public seriously may be far greater than the risk of inaction.

Summary

There are many limitations to risk assessment, such as the various interpretations of acceptable models for carcinogenesis, and levels of uncertainty. Uncertainty can be demonstrated in the animal-to-human extrapolations of risk, the lack of data on multiagent exposure, and collective exposures that do not translate easily to individual risks. Thus, the definition of an *acceptable risk* is still controversial, and there will always be conflict between the forces of economics and the assurance of public and environmental health and safety for the most sensitive ecological species or individuals.

Risk analysis must continue to use sound science in order to understand the integral mechanisms of human toxicology, how toxic chemicals impact ecosystem health, and how both areas are interconnected.

Bibliography

Agency for Toxic Substance Disease Registry, Division of Toxicology. *Health Assessment Guidelines.* Atlanta, GA: ATSDR, 1992.

AIHA Roundtable. *The Changing Face of Risk Assessment: Revolution or Evolution?* Washington, DC: AIHCE, 1996.

American Society for Testing and Materials, "Standard Guide for Risk Based Corrective Action (RBCA) Applied at Petroleum Sites." ASTM Standard E–1739–95el. West Conshohocken, PA: ASTM, 1997.

Doull, John. *Principles of Risk Assessment: Overview of the Risk Assessment Process.* Conference on Chemical Risk Assessment: Science, Policy and Practice. Cincinnati, OH: ACGIH, 1992.

Gochfeld, Michael. "Environmental Risk Assessment." In *Public Health and Preventive Medicine.* East Norwalk, CT: Appleton-Century-Crofts, 1992.

Gold, Lois, Neela Manley, and Bruce Ames. *Quantitative and Qualitative Extrapolation of Carcinogens Between Species.* Conference on Chemical Risk in the DOD: Science, Policy and Practice. Cincinnati, OH: ACGIH, 1992.

"Guidelines for Carcinogen Risk Assessment." *Federal Register* 51 (24 September 1986): 33992.

Hattis, Dale and John Froines. *Uncertainties in Risk Assessment.* Conference on Chemical Risk Assessment in the DOD: Science, Policy and Practice. Cincinnati, OH: ACGIH, 1992.

Herrick, Robert. *Exposure Assessment in Risk Assessment.* Conference on Chemical Risk Assessment in the DOD: Science, Policy and Practice. Cincinnati, OH: ACGIH, 1992.

Johnson, Barry. *Principles of Chemical Risk Assessment.* Conference on Chemical Risk Assessment in the DOD: Science, Policy and Practice. Cincinnati, OH: ACGIH, 1992.

Kodell, Ralph. *Risk Assessment for Non-Carcinogenic Chemical Effects.* Conference on Chemical Risk Assessment in the DOD: Science, Policy and Practice. Cincinnati, OH: ACGIH, 1992.

Krisnan, Kanaan, Rory Conolly, and Melvin Andersen. *Biologically Based Models in Risk Assessment.* Conference on Chemical Risk Assessment in the DOD: Science, Policy and Practice. Cincinnati, OH: ACGIH, 1992.

Lu, Frank. "Toxicolological Evaluation: Assessment of Safety/Risk." In *Basic Toxicology.* Washington, DC: Taylor and Francis, 1996.

Murdock, Duncan, Daniel Krewski, and John Wargo. *Cancer Risk Assessment with Inter-*

mittent Exposure. Conference on Chemical Risk Assessment in the DOD: Science, Policy and Practice. Cincinnati, OH: ACGIH, 1992.

Nelson, B. K. "Exposure Interactions in Occupational/Environmental Toxicology." *Appl. Occup. and Envir. Hyg.,* 12(5): 356-361, 1997.

Nelson, Deborah Imel. "Risk Assessment in the Workplace." In *The Occupational Environment–Its Evaluation and Control.* Fairfax, VA: AIHA, 1997.

Nicholson, William. "Quantitative Risk Assessment for Carcinogens." In *Environmental and Occupational Medicine.* William Rom, Ed. Boston, MA: Little and Brown, 1992.

Oregon Department of Environmental Quality. *Guidance for Ecological Risk Assessments.* Portland, OR: DEQ, 1997.

Putzrath, Resha. *Use of Risk Assessment in Evaluating Remediation of PCBs.* Conference on Chemical Risk Assessment in the DOD: Science, Policy and Practice. Cincinnati, OH: ACGIH, 1992.

Staynor, Leslie. *Methodological Studies for Quantitative Risk Assessments.* Conference on Chemical Risk Assessment in the DOD: Science, Policy and Practice. Cincinnati, OH: ACGIH, 1992.

Werner, Michael. *Site Risk Assessment: A Primer.* West Chester, PA: Roy F. Weston, Inc., 1996

Yost, Lisa. *Risk Assessment Strategies.* Responsible Environmental Management Conference. Portland, OR, 1997.

Tony Uliano, Jr. *is currently the Manager of Environmental Health and Safety at Oregon Health Sciences University in Portland, Oregon. He has been involved in hazardous materials management from both occupational and environmental perspectives for over 22 years. His diverse experiences were gained during his career as an Industrial Hygienist. Mr. Uliano was an OSHA compliance officer for 10 years and worked at Department of Veterans Affairs medical/research facilities for over 12 years. His major areas of specialty include hazardous waste management, emergency response, indoor environmental air quality, and field investigations; as well as environmental health and safety compliance. In addition to his day job, he performs consulting work and teaches a variety of hazardous materials and environmental science–related courses at several colleges and universities in the Portland area. Acknowledgements go to his wife Andrea for her continuous encouragement and support, and to his daughters Megan and Caitlin for being such an inspiration in his life. Thanks also go to Adriane P. Borgias for her patience and valuable editorial assitance.*

Part II

Safety Principles

CHAPTER 7

Safety Overview

Guy S. Camomilli, CSP, CHSP, CHMM

Introduction

An in-depth discussion about safely managing hazardous materials would take more than a single book, let alone a single chapter. Therefore, this chapter will attempt to deal broadly with the principles of safety and health associated with managing hazardous materials and substances. First, it is necessary to define the categories and subcategories of hazardous materials.

Hazardous materials naturally fall into three categories based on the following properties: chemical, physical, and biological; but remember that many hazardous substances share more than one of these properties. For example, carbon monoxide is a chemical toxicant, but is also a physical hazard when it is stored under high pressure.

Chemical Hazards

Chemical hazards are generally divided into toxic, reactive, corrosive, and flammable subcategories. *Toxic* chemicals produce reactions which are either acute or chronic. *Acutely toxic* materials are substances which produce an immediate reaction which is itself the entire reaction. Acute reactions are characterized by rapid onset and short duration of symptoms. Damage to the body from acute reactions may be reversible or irreversible.

A *chronic toxin* is one which exhibits no symptoms or only mild symptoms at the time of exposure, but may build after a series of exposures. A period of latency may follow, but chronic toxins will typically produce unhealthy effects some time later. Chronic reactions may be triggered by a build-up of the material (*e.g.,* lead) in body tissues, or by the immune system becoming sensitized to specific toxins, such as organic solvents.

Toxic chemicals are subdivided into irritants, sensitizers, tumorogens, mutagens, and teratogens. *Sensitizers* are substances which change the body's proteins. Once these proteins are altered, the body fails to recognize them as its own and reacts by stimulating the immune system to produce antibodies. The antibodies are not immediately produced but are built up after a period of exposure. Then, when next exposed to the same substance, the body tries to fight off what it perceives as an infection. An allergic reaction occurs as the antibodies try to destroy the altered proteins. This reaction can cause many harmful symptoms, including the inability to breathe.

Tumorogens produce tumors. These tumors may be *benign* (noncancerous) or *malignant* (cancerous). Substances which produce malignant tumors are known as carcinogens. Tumorogens are sometimes erroneously equated with carcinogens. Tumorogens alter the genetic code in body tissues and cause the abnormal growth of body tissues. This abnormal growth is a tumor and is usually seen as a swelling or enlargement. Carcinogens are substances which produce *malignant* tumors. Thus, all carcinogens are tumorogens, but not all tumorogens are carcinogens. There are no *realistic* threshold limit values (TLVs) for carcinogens.

Mutagens are substances which alter the genetic code in the gametes (sperm and/or eggs) of the body. The altered genetic code changes the appearance or function of the affected part of the body. This change is passed along to following generations as a permanent alteration of the genetic structure.

Teratogens produce genetic changes or tumors in developing fetuses. Like mutagens, the altered genetic code produces some change in the appearance or function of the body. However, unlike mutagens, teratogens do not affect the genetic code of the gametes and so are not passed on to following generations.

Reactive chemicals produce a violent reaction when exposed to or mixed with another substance, sometimes even water or air. The usual textbook example of a reactive chemical is pure sodium metal which is reactive in water.

Corrosive chemicals are materials that disintegrate body tissues. They particularly affect the water and fatty tissues of the body, and are capable of causing rapid and deep destruction of tissue. Examples of corrosives are lye and sulfuric acid.

Flammable liquid substances are defined differently by the standards of various Federal agencies and industry organizations. Which standard (and which definition) applies to a specific situation will depend on which agency has jurisdiction (*i.e.,* Occupational Safety and Health Administration [OSHA], National Fire Protection Association [NFPA], Environmental Protection Agency [EPA], Department of Transportation [DOT], *etc.*).

Physical Hazards

Materials which are in the category of physical hazards fall into the following groups: ionizing radiation, thermal, and pneumatic high pressure hazards. Such *health physics* hazards as ionizing gamma radiation, nonionizing radiation (*e.g.,* lasers), and electromagnetic fields are not "materials" *per se*, and will not be discussed in this chapter.

Ionizing radiation causes the body's molecules to break apart and form electrically charged particles called *ions*. The ions bond readily with other body chemicals and form toxicants such as hydrogen peroxide. Ionizing radiation can be received by the emission of alpha or beta particles from a radioactive source, or from gamma waves, or X-rays. Particles and waves are emitted from nuclear substances as they degrade to other elements. Proton particles can also damage human body tissues.

Thermal hazards can be subdivided into substances which are cold or hot. These divisions are *not* related to heat or cold stress from exposure to the environment. Temperature tolerance limits vary with the part of the body which is exposed and the duration of exposure. The guideline temperatures given below are for exposure of the bare hand. The duration of these exposures is the human reflex time. The temperature tolerances will change when the duration of exposure is longer. The following temperatures are only guidelines.

Cold substances are considered hazardous to the touch if they are 32°F (0°C) or colder. This is because moist skin will stick to the surface of frozen material. Extremely cold substances are known as *cryogenics* and are defined as substances which are –148°F (–100°C) or colder.

Hot substances are considered thermally hazardous if they are 120°F (49°C) or hotter to the touch.

Pneumatic and hydraulic high-pressure hazards come from pressurized gases and liquids. In addition to the chemical hazards associated with some gases, compressed gases contain considerable potential energy and can be serious physical hazards. Even gases stored at very low pressures can present a danger; and small volumes can be hazardous at high pressures. Vessels or systems which are overpressurized may leak, or they may suddenly and catastrophically fail, releasing a violent shock wave of energy.

Any fluid moving through an opening produces thrust. If a valve or fitting has broken off, the container may become a missile or pressure lines may whip violently. Pressure vessels have been known to fly through cement block walls and chain link fences.

Injury can also result from the injection of gases or fluids into the body. Injection injuries can come from undetected, pin-hole sized, high-pressure leaks. Highly pressurized streams of fluid may cause injuries at pressures as low as 650 pounds per square inch. Such streams of fluid often penetrate several layers of tissue and are very difficult to remove. To minimize pneumatic and hydraulic high-pressure hazards, regulators for compressed *shop* air should be set at 30 pounds per square inch or less. Solid lines should be used whereever possible; flexible pressurized lines should be weighted down with sand bags, or otherwise secured, to prevent whipping if they become detached.

The ensuing OSHA standards discussed in this chapter apply either specifically, or generally to the safe use, handling, and/or control of hazardous materials. Every Certified Hazardous Materials Manager (CHMM) should study each standard in greater detail before assessing materials and/or their hazards for practical applications.

Occupational Safety and Health Act of 1970

The Occupational Safety and Health Act (OSH) Act of 1970 (PL 91–596), was passed by the 91st Congress on December 29, 1970. It was amended by PL 101–552, Section 3101, on November 5, 1990.

The OSH Act intended

> [t]o assure safe and healthful working conditions for working men and women; by authorizing enforcement of the standards developed under the Act; by assisting and encouraging the States in their efforts to assure safe and healthful working conditions; by providing for research, information, education, and training in the field of occupational safety and health; and for other purposes.

The 91st Congress intended to reduce the number of occupational safety and health hazards. The law sought to motivate employers and employees to be responsible for achieving and maintaining safe and healthful working conditions.

Congress also adopted the idea of *national consensus standards* as a way to establish nationally recognized standards of safety and health. They then created the *general duty clause* in Section 5(a)1 of the OSH Act. The general duty clause declared that each employer has a duty to furnish to his employees a workplace free from recognized hazards that are likely to cause serious physical harm or death. The implication of the clause is that employers must use common sense in providing a safe and healthful workplace.

Several conditions must be met before OSHA can invoke the general duty clause: (1) there must be a hazard which caused or is likely to cause serious physical harm or death; (2) the employer must have recognized the hazard and failed to remedy the hazardous condition; and (3) there must have been a feasible way to correct the hazard. If all of the above conditions are met, OSHA can invoke the general duty clause.

The prudent employer utilizes *best practices*. Best practices means the employer provides the best possible engineering and administrative means (over and above the specific requirements of law) to control hazards in the workplace. In other words, the employer does what is necessary to best protect the employee. This concept is difficult to market to many employers because *prevention* (nonoccurrence) is neither quantifiable nor easily proven. The employer who looks only at short-term costs will not recognize best practices as a way to prevent greater long-term expenses. The employer who truly holds at heart the safety and well-being of the employees will reap the additional long-term benefits of employee dedication, productivity, efficiency and safety.

In addition to the Federal OSHA standards, 23 states and 2 territories have their own OSHA-approved occupational safety and health plans. The provisions of each State plan must be *at least* as stringent as the OSHA requirements. The hazardous materials manager must evaluate all requirements and invoke the most judicious standards and/or best practices for his area of responsibility.

General Industry Standards: 29 CFR 1910—Occupational Safety and Health Standards

Subpart B—Adoption and Extension of Established Federal Standards

29 CFR 1910.19: Special Provisions for Air Contaminants. This standard makes all of 29 CFR 1910 applicable to various specialized industries: construction, shipyard employment, longshoring, and work in marine terminals.

Subpart C—General Safety and Health Provisions

29 CFR 1910.20: Access to Employee Exposure and Medical Records. The provisions of this standard give employees the right of access to their exposure and medical records. Employees must be provided access to their records within 15 working days of a written request. The records are to be provided at no cost to the employee.

Subpart G—Occupational Health and Environmental Control

20 CFR 1910.94: Ventilation. The ventilation standard encompasses the application of many requirements for employee protection from airborne materials, and deals with topics such as abrasive blasting, grinding, polishing, buffing, spray finishing, and open surface tanks. The standard requires that certain breathable dusts and chemical vapors be kept below the levels specified in the 29 CFR 1910.1000 series of standards. Exposure to abrasives and surface coatings of materials must be kept to a minimum. The standard allows Personal Protective Equipment (PPE) such as safety glasses, exhaust ventilation, respiratory protection, Mine Safety and Health Administration (MSHA) approved respirators and cartridges, gloves, *etc.*, to be used in order to minimize the employees' exposure to hazardous materials.

29 CFR 1910.96: Ionizing Radiation. The ionizing radiation standard establishes radiation dose limits for employees during any one calendar quarter. This standard includes limits for radiation doses to the whole body and for employees under the age of 18. Employers are required to maintain exposure records on their employees. The standard also limits the exposure of employees to airborne radioactive material.

Employers are required to provide radiation surveys using personal monitoring equipment, and to verify that employees are in compliance with dose limits. Caution signs, labels, and signals must be posted to help protect against inadvertent exposure. The employer must also establish evacuation procedures, including emergency evacuation signals.

Employees working in or frequenting radiation areas must be informed of the presence of radioactive materials in the workplace. The employer must also provide his employees with instruction on the health hazards that come from exposure to radioactive materials and radiation. The instruction should include training on how to minimize exposure and how to take precautions against overexposure. This standard also requires the employer to post the OSHA radiation standard in a place where employees will see it on the way into and out of the workplace.

The ionizing radiation standard requires employers to notify the proper authorities in the event of a radiological incident involving the overexposure of employees to a radioactive source. Employers must keep exposure records and must inform individual employees about their individual personal exposure levels at least once a year.

Subpart H—Hazardous Materials

29 CFR 1910.101: Compressed Gases (General Requirements). Compressed gases offer unique material hazards. In addition to the obvious physical hazards, compressed gases may also possess the hazardous properties of some liquids or solids. They may be toxic, flammable, corrosive, reactive, *etc.* OSHA recognizes that the Compressed Gas Association (CGA) and Department of Transportation (DOT) provide an industry consensus on the ways that compressed gas hazards can be effectively controlled. OSHA refers to and incorporates this industry consensus into the 29 CFR 1910.101 standard.

Under this standard, employers are required to ensure that any compressed gas containers under their control are in a safe condition. *Safe conditions* are listed in the DOT Hazardous Materials Regulations (49 CFR 171–179) and the CGA Pamphlets C–6–1993, "Standards for Visual Inspection of Compressed Gas Cylinders" and C–8–1997, "Standards for Requalification of DOT–34T, CTC–34T, and TC–34HTM Seamless Steel Cylinders." Handling, storage, and use of all compressed gases in cylinders, portable tanks, rail tankcars, or motor vehicle cargo tanks, must be accomplished according to CGA's Pamphlet P–1–1991, "Safe Handling of Compressed Gases in Containers." The safety relief equipment for

compressed gas containers must have pressure relief devices installed on them and must be maintained as is outlined in CGA's Pamphlets S–1.1–1994, "Pressure Relief Device Standards Part 1–Cylinders for Compressed Gases" and S–1.2–1995, "Pressure Relief Device Standards, Part 2–Cargo and Portable Tanks or Compressed Gases."

The following standards are referred to here because they apply to the specific hazardous gases named in their headings. Each standard outlines the safe handling, transportation, and storage of its named substance. The OSH Act requires compliance to the standards that are incorporated by reference, such as those from the CGA, American Society of Mechanical Engineers (ASME), or DOT:

*   29 CFR 1910.102 Acetylene

*   29 CFR 1910.103 Hydrogen

*   29 CFR 1910.104 Oxygen

*   29 CFR 1910.105 Nitrous oxide

29 CFR 1910.107: Spray Finishing Using Flammable and Combustible Liquids. Spray finishing uses a liquid or powdered material to provide a coating on an article. The 29 CFR 1910.107 standard regulates many aspects of spray finishing, such as (1) the construction of spray booths, (2) the use of electrical equipment, and (3) other potential sources of ignition in the spray finishing process. This standard also covers:

*   Ventilation requirements for spray finishing

*   Storage and handling of flammable liquids

*   Locations in which spraying is permitted

*   Cleaning of tools and equipment

*   Disposal of coating residues

*   Control of contaminated employee clothing

*   Use of cleaning solvents

*   Combining hazardous materials

*   Requirements for fire protection during spray finishing work

29 CFR 1910.108: Dip Tanks Containing Flammable or Combustible Liquids. Any tank or other container which contains a flammable or combustible liquid and in which articles or materials are immersed for any process, is considered by this standard to be a *dip tank*. The 29 CFR 1910.108 standard covers:

- Dip tank ventilation

- Dip tank construction

- Storage and handling of dip tank liquids

- Potential sources for ignition of dip tank vapors (including electrical sources)

- Operation and maintenance of dip tanks and their surrounding facilities

- Requirements for extinguishing fires

- Requirements for special dip tank processes

Dip tanks containing flammable or combustible liquids produce vapors. Vapors can be dangerous, and so dip tank areas must be limited to the smallest practical space. Tanks and drainboards must be constructed of noncombustible materials, be properly supported, and be equipped with bottom drains and automatic fire extinguishing hardware. Only the proper National Electrical Code (NEC) Class, Division, and Group rated electrical services are allowed to be installed in the vapor areas of dip tanks. These fixtures are rated by the National Fire Protection Association's (NFPA) NEC to prevent any inadvertent ignition of the vapors by an electrical spark. Only the minimum amounts of combustible materials are allowed to be stored at the work site, and the storage of combustible debris is prohibited.

The 29 CFR 1910.108 standard requires periodic inspections. Warning signs must be posted, and there must be easy access to fire extinguishers. The extinguishers must be suitable for use on burning flammable and combustible liquids.

29 CFR 1910.109: Explosives and Blasting Agents. The 29 CFR 1910.109 standard outlines the proper storage, handling, and transportation of explosives, including requirements for ammonium nitrate, and small arms ammunition. The proper packaging, marking, transportation, use, loading, and initiation of explosive charges and small arms ammunition, and construction of storage magazines, are also covered.

29 CFR 1910.110: Storage and Handling of Liquefied Petroleum Gases. This standard regulates many aspects of the commercial use and

handling of liquid petroleum (LP) gases. The requirements for LP gas systems, equipment, and facilities are discussed, as well as the specific requirements for valves, accessories, piping, gauges, hoses, and safety devices. The standard covers how to transfer LP gases into and out of certain vehicles and how to properly mark LP gas containers. There are also safety requirements concerning the location of LP gas tanks and cylinder systems.

Many liquid petroleum gases are odorless in their natural states. Therefore, the 29 CFR 1910.110 standard requires that all LP gases be *odorized* so that leaks may be quickly detected. The standard also addresses the requirements for LP gas systems on commercial vehicles and at LP gas service stations.

29 CFR 1910.111: Storage and Handling of Anhydrous Ammonia. The 29 CFR 1910.111 standard imposes specific requirements on the design, construction, marking, location, testing, requalification, installation, and operation of anhydrous ammonia systems. Provisions are included in the standard for:

- Refrigerated ammonia storage systems

- Any associated pumps and compressors

- Safety relief devices

- Gauges, fittings, and valves

- Hoses, piping and tubing

This standard also has components which outline the requirements for respiratory protection, quick-drench facilities, and how to safely transfer ammonia liquid.

29 CFR 1910.119: Process Safety Management of Highly Hazardous Chemicals. This standard applies to processes (nonretail) that use chemicals at or above the threshold quantities listed in Appendix A of the standard. Any process that uses 10,000 pounds or more of a flammable liquid or gas stored below its natural boiling point (without using chilling or refrigeration) is covered by this standard. The standard does not apply to materials used as fuels.

The 29 CFR 1910.119 standard requires employees to participate in the employer's development of Process Hazard Analyses (PHAs). PHAs are conducted to identify and define the hazards posed by the processes that use hazardous chemicals. PHAs must go into enough detail to adequately deal with the complexity of the system. PHAs are required to use specific methods in order to

- Identify and evaluate hazards

- Pinpoint previous incidents that have catastrophic potential

- Select engineering and administrative safeguards against hazards

- Outline detection methods

- Describe the possible effects on employees if the workplace safeguards fail

The standard requires PHAs to be updated at least every five years. Operating procedures must conform to current practices. At least once a year, the employer is required to certify that its operating procedures are both current and accurate. Operating procedures are required to

- Reflect the entire range of routine operations

- Outline emergency operations

- Discuss the consequences of deviation from written procedures

- List the precautions that will prevent personnel exposure

- Warn of special or unique hazards

Written procedures must also establish the protocol for making changes to processes, equipment, and materials used. The operating procedures must also include an emergency action plan and procedures for handling small releases of chemicals.

Both initial and refresher training is required by the standard. *Initial training* is required before an employee begins participating in operations that use highly hazardous chemicals, and *refresher training* is required a minimum of every three years. All training must be documented by the employer. (Test results are considered evidence of training and understanding.) Similar requirements apply to any contractors and maintenance personnel of the employer who work on or around the process equipment.

Each process-related incident that resulted, or could have resulted, in a catastrophic release must be investigated within 48 hours of occurrence. Employers ought to keep a record of these operational mishaps. The record can be analyzed for trends that reveal weak spots in the PHAs and operating procedures. Operating procedures can then be improved in response to these mishaps and close calls. Compliance audits are required by the standard in order to ensure that procedures are adequately safe.

The 29 CFR 1910.119 standard includes appendices containing examples of safe processes, voluntary guidelines, and recommendations to help guide the employer and/or hazardous materials manager toward compliance.

29 CFR 1910.120: Hazardous Waste Operations and Emergency Response. This standard covers hazardous emergency situations for *all* toxic substances in general. The emergency responses are not limited to hazardous waste operations, hazardous wastes, or substances on any Comprehensive Environmental Response, Compensation and Liability Act (CERCLA) or Resource Conservation and Recovery Act (RCRA) site. Once any toxic chemical is spilled, it becomes a hazardous waste by definition and is covered under the provisions of this standard. Training and other provisions to protect workers are necessary whenever exposure to toxicants is possible. This is particularly true during emergencies.

OSHA believes that "[t]he treatment and disposal of hazardous wastes under RCRA and CERCLA creates a significant risk to the safety and health of employees who work in treatment and disposal operations." The most significant risks to employees come from exposure to hazardous wastes by means of absorption through the skin and inhalation.

The standard emphasizes the necessity of limiting an employee's exposure to any health hazards on the worksite. The requirements include:

- A safety and health program which incorporates the employer's organizational structure

- A comprehensive plan to develop procedures, and identify and control hazards

- Standard operating procedures for safety and health

- A safety and health plan that integrates the general program with procedures for specific sites

- A safety and health training program

- A medical surveillance program

Hazardous waste sites are evaluated in order to identify specific hazards and determine the appropriate procedures for controlling these risks to safety. The preliminary evaluation determines the potential hazards and recommends the appropriate protection employees must have before entering the site. The evaluators must be especially careful to identify any hazards which may be Immediately Dangerous to Life and Health (IDLH), such as confined spaces or hazardous atmospheres. They must pay attention to *biological indicators*, such as dead animals or vegetation. Based on the preliminary evaluation, Personal Protective Equipment (PPE) is selected, and the appropriate site-control procedures must be implemented.

Site evaluation and characterization is a continuous process, and the assessments are used to develop and *maintain* a valid safety and health plan. *All* site personnel should be constantly alert to changing site conditions and respond to any new information on the hazards around them.

Employees at hazardous waste operations face serious health and safety risks. Frequent training is needed to reinforce the safe work practices necessary to avoid exposure to these hazards. The employer must provide training that is consistent with the worker's job function and responsibilities. The training must be provided frequently enough to strengthen the initial training and to update employees on any new policies or procedures.

All employees who may be exposed to hazardous substances (including supervisors and managers) must be trained to the level their job requires *before* they are permitted to engage in, supervise, or manage hazardous waste operations. Employees whose routine activities may potentially expose them to hazardous substances (including supervisors and managers) are required to have a minimum of 40 hours of off-site classroom instruction, and three days of actual field experience. The field experience must be ac-

complished under the direct guidance of a trained and experienced supervisor. Employees who are on-site, but are not potentially exposed over the Permissible Exposure Limit (PEL), are required to have a minimum of 24 hours of off-site instruction and one day of field experience. All affected employees must complete eight hours of refresher training annually.

Sites that generate small quantities of hazardous materials are conditionally exempt to certain provisions of the 29 CFR 1910.120 standard.

Instructors and supervisors who train employees for hazardous waste operations must have either completed a training program for the subjects they teach, or earned academic credentials and have experience teaching those subjects. Trainers must demonstrate that they are competent to instruct and that they possess sufficient knowledge of the subject matter.

Some site conditions require a medical surveillance program. The program must include free medical examinations and consultations. The employer is required to keep records of these examinations and consultations, and to provide the results to the employees.

The employer needs to know the hazards faced by employees in order to develop and implement effective control measures. A written safety and health program forces the systematic identification of hazards in the workplace, makes employees aware of these hazards, and identifies the proper response employees should have to them. The more accurate and detailed the information available about a site, the more protective measures can be tailored to the actual hazards that the employees may encounter.

Employers use engineering controls, work practices, and/or personal protective equipment to protect their employees. These controls and/or equipment must keep employees below the permissible exposure limits. Specific methods for hazardous materials exposure determination and limits are provided in 29 CFR 1910, Subparts G and Z, and in the American Conference of Governmental Industrial Hygienist's *Documentation of the Threshold Limit Values* (ACGIH's TLVs). At the time of this writing, the latest edition was published in 1991, but three supplemental updates have been issued. Employers who use the ACGIH's TLVs must be certain to use the most up-to-date values.

In situations where an employee may be exposed to hazardous concentrations of chemicals, the conditions of the site must be monitored. Before entering into potentially hazardous atmospheres, the employee or supervisor must check for the presence of any conditions which may be Immediately Dangerous to Life and Health (IDLH), including flammable and oxygen-deficient atmospheres. The employee or supervisor must continue periodic monitoring of site conditions when the possibility of an IDLH condition exists. If exposure levels may possibly have risen over the PEL or over the previously documented level of the location being monitored, then the employee or supervisor must continue periodic monitoring while the employee is exposed.

DOT, OSHA, EPA, and State regulations describe the correct way to handle drums and other containers during cleanup operations. The people in charge of cleanup operations should pay particular attention to container inspection, labeling, and handling; fire safety; and protection of the environment.

Every employer must develop a decontamination procedure in order to minimize the employees' contact with hazardous substances. All employees who enter contaminated areas are required to decontaminate or dispose of their clothing and equipment *before* leaving the contaminated area. The clothing should be laundered by a commercial laundry or other cleaning establishment familiar with the potentially harmful effects of the contaminated clothing. The employers may also need to provide showers and change rooms.

Any employer who assigns employees to the danger area when an emergency occurs must develop an emergency response plan. The plan must address:

- Preemergency planning
- Evacuation routes and procedures
- Emergency recognition and prevention
- Personnel assignments
- Lines of authority
- Communication
- Decontamination procedures
- First aid and emergency medical treatment
- Safe distances

- Places of refuge
- Security
- Critique of the response and follow-up

The plan must contain the topography, layout, and weather conditions at the site, and the procedures for reporting incidents to the Federal, State, and local authorities. The plan must be compatible with the Federal, State, and local plans, and must be reviewed and rehearsed regularly.

Employers who provide temporary storage or disposal services under the Resource Conservation and Recovery Act must comply with special requirements in the standard. There are also special provisions for professionals who respond to emergencies involving hazardous materials.

Appendices A, B, and C of the standard provide tests and guidelines which may be used to assess compliance, and Appendix E provides training and curriculum guidelines for assistance in developing training for specific sites.

Subpart I—Personal Protective Equipment

29 CFR 1910.132: General Requirements. The general requirements for Personal Protective Equipment (PPE) include protection for the eyes, face, head, and extremities. The requirements also apply to protective clothing, respiratory devices, protective shields, and barriers. The 29 CFR 1910.132 standard requires PPE to be used whenever employees are exposed to hazards from processes, environment, chemicals, radiation, or mechanical irritants which could cause bodily injury or impairment through absorption, inhalation, or physical contact. It requires the employer to make sure that the PPE selected is adequately designed, maintained, and kept in sanitary condition.

The employer must complete a hazard assessment in order to identify the proper PPE for use. The assessment must be documented in writing. The certification must specify the workplace evaluated, the date of the evaluation, and the person who performed the evaluation.

Training is vital to the success of any PPE program. Every employee who is required to wear

PPE must be trained in the care and use of each article of the equipment. The employee must also learn the conditions which make PPE necessary; the PPE's selection and limitations; how to properly don, doff, adjust, and wear the equipment; and disposal of the PPE. The employer must ensure that each employee *demonstrates* an understanding of each of the training requirements, as well as the ability to use the PPE, before allowing the employee to perform work which requires PPE. If the employer has reason to believe that any person does not understand or possess the skill required to properly use the PPE, the employer must retrain that employee. Retraining is also required when changes in the workplace or PPE make the prior training obsolete. The employer must keep written records (containing the employee's name, the date, and the subject of certification) certifying that each employee has received and demonstrated an understanding of the required training.

29 CFR 1910.133: Eye and Face Protection. In addition to the hazards inherent in certain materials, the side effects of some processes (*i.e.*, splashes or flying particles) pose particular risks to the eyes and face. Because of the variety of ways in which eyes and face can be injured, it is important to select the appropriate eye and face protection for the task. All protective eyewear and facewear must comply with the American National Standards Institute (ANSI) standard Z87.1–1989, "Practice for Occupational and Educational Eye and Face Protection." However, the ANSI Z87.1 inscription on the eye protection does not mean that the proper PPE was selected. Glasses with side shields are appropriate to protect the eyes from flying objects, but goggles or full-face respirators which provide an air, liquid, and dust-tight seal are needed wherever an airborne presence, splashing, or spattering of hazardous materials may occur. Employees who need prescription lenses must wear eye protection that either incorporates the prescription into the protective gear, or fits over the employee's regular prescription eye wear without disturbing either the prescription lenses or the PPE.

29 CFR 1910.134: Respiratory Protection. The respiratory protection standard was intended to prevent occupational diseases caused by breathing contaminated air in the workplace. The standard requires that, where feasible, engineering controls or substitution with a less hazardous substance be the first line of employee protection. When engineering controls cannot reasonably be used, employers may use respiratory equipment to protect their employees' health. The employers must provide the respiratory protection free of charge, and they are responsible for implementing respiratory protection programs. OSHA requires employees to use the respiratory protection provided in accordance with the instruction and training given by the employer.

Training programs must present the written standard operating procedures for selecting and using respirators. Employees should select and use the equipment on the basis of hazards they will encounter. Their selections must be in compliance with the guidelines of the ANSI standard Z88.2–1992, for respiratory protection. Only those respirators that are jointly approved by the Mine Safety and Health Administration and the National Institute for Occupational Safety and Health may be used. The standard also requires employers to train and instruct their employees about the limitations, cleaning, disinfecting, inspecting, and storing of respirators.

The 29 CFR 1910.134 standard requires initial and annual medical examinations. The examinations must be performed by a Physician or other Licensed Health Care Professional (PLHCP), and must be well documented.

The employer must develop standard procedures for the use and maintenance of respirators. These procedures should outline how to correctly select, use, and care for respirators, and how to safely use respirators in dangerous atmospheres. The employers should also record their plans for hands-on, on-the-job training for employees who are required to use respirators, and for frequent random inspections to ensure that respirators are being properly used and maintained.

Employees are prohibited from wearing contact lenses when using a respirator in a contaminated atmosphere.

The air supplies for a self-contained breathing apparatus and for supplied-air respirators must meet the United States Pharmacopoeia requirements, and the requirements for Grade D breathing air defined by the Compressed Gas Association (CGA) G–7.1–1997, "Commodity Specification For Air." Airline couplings must be incompatible with the couplings of other gas lines in order to prevent the users from connecting respirator lines to supplies of nonrespirable gases by mistake.

Subpart J—General Environmental Controls

29 CFR 1910.146: Permit-Required Confined Spaces. The Permit-Required Confined Spaces standard was formulated to protect employees in general industry from exposure to the hazards which may be found in confined spaces. This standard is a re-creation of the previous confined space standard. The main difference is that confined spaces have been separated into two categories: confined spaces and permit-required confined spaces. **Confined spaces** must

- Be large enough for an employee to enter

- Have a limited or restricted means of entry or exit

- Not have been designed for continuous occupancy

A *permit-required confined space* is a confined space that

- Contains or has the potential to contain a hazardous atmosphere

- Contains a material that has the potential to engulf an entrant

- Has inwardly converging walls or a floor which slopes downward and tapers to a smaller cross section so that an entrant could be trapped or asphyxiated inside

- Contains any other recognized serious safety or health hazard

It is a good practice to periodically reevaluate the workplace for any changes, additions, or eradications of confined spaces. All assessments of confined spaces must be documented, and include information on the location(s) evaluated, the date of the evaluation, and the name and signature of the person performing the evaluation.

The employer has several duties under the 29 CFR 1910.146 standard. First, the workplace must be evaluated to determine if any areas meet the criteria of confined spaces. If any spaces are found that meet the criteria, danger signs or other effective means must be used to warn personnel not to enter the space. If an employer decides that its employees will *not* enter the confined space, the employer must take positive measures to ensure that no employee will be able to enter. If the employer decides that it is necessary for employees to enter the space, then the employer must develop and implement a written program that satisfies the requirements of the standard.

Many elements go into a program covering permit-required confined spaces, beyond controlling entry and identifying and evaluating the hazards. For example, the employer is required to

- Furnish all equipment necessary for the employee to enter safely

- Monitor the permit-required confined space when anyone is inside

- Provide at least one attendant throughout the duration of the entry

- Assign specific responsibilities to each person who has an active role in the entry

- Develop and implement emergency rescue procedures

- Develop and implement a detailed system for granting confined space entry permits

- Coordinate entries with other employers at the site

- Review entry operations and permit requirements when conditions change

Permits are required to display the following:

- The responsible supervisor's signature

- Date and authorized duration of the permit

- The permit space to be entered

- Reason for entry

- Names of the authorized entrants

- Names of attendants

- Name of the entry supervisor

- Any hazards of the space being entered

- Measures used to isolate or eliminate the hazards in the confined space

- Acceptable entry conditions
- Test results
- Available rescue and emergency services
- Communication procedures
- Required equipment
- Hot-work permits and any other pertinent information

Because the conditions in confined spaces can change, employers are required to reevaluate confined spaces if changes occur, in order to determine if the space must be reclassified.

When an employer makes arrangements for the employees of another employer to enter a confined space, the host employer is responsible for informing the contractor about the presence of the confined space(s) under its control. Additionally, the host employer must notify the subcontractor that a permit system for entry into confined spaces is in effect and that entries must be coordinated prior to their execution. The standard requires that personnel be debriefed after their entry work has been completed. Other information which must be made known are the hazards associated with the confined spaces, and the precautions and/or procedures implemented by the host.

As with most hazardous operations, training is crucial to guaranteeing that the risks are reduced to acceptable levels. Training must ensure that employees have mastered the procedures for entering a permit-required confined space. The training and evidence of mastery must be sufficient to do the job, and must be documented.

The requirements for entrants, attendants, supervisors, and rescue personnel are outlined in the 29 CFR 1910.146 standard, and Appendices A through E provide information and guidelines to assist in compliance.

29 CFR 1910.147:   The Control of Hazardous Energy (Lockout/Tagout). The lockout/tagout standard is covered here because it complements other measures which control exposure to materials which contain chemical and/or physical energy. This standard fits hand in glove with the process safety standard 29 CFR 1910.119 for general industry during servicing and/or maintenance of machines and equipment. It requires employers to set up a program to positively lock out and tag out devices, in order to prevent employees from being exposed to uncontrolled hazardous energy.

OSHA classifies employees by their relationship to the hazardous energy sources. *Affected employees* are those who operate or use the machinery, processes, or systems that are being maintained or serviced; *authorized employees* are those who actually perform the service or maintenance. Although OSHA does not name another category, it mentions the group of *all other employees*, who may have limited or indirect contact with the machinery, processes, or systems that are locked or tagged out.

The program requires the employer to control hazardous energy, train employees, and provide periodic inspections to ensure that starting the machine, activating the process or system, or releasing energy cannot injure an employee. At least once a year, the program must be inspected in order to certify the employer's compliance with the standard.

A *tagout* system may be used in place of a *lockout and tagout* system *only* when the system cannot be locked out, and when it can be *proven* that the tagout is as effective in preventing injury as the lockout/tagout system.

Again, training is vital to effectively protecting employees. The degree of training required depends on the employee's role: affected, authorized, or other. The employer must train employees on the

- Purpose and function of the energy control program
- Recognition of hazardous energy sources
- Type and magnitude of the energy present in the workplace
- Means required for energy isolation and control
- Purpose and use of the control procedure
- Prohibition relative to starting or energizing machinery, or activating the process or system which is locked and/or tagged out

The employer must also verify that the employees possess the knowledge and skills necessary to follow the lockout/tagout procedures.

Other requirements of the standard outline how and by whom lockout/tagout operations may be performed. Lockout and/or tagout will only be performed by the maintenance (authorized) employee(s). Prior to beginning the lockout/tagout of the machinery, process, or system, the authorized employee must notify all affected employees of the coming outage. Next, the machine must be shut off using the approved methods. All energy-isolating devices must be installed and all potential energy must be relieved or otherwise rendered safe. Then the authorized employee must verify that the machine, process, or system cannot transmit any hazardous energy. After the maintenance or repairs are completed and before energy is restored, the authorized employee must inspect the area to make sure that the machine, process, or system can be restored to operational status. Also, the authorized employee must make certain that all other employees are a safe distance from the machine, process, system, and/or energy source when it is made operational. The employee who placed the lock or tag will then remove the device(s). After the devices are removed, the authorized employee must notify all the affected employees that the machinery, process, or system is no longer locked or tagged out.

If a subcontractor's employees use this equipment, then they must advise the host employees of their lockout/tagout procedures and *vice versa*.

Where work continues across shifts, arrangements must be made between the authorized employees of all shifts for the continuity of protection.

Subpart Z—Toxic and Hazardous Substances

OSHA's objective in writing standards that dealt with toxic materials or harmful physical agents was to "set the standard which most adequately assures, to the extent feasible, on the basis of the best available evidence, that no employee will suffer material impairment of health or functional capacity even if such employee has regular exposure to the hazard dealt with by such standard for the period of his working life."

OSHA required standards to be based upon research, demonstrations, experiments, and other appropriate information, and to attain the highest degree of health and safety protection for the employee. Whenever practical, the standards were to be expressed in terms of *objective criteria* and of *the performance desired*.

OSHA made the 29 CFR 1910.1000 series of standards the foundation and framework on which the rest of the standards were built. OSHA recognized that many employee deaths and injuries were associated with exposure to hazardous materials. The root causes appeared to be: (1) the manufacturers' and distributors' failure to identify the health and physical hazards and the safe exposure levels of their products, (2) the employers' failure to investigate the hazards and safe exposure levels of the materials they used in the workplace, and (3) the employees' failure to perceive the hazards associated with the materials that they were exposed to in the workplace.

The objectives of the 29 CFR 1910.1000 series of standards were to: identify selected chemicals used in industry and determine their safe exposure limits (Personal Exposure Limits [PELs]), require employers to notify their employees about each hazard in the workplace, and require employees to learn how to recognize hazards and how to protect themselves.

29 CFR 1910.1000: Air Contaminants. With air contaminants, as with all other toxicants, *the dose makes the poison*. The 29 CFR 1910.1000 standard defines the limits to which an employee may be exposed to any chemical listed in the standard's tables. The limits for the listed chemicals are defined in two ways. The first definition is the eight-hour *time-weighted average* (TWA). The TWA is the measured concentration of the chemical multiplied by the duration of the exposure. The total value of this calculation must not exceed the limit for an eight-hour constant exposure at the PEL. The formulas for calculating a TWA are provided in the standard. The second way to calculate exposure limits is by determining the highest concentration of air contaminants to which any employee may be exposed. This is called the *ceiling*. Employees may not be exposed to chemicals at the ceiling level for more than 15 minutes at a time, and no more than four times daily. These 15-minute exposures must be at least 60 minutes apart. Employees may never be exposed to levels above the ceiling value. The 29 CFR 1910.1000 standard requires the employer to use *engineering controls* (where feasible) to protect employees from exposure to hazardous chemicals. *Administrative controls* and PPE are allowed only when engineering controls are not feasible.

29 CFR 1910.1030: Bloodborne Pathogens. The scope of this standard is not based on a few specified industries. It is based upon the reasonable possibility that any employee can be exposed to bloodborne pathogens. The risk of exposure to bloodborne pathogens is not exclusive to the healthcare industry. The employee is at risk of exposure anytime he or she can come into contact with another person's blood OSHA believes that there is a reason for national concern for the occupational safety and health of employees exposed to bloodborne pathogens.

In 1983, OSHA issued a set of voluntary guidelines designed to reduce the risk of occupational exposure to the Hepatitis B Virus (HBV). The guidelines included recommended work practices and recommendations for use of immune globulins and the hepatitis B vaccine. Despite this measure, significant risks remained. OSHA concluded that an occupational health standard would reduce these risks more effectively.

The 29 CFR 1910.1030 standard covers *all* bloodborne pathogens, not just Hepatitis B Virus (HBV) and Human Immunodeficiency Virus (HIV).

OSHA realizes that assessing the risk of exposure to bloodborne diseases is different from quantifying the risks that come from exposure to toxic chemicals. The risks associated with chemicals generally are acute dose reactions or chronic health effects. The effects of exposure to bloodborne pathogens do not depend upon cumulative dose. Each time an employee is exposed, infection either occurs or it does not occur. OSHA concluded that the best way to reduce the risk of the transmission of bloodborne diseases is by reducing exposure.

Exposure to blood or other potentially infectious materials place workers at risk. OSHA notes that "[a]ll employees who are exposed to blood and other potentially infectious materials may be at risk of infection." The agency requires employers to conduct an exposure determination, in order to identify any occupational exposure to bloodborne pathogens (traditional and nontraditional). OSHA's intent is to identify and protect all employees who are reasonably at risk of skin, eye, mucous membrane, or parenteral contact with blood or other po-tentially infectious materials during the per-formance of their job duties (occupational exposure).

Every employer needs to know who among the employees is potentially exposed to bloodborne pathogens so that the proper training, engineering, work practice controls, personal protective equipment, and the other provisions required by the standard, can be provided to the employee. To learn who is potentially exposed, the employer should examine the tasks and procedures carried out by employees, and ascertain any exposure that can reasonably be anticipated to occur. For example, it would be reasonable to anticipate that someone who provides first aid at a construction site will eventually be exposed to blood in the course of his or her work duties. The employer must identify those tasks and procedures where occupational exposure may occur, and identify the workers whose duties include those tasks and procedures. The Exposure Control Plan is a key provision of the standard. It requires employers to identify the individuals who will receive training, personal protective equipment, vaccination, and other protection required by standard.

Hazardous materials managers who provide professional guidance for the control of human blood and/or blood products (such as professions in the traditional health-related industries) should supplement the material in this chapter with readings from James Tweedy's *Healthcare Hazard Control and Safety Management*.

OSHA names 24 industry sectors where the exposure to bloodborne pathogens is significant, but the 29 CFR 1910.1030 standard encompasses *all* workplaces in which employees might be exposed to blood or other potentially infectious materials during the performance of their duties. Although most of the 24 named sectors are directly related to patient care, 11 are not. They are:

- Personnel services

- Funeral homes and crematories

- Research laboratories

- Linen services

- Medical and dental equipment repair
- Law enforcement
- Fire and rescue
- Correctional institutions
- Schools
- Lifesaving
- Regulated waste removal

OSHA's plan for reducing an employee's exposure to bloodborne pathogens is based on **universal precautions** of infection control. The carriers of any disease are not usually identifiable, and contaminated materials and articles are not always properly identified. As a result, the exposed worker could be unaware of a grave hazard. Universal precautions require employees to assume that all human blood and body fluids contain potentially infectious bloodborne pathogens, and to handle those articles and substances as if they were pathogenic

The 29 CFR 1910.1030 standard is a **performance standard**. It requires the employer to *effectively* implement a program to protect employees from bloodborne pathogens. The employer must develop measures to abate this hazard, that best suit the work place and protect the worker from contact with blood, body fluids, and other potentially infectious materials.

Employers must develop an exposure control program that identifies the tasks and/or positions at risk from bloodborne pathogens, and then document the implementation of the measures that will reduce the potential risk. They must then develop procedures to evaluate the circumstances surrounding exposure incidents. Careful documentation of the circumstances leading to an exposure can help managers more efficiently identify future problems. The standard also requires employers to offer the HBV vaccine and to provide evaluation and treatment to employees following an exposure.

Training is an integral part of reducing risks. Training that accurately informs the employees about risks is an indispensable link in hazard abatement. The training requirement of the 29 CFR 1910.1030 standard ensures the effectiveness of the other provisions of the standard.

Work practices and equipment design can also lessen exposure to bloodborne pathogens. If necessary, an employer should alter the way a task is performed or change the way equipment is designed in order to prevent occupational exposure. Workers who are exposed to blood, body fluids, or tissues can be protected from the risk of infection by using appropriate clothing, masks, gloves, and other protective equipment. When this is coupled with safe work practices, operating procedures, and education, the employee will have the best protection possible from bloodborne diseases.

Many employees are exposed to bloodborne disease because of broken skin, or through their mucous membranes. Personal protective equipment (PPE) is a direct line of defense for all workers who come into contact with blood or other potentially infectious materials. Minute breaks in the skin and airborne particles (*e.g.*, from a cough or sneeze) often cannot be seen. Since PPE isolates such portals of entry from potentially infectious materials, it is an effective means of preventing infections when properly used.

Communicating hazards to employees and providing training and information is critical to the implementation of the 29 CFR 1910.1030 standard. PPE and proper work practices will not protect personnel unless the workers are instructed in their effective use.

Safe work practices are essential to preventing exposure to biohazards. These controls will reduce risk by requiring employers to make sure that employees are performing their tasks in the safest manner possible, consistent with universal precautions. Examples of work practice controls would be prohibiting the recapping of syringe needles, forbidding mouth-pipetting, and requiring hands to be washed after removing personal protective equipment.

Engineering controls reduce the risk of employee exposure in the workplace by either removing the hazard or isolating the worker from exposure. Work practices and engineering controls work in tandem. It is often necessary to employ work practice controls to assure that engineering controls operate effectively. For example, a sharps disposal container provides no protection if an employee persists in recapping needles by hand or disposing of them in the wastebasket. Proper work practices and engineering controls must both be utilized to ensure safe, acceptable sharps disposal.

29 CFR 1910.1200: Hazard Communication. The 1910.1200 standard is the essence of the hazardous materials regulations. This standard states that employees have the *right-to-know* about the hazards present in their workplace, and outlines the specific methods of how employees will be informed of those hazards.

Beside the specific items of compliance in the standard, should your workplace be inspected, compliance officers will be assessing the adequacy of the program. A list of the items which compliance officers may ask to see is included in Appendix E of the standard.

Any chemicals which fit the standard's definition of a carcinogen, corrosive, highly toxic, irritant, sensitizer, toxic, hepatotoxin, nephrotoxin, neurotoxin, agent acting on the blood or hematopoietic system, agent damaging the lung, reproductive toxin, cutaneous hazard, or eye hazard are regulated by the 29 CFR 1910.1200 standard. These definitions can be found in Appendix A.

Employers must implement a hazard communication program, which includes container labeling, employee training, Material Safety Data Sheets (MSDSs), and other forms of information and warning. It also should incorporate the guidelines of the American Conference of Governmental Industrial Hygienists (ACGIH) Threshold Limit Values (TLVs) as exposure limits for chemicals.

Employers must also develop a written hazard communication program which contains a list of the hazardous chemicals known to be present in the workplace. Each chemical must have an MSDS on file. The chemicals on the list and the MSDSs on file must match. The written program should also outline the methods used to inform employees of the hazards of nonroutine tasks.

If more than one employer shares a workplace, each employer is required to provide the others with information about on-site access to MSDSs, the precautionary measures to be taken in order to protect employees during normal operating conditions and in foreseeable emergencies, and of the labeling system used in the workplace.

The manufacturer, importer, or distributor of a chemical is responsible for ensuring that the containers of hazardous chemicals leaving the workplace are labeled, tagged, or marked with hazard warnings and the identity of the hazardous chemical(s), and the manufacturer's name and address. The chemical manufacturer must also supply the employer with an MSDS.

The employer's responsibilities include ensuring that each container of hazardous chemicals remain labeled, tagged, or marked with the hazardous chemical identity, appropriate hazard warnings, and the specific information regarding the physical and health hazards of the hazardous chemical. The employer must also maintain the list of hazardous materials and the MSDSs in good order.

Material Safety Data Sheets may be in any language (in support of the language spoken by the employees) but must also be in English. All MSDSs must contain:

- The identity used on the label

- The chemical and common names

- The physical and chemical characteristics, and hazards of the chemical

- The health hazards and medical symptoms of the chemical

- The chemical's primary route of exposure

- The OSHA PEL, ACGIH TLV

- Whether the chemical is listed in the National Toxicology Program (NTP) Annual Report on Carcinogens or has been found to be a potential carcinogen by the International Agency for Research on Cancer (IARC)

- Any precautions for safe handling procedures, cleanup of spills, and leaks

- Appropriate engineering controls, work practices, and personal protective equipment

- Emergency and first aid procedures

- The date of preparation of the MSDS or the date of the last change

- The name, address and telephone number of the chemical manufacturer, importer, employer or other responsible party preparing or distributing the Material Safety Data Sheet (the objective being to provide an expert who can give additional information on the hazardous chemical and appropriate emergency procedures, if necessary)

Appendix E, although nonmandatory, provides an excellent primer for compliance with the 29 CFR 1910.1200 standard, and a concise, no-nonsense explanation of the intent of the regulation.

29 CFR 1910.1201: Retention of DOT Markings, Placards and Labels. The 29 CFR 1910.1201 standard simply states that employers must preserve the shipping labels of the shipping packages of hazardous materials, and that the labels must be visible. For the purposes of the standard, the term *hazardous material* has the same definition as the one contained in 49 CFR 171–180.

29 CFR 1910.1450: Occupational Exposure to Hazardous Chemicals in Laboratories. After the 29 CFR 1910.1200 standard was promulgated, OSHA recognized that compliance with the Hazard Communication Standard (HCS), although effective for general industry, was not effective for all types of laboratories. In particular, the HCS was effective in large production-type laboratories. Although many of the principles of the HCS were effective for smaller laboratories, some were not effective for small chemical research laboratories. In order to encourage safe operations in all types of laboratories, there had to be a break with the HCS and a distinction made between production laboratories and research laboratories. To make the distinction, OSHA said:

> *Laboratory* means a facility where the *laboratory use of hazardous chemicals* occurs. It is a workplace where relatively small quantities of hazardous chemicals are used on a non-production basis. *Laboratory scale* means work with substances in which the containers used for reactions, transfers, and other handling of substances are designed to be easily and safely manipulated by one person. *Laboratory scale* excludes those workplaces whose function is to produce commercial quantities of materials. *Laboratory use of hazardous chemicals* means handling or use of such chemicals in which all of the following conditions are met:
>
> i)   Chemical manipulations are carried out on a *laboratory scale*
>
> ii)   Multiple chemical procedures or chemicals are used;
>
> iii)   The procedures involved are not part of a production process, nor in any way simulate a production process; and

> iv)   *Protective laboratory practices and equipment* are available and in common use to minimize the potential for employee exposure to hazardous chemicals.

Laboratories that do not fall under the 29 CFR 1910.1450 standard must comply with the requirements of the HCS.

Employers whose laboratories must comply with the 29 CFR 1910.1450 standard must measure the employees' exposure to any regulated substance which requires monitoring if the exposure level for that substance routinely exceeds the action level or the PEL. If the exposure level is over the action level or PEL, the employer must comply with the requirements for the monitoring of exposure prescribed for that substance.

Employers who come under the 29 CFR 1910.1450 standard must develop and implement a written Chemical Hygiene Plan (CHP). The plan must ensure that employees are protected from the health hazards associated with hazardous chemicals, and that employee exposures to hazardous chemical substances do not exceed the permissible exposure limits. (The Permissible Exposure Limits may be found in 29 CFR 1910, Subpart Z.) The plan must also be readily available to employees, and must include the specific measures that the employer will take to protect its laboratory employees. The employer must review and evaluate the effectiveness of the Chemical Hygiene Plan annually and update it as necessary.

The 29 CFR 1910.1450 Standard's training requirements include instruction on how the presence or release of hazardous chemicals may be detected, the physical and health hazards of the chemicals in the workplace, the actions employees can take to protect themselves from hazards, and the details of the CHP.

All employees who work with hazardous chemicals have the right to receive medical attention, examinations, and/or consultations by a licensed physician if the employee develops the symptoms of hazardous chemical exposure as a result of employment. If exposure monitoring indicates that the employee may have been exposed above the action level or PEL, or if there is a spill or other event which may have exposed the employee, these same medical services must be provided. All records must be kept, transferred, and made available in accordance with 29 CFR 1910.1020.

The employer must also make sure that the labels on incoming materials are preserved, that MSDSs are maintained on all hazardous chemicals, and that the MSDSs are easily accessible by laboratory employees.

The National Research Council Recommendations Concerning Chemical Hygiene in Laboratories (29 CFR 1910.1450, Appendix A) are nonmandatory; however they provide an excellent reference for the development of a CHP. The references in the Appendix were extracted from *Prudent Practices for Handling Hazardous Chemicals in Laboratories* (referred to as *Prudent Practices*), published in 1981 by the National Research Council, and are listed in such a way that the recommendations correspond to specific sections of the standard.

Construction Industry Standards: 29 CFR 1926—Safety and Health Regulations for Construction

Subpart C—General Safety and Health Provisions

29 CFR 1926.21: Safety Training and Education. The 29 CFR 1926.21 standard points out that "[e]mployees required to handle or use poisons, caustics, and other harmful substances shall be instructed regarding the safe handling and use, and be made aware of the potential hazards, personal hygiene, and personal protective measures required." Later in the standard it states, "Employees required to handle or use flammable liquids, gases, or toxic materials shall be instructed in the safe handling and use of these materials."

29 CFR 1926.55: Gases, Vapors, Fumes, Dusts, and Mists. The 29 CFR 1926.55 standard prohibits employees to be exposed to any material or substance at a concentration above those specified in the ACGIH *Documentation of the Threshold Limit Values.* Here too, the first measure of hazard control is engineering. Only when engineering controls are not feasible, may employers use other protective measures. In any case, full compliance with the regulation and protection of the employee must be achieved. This section also contains references to the specific substance rules in the General Industry Standard. For example, "Whenever any employee is exposed to formaldehyde, the requirements of 29 CFR 1910.1048 of this title shall apply."

29 CFR 1926.57: Ventilation. The terms and applications of the 29 CFR 1926.57 rule are different from the 29 CFR 1910.94 General Industry Standard. However, the principles of the two standards are the same.

29 CFR 126.59: Hazard Communication and 29 CFR 1926.61—Retention of DOT markings, placards and labels. OSHA chose to apply the General Industry 29 CFR 1910.1200 standard to the construction industry. Both the 29 CFR 1926.59 and the 29 CFR 1926.61 standards state: "Note: The requirements applicable to construction work under this section are identical to those set forth at 29 CFR 1910.1200 of this chapter."

29 CFR 1926.64: Process Safety Management of Highly Hazardous Chemicals. With few exceptions, the General Industry and Construction Industry rules for process safety, hazardous waste operations, PPE, and respiratory protection are alike in principle.

Bibliography

American Conference of Governmental Industrial Hygienists. *Documentation of the Threshold Limit Values.* Cincinnati, OH: ACGIH, 1991.

Bauer, R. L. *Safety and Health for Engineers.* 16th ed. New York, NY: Van Nostrand Reinhold, 1990.

Klaassen, C. D. *Casarett and Doul's Toxicology, The Basic Science of Poisons.* 5th Ed. New York NY: McGraw-Hill, Health Profession Division, 1996.

National Research Council. *Prudent Practices for Handling Hazardous Chemicals in Laboratories.* Washington, DC: National Academy Press, 1981.

Moeller, D. W. *Environmental Health.* Cambridge, MA: Harvard University Press, 1997.

Parker, James F. and Vita R. West, Eds. *Bioastronautics Data Book.* 2nd Ed. NASA SP–3006. Washington DC: National Aeronautics and Space Administration, Scientific and Technical Information Office, 1973.

Plog, B. A., *Fundamentals of Industrial Hygiene.* 4th Ed. Itasca, IL: National Safety Council, 1996.

Tweedy, J. T. *Healthcare Hazard Control and Safety Management.* Delray Beach, FL: GR/St. Lucie Press, 1997.

Internet Resource

<http://www.osha-slc.gov/> (Occupational Safety and Health Administration. *OSHA Computerized Information System.* 1997)

Guy S. Camomilli *is employed as the Manager of Safety, Quality, and Environmental Compliance with Dynamac International, Inc. He works at the Kennedy Space Center on the Life Sciences Support Contract. Mr. Camomilli is a Certified Safety Professional, a Certified Hazardous Materials Manager, and a Certified Health Care Professional. He has 16 years experience in hazardous materials management, predominantly in safety and environmental compliance management of life sciences research, biomedical procedures, and space life sciences.*

CHAPTER 8

OSHA and the Hazardous Material Manager

Steven G. Weems, MPH, CIH, CHMM

Today, over 95 million employees are guaranteed a safe and healthful workplace by Federal law. Prior to 1970, industry saw enormous losses in manpower due to death, dismemberment, and illness. More time was lost to job-related accidents than to strikes. To remedy this situation, Congress passed the Occupational Safety and Health Act of 1970 (OSH Act) and created the Occupational Safety and Health Administration (OSHA).

What Is OSHA?

The OSH Act

The Occupational Safety and Health Administration came into existence as a result of the OSH Act of 1970 (PL 91–596; 29 December 1970). The purpose of the law was "to assure safe and healthful working conditions for working men and women; by authorizing enforcement of the standards developed under the Act; by assisting and encouraging the States in their efforts to assure safe and healthful working conditions; by providing for research, information, education, and training in the field of occupational safety and health; and for other purposes."

Duties. The OSH Act spells out in Section 5(a) the duties of the employer:

- The employer shall furnish to each of his employees employment and a place of employment which are free from recognized hazards that are causing or are likely to cause death or serious physical harm to his employees[this is the 5(a)(1) General Duty clause]

- [The employer] shall comply with occupational safety and health standards promulgated under this Act

The employee's duties are spelled out in Section 5(b): "Each employee shall comply with occupational safety and health standards and all rules, regulations, and orders issued pursuant to this Act which are applicable to his own actions and conduct."

Research. OSHA adopted by reference industry's consensus standards in 1970. The research arm under the OSH Act is NIOSH (National Institute for Occupational Safety and Health). NIOSH is not under control of OSHA, or even the Department of Labor; NIOSH is under the United States (US) Department of Health, Education, and Welfare (now the Department of Health and Human Services). This separation was intended to keep the law enforcer, OSHA, from also establishing worker exposure limits. NIOSH has no compliance authority like OSHA, but does have right of access into a work site to establish industry's exposure data.

NIOSH conducts research on various safety and health problems, provides technical assistance to OSHA, and recommends standards for OSHA to adopt. While conducting its research, NIOSH may conduct workplace investigations, gather testimony from employers and employees, and require employers to measure and report the exposure of employees to potentially hazardous materials. NIOSH also may require employers to provide medical examinations and tests in order to determine the incidence of occupational illness among employees. When such examinations and tests are required by NIOSH for research purposes, they may be paid for by NIOSH rather than the employer.

Informational Resources. OSHA maintains a large library of pamphlets and books on a large variety of safety and health topics. The standards are public domain materials and can be freely copied, unless the book or article has been copyrighted. Most of these publications are available from OSHA or can be downloaded from the Internet (refer to the "Help" section in this chapter).

Training and Education. As stated in the OSH Act, OSHA will provide training and education. The OSHA Training Institute (OTI) at Des Plaines, Illinois, provides a formal classroom for use by the OSHA Compliance Safety and Health Officer (CSHO), State consultation members, and (on certain subjects) the general public. Periodically, the institute will export some of their classes. These courses cover areas such as electrical hazards, machine guarding, ventilation, noise, respiratory protection, and laboratory safety. The OTI facility also has a laboratory, library, and audiovisual support. The instructors are former field compliance officers. OSHA also allows various industries to conduct some courses for a small fee.

OSHA currently has 73 Area Offices that are also fully functioning service centers. These offices offer a wide variety of informational services such as speakers, publications, audiovisual aids on workplace hazards, and technical advice. The laboratories contain various demonstrations and equipment, such as power presses, woodworking and welding shops, a complete industrial ventilation unit, and a sound demonstration laboratory.

Inspections!

The OSH Act gave OSHA the responsibility to conduct inspections of the workplace. As stated in the preamble, OSHA is "to assure safe and healthful working conditions for men and women; by authorizing enforcement of standards developed under the Act . . . ." The type of workplace you work in determines how or whether OSHA has jurisdiction. Employer's compliance issues are monitored by Federal officers, unless the specific state has an approved plan to conduct its own resident inspections. These are called *State Plan* states. About half of the states have their own approved plans. For approval of the State Plan, OSHA requires that the State's standards must be at least as strict as the Federal OSH Act. The requirements for a State Plan state are found in 29 CFR 1902.

Federal

Under the Act, the heads of Federal agencies are responsible for providing safe and healthful working conditions for their employees. An Executive Order requires agencies to comply with standards consistent with the OSHA standards for private sector employers. OSHA conducts Federal workplace inspections in response to employees' complaint reports, and as part of a special program that identifies Federal workplaces with higher-than-average rates of injury and illness.

OSHA's Federal sector authority is different from that in the private sector in several ways. The most significant difference is that OSHA cannot propose monetary penalties against another Federal agency for failure to comply with OSHA standards. Instead, compliance issues that cannot be resolved at the local level are raised to higher organizational levels until they are resolved. Another significant difference is that OSHA does not have the authority to protect the Federal employee who becomes a "whistle-blower." However, the Whistleblower Protection Act of 1989 affords present and former federal employees (other than US Postal Service employees and certain intelligence agencies) an opportunity to file their reports of reprisal with the Office of Special Counsel, US Merit Systems Protection Board.

State and Local Government

Federal OSHA has no jurisdiction over state and local government employees. However, if compliance is done as part of a *State Plan* state's program, then the state's compliance officers cover these employees.

Private Sector

Any private sector employer can be inspected under certain situations. There are several reasons for inspections. These are (in order of priority) imminent danger, catastrophes and fatal accidents, employee complaints, programmed high hazard inspections, other programmed inspections, and follow-up inspections.

Imminent Danger. An *imminent danger* is any condition where there is the reasonable certainty that an immediate danger exists that may cause death or serious physical harm before the danger can be eliminated through normal enforcement procedures.

Catastrophic and Fatal Accidents. *Catastrophic injuries* occur when three or more employees are hospitalized as the result of a single incident. The employer must report catastrophic injuries and any fatalities to OSHA within 8 hours of the occurrence. OSHA will investigate for possible violations of standards and to prevent the recurrence of similar accidents. There are strong penalties for failure to report these accidents to OSHA.

Employee Complaints. All employees have the right to request an OSHA inspection if they feel they are in imminent danger, threatened by physical harm, or that OSHA standards are not being met. OSHA keeps the complaints in strictest confidence and the employer cannot take actions against any employee for making such a complaint, even if he finds out the identity of the employee. OSHA will notify the complainant of the results, if requested. Complaints can be formal or informal. Formal complaints are a signed statement containing the allegations, and a compliance officer will visit the site of alleged offense. Informal complaints are transferred to a form letter and mailed or faxed to the employer. The employer is asked to investigate the allegations and inform OSHA whether or not they are legitimate. If the complaint is legitimate, the employer must explain how and when corrections were made. Compliance officers rarely visit on an informal complaint, unless no reply or a false reply is returned.

That Knock on the Door

Contrary to popular belief, OSHA's job is not to harass businesses or to drive them to bankruptcy. It does not matter how compliant or well run your company is, the "dreaded knock on the door" happens. Preparedness is the best defense. It is illegal for anyone but OSHA to reveal the news of an upcoming visit. OSHA can reveal upcoming visits if it notifies both the employer and the employees.

All inspections have the following in common: (1) inspector's research, (2) inspector's credentials, (3) the opening conference, (4) the inspection tour, and (5) the closing conference.

Inspector's Research. This is the preparation stage. The OSHA Compliance Safety and Health Officer (CSHO) assigned to the case becomes familiar with the company's history of past inspections and complaints (if any), the nature of the business, and the particular standards involved. After researching, the CSHO obtains any necessary sampling media, selects and calibrates the appropriate equipment, and travels to the site.

Inspector's Credentials. The inspection begins when the CSHO arrives at the site and displays official credentials. The CSHO will ask to meet an appropriate employer representative (the representative should insist upon seeing the inspector's credentials, because there have been impostors in the past). The inspector's credentials can be verified by calling the nearest OSHA office.

The Opening Conference. In the opening conference, the CSHO explains why the company has been selected (*i.e.*, imminent danger, catastrophes or fatalities, programmed high hazard, or follow-up). The CSHO will also determine whether the company falls under an exemption program or is covered by an ongoing OSHA consultation (refer to the "Consultation" section of this chapter). The CSHO will give the owner/operator copies of the applicable standards and the complaint (if any). The complaint will not contain the identity of the complainant if the complainant wishes to remain anonymous.

The Inspection Tour. The CSHO, accompanied by an employer representative, will tour the site. In an organized labor plant, the unions are allowed to designate representatives to accompany the CSHO in the inspection if they so desire. If the inspection was prompted by a complaint, then only the affected areas are visited. The CSHO may also conduct confidential interviews with employees. The CSHO will also review recordkeeping and injury and illness (OSHA Form 200) logs. If sampling is required, the CSHO may return to acquire full-shift noise or air contaminant samples. If the CSHO sees additional hazards that are not contained in the original scope of the visit, these areas may also be included in the inspection.

The Closing Conference. During the closing conference the CSHO will discuss the results of the inspection and give the employer an opportunity to ask questions about the findings. If any "on-the-spot" corrections have been made, it is in the employer's best interest to point them out

during the closing conference. The compliance officer will explain appeal procedures, informal conferences, and how to contest any citations. Next, the CSHO's supervisor will review the complete inspection file. The results of any sampling that has been carried out will not be known until later, and the supervisor will review these results as well.

Citations

The prospect of citations as a result of the inspection can be compared to a scene in the movie, *Jaws.* Quint describes a shark attack saying, "sometimes, the shark, he goes away, sometimes the shark, he doesn't go away." OSHA has several classes of violations. In order of seriousness, they are: *de minimus*, regulatory, other than serious, serious, willful, repeat, and failure to correct prior violations. These violations are found in 29 CFR 1903.

De Minimus. *De minimus* violations are small infractions and don't carry a monetary penalty.

Regulatory. Like *de minimus*, these are small recordkeeping infractions. They normally don't carry a monetary penalty, but if one is imposed, its maximum is $1,000. Examples of regulatory infractions include: no OSHA poster or failure to post the OSHA Form 200 during the month of February.

Other Than Serious. These infractions would probably not result in death or physical harm. Fines may be imposed of up to $7,000 per incident.

Serious. Serious violations could cause death or serious physical injury. The employer knew or should have known that these hazards existed. Fines of up to $7,000 per violation may be imposed.

Willful. These are violations that the employer intentionally and knowingly commits. Fines for willful violations range from $5,000 to $70,000 per violation. In addition, if the violation resulted in the death of an employee, the fine could be as much as $250,000 per individual and $500,000 for the corporation. There may also be jail time of up to six months.

Repeat. Repeat offenses are similar violations that are noted on a follow-up visit. Fines may reach

$70,000 per violation. An infraction is a repeat violation only after the original citation has become final and is not contested.

Failure to Correct Prior Violation. Failure to correct these violations may result in fines of up to $7,000 per day past the prescribed abatement date.

Additional Violations. Additional violations may include: falsifying records; not having the OSHA poster; and assaulting, intimidating, or interfering with compliance officers in the performance of their duties. Fines range from $10,000 plus six months jail, $7,000, and $5,000 plus three years jail, respectively.

The Standards

All the OSHA standards are found in 29 CFR (the US Department of Labor) 1900 series. With two exceptions, parts 1900–1909 are OSHA and OSHA Consultation procedures.

The standard that affects most employers is the General Industry Standard.

General Industry

The General Industry Standards are found in 29 CFR 1910. There are two additional parts: 29 CFR 1903 (OSHA poster), and 29 CFR 1904 (OSHA Form 200). The major areas cited are:

- Subpart D, Walking–Working Surfaces (29 CFR 1910.21–30)

- Subpart E, Means of Egress (29 CFR 1910.35–38)

- Subpart F, Powered Platforms (29 CFR 1910.67–68)

- Subpart G, Occupational Health and Environmental Control (29 CFR 1910.96)

- Subpart H, Hazardous Materials (29 CFR 1910.101–120)

- Subpart I, Personal Protective Equipment (29 CFR 1910.132–38)

- Subpart J, General Environmental Controls (29 CFR 1910.141–147)

- Subpart K, Material and First Aid (29 CFR 1910.151)

- Subpart L, Fire Protection (29 CFR 1910.155–165)

- Subpart M, Compressed Gas (29 CFR 1910.169)

- Subpart N, Materials Handling (29 CFR 1910.176–184)

- Subpart O, Machine Guarding (29 CFR 1910.211–219)

- Subpart P, Hand and Portable Powered Tools (29 CFR 1910.241–244)

- Subpart Q, Welding (29 CFR 1910.251–255)

- Subpart R, Special Industries (29 CFR 1910.261–272) such as paper, textiles, bakery equipment, logging, and telecommunications.

- Subpart S, Electrical (29 CFR 1910.301–399)

- Subpart T, Commercial Diving (29 CFR 1910.401–441)

- Subpart Z, Toxic and Hazardous Substances (29 CFR 1910.1000–1500)

Of the standards listed, the most frequently cited are:

- Hazard Communication (29 CFR 1910.1200)

- Lead (29 CFR 1910.1025)

- Respiratory Protection (29 CFR 1910.134)

- Noise (29 CFR 1910.95)

- Machine Guarding (anything in Subpart O)

- Lockout/Tagout (29 CFR 1910.147)

- Air Contaminants (29 CFR 1910.1000)

- Confined Space (29 CFR 1910.146)

Construction

The construction standards basically mirror the General Industry standards, but have a different numbering system. For example, Hazard Communication is 29 CFR 1926.59 and the lead standard is 29 CFR 1926.62. Another difference is in the lead standard itself. Respirators and personal protective equipment (PPE) are required unless the employer has monitoring data showing that the employee's exposure does not exceed the permissible exposure limits.

Maritime

The maritime standards are divided into four major areas: Shipyards (29 CFR 1915), Marine Terminals (29 CFR 1917), Longshoring (29 CFR 1918), and Gear Certification (29 CFR 1919). The standards are specific to their own area and the maritime safety specialist must become familiar with them all.

Others

There are a whole host of other parts in the 1900 series of standards. Three of the more important parts are 29 CFR 1924, Workshops and Rehab Facilities; 29 CFR 1925, Federal Service Contracts; and 29 CFR 1928, Agriculture.

The General Duty Clause

The general duty clause of the Act states that each employer "shall furnish . . . a place of employment which is free from recognized hazards that are causing or are likely to cause death or serious physical harm to his employees." This is the catchall clause that applies to all employers.

OSHA does not have a standard for everything. It uses the recommendations of industry, research, and professional associations. For example, OSHA uses the American Conference of Governmental Industrial Hygienists' (ACGIH) Threshold Limit Values (TLV) if it does not have an established exposure limit.

Consultation

Every state and territory has an OSHA consultation program, sometimes referred to as the 7(c)(1) program. The consultation program is conducted under the guidelines in the OSH Act and 29 CFR 1908.

Consultants are trained by OSHA and often sit in on the compliance classes. Many of these consultants are former compliance officers, hold advanced degrees, and many have Certified Safety Professionals (CSP) and Certified Industrial Hygienists (CIH) certifications. This consultation service is available to the smaller employers (250 employees or less at one site, 500 total at all sites). The service is free, but there is a catch: the employer must agree to correct all serious hazards by an agreed-upon abatement date. Failure to abate the hazard will result in the case being turned over to OSHA. This drastic measure rarely occurs, and then only after all means of getting the employer to abate the identified hazards have been exhausted. In situations beyond the company's control, the abatement date can be extended if the employer requests it.

The consultation services may be run from the state's Department of Labor, or at a State university. In large states there may be several consultation offices. You can contact your state's consultation project by contacting the OSHA area office in your state, or by calling the OSHA number on the poster that is displayed in the work area.

The consultation can be wall-to-wall or specific in nature (i.e., noise only). The information will be kept in confidence between the company and the consultation project. Ordinarily, the project does not discuss consultations with OSHA. The only time OSHA will find out about a consultation visit is if

- The company tells OSHA
- OSHA sees the report posted in a common area
- The consultation file is subpoenaed for a court case
- The file is turned over to OSHA for failure to abate

In addition to verifying whether the company is or is not in compliance, the consultant will explain the various standards and offer suggestions on how to fix identified problems. Sometimes the consultant will have model programs that the employer may use. Many of the projects have references to products that have been used with good results. The consultant cannot recommend a specific manufacturer or brand name, but may have a variety from which the employer may choose.

Help and Information

Help

Confused? Many are, and they get help by calling their local OSHA office. OSHA personnel are helpful and ready to explain a standard or to send some of their literature. You don't have to give your name, company, or location, and OSHA does not trace the phone call so it can visit the plant the next day. You can also contact your state's consultation service.

The Safety and Health Information Superhighway

Good news! OSHA is on the Internet. Dial up

<http://www.osha.gov>

to find out the latest standards, complete with the preambles. You can download these standards and print them. Also available are the Field Inspection Reference Manual (FIRM), Industrial Hygiene Tech Manual, Most Frequently Asked Questions (FAQs), and interpretations. OSHA's Salt Lake City Laboratory can also be accessed at

<http://www.osha.slc.gov>

if you have sampling questions.

OSHA makes the above items available on CD-ROM. Many commercial companies make versions of these standards in CD-ROM and binder formats. Some offer updates quarterly by replacement CDs or paper inserts to the binders. The updates are also available through the government printing office, the publisher, or many of the safety products distributors.

The *Federal Register* can be consulted if you want the latest news from the Federal government.

Unfortunately, each volume contains everything that came out of any branch of the Federal government on that day. That's a lot of daily reading!

Summary

OSHA is not a mindless bureaucracy. It can assist you in staying current with the ever-changing field of safety and health. Use it like another map in your quest for safety. By learning about OSHA, its mission, and how it operates, you will become a better safety manager.

Bibliography

91st Congress. *Occupational Safety and Health Act.* PL 91–569, S2193, 29 December 1970; as amended by PL 10–552, Section 3101, 5 November 1990; as amended by PL 105–198, 16 July 1998.

Mintz, Benjamin W. "Occupational Safety and Health: The Federal Regulatory Program—A History." In *Fundamentals of Industrial Hygiene*, 3rd ed. Barbara A. Plog, MPH, CIH, CSP, Ed. Itasca, IL: National Safety Council, 1988.

Occupational Safety and Health Administration. *OSHA Handbook for Small Businesses.* Pamphlet Number 2209. Washington, DC: OSHA, 1990.

Occupational Safety and Health Administration. *All About OSHA.* Pamphlet Number 2056. Washington, DC: OSHA, 1995.

Robbins, Chain M. "Governmental Regulations." In: *Fundamentals of Industrial Hygiene*, 3rd ed. Barbara A. Plog, MPH, CIH, CSP, Ed. Itasca, IL: National Safety Council, 1988.

***Steven G. Weems** is currently a Senior Health Consultant for the Department of Environmental and Industrial Programs, at Safe State. Mr. Weems joined Safe State, Alabama's OSHA consultation program, in July 1989 as an Industrial Hygienist. Prior to joining Safe State, Mr. Weems was the Safety Administrator of Southern Research Institute in Birmingham, Alabama. He is a veteran of 12 years in the United States Army and currently Lieutenant Colonel in the*

Alabama Army National Guard. Mr. Weems earned board certification in the comprehensive practice of Industrial Hygiene in July 1994. He also became a Certified Hazardous Materials Manager in December 1993. Mr. Weems has extensive experience in evaluating workplace hazards and prescribing controls in a variety of industries and services. He serves as an instructor for numerous training courses. He has a Master's degree in Public Health from the University of Alabama at Birmingham (1992) and a Bachelor of Science degree in Chemistry from Auburn University (1976).

CHAPTER 9

Personal Protective Equipment

Michael H. Ziskin, CHCM, CHMM
Dawn Han, MS, CIH

Introduction

Personal protective equipment (PPE) includes all clothing and other protective devices designed to be worn to protect workers from workplace hazards. PPE is generally needed for physical or health hazards that cannot be eliminated through engineering or administrative control techniques. The effective use of PPE is dependent upon the commitment of the organization's management, the effectiveness of the risk assessment process, the objective selection of equipment, the training and demonstrated competency of the wearer, as well as the budget to maintain, care for, repair, replace, and properly dispose of the equipment. In this chapter on PPE, we will review related regulatory requirements, hazard assessment, types of different equipment, PPE selection process, various PPE-related factors, management approaches, new trends, and resources.

Personal protective equipment is anything that a worker can wear, carry, or use to protective himself or herself against some of the hazard(s) that may be encountered while doing work. In reality, the workplace can require complex choices about PPE use. This chapter covers a broad range of PPE, including primary PPE items such as:

- Full and partial body protective garments
- Protective gloves and other handwear
- Protective footwear
- Protective headwear
- Protective face and eyewear
- Respirators
- Hearing protectors or hearing protection devices

Other types of clothing and equipment, often considered as PPE, are described in this chapter but not discussed in as much detail as the PPE above. These items include fall protection and cooling devices.

This chapter also covers selection and use of PPE with information relevant to several different environments, including the following types of protection:

- Physical
- Environmental
- Chemical
- Biological
- Thermal
- Electrical
- Radiation

Because of the broad range of PPE types and applications, this chapter will focus on those areas in which there is less information available to industry for selection of *appropriate* PPE. See Figure 1 for examples of personal protective equipment.

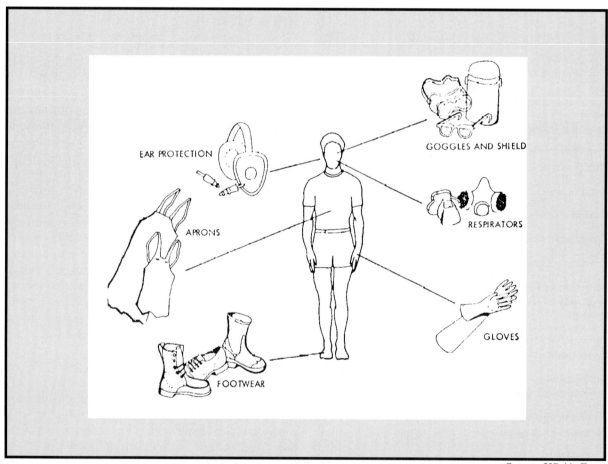

Source: US Air Force

Figure 1. Examples of Personal Protective Equipment

PPE and Its Role in Providing Worker Protection

Personal protective equipment includes items of clothing and equipment which are used by themselves or in combination with other protective clothing and equipment to isolate the individual wearer from a particular hazard or a number of hazards. PPE is also used to protect the environment from the individual, such as in the case of cleanroom apparel and medical devices for infection control.

PPE is considered the *last line of defense* against particular hazards when it is not possible to prevent worker exposure by using engineering or administrative controls. *Engineering controls* should be first used to eliminate a hazard from the workplace by modifying the work environment or process to prevent any contact of workers with the hazard. An example of an engineering control is the replacement of a manual task involving potential hazard exposure with an automated process. In the absence of engineering controls, *administrative controls* should then be used to prevent worker contact with the hazard. An example of an administrative control is to establish a procedure which dictates that workers be out of an area when the hazards are present or to limit the time of exposure to those hazards. Finally, when neither engineering nor administrative controls are possible, PPE should be used.

While PPE is designed for protection of personnel against various hazards, *PPE cannot provide protection to the wearer against all hazards under all conditions.* Workers should not rely on PPE exclusively for protection against hazards, but should use PPE in conjunction with mechanical guards, engineering controls, and sound manufacturing practices in the workplace setting. The use of PPE in itself may create additional hazards or stressors for the wearer such as heat stress or reduced mobility, dexterity, and tactility; and impaired vision or hearing.

Regulatory Overview—OSHA 29 CFR 1910

In the United States, the majority of PPE-related regulations are promulgated by the Occupational Safety and Health Administration (OSHA). The Administration provides regulations that address selection and use of PPE in Title 29, *Code of Federal Regulations* (CFR) Subpart I, Parts 1910.132 through 1910.140. The standard was first adopted in 1971, and was revised in 1994. *States with their own OSHA plans must meet or exceed these requirements.*

The PPE standards that are included in Subpart I are shown in Table 1.

OSHA released the new revision to the Respiratory Protection Standard in January 1998, replacing the standard with the same name and number that was released in 1971. The old standard was renamed 29 CFR 1910.139 and pertains to respiratory protection for tuberculosis only.

Table 1. Overview of OSHA PPE Standards

29 CFR Section	Topic
1910.132	Provides general requirements for PPE
1910.133	Pertains to eye and face protection
1910.134	Pertains to respiratory protection
1910.135	Pertains to head protection
1910.136	Pertains to foot protection
1910.137	Pertains to electrical protective equipment
1910.138	Pertains to hand protection
1910.139	Pertains to respiratory protection for *Mycobacterium tuberculosis*
1910.140	Provides a list of standards organizations relative to PPE

The OSHA standards require that workers use eye, face, respiratory, head, foot, hand and electrical protective equipment where there are hazards that cannot be sufficiently controlled by other means. This would include the use of engineering controls (for example, ventilation) or changing or modifying the way in which specific tasks are performed (such as avoiding heavy lifting). It also requires that the equipment for use by workers must be reliable, clean, and in good working condition. In most cases, employers are required to provide workers with PPE. The employer is required by OSHA: to perform a hazard assessment at each workplace to determine whether PPE is needed; document the assessment in a written certification; select the appropriate PPE if needed; and train employees about PPE usage, proper care, maintenance, useful life, and disposal.

OSHA requires that the employer ensure that the PPE used to protect the worker meets minimum performance requirements as specified. For example, OSHA specifies that employers shall provide National Institute for Occupational Safety (NIOSH) certified respiratory protection as specified in the OSHA respiratory protection standard. OSHA also requires eye, face, hearing, and foot protection to meet specific American National Standards Institute (ANSI) standards.

OSHA can also recommend the employer to follow nonmandatory standards. For example, OSHA's Hazardous Waste Operations and Emergency Response (HAZWOPER) standard recommends that chemical protection needed for response to hazardous materials emergencies meet the National Fire Protection Association's (NFPA) 1991 or 1992 standards. The NFPA standards and

Table 2. Other OSHA PPE-Related Standards

29 CFR Section	Topic	29 CFR Section	Topic
1910.66	Working with chemicals on elevated work surface	1910.1013	beta-Propiolactone
		1910.1014	2-Acetylaminofluorene
1910.94	Work around open-surfaced tanks	1910.1015	4-Dimethylaminoazobenzene
1910.95	Occupational noise and hearing protectors	1910.1016	N-Nitrosodimethylamine
		1910.1017	Vinyl chloride
1910.119	Process safety management of highly hazardous chemicals	1910.1018	Inorganic arsenic
		1910.1025	Lead
1910.120	Hazardous waste operations and emergency response (HAZWOPER)	1910.1027	Cadmium
		1910.1028	Benzene
1910.146	Permit-required confined spaces	1910.1029	Coke oven emissions
1910.156	Fire brigades	1910.1030	Bloodborne pathogens
1910.183	Helicopters	1910.1043	Cotton dust
1910.242	Hand and portable powered tools and equipment, general	1910.1044	1,2-Dibromo-3-chloropropane (BCP)
		1910.1045	Acrylonitrile
1910.252	Welding, general requirements	1910.1047	Ethylene oxide
1910.331	Electrical work	1910.1048	Formaldehyde
1926.102	Subpart E: Lasers	1910.1050	Methylenedianiline
1910.250	Subpart Q: Welding, cutting and brazing	1910.1200	Hazard Communication
1910.262	Acid and caustics in textile production	1910.1450	Occupational exposure to hazardous chemicals in laboratories
1910.1001	Asbestos		
1910.1003	13 Carcinogens (4-Nitrobiphenyl, *etc.*)	1915 Subpart I	PPE for shipyard employment
1910.1004	alpha-Naphthlamine		
1910.1006	Methylchloromethyl ether	1917 Subpart E	PPE for marine terminals
1910.1007	3,3'-Dichlorobenzidine		
1910.1008	Bis-Chloromethyl ether	1918 Subpart J	PPE for longshoring
1910.1009	beta-Naphthylamine		
1910.1011	4-Aminodiphenyl	1926 Subpart E	PPE for the construction industry
1910.1012	Ethyleneimine		

other related standards guidelines will be discussed later in this chapter.

Other OSHA Standards

In addition to the general personal protective equipment and respiratory protection regulations from OSHA, there are other standards that relate to PPE applications when exposed to specified hazards. For example, the standard for laser protection specifies *suitable* safety goggles that will offer adequate protection for the specific wavelength of the laser and optical density for the energy involved, while the welding standard provides shade requirements for optical protection during welding operations. Table 2 contains a partial list of OSHA standards that relate to PPE.

Other Related Standards and Guidelines

There are other PPE-related standards and guidelines available from nationally recognized organizations, in addition to the OSHA regulatory requirements. These organizations are shown in Table 3.

Standards and guidelines from these organizations may be incorporated by reference into the regulations. For example, OSHA specifically references the American National Standards Institute (ANSI) for eye and face protection and other personal protective equipment. OSHA also references the NFPA chemical protective clothing standards (NFPA 1991 and NFPA 1992) in 29 CFR 1910.120 Hazardous Waste Operations and Emergency Response Standard nonmandatory appendices.

General Employer Selection Responsibilities under OSHA 29 CFR 1910.132

Under the OSHA regulations, the employer is responsible for conducting a hazard assessment of the workplace to determine if hazards requiring PPE are present or are likely to be present. If hazards are present, the employer must

- Select and have the affected employees use the types of PPE that will protect them from the hazards identified in the hazard assessment

- Communicate selection decisions to each affected employee

Table 3. Other Organizations with PPE-Related Standards and Guidelines

Organization or Agency	Acronym
American National Standards Institute	ANSI
American Industrial Hygiene Association	AIHA
American Conference of Governmental Industrial Hygienists	ACGIH
American Society for Testing and Materials	ASTM
International Organization for Standardization	ISO
National Fire Protection Association	NFPA
National Institute for Occupational Safety and Health	NIOSH
US Coast Guard	USCG
US Department of Defense	DOD
US Department of Energy	DOE
US Department of Justice	DOJ
US Environmental Protection Agency	EPA
US Food and Drug Administration	FDA

- Select PPE that properly fits each affected employee

The employer must verify that the required workplace hazard assessment has been performed through a written certification that identifies

- The workplace evaluated
- The person certifying that the evaluation has been performed
- The date(s) of the hazard assessment
- Signed documentation of certification of the hazard assessment

For selection of PPE for protection against respiratory and electrical hazards, the employer should refer to OSHA's Respiratory Protection Standard at 29 CFR 1910.134 and the Electrical Protection Standard at 29 CFR 1910.137.

General Employer Training Responsibilities under OSHA 29 CFR 1910.132

The employer must train the employee who is required to use PPE so that he or she knows

- When PPE is necessary
- What PPE is necessary
- How to properly don, doff, adjust, and wear PPE
- The limitations of PPE
- The proper care, maintenance, useful life, and disposal of PPE

The employer must have each affected employee demonstrate an understanding of the required training and the ability to use PPE properly before being allowed to perform work requiring the use of PPE. If the employer has reason to believe that any affected employee who has already been trained does not have the required understanding, then the employer must retrain each such employee. The employer must conduct retraining under circumstances that include, but are not limited to, situations in which

- Changes in the workplace render previous training obsolete

- Changes in the types of PPE to be used render previous training obsolete
- Inadequacies in an affected employee's knowledge or use of assigned PPE indicate that the employee has not retained the requisite understanding or skill

The employer must verify that each affected employee has received and understood the required training through a written certification that

- Lists the name of each employee trained
- Indicates the date(s) of training
- Identifies the subject of the certification

For additional training requirements against respiratory and electrical hazards, the employer should refer to 29 CFR 1910.134 for respiratory protection and 29 CFR 1910.137 for electrical protection.

Overview of Requirements in OSHA 29 CFR 1910.133—1910.140

OSHA 29 CFR 1910.133 through 1910.138 cover PPE in varying levels of detail but do not specifically address all hazards or types of PPE. For example, no specific requirements are provided for overall skin or body protection. Some of the specific requirements are summarized below.

Eye and Face Protection. 29 CFR 1910.133 on eye and face protection requires

- The use of eye and face protection for specific hazards, including:
  - flying particles
  - molten metal
  - liquid chemicals
  - acids or caustic liquids
  - chemical gases or vapors
  - potentially injurious light radiation
- Side protection for flying object hazards, provision for prescription lenses, PPE marking, and use of filter lenses for protection against injurious light radiation

- Compliance of protective eye and face devices with ANSI standard Z87.1–1989, "American National Standard Practice for Occupational and Educational Eye and Face Protection." (Devices purchased before July 4, 1994 must meet ANSI Z87.1–1968).

Respiratory Protection. 29 CFR 1910.134 on respiratory protection was updated on January 8, 1998 and requires

- The use of engineering controls where feasible and the use of respirators when necessary to protect employee health against occupational diseases caused by contaminated air

- The establishment of a respiratory protective program.

- That an individual be named as the administrator for the respiratory protection program

- The selection of respirators based on an evaluation of respiratory hazards and relevant workplace and user factors which effect respirator performance and reliability

- Medical exams for employees who must wear respirators

- Fit testing of employees who must wear respirators

- That employers implement procedures for use of respirators that provide

  - proper facepiece seal protection

  - continuing respirator effectiveness

  - protection in Immediately Dangerous to Life and Health (IDLH) atmospheres

  - protection during interior structural fire fighting

- Specific maintenance and care of respirators, including:

  - inspection for defects

  - cleaning and disinfection

  - repair

  - storage

- Meeting minimum air quality standards

- Identification of respirator filters, cartridges, and canisters

- Employers to provide effective training to employees who must use respirators

- Evaluation of the workplace to ensure that provisions of the respirator program are being carried out

- Recordkeeping of medical examinations and fit testing

Head Protection. 29 CFR 1910.135 on head protection requires

- The use of protective helmets in areas where potential exists for head injury from falling objects

- The use of protective helmets designed to reduce electric shock when employees are near exposed electrical conductors that could contact the head

- That protective helmets comply with ANSI standard Z89.1–1986, "American National Standard for Personnel Protection–Protective Headwear for Industrial Workers–Requirements." (Protective helmets purchased before July 5, 1994 must meet ANSI Z89.1–1969. A new edition of this standard exists as ANSI Z89.1–1997).

Foot Protection. 29 CFR 1910.136 on foot protection requires

- The use of protective footwear in areas where the potential for foot injury exists from:

  - falling or rolling objects

  - objects piercing the sole

  - exposure to electrical hazards

- That protective footwear comply with ANSI standard Z41–1991, "American National Standard for Personal Protection–Protective Footwear." (Protective footwear purchased before July 5, 1994 must meet ANSI Z41.1–1967).

Electrical Protective Equipment. 29 CFR 1910.137 regulates electrical protective equipment by:

- Addressing insulating electrical protective equipment made from rubber. Including:

  - blankets

  - mattings

  - covers

  - line hose

– gloves

– sleeves

- Setting specific design requirements for electrical protective equipment that include

  – manufacture and marking

  – electrical requirements

  – workmanship and finish

- Setting requirements for in-service care and use for electrical protective equipment

Hand Protection. 29 CFR 1910.138 on hand protection requires

- Employers to select *appropriate* hand protection when employees' hands are exposed to hazards from

  – skin absorption of harmful substances

  – severe cuts or lacerations

  – severe abrasions

  – punctures

  – chemical burns

  – thermal burns

  – harmful temperature extremes

- Employers to base selection of hand protection on

  – performance of hand protection relative to task(s) to be performed

  – conditions present

  – duration of use

  – identified hazards and potential hazards

29 CFR 1910.140 lists the American National Standards Institute (ANSI) as the organization recognized for obtaining referenced standards.

Limitations and Shortcomings of Regulations

Regulations specifying the selection and use of PPE are often general in scope, limited to specific applications, or do not provide specific guidance. Only in a few areas are regulations specific in recommending particular types of PPE in terms

of design, performance properties, and service life. The majority of OSHA and other governmental regulations simply specify the use of general types of PPE (*e.g.*, they only require the employee use of *protective clothing*) without indicating a particular configuration and required performance.

The principle exceptions to generic requirements exist when there are national standards such as:

- Protective footwear (ANSI Z41.1)

- Protective face and eyewear (ANSI Z87.1)

- Protective headwear (ANSI Z89.1)

- Respirators (ANSI Z88.2 and 42 CFR 84)

There are no national standards for protective garments or gloves except for very specific applications (*e.g.*, emergency response). Employers and end users, therefore, are faced with a variety of choices for PPE selection for meeting regulatory requirements and must use a risk assessment and determination of protection needs (based on PPE design, performance, and service life) in order to decide on the appropriate PPE.

Hazard Assessment

Overview

For each workplace, employers are required to perform a *hazard assessment* to determine if hazards requiring the use of PPE are present or are likely to be present. The hazard assessment usually involves a walk-through survey of the job area to identify sources of hazards, and consideration of the following categories of hazards: impact, penetration, compression, chemical, heat, harmful dust, light radiation, falling, *etc.* During the walk-through, the sources or potential problem areas for these hazards should be identified. The data and information for use in the assessment of hazards are then recognized and an estimate of the potential for injuries is made. Each of the basic hazards is then reviewed and determinations made as to the type, level of risk, and seriousness of potential injury from each of the hazards. The hazard assessment should be documented in writing, and a reassessment of the hazards should be performed as necessary. A flow chart summarizing the PPE hazard assessment process can be found in Figure 2.

OSHA Requirements for Performing a Hazard Assessment

OSHA 29 CFR 1910.132 requires that employers conduct a hazard assessment to determine the need for and then to select PPE. Appendix B to Subpart I of OSHA 29 CFR provides non-mandatory guidelines for conducting hazard assessments and for selecting personal protective equipment. The following subjects are addressed:

- Controlling hazards
- Assessment guidelines
- Selection guidelines
- Fitting the devices
- Devices with adjustable features
- Reassessment of hazards
- Selection guidelines for eye and face protection
- Selection guidelines for head protection
- Selection guidelines for foot protection
- Selection guidelines for hand protection
- Cleaning and maintenance

The principal source of information for conducting a hazard assessment comes from an inspection of the workplace with actual observation of specific tasks being carried out. Additional information can be obtained by interviewing the affected employees and asking about

- The types and frequency of hazards encountered in the tasks
- Specific instances in which hazards have been encountered in the past
- The past effectiveness or ineffectiveness of any PPE used in the tasks

Another source of information is a review of the log and summary of all occupational illnesses and injuries at the workplace. This may be accomplished by examining the OSHA Occupational Injury and Illness log or equivalent forms. The specific hazard and nature of any accidents or exposures should be evaluated to determine the

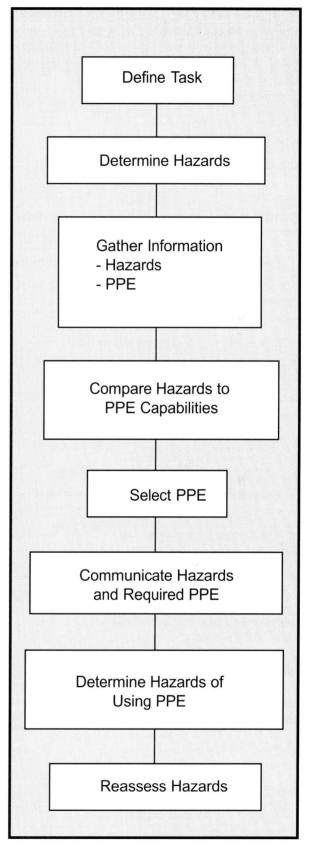

Figure 2. Flow Diagram for PPE Hazard Assessment

possible preventative role of using PPE or improving PPE if involved. In some cases, it will be necessary to measure hazard levels using special instrumentation such as portable sampling devices to measure airborne concentrations of chemicals or noise-monitoring equipment.

Recommended Risk Assessment Approach

The risk assessment based-approach presented in this chapter for selecting PPE uses the following steps:

- Conduct a hazard assessment of the workplace

- Determine the risk of exposure, and rank protection needs

- Evaluate available PPE designs, performance, and applications against protection needs

- Specify appropriate PPE

Conducting the Hazard Assessment. The hazard assessment consists of the following steps:

1) Define the workplace and tasks to be evaluated

2) Identify the hazards in the workplace

3) Determine areas of the body or body systems that are affected by the hazards

4) Estimate the potential for employee contact with hazards in the workplace

5) Estimate the consequences of employee contact with hazards

The specific steps of this process are described in more detail below.

Define Each Workplace and Tasks to Be Evaluated. The *workplace* includes the area that encompasses the range of hazards that may be encountered. Examples of a workplace include

- The specific work locations for a particular employee

- A laboratory

- A part of a production process

Work tasks should be defined as those worker activities that

- Involve unique hazards

- Are accomplished by a single individual or group of individuals within a given period of time

Identify the Hazards Associated with Each Work Task. General hazard categories include

- Physical

- Environmental

- Chemical

- Biological

- Thermal

- Electrical

- Radiation

- Person–position

- Person–equipment

Determine Each Affected Body Area or Body System. For each hazard, determine which portion of the body can be affected by the hazard. General body areas and body systems typically affected by workplace hazards include

- Head

- Eyes and face

- Hands

- Arms

- Feet

- Legs

- Trunk or torso

- Entire body

- Respiratory system

- Hearing

Estimate the Likelihood of Employee Exposure to Identified Hazards. For every identified hazard affecting a specific portion of the body (or the whole body), indicate the likelihood of exposure. One method of doing this is to use a rating scale of 0 to 5 based on both the risk and the frequency of exposure:

0:    Exposure cannot occur

1:    Exposure very unlikely

2:    Exposure possible, but unlikely

3:    Exposure likely

4:    Multiple exposures likely

5:    Continuous exposure likely

Estimate the Possible Consequences of Exposure to Identified Hazards.

For every identified hazard affecting a specific portion of the body (or the whole body), indicate the consequences of exposure. A rating scale of 0 to 5, based on the *worst case* effects on the potentially exposed worker, can be used:

0:    No effect

1:    Temporary effect on employee (such as discomfort) with no long-term consequences

2:    Exposure results in temporary, treatable injury

3:    Exposure results in serious injury with loss of work time

4:    Exposure results in permanent debilitating injury

5:    Exposure results in likely death

A completed hazard assessment will provide a list of hazards and show which parts of the worker body may be affected, how likely exposure will occur, and what the probable consequences of exposure might be.

Determining Relative Risk and Establishing Protection Needs.

The risk of exposure is determined by the hazard assessment. Protection needs are ranked by the relative risk. For determining relative risk, establish a risk assessment form for listing

- Hazards identified for the specific workplace/task

- Body areas or body systems affected by the respective hazard

- The rating associated with the likelihood or frequency of the respective hazard

- The rating associated with the severity or consequences of the respective hazard

Using risk determinations from the risk assessment form, rank all hazards associated with the workplace/tasks. Those hazards with the highest amount of risk should be assigned higher priority for prevention or minimization. Those hazards with zero risk or low risk should be assigned lower priority for prevention or minimization.

Examine possible engineering or administrative controls for those hazards with the highest risk. Engineering controls can encompass changes in the task and process, use of protective shields, or other designed measures that eliminate or reduce possible exposure to hazards. Administrative controls can include changes in tasks or work practices to eliminate or limit employee exposure time to a hazard. If engineering or administrative controls are not possible, examine different types of PPE for elimination or reduction of exposure to hazards.

From information about affected body area or body systems, decide which type of PPE can be used to eliminate or minimize the hazard. In many cases, the type of PPE to be used will be obvious and limited to a single general type (for example, inhalation hazards can be protected against with a respirator). In other cases there may be several types of PPE, which can be used to provide the needed protection.

General Approach for Evaluating PPE Designs, Design Features, Performance, and Applications

The evaluation of PPE designs, design features, performance, and applications encompasses the following steps:

- Understand and choose the types of PPE available for protection

- Understand and choose relevant performance properties of PPE to consider during PPE selection

The types of PPE available in the marketplace and thus the choices available to the end user are rapidly increasing. PPE exists in a variety of designs, materials, and methods of construction, each having advantages and disadvantages for specific protection applications. End users should have an understanding of the different types of PPE and their features in order to make appropriate selections.

> **! WARNING: PPE that are similarly designed may offer different levels of performance. Examine PPE performance in addition to design and features.**

PPE must be properly sized to provide adequate protection.

> **! WARNING: Improperly sized or ill-fitting PPE may reduce or eliminate protective qualities of PPE.**

PPE may be classified by

- Design
- Performance
- Intended service life

Design. The classification of PPE by its design usually reflects either how the item is configured or the part of the body area or body systems that it protects. For example, footwear by design provides protection to the wearer's feet.

The classification of PPE by design may also provide an indication of specific design features that differentiate PPE items of the same type. For example, closed-circuit self-contained breathing apparatus are configured with significant design differences when compared to open-circuit self-contained breathing apparatus. Some designs of PPE may offer varying protection against hazards in different parts of the PPE item. For example, the palm material in a glove palm may provide a better grip surface than the glove's back material.

The types of PPE can be generally categorized as:

- Full body garments
- Partial body garments
- Gloves
- Footwear
- Head protection
- Face and eye protection

- Hearing protection
- Respirators

> **! WARNING: PPE coverage of a specific body area, in and of itself, does not guarantee protection of that body area.**

Performance. The classification of PPE by performance indicates the actual level of performance to be provided by the item of PPE. This may include a general area of performance or a more specific area of performance. For example, while two items of PPE might be considered to be chemical protective clothing, one item may provide an effective barrier to liquids but not to vapors while the other item provides an effective barrier to both liquids and vapors. The classification of PPE by performance is best demonstrated by actual testing or evaluations of PPE with a standard test.

> **! WARNING: Intended or manufacturer-claimed performance does not always match actual performance.**

Service Life. The classification of PPE by expected service life is based on the useful life of the PPE item. PPE may be designed to be

- Reusable
- Used a limited number of times (limited use)
- Disposable after a single use

The classification of PPE by expected service life is based on:

- Durability
- Life-cycle cost
- Ease of reservicing

Durability is determined by evaluating how the item of PPE maintains its original performance properties following the number of expected uses. The *life-cycle cost* of PPE is the total cost for using an item of PPE and is usually represented as the cost per use for a PPE item. The following costs should be considered in determining the life-cycle cost:

- Purchase cost
- Labor cost for selection/procurement of PPE

- Labor cost for inspecting PPE

- Labor and facility cost for storing PPE

- Labor and materials cost for cleaning, de-contaminating, maintaining, and repairing PPE

- Labor and fees for retirement and disposal of used PPE

The total life-cycle cost is determined by adding the separate costs involved in the PPE life cycle and dividing by the number of PPE items and number of uses per item.

If an item of PPE cannot be *reserviced* to bring it to an acceptable level of performance, then it cannot be reused.

> **! WARNING: Expected service life does not always equal actual use life.**

Additional detail for selection factors for different types of PPE is provided later in this chapter.

General Approach for Specifying PPE. Specifying appropriate PPE entails one or more of the following:

- Referencing appropriate standards

- Developing a comprehensive product design and performance

- Establishing acceptance criteria

This part of the process is covered in more detail later in this chapter.

Types of Protective Equipment

Personal protective equipment includes all items worn by a worker that are designed to create a barrier against workplace hazards. Such items would include respirators, protective clothing, and other devices for head, eye and face, hearing, foot, hand, and fall protection, *etc.*

General Use *vs.* Emergency Applications

Personal protective equipment is used in a variety of applications. These applications include protecting workers from sharp objects, flying projectiles, hot surfaces, chemical splash, falling, *etc.* In addition, PPE is also used to protect workers from simultaneous hazards (multiple hazards), present in many industrial operations, hazardous waste site operations, confined space operations, and emergency response.

Each application for PPE may present unique challenges dictating different designs and levels of performance. Many industrial operations are associated with hazards that are predictable such as known chemicals, identified confined spaces, *etc.* However, for hazardous waste operations or emergency response situations, a large number of complex issues may need to be considered in both selecting and using the appropriate protection. Unlike many other more predictable work environments, hazardous waste site operations and emergency response involve widely ranging conditions and a great deal of uncertainty. Any number of hazards due to the type(s) of operations, environmental conditions, varying chemical mixtures and concentrations may be encountered. Differences in the performance offered by the various types of PPE could make the difference between adequate protection and hazardous acute or chronic exposure.

Respiratory Protection

The basic purpose of any respirator is to protect the respiratory system from inhalation of hazardous atmospheres. Airborne hazards include dusts, fumes, gases, vapors, mists, aerosols, and smoke, *etc.* Every effort must be made to eliminate airborne respiratory hazards through engineering and administrative control methods. However, when such controls cannot eliminate a hazard or reduce its danger to an acceptable level, adequate respiratory protection is required. Any user of any type respiratory equipment must be in good health and demonstrate competence in the use of the equipment and in the equipment's limitations.

Respiratory protective devices can be categorized into two major types: *air-purifying respirators* and *atmosphere-supplying respirators*. Air-purifying respirators remove contaminants from the ambient air, while an atmosphere-supplying respirator provides air from a source other than the surrounding atmosphere. Respirators are qualified by their purpose: for entry and escape, or for escape only.

Respirators are further differentiated by the types of environments they can be used in:

- Not for oxygen-deficient atmospheres (atmospheres containing less than 19.5% oxygen)

- Not for Immediately Dangerous to Life or Health atmospheres (*i.e.,* hazardous atmospheres which may produce physical discomfort immediately, chronic poisoning after exposure, or acute physiological symptoms after prolonged exposure)

Respirators which rely on finite air supplies or filtering capabilities are also classified by their service time ranging from three minutes to four hours as defined in 42 CFR 84, "Approval Of Respiratory Protective Devices."

Air-Purifying Respirators

Air-purifying respirators remove certain contaminants from air by either mechanical or chemical means. Prior to inhalation, ambient air is passed through a filter, cartridge, or canister packed with the appropriate materials to remove or neutralize the contaminants. Air-purifying respirators are of two types: those that are powered with an external power source (powered air-purifying respirator), and those that operate solely through the breathing effort of the wearer.

Filter Respirators. Filter (mechanical) respirators offer protection against airborne particulate hazards such as dust, mists, metal fumes, and radionuclides. They are equipped with a facepiece; either half or full face. Directly attached to the facepiece is a mechanical filter, made of an appropriate fibrous material that physically traps the airborne particles and delivers purified air to the user. The specific type of mechanical filter that matches the airborne hazard must be selected, as there are different types of mechanical filters, each capable of removing a certain type and size of particulate matter from the air.

Chemical Cartridge Respirators. These respirators are capable of removing low concentrations of hazardous vapors and gases from breathing air. Cartridges usually attach directly to the respirator facepiece. The removal of air contaminants is accomplished either by adsorption, or chemically

by neutralization of the particular contaminants. When chemical cartridges are opened, they begin to absorb humidity and air contaminants, which will cause their efficiency and service life to decrease. Cartridges should be discarded after use but should not be used for longer than the calculated use time. Cartridge change-outs can be determined by end-of-service-life indicators (ESLIs). However, manufacturers may not have an ESLI for a specific cartridge. In this case, *cartridge use times* must be determined by the employer, prior to being used by the employee. Cartridge use time is based upon many factors such as gas/vapor concentration, use duration, breathing rate, and environmental conditions. A qualified health and safety professional will determine use time as part of the respirator selection process. Many manufacturers of respiratory protection devices as well as OSHA, provide information on calculating use time and cartridge change-out schedule.

Atmosphere-Supplying Respirators

Supplied-air respirators provide breathing air to a facepiece. Two main types of supplied-air respirators are available: Self-Contained Breathing Apparatus (SCBA) and Airline Systems. OSHA 29 CFR 1910.134 requires the employer to provide employees using atmosphere-supplying respirators (supplied-air and SCBA) with breathing gases of high purity.

SCBA. The main SCBA components consist of an air or oxygen supply designed into a harness/backpack-type assembly with attached regulators, hoses, and a facepiece. SCBAs operate by maintaining a positive pressure of breathing air in the user's facepiece or by providing a continuous flow of breathing air to the facepiece. Positive pressure SCBAs represent the highest level of respiratory protection available and are used for Immediately Dangerous to Life and Health (IDLH) conditions. SCBAs usually can provide the user from 20 to 40 minutes of actual breathing time, even though the manufacturer may rate the use time longer. Some forms of SCBAs, especially those designed for mine rescue operations, may be used for extended periods (up to a few hours).

SCBAs can be of either open-circuit or closed-circuit design. In open-circuit devices, the expired air from the user is exhausted directly and is not reused. In closed-circuit equipment, the exhaled air passes through an adsorbent to remove carbon dioxide and water, and breathable air is regenerated. SCBAs provide breathing air that is filtered to a quality as specified by the Compressed Gas Association as Grade D air. OSHA specifies that this quality of air be provided to employees and that the employer document this carefully.

Airline Respirators. Airline respirators are available in pressure demand and continuous flow configurations. This system includes a facepiece component, air line, regulator assembly, escape bottle (for emergency egress, such as in confined space operations) and a breathing supply obtained from compressed gas cylinders or an air compressor system. The air quality must be as mentioned previously, Grade D. The user may wear a standard full or half facepiece, helmet, hood, or complete suit. A pressure demand airline respirator is very similar in operation to a pressure demand open circuit SCBA, except that the air is supplied through a small diameter hose from a stationary source of compressed air rather than from a cylinder of air worn on the user's back. Some important considerations in the use of airline respirators are to ensure that a clean supply of air is available and the air lines do not tangle, puncture, or degrade due to chemical or thermal exposure. In addition, airline systems should not pose an unsafe working condition to the user by restricting their movement or the performance of their work.

For both air-purifying and atmosphere-supplying respirators, OSHA requires frequent inspection, maintenance, and cleaning. SCBAs must be inspected monthly, with written documentation maintained, and the air cylinders hydrostatically tested either every three or five years, depending upon the material of construction of the cylinder.

General Designs and Features

All respirators are equipped with respiratory inlet covers or facepieces to provide a barrier from the hazardous atmosphere and for *connecting* the wearer's respiratory system with the respirator.

The two types of respiratory inlet covers include

- Tight-fitting (facepieces)
  - quarter masks
  - half masks
  - full facepieces
- Loose-fitting
  - helmets
  - hoods
  - blouses
  - suits

General respirator designs include

- Air-purifying respirators (APR)
  - disposable respirators

! Warning: Many disposable respirators do not provide an adequate seal on the user's face to prevent inward penetration of atmospheric contaminants and may not easily be evaluated by fit testing.

  - particulate filter respirators
  - cartridge or canister respirators (gas mask)
  - cartridge or canister respirators (gas mask) with particulate filter
- Powered air-purifying respirators (PAPR)
- Supplied-air respirators (SAR)
  - demand supplied-air respirators
  - continuous flow supplied-air respirators
  - pressure demand supplied-air respirators
- Combination supplied-air/air-purifying respirators
  - continuous SAR/APR
  - pressure demand SAR/APR
- Self-contained breathing apparatus (SCBA)
  - demand self-contained breathing apparatus
  - continuous flow self-contained breathing apparatus
  - pressure demand self-contained breathing apparatus

- Combination supplied-air respirators with auxiliary self-contained air supply (SCBA/SAR)

    - demand SCBA/SAR

    - continuous SCBA/SAR

    - pressure demand SCBA/SAR

Respirators may be either negative pressure or positive pressure respirators. All nonpowered air-purifying respirators are negative pressure respirators. Other negative pressure respirators include:

- Demand supplied-air respirators

- Demand self-contained breathing apparatus

- Combination continuous or pressure demand supplied-air/air-purifying respirator (SAR/APR)

Positive pressure respirators include:

- Powered air-purifying respirators (PAPR)

- Continuous flow supplied-air respirators

- Pressure demand supplied-air respirators

- Continuous flow self-contained breathing apparatus

- Pressure demand self-contained breathing apparatus

- combination pressure demand supplied-air respirator with auxiliary self-contained air supply (SCBA/SAR)

Respirators have several design and performance features associated with each type, such as the mask material, number of mask sizes, and rating service time.

Respirator Selection Approach

There are a number of regulatory requirements and guidelines that govern the selection of appropriate respiratory protection. In the United States, respirators must be certified by the National Institute for Occupational Safety and Health (NIOSH) to the respective requirements in 42 CFR 84. The selection of general respirator types is specified in OSHA 29 CFR 1910.134 (January 8, 1998). These regulations update previous selection practices specified by the regulation from ANSI Z88.2–1992, "American National Standard for Respiratory Protection."

OSHA 29 CFR 1910, Subpart Z provides for specific selection of respirators for protection against the referenced substances. *Recommended Practices* documents, and specific respirator selection guidance, are available. The guidelines for respirator selection are contained in their *Guide to Industrial Respiratory Protection* (DHHS/NIOSH Publication No. 87–116, 1987).

As discussed earlier, in order to select the appropriate respiratory protection, a hazard assessment must be conducted. The information needed to conduct the specific respiratory hazard assessment includes

- Identification of atmospheric contaminant(s)

- Determination of specific regulations or guidelines which may be available for identified contaminant(s)

- Measurement of concentration(s) for specific contaminant(s)

- Determination of Immediately Dangerous to Life and Health (IDLH) concentrations for contaminant(s). (An IDLH atmosphere poses an immediate threat to life, would cause irreversible adverse health effects, or would impair an individual's ability to escape)

- Measurement of oxygen concentration in atmosphere. (Oxygen deficiency exists when an atmosphere has an oxygen content below 19.5% by volume.)

- Determination if respirator use is for work or escape

- Determination of chemical and physical state of contaminant(s)

    - gas or vapor

    - particulates (dusts, aerosols, mists, fumes)

The general respirator risk assessment allows for specific types of respirators for IDLH environments and only general respirator types for selection in non-IDLH environments. A more detailed analysis is required to allow decisions between different specific respirator types and features. This analysis consists of

- Determining specific exposure limits and characteristics of the contaminants

- Evaluating workplace factors which affect respirator selection

- Reviewing respirator features related to protection

Specific respirators must be selected for special environments.

Fire Fighting. For fire fighting, select a full face-piece pressure demand self-contained breathing apparatus (SCBA) which meets the requirements of NFPA 1981.

Chemical Emergency Response. For chemical emergency response or hazardous waste site cleanup requiring Level A or B protection, select

- Full facepiece pressure demand self-contained breathing apparatus (SCBA)

- Combination full facepiece pressure demand supplied-air respirator with auxiliary self-contained air supply (SCBA/SAR)

Airline or Air-Supplied Suits. For airline or air-supplied suits (without internal respiratory inlet covering), select suits which have been approved by the requirements specified by the United States Department of Energy.

Abrasive Blasting. For operations involving abrasive blasting, select respirators approved for abrasive blasting:

- Powered air-purifying respirators

- Type AE, BE, or CE supplied-air respirators

Biological Airborne Pathogens. For protection against biological airborne pathogens (*Mycobacterium tuberculosis* [TB]), choose particulate filter facepiece air-purifying respirators equipped with a high-efficiency particulate air (HEPA) filters. Wear specified respirators when

- Employees enter rooms housing individuals with suspected or confirmed infectious TB diseases

- Employees perform high-hazard procedures on individuals who have suspected or confirmed TB diseases

- Emergency medical response personnel or others must transport, in a closed vehicle, an individual with suspected or confirmed TB diseases

When selecting a respirator, it is also important to consider the potential hazards associated with that respirator:

1) Do respirator materials (especially those in the facepiece) irritate or sensitize the wearer's skin?

2) Is the respirator likely to retain contamination even after cleaning?

3) Should the respirator be protected from exposure to liquids and other contaminants?

4) Does the respirator have a design with loose bands, straps, or material that can be caught in moving machinery?

5) Does the respirator (full facepiece or helmet/hood configurations) provide clear and unobstructed vision for performing required tasks?

6) Is the respirator available in a sufficient number of sizes or can the respirator be adjusted to fit personnel? Has each individual who must wear a respirator been fit-tested?

7) Is the respirator difficult to use and reservice?

8) Is the respirator uncomfortable for the wearer under use conditions?

9) Is the respirator reliable for meeting the intended service life?

Protective Clothing

Protective clothing includes specific components of a protective ensemble (*e.g.*, garments, gloves, boots, *etc.*) that provides dermal protection. Dermatological disorders are primarily a result of unprotected exposures to harmful chemical, biologic, and physical agents. Most of these exposure risks can be prevented or reduced through the proper selection and use of protective clothing if engineered or administrative controls are not effective or are inapplicable or unavailable. Personal hygiene practices should always follow use of protective equipment.

Chemical Protective Clothing

The types of chemical protective clothing (CPC) range from basic work clothes to total encapsulating chemical protective ensembles, with a wide variety of designs in between. Basic items for providing splash protection for specific areas of the body, including the head, torso, and appendages, are available in a wide variety of materials. These materials will be discussed further on in this section.

Chemical protective clothing is used in a variety of operations such as cleanrooms and laboratories, in the manufacture of electronic devices, and for hazardous waste operations and emergency response. When specific chemical protection is required, the specifier of the CPC should research the chemical protection information from the CPC manufacturers. Each manufacturer has its own laboratory data for use in selecting CPC. This information is specific to the manufacturer and should not be used as a reference for other manufacturers' products. It should also be noted that laboratory data from any manufacturer might not directly relate to the use environment of concern to the specifier. Depending on the CPC materials, garment design, and construction, these suits can protect the wearer against a wide range of chemicals, including acids, solvents, oxidizers, alkalis, *etc.*

Chemical protective clothing is generally classified into different types: gas/vapor-resistant and splash/particulate-resistant. The gas/vapor-resistant clothing is generally configured as totally encapsulating suits, providing head to toe coverage to protect the wearer. These are the large *moon suits* that have special seams and zippers to prevent chemicals from leaking into the suit. The suits have a face shield which is made a part of the hood. *Splash/particulate-protective clothing* provides good protection and is used when less skin protection is needed. Splash/particulate-resistant clothing is available in various designs–some with hood, some with elastic wrists and ankles, *etc.* The hood can either be part of the suit or detached.

All CPC can be either reusable or disposable. *Reusable CPC* would provide the same protection for the second use as it was intended to provide initially. If the CPC cannot meet these criteria, it is considered disposable. Thus performance, not cost, should drive reusability. Unfortunately, because the state of the art in proving reusability of CPC is so poor, many CPC use scenarios require CPC to be disposed of after one or a limited number of uses. All CPC is vulnerable to chemical attack, environmental conditions, and physical abuse. The true useable life of CPC is very difficult to predict, so in many situations it is simply better to err on the side of safety and replace the CPC prior to evidence of failure.

CPC materials and quality of construction also influence the performance of CPC as a barrier to chemicals. CPC is available in a variety of materials and designs. Materials used for protective garments include natural rubber, neoprene, nitrile, polyethylene, chlorinated polyethylene, polyvinyl chloride (PVC), Saran-coated Tyvek or Saranex®, polyurethane, butyl polymers, treated woven fabrics, *etc.* These materials can be supported on cotton, nylon, polyester, and other materials. The most appropriate clothing material will depend on the chemicals present and the task to be accomplished. Ideally, the chosen material should resist permeation, degradation, and penetration. The manufacturers' literature usually provides charts indicating the resistance that various clothing materials have to permeation, subsequent breakthrough, or degradation by certain chemicals. However, no single material can protect against all chemicals or any combination of chemicals, and few currently available materials are effective as barriers to any prolonged high level of chemical exposure beyond 60 minutes. Also, for a given clothing material type, chemical resistance can vary significantly from product to product. And for certain chemicals or combinations of chemicals, there may be no commercially available CPC material that will provide more than an hour's worth of protection following contact.

In recent years, public concerns have been raised regarding civilian exposure to toxic chemical warfare agents with terrorist attacks and activities involving stockpiled weapons, *etc.* Aside from deliberate attacks, there is the potential for exposure from unknown military ordnance dumps. The possibility of the presence of any chemical agent (nerve agents, blister agents, blood agents, choking agents, irritants, and biological agents) may warrant the highest level of personal protection available for initial response efforts. The compatible protective apparel material may vary depending on the specific agent, but the highest precautionary level is imperative until the threat is clearly identified. Also, it is impossible to guarantee 100 percent protection against a chemical agent incident.

The design and construction of CPC is also very important for CPC performances. Here are some examples of design and construction factors:

- Stitched seams of clothing may be highly penetrable by chemicals if they are not overlaid with tape or sealed with a coating.

- Lot-to-lot variations do occur, and may have a significant effect on the barrier effectiveness of the CPC.

- Pinholes may exist in elastomeric or plastic products due to deficiencies or poor quality control in the formulation or in the manufacturing processes.

- Thickness may vary from point to point on the clothing item.

- Garment closures differ significantly from one manufacturer to another and within one manufacturer's product line.

The most important criterion in the selection and use of CPC is the effectiveness in protecting against the chemical(s) of interest. However, CPC testing methods are not universal. And manufacturers' testing standards vary substantially. The American Society for Testing and Materials (ASTM) has a standard for permeation testing of protective garment materials (ASTM, F739–96). The ASTM method determines two critical properties of the CPC material, breakthrough time and steady-state permeation rate. The ideal CPC material should have a very long breakthrough time and a very low steady-state permeation rate.

Thermal Protection

The thermal environment people work in has a great impact on workers' physical conditions as well as work performances. Extreme environmental temperatures can cause heat or cold stress. Common heat disorders may include heat cramps, heat exhaustion, and heat stroke. Cold stress can cause hypothermia and frostbite. Besides good work practices for preventing heat or cold stress, protective clothing can make a big difference in protecting workers from extreme cold or heat.

In hot environments, it is important to ensure that clothing keeps workers cool and prevents dehydration. Clothing that is light in color and fabric, and fits loosely to allow sweat to evaporate is recommended. Cooling vests are also available which can absorb the extra body heat, keeping worker's internal temperatures within a safe range. Studies done by the United States Navy have shown that crewmen using cooling vests can safely work in hot environments for twice as long as those without them. There are two basic types of cooling vests available: passive and active. *Passive devices* generally involve water-based or liquid chemical solutions packaged in sealed sectional plastic strips that can move with a worker. Once frozen, the packs are inserted into insulated vests. *Active cooling devices* usually circulate cool water over the body using a power-generated pump. There are tubes that carry chilled water through the vest and recirculate water warmed by the body to an ice-water reservoir where it is recooled.

In cold environments, one of the ways to stay warm is to dress in layers. An outer layer of a waterproof windbreaker material may be worn—acting as a buffer against the wind. A light, airy, dry shirt should be worn next to the skin. It may also be important to have ventilation in the cloth next to the skin so the body can sweat freely. Hands, fingers, feet, ears, and nose should be protected from the cold, since frostbite tends to attack extremities first.

When proper heat or cold stress protection is used, it should help make most environments more tolerable, which is essential for worker protection and productivity.

Head Protection

Hazards associated with the head are impact, falling or flying objects, electrical shock, burns, and scalping, *etc.* Under the OSHA requirements for head protection (29 CFR 1910.135), the employer must determine the needs for protecting workers from head injuries and ensure that the appropriate protection is provided and worn. The protective helmet provided by the employer must provide protection against impact and penetration of falling and flying objects, protection from limited electric shock and burn, as well as meet criteria and requirements contained in ANSI Z89.1–1997 for helmets.

Protective helmets are made in the following types and classes:

- Type I – crown impact
- Type II – crown and lateral impact

Three classes of *hard hats* are recognized:

- Class G–general, limited voltage protection (proof-tested at 2,200 volts)
- Class E–electrical, high-voltage protection (proof-tested at 20,000 volts)
- Class C–conductive, no voltage protection

Other head protection–related items may include *helmet liner* (for insulating against cold), hood (for protecting against chemical splashes, particulate, and rain), and *protective hair covering*.

Eye and Face Protection

Eye and face injuries are typically caused by the following hazards:

- Flying objects, such as dust, metal, or wood chips
- Splashes of toxic or corrosive chemicals, hot liquids, and molten metals
- Fumes, or gases or mists generated by toxic or corrosive chemicals
- Radiant energy and/or intense heat
- Lasers

Thus, eye and face protective equipment are designed for protecting the users from glare, liquids, flying objects, ultraviolet radiation, or a combination of the above hazards. The OSHA requirements for eye and face protection, which require the employer to provide eye and face protection where there is a reasonable likelihood that injuries could be prevented by use of the

protective equipment, are listed in 29 CFR 1910.133. The design, construction, testing, and use of eye and face protective devices must be conducted according to the ANSI standard Z87.1–1989. The OSHA 29 CFR 1910.133 standard also has a chart identifying lens filtering properties.

Eye and face protective equipment usually includes safety glasses, safety goggles, and face shields. Sometimes, different forms of protection need to be combined to provide better protection.

Safety Glasses. Safety glasses (some are equipped with side shields) are designed for protecting the eyes from flying objects and moderate impacts encountered in operations such as grinding, spot welding, woodworking, scaling, and machine operation, *etc.* Safety glasses normally have hard plastic, clear hardened glass or wire mesh lenses and frames made of plastic, plastic coated wire, wire, or fiber. Safety glasses with metal frames should not be used if there is potential of electrical contact or in an explosive atmosphere where non-sparking tools are required. A face shield needs to be used over the safety glasses if a potential chemical splash hazard exists.

Safety goggles. Safety goggles protect the eyes from dust, splashes, flying objects, and sparks from any direction. Some goggles have a cup over each eye, whereas others have a frame that extends over both eyes. Goggles have a variety of designs and forms such as: chemical goggles, leather mask dust goggles, miner's goggles, melter's goggles, welder's goggles, and chipper's goggles. Goggles are appropriate during dusty operations, such as sanding. Welding jobs require their own special helmets, either with stationary or lift front windows, to protect against burns as well as sparks and molten splashes.

Face shields. Face shields protect the face and neck from splashes, heat, glare, and flying objects. They usually have detachable windows made of plastic, wire mesh, or both. Face shields can be attached to a hard hat. They are also available with infrared absorbers to provide protection against radiation. In potentially explosive atmospheres, nonsparking shields must be used. Face shields do *not* take the place of safety glasses. If safety glasses are required in an area where a face shield is to be worn, the safety glasses are worn under the face shield.

Hearing Protection

Excessive noise can create stress on workers, damage their ability to hear, and reduce their job performance. The OSHA regulations for noise protection (29 CFR 1910.95) require each employer to administer a hearing conservation program whenever employee noise exposures equal or exceed 85 decibels for an 8 hour time weighted average. Hearing protective devices are needed when engineering and administrative controls can not reduce noise levels to the permissible noise exposure levels required by OSHA.

The general types of hearing protection include the insert type (earplugs) and the ear muff type. Figure 3 provides examples of various types of hearing protection.

Earplugs. Earplugs are the most common types of hearing protection devices used in industry. They are designed to be inserted into the ear canal.

Earplugs come in many different varieties and are made of soft materials. Some devices are made in several sizes to accommodate different-sized ear canals. Others come in one size that can be adapted by their natural expansion when inserted in the canal, or by removing one or more flanges on the unit. Some earplugs are disposable; they are normally made of expandable foam. Other earplugs are reusable; they are usually made of a flexible rubber or silicone. They may be flanged or cone-shaped, and are often jointed by a cord so that they are not easily lost.

Earmuffs. Earmuffs consist of a pair of padded plastic shells connected by a flexible band. The shell interior is lined with foam rubber, plastic, or a liquid-filled cushion. Some devices fit only in one position, others are multipositioned and can be worn with the headband over the head, behind the head, or under the chin. Earmuffs can offer good protection if the cushions fit tightly around the ears. Ear muffs cost more initially, but they are cleanable, and replacement parts are available.

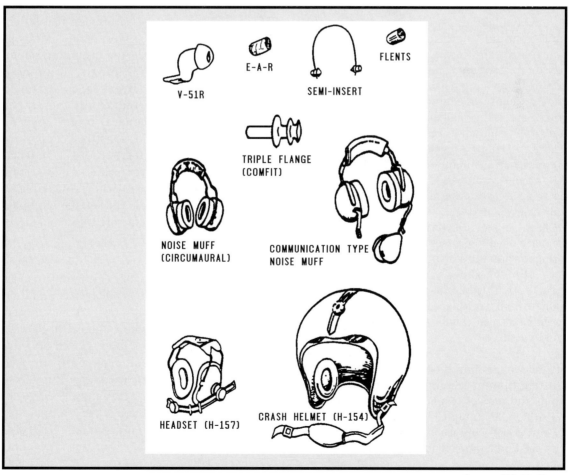

Source: US Air Force

Figure 3. Hearing Protection

Each type of hearing protective device has a noise reduction rating (NRR) which tells how much the protector lessens the noise. The NRR is a key factor for selecting the hearing protective devices. However the selection process is also affected by many other factors, among them comfort, durability, ease of use, hygiene, cost, range or sizes available, care and cleaning requirements, *etc.* All of these are very important factors.

Foot Protection

Foot protection such as footguards, toe caps, metatarsal guards, safety shoes, or boots are needed for the protection of feet from falling or rolling objects, sharp objects, molten metal, hot surfaces, contaminated materials, and wet slippery surfaces, *etc.* OSHA 29 CFR 1910.132 has general requirements for the employer to protect workers from hazards that have a potential for causing injury to the foot. OSHA 1910.136 has detailed requirements regarding occupational foot protection.

Boots are available in a wide variety of lengths ranging from hip length to those that cover only the bottom of the foot. Boots constructed of chemical-resistant material are used for protecting feet from contact with chemicals. Chemical protective boots are made of a few different polymers since the boot heel and sole require a high degree of abrasion resistance. The common polymers and rubbers used in chemically resistant boot construction include butyl rubber, nitrile, neoprene rubber, and PVC materials. Disposable shoes or boot covers are also available and made from a variety of materials. Boots constructed with some steel materials (*e.g.*, toes, shanks, and insoles) are used to protect feet from compression, crushing, or puncture by falling, moving, or sharp objects. Safety shoes and boots may also be constructed with fiberglass toes, as a substitute for steel toes in workplaces involving unprotected electrical sources (such as the third rail at a railroad yard).

Safety footwear is classified according to its ability to meet minimum requirements for both compression and impact tests. ANSI standard ANSI Z41–1001 is the standard for Personal Protective Footwear. Boots constructed from nonconductive, spark-resistant materials or coatings are used to protect the wearer against electrical hazards and prevent ignition of combustible gases or vapors.

Hand Protection

Hazards to hands may be presented by chemicals, sharp objects, hot surfaces, moving parts of machines or tools, and heavy equipment, *etc.* Besides the general requirements, OSHA has hand protection requirements in 29 CFR 1910.138.

Gloves. Gloves can be used to protect hands from potential hazards such as burns, cuts, electrical shock, amputation, and absorption of chemicals, *etc.* There are gloves available that can protect the wearers from any of the individual hazards or any combination thereof. Gloves may be integral, attached, or separate from other protective clothing. Overgloves are sometimes used to provide supplemental protection to the wearer and protect more expensive undergarments from abrasions, tears, and contamination. Disposable gloves are often used to reduce decontamination needs in contamination protection situations. Cloth gloves are often used for light duty materials handling for protection against cuts or abrasions. They may also be used to provide insulation for moderately hot or cold environments. Many glove materials are available for providing chemical protection, such as PVC, nitrile, neoprene, and natural rubber, *etc.* Breakthrough time and permeation rate are important factors when choosing gloves for chemical protection. When dealing with glove selections, it is important to match the gloves selected to the specific hazards. Comfort, dexterity, glove thickness, glove interface with chemical protective clothing, available sizes, use duration, frequency and degree of exposure, *etc.* are also important factors to be considered.

The health care industry, research laboratories, food service workers, *etc.* commonly use latex gloves. Latex gloves are also used as inner gloves by environmental professionals and emergency response personnel when chemical gloves are required. Latex gloves have proved effective in preventing transmission of many infectious diseases to health care workers. But for some workers, exposures to latex may cause allergic reactions and result in skin rashes, sinus symptoms, asthma, and shock. NIOSH has published an *Alert* publication that recommends employers adopt policies to protect workers from latex exposure and allergy in the workplace.

Fall Protection

Fall protection is needed in many work areas or job activities such as excavations, hoist areas, holes, unprotected sides and edges, roofing work, wall openings, building construction, and other walking/working surfaces. In 1995, two major fall protection–related construction standards took effect: OSHA's Subpart M standard, which relates directly to equipment and systems used for fall protection in the construction industry, and Subpart L, which deals specifically with scaffolds. Residential construction activities are also impacted by these regulations. Other available OSHA fall protection standards include 29 CFR 1926.104 for safety belts, lifelines and lanyards and 29 CFR 1926.105 for safety nets. As required by the OSHA Fall Protection Standard, employers must protect their employees from fall hazards and falling objects whenever an affected employee is 6 feet or more above a lower level.

Fall protection generally can be provided through the use of guardrail systems, safety net systems, personal fall arrest systems, positioning device systems, and warning line systems, *etc.* Canopies and toeboards are methods for protection from falling objects. Employers can select fall protection measures compatible with the type of work being performed. Specific fall protection system criteria and practices required by the fall protection standard must be met when using a system. For construction employees, OSHA's Subpart M standard for construction (29 CFR 1926.500–502) requires the use of full body harnesses as of January 1, 1998 as part of personal fall arrest systems. Locking snap hooks must be used to replace nonlocking snap hooks that took effect at the same time.

Selection of Protective Equipment: Anticipation, Recognition, Evaluation, and Control

The association of potential hazards with modern technology and the increasing use of new biological and chemical reagents requires a positive approach to the anticipation, recognition, evaluation, and control of health hazards, as well as physical stresses in the work place. Each employer and

his or her supervisors must be diligent in the recognition of hazards associated with his/her activities. This requires a review of health and physical hazards associated with activities or operations to be conducted. Employers must evaluate the hazards associated with each procedure and process, to determine where exposures exist or may occur; evaluate actual exposure through air sampling, analysis, or other sampling techniques as necessary; conduct medical surveillance programs on employees exposed to hazardous materials as necessary; and control hazards through substitution or elimination of materials, modification of equipment, addition of ventilation, use of personal protective equipment or administrative changes, *etc.*

Risk Assessments and PPE Selection

As discussed earlier, in order to select the PPE needed for a specific task, it is very important that the needs be assessed for the particular job, performance criteria be developed, and a hazard/risk assessment be conducted. The OSHA PPE standard requires the employer to perform a hazard assessment at the workplace to determine if hazards that require the use of head, eye, face, hand, or foot protection are present or are likely to be present. 29 CFR 1910.132 includes a non-mandatory Appendix B that offers compliance guidelines for hazard assessment and PPE selection.

The hazard assessment usually involves a walk-through survey of the job area to identify sources of hazards, and consideration of the following categories of hazards: impact, penetration, compression, chemical, heat, harmful dust, light radiation, and falling, *etc.* During the walk-through, the sources or potential problem areas for these hazards should also be identified. The data and information for use in the assessment of hazards are then organized, and an estimate of the potential for injuries is made. Each of the basic hazards is then reviewed and a determination is made as to the type, level of risk, and seriousness of potential injury from each of the hazards.

If hazards or the likelihood of hazards are found, suitable PPE needs to be selected for protecting the affected employees from these hazards. The PPE selection process usually involves determining the following conditions:

- Potential hazard(s)
- Adverse effects of unprotected exposure
- Other control options that can be used
- Requisite performance characteristics
- Decontamination requirements
- Physiological, psychological, and physical stresses associated with the work task and PPE usage
- Ergonomic constraints
- Cost of various options
- Making the selection

Once it is determined that certain types of hazards exist or potentially exist and the level of risk, and the seriousness of any potential injury analyzed, then the environmental health and safety professional making the PPE selection should become familiar with the types of PPE that are available for the specific hazards. The hazards associated with the environment they exist in are compared with the capabilities of the available PPE. The PPE that ensures a level of protection greater than the minimum required to protect employees from the hazards should be selected. The minimum level of protection required will depend on the performance standard required or selected, such as the Threshold Limit Value (TLV) or Permissible Exposure Limits (PEL), break-through time, *etc.* Besides performance criteria, the selection may also be affected by the need for decontamination, ergonomic constraints presented, the cost, *etc.* Once the individual PPE items are selected, they should be assembled according to previously established ensemble configuration criteria, and fitted to the user. It is also very important that the users are instructed in the use and maintenance of the PPE, and are made aware of all warning labels and limitations of their personal protective equipment.

In summary, the risk-based PPE selection process should involve: defining work task, determining workplace hazards (type and source of hazards) by using a job safety analysis, analyzing potential risks, determining type of PPE needed for the specific hazard, and comparison of PPE capabilities with performance standard or action level criteria. If the performance standard can be met by the PPE, and other miscellaneous factors

including economic constraints are satisfied, then the PPE is selected for the ensemble, configuration, and fitting phases.

The Selection Process

Respiratory Protection. The selection of a specific respirator must be made by individuals knowledgeable about the limitations associated with each class of respirators and acquainted with the actual workplace environment, including the job task(s) to be performed. Many factors may influence the respirator selection, such as: adequate warning, type of hazard, concentration of contaminant, acuteness of hazard, time spent in contaminated atmosphere, nature of the working environment, activity of the wearer, mobility of the wearer, and whether the use is for routine or emergency application, *etc.*

NIOSH Certification. In 1995, NIOSH established new testing procedures and performance requirements for nonpowered, air-purifying respirators in its new certification regulation (42 CFR 84). NIOSH has been approving respiratory products using the new rules since then. The 42 CFR 84 approval procedure uses a letter and number system to differentiate products. There are three types of filters: N–series for protection against solid and water-based particulates; R–series for protection against any particulates, including oil-based materials with eight-hour maximum usage; and P–series for protection against any particulates, including oil-based materials, with no specific time limit. Manufacturers will be able to sell only those products certified under the new procedures after June 10, 1998. Distributors and employers will be able to exhaust their old supplies before switching over to the new ones.

Personal Protective Clothing and Equipment Selection. Besides the selection of respiratory equipment, the PPE selection process involves the selection of other individual items, based on needs, including but not limited to protective equipment for: eye and face, head, foot, hand, torso, and hearing. Many of the factors that affect the selection of individual items were discussed in earlier sections of this chapter. The factors affecting chemical protective clothing (suits, gloves, boots) selections for skin protection are emphasized here.

Several factors have to be considered when making a chemical protective clothing selection. These factors include: body coverage needed, physical properties for avoiding chemical penetration, prior use experience, permeation resistance, protection period required, chemical toxicity, severity of potential chemical contact, temperature of chemicals, multiple chemical exposures, degradation conditions, and decontamination methodology, *etc*.

Whenever exposure to vapors of a chemical is considered unacceptable, permeation resistance data should be used. The permeation resistance performance for protective clothing is particularly important for chemicals that pose known health hazards through irritation, reaction with, or absorption into the skin. Whenever the workplace conditions indicate the likelihood of chemical exposure, such as direct contact with solids or particulate-based chemicals, direct liquid contact through immersion, or contact with liquid through splashing, spraying, or misting, and vapor contact, permeation data should also be considered.

The most common practice for using permeation resistance data is to compare the breakthrough time with the intended period for using the relevant PPE garments. Many users select a chemical protective garment when the garment materials of interest show breakthrough times greater than the longest anticipated use period. However, according to PPE experts, using this method alone can provide an inadequate PPE selection, resulting in both underprotection and overprotection.

It is recommended that the consideration to use permeation data be based on a risk assessment which identifies the hazards of the chemicals in the work environment, combined with estimates for likelihood of exposure and the possible consequences of exposure. However, unlike respiratory protection, there is little data available for defining acceptable exposure limits for skin to chemicals, which makes it more difficult to define an acceptable risk for selecting PPE. The two available measures may be the *toxic dose low* (TD$_{LO}$) (the lowest concentration of a substance introduced by any route, other than inhalation, over any given period of time and reported to produce any toxic effect in humans or to produce teratogenic or reproductive effects in animals) and the *toxic concentration low* (TC$_{LO}$)–the lowest

concentration of a substance in air to which humans or animal have been exposed for any given period of time that has produced any toxic effect in humans or to produce teratogenic or reproductive effects in animals.

One alternative approach recommended by some experts for selecting protective clothing on the basis of chemical permeation resistance data is determining the potential of exposure to chemicals through the permeation rate or minimum detectable permeation rate. The cumulative permeation through the protective clothing is based on a number of assumptions and is calculated using the permeation rate of the chemical together with the exposed clothing surface area and exposure time.

Simultaneous Hazards. Simultaneous hazards may be present under certain conditions. The PPE selection process under these conditions may be even more complicated. The selected PPE should not only protect against each individual hazard, but also against the combination of those simultaneous hazards, if any synergistic effects exist.

Levels of Protection. The components of personal protective equipment may be assembled into a protective ensemble that not only protects the worker from site-specific hazards but also minimizes the hazards and drawbacks of the PPE ensemble itself. The Environmental Protection Agency (EPA) has defined four levels of protection: Level A, B, C, and D. These levels are widely referenced especially for hazardous waste site operations and emergency response applications.

Level A is required where there is the greatest potential for exposure to skin, respiratory, and eye hazards. Level A includes respiratory protection with positive pressure, full facepiece self-contained breathing apparatus (SCBA), or positive pressure supplied-air respirator with escape SCBA; totally encapsulated chemical-and-vapor-protective suits; inner and outer chemical-resistant gloves; and disposable protective suits, gloves, and boots.

Level B protection includes the highest level of respiratory protection but a lesser level of skin protection. At most abandoned outdoor hazardous waste sites, ambient atmospheric vapors or gas levels are not high enough to warrant Level A protection and Level B often is adequate. Level B includes respiratory protection with positive pressure, full facepiece self-contained breathing apparatus (SCBA) or positive pressure supplied-air respirator with escape SCBA; inner and outer chemical-resistant gloves; faceshield; hooded chemical-resistant clothing, coveralls, and outer chemical-resistant boots.

Level C is required where the concentration and type of airborne substances are known and the criteria for using air-purifying respirators are met. Typical Level C equipment includes full face air-purifying respirators, inner and outer chemical-resistant gloves, hard hat, escape mask, and disposable chemical-resistant outer boots.

Level D protection is the minimum protection required. Level D may be sufficient when no contaminants are present or work operations preclude splashes, immersion, or the other potential for unexpected inhalation or contact with hazardous chemicals. Appropriate Level D protective equipment may include gloves, coveralls, safety glasses, faceshield, and chemical-resistant, steel-toe boots or shoes.

Figure 4 provides examples of the levels of protection. *Note that these levels of protection do not provide specific information to the user regarding the specific ensemble components. The specific ensemble components must be specified based upon a site- or workplace-specific risk assessment.*

Validating the Selection Process

Once the PPE is selected, the selection process should be validated. PPE should be fitted to the specific user. For example, respirator fit-testing needs to be performed to decide what size of a particular air-purifying respirator should be assigned to a user. Medical surveillance may be needed for a particular PPE user.

Documenting the Selection Process

In the OSHA PPE standard at 29 CFR 1910.132, the employer is required to provide written certification that the workplace hazard assessment has been performed. The certification shall identify the workplace evaluated, the person certifying that the evaluation has been performed,

and the date(s) of the hazard assessment. It is recommended that the whole PPE selection process be documented to include not only the hazard assessment performed but also the PPE selection criteria used, the PPE selected and assigned (including personnel and type of task), and the PPE training performed (what kind of training and date of training).

Various Factors Affecting PPE Performances

The performances of PPE are affected by many factors, including physical environment conditions, ergonomics and human factors, the types of PPE selected, *etc.*

Adverse physical environments such as extreme temperature conditions can not only cause heat or cold stress to the PPE users, but may also affect the performance of the PPE itself, potentially exposing the worker unexpectedly. For example, low temperatures may fog respirator lenses, impeding the worker's ability to see what he or she is doing. In addition, many elastomeric materials (PVC, butyl, neoprene, and viton rubber) are not designed for use in extremely cold or hot environments and will crack, puncture, and degrade more readily than at normal room temperature. The PPE user should be made aware of the limitations of PPE they are assigned, and use them only under the assumed conditions and tasks the PPE was originally selected for.

In certain cases, the use of personal protective equipment could increase workers' discomfort, cause heat stress, and decrease mobility, communications, dexterity, tactility, strength, and endurance, and affect productivity. For example, when workers use more than one layer of gloves, the ability to perform simple motor functions such as trying to grip an object becomes increasingly difficult and muscle fatigue occurs more rapidly. These *ergonomic factors* should be considered when PPE is selected and used. The use of protective clothing may also affect the quality of a worker's performance and increase the chance for human error.

Wearing personal protective equipment can impose some physiological and psychological stress

LEVEL A PROTECTION
Totally encapsulating
vapor-tight suit with full-
facepiece SCBA or
supplied-air respirator

LEVEL B PROTECTION
Totally encapsulating
suit does not have to be
vapor-tight. Same level of
respiratory protection as
Level A.

LEVEL C PROTECTION
Full-face canister air-
purifying respirator.
Chemical protection suit
with full body coverage

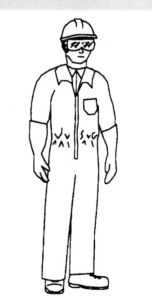

LEVEL D PROTECTION
Basic work uniform; *i.e.,*
longsleeve coveralls,
gloves, hardhat, boots,
faceshield or goggles.

Source: US Air Force

Figure 4. Levels of Protection

on the user. The weight of the equipment, for example, increases the energy requirement for a given task. The use of PPE may also affect the human response and endurance, especially in hot environments. Various human factors may not only affect the PPE performance but may also affect the safety of the PPE user. These factors may include training, experience, stress, height, weight, false sense of security, *etc*. The mobility, tactility, dexterity, and visibility can also be limited due to PPE usage.

The following practices are recommended for reducing adverse effects caused by various factors:

- Select the lightest-weight protective ensembles and respiratory protective devices that adequately protect the worker. This will minimize the physiological demands placed on the worker that are associated with carrying the weight of this equipment. The size of the PPE should fit the user

- If available, select protective clothing made of material that will allow evaporation of water vapor, while providing skin protection from the contaminant

- When PPE is used during intensive work, reduce the work rate by adjusting work/rest schedules, using automated procedures and/ or mechanical assistance where possible, and minimize the work intensity

- Educate workers on symptoms and prevention of heat illness, and schedule periodic fluid replacement breaks

- Reduce heat or cold stress by scheduling work accordingly and using engineering control methods

Behavior Modification and the Human Factors

Adequate education for PPE users is essential to its effective use. The OSHA PPE standard requires that employees be trained to know when PPE is necessary, what type is necessary, how it is to be worn, and what its limitations are, as well as to know its proper care, maintenance, useful life, and disposal procedures. This must be accomplished before any work that requires the use of PPE is assigned. Retraining may be required if the workplace changes, if the types of PPE to be used change, or if inadequacies are identified which indicate that employees have not retained the necessary understanding or skills. By educating the PPE user, one may be able to improve or control the human factors and achieve a better PPE performance goal.

Management Approaches to Personal Protection Strategies

Personal protective equipment should not be used as a substitute for engineering controls, work practice, and/or administrative controls. The workplace should be designed, engineered, modified, improved, or controlled by reasonable means to eliminate hazards requiring PPE. PPE should be used in conjunction with these controls— as part of the *big picture* to provide for employee safety and health in workplace. Management must commit its financial, human, and equipment resources towards establishing and maintaining an effective program within which the employees are actively participating. In many organizations, personal protective equipment programs are an integral part of the daily implementation of the corporate environmental health and safety (EHS) program. The corporate EHS program encompasses all strategies (including personal protective equipment) to maintain a health and safe workplace. The following section discusses overall workplace health and safety strategies as part of an organization's EHS program.

Environmental Health and Safety Management Program

In order to provide work environments that maintain and protect employee health and safety, organizations must develop and maintain a comprehensive and proactive program. Organizations must be committed to developing and maintaining an environmental, health, and safety (EHS) program that prevents injury and controls loss by minimizing occupational injuries and illnesses; monitoring the workplace environment (*e.g.*, dust, vapor, chemical hazard, radiation, and noise); and using this data as the basis for engineering and work practice controls as well as personal protective equipment selection.

To be effective, this effort must have the strong commitment and support of management. The following elements are important components of an effective EHS management plan:

- Management leadership
- Assignment of responsibility and accountability
- Design and maintenance of safe working conditions
- Accident investigation and analysis
- Safe operating rules and procedures
- Safety training and education
- Recognition, evaluation, and control of occupational hazards
- EHS administration

Management Leadership. The organization must realize that the attitude and commitment of upper management are invariably reflected in the attitude of the supervisory force towards environment, safety, and health. Therefore, management must demonstrate a sincere interest in the EHS program if employee cooperation and participation are to be achieved. Management must issue clear statements of policy for the guidance of supervisors and employees. If safety is a shared responsibility, then management must do its share.

Assignment of Responsibility. While the prime responsibility for the safety and health of all employees is, by law, vested in the Chief Executive, the day-to-day responsibility rests with supervisors. This means that all supervisors have the responsibility to monitor the EHS performance of all employees that report to them. Each employee is responsible for following all of the safety rules and requirements that are applicable to his or her assigned responsibilities. Management should encourage employee participation in the development of EHS promotional programs, goals, and objectives. Employees have the responsibility to make others aware of environmental, health, and safety problems.

Design and Maintenance of Safe Working Conditions. The supervisor or worker is responsible for checking out all new equipment and processes as well as initiating the development of standard operating procedures (including relevant safety precautions) before authorizing use of such equipment. Workplace inspections are essential not only for the maintenance of safe work conditions and operating practices, but also for the discovery of causes of accidents and harmful exposures. Comprehensive inspections of the entire facility should be conducted periodically and at least annually. Self-audits should be conducted daily—before and at the end of the workday. Written reports should be prepared after inspections, and appropriate corrective action should be initiated to correct all deficiencies that are identified. Follow-up inspections by the person responsible for employee safety (usually the Safety Officer) should be conducted to evaluate the corrective action taken.

Accident Investigation and Analysis. Employees must be encouraged to report all near-misses, accidents, injuries, and occupational illnesses. These events should be investigated to determine the cause, and to establish corrective measures to prevent reoccurrence. It is essential that an accident investigation and injury report be completed for all work-related injuries or illnesses. Each supervisor is responsible for investigating all incidents, accidents, and injuries that occur within their area(s) of supervision.

Safe Operating Rules and Procedures. Safe operating rules and procedures must be established as a guide to all employees for the prevention of accidents, injuries, illness, or exposure to harmful substances. These rules and procedures need to be tested and proven to work and reviewed periodically by employees with their supervisors. It may be necessary, on occasion, to establish new safety rules to cover new equipment or operations, or to revise existing rules to improve conditions or procedures. When any safe operating rule or procedure is violated or disregarded, appropriate disciplinary action must be taken, immediately. Employees must be advised of and required to comply with these rules and procedures. This would apply to the management of toxic chemicals, infectious agents, radiation exposures, *etc.*

Safety Training and Education. OSHA has come to the realization that adequate protection of the safety and health of employees goes beyond the provision of personal protective equipment and safe working conditions. Human factors must be considered and action must be taken to encourage proper work practices. Education, training, and safety promotional programs are the means of developing a spirit of safety consciousness and good attitudes in the workplace. OSHA and EPA include a worker training section as part of every standard that they promulgate.

All new and transferred employees must be oriented and trained in their jobs. This basic orientation is usually followed by frequent on-the-job safety instructions by supervisors during their first few days on the job. Potential hazards, if any, in the area are identified, and precautions to avoid injury are defined. Information on fire and emergency procedures, reporting of injuries and accidents, and available health care, should be covered as well. Attention to the integration of safe and healthy work practices into daily tasks is an important factor in this follow-up training. Safety Teams composed of representatives of senior management, employees, and staff should conduct safety reviews periodically to assess the *environment of safety*.

Developing and Managing PPE Programs

Written programs for PPE, respiratory protection, fall protection, confined space entry and rescue, as well as a hearing conservation program should be developed if PPE is needed in the workplace. These programs usually include policy statements, procedures, and guidelines. A risk assessment methodology for selecting the PPE; an evaluation of other control options to protect the worker; PPE selection criteria and procedures; PPE performance criteria; user training requirements; PPE storage, maintenance, and decontamination requirements; and an auditing or program reevaluation procedures are normally included in the programs. By implementing formal written programs, the chance for error is reduced, worker protection is increased, and a companywide, consistent approach for the selection and use of PPE is established. The PPE program should be reviewed at lease annually.

Corporate management and employees should make diligent and dedicated efforts to ensure corporate respiratory and PPE programs meet their standard goals. The person responsible for administering the program must have the appropriate technical and professional background. It is important that one member of management have final responsibility as well as the necessary management support for the company's respiratory and PPE program; with appropriate staff specialized in all needed areas of respirator and PPE maintenance and use.

PPE Inspection, Cleaning, Maintenance, Storage and Repair. PPE maintenance should be made an integral part of the overall PPE program. Manufacturer's instructions for inspection, cleaning, and maintenance of PPE should be followed to ensure that the PPE continues to function properly. Wearing poorly maintained or malfunctioning PPE may be even more dangerous than not using it at all. The worker wearing a defective device may falsely assume that protection is being provided. Emergency escape and rescue devices are particularly vulnerable to inadequate inspection and maintenance. Although they generally are used infrequently, they are used in the most hazardous and demanding circumstances. The possible consequences of wearing a defective emergency escape and rescue device may be lethal. An adequate PPE maintenance program tailored to the type of workplace and the type of hazards is very important.

PPE Selection Reevaluation. Normally, the EPA levels of protection (A, B, C or D) are used as a starting point for ensemble creation. Each ensemble is then tailored to a specific situation in the selection process in order to provide the most appropriate level of protection. As the amount of information about the specific operation or site increases, the overall level of protection and PPE selected are validated to make sure proper and effective selections were made. The level of protection may also need to be upgraded or downgraded based on the new evaluation.

Reasons to upgrade PPE may include

- Known or suspected presence of additional hazards

- Changes in work task that will increase exposure or potential exposure to hazardous materials

- Request of the individual performing the task

Reasons to downgrade PPE may include

- New information indicating that the situation is less hazardous than was originally thought

- Change in work conditions or work task that decreases the hazard

Trends in Biological and Chemical Personal Protection

The trends in personal protection can be described based upon an assessment of the gaps that exist within the collective knowledge of the allied health and safety professions. Recently a national workgroup, spearheaded by NIOSH, established priorities to focus research in a variety of areas of personal protection. This workgroup, the National Organization Research Agenda (NORA), has identified some key areas for study. More specifically, biological and chemical protection research efforts are needed to close the gaps in the areas discussed below.

Dynamics of Biological and Chemical Dermal Exposure

- Research the need for conducting dermal exposure risk assessments and surveillance as part of Biological and Chemical Protective Clothing (BCPC) use.

- Develop state-of-the-art dermal monitoring equipment or techniques.

- Advance biological monitoring practices.

- Establish task-based worker behavior assessment where BCPC is used.

- Study the adverse impact of BCPC on the wearer; *e.g.*, biocompatibility (dermatitis, allergy, sensitization), physical impairments, and ergonomic limitations.

- Investigate the impact of dermal expiration on BCPC performance; *e.g.*, synergetic and inhibition factors that affect the BCPC performance and the establishment of surveillance procedures to monitor this phenomenon.

- Develop approaches and parameters for determining acute and chronic dermal ex-

posure (for the purpose of setting permissible exposure levels).

Improvement of Laboratory and Field Testing Methodologies

- Develop standardized tests for key BCPC performance areas not addressed by industry.

- Identify improvements in laboratory testing methodology, for easier end-user interpretation of results.

- Develop surveillance system for monitoring the selection, use, limitations, and failure of specific BCPC.

- Establish research to validate BCPC and use of end-of-service-life indicators (ESLI).

- Advance real-time biological and chemical monitoring technologies, such as microsensors, colorimetric techniques, and other analytical techniques, as well as field detectors.

Improvement of BCPC Human Factors and Ergonomics

- Investigate procedures for reporting BCPC sizing and techniques for determining correct fit.

- Establish techniques for measuring BCPC comfort and impact on worker productivity and acceptance.

- Develop specific workplace evaluation procedures for assessing the working environment with respect to its effect on BCPC use.

- Investigate BCPC use applications and examine ergonomic factors.

Development of BCPC Decision Logic

- Develop a BCPC management system and practices for its implementation.

- Establish decision logic for BCPC selection that accounts for a range and combination of biological/chemical hazards.

- Integrate ergonomics and human-factor engineering in establishing worker BCPC fit, comfort, functionality, task requirements, physical environment, and contamination avoidance.

- Incorporate feedback for decision logic from dermal exposure and risk assessment data.

- Increase cooperation among small business, trade associations, government agencies, and academia for BCPC use; additionally, establish a partnership among these organizations for ongoing BCPC applications.

Promotion of Education and Training on Selection, Use, and Limitations of BCPC

- Identify ways of reaching users, especially small businesses through trade associations, unions, government public services, and volunteers.

- Establish methods for communicating hazards to the users using available training techniques; *e.g.*, adult education, distance learning, Internet, *etc.*

- Develop approaches for promoting student learning of BCPC selection and use.

- Establish a model motivation program on proper use and care of BCPC, using organizational reward programs and occupational health and safety programs.

- Evaluate approaches for modifying worker behavior with respect to BCPC use.

- Create a surveillance program for recording problems and successes with use of BCPC.

Investigation of Physiological and Psychological Factors Associated with the Physical Environment

- Investigate practices for reducing the impact of heat stress from BCPC use in specific work applications.

- Determine BCPC factors for protection from extreme cold *(frostbite and hypothermia).*

- Examine increasing effects of nonionizing radiation for impact on workers and BCPC.

- Establish techniques to determine appropriate levels of exertion while wearing BCPC.

- Establish techniques to determine criteria for evaluating BCPC user phobias.

- Evaluate in-service procedures used to evaluate physiological and psychological factors of BCPC users prior to, during, and after mission/task performance.

Encouragement of Engineering Technology

- Identify areas where new BCPC material would provide benefit for different industries.

- Examine aspects of BCPC design which maximize protection and minimize physical stresses on the wearer.

- Establish BCPC human factors and ergonomic design principles for improvement of BCPC.

- Determine procedure for evaluating new BCPC materials with respect to potentially adverse effects before marketing.

Development of BCPC Decontamination and Disposal Methods

- Conduct research to determine when decontamination is possible and effective.

- Establish techniques for evaluating effectiveness of decontamination techniques.

- Investigate novel decontamination approaches for BCPC use.

- Develop decision logic for BCPC decontamination and disposal.

Bibliography

American National Standards Institute. *Industrial Head Protection.* ANSI Z89.1–1997. New York, NY: ANSI, 1997.

American National Standards Institute. *Personal Protection—ProtectiveFootwear.* ANSI Z41–1991. New York NY: ANSI, 1991.

American National Standards Institute. *Practice for Occupational and Educational Eye and Face Protection.* ANSI Z87.1–1989. New York, NY: ANSI, 1989.

American Society for Testing and Materials. *Standard Test Method for Resistance of Protective Clothing Materials to Permeation by Liquids or Gases Under Conditions of Continuous*

Contact. ASTM F739–96. West Conshohocken, PA: ASTM, 1996.

Collert, J. *Environmental Health and Safety CFR Training Requirements.* Rockville, MD: Government Institutes, Inc., 1995.

Kavianian, H. R., *et al. Occupational and Environmental Safety Engineering and Management.* New York, NY: Van Nostrand Reinhold, 1990.

National Institute for Occupational Safety and Health. *NIOSH Guide to Industrial Respiratory Protection.* NIOSH #87-116. Washington, DC: NIOSH, 1987.

National Institute for Occupational Safety and Health. *NIOSH Guide to the Selection and Use of Particulate Respirators Certified Under 42 CFR Part 84.* NIOSH #96-101. NTIS No: PB 96-191-937/A03. Washington, DC: NIOSH, 1996.

National Institute for Occupational Safety and Health. *NIOSH Respirator Decision Logic.* NIOSH # 87–108. Washington, DC: NIOSH, 1987.

National Institute for Occupational Safety and Health. *Occupational Safety and Health Guidance Manual for Hazardous Waste Site Activities.* NIOSH 85-115. Washington, DC: NIOSH, 1985

"Occupational Exposure To Bloodborne Pathogens, Final Rule." *Federal Register.* 56 (6 December 1991), pp. 64004–64122.

Occupational Safety and Health Administration. *Fall Protection In Construction.* OSHA 3146. Washington, DC: OSHA, 1998.

Occupational Safety and Health Administration. *Protect Yourself with Personal Protective Equipment.* Fact Sheet OSHA 92–08. Washington, DC : OHSA, 1992.

"Occupational Safety and Health Standards." *Code of Federal Regulations.* Title 29. Part 1910.

Raouf, A. and B. S. Dhillon. *Safety Assessment.* Boca Raton, FL: CRC Press/Lewis Publishers, 1994.

Roland, H. E. and R. B. Moriarty. *System Safety Engineering and Management.* 2nd ed. New York, NY: John Wiley and Sons, 1990.

"Safety and Health Regulations for Construction." *Code of Federal Regulations.* Title 29. Part 1926.

Schwope, A. D., *et al. Guidelines for the Selection of Chemical Protective Clothing,* 3rd ed. Cincinnati, OH: ACGIH, 1987.

Ziskin, Michael, *et al. Chemical Protection: Practical Selection and Use in Hazardous Waste Operations and Emergency Response.* Supplemental materials for Professional Development Course No.4, American Industrial Hygiene Conference and Exposition. Anaheim, CA: AIHA, 1994.

Ziskin M., J. Stull , and J. Zvetan. *Risk-based Assessment for the Selection of Personal Protective Equipment.* American Industrial Hygiene Conference and Exposition–Professional Development Course. Atlanta, GA: AIHA, 1998.

Michael H. Ziskin is the founder and executive vice president of Field Safety Corporation, an environmental, health and safety consulting firm providing management, training, and field services. He has completed undergraduate and graduate studies in Environmental Health Science and Industrial Hygiene. Mr. Ziskin holds certifications as a Certified Hazardous Materials Manager (CHMM) and as a Hazard Control Manager (CHCM). He has received specialized training from various governmental and industrial organizations including: The USEPA, The National Institute for Occupational Health and Safety, and E. I. DuPont de Nemours and Company. Mr. Ziskin has 23 years of experience in the environmental and health and safety industries. He has managed and directed projects including: air pollution assessments, hazardous waste site investigations, remedial actions, emergency responses, and industrial facility compliance audits. Mr. Ziskin has designed and implemented a variety of corporate environmental, industrial hygiene and safety programs for businesses ranging in size from small industrial shops to Fortune 100 corporations. Mr. Ziskin is a nationally recognized authority in personal protection and has been appointed to AIHA, NFPA, AESF, and ASTM Committees involved with occupational health and safety issues. He is the immediate past chairman of the Personal Protective Clothing and Equipment Committee of the AIHA, a Principal on the National Fire Protection Association Technical Committee on Hazardous Materials Protective Clothing and Equipment, and the AIHA representative on Personal Protection for the NIOSH National Organization Research Agenda. He is an instructor and committee member of the Wolcott State Fire School. Mr. Ziskin is a member of the Mid Fairfield County Haz Mat Team, the Greens Farms Volunteer Fire Department, and the local emergency planning commission. He is also an adjunct Professor at the University of New Haven, teaching Hazardous Materials Management courses in the Graduate Environmental Program. Mr. Ziskin is a Fellow of the Institute of Hazardous Materials Management. Mr. Ziskin's publications include EPA reports, field studies, training manuals, instructional aids, and lesson plans. He has presented over thirty papers to professional and trade organizations on topics such as health and safety training, emergency response planning, managing hazardous materials, identification of unknown materials, radioactive disposal practices, and personal protective equipment selection, uses, and limitations.

Dawn Han *has eight years experience in environmental and occupational health and safety, and Industrial Hygiene, with specialized expertise in health and safety hazard analysis, data validation and quality control and assurance, pollution control, ventilation system design, and indoor air quality. She has an MS in Industrial Hygiene and a BS in HVAC Engineering. Ms. Han is certified in the comprehensive practice of industrial hygiene. She is also certified in Hazardous Waste Site Safety Supervision, First Aid, and CPR. Ms Han is licensed as an Asbestos Inspector, and is an Asbestos Project Designer. She has been responsible for the development of site-specific health and safety plans, conducting health hazard studies, performing on-site OSHA compliant inspections and audits, and technical research. She also participates in the development of corporate environmental health and safety programs, and in the performance of comprehensive Industrial Hygiene and safety services. Ms. Han has performed hundreds of hours of worker exposure monitoring for numerous air contaminants including metals, corrosives, carcinogens, and other respiratory toxicants. She has performed numerous worker exposure studies and risk assessments, especially in the metal-finishing industry, hardware, defense and tool-making manufacturing industries. She has also provided her expertise as a third-party auditor in reviewing exposure cases for worker compensation and legal cases. Ms. Han has previously served as an Industrial Hygienist for an environmental laboratory. In this capacity she directed the company's health and safety program and provided technical oversight for the company's field sampling programs at RCRA and CERCLA sites. She has also conducted a variety of occupational hygiene programs, and has managed asbestos abatement, underground storage tank removals, and waste disposal projects for the University of Cincinnati. Ms. Han is currently co-authoring a book for one of the largest safety organizations in the United States.*

Industrial Toxicology

Robert Roy, PhD, DABT
Robert Skoglund, PhD, DABT, CIH

Introduction

Toxicology is the study of the adverse effects of chemical, physical and biological agents on living organisms. *Toxicity* is the ability of a chemical, physical, or biological agent to cause damage to biological material. Hazard has also been used to describe the ability of substances to cause damage to biological material. For example, in the preamble of the Final Rule for the Occupational Safety and Health Act (OSH Act) Hazard Communication Standard, *hazard* is defined as an inherent property of the chemical that would exist no matter what quantity was present. Toxic agents may be classified in terms of

- *Use*–industrial solvent, pesticide, food additive, pharmaceutical, *etc.*

- *Source*–plant and animal agents

- *Health effect*–carcinogen, teratogen, liver damage, lung damage, kidney damage, *etc.*

- *Physical state*–solid, liquid, gas, particulate, mist, *etc.*

- *Labeling requirements*–flammable, oxidizer, corrosive, irritant, *etc.*

- *Chemistry*–aromatic amine, halogenated hydrocarbon, inorganic, aliphatic hydrocarbon, aromatic hydrocarbon, metal, *etc.*

- *Biochemical mechanism of action*–enzyme inhibitor, binding/damage to biomolecules such as deoxyribonucleic acid (DNA) and proteins, *etc.*

- *Poisoning potential*–relatively harmless, slightly toxic, moderately toxic, highly toxic, and extremely toxic

Risk is the likelihood (probability) that an adverse effect (injury or harm) will occur in a given situation. OSHA defines risk as a function of both hazard and the amount of exposure (see 59 FR 6126). It is a function of the *toxicity* (hazard) of the agent and *exposure* to the agent:

Risk = f (Toxicity, Exposure)   (1)

or, expressed in a very useful manner:

Risk = Toxicity x Exposure   (2)

Toxicity is an inherent property of all substances. *The toxicity of a chemical cannot be altered or eliminated.*

Exposure can be defined as the amount of an agent available for absorption into the body. Unlike toxicity, exposure to chemical substances can be significantly reduced or completely eliminated. Methods of reducing or eliminating exposure to chemicals include

- Substituting a less toxic chemical for one that has a higher relative toxicity

- Using administrative controls (*e.g.*, limiting access to certain areas)

- Using appropriate ventilation (*e.g.*, passive and mechanical)

- Using appropriate personal protective equipment (*e.g.*, respiratory protection, eye protection, gloves)

Since the toxicity of a chemical (or chemical mixture) cannot be altered, one can see by using Equation (2) that the risk of adverse health effects from the chemical or mixture is directly proportional to the amount of exposure (*i.e.*, as exposure increases, risk increases, and *vice versa*).

Safety is the probability that adverse effects (harm) will *not occur* under specified conditions (the inverse of risk).

Factors Affecting Responses to Toxic Agents

If a chemical agent reaches an appropriate site (or sites) in the body at a high enough concentration and for a sufficient length of time, it will produce adverse (toxic) effect(s). Many factors affect the responses (*e.g.*, adverse effects) that may occur in a given situation. These factors are:

1) **Chemical and Physical Properties of the Substance.**

- The **physical state** (*e.g.*, liquid, gas, fume, mist, dust, vapor, *etc.*) can influence exposure potential, route of exposure, and amount potentially absorbed.

- The **solubility** in biological fluid/material such as blood and lipids (fat) may affect its absorption into the body, its movement (distribution) within the body, and its potential for storage in the body (*e.g.*, in fat or bone).

2) **Exposure Situation.**

- *Duration of exposure* includes acute (*e.g.*, usually a single dose), subchronic (*e.g.*, days to years), and chronic (*e.g.*, years to a lifetime).

- *Frequency of exposure* influences the amount of substance available for absorption into the body. It may be expressed as number of exposures per time period.

- *Routes of exposure* include the gastrointestinal tract (ingestion), lungs (inhalation), and skin (dermal, topical, or percutaneous). The order of effectiveness of the exposure routes into the body (moving from most effective to least effective) is: intravenous, inhalation, intraperitoneal, subcutaneous, intramuscular, oral, dermal. In the workplace, the major routes of exposure to chemical agents are inhalation, dermal, and oral

- *Dosage* (dose) is the most critical factor in determining whether a toxic response will take place. It is defined as the unit of chemical per unit of biological system (*e.g.*, mg/kg body weight, mg/body surface area, ml/kg body weight, *etc.*).

3) Individual factors.

- *Age* of the exposed person(s) (*e.g.*, neonates have a less developed blood/brain barrier, thus certain chemicals may easily pass into the central nervous system).

- *Genetic background* is very important in determining the variations seen between individuals (*e.g.*, may determine whether allergic reactions occur).

- *General health status* (*e.g.*, persons with preexisting damage to organs such as the liver, kidney, or lungs may be at an increased risk for damage if they are exposed to chemicals which specifically damage these organs).

Toxicokinetics

Toxicokinetics is the movement of substances (chemicals) within the body. This movement is usually divided into four interrelated processes termed absorption, distribution, biotransformation (metabolism), and excretion.

- *Absorption* is the process by which chemicals cross cell membranes and enter the bloodstream. Major sites of absorption are the skin, lungs, gastrointestinal (GI) tract, and parenteral (*e.g.*, intravenous, subcutaneous). The *degree of ionization* and *lipid (fat) solubility* affects how easily a chemical can move through cell membranes. A chemical's degree of ionization depends on its pK_a (the pH at which a chemical is 50% ionized) and the pH of the solution. For example, at low pH (*e.g.*, pH = 2-3), a weak organic acid such as benzoic acid (pK_a = 4.0) is usually ***non-ionized*** ($RCOO^- + H^+ \rightarrow RCOOH$), and an organic base such as aniline (pK_a = 5.0) is usually ***ionized*** ($RNH_2 + H^+ \rightarrow RNH_3^+$). In general, more of the non-ionized form of a chemical will be absorbed than of the ionized form. Also, the greater the lipid solubility of a chemical the greater its tendency to be absorbed.

- *Distribution* is the movement of chemicals throughout the body after they have been absorbed. It is usually rapid. Anatomical barriers such as the blood/brain barrier, which limits the passage of chemicals to the central nervous system, can impede distribution. The placenta is not as good a barrier as was once thought. Many chemicals and drugs can cross the placenta and reach the fetus. Some chemicals can be stored in body fat (*e.g.*, lipid-soluble compounds such as dichlorodiphenyltrichloroethane [DDT] and polychlorinated biphenyls [PCBs]) and in bone (*e.g.*, lead, fluoride, and strontium).

- *Biotransformation* (metabolism) is the process by which living organisms can chemically change a substance. Biotransformation reactions take place primarily in the liver, but can also occur in other organs such as the kidneys and lungs. The principal biotransformation enzymes of the body are called the cytochrome P-450 monooxygenases (P-450). Biotransformation reactions convert lipid-soluble chemicals into more polar (water-soluble) metabolites and increase the excretion of the chemicals from the body. The biotransformation of chemicals can produce metabolites that have little or no therapeutic or toxicologic activity (called detoxification reactions). Also, biotransformation reactions can produce metabolites that are capable of causing far more cell and tissue damage than the parent chemical alone would have caused. The biotransformation of a relatively nontoxic chemical into one or more toxic metabolites is called ***bioactivation***. Examples of occupational chemicals that undergo bioactivation into highly toxic metabolites include:

Chloroform	$\rightarrow$	Phosgene
Carbon Tetrachloride	$\rightarrow$	Trichloromethyl radical
n-Hexane	$\rightarrow$	2,5-Hexanedione
Methyl *n*-Butyl Ketone	$\rightarrow$	2,5-Hexanedione
Vinyl Chloride	$\rightarrow$	Chloroethylene Oxide

- *Excretion* is the elimination of chemicals from the body. Chemicals may be excreted either unchanged (in the chemical form in which they were absorbed) or as metabolites (the products of biotransformation). The primary organs involved in the excretion of chemicals are the kidneys (urine), liver (bile/feces), and lungs. Renal excretion (kidney) is the primary means by which the polar chemicals and polar metabolites of lipid-soluble chemicals are eliminated from the body. Certain chemicals and their metabolites are eliminated by means of biliary excretion (excretion into the bile from the liver). Volatile chemicals and metabolites can be excreted rapidly and efficiently through the lungs in a process called *pulmonary excretion.*

The Dose-Response Relationship

Dose-response is the quantitative (measurable) relationship between the dose of a chemical (*e.g.*, mg chemical/kg body weight) and an effect (response) caused by the chemical. The dose-response relationship is the most fundamental concept in toxicology.

The dose-response curve (Figure 1) is a graphical presentation of the relationship between the degree of exposure to a chemical (dose) and the observed biological effect or response. One can usually identify both very sensitive (hypersensitive) and resistant populations on a dose-response curve.

The following data can be calculated (estimated) from dose-response curves:

- The ***median lethal dose (LD$_{50}$)*** is the single dose of a chemical that can be expected to cause death in 50% of the exposed population. The LD$_{50}$ is not a biological constant. There are many factors that can influence the estimation of the LD$_{50}$, such as the route of administration of the chemical and the species and strain of animals used.

- The ***median effective dose (ED$_{50}$)*** is the single dose of a substance that can be expected to cause a particular effect (other than lethality) to occur in 50% of the exposed population.

- The ***median lethal concentration (LC$_{50}$)*** is similar to the LD$_{50}$ except that the chemical is quantified as an exposure concentration

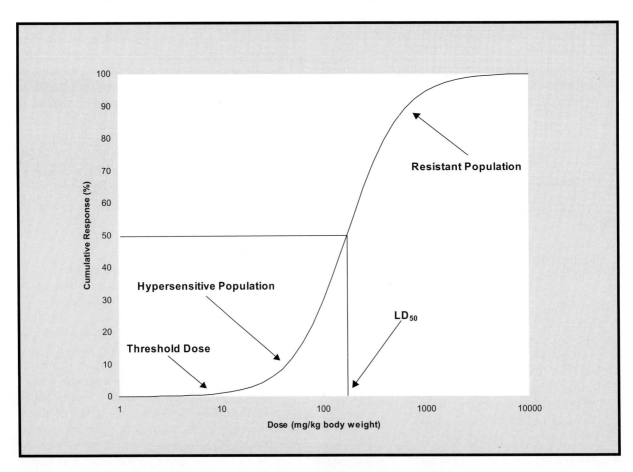

Figure 1. Hypothetical Dose-Response Curve

(*e.g.*, concentration in air such as ppm or mg/m^3) rather than a dose (*e.g.*, mg/kg-body weight)

- The ***threshold dose*** is the lowest dose of a chemical at which a specified measurable effect is observed and below which it is not observed (*e.g.*, No Observed Effect Level [NOEL])

The dose-response data have a variety of uses. They can be used to

- Establish a causal relationship between a substance and a particular effect

- Provide an indication of the range of effective dosages (for both toxic and therapeutic effects)

- Provide an indication of the threshold or no-effect-level dosage

- Allow for a quantitative comparison of the toxicities of two substances

- Determine the regulatory classifications (such as Highly Toxic, Very Toxic, Toxic; Packing Group; and Toxicity Category [based on oral LD_{50}, dermal LD_{50}, and inhalation LC_{50}]) of a chemical according to various regulatory and authoritative organizations such as the Occupational Safety and Health Administration (OSHA), Consumer Product Safety Commission (CPSC), Environmental Protection Agency (EPA), Department of Transportation (DOT), and the American National Standards Institute (ANSI).

Chemical Interactions

People are never exposed to only one chemical at a time. We are always exposed to mixtures of chemicals whether we are in the workplace, at home, or in the ambient outdoor environment. Most chemicals do not cause a similar degree of injury in all the tissues or organs they encounter. Major adverse effects usually occur in the ***target organs***. For example, benzene usually affects the bone marrow, *n*-hexane affects the peripheral nervous system, and paraquat affects the lungs. The majority of the available toxicological data comes from controlled studies where test subjects have been exposed to only a single substance. There are little data from studies where test subjects have been exposed to two or more substances simultaneously. Based on the available data and other observations, toxicologists have identified five ways in which chemicals may interact with each other in the body to produce responses. These interactions are called

- ***Independent effect***—Substances exert their own toxicity independently of each other.

- ***Additive effect***—The combined effect of exposure to two chemicals that have both the same mechanism of action and target organ, is equal to the sum of the effects of exposure to each chemical when given alone (*e.g.*, 3 + 5 = 8). In the absence of any data to the contrary, chemicals are assumed to interact in an additive manner (example: Two different organophosphate insecticides and inhibition of acetylcholinesterase at neuromuscular junctions).

- ***Synergistic effect***—The combined effect of exposure to two chemicals is much greater than the sum of the effects of each substance when given alone (*e.g.*, 3 + 5 = 30); each substance magnifies the toxicity of the other.

- ***Potentiating effect***—One substance, having very low or no significant toxicity, enhances the toxicity of another (*e.g.*, 0 + 5 = 15); the result is a more severe injury than that which the toxic substance would have produced by itself.

- ***Antagonistic effect***—An exposure where two chemicals together interfere with each other's toxic actions (*e.g.*, 4 + 6 = 8), or one chemical interferes with the toxic action of the other chemical, such as in antidotal therapy (*e.g.*, 0 + 4 = 2).

General Classification of Toxic Effects

Chemical, physical, and biological agents can produce many general types of toxic effects following exposure:

- ***Allergic (hypersensitive, sensitization) reactions*** are generally mediated by two mechanisms: those that are mediated via the production of antibodies to the allergen (antigen), (reaction Types I, II, and III), and those mediated via specialized immune cells (Type IV). An example of a Type I reaction is the occupational asthma produced in susceptible individuals following an exposure to certain diisocyanates such as toluene diisocyanate (TDI). A Type IV reaction would be like the allergic contact dermatitis produced in susceptible individuals following an exposure to nickel salts and chromates.

- *Immediate or acute toxicity* occurs rapidly after a single exposure. Examples of acute toxicity include: eye and skin burns from corrosives, or eye and skin irritation; asphyxiation by simple asphyxiants (*e.g.*, propane and methane) which displace oxygen from inspired air; and chemical asphyxiants (*e.g.*, carbon monoxide and cyanide ion) which interfere with oxygen transport or utilization by cells.

- *Delayed or chronic toxicity* manifests itself after a period of latency (may be many years). Chronic toxicity may also occur as a result of long-term exposure to low levels of various chemicals. Examples include cancer, peripheral and central nervous system damage, and liver and kidney disease.

- *Reversible effects* are those adverse effects that wear off (or reverse), given sufficient time after the exposure ceases. Examples include central nervous system depression, skin and eye irritation, upper respiratory tract (*e.g.*, nose and throat) irritation, certain types of liver/kidney damage, and certain types of nervous system damage.

- *Irreversible effects* are those adverse effects that do not reverse after the exposure ceases. The damage is permanent. Examples include neuronal damage in the central nervous system, liver and lung fibrosis, skin and eye corrosion, and birth defects.

- *Local toxicity* occurs at the site of chemical contact. The chemical need not be absorbed to cause this reaction. Examples include skin and eye irritation and corrosion, and upper respiratory tract irritation.

- *Systemic toxicity* occurs at a site or sites distant from the site of chemical absorption.

Examples include liver damage following inhalation of carbon tetrachloride, peripheral nervous system damage following inhalation of *n*-hexane, bone marrow damage following inhalation of benzene, and nervous system damage following dermal exposure to methyl parathion (insecticide). *Most chemicals produce systemic toxicity following absorption.*

Adverse Health Effects, Associated Chemicals, and Other Agents

Some very important adverse health effects and their major symptoms/signs are:

- *Dermal irritation* is a localized, nonimmune, inflammatory response of the skin that is characterized by reversible redness, swelling and pain at the site of contact.

- *Eye irritation* is a localized, nonimmune, inflammatory response of the eye that is characterized by reversible redness, swelling, pain and tearing of the eyes.

- *Dermal and eye corrosion* is an irreversible destruction of tissue with pain, ulcerations, and scarring.

- *Dermal sensitization* is a skin reaction triggered by an immune response to allergens. This reaction is characterized by redness, swelling, crusting/scaling, and vesicle formation.

- *Respiratory sensitization* is a pulmonary (lung) reaction triggered by an immune response to allergens. Pulmonary reactions are characterized by coughing, labored breathing, tightness in the chest, and shortness of breath.

- *Upper respiratory tract irritation* is a reversible inflammatory reaction that occurs in the nose and throat. This reaction is characterized by sneezing, nasal discharge, coughing, hoarseness, the productions of phlegm, and nasal inflammation (rhinitis).

- *Hepatotoxicity* can be caused by cell death (necrosis), accumulation of lipids, damage to the bile ducts, inflammation (hepatitis), and the presence of fibrotic tissue in the liver. Hepatotoxicity may be characterized by malaise, abdominal pain, nausea/vomiting, jaundice, loose stools, GI bleeding, liver enlargement, and an increase in the presence of liver enzymes in the plasma.

- *Nephrotoxicity* can be caused by damage to various portions of nephrons (the functional units of the kidney that produce urine

Table 1. Adverse Health Effects, and Some of the Chemicals/Agents Associated with Them

Adverse Effect	Chemicals/Agents
Dermal irritation	Acetic acid (<10%), hydrochloric acid (<5%), chloroform, toluene, trichloroethylene, methanol
Eye irritation	Acetone, propylene glycol, sodium hydroxide (<1%), methyl ethyl ketone, xylene, toluene
Dermal and eye corrosion	Acetic acid (>50%), ammonium hydroxide, hydrofluoric acid, hydrochloric acid (>10%), phenol (>5%), sulfuric acid (>10%)
Dermal sensitization	Nickel salts, chromates, epoxy resins, formaldehyde, p-phenylenediamine
Respiratory sensitization	Trimellitic anhydride, toluene diisocyanate (TDI), ethylenediamine, diphenylmethane diisocyanate (MDI), avian proteins, Western red cedar, various bacterial and fungal antigens
Upper respiratory tract irritation	Acrolein, ammonia, hydrogen chloride, formaldehyde, sulfur dioxide, chlorine
Hepatotoxicity (Liver)	Carbon tetrachloride; dimethylnitrosamine; allyl alcohol; dioxane; 1,2,3-trichloropropane; 1,1,2,2-tetrachloroethane; ethanol
Nephrotoxicity (Kidneys)	1,3-dichloropropene; hexachlorobutadiene; cadmium; lead; mercury; chloroform; 1,2-dichloroethane; ethylene glycol; trichloroethylene; tetrachloroethylene; nickel
Neurotoxicity (CNS damage)	Carbon disulfide, lead, manganese, carbon monoxide, toluene
Neurotoxicity (PNS damage)	n-hexane, acrylamide, methyl n-butyl ketone, lead
Neurotoxicity (CNS depression)	Numerous volatile organic compounds (VOCs) including: acetone, benzene, carbon tetrachloride, ethylbenzene, ethyl ether, Stoddard solvent, toluene, trichloroethylene
Pulmonary toxicity (Lungs)	Silica, coal dust, asbestos, ozone, phosgene, paraquat, nitrogen dioxide, beryllium, vanadium, cadmium oxide
Hematotoxicity (Blood)	Benzene, carbon monoxide, cyanide (alkali), arsine gas, aniline, p-nitroaniline, lead, stibine, nitrates/nitrites
Male reproductive toxicity	Chlordecone, dibromochloropropane (DBCP), lead, ethylene glycol monomethyl ether, ethylene glycol monoethyl ether, ionizing radiation
Female reproductive toxicity	Lead, carbon monoxide, ethylene oxide, ionizing radiation, toluene, polychlorinated biphenyls (PCBs)
Developmental toxicity	Mercury (organic), carbon monoxide, lithium, arsenic, PCBs
Cancer	Benzene; vinyl chloride; crystalline silica; arsenic; benzidine; 1,3-butadiene; N-nitrosodimethylamine; acrylonitrile; chromium (VI); asbestos; benzo[a]pyrene

including the glomerulus and the proximal tubule) and may be characterized by decreased urine output, increased blood-urea-nitrogen (BUN), electrolyte imbalances, and the presence of protein and glucose in the urine.

- *Central nervous system (CNS) toxicity* affects the brain and spinal cord. CNS toxicity may be characterized by numerous symptoms, both reversible and irreversible, which include depression, irritability, memory disturbances, personality changes, and attention deficits.

- *Peripheral nervous system (PNS) toxicity* affects the sensory and motor nerves of the extremities and often is reversible. PNS toxicity is characterized by numbness and tingling in the hands and feet, loss of tactile sensitivity, fatigue, muscle weakness, loss of coordination, and tremor.

- *CNS depression (narcosis)* is a reversible effect characterized by drowsiness, headache, nausea, slurred speech, difficulty in concentrating, dizziness, loss of coordination, and possibly coma and death in extreme situations.

- *Pulmonary toxicity* can be caused by obstruction of the airways (*e.g.*, by swelling or constriction), damage to the area of gas exchange (alveolar area) in the lungs, and various types of immune reactions. It may be characterized by coughing, shortness of breath, fluid build-up (edema), chest pain, fatigue, radiographic changes, and changes in lung function tests (spirometry).

- *Hematotoxicity* damages the formed elements of the blood (*e.g.*, red blood cells, white blood cells, platelets, *etc.*) and/or damages the bone marrow (area of blood cell formation) and may be characterized by fever, chills, malaise, fatigue, susceptibility to infections, shortness of breath, and an ashen appearance.

- *Male reproductive toxicity* affects the male reproductive system and may be characterized by changes in sexual behavior, the onset of puberty, fertility, sperm production, and hormone levels.

- *Female reproductive toxicity* affects the female reproductive system and may be characterized by changes in sexual behavior, menstrual cycle, fertility (the ability to conceive), length of gestation, lactation, and hormone levels.

- *Developmental toxicity* affects the developing organism (due to the exposure of either parent) and may be characterized by spontaneous abortion, birth defects, altered growth (*e.g.*, low body weight, *etc.*), and alterations in learning and memory.

- *Cancer* is the unrestrained proliferation of immature or abnormal cells that leads to tumor formation and, eventually, to the inhibition of the normal function of the organ or tissue.

See Table 1 for examples of agents which cause these adverse effects.

Exposure Limits for Airborne Chemicals in the Workplace

The exposure concentration limits for airborne chemicals in the workplace are called the *occupational exposure limits* (OELs). The OEL for an airborne substance is the acceptable limit on both the concentration and the duration of an employee's exposure to the substance. OELs are exposure limits to which it is believed that nearly all workers could be repeatedly exposed without an adverse effect. OELs are determined based on data from numerous sources including industrial experience with the chemical, controlled exposure studies involving humans and/or experimental animals, and analogy to chemicals that have a similar structure or are from the same chemical class. Three types of OELs are recognized: (1) the 8-hour time-weighted average (TWA) concentration, (2) the short-term (usually 15 minutes) time-weighted average concentration, and (3) the ceiling (instantaneous) limit.

OELs have been set up by many governmental and other authoritative organizations worldwide. The organizations in the United States that establish OELs include the Occupational Safety and Health Administration (OSHA), the American Conference of Governmental Industrial Hygienists (ACGIH), the National Institute for Occupational Safety and Health (NIOSH), and the American Industrial Hygiene Association (AIHA).

Bibliography

Clayton, G. and Clayton, F., Eds. *Patty's Industrial Hygiene and Toxicology*, 4th ed. New York, NY: John Wiley and Sons, 1994.

Frazier, L. M. and Hage, M. L., Eds. *Reproductive Hazards of the Workplace.* New York, NY: Van Nostrand Reinhold, 1997.

Hathaway, G. J., N. H. Proctor, and J. P. Hughes. *Proctor and Hughes' Chemical Hazards of the Workplace.* 4th ed. New York, NY: Van Nostrand Reinhold, 1996.

Klaassen, C. D., Ed. *Casarett and Doull's Toxicology–The Basic Science of Poisons),* 5th ed. New York, NY: McGraw-Hill, 1996.

Lu, F. C. *Basic Toxicology: Fundamentals, Target Organs, and Risk Assessment,* 3rd ed. London: Taylor & Francis, 1996.

Raffle, P. A. B., P. H.Adams, P. J. Baxter, and W. R. Lee. *Hunter's Diseases of the Occupations,* 8th Ed. London: Edward Arnold Publishers, 1994.

Stacey, N. H., Ed. *Occupational Toxicology.* London: Taylor & Francis, 1993.

Sullivan, J. and G. Krieger, Eds. *Hazardous Materials Toxicology–Clinical Principles of Environmental Health.* Baltimore, MD: Williams and Wilkins, 1992.

Dr. Robert Roy is a Toxicology Specialist and Team Leader of Industrial and Consumer Markets Toxicology in Corporate Toxicology at the 3M Company in St. Paul, Minnesota. Dr. Roy has over 10 years of experience in the fields of industrial and occupational toxicology, chemical hazard communication, and regulatory toxicology. Dr. Roy holds appointments as an adjunct assistant professor and graduate faculty member in the Toxicology Graduate Program and the School of Public Health (Environmental and Occupational Health) at the University of Minnesota. Dr. Roy is a member of the Society of Toxicology, the Society for Chemical Hazard Communication, and the American Industrial Hygiene Association.

Dr. Robert Skoglund is a Toxicology Specialist with Corporate Toxicology at the 3M Company in St. Paul, Minnesota and is a member of the Graduate School faculty at the University of Minnesota. Dr. Skoglund has over 15 years of experience in the chemical industry. At 3M he has responsiblity for toxicology and hazard communication in the Asia-Pacific region. His research interests and expertise are in the areas of environmental and occupational contaminant toxicology; transport and fate of chemical; hazard determination and risk assessments of chemical; as well as the toxicological aspects of product stewardship, hazard communication and chemical regulations. Dr. Skoglund is certified in toxicology and industrial hygiene, and as a hazardous materials first responder (technical level). He is a member of the American Academy of Industrial Hygiene, the Society of Chemical Hazard Communication, and the Society of Toxicology.

CHAPTER 11

Process Safety Management of Highly Hazardous Chemicals

Ken B. Baier, CSP, CEA
Alan R. Hohl, CSP, CEA, CHMM
John S. Kirar, CSP, CEA, CHMM

Introduction

Purpose

The purpose of the Occupational Safety and Health Administration (OSHA) Process Safety Management (PSM) Rule, promulgated in 1992, is to help employers prevent or mitigate episodic chemical releases of highly hazardous chemicals, and to protect workers, the public, and the environment from catastrophic accidents or events. The Rule's objective is to emphasize the management of hazards that are associated with highly hazardous chemicals. The Rule also aims to establish a comprehensive management program of identification, evaluation, and prevention of chemical releases that could result from failures in processes, procedures, or equipment. These objectives are achieved by first building safety into a process, and then by maintaining and updating hazard controls as needed throughout the life of the process. Through these measures, OSHA hopes

155

to keep industrial facilities operating safely throughout their life cycle. PSM is a vital part of the ongoing effort to prevent catastrophic accidents involving hazardous process materials and energies. Reducing the risks associated with covered processes requires the application of management principles and analytic techniques. Covered processes are the on-site manufacture, use, handling, storage, and movement of chemicals.

The PSM Rule is a performance-based rule consisting of 14 elements. The Rule does not prescribe *how* each of the 14 elements is to be implemented in the management of facilities, technology, and personnel, but instead identifies the minimum requirements that must be met by employers. The 14 elements of PSM are:

- Employee participation
- Process safety information
- Process hazard analysis (PHA)
- Operating procedures
- Training
- Contractor safety
- Pre-startup safety review
- Mechanical integrity
- Hot work permit (authorization for nonroutine work)
- Management of change
- Incident investigation
- Emergency planning and response
- Compliance audits
- Trade secrets

We will discuss each of these elements later in this chapter.

Application

The PSM Rule applies to processes rather than facilities. The following criteria are used to determine whether or not an employer must implement the PSM Rule:

- The Rule applies to processes involving chemicals at or above threshold quantities (TQs), and processes that involve flammable liquids or gases in quantities of 10,000 pounds or more onsite in one location.

- Hydrocarbon fuels are included if the fuel is part of a process covered by the PSM Rule. Hydrocarbon fuels used solely for workplace consumption are excluded.

- The Rule does not apply to retail facilities, oil or gas well drilling or servicing operations, or normally unoccupied remote facilities. It also does not apply to flammable liquids kept below their normal boiling point without benefit of chilling or refrigeration.

Manufacturers of explosives and pyrotechnics must meet the requirements of the Rule.

Important Terms

Boiling point—The boiling point of a liquid at a pressure of 14.7 pounds per square inch absolute (psia).

Catastrophic release—A major uncontrolled emission, fire, or explosion involving one or more highly hazardous chemicals that presents a serious danger to employees in the workplace.

Covered process—Any activity involving a highly hazardous chemical in excess of a TQ amount. Covered process includes any use, storage, manufacturing, handling, or onsite movement of such chemicals, or combination of these activities.

Facility—The buildings, containers, or equipment which contain a process.

Highly hazardous chemical—A substance possessing toxic, reactive, flammable, or explosive properties specified by the Application section above.

Hot work—Work involving electric or gas welding, cutting, brazing, or similar flame- or spark-producing operations.

Replacement in kind—A replacement which satisfies the design specification.

Trade secret—Any confidential formula, pattern, process, device, information, or compilation of information that is used in a particular company, and that gives the company an opportunity to gain an advantage over competitors who do not know or use it.

Employee Participation

The first element of the PSM program is employee participation. Employers are required to include their employees fully in the PSM program. An effective employee participation program provides a cooperative, participatory environment and the necessary flow of information between management and employees. This element requires employee participation during all phases of the program.

At the minimum, employers must have a written plan (action plan) outlining employee participation in the conduct and development of PHAs and in the development of other elements of the PSM Rule. The employers should consult with employees and their representatives about process safety matters, and employees, contractors, and their representatives must have access to PHA information.

Action Plan

The employee participation policy should include explicit details on employee involvement. Key areas of the policy should incorporate proof of active participation, consultation with affected employees, an anonymous means to submit complaints or suggestions, a description of how volunteers are selected (must be nonbiased), and a description of how participation is carried out in the organization.

Employee Consultation

Broad and active employee participation in all elements of the PSM program is vital to the overall program. Participation should include hourly, exempt, nonexempt, and contract employees working together to make their workplace safer. Industry has used different methods to comply with the requirements of this element. These methods include (1) management/employee safety committees, (2) management/union representative safety committees, (3) involving hourly operations and maintenance employees in PHAs and develop-

ment of safe work practices and procedures, (4) suggestion programs, and (5) safety audits conducted by employees.

Access to Process Hazard Analyses

Employers are required to provide employees, employee representatives, and contractors with access to PHAs and to all other information required under the PSM Rule. The information should be readily accessible at all times. Employees, employee representatives, and contractors should be told where to find the PSM information.

Process Safety Information

Written process safety information is essential for an effective PSM program and for conducting PHAs. Employers must collect and document the complete, accurate process safety information before any PHA is performed on a process. The employer is required to compile safety information on process chemicals (Material Safety Data Sheets [MSDSs] may be used to comply), process technology, and process equipment (employers must document that existing equipment is designed, maintained, inspected, tested, and operated in a safe manner). Process safety information should be updated as part of the Management of Change element of the PSM Rule and should be maintained for the life of a process. Process safety information must be sufficient to allow the accurate assessment of toxic, reactive, fire, and explosion hazards; the effects of process chemicals on equipment and instruments; the potential for overpressures or runaway reactions; and the existence of incompatible process materials. Table 1 on the next page lists the minimum required process safety information.

Hazards of Highly Hazardous Chemicals in Process

A *hazard* can be defined as an inherent characteristic of a material, system, process, or plant that must be controlled in order to avoid undesirable consequences. Typical hazards associated with processes involving highly hazardous chemicals include: combustible/ flammable, explosive, toxic, simple and chemical asphyxiant, corrosive, chemical reactant, thermal, potential energy,

Table 1. Process Safety Information

Pertaining to:	Information Required
Highly Hazardous Chemicals in Process	• Toxicity Information • Exposure Limits (PEL, TLV, IDLH, *etc.*) • Physical Data • Reactivity Data • Corrosivity Data • Thermal & Chemical Stability Data • Compatibility Data
Technology of Process	• Block Flow/Simplified Process Flow Diagram • Process Chemistry • Maximum Intended Inventory • Safe Upper and Lower Limits (temperatures, pressures, flows, compositions, *etc.*) • Consequences of Deviations Outside Safe Operating Limits
Equipment in Process	• Materials of Construction • Piping and Instrument Diagrams (P&IDs) • Electrical Classification • Relief System Design and Design Basis • Ventilation System Design • Design Codes/Standards • Material/Energy Balances • Safety Systems

kinetic energy, electrical energy, and pressure source hazards. The primary purpose of PSM is the prevention of a catastrophic release of highly hazardous chemicals. A ***highly hazardous chemical*** is defined as a substance possessing toxic, reactive, flammable, or explosive properties that is listed in the PSM Rule. According to PSM regulations, a hazard can be a combination of a highly hazardous chemical, an operating environment (process), and certain unplanned events (accident initiators) that could result in undesirable consequences.

Toxic Chemical Hazards. Toxic chemical hazards involve materials that can have a harmful biological effect on surrounding organisms. Toxic chemical hazards include acute and chronic health effects on employees and the public. Toxic effects can also disable operating personnel and/or response personnel. Harmful effects include cancer, nervous system damage, birth defects, lung damage, skin damage, chromosome damage, liver damage, and kidney damage. The primary routes

of entry for toxic chemicals are inhalation, ingestion, injection, and dermal absorption.

In order to identify and evaluate toxic hazards, the safety manager must know the material properties; process conditions, concentrations, and quantities; chemical reactions that could produce other toxic materials; industrial hygiene exposure limits; and release and dispersion characteristics of the chemicals involved. Common industrial hygiene exposure limits, also known as *exposure guidelines*, *levels of concern*, or *toxic endpoints* are published as: American Conference of Governmental Industrial Hygienists (ACGIH) Threshold Limit Values (TLVs), OSHA permissible exposure limits (PELs), National Institute for Occupational Safety and Health (NIOSH) Immediately Dangerous to Life and Health (IDLH) Guidelines, American Industrial Hygiene Association (AIHA) Emergency Response Planning Guidelines (ERPG) and Workplace Environmental Exposure Limits (WEELs), the National Academy of Sciences (NAS) Emergency Exposure Guidance Limits (EEGL) and Short-Term Public Emergency Guidance Levels (SPEGL), and EPA Levels of Concern (LOC).

The data on toxic chemicals that are used in PSM-covered processes can be obtained from a multitude of sources including: MSDSs that meet the requirements of 29 CFR 1910.1200(g); reference textbooks such as Sax's *Dangerous Properties of Industrial Materials*, Lewis's *Hazardous Chemicals Desk Reference*, Meyer's *Chemistry of Hazardous Materials*, National Fire Protection Association (NFPA) Standard NFPA 49, *Hazardous Chemicals Data*; the NIOSH *Pocket Guide to Chemical Hazards*; AIHA *Emergency Response Planning Guidelines and Workplace Environmental Exposure Limits Guidelines*; and ACGIH *Threshold Limit Values (TLVs) for Chemical Substances and Physical Agents and Biological Exposure Indices (BEIs)*.

Reactive Chemical Hazards. When one or more substances have the potential to chemically combine or self-react with dangerous side effects, a reactive chemical hazard exists. Dangerous side effects may be increased process pressures and temperatures; ignition of combustible, flammable, or explosive material; and generation of toxic, asphyxiating, or corrosive materials. These side effects can lead to equipment damage and subsequent chemical release with adverse effects on human health and the environment. Out-of-control chemical reactions in processes are typically caused by process irregularities involving pressures and temperatures, the inadvertent mixing of incompatible materials, material contamination, and material handling and storage errors.

The chemical characteristics to look for when identifying and evaluating reactive chemical hazards are: the character of reactions between materials or classes of materials associated with a process; reaction energies; reaction rates; material decomposition temperatures; material stability; flashpoints; concentrations; and process temperatures and pressures. To identify reactive chemical hazards for a process, the hazard evaluation team should prepare a chemical interaction matrix that lists all the chemicals or chemical families used or produced by the process. This interaction matrix is used to screen for materials which may pose any unusual and/or dangerous conditions when inadvertently combined. Such a matrix will help to determine whether or not incompatibilities exist near or within the process.

In addition to the previously mentioned information sources, data on reactive chemical hazards can be found in the Federal Emergency Management Agency's *Handbook of Chemical Hazard Analysis Procedures*; NFPA 491M, *Manual for Hazardous Chemical Reactions*; and Lab Safety's *Pocket Guide to Chemical Compatibility*.

Flammable Chemical Hazards. Flammable chemical hazards exist when there is the potential for one or more materials to rapidly react with an oxidant. Such reactions release energy in the form of heat and light. A flammable liquid has a flash point below 100°F, and has a vapor pressure not exceeding 40 psia at 100°F. Flammable liquids are further subdivided by NFPA 30, *Flammable and Combustible Liquid Code*. The PSM standard also addresses flammable gases, which are defined as: gases that, at ambient temperature and pressure, form a flammable mixture with air at a concentration of 13% or less by volume, and have a flammable range of 1 atmosphere (atm) with air of at least 12% regardless of the lower limit. Fires involving flammable gases or liquids include flash fires, pool fires, fireballs, and jet fires. Consequences of these fires include burns, smoke inhalation, undesired chemical reactions, thermal radiation and damage to process equipment, exposure to toxic by-products from combustion, and environmental impacts.

The chemical data needed to identify and evaluate flammable chemical hazards include: material flashpoints, flammability limits, autoignition temperatures, ignition energies, burning velocities, pyrophoric properties, and process conditions, including temperature, pressure, concentration, and quantity involved.

Information sources available for flammable gases and liquids include: NFPA 325, *Fire Hazard Properties of Flammable Liquids, Gases, and Volatile Solids*; and NFPA 704, *Identification of the Fire Hazards of Materials*.

Explosive Chemical Hazards. Explosive chemical hazards exist when there is a potential for one or more substances to release energy over a short period of time. The behavior of explosions varies depending on temperature, pressure, quantity, properties, and composition of chemicals; characteristics of ignition sources; geometry of surroundings (*i.e.*, confined or unconfined explosion); delayed ignition; and atmospheric mixing prior to ignition. Types of explosions include unconfined vapor cloud explosions, physical explosions, Boiling Liquid Expanding Vapor Explosions (BLEVEs), confined explosions, and chemical explosions (*i.e.*, runaway reactions). Each of these types of explosions can produce shock wave (overpressure), projectile, and thermal radiation effects resulting in blast injuries, fragmentation injuries, equipment damage, dispersion of toxic materials, damage to nearby processes, and human health and environmental impacts. Flammable gases and liquids, highly reactive chemicals, strong oxidizers, cryogenic liquids, and compressed or liquefied gases are examples of potentially explosive materials.

The chemical data one needs in order to identify and evaluate explosive chemical hazards include: flammability characteristics, thermodynamic properties, minimum ignition energies, shock sensitivity, ignition source characteristics, dispersion characteristics, and process conditions including temperature, pressure, concentration, and quantity involved.

Technology of the Process

Information on process technology should describe the process chemistry, maximum inventories of process chemicals, and the limits on process parameters. This information can be used to support a qualitative estimate of the consequences of deviations or process upsets outside the established limits. Diagrams that illustrate the flow of the process should indicate:

- Equipment sizes and ratings
- Process parameters for each mode of operation
- Limits on chemical levels in process vessels
- Flow and pressure data for pumps and components
- Process temperature and pressure limits for all equipment

Such diagrams may also include the settings for pressure relief valves and alarms, monitoring and surveillance equipment, and batch size information. Therefore, detailed process flow diagrams may contain all of the required process technology information except process chemistry information. Other process information can be obtained from Piping and Instrumentation Diagrams (P&IDs) and written and detailed process descriptions. Process technology information may be developed in conjunction with the PHA.

Equipment in the Process

Process equipment information should describe all of the hardware utilized in the process and provide the *as-built* design, including all codes, standards, or other good engineering practices that the equipment meets. The description must outline the materials of construction; electrical classification; and the design of pressure relief, ventilation, monitoring and surveillance equipment, and other safety systems. The process safety information should contain a functional description of the safety systems in a process in order to communicate its protective and mitigative features in the event of an emergency. P&IDs should be used to show the relationship between equipment and instrumentation. The information on new processes must include material and energy balances. For older processes, where the design basis is unknown or the standards, codes, and practices are not in general use, documentation must be developed to demonstrate that the equipment is still safe to use.

Safe operation of a process can be demonstrated by (1) conducting engineering analyses or empirical testing to show that a level of protection exists and is equivalent to current codes and standards, (2) changing the design parameters of the process in order to comply with current codes

and standards, or (3) using a PHA to demonstrate that the continued use of existing equipment does not significantly increase the probability of a catastrophic release compared to equipment designed to current codes and standards.

Process Hazard Analysis

A PHA is a thorough, systematic approach for identifying, evaluating, and controlling process hazards; it is the major element of the PSM program. PHAs can be used to identify the causes and consequences in the potential accident scenarios that are associated with equipment, instrumentation, utilities, human performance, external events, and natural phenomena. The objective of a PHA is to determine areas of risk where preventive and mitigative measures may be warranted in order to better control the identified hazards. PHAs can help identify accident scenarios that might lead to injuries or fatalities, property or equipment damage, human exposure to highly hazardous chemicals, environmental impacts, or other adverse consequences. The minimum criteria for implementing a PHA include:

- Setting priorities and conducting analyses according to a defined schedule
- Using the appropriate methodologies to identify and evaluate process hazards
- Addressing process hazards, previous incidents that had catastrophic potential, engineering and administrative controls that apply to the hazards, consequences of the failure of controls, facility site, human factors, and a qualitative evaluation of the possible safety and health effects on employees of a failure of hazard controls
- Performing a PHA using a team with expertise in the process technology, the process operations, and the PHA methodology
- Establishing a system to promptly address findings and recommendations that includes resolution, documentation of corrective actions, schedule for completing actions, and communication of actions to employees who are affected by the process
- Updating and revalidating PHAs at least every five years

- Retaining PHAs and updates for the life of the process

Hazard Identification

Hazard identification is the process of pinpointing specific undesirable consequences and the materials, systems, processes, and plant characteristics that could produce such consequences.

A preliminary list of process hazards can be compiled by using one or more of the following methods: (1) hazard identification matrix, (2) hazard screening worksheet, (3) hazard identification checklist, or (4) an interaction matrix. (An interaction matrix identifies the interactions between process materials, process conditions, materials of construction, contaminants, human exposure limits, and environmental exposure.) The preliminary results of the hazard identification usually include lists of the process chemicals (toxic, reactive, flammable, and explosive); potentially hazardous reactions with chemicals; system and process equipment hazards; and the conditions that could lead to a runaway reaction. A safety analyst can then use the hazard identification results to define the scope of the hazard analysis and to select an appropriate method for analyzing the hazards.

Hazard Analysis Methodologies

According to the PSM Rule, employers must use one or more of the following methods to analyze the hazards that may be present in the workplace. The method(s) should be appropriate for evaluating the hazards of the process that is under consideration:

- What-If
- Checklist
- What-If/Checklist
- Hazard and Operability Study (HAZOP)
- Failure Mode and Effects Analysis (FMEA)
- Fault Tree Analysis (FTA)
- An appropriate equivalent methodology

Several factors influence the selection of an appropriate method of hazard analysis. These factors include: the reason for performing the hazard analysis, the type of results that are needed, the type and extent of information that is available, the characteristics of the process that is being analyzed, the perceived risk that is associated with the process, and the resources that are available.

A PHA usually is performed in order to comply with corporate policy, function as a risk management tool, or comply with regulations. The result(s) of a PHA could be a simple list of hazards, a list of potential accident scenarios, a list of alternatives for reducing risks, a list of areas that require further analysis, and a means of assigning priorities to results.

The safety analyst will need to know the life cycle of the process that is under analysis, and where in this life cycle the process currently is located. The analyst will also need accurate documentation on the process prior to performing a PHA. The documentation should outline characteristics such as the type of process, the size and complexity of the process, the type of operation(s) involved, the inherent hazards associated with the process, and accident events of concern.

The less exhaustive What-If or Checklist methods are appropriate when the process is perceived to be low-risk and the employer is experienced with the operation of the process. If, however, the process is perceived as a high risk, a more rigorous method of analysis may be warranted. When deciding on a method of analysis, the analyst should consider the availability of skilled and knowledgeable personnel, the schedule for performing the analysis, and the financial resources that are available.

The safety analyst should refer to industry references such as the Center for Chemical Process Safety's *Guidelines for Hazard Evaluation Procedures, Second Edition with Worked Examples* for detailed information about the appropriate PHA methodologies.

What-If. The What-If Analysis is an unstructured brainstorming method in which a team of people who are familiar with the process under analysis asks questions about possible problems. The questions are developed based on the experience of the team and are applied to P&IDs, process flow diagrams, and descriptions of the process. The questions should apply to conditions which can upset the process as well as failures of the equipment or components and variations in the process. The What-If Analysis can be used on any design or procedure, can be performed at any stage of the process's life, can be led by a relatively inexperienced safety analyst, and produces qualitative results.

The process encourages the hazard evaluation team to ask questions that start with "What if." All questions or concerns arising from the analysis should be written down and then divided into specific areas of investigation, such as electrical safety, fire protection, or occupational safety. One or more experts in each area can then address the set of questions relating to his or her field. What-If Analyses produce lists of questions and answers about the process being analyzed. They may also produce organized lists of hazardous situations (potential accident scenarios), consequences, safeguards, and possible ways of reducing risks.

Checklist. The Checklist analysis is a structured question method that is versatile, easy to use, and can be applied at any stage in the life of a process. The Checklist can be used by an individual analyst or by a hazard evaluation team, and requires no formal training. Checklists provide standard evaluations of process hazards. They can focus on specific areas of concern, such as fire protection, electrical systems, and pressure systems. Checklists are often created by organizing the information from current relevant codes, standards, and regulations. Both the regulations and the Checklists should be reviewed and updated regularly. The Checklist analysis will usually produce a list of hazards associated with the process, and identify deficiencies in the process that could cause accidents. The completed checklist can be used by the analyst or managers to identify alternatives that can improve safety in the workplace. Generic hazard checklists are often combined with other hazard analysis techniques.

What-If/Checklist. This analysis method combines the What-If and Checklist techniques. It can identify hazards, predict the general types of accidents that can occur in a process, evaluate the effects of these accidents, and determine which controls would be adequate to prevent or mitigate these accidents.

Hazard and Operability Study. The Hazard and Operability (HAZOP) analysis method is a systematic way to examine how variations in a

process affect a system. Process variations occur through equipment failures, human errors, and process upsets such as localized chemical reactions. The technique can be used for systems with continuous processes as well as batch processes. The analyst who wants to use the HAZOP analysis method must have detailed information about the design and operation of the process. The HAZOP can be used effectively during the design phase of a process or for existing processes.

In the HAZOP analysis, a hazard evaluation team composed of experts from different areas systematically examines every part of the process to discover how deviations from the design of the process can occur. The leader of the hazard evaluation team guides the team through the design of the process, using a fixed set of *guide words*. These guide words are applied at crucial points or *nodes* of the process. The guide words include *no, more, less, as well as, part of, reverse,* and *other than*. They are combined with a condition, such as *flow* or *pressure*, to define the deviation. For example, "What is the effect of *low flow* on the process?" Typical deviations include leaks or ruptures, loss of containment, ignition sources, and chemical reactions. The HAZOP analysis should identify hazards and operating problems, and enable the HAZOP team to recommend design or procedural changes that will improve the safety of the process. The results of the HAZOP analysis are documented in a tabular format with a separate table for each segment of the process under study.

Failure Mode and Effects Analysis. The Failure Mode and Effects Analysis (FMEA) tabulates the ways in which equipment and components can fail, and the effects of these failures on a system, process, or plant. Failure modes describe the ways in which equipment can fail (such as open, closed, on, off, leaks, *etc.*). The analyst lists all of the components of the system under review and all the failure modes for these components. The FMEA identifies the individual failure modes that can either cause or contribute to an accident. This method of analysis does not address multiple failures. An FMEA analysis should produce a qualitative, systematic list of equipment and components, a list of associated failure modes, and a list of the effects of the failure modes on the system. The list of effects should include a worst-case estimate of the consequences of each failure mode. The information produced by an FMEA

- Include safety precautions and information on safety implications

- Describe the safety systems as they are defined in the process safety information

- Walk down and/or monitor operations to ensure that the actual operations match the written operating procedures

Accessibility of Procedures

The employer must keep the written procedures in a place that is readily accessible to employees who work in or maintain a process. *Readily accessible* means that the employees know where the procedures are located and are encouraged to refer to the procedures if they have questions or concerns during the conduct of operations.

Review and Update

Appropriate personnel should occasionally review the procedures in order to make sure that the operating procedure matches the actual operations and that the appropriate safety features and safety implications have been identified. The review of the procedure ensures that the procedure contains all of the elements that are required by OSHA. The review also guarantees that the procedure is accurate and the instructions are clear and understandable.

The employer is responsible to make sure that any changes to the process are incorporated into the written procedure. The Management of Change element of the PSM program requires the employer to update any changes to a procedure before operations resume.

Development and Implementation of Safe Work Practices

Employers and managers must show by their attitude and actions that safety is important. The employer must demonstrate to employees and contractors that stopping work whenever they feel conditions are unsafe is the right thing to do, and that management will support the employee's decision if they do stop work. Management has to be committed to promptly resolving any unsafe work conditions and to encouraging the employees to ensure the safety of operations.

Training

Employees must be trained about the nature and causes of problems that arise from process operations. It is in the employer's best interests to increase employee awareness about the hazards peculiar to a process. The process owner or operator is required to train and periodically retrain the employees who are involved in operating processes on the hazards of the process and the routine and nonroutine tasks that are necessary to safely perform their jobs.

By developing a training program, the employer shows his commitment to providing a safe working environment. Consulting with employees during development of the training program helps build an effective training program because of their direct experience with the processes. All applicable elements of the PSM program should be discussed during the training session.

The employer must determine the initial and refresher training requirements for employees, and implement a process for documenting the training.

Initial Training

The initial training for employees who are new to a process must emphasize the importance of safe work practices. The trainer should discuss the safe work practices that are appropriate to all operating phases of the process. The trainer should describe the hazardous chemicals that are present in the process, the safety and health hazards that are inherent in the process, and the controls that have been implemented to prevent exposure to the hazardous chemicals. During the initial training, the employee should learn the operating limits of the process, the consequences of deviating from these limits, and the ways to avoid deviation. Emergency drills (how to handle upset conditions and emergencies) should be discussed in detail. Every employee should be given the history of any accidents or near-misses that have occurred, along with the lessons that were learned from the incident and the actions that have been taken to prevent similar incidents. This history helps to emphasize the importance the employer places on safe work practices.

Refresher Training

Refresher training shall be provided to each employee at least every three years. Employers and employees should cooperate to determine the necessary frequency for refresher training. This training should review all of the areas that are addressed in the initial training. Any changes to the process and operating procedures should be covered, including why these changes were necessary.

Training Documentation

Employers are required to certify that each employee has received and understood the training. A written test is one method of measuring comprehension, but it is not the only acceptable method. A demonstration of the necessary skills is another acceptable method, but the results need to be well documented. Training records should show the date of the employee's most recent training, the type of training provided (classroom, on-the-job, or a combination), and the type of competency test used to measure comprehension.

Contractors

Employers are required to hire and use only those contractors who accomplish their tasks without compromising the safety and health of other employees at a facility.

Application

Contractors are affected by these requirements if they perform maintenance, repair, turnaround, major renovation, or specialty work on or adjacent to covered processes. Contractors who provide incidental, nonsafety-related services (*e.g.*, janitorial, food and drink service, laundry, delivery services) are not subject to these regulations.

Employer Responsibilities

Employers should evaluate a contractor's safety performance prior to making a selection, and should also periodically reevaluate the safety performance of the contractor. The employer is also responsible to inform the contractor of potential hazards that are related to the contractor's work, as well as the facility's safety rules and the provisions of the emergency response plan. The employer is required to train contract employees in the procedures that will allow them to safely perform their job responsibilities at the site. This training can be administered by the employer or the contractor. A contract employee injury and illness log must be maintained by the employer.

There are different methods the employer can use to make sure that contract employees abide by the safety rules of the facility. These include:

- Purchase requisition clauses requiring pre-award submission and approval of the contractor safety program.

- Identify a technical representative that interfaces with the contractor and is responsible for ensuring that all contract employees receive a safety orientation before they are allowed unescorted access to facilities.

- Periodically evaluate the performance of contract employers in fulfilling their obligation.

- Maintain contractor employee injury and illness logs related to contractors' work in process areas.

- Establish a permit system or work authorization system for activities on or near covered processes.

- Document that each contract employee has received and understood the required training.

Contract Employer Responsibilities

Contract employers are required to certify that their employees are trained in all applicable work practices and safety rules of the facility, and that they know about the potential hazards of the process they will be working on. Any unique hazard that is presented by the contractor's work and any hazards that are revealed by the contractor's work shall be presented to the employer by the contractor.

Pre-Startup Safety Review

The pre-startup safety review is designed to guarantee that safety requirements have been fulfilled prior to the introduction of highly hazardous chemicals to a process. The review verifies that

- New or modified facilities and equipment are built and installed in accordance with design requirements

- Process procedures and operator training are adequate and completed prior to the introduction of hazardous materials into the process

- Adequate safety reviews are conducted

- PHA safety recommendations have been implemented prior to startup

This element provides the employer with an opportunity to prove that all of the safety requirements have been implemented before workers are exposed to highly hazardous chemicals during operations.

Requirements for Performing

The employer must perform a pre-startup safety review for new facilities and for modified facilities when the modification requires a change in the process safety information. A pre-startup safety review should also be completed before restarting a process that was shut down for safety-related reasons.

Items Covered in Review

The pre-startup safety review must confirm that

- Facilities and equipment are in accordance with design specifications

- Adequate safety, operating, maintenance, and emergency procedures are in place

- A PHA has been performed for new facilities, and recommendations were resolved or implemented

- Modified facilities meet the requirements of the management of change element

- Training of each employee involved in the process has been completed

Depending on the complexity of the new or modified process, different methods may be used to conduct the pre-startup safety review. A qualified team should be assembled to conduct each pre-startup safety review. Then they can use a formal checklist, interviews with key personnel, a physical examination of the process and its required safety features, a review of the employee training, and reviews of the documentation (procedures, drawings, *etc.*) to satisfy the review stipulations. Records of the pre-startup safety review should indicate which method or combination of methods was used to fulfill the review requirements.

Mechanical Integrity

The employer should have a philosophy of equipment integrity that ensures that all process equipment and instruments are designed, constructed, installed, and maintained so as to minimize the risk of hazardous releases. A proactive approach, encompassing all process equipment from installation through retirement, is necessary to ensure the effectiveness of the safety program. The equipment that is in contact with highly hazardous chemicals is the first line of defense against any uncontrolled catastrophic releases of these chemicals.

An effective mechanical integrity program begins with compiling a list of the process equipment and instrumentation included in the program. The process equipment and instruments are assigned priorities based on their importance to the process. The list is then expanded to include the important mitigation system components (*e.g.*, fire protection system components, dikes, drainage systems, *etc.*). The records that fulfill the mechanical integrity element should indicate

- The proper application to specific types of process equipment

- The development of written maintenance procedures

- Training of maintenance personnel

- The performance of periodic inspections and tests

- Prompt identification and correction of equipment deficiencies

- Implementation of a Quality Assurance (QA) program for fabrication, installation, and materials

Application

The mechanical integrity program applies to pressure vessels and storage tanks, piping systems, relief and vent systems and devices, emergency shutdown systems, controls (including monitoring devices, sensors, alarms, and interlocks), pumps, and other equipment and systems. Each facility has to determine which equipment prevents, detects, controls and/or mitigates releases of highly hazardous chemicals. Equipment such as fire protection equipment, ammonia detectors, ventilation systems and electrical systems are subject to this element if the PHA has identified this equipment as necessary to detect, control, or mitigate releases.

Written Procedures

Written procedures help ensure the integrity of process equipment and instruments. The procedures facilitate the tests, inspections, and preventive maintenance activities that must be conducted properly and consistently, even when different employees may be involved. Some components of the maintenance procedures are:

- Identification of safe work practices when working around highly hazardous chemicals

- Preventive maintenance plans

- A maintenance work control system

- Use of manufacturers' manuals

- Incorporation of job safety analyses for maintenance tasks

- The use of detailed maintenance procedures for accomplishing the task

Training for Process Maintenance

Employees who are responsible for maintaining the process equipment should be trained in an overview of the process, the hazards of the process, and the procedures and equipment applicable to their job. The training that is provided to personnel who maintain process equipment and instruments is different from the training provided to the personnel who operate a process. Formal classroom training, reviews of the mechanical integrity of the procedure, and on-the-job training with qualified personnel are all acceptable methods of training maintenance employees.

Inspection and Testing

All process equipment must be inspected and tested in accordance with the manufacturer's recommendations and good engineering practices, or more frequently if experience warrants it. The employers shall certify that inspections and tests have been performed and shall document the test procedures and results. Employers must correct any deficiencies or malfunctions before further use of the equipment (or in a safe and timely manner when necessary means are taken to ensure safe operation). Employers must also ascertain that fabricated equipment is suitable for use in the process, and is installed properly (maintenance materials and spare parts are included in this requirement).

Equipment Deficiencies

When equipment deficiencies are found (as defined by the process safety information), the employer is responsible to promptly correct the deficiency before allowing further use of the equipment. The safety manager can correct the deficiency immediately, remove the equipment from service, follow preapproved temporary operating procedures, or conduct a management of change review to guarantee that safe operation can be continued until the deficiency is corrected.

Quality Assurance

This subsection of the mechanical integrity element requires the employer to implement a quality assurance (QA) program for equipment fabrication, equipment installation, and the use of maintenance materials, spare parts, and equipment. The QA program should focus on reducing the possibility of unwanted chemical releases from process equipment and guaranteeing that the features that are designed to control or mitigate releases will function when required. The QA program should verify that the proper materials of construction are used in fabrication, that inspection procedures are proper, that as-built drawings are developed, and that there are certifications of coded vessels and other equipment. The QA program for the installation of equipment certifies that the equipment is installed properly, is consistent with design specifications and manufacturer's instructions, and that qualified craftsmen were used to perform the installation work. For the maintenance procedures, the QA program ensures that maintenance materials, spare parts, and equipment are suitable for the task.

Hot Work Permit

This element of the PSM Rule requires employers to uniformly control all nonroutine work that is conducted on or near covered processes and that might initiate or promote a release of highly hazardous chemicals. Nonroutine work authorizations include hot work permits, radiation work permits, and confined space entry permits. Hot work permits cover welding, cutting, and other spark-producing operations. Routine work is covered by approved operating procedures and training. The hot work permit element outlines the requirements of the permit and its issuance.

Issuance

Employers shall issue a permit for all hot work (temporary or permanent) conducted on or near a covered process. The permit documents the fire protection provisions, indicates the authorized date(s), and identifies the item which requires the hot work. The permit is kept on file until the hot work is completed.

All nonroutine work authorizations should be issued under consistent guidelines and should communicate the known hazards to those who perform the work and to the affected operating personnel. Authorization guidelines should describe the steps that the maintenance supervisor, contractor representative, or others must take to obtain clearance before starting their work. The authorization procedure should also outline clear steps for closing the special work so the personnel know that equipment can be returned to normal. In addition, supervisors shall ensure that all combustibles present in the hot work area are protected from ignition prior to welding or cutting. Move the combustibles or shield them during the time that welding is scheduled.

Permit Requirements

Hot work permits are required to document compliance with the fire prevention and protection requirements of 29 CFR 1910.252(a). According to the Rule, the employer must

- Establish areas and procedures for safe welding and cutting based on fire potential
- Designate an individual who will be responsible for authorizing cutting and welding in process areas
- Certify that welders, cutters, and their supervisor(s) are trained in the safe operation of their equipment
- Advise any subcontractors about the hot work permit program

Some of the items that should be identified on the hot work permit are:

- The date the hot work is authorized
- The item which requires the hot work
- Identification of openings, cracks, and holes where sparks may drop onto combustible materials
- Description of fire extinguishers that are available to handle fires if they occur
- Assignment of fire watchers in locations where more than a minor fire could develop
- A description of the precautions associated with combustible materials
- A warning prohibiting welding or cutting in unauthorized areas, in buildings with impaired sprinkler systems, in explosive atmospheres, and in storage areas with large quantities of readily ignitable materials
- Requiring relocation of combustibles where practicable, and covering with flameproof covers where removal is not practicable
- Requiring that any ducts or conveyors by which sparks may reach distant combustibles be shutdown

Management of Change

The management of change element requires all modifications of equipment, procedures, raw materials, and processing conditions other than *replacement in kind* to be identified and reviewed prior to implementing the change. Process changes that should be included under this element include

- Changes in process technology
- The addition or removal of process equipment or piping
- Changes in process parameters
- Changes in utilities
- Changes in procedures
- Changes in facilities
- Personnel changes (according to OSHA, personnel changes are implied in the term *facilities*)

This element is a critical part of the PSM Rule because it integrates many of the other elements of the Rule.

The management of change element requires written procedures to manage change and procedure considerations. For the management of change program to be effective, personnel at all levels should be trained to recognize and understand the ramifications of proposed changes to the safe operation of a chemical process, including the interdependencies and relationships among facility functions, processes, and activities.

Written Procedures

All changes should be identified and reviewed prior to implementation, and the impact of design, operational, and procedural changes on process safety should be examined. Written management of change procedures should address employee responsibilities, the basis for the change, and the impacts of the change. In addition, the written procedure should designate who is responsible for requesting changes to existing systems, processes, and procedures; who can approve changes (and the criteria for approving changes); and who is responsible for implementing the approved changes.

Procedure Considerations

An effective management of change program includes an assessment procedure to determine the importance of the change to process safety, and a procedure to manage each proposed change. The procedure describing the management of change program enables management to

- Determine the level of effort required for review and implementation of a change
- Update process safety information and PHAs, and address any resulting recommendations
- Modify existing operating and maintenance procedures
- Inform and retrain affected employees and subcontractors
- Update emergency plans
- Develop a schedule and a list of required authorizations
- Update pre-startup procedural changes
- Modify pre-startup inspection and testing procedures
- Verify mechanical and system integrity prior to startup

Incident Investigation

The incident investigation element aims to prevent the recurrence of incidents that have the same nature or the same root cause. Employers must investigate every incident that results in, or could reasonably have resulted in, a catastrophic release of a covered chemical in the workplace. The investigation shall be started no later than 48 hours following the incident. An investigation team is established and must have at least one person who is knowledgeable in the process. A contract employee is included on the team if the incident involved a contractor's work. The team investigates the incident and prepares a report on the incident. The report should be reviewed with all of the affected personnel who work in the facility. The employer should have an agendum in place to address and respond to the report's findings and recommendations. The incident investigation team's recommendations must be implemented, unless it can be proven that an alternative will address the concerns as effectively and efficiently as the team recommendation. Incident investigation reports are to be retained for five years.

The incident investigation reports must record the date the incident occurred, the date the inspection began, a description of the incident, the factors contributing to the incident, and recommendations resulting from the investigation. A *lessons learned* program should be developed, to share the results of the incident investigation with other operations personnel at the site and with operations personnel at other company locations.

A thorough analysis of all incidents is important, since the analysis will generally reveal a number of deeper factors which allowed or even encouraged an employee's action. Such factors may include inadequate equipment, or a work practice that is difficult for the employee to carry out safely, or a supervisor allowing or pressuring the employee to take shortcuts in the interest of production. An effective analysis will address each of the causal factors in an accident or *near-miss* incident.

Emergency Planning and Response

This element outlines the actions employees are required to take when there is a release of highly hazardous chemicals. Emergency planning and response are required under the PSM Rule in order to mitigate the consequences of catastrophic releases. Emergency plans form the last line of defense in protecting workers from such events.

Employers shall establish and implement an emergency action plan in accordance with 29 CFR 1910.38(a). (29 CFR 1910.120[a], [p], and [q] may also be applicable–OSHA's Hazardous Waste Operations Rules, or HAZWOPER). The emergency action plan shall address both small and large releases.

The emergency plan should describe escape routes and procedures and the best means for reporting emergencies. The duties and procedures of employees who remain to operate critical equipment and those employees who perform rescue and medical duties need to be described in the emergency plan. The names of persons or locations to contact for more action plan information, and employee alarm systems, must also be provided in the plan. The plan should establish a procedure for accounting for employees after an evacuation.

The emergency plan shall be reviewed with each employee covered by the plan. The plan should be reviewed when the plan is initially developed, whenever employees' responsibilities or designated actions in the plan change, and whenever the plan is changed.

Training requirements differ according to the responsibilities of employees. Employees who are likely to discover hazardous substance releases are trained and should demonstrate competency in the provisions listed in the awareness level for first responders. Employees who will be required to take defensive action in containing and controlling a release as part of the response are trained as and should demonstrate knowledge of the operations required of a first responder. Employees who will be responsible for taking offensive action in containing and controlling a release as part of the response are trained as and should demonstrate the competencies required of a hazardous materials (HAZMAT) technician.

Compliance Audits

The compliance audits element requires employers to self-evaluate the effectiveness of their PSM programs by identifying any deficiencies and taking corrective and preventive actions. This element ensures that the program is operating in an integrated and effective manner. The audit team provides an independent assessment of the degree of compliance with the PSM Rule, even though one member may by necessity be involved in the operation of the process.

Compliance audits have two major objectives. First, they assess whether the management system in place adequately addresses all elements of the PSM Rule. Second, the audits assess whether the management system has been adequately implemented for every facility or process.

Employers must conduct a PSM compliance audit at least every three years. The audit evaluates all elements of the PSM program. The audit team must include at least one person who is knowledgeable in the process. The team will submit a report of its findings, and the employer must promptly respond to each finding and document that any deficiencies have been corrected. The employer must keep the two most recent audit reports on file.

Trade Secrets

Regardless of whether a process has trade secret status, the PSM Rule requires employers to provide all necessary information to

- Personnel who are responsible for compiling the process safety information

- Any personnel who are assisting in the development of the PHA

- Personnel who are responsible for developing the operating procedures

- Personnel who are involved in incident investigations, emergency planning and response, and compliance audits

A trade secret is any confidential formula, pattern, process, device, information, or compilation of information that is used in an employer's business, and that gives the employer an opportunity to obtain an advantage over competitors who do not know it or use it. Appendix D of 29 CFR 1910.1200 sets out the criteria to be used when evaluating trade secrets. Employers may require confidentiality agreements with personnel who receive trade secret information.

Summary

The PSM Rule was promulgated to help employers prevent or mitigate the release of highly hazardous chemicals and to protect workers, the public, and the environment from catastrophic accidents or events. The Rule applies to individual processes rather than to plants. Processes involving chemicals at or above TQs, and processes which involve flammable liquids or gases in quantities greater than or equal to 10,000 pounds on-site in one location, are required to implement the PSM Rule. Manufacturers of explosives and pyrotechnics must also adhere to the requirements of the Rule. The PSM Rule provides exclusions for

hydrocarbon fuels that are used solely as a fuel for workplace consumption, for retail facilities, for oil or gas well drilling or servicing operations, for facilities that normally are unoccupied, and for flammable liquids that are kept below their normal boiling point without benefit of chilling or refrigeration.

The PSM program provides a way for employers to build safety into their processes and then keep the processes operating safely throughout their life cycles by identifying hazards and providing the necessary controls. The program requires management principles and analytic techniques to be applied in order to reduce the risks associated with covered processes during the manufacture, use, handling, storage, and movement of highly hazardous chemicals. Employers must hold worker safety and health to be a fundamental value of their organization and apply this commitment to safety and health protection with vigor. The management can accomplish this by doing the following:

- Clearly state a work-site policy on safe and healthful work practices and working conditions.

- Establish and communicate the goal of the safety and health program and the objectives for meeting that goal.

- Provide visible top management involvement in implementing the program.

- Provide for and encourage employee involvement in the structure and operation of the program, and in decisions that affect their safety and health.

- Assign and communicate responsibility for all aspects of the program.

- Provide adequate authority and resources to responsible parties.

- Hold managers, supervisors, and employees accountable for meeting their responsibilities.

- Review program operations at least annually to evaluate their success in meeting the goal and objectives.

- Provide a reliable system wherein employees, without fear of reprisal, can notify management personnel about conditions that appear hazardous; provide employees with timely and appropriate responses to such notices; and encourage employees to use the system.

The hazard evaluation and control techniques described in PSM are also useful for minimizing risks in operations involving less than TQ amounts of listed chemicals, or other chemicals not listed in Appendix A of 29 CFR 1910.119. The principles of PSM should be utilized at all facilities where highly hazardous chemicals are used in any other than an incidental manner.

Bibliography

American Conference of Governmental Industrial Hygienists. *Threshold Limit Values (TLVs) for Chemical Substances and Physical Agents and Biological Exposure Indices (BEIs).* Cincinnati, OH: ACGIH, 1996.

American Industrial Hygiene Association. *Emergency Response Planning Guidelines and Workplace Environmental Exposure Level Guides Handbook.* Fairfax, VA: AIHA, 1996.

American Institute of Chemical Engineers. *Guidelines for Chemical Process Quantitative Risk Analysis.* New York, NY: AIChE–Center for Chemical Process Safety, 1989.

American Institute of Chemical Engineers. *Guideline for Hazard Evaluation Procedures, Second Edition with Worked Examples.* New York, NY: AIChE–Center for Chemical Process Safety, 1992.

"Compliance Guidelines and Recommendations for Process Safety Management ." *Code of Federal Regulations.* Title 29, Pt. 1910.119, Appendix C.

Cote, Arthur E. *Fire Protection Handbook,* 18th ed. Quincy, MA: NFPA, 1997.

"Definition of 'Trade Secret.'" *Code of Federal Regulations.* Title 29, Pt. 1910.1200, Appendix D.

Department of Energy. *Chemical Process Hazards Analysis,* DOE-HDBK-1100-96. Washington, DC: Office of Scientific and Technical Information, 1996.

Department of Energy. *Process Safety Management for Highly Hazardous Chemicals.* DOE–HDBK–1101–96. Washington, DC: Office of Scientific and Technical Information, 1996.

Department of Health and Human Services. Center for Disease Control and Prevention. National Institute for Occupational Safety and Health. *Pocket Guide to Chemical Hazards.* Atlanta, GA: Government Printing Office, 1994.

"Employee Action Plans and Fire Prevention Plans." *Code of Federal Regulations.* Title 29, Pt. 1910.38(a).

Federal Emergency Management Agency. *Handbook of Chemical Hazard Analysis Procedures.* Washington, DC: Government Printing Office, 1987.

"General Requirements for Welding, Cutting and Brazing." *Code of Federal Regulations.* Title 29, Pt. 1910.252(a), Subpart Q.

Government Institutes, Inc. *EPA Technology Handbook.* Rockville, MD: Government Institutes, Inc., 1992.

Government Institutes, Inc. *Process Safety Management Standard Inspection Manual.* Rockville, MD: Government Institutes, Inc., 1993.

"Hazardous Waste Operations and Emergency Response." *Code of Federal Regulations.* Title 29, Pt. 1910.120 (a), (p), and (q).

Lab Safety Supply. *Pocket Guide to Chemical Compatibility.* Janesville, WI: Lab Safety Supply, 1995.

Lewis, Richard J. *Sax's Dangerous Properties of Industrial Materials,* 5th ed. New York, NY: Van Nostrand Reinhold, 1992.

Lewis, Richard J., Sr., *Hazardous Chemicals Desk Reference,* 4th ed. New York, NY: Van Nostrand Reinhold, 1997.

"List of Highly Hazardous Chemicals." *Code of Federal Regulations.* Title 29, Pt. 1910.119, Appendix A.

Meyer, Eugene. *Chemistry of Hazardous Materials,* 2nd ed. Englewood Cliffs, NJ: Brady Prentice Hall, Career and Technology, 1990.

National Fire Protection Association. *Hazardous Chemicals Data*, NFPA 49. Quincy MA :NFPA, 1994.

National Fire Protection Association. *Fire Hazard Properties of Flammable Liquids, Gases, and Volatile Solids*, NFPA 325. Quincy, MA: NFPA, 1994.

National Fire Protection Association. *Identification of Fire Hazards of Materials*, NFPA 704. Quincy, MA: NFPA, 1996.

National Fire Protection Association. *Guide for Hazardous Chemical Reactions*, NFPA 491. Quincy, MA: NFPA, 1997.

Plog, Barbara A., Ed. *Fundamentals of Industrial Hygiene*, 4th ed. Itasca, IL: National Safety Council, 1996.

Process Safety Institute. *Performing Hazard Assessments to Comply with EPA's RMP Rule*. Resource Materials for Course #302. Knoxville, TN: ABS Groups, Inc. (formerly JBF Associates Inc.), 1997.

"The Process Safety Management Approach for Compliance." *OSHA Regulation 29 CFR 1910 .119* presented by the American Institute of Chemical Engineers in conjunction with The Process Safety Institute and JBF Associates, Inc. Knoxville, TN, 1994.

"Process Safety Management of Highly Hazardous Chemicals." *Code of Federal Regulations*. Title 29, Pt. 1910.119.

Internet Resources

<http://www.abs-jbfa.com> (ABS Group, Inc. Home Page; complete references, courses, links)

<http://atsdr1.atsdr.cdc.gov:8080/ > (Agency for Toxic Substances and Disease Registry)

<http://tis.eh.doe.gov:80/web/chem_safety/> (DOE Chemical Safety Program)

<http://www.epa.gov/swercepp> (Chemical Emergency Preparedness and Prevention Office)

<http://www.thompson.com/tpg/enviro/chem/chem.html> (Chemical Process Safety Report)

<http://www.access.gpo.gov/nara/cfr> (Government Printing Office. Sources for Code of Federal Regulations on the Internet)

<http://ull.chemistry.uakron.edu/erd/> (Hazardous Chemical Database)

<http://www.sba.gov> (Small Business Administration; important links to help for businesses)

Ken B. Baier, Alan R. Hohl, and John S. Kirar are Vice Presidents of Engineered Safety Source Company, providing safety engineering and environmental management services to government, environmental, and commercial clients. The authors work as nuclear safety engineers at the Department of Energy's Rocky Flats Environmental Technology Site in Golden, Colorado.

Mr. Baier has 20 years experience in safety engineering and risk management, systems engineering, project management, product assurance, and environmental regulatory management. He has worked on multiple Department of Defense, National Aeronautics and Space Administration, and Department of Energy projects, managing the risk to workers, the public, and the environment.

Mr. Hohl has 25 years experience in safety engineering and risk management, project management, waste management, and environmental regulatory compliance. He has worked on a wide variety of Department of Defense, National Aeronautics and Space Administration, and Department of Energy projects, managing the risk to workers, the public, and the environment, as well as to the clients' facilties and hardware.

Mr. Kirar has 20 years experience in safety engineering and risk management, systems engineering, project management, environmental health and safety management, and regulatry compliance. He has worked in the hazardous matrerials management field for seven years evaluating public, worker, and environmental risks associated with the release of chemicals and radiological contaminants.

CHAPTER 12

Laboratory Safety and Resource Management

Mignon J. Clarke, MS, CHMM

Introduction

Safety and resource management poses a variety of challenges for today's laboratory manager. Managers must be creative in providing the safest and most efficient work environments while remaining within the confines of today's regulatory requirements and shrinking budgets.

The intent of this chapter is to provide the reader with insight into the safety and environmental issues unique to the laboratory. The first part of the chapter addresses laboratory safety and regulatory requirements, while the second half concentrates on the management of laboratory resources. Upon completion of this chapter the reader will better understand the role each plays in the daily operation of today's modern laboratory.

The Modern Laboratory

Laboratories, for the purpose of this chapter, are defined as facilities that use chemicals to perform a variety of functions including chemical analysis, research, and education. The activities conducted within them are extremely varied, and frequently changed. However, the use of chemicals is the common thread among all laboratory facilities. Facilities including: (1) academic laboratories such as those found in universities, high schools, educational or scientific organizations, (2) commercial testing laboratories including labs that analyze hazardous waste samples, (3) research and development laboratories such as government labs, industrial labs, and commercial labs, and (4) medical laboratories, including dental and hospital labs, are all included in this definition.

Laboratory Safety

Laboratory Safety Programs

Safety is a major concern in all occupational settings; however, because of the many types of chemicals and varying activities, the laboratory presents a particular set of challenges not found in other workplaces. The key to meeting these challenges depends on the development, implementation, and maintenance of a comprehensive proactive safety program. An effective program will require the participation of all laboratory personnel and the utilization of laboratory resources.

One of the main points of consideration when developing a laboratory safety program is assigning accountability and responsibility. The program must identify various key positions and assign individuals to them. Laboratory personnel must understand their individual responsibilities in maintaining safe work environments. This is accomplished by providing administrative resources and mechanisms to empower employees to mitigate or correct any unsafe acts or conditions. Ultimately, the responsibility for ensuring a safe work environment rests with the laboratory employee.

Training is another point of consideration when developing a safety program. Laboratory workers must be able to identify hazards and risks involved in the work they perform. Individuals trained in handling hazardous materials are better equipped to minimize the risk of exposure to themselves as well as their coworkers. Personnel who are potentially exposed to hazards in the laboratory should be provided with written materials on the nature of the hazard. The method of training, when not otherwise specified, should include: formal classroom training, informal training in the form of routine safety meetings, and on-the-job training. The training should always be conducted by a qualified safety person and properly documented.

The safety program should include, but not be limited to, the following issues:

- Hazard communication
- Fire safety training
- Respiratory protection
- Handling radiological materials
- Emergency response and evacuation
- First aid and CPR
- General laboratory safety
- Chemical hygiene plan
- Engineering controls
- Personal protective equipment

General Safety

All laboratories are subject to general rules of safety when handling chemicals. These rules should be incorporated into daily laboratory activities.

- *Accidents–eye contact:* promptly flush eyes with water for at least 15 minutes

- *Ingestion:* encourage the victim to drink large amounts of water

- *Skin contact:* flush affected area with water

Note: Always refer to the Material Safety Data Sheet (MSDS) for specific actions, and always seek medical attention.

- Promptly clean up spills using appropriate materials and protective equipment.

- Avoid routine exposures through safe handling procedures.

- Do not smell or taste chemicals.

- Inspect gloves and protective equipment before and after each use.

- Ensure that ventilation equipment is appropriate for chemicals and procedure being used.

- Do not eat, drink, smoke, or apply cosmetics in areas where laboratory chemicals are present.

- Avoid practical jokes or other behavior which may confuse, startle, or distract another worker.

- Confine long hair and loose clothing while working in the laboratory.

- Always wear the proper protective equipment when working in the laboratory.

- Do not work alone.

- Dispose of chemicals in accordance with established procedures.

Personal Protective Equipment

Activities that involve the use of hazardous materials may require the use of personal protective equipment (PPE) to prevent harmful chemicals from being absorbed through the skin, inhaled, ingested, or injected into the body. Laboratories should develop and implement a formal plan to address these concerns.

The two basic objectives of any PPE program should be to protect the wearer from safety and health hazards, and to prevent injury to the wearer from incorrect use and or malfunction of the PPE. To accomplish this, the program should include hazard identification, medical monitoring, environmental surveillance selection, use maintenance, and decontamination of PPE. The program must also include policy statements, procedures, guidelines, and a method for review and evaluation.

Before PPE is prescribed to mitigate hazardous conditions, engineering and administrative controls should be evaluated. Engineering controls form the first line of defense against exposure to toxic or hazardous agents in the laboratory. In laboratories this may consist primarily of ventilation equipment. The effectiveness and efficiency of the system will depend on regular preventative maintenance and routine measurements. The second line of defense for protecting workers is the use of administrative controls. These controls include training, labeling, and posting, and the development of standard operating procedures.

© *Lab Safety Supply, Inc., Janesville, WI. Reproduced with permission. Related information available at www.LabSafety.com*

The selection process for PPE must also take into account the hazards created by the PPE itself. Individuals using PPE may experience any one or combination of the following: reduced visibility, reduced dexterity, reduced ability to communicate, and increased physical and emotional stress. Therefore, decisions regarding the level of protection must also consider the increased safety hazard involved in donning any article of PPE. Table 1 describes the type of PPE used in the laboratory.

Chemical Compatibility and Storage

Chemical compatibility is always a concern when storing chemicals in the workplace. Compatibility refers to the reaction of certain chemicals when they come in contact with other chemicals. Some of the reactions that can occur include heat, vapor, or gas generation; polymerization; corrosion; explosions; and fires. Therefore, when planning to store chemicals it is important to minimize these potential reactions.

Spills often result in the mixing of incompatible materials. They can occur from corrosion of metal containers or the storage of materials in containers that are not compatible with its contents.

The other possibility for combining incompatible materials happens when personnel intentionally mix materials or wastes. This activity can be especially dangerous in the case of wastes. Waste streams often change composition, and depending on how well the containers are managed can contain many unknown constituents.

Incompatible materials should be stored in a manner that reduces the possibility of mixing in the event of a spill or other accidental releases.

Environmental Regulations Relative to Laboratories

The environmental regulatory requirements for laboratories can be very cumbersome and sometimes confusing. Current environmental regulations control the release of pollutants to the air, water, and land. Some of the environmental regulations that may affect the laboratory are discussed below.

Table 1.
Laboratory Personal Protective Clothing

Body Part Protected	Type of Clothing	Type of Protection	Use Consideration
Body	Aprons, Sleeve protectors	Splash for arms, legs, chest	Sampling, labeling activities
Eyes, Face	Faceshield, Glasses, Goggles	Splashes; impact-resistant eyewear protects against projectiles	Sampling, labeling activities
Ears (Hearing)	Earplugs, Ear muffs	Protects against physiological and psychological disturbance	Must meet OSHA requirements; could introduce contaminants to the ear
Hands	Gloves	Protects hands from chemical contact	Wear sleeves over glove cuffs to prevent splashed liquid from entering glove
Foot	Safety boots or shoes, chemical-resistant boots or shoes	Protects feet from chemical contact, puncturing, and crushing	Must meet OSHA requirements

Clean Air Act

The Clean Air Act (CAA) addresses air pollutants released from facilities into the atmosphere. Laboratory facilities that practice incineration of animal carcasses and other wastes (including hydrocarbons, radioactive materials, biological hazards, or particulate matter) may be required to obtain an air quality permit.

Clean Water Act

The Clean Water Act (CWA) affects those laboratories that discharge wastewater effluents into sewers or bodies of water such as lakes, rivers, and streams. This act also regulates discharges from facilities to sewer systems that are connected to treatment works, or publicly owned treatment works (POTW) which have National Pollutant Discharge Elimination System (NPDES) permits.

Toxic Substances Control Act

TSCA, or the Toxic Substances Control Act, has limited applicability for most laboratories. It regulates the manufacture and use of chemicals. TSCA authorizes the EPA to (1) obtain data from industry regarding the production, use, and health effects of chemical substances and mixtures, and (2) regulate the manufacture, processing, and distribution in commerce, as well as use and disposal, of a chemical substance or mixture.

Resource Conservation and Recovery Act

The Resource Conservation and Recovery Act (RCRA) governs the disposal of solid and hazardous waste. It can be considered as a *cradle-to-grave* approach to managing hazardous waste. **Cradle-to-grave** is a term used to describe the generator's responsibility for tracking hazardous waste from generation through final disposition. This law makes it possible for a hazardous waste generator to be made liable for waste even after it has been removed from their facility. Under this law many laboratories generating hazardous waste are now required to dispose of their waste at a RCRA-permitted treatment, storage, and disposal facility (TSDF).

Comprehensive Environmental Response, Compensation and Liability Act

The Comprehensive Environmental Response, Compensation and Liability Act (CERCLA), or Superfund, regulates past disposal sites and the release of regulated hazardous materials to the environment. The law includes a protocol that must be followed in the event of a release of a regulated material. Title 40 of the *Code of Federal Regulations*, Part 302 (40 CFR 302) explains the reporting steps and minimum reportable quantities (RQs) for each regulated material. The relevance of this law to the laboratory is the potential for the release of a regulated material (waste or nonwaste) from the laboratory facility.

OSHA Health Standards

General OSHA Overview

The protection of workers in the workplace is covered by a set of regulations enforced by the Occupational Safety and Health Administration (OSHA). These regulations, found in 29 CFR 1900–1999, were established to ensure that each working person was provided with a safe work environment.

OSHA health standards have four categories:

1)   **Design Standards.** These standards prescribe specific design criteria for issues such as ventilation and personal protective equipment. They are also referred to as *general industrial hygiene standards*.

2)   **Performance Standards.** These standards, also known as generic *performance-oriented (or based) standards*, state the objective that must be obtained, and leave the method of achievement up to the employer.

3)   **Vertical Standards.** Apply to a particular industry. Examples include standards for pulp, paper, and paperboard mills.

4)   **Horizontal Standards.** Apply to all workplaces and relate to broad areas such as Sanitation or Walking on Surfaces.

Under the OSHA Act, OSHA is authorized to conduct inspections, and when alleged violations of safety and health standards are found, to issue citations and to assess penalties.

Inspection schedules are conducted on a priority system. The first priority is inspections conducted in response to fatalities and multiple (five or more) hospitalization incidents and imminent-danger situations; second, inspections in response to employee complaints; third, random inspections of high-hazard industries; and fourth, follow-up inspections.

It is important to mention that OSHA compliance officers may enter the employer's premises without delay to conduct inspections. In addition, inspectors have the right to inspect records of injuries and illnesses, including certain medical records.

Since its inception, OSHA has developed a number of regulations that pertain to the management of hazardous materials in the workplace. Table 2 lists some OSHA health standards.

Table 2. OSHA Health Standards

Reference	Standard
29 CFR 1910.132	General Requirements for Personal Protective Equipment
29 CFR 1910.1000	General
29 CFR 1910.1000 through 1910.1045	Z Tables
29 CFR 1910.133(a)	Eye and Face
29 CFR 1910.95	Noise Exposure
29 CFR 1910.134	Respiratory Protection
29 CFR 1910.135	Head
29 CFR 1910.136	Foot
29 CFR 1910.120	Hazardous Waste Operations and Emergency Response
29 CFR 1910.1200	Hazard Communication
29 CFR 1910.1450	Laboratory Standard

Hazard Communication Standard

The hazard communication standard, sometimes referred to as the "Worker *Right-to-Know*" law, protects workers by requiring a written Hazard Communication Plan and a program to explain how the plan will be implemented. The program for employees must address container labeling requirements, the use of Material Safety Data Sheets (MSDS), employee training, and a list of the hazardous chemicals in the workplace. The program must also include methods the employer will use to inform workers of workplace hazards.

Laboratory Standard

This standard requires laboratories using hazardous chemicals to take additional measures to inform workers of chemical hazards they may be exposed to under normal conditions or in a foreseeable emergency. The law affects all employers engaged in the laboratory use of toxic substances. This standard will be discussed further in the next section.

OSHA Laboratory Standard

Laboratory operations are inherently more difficult to regulate than other industries because of the variability of operations and types of chemicals used; as a result, OSHA enacted the laboratory standard (29 CFR 1910.1450). The standard provides more protection to workers who may be exposed to small quantities of chemicals used on a short-term basis as opposed to other workers in chemical manufacturing operations who may be exposed to a limited number of chemicals over a longer period of time.

The focus of the standard is not to establish new exposure limits, but to encourage worker awareness of potential risks, improve work practices, ensure the appropriate use of personal protective equipment, and the use of engineering controls.

The applicability of this standard depends on meeting two key definitions. The first refers to the use of laboratory chemicals. As defined by the standard, *laboratory use of hazardous chemicals* means handling or use of such chemicals when (1) chemical manipulations are carried out on a *laboratory scale*, (2) multiple chemical procedures or chemicals are used, (3) the procedures involved are not part of a production process, nor in any way simulate a production process, and (4) protective laboratory practices and equipment are available and in common use to minimize the potential for employee exposure to hazardous chemicals.

The second condition of applicability of the laboratory standard relates to the use of hazardous chemicals. Laboratory operations that use hazardous chemicals but provide no potential for employee exposure are not subject to this standard. Two examples include: (1) procedures using chemically-impregnated test media such as *Dip-and-Read* tests where a reagent strip is dipped into the specimen to be tested and the results are interpreted by comparing the color reaction to a color chart supplied by the manufacturer of the test strip, and (2) commercially prepared kits such as those used in performing pregnancy tests, in which all of the reagents needed to conduct the test are contained in the kit.

The laboratory standard, like the OSHA Hazard Communication Standard (29 CFR 1910.1200), ensures that employees are apprised of the hazards of chemicals present in their work area.

Chemical Hygiene Plan

A major component of the laboratory standard is the development and implementation of a chemical hygiene plan. The **chemical hygiene plan** is a comprehensive plan that is capable of (1) protecting employees from health hazards associated with hazardous chemicals in that laboratory and (2) keeping exposures below specified OSHA permissible exposure limits.

Developing a Chemical Hygiene Plan

Developing an accurate and up-to-date plan requires the laboratory to carefully examine and define what hazards exist in the laboratory, what programs are necessary to control the hazards, and what methods can be used to determine their effectiveness.

This information may be found in existing programs. For example, methods for protecting employees from exposure to chemicals may already be addressed in the personal protective equipment program, equipment preventative maintenance program, and/or established standard operating procedures.

The chemical hygiene plan ties information presented in other programs together in an all-encompassing; readily available plan specific to work conducted in the laboratory.

As noted in the regulation, the chemical hygiene plan must include the following items:

- Standard operating procedures (SOPs) relevant to health and safety to be followed when handling hazardous chemicals

- Criteria that the employer will use to determine and implement control measures to reduce employee exposure to hazardous chemicals

- Requirements to ensure proper selection and performance of chemical fume hoods and personal protective equipment

- Provisions for employee information and training

- Defined circumstances for obtaining prior approval for particular laboratory operations

- Provisions for medical examinations and consultations

- Designation of personnel responsible for implementation of the chemical hygiene plan

- Provisions for additional employee protection for work with particularly hazardous substances, including carcinogens and reproductive toxins

The first step in developing a chemical hygiene plan is to identify activities that require the handling of chemicals in your laboratory. This will also require gathering Material Safety Data Sheets (MSDS) to determine the requirements for safe handling of any chemicals associated with these activities. You will need to determine whether standard operating procedures (SOPs) exist for those activities, and the date they were last reviewed. If you don't have SOPs for the activities, or the current SOPs are outdated, new SOPs will have to be written. Depending on the laboratory's organizational structure, you may be able to modify an SOP from another laboratory that conducts similar activities.

All SOPs should be reviewed by various levels of management in order to evaluate their technical adequacy and administrative accuracy. Be sure to assign a tracking number to the SOP so that it may be easily referenced and retrieved. Final versions should be dated and have approval signatures from appropriate laboratory personnel.

The second step in developing a chemical hygiene plan will include gathering existing safety- and health- related plans. Examples include fire protection, spills and accidents, waste disposal, and monitoring plans. This information will be used to write the chemical hygiene plan, or can be referenced or added to the plan as appendices.

Once all appropriate information is gathered you can begin to write the formal plan. Table 3 is an example of an outline of the chemical hygiene plan.

Table 3. Outline of the Chemical Hygiene Plan

Section	Topic
1	Basic rules and procedures
2	Chemical procurement, distribution, and storage
3	Environmental monitoring
4	Housekeeping, maintenance, and inspections
5	Medical Program
6	Protective apparel and equipment
7	Records
8	Signs and labels
9	Spills and accidents
10	Training and information
11	Waste disposal

Implementing the Chemical Hygiene Plan

Successful implementation of the chemical hygiene plan will depend on several factors. The most influential will in most cases be managerial support. Management must maintain an atmosphere where safety is the top priority. In addition the chemical hygiene plan, like any other safety-related program, must have delegated resources and policy enforcement.

Maintaining an atmosphere conducive to promoting safe work habits may include motivating and shaping employee attitudes about safety. This can be accomplished through employee empowerment, and through incentive and recognition programs.

Resource Management

Pollution Prevention

Most environmentalists would agree that the overall emphasis of pollution prevention appears to center on the elimination or reduction of waste. Pollution prevention as it applies to laboratory activities must incorporate both source reduction and end treatment strategies. There are three basic benefits received from implementing a pollution prevention program in your laboratory: (1) economic savings, (2) improved public image, and (3) enhanced personnel morale.

Resource management is the main tool used to enhance pollution prevention in the laboratory. In its broadest definition, resource management involves the conservation of chemical resources through proactive management of chemical stores and disposal practices.

Chemical Procurement

A common approach to purchasing materials is to buy at the lowest per-unit cost. Unfortunately, this practice usually leads to purchasing materials in larger quantities. In many instances only a small quantity of the chemical is used, while the excess is kept in stock for future use. As a result these excess chemicals can remain in storage beyond the life of the initial project, creating various safety and housekeeping problems. The lack of necessary storage space for large quantities of chemicals often hampers efforts to provide proper storage conditions.

To avoid build-up of surplus chemicals, laboratories should only purchase quantities that will be used in the foreseeable future. One way to ensure immediate use is through planning. Each project manager or researcher should carefully consider how much material will actually be needed to complete his or her project.

Whenever feasible, toxic chemicals should be substituted for less toxic ones. Purchasing chemicals that are lower in toxicity will in most cases reduce the disposal cost of the chemical.

Inventory Management

Proactive chemical inventory management is also an important tool in reducing waste in the laboratory. This involves tracking chemicals from purchase to disposal. There are several advantages to tracking chemicals in a laboratory facility. First, tracking chemicals will assist procurement personnel in identifying usage patterns for community stock chemicals. As a result, chemical purchases can be adjusted to meet laboratory consumption. Second, a tracking system can identify chemicals that are not being used in a timely manner. Third, information can be placed in a database that may be used to satisfy Federal, State, and local reporting requirements.

Excess chemicals are inherent to all laboratory operations. The best way to avoid this is to plan ahead; however, when excess chemicals are encountered, the laboratory should explore alternatives other than treatment for disposal.

Waste Management

Until recently, the disposal of laboratory waste consisted of pouring waste down the laboratory drain or throwing it in the dumpster. A common engineering practice was to include a limestone effluent trap to neutralize any acids that might be present.

In 1976 the United States Congress passed the Resource Conservation and Recovery Act (RCRA). Under RCRA, laboratories generating hazardous waste were now required to dispose of their waste at a RCRA-permitted treatment, storage, and disposal facility (TSDF).

Laboratories have a unique set of waste management issues that are inherent to their operations. Unlike industrial operations that typically have a small number of larger-volume waste streams, laboratories typically have smaller volumes of many different waste streams.

Laboratories are capable of producing a variety of wastes. The basic categories include air emissions, wastewater, solid and hazardous wastes, biological wastes, and radioactive wastes. However, the majority of waste that is generated from a laboratory facility will more than likely be solid

and hazardous waste, such as unused chemicals, sample residues, spent solvents, and analytical process waste. Table 4 describes common laboratory hazardous wastes.

Table 4. Laboratory Hazardous Wastes

Waste Stream	Source
Spent Solvents	Cleaning, extraction, analytical processes
Unused Reagents	Unused chemicals that are no longer needed; do not meet specifications; contaminated; exceeded storage life; otherwise unusable
Reaction Products	Unknown and known composition, typically produced by academic and research labs
Testing Samples	Sample residues not consumed in analytical procedures
Contaminated Materials	May include glassware, protective clothing, and other miscellaneous solid waste

If hazardous wastes are generated by your laboratory, it is important to dispose of your waste in accordance with all applicable Federal and State environmental regulations.

Summary

Laboratories are defined as facilities that use chemicals to perform a variety of functions including chemical analysis, research, and education. The activities conducted within them are extremely varied, and frequently changed. Unlike industrial activities, which typically deal with a small variety of chemicals on a larger scale, laboratories typically use a large variety of chemicals on a much smaller scale. This makes the job of protecting laboratory workers that much harder. In addition to protecting workers, today's laboratory manager must optimize resources under ever-shrinking budgets. Worker safety and resource management are related on several levels.

For example, substituting a product with less toxic substance not only demonstrates sound resource management, but it may also reduce the workers' exposure to toxic substances, possibly eliminating a serious safety concern. There are many examples of how each influences the other; therefore, both issues should be considered in all daily activities.

Bibliography

American Chemical Society, Department of Government Relations. *Less is Better: Laboratory Chemical Management for Waste Reduction*, 2nd ed. Washington, DC: ACS, 1993. (Also accessible via the web at: <www.acs.org/govt/pubs/5s+45b.htm>).

American Chemical Society Department of Government Relations. *The Waste Management Manual: For Laboratory Personnel.* Washington, DC: ACS, 1990.

Ashbrook, Peter C. *Safe Laboratories: Principles and Practices for Design and Remodeling.* Boca Raton, FL: Lewis Publishers Inc., 1990.

Carson, H. Tom. *Handbook on Hazardous Materials Management.* Rockville, MD: Institute of Hazardous Materials Management, 1992.

Furr, A. Keith. *CRC Handbook of Laboratory Safety*, 4th ed. Boca Raton, FL: CRC Press, 1995.

Kaufman, James A. *Waste Disposal in Academic Institutions.* Boca Raton, FL: Lewis Publishers Inc., 1990.

National Safety Council. *Principles of Occupational Safety and Health.* Itasca, IL: NSC, 1998.

"Occupational Exposure to Hazardous Chemicals in Laboratories." *Code of Federal Regulations.* Title 29, Pt. 1910.1450.

Stricoff, R. Scott. *Laboratory Health and Safety: A Guide for the Preparation of a Chemical Hygiene Plan.* New York, NY: Wiley-Interscience Publications, 1990.

Wentz, Charles A. *Hazardous Waste Management.* New York, NY: McGraw-Hill, 1989.

Mignon J. Clarke *has over nine years of combined technical, analytical, and research support experience resulting from multidisciplined work environments. She currently serves as an environmental specialist for the Department of the Army, Environmental Management Division, at Fort Benning, Georgia. Her responsibilities include reviewing environmental protection regulatory developments in order to identify impacts to future installation projects, particpating in environmental compliance inspection activities, and reviewing permit applications and reports for compliance with environmental protection regulations. Ms. Clarke has an extensive background in laboratory health and safety, waste minimization, and analytical, testing procedures. Her previous professional experience has included such titles as chemical technician, process chemist/ compliance coordinator, and environmental, safety, and health professional.*

CHAPTER 13

Bloodborne Pathogen Program

Frank Pfeifer, RS, CHMM
Martha Boss, CIH, CSP

Introduction

Part 1910.1030 of Title 29 of the *Code of Federal Regulations*, was published in the *Federal Register* on December 6, 1991, (29 CFR 1910.1030) and became effective on March 6, 1992. This standard originally sought to protect workers in hospitals, funeral homes, nursing homes, clinics, law enforcement agencies, emergency responders, human immunodeficiency virus/hepatitis B virus (HIV/HBV) research laboratories, and dentists. In reality, however, *all workers* who potentially could be *occupationally* exposed to bloodborne pathogens in the workplace environs are covered.

> ***Bloodborne Pathogens*** means pathogenic microorganisms that are present in human blood and can cause disease in humans. These pathogens include *but are not limited to,* hepatitis B virus (HBV) and human immunodeficiency virus (HIV).

The above-mentioned phrase, "but are not limited to," is extremely important and often ignored. Remember, this standard applies to all bloodborne pathogens, not just HBV and HIV. Hepatitis C is increasingly being considered a workplace health hazard.

This Occupational Safety and Health Administration (OSHA) standard gave the Center for Disease Control and Prevention's (CDC's) universal precautions the force of law. This standard does not apply to hospital patients, or any other exposure situation outside of the workplace. *Good Samaritan* acts that result in exposure to blood or other potentially infectious materials from assisting a fellow employee (*i.e.*, assisting a co-worker with nosebleed, giving Cardiopulmonary Resuscitation [CPR] or first aid) are not included in the Bloodborne Standard, unless such duties could be expected to follow from the job's task descriptions. OSHA, however, encourages employers to offer postexposure evaluation and follow-up in such cases.

Exposure Control

OSHA's rule applies to all persons *occupationally exposed* to blood or other potentially infectious materials. Blood means human blood, blood products, or blood components (see Figure 1).

Other potentially infectious materials (OPIM) include

- Human body fluids: semen, vaginal secretions, cerebrospinal fluid, synovial fluid, pleural fluid, pericardial fluid, peritoneal fluid, amniotic fluid.

- Saliva in dental procedures.

- Any body fluid visibly contaminated with blood.

- All body fluids in situations where differentiation between body fluids is difficult. This would include waste mixtures such as vomit, toilet overflow, sewage, and leachate associated with refuse containers.

Written Exposure Control Plan

The standard requires the development of a written exposure control plan. Refer to the standard for specific points, but at a minimum, the exposure control plan must include:

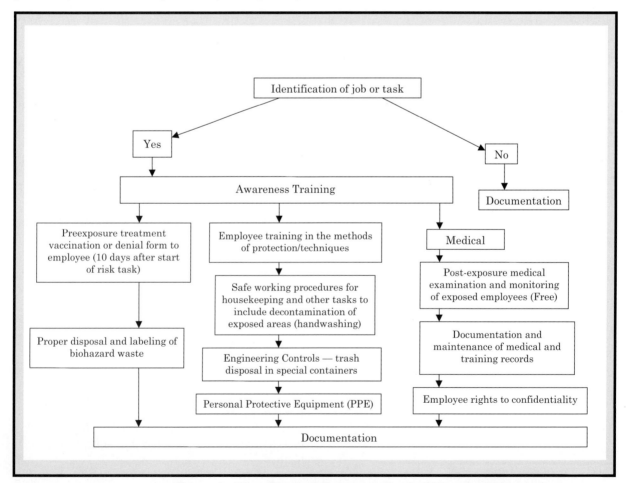

Figure 1. Flow Chart of Exposure Control Plan Development

- The exposure determination (without regard for personal protective equipment [PPE])

- The procedures for evaluating the circumstances surrounding an exposure incident

- The schedule and method for implementing sections of the standard covering the methods of compliance

- Hepatitis B vaccination and postexposure follow-up

- Communication of hazards to employees, and recordkeeping

- Schedule for annual plan review

Bloodborne Pathogenic Diseases

Human Immunodeficiency Virus

Acquired immune deficiency syndrome (AIDS) is caused by a virus. Technically, the viral infection is human immunodeficiency virus (HIV) and the resultant illness is AIDS. The HIV attacks the immune system and thus, the symptoms of AIDS emerge. AIDS is ultimately a fatal disease with no known cure. Each year, 35,000 people in the United States are infected with HIV, and many do not even know they are infected! HIV is a very fragile organism outside of the human body. HIV requires narrow temperature–time period environments to survive.

Hepatitis B Virus

Hepatitis B (HBV) is a serious public health problem that affects people of all ages in the United States and around the world. Each year, more than 240,000 people in the United States contract hepatitis B. The disease is caused by a highly infectious virus that attacks the liver. HBV infection can lead to severe illness, liver damage, and, in some cases, death. Each year, approximately 5,000 people in the United States die of liver failure related to hepatitis B, and another 1,500 die of liver cancer related to hepatitis B. Hepatitis B is the most common cause of liver cancer worldwide, and liver cancer is one of the three most common cancers in the world. The best

way to be protected against hepatitis B is prevention. The best prevention is to be vaccinated with hepatitis B vaccine, a vaccine that has been proven safe and effective.

People who do not clear the virus from their blood have not fully recovered, and are called *hepatitis B carriers*. There are over one million carriers in the United States today. An HBV carrier is someone who has had hepatitis B virus in his/her blood for more than six months. While about 10 percent of adults who acquire HBV infection become carriers, children have a greater risk. The younger the child is at the time of infection, the greater the risk that the child will become a lifelong carrier. Many babies born to carrier mothers will also become carriers of HBV unless the babies are given special shots in the hospital and during the six months after birth to protect them from the infection. A carrier usually has no signs or symptoms of HBV but remains infected with the virus for years or for a lifetime and is capable of passing the disease on to others. Sometimes HBV carriers will spontaneously clear the infection from their bodies, but most will not. Although most carriers have no serious problems with hepatitis B and lead normal healthy lives, some carriers develop liver problems later. Hepatitis B carriers are at significantly higher risk than the general population for liver failure or liver cancer.

Hepatitis C Virus

Hepatitis C (HCV) is a viral illness that affects the liver. In 1990 an antibody to the hepatitis C virus was identified, but before that the illness was known in the United States as, *non-A, non-B hepatitis*. Hepatitis C is spread by blood-to-blood contact, and is therefore a bloodborne pathogen. Some of the most common ways of spreading the virus are

- Transfusion of blood products

- Intravenous drug use

- Tattooing

- Body piercing

- Sharing needles

Menstrual blood can also contain the virus.

Additional Pathogens

There are other pathogen diseases that should also be considered in a Bloodborne Pathogen Program or a Biological Protection Program, besides the usually covered HIV and HBV. Examples of other Bloodborne Pathogens include HCV, syphilis, and malaria. HCV has been previously referred to as non-A and non-B hepatitis in other literature. A related disease is tuberculosis, which, while not considered a bloodborne disease, is transmitted by contact with body fluids and liquids (*i.e.*, OPIM). If a risk in a population is to be covered by a protective program, tuberculosis should be considered for addition to the Bloodborne Pathogen Program. Separate regulatory requirements are in effect for tuberculosis.

Training

In determining exposure, you may consult the standard as to employees who are always at risk, versus employees who are at risk only in extraordinary situations. However, in point of fact, all employees with defined potential exposure risk will need information and training. Through awareness training, assessment of actual risk and the exposure determination may be made from a more educated platform. Be sure to document the decision logic for inclusion or exclusion of employees from the Bloodborne Pathogen Program.

Additional training is needed for those who have been determined to be exposed and at risk of acquiring bloodborne pathogen infections. When existing tasks are modified or new tasks that involve occupational exposure to bloodborne pathogens affect the employee's exposure, decision logic as to employee inclusion in the program may also have to be altered.

Training must be accomplished by trainers with knowledge of both the standard and the implications of the standard given current employee exposure. Thus, training must be site- or facility-specific. Information as to routes of exposure and preventative measures, including postexposure follow-up, must include the epidemiology and symptoms of bloodborne diseases. Given the potential to encounter labeled biohazards and unforeseen biohazards when handling waste, the emphasis during training should include discussions of *what-if* situations.

Training Records

Accurate training records must be kept for 3 years and include the following:

1) Training dates

2) Content or a summary of the training

3) Name(s) and qualification(s) of trainer(s)

4) Names and job titles of trainees

Upon request, both medical and training records must be made available to the Director of the National Institute for Occupational Safety and Health (NIOSH) and to the Assistant Secretary of Labor for Occupational Safety and Health. Training records must be available to employees or employee representatives upon request.

Vaccination

Hepatitis B Virus

Fortunately, a vaccine is available to protect employees from hepatitis B (HBV) prior to exposure and also to provide prophylactic treatment postexposure. Employees who decline the vaccination must sign a declination form. The employee may request and obtain the vaccination at a later date and at no cost, if he/she continues to be exposed.

The hepatitis B vaccine and vaccination series must be offered within 10 working days of initial assignment to employees who have occupational exposure to blood or other potentially infectious materials. Any booster doses of the hepatitis B vaccine recommended by the United States Public Health Service also must be offered.

A physician may determine and provide recommendations as to employees who should be precluded from receiving the vaccine, despite proven occupational exposure potential. The employer must obtain and provide the employee with a copy of the health care professional's written opinion stating whether a hepatitis B vaccination is indicated for the employee and whether the employee has received such vaccination.

HBV Vaccination Declination Statement

In the event an employee elects not to receive the HBV vaccine, he or she should sign and date a declination statement to this effect. An example declination statement is:

> I understand that due to my occupational exposure to blood and other infectious materials I may be at risk of acquiring hepatitis B virus (HBV) infection. I have been given the opportunity to be vaccinated with the HBV vaccine at no charge to myself. I decline the HBV vaccine at this time. However, I understand that by declining the HBV vaccine, I may continue to be at risk of acquiring HBV. If in the future I continue to have occupational exposure to blood or other potentially infectious materials and I later want the HBV vaccine, I can receive the vaccination series at no charge to me.

If a declination statement is made and accepted, the employer must also document that the employee had appropriate training to make an informed decision. This training need not be a full program training sequence; however, minimum awareness training is required.

Engineering and Work Practice Controls

Engineering and work practice controls are the primary methods used to prevent occupational transmission of HBV and HIV. Personal protective equipment (PPE) is necessary when occupational exposure to bloodborne pathogens remains even after instituting these controls. Any booster doses of the hepatitis B vaccine recommended by the United States Public Health Service also must be provided.

Proper work practice controls alter the manner in which a task is performed. In work areas where a reasonable likelihood of occupational exposure exists, work practice controls include restricting eating, drinking, smoking, applying cosmetics, and handling contact lenses, as well as prohibiting mouth pipetting. Foods, including stored food and drink, should be excluded entirely from areas

where bloodborne pathogens may be present. This preclusion includes storage of foods and drinks in sample refrigerators.

Handwashing is necessary, and handwashing facilities must be provided. All personnel, upon doffing gloves and other PPE, must immediately wash their hands. In the event that an exposure incident has occurred, gloves and other PPE should be disinfected prior to doffing. If the PPE has already been breached, measures to prevent further worker exposure as PPE is doffed must be determined on a situation-specific basis.

Personal Protective Equipment

Personal protective equipment is considered appropriate only when blood or other potentially infectious materials are not permitted to pass through or reach employees' work clothes, street clothes, undergarments, skin, eyes, mouth, or other mucous membranes under normal conditions of use and for the duration of time used. Thus, PPE in these cases provides a limited barrier between the user and infectious materials. As with any barrier, breakthrough, molecular penetration, and mechanical degradation of PPE will all lessen the barrier's effectiveness.

Gloves, while providing an effective barrier form of PPE, should not be thought of as a foolproof mode of protection. Proper donning and doffing of gloves will lessen worker exposure to surficial contamination. The practice of reusing gloves may not be appropriate in these situations, and glove reuse must be discussed thoroughly during training exercises. *If gloves cannot be reused without increased exposure to the workers, gloves should be disposed of during the initial doffing interval.* Some people have an allergy to latex, and thus alternative gloves should be considered in a PPE program.

Housekeeping

Normal housekeeping may not be sufficient to deal with bloodborne hazards. In all cases, any housekeeping that provides a way to handle materials without directly touching contaminated surfaces should be encouraged. The use of mechanical means such as tongs, forceps, or a brush and a dustpan to pick up contaminated broken glassware are examples of good house-keeping procedures. Scientific and lay publications provide varying time-period discussions as to the continued virulence of bloodborne pathogenic materials. Currently, the HIV virus is thought to exist in virulent numbers or status for only 30 minutes outside of the human body. HBV can, however, remain infective on dry surfaces for 7 days, and in raw sewage for up to 21 days within a *wet* suspension. For housekeeping purposes, you may not know the time or conditions under which the initial spill occurred. So during housekeeping, as in any other exposure situation, assume the materials remain virulent and take required precautions. Reducing splash during cleanup is desirable, and may be accomplished by placing a disposable cloth over the spill and then carefully pouring the disinfectant over the cloth.

Disinfection

All surfaces, reused PPE, tools, and other devices that come in contact with bloodborne pathogens must be disinfected as soon as possible. A solution of 5.2% sodium hypochlorite (unscented household bleach) diluted between 1:10 and 1:100 with water may be used (read the label to confirm). The recommended dilution is 1:10 (1/4 cup per gallon of water) whenever practicable. The resident time for this solution is also important and varies between 10 to 30 minutes depending on the original solution concentration and the porosity of the materials being disinfected.

For surfaces, such as the neoprene in respirator harnesses, chlorine is not a recommended disinfecting agent in most cases—the chlorine degrades the neoprene. When neoprene respirator parts must be disinfected, consult the manufacturer as to appropriate disinfection solutions.

Remember that not all disinfecting agents are appropriate or effective in these situations. To obtain information specific to your disinfection needs, contact the National Antimicrobial Information Network (NAIN): (800)447-6349. Remember also that disinfection is not sterilization of the surfaces—you are reducing the number of infectious agents, not entirely eliminating all infectious organisms.

Labeling

The standard requires that fluorescent orange or orange-red warning labels be attached to containers of regulated waste, to refrigerators and freezers containing blood and other potentially infectious materials, and to other containers used to store, transport, or ship blood or other potentially infectious materials. The warning label must be fluorescent orange or orange-red, contain the biohazard symbol and the word BIOHAZARD, in a contrasting color, and be attached to each object by string, wire, adhesive, or another method to prevent loss or unintentional removal of the label. The labels are not required when

1) Red bags or red containers are used

2) Containers of blood, blood components, or blood products are labeled as to their contents and have been released for transfusion or other clinical use

3) Individual containers of blood or other potentially infectious materials are placed in a labeled container during storage, transport, shipment, or disposal

If an Exposure Incident Occurs

The standard requires that the postexposure medical evaluation and follow-up be made available immediately for employees who have had an exposure incident. At a minimum, the evaluation and follow-up must include the following elements:

- Document the routes of exposure and how exposure occurred.

- Identify and document the source individual, unless the employer can establish that identification is infeasible or prohibited by state or local law. Keep in mind that local regulations and legal interpretations should be consulted in the development of the program, rather than making a random last-minute decision to question the source individual. If the source individual is known to be infected with either HIV or HBV, testing need not be repeated for known pathogens.

– obtain consent, and test source individual's blood as soon as possible.

♦ provide the exposed employee with the source individual's test results and information about applicable disclosure laws and regulations concerning the source identity and infectious status.

♦ if the employee does not give consent for HIV serological testing during the collection of blood for baseline testing, preserve the baseline blood sample for at least 90 days.

• Provide HBV, HCV, and HIV serological testing, counseling, and safe and effective postexposure prophylaxis.

• The employer must give the health care professional responsible for the employee's hepatitis B vaccination and postexposure evaluation and follow-up a copy of the OSHA standard. The employer also must provide to the health care professional evaluating the employee after an exposure incident a description of the employee's job duties relevant to the exposure incident, documentation of the route(s) of exposure, circumstances of exposure, and results of the source individual's blood tests, if available, and all relevant employee medical records, including vaccination status.

• Within 15 days after evaluation of the exposed employee, the employer must provide the employee with a copy of the health care professional's written opinion. The written opinion is limited to whether the vaccine is indicated and has been received. The written opinion for postexposure evaluation must document that the employee has been informed of the results of the medical evaluation and of any medical conditions resulting from the exposure incident that may require further evaluation or treatment. All other diagnoses must remain confidential and not be included in the written report.

Recordkeeping

Occupational Exposure

The employer also must preserve and maintain for each employee an accurate record of occupational exposure according to OSHA's rule governing access to employee exposure and medical records (29 CFR 1910.20).

Medical records must include the following information:

1) Employee's name and social security number

2) Employee's hepatitis B vaccination status including vaccination dates and any medical records related to the employee's ability to receive vaccinations

3) Results of examinations, medical testing, and postexposure evaluation and follow-up procedures

4) Health care professional's written opinion

5) A copy of the information provided to the health care professional

Medical Records

Medical records must be kept confidential and maintained for at least the duration of employment plus 30 years. An employee's medical records can be obtained by that employee or anyone having that employee's written consent. Also, if the employer ceases to do business, medical and training records must be transferred to the successor employer. If there is no successor employer, the employer must notify the Director, National Institute of Occupational Safety and Health (NIOSH), United States Department of Health and Human Services, for specific directions regarding disposition of the records at least 3 months prior to intended disposal.

Summary

The Bloodborne Pathogen Standard requires that a written program be developed and implemented by employers when their employees may be exposed to these pathogens at work. Before a program can be implemented, a task/job risk identification or hazard assessment must be documented. Awareness training should then be offered to all affected personnel within the work force, including any decision-makers who oversee program implementation. For those personnel judged to be at risk, additional training is needed and must include engineering controls, housekeeping, personal protective equipment (PPE), waste disposal, medical procedures, confidentiality, and documentation requirements. Some states have regulations pending that will add additional requirements for their jurisdictions, so be sure to check with local and state regulatory agencies for additional requirements. There are many resources for more information on specific information on Bloodborne Pathogen Programs, in book form and on the Internet. Some are listed in the following section.

Additional Resources

Many resources are available in printed form and electronically, on the Internet. OSHA, CDC, EPA, and universities have web sites that provide example programs and other valuable information as to requirements and workplace implementation strategies. Often, grant monies are used to fund phone-number information resources and web pages. When grant resources are altered (*i.e.*, grants are placed with new institutions), the listed phone numbers may change unexpectedly. Be very specific about the information you are seeking, and query the information source as to how they disseminate information. What information do they disseminate? For example if you are looking for disinfectants and ask for "recommended disinfectants," you may receive the reply that "they do not recommend disinfectants." However, if you ask for a copy of the EPA's "List of Recommended Disinfectants," you will be provided with a list of recommended disinfectants and with guidance to read the labels.

National Antimicrobial Information Network
(NAIN): (800) 447-6349

CDC's HIV and AIDS Hotline:
(800) 342-2437

OSHA's Emergency Number:
(800) 321-6742

OSHA's Technical Support:
(202) 693-2300

Internet Resources

<http://www.cdc.gov/> (Centers for Disease Control and Prevention Homepage)

<http://www.epa.gov/epahome/Offices.html> (Environmental Protection Agency Regional Offices)

<http://www.osha.gov> (Occupational Safety and Health Administration Homepage)

<http://www.osha-slc.gov/SLTC/> (Occupational Safety and Health Administration Technical Links)

Frank Pfeifer has a degree in Environmental Studies from the University of Kansas. Frank has worked in several environmental positions since graduation in 1976. Some of these include: Water and Wastewater Plant Operator, Sanitarian (Health Inspector at city, county, and State levels), Food and Drug Inspector, Asbestos Course Auditor under an EPA Grant, Program Specialist for the Nebraska Department of Health (Water and Asbestos Programs), Industrial Hygienist Consultant for a regional engineering firm, and writer. Currently, Mr. Pfeifer is a Safety/ Environmental Engineer for 3M in Valley, Nebraska. In the last year, Mr. Pfeifer's 3M plant was recommended for OSHA's Voluntary Protection Program Star Award and has also received ISO 14001 certification. Mr. Pfeifer is the interim President for the Midwest Plains Chapter of ACHMM.

Martha Boss is a Certified Industrial Hygienist and a Certified Safety Professional specializing in hazardous materials management, hazardous waste abatement and remediation, OSHA process safety, and EPA risk management program development, risk assessment, biological risk evaluations, due diligence, and OSHA-type compliance and ventilation design. Ms. Boss has been active in these endeavors for over 10 years, during which time she has overseen hazardous materials management at major Environmental Protection Agency, CERCLA, and Department of Defense and Department of Energy sites throughout the nation. These experiences have resulted in published works in Applied Toxicology, Environmental Protection *and* Compliance *magazines. Martha wishes to acknowledge her former colleagues at the US Army Corps of Engineers, who continue to provide Industrial Hygiene and safety oversight at challenging sites throughout the world.*

Part III

The Right to Know

Part III

CHAPTER 14

Release Reporting and Emergency Notification

Alan A. Eckmyre, CHMM

Introduction

Several Federal environmental laws require that a *release of hazardous substances to the environment* above certain threshold amounts—Reportable Quantities or RQs—be reported in a timely manner. Failure to report certain releases can result in substantial penalties to be imposed on both the facility and its employees.

Hazardous substances are

- Any elements, compounds, mixtures, solutions, or substances designated by the United States Environmental Protection Agency (EPA) under Section 311 of the Clean Water Act (CWA), (40 CFR 116.4) or under Section 102 of the Comprehensive Environmental Response, Compensation, and Liability Act (CERCLA), (40 CFR 302.4)

- Any toxic pollutants listed under Section 307(a) of the CWA

- Any hazardous substances regulated under Section 311(b)(2)(A) of the CWA (40 CFR 110, 117, and 122)

- Any listed or characteristic Resource Conservation and Recovery Act (RCRA) hazardous wastes (40 CFR 262)

- Any hazardous air pollutants listed under Section 112(r) of the Clean Air Act (CAA) (40 CFR 68)

- Any imminently hazardous chemical substances or mixtures regulated under Section 7 of the Toxic Substances Control Act (TSCA) (40 CFR 761.120, *et seq.*)

Under TSCA, any release, leak, or spill of a hazardous chemical that *seriously threatens humans with cancer, birth defects, mutation, death, or serious or prolonged incapacitation, or seriously threatens the environment with large-scale or ecologically significant population destruction* must be reported immediately to the appropriate EPA Regional Office by telephone. A written follow-up report is required within 15 days of the oral report. If the incident has been reported under the CWA, the facility does not have to submit the *substantial risk report* for the incident (40 CFR 761.120, *et seq.*).

The Emergency Planning and Community Right-to-Know Act (EPCRA) also establishes emergency reporting requirements for *extremely hazardous substances* (40 CFR 355, Appendix A). All of these substances are also CWA and CERCLA *hazardous substances*.

A *reportable quantity (RQ)* is the amount of a hazardous substance which, when released to the environment, must be reported to governmental authorities under the CWA, CERCLA, SARA Title III, or RCRA. Many states also require release reporting in addition to that called for by the Federal statutes. Consequently, a complicated maze of notification requirements exists that makes it difficult for responsible parties to determine which requirements are applicable.

The initial release notification usually is required immediately, or within 24 hours of knowledge of the release. In some cases, follow-up written reports are also required. Therefore it is important to establish in advance a release response program that addresses all appropriate release reporting requirements at a particular facility.

In order to assist individuals with notification obligations, a summary of reporting requirements

contained in the CWA Section 311, CERCLA Section 103, Superfund Amendments and Reauthorization Act (SARA) Title III Section 304, and RCRA Subtitle I is provided in Table 1. The table provides a summary of the various reporting requirements, including type of releases subject to reporting, substances subject to reporting, quantities subject to reporting, parties responsible for reporting, when reporting is required, to whom to report, and penalties for failure to notify.

EPA considers a *release* to be virtually all conceivable contacts with the environment, including any spilling, leaking, pumping, pouring, emitting, emptying, discharging, injecting, escaping, leaching, dumping, or disposing into the environment. The abandonment or discarding of barrels, containers, and other closed receptacles containing hazardous substances is also considered a release to the environment. However, EPA has indicated that certain administrative exemptions from reporting hazardous substances contact with the environment may be appropriate.

Some releases are excluded, including

1) Releases solely in the workplace

2) Exhaust emissions from vehicles, aircraft, vessels, pumping station engines, *etc.*

3) Normal applications of fertilizer

4) Releases of source, by-product, or special nuclear material subject to Section 170 of the Atomic Energy Act (AEA) or Sections 102(a)(1) or 302(a) of the Uranium Mill Tailings Radiation Control Act (UMTRCA)

5) Federally permitted releases

EPA has not formally defined what constitutes a release solely in the workplace. However, in the final rule setting RQs for radionuclides (54 FR 22524; May 24, 1989) EPA makes clear that this exemption applies only to releases that occur within a closed space with no emissions to the ambient environment. Therefore, spills onto concrete floor of an enclosed building or plant would qualify as a release solely within the workplace as long as the hazardous substances do not leave the building or structure by penetrating the floor or any other route. However, for volatile materials with significant vapor pressure at the release temperature, it is assumed that if an RQ of the material goes off-site immediately then this type of volatile substance is a reportable incident. In addition, the stockpiling of a hazardous substance in any unenclosed containment structure—surface impoundment, lagoon, tank, or other holding device that has an open side with the contained materials directly exposed to the ambient environment—is a release to the environment.

The general assumption underlying the exemptions from reporting requirements for Federally permitted releases is that such releases have been

evaluated through the permit process and are not considered to be harmful to human health and the environment. Section 101(10) of CERCLA identifies specific types of releases considered to be Federally permitted, including

- Releases of substances and quantities specified in National Pollutant Discharge Elimination System (NPDES) permits, permit applications, or permit administrative records under the CWA

Table 1. Release Reporting Requirements Summary

Reporting Requirements	CWA Section 311	CERCLA Section 103	SARA Title III Section 304	RCRA Subtitle I (UST Program)
Type of release subject to reporting	Discharge to surface water	Releases into the environment (outside enclosed building or structure)	Releases into the environment with potential to result in exposure to persons offsite	Ground water, surface water, or surface soils
Hazardous substances subject to reporting	Oil and hazardous substances listed in 40 CFR 116.4	CERCLA hazardous substances listed in 40 CFR 302.4	Extremely hazardous substances listed in 40 CFR 355.20 and CERCLA hazardous substances	Petroleum and CERCLA hazardous substances (excluding RCRA hazardous wastes)
Trigger amount for reporting	All discharges of oil that form a sheen and discharges of hazardous substances that equal or exceed the CERCLA RQ	Releases that equal or exceed the CERCLA RQ	Releases that equal or exceed the CERCLA RQ, or one pound for EHSs that are not CERCLA hazardous substances	Any suspected or confirmed release, spills that equal or exceed the CERCLA RQ, and spills of petroleum that exceed 25 gallons or that result in a sheen
Facilities subject to reporting	All facilities and vessels	All facilities and vessels, except where consumer products are in consumer use	All facilities that produce, use, or store an OSHA hazardous chemical	Facilities with an underground storage tank system, the volume of which is 10 percent or more beneath the surface of the ground
Parties responsible for reporting	Person in charge	Person in charge	Owner/operator	Owner/operator
When report is required	Immediately	Immediately	Immediately, with written follow-up report	Within 24 hours
To whom to report	National Response Center	National Response Center	Relevant SERCs[1] and LEPCs[2]; for transportation-related releases, the 911 emergency telephone number or the operator	Implementing agency (usually a State agency)
Maximum penalties	$10,000 and/or prison sentence of one year	$50,000 and/or prison sentence of three years	$50,000 and/or prison sentence of three years	$25,000 per day for each day of noncompliance

[1]State Emergency Response Commissions
[2]Local Emergency Planning Committees

- Discharges complying with permits for dredge or fill materials under Section 404 of the CWA

- Releases in compliance with RCRA final permits

- Releases in compliance with enforceable permits under the Marine Protection, Research, and Sanctuaries Act (MPRSA)

- Underground injection of fluids permitted under the Safe Drinking Water Act (SDWA)

- Injection of fluids authorized by State law regulating underground injection of fluids used in petroleum product production or recovery

- Air releases complying with permit or control regulations under specific provisions of the Clean Air Act (CAA)

- Releases to a publicly owned treatment works (POTW) in compliance with a pretreatment standard and program submitted to EPA for review (must be in compliance with local limits that take into account site-specific characteristics)

- Releases of source, special nuclear, or by-product material in compliance with a license, permit, or order issued pursuant to the AEA

In general, in order to qualify as *Federally permitted*, the hazardous substances, quantities released, and activities causing the release must be within the scope of a permit. If a release exceeds permitted levels, the excess is not in compliance with the permit and cannot be Federally permitted. Therefore, if the amount of a release that exceeds a permit level is equal to or exceeds an RQ, the release must be reported.

The following discussion provides a general summary of the major Federal requirements for release notification and reporting. To be in full compliance with release notification requirements, the discharger should contact the proper authorities and provide them with the appropriate information on the required form. Since the response to a substance release is primarily a function of local government response teams, it is important to review local rules and ordinances for any additional notification requirements.

Clean Water Act Reporting Requirements

The CWA Section 311 reporting requirements apply to releases of CWA hazardous substances and oil from vessels and facilities into waters of the United States and the contiguous zone. CWA hazardous substances are listed in 40 CFR 116.4. For purposes of CWA reporting, *oil* is defined as *oil of any kind or in any form, including, but not limited to, petroleum, sludge, oil refuse, and oil mixed with wastes other than dredged spoil* (CWA Section 311[a][1]).

For a CWA hazardous substance, notification is necessary when a quantity at or above the reportable quantity (RQ) is discharged (released) to surface water within a 24-hour period.

Notification is required for oil discharges that

- Cause a sheen

- Violate applicable water quality standards

- Cause a sludge or emulsion to be deposited beneath the water surface or upon the shoreline (40 CFR 110).

Discharges in compliance with NPDES permits are exempt from the CWA notification requirements. However, if a discharge of oil or CWA hazardous substance violates an NPDES permit, the reporting requirements must be met.

In the event of a release of a hazardous substance, if the National Response Center (NRC) cannot be contacted immediately, reports may be made to the Coast Guard or EPA predesignated On-Scene Coordinator (OSC) for the geographic area where the discharge occurs. All reports are then promptly relayed by the OSC to NRC. If it is not possible to notify NRC or OSC, reports may be made immediately to the nearest Coast Guard unit, provided the person in charge of the vessel or onshore or offshore facility notifies NRC as soon as possible.

Oil and hazardous substance releases require that oral reporting be made immediately upon knowledge of a reportable release. A written follow-up report is also required within 30 days and any additional written follow-up within 60 days. If NPDES permit limits are exceeded, an oral report is required within 24 hours of any noncompliance, with written follow-up within 5 days. Follow-up reports are required to be developed and should be forwarded to EPA regional office, State Emergency Response Commission, Local Emergency Planning Committee or fire department, within reporting deadlines.

Regulations specify the contents of the initial oral report for oil and hazardous substance release. The report must include

- Name of person reporting

- Type and amount of substance

- Location, date, time, and duration of release

- Basis for hazard classification

- Reportable quantity; whether release is to ground, water, or air

- Remedial action taken

- Identity of other regulatory authorities

Furthermore, if there is potential for injury, additional newspaper notice is required to notify potentially injured parties. Table 2 provides a summary of emergency notification procedure content to satisfy notification-reporting requirements.

The follow-up written report must include

- Initial oral report contents

Table 2. Emergency Notification Procedure Content

Emergency Notification Procedure Content

1) Chemical Name (CAS# if available)

2) Chemical an Extremely Hazardous Substance?

3) Estimate of Quantity Released (refine later during follow-up report)

4) Time and Duration of Release

5) Endangered Media (air, water, and/or land)

6) Known Health Risks and Medical Advice

7) Proper Precautions (evacuation, shelter-in-place, personal protective equipment [PPE])

8) Name and Phone Number for Return Calls

9) Actions Taken and Responders On-Scene or Enroute

10) On-Scene Incident Commander and How to Contact

11) Follow-up Schedule Based upon Incident (within 15 days)

Notes:

- Requirements imply that the person constructing the Emergency Notification Procedure is familiar with the Incident Command System.

- A follow-up report to the same agencies is required, detailing the incident, response, cleanup, effects of the spill, and measures taken to prevent the recurrence of a similar incident.

- Name and telephone number of person in charge
- NRC or EPA case number
- Dun and Bradstreet facility or company number
- Source and frequency of release
- Population density
- Sensitive populations
- Sensitive ecosystems within one mile
- Releases over past year
- Basis for stating that release is continuous and stable
- Port of registration if release is from a vessel

Additional follow-up reports should include

- Name of person reporting
- Case number
- Information verifying contents of previous reports

Written follow-up report contents for any NPDES permit limit exceedance include

- Description of noncompliance
- Cause of noncompliance
- The period of noncompliance, including dates and times
- If the noncompliance has not been corrected, the anticipated time it is expected to continue
- Steps to reduce, eliminate, and prevent reoccurrence of the noncompliance (40 CFR 122.41, 403.12, and 403.16)

The following civil penalties exist for reporting requirement violations under the CWA:

- Possible $10,000 fine and up to 1 year imprisonment for failure to notify. (40 CFR 117.22[a])
- $5,000 fine per each release or discharge in a 24-hour period exceeding an RQ (For most petroleum products, the RQ is one barrel, with the amount in gallons or liters.). (40 CFR 117.22[b])
- Additional civil penalty of up to $5,000, or up to $250,000 if the discharge is the result of

willful negligence or willful misconduct (40 CFR 117.22[b])

Comprehensive Environmental Response, Compensation, and Liability Act Reporting Requirements

Under CERCLA, notification is necessary if an amount of a CERCLA hazardous substance greater than or equal to its reportable quantity (RQ) is released into the environment from a vessel or facility within a 24-hour period.

CERCLA *hazardous substances* include

- Substances designated under Sections 307(a) and 311(b)(4) of the CWA
- Hhazardous air pollutants listed under Section 112 of the CAA
- RCRA hazardous wastes
- Chemicals or mixtures for which EPA has taken action under Section 7 of the Toxic Substances Control Act (TSCA) (CERCLA Section 101[14])

The statute also gives EPA the authority to designate additional hazardous substances.

Petroleum Exclusion

CERCLA exempts the following petroleum products from regulation as hazardous substances:

- Petroleum, including crude oil or any fraction thereof which is not otherwise specifically listed or designated as a hazardous substance
- Natural gas, natural gas liquids, liquefied natural gas
- Synthetic gas usable for fuel (or mixtures of natural gas and such synthetic gas) (CERCLA Section 101[14])

According to EPA's interpretation, the petroleum exclusion applies to materials such as crude oil, petroleum feedstocks, and refined petroleum products, even if a CERCLA hazardous substance is a constituent of such products or is normally added to them during refining. However, the exclusion does not apply to hazardous substances that are added to petroleum products after the refining process or contamination that occurs during use, such as with used oil.

In order for notification to be required, all of the following criteria must be met:

- A release of a CERCLA hazardous substance must occur from a vessel or facility. A *release* is defined as *any spilling, leaking, pumping, pouring, emitting, emptying, discharging, injecting, escaping, leaching, dumping, or disposing into the environment.* (40 CFR 302.3)

- The release must occur *into the environment*. A release is considered to have entered the environment if it is not completely contained within a building or structure, even if it remains on the plant or facility grounds.

- The hazardous substance must be released in a quantity that equals or exceeds the RQ for that substance over a 24-hour period. RQs are listed in 40 CFR, Table 302.4.

Determination of whether an RQ has been released is more complicated for mixtures. When the hazardous substances in the mixture and their concentrations are known, releases of the mixture must be reported as soon as any hazardous substance is released in an amount equal to or greater than its RQ.

If the concentrations of hazardous substances are unknown, reporting must occur when the amount of the entire mixture released reaches the RQ for the component having the lowest RQ.

Several types of incidents are exempt from CERCLA notification requirements, including

- Federally permitted releases.

- Proper application of pesticide products registered under the FIFRA. (Accidents, spills, improper application and improper disposal of pesticides must be reported.)

- Releases of solid particles of antimony, arsenic, beryllium, cadmium, chromium, copper, lead, nickel, selenium, silver, thallium, or zinc, when the mean diameter of particles is larger than 100 micrometers (0.004 inches).

- Releases from consumer products in consumer use.

- Emissions from the engine exhaust of a motor vehicle, rolling stock, aircraft, vessel, or pipeline pumping-station engine.

- Certain releases of source, by-product, or special nuclear material from a nuclear incident

- Normal application of fertilizer.

In addition, EPA has reduced the reporting requirements for *continuous releases.*

Following initial notification, additional reporting for continuous releases is only required annually or when there is a statistically significant increase in the quantity of the release.

CERCLA Section 103(f)(2) provides relief from the reporting requirements of Section 103(a) for a release of a hazardous substance that is continuous, stable in quantity and rate, and either is a release from a facility for which notification of known, suspected, or likely releases of hazardous substances has been given under Section 103(c) or is a release for which notification has been given under Section 103(a) for a period sufficient to establish the continuity, quantity, and regularity of such release. Section 103(f)(2) further provides that in such cases, notification shall be given annually or at such time as there is any statistically significant increase (SSI) in the quantity of hazardous substance released.

The *person in charge* of a facility or vessel must make notification of hazardous substance releases to the National Response Center (NRC) as soon as he/she learns that a reportable release has occurred. The person in charge is not defined by CERCLA. Therefore, each company must designate a person to be responsible for reporting.

In the event of a reportable release of a CERCLA hazardous substance, an initial oral report and follow-up report by the facility emergency coordinator are required to be received by the National Response Center (NRC), State Emergency Response Commission, Local Emergency Planning Committee or fire department. The facility emergency coordinator reports immediately following any release of a reportable quantity. Subsequently, a written follow-up report within 30 days and additional written follow-up report within 60 days are required by reporting deadlines (40 CFR 302, *et seq.*; Appendices A and B).

The content of emergency notification information under CERCLA is very similar to that required under the CWA discussed previously. The initial report must include

- Name of person reporting
- Type and amount of substance
- Location
- Date
- Time and duration of release
- Basis for hazard classification
- Reportable quantity
- Whether release is to ground, water, or air
- Remedial action taken
- Identity of other regulatory authorities

If there is potential for injury, additional newspaper notice also is required to notify potentially injured parties.

The first follow-up written report must include

- Initial report contents
- Name and telephone number of person in charge
- NRC or EPA case number
- Dun and Bradstreet facility or company number
- Source and frequency of release
- Population density
- Sensitive populations and sensitive ecosystems within one mile
- Releases over the past year
- Basis for stating that release is continuous and stable
- Port of registration, if release is from a vessel

Additional follow-up reports should include

- Name of person reporting
- Case number
- Information verifying contents of previous reports

The following criminal penalties exist for reporting requirement violations under the CERCLA:

- Possible fines according to Title 18 of the United States Criminal Code and up to 3 years imprisonment/first offense or 5 years imprisonment/subsequent offenses for failure to notify, submitting false or misleading information, or destroying or falsifying evidence (Section 103[b][2])

Superfund Amendments and Reauthorization Act Title III Reporting Requirements

SARA Title III Section 304 requires notification of State and local response authorities when releases of CERCLA hazardous substances and SARA Title III extremely hazardous substances (EHSs) extend beyond the site boundary. (On-site releases that do not migrate off-site are exempt from this requirement). This reporting requirement is in addition to CERCLA reporting requirements. Therefore, a release may require notification of both the National Response Center (under CERCLA) and the State and local emergency planning bodies (under SARA Title III).

The applicability of the SARA Title III notification requirements is exactly the same as CERCLA, with 4 exceptions:

1) The SARA Title III reporting requirements apply to CERCLA hazardous substances and extremely hazardous substances (EHSs) listed under SARA Title III. Lists of these substances and their RQs are contained in 40 CFR, Table 302.4 and 40 CFR 355, Appendix, respectively.

2) The SARA Title III requirements apply only to facilities—vessels are excluded.

3) SARA Title III does not contain a petroleum exclusion. Therefore, if a petroleum contains a listed EHS or CERCLA hazardous substance, which is released above its RQ, notification is required.

4) Releases that only result in exposure to persons solely within the boundaries of the facility are exempt from SARA Title III reporting requirements. CERCLA requires notification of a release regardless of the extent of its impact.

For releases of EHSs and CERCLA hazardous substances that equal or exceed their RQs, facility owners/operators must immediately notify the State Emergency Response Commission (SERC) and the Local Emergency Planning Committee (LEPC). If the release involves a CERCLA hazardous substance, the NRC must be notified as well to comply with CERCLA reporting requirements. The initial SARA Title III notification must be followed by a written report that describes the release, response actions, known or anticipated health risks, and recommendations for medical attention of exposed individuals.

For transportation-related releases, and SARA Title III LEPC notification requirement, the notification requirements may be met by providing the required information to the 911 operator, or, in the absence of a 911 emergency operator, to the telephone operator.

Regulations specify that, in the event of a reportable release of SARA Title III extremely hazardous substances (EHSs), an initial report and follow-up report must be received by the National Response Center (NRC), State Emergency Response Commission, Local Emergency Planning Committee or fire department. The facility emergency coordinator reports immediately following releases of a reportable quantity and written follow-up within 30 days and additional written follow-up within 60 days are required by reporting deadlines (40 CFR 355.40, et seq.).

Table 2 provides a summary of the contents of the emergency notification that must be provided to appropriate agencies to satisfy requirements for notification. This information is similar to the information required under both CERCLA and CWA notification requirements previously discussed. Similarly, if there is potential for injury, additional newspaper notice is required to notify potentially injured parties.

The first follow-up written report must include

- Initial report contents
- Name and telephone number of person in charge

- NRC or EPA case number
- Facility or company Dun and Bradstreet number
- Source and frequency of release
- Population density
- Sensitive populations and sensitive eco-systems within one mile
- Releases over past the year
- Basis for stating that release is continuous and stable
- Port of registration if release is from a vessel

Additional follow-up reports should include

- Name of person reporting
- Case number
- Information verifying contents of previous reports

The following civil and criminal penalties exist for reporting requirement violations under the SARA Title III (EPCRA):

- Civil penalties up to $25,000 fine for failing to provide emergency notification
- Criminal penalties of up to $25,000 and 2 years imprisonment/first offense or $50,000 and 5 years imprisonment/subsequent offenses for willfully failing to provide emergency notification (40 CFR 355.50)

Resource Conservation and Recovery Act Subtitle I Reporting Requirements

Notification is required for releases from underground storage tank (UST) systems under RCRA Subtitle I. The RCRA regulations define a *release* as any *spilling, leaking, emitting, discharging, escaping, leaching, or disposal from an UST into groundwater, surface water, or subsurface soils.* *Regulated substances* for which notification is necessary are petroleum (including crude oil, crude oil fractions, and petroleum-based substances, and CERCLA hazardous substances [excluding RCRA hazardous wastes]).

UST systems regulated by the notification requirements include underground storage tanks for which more than 10% of the volume of its tank and associated piping is underground and that contain regulated substances. The notification requirements also apply to associated underground piping and ancillary equipment, and containment systems. Farm or residential tanks (1,000 gallon capacity or less), tanks used to store heating oil for consumptive use on the premises where it is stored, septic tanks, and certain pipeline facilities are excluded from the requirements.

Notification is required for the following situations:

- Suspected releases discovered through release detection methods, unusual operational conditions, or evidence of contamination

- Confirmed releases

- Spills or overfills of a hazardous substance that equal or exceed its RQ, or that are less than the RQ if the owner/operator is unable to clean up the release within 24 hours

- Spills or releases of 25 gallons of petroleum that the owner/operator is unable to clean up within 24 hours

- Spills or overfill of petroleum resulting in a sheen on nearby surface water

When a regulated release occurs, the owner/operator of the UST system must report to the implementing agency (usually a State agency) by telephone or electronic mail within 24 hours, unless otherwise specified by the implementing agency. The NRC must also be immediately notified if a CERCLA hazardous substance is released in a quantity greater than or equal to its RQ, or if a petroleum release has created a sheen on a nearby surface water. A written report summarizing response actions and plans for future remediation must be submitted to the implementing agency within 20 days after the release, unless otherwise specified (40 CFR 280.50).

EPA may issue compliance orders or file civil actions against owner/operators who fail to comply with the notification requirements. Failure to comply with a compliance order may result in penalties of up to $25,000 for each day of noncompliance.

Complying with Reporting Requirements

As a best management practice to ensure proper notification of the appropriate authorities, it is recommended that individuals/companies

- Determine which statutes regulate the hazardous substances that are on-site.

- Maintain records that document where all regulated substances are located at the facility.

- Create a list of RQs for substances managed on-site, and post it at all locations where hazardous substances are present.

- Determine the constituents of all mixtures managed on-site, the concentrations of regulated constituents in the mixtures, and their RQs.

- Determine the RQs for all waste streams.

- Designate a person or alternate responsible for reporting releases, and ensure that the designated individual is familiar with proper notification requirements. That person or designated alternate is the only person to notify the NRC and SERC. Calling 911 in the event of an emergency should be at a supervisor's discretion.

- Post by the telephone the numbers of all agencies to whom reporting may be necessary.

- Check and reconcile product inventory routinely.

- If a release occurs, keep detailed records of any actions taken or planned.

- If there is any doubt about whether reporting is required, take a conservative approach and consider reporting anyway to avoid potential violations.

The advice of counsel should be obtained in the event of a spill, to avoid any penalties for failure to properly report the release. Failure to report releases under the various applicable Federal and State laws and regulations may result in substantial civil and criminal penalties. Therefore preplanning, development, and implementation and testing of notification plans and procedures detailing incident, response, and mitigative measures are imperative.

It is implicit in the requirement to notify *immediately* that the notifier knows that the release occurred. However, EPA has stated that *if the facility owner/operator should have known of the release, then the fact that he or she was unaware of the release will not relieve the owner/operator from the duty to provide release notification.*

Significant fines have been levied involving incidents where responsible parties waited until they could determine if an RQ had been released. If the possibility that a release may be of a RQ, or if it is of a substance with an RQ and the amount is not immediately known, strong consideration must be given to initiating the notification process. However, there are no penalties for notifying the agencies (NRC, SERC, and LEPC) and then showing in the follow-up report that an RQ was not released.

Fundamentally, requirements for initial notification can be met by calling

- 911 or operator—this obtains emergency assistance and satisfies the requirement to notify the LEPC

- (800) 424–8802—this satisfies the requirement to notify the National Response Center (Duty Officer of the US Coast Guard—the ultimate Incident Commander in the event of a major incident)

- The State Emergency Response Commission

- Any other affected LEPCs or SERCs (if near a State or LEPC boundary)

- CHEMTREC—if it is a transportation incident (800) 424–9300

- The local poison control center, if persons have been exposed

Calling CHEMTREC, or using any of the emergency guides is only a first resort. CHEMTREC and other contracted response agencies will only read the Material Safety Data Sheet. They are not able to give additional information. The poison control center will give medical advice that may be of assistance in making decisions such as evacuation or shelter-in-place.

With today's advances in communication technology, a facility with significant release risks might decide to have a one-to-many broadcast call preprogrammed to all required agencies simultaneously. The call must be initiated at the earliest instant that the sort of information contained in Table 2 is assembled, but well within the deadlines imposed by regulations discussed above.

An emergency notification procedure should be included as a component of any environmental risk management program, to ensure that a facility has an effective notification process. A simple but well-designed emergency notification procedure can ensure compliance and assist in preventing or minimizing a hazardous materials incident, thus avoiding or reducing injuries, loss of life, environmental damage, and creation of a public relations disaster. Having developed and implemented a procedure will help ensure appropriate response to the emergency notification requirements of a reportable accidental spill or release of hazardous substances into air, soil, or water under the pressure of a pending notification deadline.

For additional information, questions concerning SARA Title III or Section 112(r) of the CAA can be addressed to the Emergency Planning and Community Right-to-Know Information Hotline

USEPA (5104)
410 M Street, SW
Washington DC, 20460
(800) 424–9346 or (703) 412–9810
TDD: (800) 553–7672)
9:00 am to 6:00 pm, EST, M–F

A valuable desk reference is

Environmental Protection Agency, *Title III List of Lists—Consolidated List of Chemicals Subject to the Emergency Planning and Community Right-to-Know Act (EPCRA) and Section 112 (r) of the Clean Air Act, as Amended.* EPA 550–B–96-015.

It is available through the United States EPA:

National Center for Environmental Publications and Information (NCEPI)
PO Box 42419
Cincinnati, OH 45242
(800) 490–9198

Bibliography

Business and Legal Reports, *Environmental Compliance*, Rev. 3, Washington, DC: BLR, 1997.

"Designation, Reportable Quantities, and Notification." *Code of Federal Regulations*. Title 40. Pt. 302.4.

"Discharge of Oil." *Code of Federal Regulations*. Title 40. Pt. 110.11.

Environmental Protection Agency. "Chemical Accident Prevention Provisions." *Code of Federal Regulations*. Title 40. Pt. 68.

"Emergency Planning and Notification." *Code of Federal Regulations*. Title 40. Pt. 355.

"Immediate Notice of Certain Hazardous Materials Incidents, Detailed Hazardous Materials Incident Report." *Code of Federal Regulations*. Title 49. Pt. 171.15–171.16.

"Polychlorinated Biphenyls (PCBs) Manufacturing, Processing, Distribution in Commerce, and Use Prohibitions." *Code of Federal Regulations*. Title 40. Pt. 761.30(a).

"Requirements for PCB Spill Cleanup." *Code of Federal Regulations*. Title 40. Pt. 761.125.

"Standards Applicable to Transporters of Hazardous Waste." *Code of Federal Regulations*. Title 40. Pt. 263.30–263.31.

"Technical Standards and Corrective Action Requirements for Owners and Operators of Underground Storage Tanks (UST)." *Code of Federal Regulations*. Title 40. Pt. 280.60–280.67.

Thomas F. P. Sullivan, Ed. *Environmental Law Handbook*, 13th ed. Rockville, MD: Government Institutes, 1995.

"Transmission and Gathering System: Incident Report." *Code of Federal Regulations*. Title 49. Pt. 191.15.

Internet Resources

<http://tis-nt.eh.doe.gov/oepa/guidance/alldocs.htm> (Department of Energy, Office of Environmental Policy and Administration)

<http://www.epa.gov/ncepihom/orderpub.html> (Environmental Protection Agency, National Service Center for Environmental Publications, Request Form)

<http://www.access.gpo.gov/nara/cfr/cfr-table-search.html> (National Archives and Records Administration, Code of Federal Regulations)

Alan A. Eckmyre has a BS in Business Administration from the New York Institute of Technology, a BE in Nuclear Engineering from the State University of New York, an MBE from Emporia State University, and an MS in Environmental Engineering from the University of Kansas. Mr. Eckmyre has been a Certified Hazardous Materials Manager since 1991 and has served as the Chair of the national Academy Publications Committee, Secretary and Director of the Heartland Chapter in Kansas, and President of the Magnolia Chapter in South Carolina. He is also a member of the American Society of Testing and Materials' Environmental Assessment Committee. Mr. Eckmyre has over 20 years of experience in nuclear and environmental engineering, consulting, business development, and program management. His project management experience includes: UST system closures, State UST ground water/soil remediation investigation and design projects, Phase I and II site assessments, Federal facility specific and programmatic environmental audits, RCRA permitting and corrective measures studies, CERCLA preliminary assessments and Remedial Investigation/Feasibility Studies, NPDES permitting, construction remediation, expert witness testimony, and managment of technical staff. He currently provides environmental engineering support to DOE–HQ EM–40 programs. Mr. Eckmyre has published a number of reports relating to field evaluation, planning, and analysis of the West Valley Demonstration Project, Tar Creek Superfund Site, Paducah Gaseous Diffusion Plant, Savannah River Federal Facility, Culvert Cliffs Nuclear Power Plant, and the Department of Energy's spent nuclear fuel and waste management programs.

CHAPTER 15

Hazard Communication Standard

Charles M. Bessey, CHMM

Introduction

Approximately one in every four workers comes into contact with hazardous chemicals on the job. In many cases, the chemicals may be no more dangerous than those in use in the typical household, but in the workplace, exposure is likely to be greater, concentrations higher, and exposure times much longer. Because worker exposure to chemicals can cause or contribute to many serious safety and health problems, the Occupational Safety and Health Administration (OSHA) issued the Hazard Communication Standard or the Worker Right-To-Know Standard.

The standard, found in the *Code of Federal Regulations* at 29 CFR 1910.1200, establishes uniform requirements to ensure that the hazards of all chemicals imported, produced, or used in the workplace are fully evaluated for possible physical or health hazards, and that this hazard information is transmitted to affected employers and exposed workers. The standard preempts inconsistent state and local laws and provides for a national law to simplify communication of chemical hazards.

All employees need to learn about the chemicals they work with and how to take precautions against the negative effects associated with them. Chemical manufacturers and importers must communicate the known hazard information they learn from their evaluations to downstream companies by means of Material Safety Data Sheets (MSDSs) and container labeling. In addition, all covered companies must have a written hazard communication program and provide this information to their employees.

Requirements

The basic requirements specifically covered in the standard include

- Hazard determination
- Material Safety Data Sheets
- Labels and labeling
- Written Hazard Communication Program
- Informing and training employees
- Trade secret provisions

Hazard Determination

The Hazard Communication Standard covers all employees who may be exposed to hazardous chemicals under normal working conditions or where a chemical emergency might occur. Manufacturers and importers must evaluate their chemicals by assessing available scientific information or optionally performing laboratory tests.

At a minimum, *hazardous chemicals* include all chemicals listed by OSHA with a Permissible Exposure Limit (PEL) or by the American Conference of Governmental Industrial Hygienists (ACGIH) with a Threshold Limit Value (TLV) or those listed in the National Toxicology Program (NTP) Annual Report on Carcinogens (latest edition) or that have been found to be a potential carcinogen in the International Agency for Research on Cancer (IARC) Monographs (latest editions), or by OSHA.

The standard does not apply to hazardous waste, tobacco or tobacco products, wood or wood products, food, drugs or cosmetics that are intended for personal use or, **articles.** Articles are manufactured items that are formed to specific shapes during manufacture. They have end-use functions that are dependent upon their shapes during end use. They do not release or otherwise result in exposure to a hazardous chemical under normal conditions of use.

The standard has limited requirements for laboratory workers; such as, they must maintain MSDSs; verify that incoming containers are labeled and that employees are trained. For most laboratories, OSHA's Occupational Exposure to Hazardous Chemicals in Laboratories, 29 CFR 1910.1450, applies depending on the criteria, as defined in the standard.

A work-site risk assessment and inventory of chemicals should be the starting point of the written program. Identification of piping system chemicals and potential in-house reactants (like welding fumes, carbon monoxide from combustion sources, and wood and metal dusts from cutting and grinding operations) should be categorized. A determination of nonroutine tasks in the work-place should be identified and evaluated for chemical exposures.

A *physical hazard* means a chemical for which there is scientifically valid evidence that it is

- Flammable liquid or solid
- Combustible liquid
- Compressed gas
- Explosive
- Organic peroxide
- Oxidizer
- Pyrophoric material (may spontaneously ignite in air at 130°F or less)
- Reactive (unstable) material
- Water reactive

A *health hazard* means a chemical for which there is statistically significant evidence, based on at least one study conducted in accordance with established scientific principles, that acute or chronic health effects may occur in exposed workers. Health hazards may not necessarily cause immediate, obvious harm or even make you sick right away. You may not always see, feel, or smell the danger.

Health hazards are categorized as

- Toxic or highly toxic
- Reproductive toxins
- Carcinogens
- Irritants
- Corrosives
- Sensitizers
- Hepatotoxins (liver)
- Nephrotoxins (kidney)
- Neurotoxins (nerve)
- Hematopoietic (blood)
- Other agents that damage the lungs, eyes, skin, and mucous membranes

If a material is a mixture of chemicals that has not been tested as a whole, the evaluator may use whatever valid data is available on the constituents of the mixture to assess the physical and health hazards. A mixture must be assumed to have the same hazards for carcinogens at 0.1% of the mixture and for other components at 1% of the mixture to have the same hazards.

If components are present in lesser amounts, but could be released in concentrations that would exceed the permissible exposure limit (PEL) or Threshold Limit Value (TLV), the evaluator must assume that those hazards exist. The procedure used to evaluate chemical hazards must be recorded in writing.

Material Safety Data Sheets

Material Safety Data Sheets must be prepared or obtained by manufacturers or importers and provided to downstream manufacturers and distributors with the initial shipment and after each MSDS update or revision. Employers must have an MSDS for every hazardous chemical they use and have copies readily available to workers during all shifts.

There is no format for Material Safety Data Sheets but they must be in English, contain no blank fields, and contain at a minimum the following information:

- Manufacturer/importer identification and phone number
- Identity used on the label
- Specific and common chemical names
- Physical and chemical characteristics
- Physical and health hazards
- Primary routes of entry
- Exposure limits (PEL-TLV)
- Whether it is a carcinogen
- Emergency and first-aid procedures
- Safe handling precautions
- Applicable control measures
- Last revision date

Labels and Labeling

Labels must be in English and identify the hazardous chemicals with the appropriate hazard warnings as well as the manufacturer. On individual stationary containers; signs, placards, batch tickets, or printed operating procedures may be used in place of labels. An employee is not required to label a portable transfer vessel where the material is intended only for use by the same worker during his normal work shift.

Written Hazard Communication Program

Employers must develop and implement a written hazard communication program for the hazardous chemicals present in the workplace. The program must describe how the employer will meet the standards requirements for labeling, the type of labeling, MSDS management, and a list of known chemicals in the workplace. It also must contain provisions for employee information and training, including nonroutine tasks and contractor information.

Training and Information

The employer must inform workers of the requirements of the standard, the operations where hazardous materials are present, and the location and availability of the written program. Employees must be trained at the time of employment and whenever new hazards are introduced to the workplace. Training must include methods of observation used to detect the presence or release of hazardous chemicals, and the protective measures an employee can take, including the use of appropriate personal protective equipment.

Employers must also define how outside contractors will be informed of potential exposure to hazardous materials in their work areas, and suggest the appropriate protective measures that are to be employed.

Trade Secrets

The standard strikes a balance between the need to protect exposed workers and the employer's need to maintain the confidentiality of trade secrets. This is achieved by providing, under specified conditions of need and confidentiality, for limited disclosure to health professionals who are furnishing medical services, or to other employee health service providers and their designated representatives.

While trade secrets can be maintained, the specific hazards must be identified and disclosed. The responsible parties must provide the specific chemical identity to a doctor or nurse if requested in an emergency. A confidentiality agreement may be requested later. In nonemergency situations, health professionals must request trade secret identifications in writing and be willing to sign a confidentiality agreement.

Bibliography

"Hazard Communication, Final Rule." *Federal Register*. 59 (13 April 1994): 17487–17479.

"Hazard Communication, Correction." *Federal Register*. 59 (22 December 1994): 65947–65948.

"Hazard Communication Standard." *Code of Federal Regulations*. Title 29. Pt. 1910.1200.

"Office of Management and Budget Control Numbers under the Paperwork Reduction Act, Final Rule." *Federal Register*. 61 (13 February 1996): 5507–5510.

"Toxic & Hazardous Substances." *Code of Federal Regulations*. Title 29. Pt. 1910, Subpart Z.

Charles M. Bessey has been with Kolene Corporation for over 32 years, and is the Supervisor of Special Projects. Kolene Corporation is the world leader in high-temperature molten salt technology, equipment, and process chemicals. His major responsibilities include serving as Chairman of the Corporate Safety Committee, which addresses internal and external safety and health concerns, accident/incident review, and liability issues.

CHAPTER 16

MSDS Management and Development

Eldon L. Burkett, PE, CSP, CHMM

Introduction

A *Material Safety Data Sheet* or *MSDS*, is a document that provides the user with important information about a chemical product. MSDSs are developed for products that contain one chemical substance, or many chemical substances combined. MSDSs are required to contain information about

- The chemical's identity, or product name
- The names of composition chemicals
- Percentages of certain listed chemicals (40 CFR 372.45)
- Health hazards
- Physical hazards
- Personal protective equipment (PPE) required for users of the product
- Engineering controls
- First aid treatment
- Emergency planning (*e.g.*, spill control)

Information contained in the MSDS may be used for compliance with the many Federal Occupational Safety and Health Administration (OSHA), the Environmental Protection Agency

(EPA), and the Department of Transportation (DOT) regulations, as well as for limiting injuries, preventing illnesses, and controlling property loss.

Management

MSDSs may be kept in notebooks, computer CD-ROMS, or on-line (*e.g.,* the Internet). MSDSs may be maintained for the entire facility or by work area within the facility.

The MSDS Format

MSDSs may be in any format and style, but must be in English. Although there is not a specific format for MSDSs, a number of organizations have developed recommended formats. For instance, the National Institute for Occupational Safety and Health (NIOSH) has proposed a four-page MSDS format, and OSHA originally issued a two-page format, Form 20, in 1983. Form 20 was later replaced by an improved two-page format, Non-Mandatory Form 183, in 1985. The American National Standards Institute (ANSI) has proposed a sixteen-section MSDS format. ANSI recommends the following format:

- Section 1: Chemical Product and Company Identification
- Section 2: Composition, Information On Ingredients
- Section 3: Hazards Identification
- Section 4: First Aid Measures
- Section 5: Fire Fighting Measures
- Section 6: Accidental Release
- Section 7: Handling and Storage
- Section 8: Exposure Controls, Personal Protection
- Section 9: Physical Properties and Chemical Properties
- Section 10: Stability and Reactivity
- Section 11: Toxicological Information

- Section 12: Ecological Information
- Section 13: Disposal Consideration
- Section 14: Transport Information
- Section 15: Regulatory Information
- Section 16: Other Information

Notification and Changes of Information

Any time a hazardous material is shipped or transported from the manufacturer, wholesaler, or retailer, an MSDS must accompany the shipment. If the hazardous chemical product arrives and no MSDS is on file or received with the shipment, the employer is required to obtain an MSDS from the manufacturer or importer. Many companies have adopted the policy of refusing chemical shipments if an MSDS is not included with the shipment. The manufacturer or importer is responsible for updating the MSDS when they become aware of any significant new information regarding the hazards of a chemical and/or ways to protect personnel against the hazard(s) that the material may pose. These changes must be completed on the MSDS within three months, and the MSDS preparation date must be updated as well. Any time new hazards are identified for a substance, it is the responsibility of the employer to retrain all affected employees on these new hazards.

A copy of a company's MSDSs must be provided to the Local Emergency Planning Committee (LEPC) and/or the fire department (40 CFR 370.20[b][3]). When a product is changed or new hazard information becomes available, a new and/or revised MSDS must be provided to the LEPC within three months. If the LEPC requests an MSDS, a facility must provide that MSDS within 30 days of the request.

Availability

MSDSs must be available for employees who work with or around hazardous chemicals. MSDSs must be kept in the workplace and be readily accessible to employees during each work shift. If employees travel to more than one geographical location, MSDSs can be kept at the primary workplace facility.

MSDS Retention

MSDSs may be used for meeting the OSHA *Access to Employee Exposure and Medical Records* requirement for identifying toxic substances in exposure and medical records. The MSDS may be discarded after a product is no longer used, or it may be replaced by a newer version if the formula is not changed. However, the original and subsequent chemical formulas (compositions) must be kept on file for more than 30 years (29 CFR1910.1020[d][1][ii][B]). Discarding an MSDS can result in lost data and documentation that could be used later to show the date that the employer was notified of new health and physical hazards arising from changes in product formulation. All MSDSs should be archived and stored for at least 30 years after the date of the product's last use in order to minimize employer liability.

Also, OSHA "does not require nor encourage employers to maintain MSDSs for nonhazardous chemicals. Consequently, an employer is free to discard MSDSs for nonhazardous chemicals." This data can and should be saved, to verify the employer's awareness of hazard information.

Application and Use

MSDSs were originally designed to provide specific safety information to emergency personnel and to employees who handle hazardous chemicals. This information is now made available to the public as well.

The chemical composition outlined in the MSDS may be used to determine whether the product is regulated under chemical lists and/or if it has a listed hazardous physical or health characteristic that is regulated by any agency. The following are *examples* of the regulations that may apply to hazardous materials.

OSHA (29 CFR)

- Compressed Gases–29 CFR 1910.101

- Dip Tanks Containing Flammable or Combustible Liquids–29 CFR 1910.108

- Flammable and Combustible Liquids–29 CFR 1910.106

- Medical Services and First Aid–29 CFR 1910.151

- Process Safety Management of Highly Hazardous Chemicals–29 CFR 1910.119

- Respiratory Protection–29 CFR 1910.134

- Spray Finishing Using Flammable and Combustible Materials–29 CFR 1910.107

- Ventilation–29 CFR 1910.94

- Air Contaminants–29 CFR 1910.1000 Series

EPA (40 CFR)

- Clean Air Act (CAA)

- Clean Water Act (CWA)

- Emergency Planning and Community Right-to-Know Act (EPCRA)

- Reportable Quantities (RQ)–Comprehensive Environmental Response, Compensation and Liability Act (CERCLA)/ Emergency Planning and Community Right-to-Know Act (EPCRA)

- Resource Conservation and Recovery Act (RCRA)

DOT (49 CFR)

- Classification of hazardous material–shipping papers with technical name(s)

- Reportable Quantities–RQ

- Chemical Storage (Incompatibles)

Development

The chemical manufacturer or importer of a hazardous product is required to develop or obtain an MSDS for their product. When products have similar hazards and contents (even if the chemical concentrations vary), the same MSDS may be used for these products. The MSDS must contain the following information:

- In the ingredient section, the chemical name (*i.e.*, International Union of Pure and Applied Chemistry [IUPAC], the Chemical Abstracts Service [CAS] Rules of Nomenclature, and/or a name which will clearly identify the ingredient) for each component must be given. If a trade secret is claimed, it must be marked as a trade secret, proprietary formulation, or the like.

- MSDSs must be written in English, although other languages may be used in addition.

- An identity (*i.e.*, cross reference, product name) must be used for the hazardous chemical list, label, and MSDS.

- The date the MSDS was prepared or last revised or amended.

- The preparer and/or distributor of the MSDS must list

  - name

  - address

  - telephone number

  Note: The DOT requires "[a] person who offers a hazardous material for transportation to provide a 24-hour emergency response telephone number (including the area code or international access code) for use in the event of an emergency involving the hazardous material" (49 CFR 172.604).

- Emergency containment and control procedures (29 CFR 1910.120).

- First aid procedures including the OSHA requirements for medical services and first aid (29 CFR 1910.151).

- If the product contains a chemical on the American Conference of Governmental Industrial Hygienists (ACGIH) Threshold Limit Values (TLV) for Chemical Substances list, the OSHA Toxic and Hazardous Substances, or any other known list of exposure limits, the limit(s) must be given. This data can be used for OSHA Respiratory Protection (29 CFR 1910.134), Ventilation (29 CFR 1910.95), and Flammable and Combustible Liquids standards (29 CFR 1910.106).

- The general applicable precautions for safe handling and use of the products, including

  - hygienic practices (*e.g.*, number of employees per shower and sinks) (29 CFR 1910.141)

  - PPE requirements including the specific type of glove, suit, apron, and respirator (29 CFR 1910.38)

  - protective measures for the repair and maintenance of contaminated equipment (the information supplements service bulletins, product information sheets, and sales literature) (29 CFR 1910.95)

  - the generally known control measures, including engineering practices and controls, work practices (administrative controls), and PPE

If no relevant information can be found for a section of the MSDS, it must be marked (*e.g.*, No Data, Not Applicable). Blank sections are *not* acceptable.

Hazard Determination

Hazard determination pertains to any chemical which is known to have physical and/or health hazards. The MSDS must address the hazards which could be present in the workplace under normal conditions of use or in a **foreseeable emergency**. A foreseeable emergency is any potential occurrence (*e.g.*, equipment failure, rupture of containers, *etc.*) which could result in an uncontrolled release of a hazardous chemical into the workplace. Most of the physical hazard characteristics (*e.g.*, flash point, flammable, oxidizer, high-pressure gas, water reactive, *etc.*) have been identified and defined under DOT regulations. The criteria for health hazards are similar to the toxic substance criteria, including those chemicals thst are listed by NIOSH–Registry of Toxic Effects of Chemical Substances (RTECS), and/or have yielded positive evidence of an acute or chronic health hazard. These health hazards should have been reported by the manufacturers and importers when filing EPA's Toxic Substances Control Act (TSCA) Premanufacture Notification (PMN) (40 CFR 720.50) for new chemical substances.

When determining health and physical hazards, the following references should be consulted:

- ACGIH–TLV (American Conference of Governmental Industrial Hygienists, "Threshold Limit Values for Chemical Substances and Physical Agents in the Work Environment")

- California Proposition 65, The Safe Drinking Water and Toxic Enforcement Act of 1986

- Consumer Product Safety Commission (16 CFR)

- DOT (49 CFR)

- EPA (40 CFR)

- IARC (International Agency for Research on Cancer) "Monographs"

- NTP (National Toxicology Program) "Annual Report on Carcinogens"

- OSHA (29 CFR)

- NIOSH Registry of Toxic Effects of Chemical Substances (RTECS)

Additional references are listed in the Bibliography section. Health data may have been disclosed for each chemical component in EPA's Premanufacture Notification (PMN) under TSCA.

Health Hazards

There is an array of potential health hazards associated with exposure to hazardous chemicals. The effects from exposure may be either acute or chronic. Please refer to the chapter entitled "Industrial Toxicology," for more information on this subject.

According to the regulations,

> evidence which is statistically significant, which is based on at least one positive study conducted in accordance with established scientific principles is considered to be sufficient to establish a hazardous effect (29 CFR 1910.1200[d][2]).

The hazardous chemical components are present in the mixture of 1 or more percent (or, in the case of carcinogens, 0.1 or more percent) and/or could release to the air concentrations of the chemicals that present health risks (29 CFR 1910.1200[d][5][iv]). These chemicals must be listed for all the health hazards as if they were 100% of the product, unless the product has been tested in whole for the specific mixture (29 CFR 1910.1200 [d][5][iii]).

Under other regulations, the concentration for reporting may be lower. For example,

- DOT–RQ can be as low as 0.002 percent (49 CFR 171.8).

- California–Proposition 65 lists carcinogens and reproductive toxicants at any concentration.

- OSHA–Formaldehyde can be as low as 0.05 percent in a mixture (29 CFR 1910.1048 [1][8][i] and [n]). For formaldehyde, the OSHA Action Level is 0.5 ppm in air and the OSHA Permissible Exposure Limit (PEL) is 0.75 ppm.

Physical Hazards

Just as a hazardous chemical may pose many health hazards, it may also pose physical hazards. Please refer to the chapter on the chemistry of hazardous materials for more information. The chemical manufacturer or importer must conduct a hazard assessment and determine that the chemical product is not hazardous (as defined by OSHA, DOT, *etc.*). Such chemical products are not regulated by any jurisdictional entity and an MSDS is not required.

Chemical Components—Trade Secrets

OSHA defines a *trade secret* as any formula, pattern, device, or compilation of information which is used in one's business, and which affords an opportunity to obtain an advantage over competitors who do not know or use it.

The composition must list each hazardous constituent by chemical name (*i.e.*, International Union of Pure and Applied Chemistry [IUPAC], the Chemical Abstracts Service [CAS] Rules of Nomenclature, and/or a name which will clearly identify each chemical). If a manufacturer claims trade secret status for a chemical product, it must be proclaimed as a trade secret, proprietary formulation, or the like. The hazardous properties and effects of the product must still be listed. There are many other factors where the chemical composition must be disclosed under DOT (shipping name, technical name) and in the MSDS (ACGIH–TLV, the OSHA–Permissible Exposure Limit [PEL]) under the Emergency Planning and Community Right-to-Know Act (EPCRA)(the Reportable Quantities [RQs])(40 CFR 372.45[b]).

For a chemical to be considered for trade secret status, the following factors must be considered:

- The extent to which the information is known outside of the business

- The extent to which it is known by employees and others involved in the business

- The extent of measures taken by the business to guard the secrecy of the information

- The value of the information to competitors

- The amount of effort or money expended by the business in developing the information

- The ease or difficulty with which the information could be properly acquired or duplicated by others

According to the EPCRA, any trade secret compound must be filed with the EPA. Provide th following information to EPA when claiming a trade secret:

1) Describe the specific measures taken to safeguard the confidentiality of the chemical identity claimed as a trade secret, and indicate whether these measures will continue in the future.

2) Report whether the information claimed as a trade secret has been disclosed to any other person (other than a member of a Local Emergency Planning Committee, officer or employee of the United States or a State or local government, or an employee) who is not bound by a confidentiality agreement to refrain from disclosing this trade secret information to others.

3) List all local, State, and Federal government entities to which this trade secret information has been disclosed. Also, indicate whether a confidentiality claim for the chemical identity has been asserted and whether the government entity denied that claim.

4) Demonstrate the validity of a trade secrecy claim by identifying the specific use of the chemical claimed as a trade secret and explaining why it is of interest to competitors.

i) Describe the specific use of the chemical claimed as a trade secret, identifying the product or process in which it is used. (If the chemical is used in a manner other than as a component of a product or in a manufacturing process, identify the activity where the chemical has been used.)

ii) Report whether the identity of the company or facility has been linked to the specific chemical claimed as a trade secret in a patent, in publications, or in other information sources that are available to the public or competitors. (The company must explain why this knowledge does not eliminate the justification for trade secrecy.)

iii) Explain, if the chemical claimed as a trade secret is unknown outside the company, how the competitors could deduce its use from disclosure of the chemical identity together with other information on the Title III submittal form.

iv) Explain why the use of the chemical claimed as a trade secret would be valuable information to competitors.

v) Indicate the nature of the harm to the business's competitive position that would likely result from disclosure of the specific chemical identity, and indicate why such harm would be substantial.

5) Indicate the extent to which the chemical claimed as a trade secret is available to the public or competitors in products, articles, or environmental releases.

6) Indicate to what extent the chemical claimed as a trade secret is available to

the public or to competitors in products, articles, or environmental releases.

7) Describe the factors which influence the cost of determining the identity of the chemical claimed as a trade secret by chemical analysis of the product, article, or waste which contains the chemical (*e.g.*, whether the chemical is in pure form or is mixed with other substances)(40 CFR 350.7[a]).

Finally, while OSHA does not require a company to report the percentage of each hazardous chemical contained in a mixture, EPA's EPCRA requires the company to report these percentages for listed chemicals.

Summary

A complete and accurate MSDS is a good tool for identifying, managing, and controlling injuries, illness, and property loss due to the effects of hazardous chemicals, while meeting OSHA, EPA, and DOT regulations.

The MSDS should provide the chemical user with: information on the physical and health hazards associated with the chemical; precautions to take when handling the product; the treatment for overexposure to the chemical; and a source to contact for more information about the product.

Bibliography

The following printed publications may be useful for preparing MSDSs and handling hazardous chemicals. Additional resources include the Internet; governmental regulations; and safety, industrial hygiene, toxicology, medical, and engineering publications.

Alden, J. L. and J. M. Kane. *Design of Industrial Ventilation Systems*. New York, NY: Industrial Press, 1982.

American Conference of Governmental Industrial Hygienists. *1999 TLV's and BEI's*. Cincinnati, OH: ACGIH, 1999.

American Conference of Governmental Industrial Hygienists, Committee on Industrial Ventilation. *Industrial Ventilation, A Manual of Recommended Practice*. Cincinnati, OH: American Conference of Governmental Industrial Hygienists, new editions issued biennially.

American Industrial Hygiene Association. *1998 Emergency Response Planning Guidelines and Workplace Environmental Exposure Levels Guides*. Fairfax, VA: AIHA, 1998.

Bollinger, N. J. and R. H. Schutz. *NIOSH Guide to Industrial Respiratory Protection*. NIOSH Publication No. 87–116. Cincinnati, OH: GPO, 1987.

Budavari, S., Ed. *Merck Index: An Encyclopedia of Chemicals and Drugs*. 12th ed. Whitehouse Station, NJ: Merck & Co., Inc., 1996.

Burgess, W. A. *Recognition of Health Hazards in Industry: A Review of Materials and Processes*. New York, NY: Wiley-Interscience, 1995.

Clayton, George D. and F. E. Clayton. *Patty's Industrial Hygiene and Toxicology: Volume 1, Parts A and B, General Principles*. New York, NY: John Wiley and Sons, 1991

Colton, C. E., L. R. Birkner, and L. M. Brosseau. *Respiratory Protection: A Manual and Guideline*. Fairfax, VA: AIHA, 1991.

Cralley, L. V. and L. J. Cralley, Eds. *In-Plant Practices for Job-Related Health Hazards Control*, New York: John Wiley & Sons, 1989.

Garrett, J. T., L. J. Cralley and L. V. Cralley, Eds. *Industrial Hygiene Management*. New York, NY: John Wiley & Sons, 1988.

Gleason, R. E., Roger P. Smith, Harold C. Hodge, and Jeanette Braddock, *Clinical Toxicology of Commercial Products*, Baltimore, MD: Williams and Wilkins, 1984.

Grant, W. Morton, and Joel S. Schuman. *Toxicology of the Eye*. Springfield, IL: Charles C. Thomas, 1993.

Harbison, Raymond D. and Harriet L Hardy, Eds. *Hamilton and Hardy's Industrial Toxicology*, 5th ed. Mineola, NY: Mosby-Year Book, 1998.

Hathaway, Gloria J., John P. Hughes and James P. Hughes. *Chemical Hazards of the Workplace.* New York, NY: John Wiley and Sons, 1996.

Hemeon, W. C. and D. Jeff Burton. *Hemeon's Plant and Process Ventilation.* Boca Raton, FL: Lewis Publishers, 1998.

International Agency for Research on Cancer. *IARC Monographs on the Evaluation of the Carcinogenic Risk of Chemicals to Man, Geneva: World Health Organization, International Agency for Research on Cancer, 1972 – Present.* (Multivolume work). Summaries are available in supplement volumes. Albany, NY: IARC.

Klaassen, C. D., Ed. *Casarette and Doull's Toxicology – The Basic Science of Poisons.* 5th ed. New York, NY: McGraw-Hill, 1996.

Levine, S. P. and W. F. Martin, Eds. *Protecting Personnel at Hazardous Waste Sites.* Boston, MA: Butterworth-Heinemann, 1994.

Lewis, R., Hawley's *Condensed Chemical Dictionary.* 13th ed. New York, NY: Van Nostrand Reinhold Co., 1998.

Lide, David R. *CRC Handbook of Chemistry and Physics,* 79th ed. Boca Raton, FL: CRC Press, 1998.

McDermott, H. J. *Handbook of Ventilation for Contaminant Control.* Stoneham, MA: Butterworth-Heinemann Publishers, 1991.

Molinelli, Richard P., Michael J. Reale, and Ralph I. Freudenthal, Eds. *Material Safety Data Sheets: The Writer's Desk Reference.* Boca Raton, FL: Hill and Garnett Publishing, Inc., 1992.

National Institute for Occupational Safety and Health. *Criteria for a Recommended Standard* (various topics). Cincinnati, OH: GPO, 1988–98.

National Institute for Occupational Safety and Health. *The Industrial Environment: Its Evaluation and Control.* Fairfax, VA: AIHA, 1991.

National Institute for Occupational Safety and Health and Occupational Safety and Health Administration. *NIOSH Pocket Guide to Chemical Hazards.* NIOSH Publication No. 94–116. Cincinnati, OH: GPO, 1994.

National Institute for Occupational Safety and Health, the Occupational Safety and Health Administration, the United States Coast Guard, and the Environmental Protection Agency. *Occupational Safety and Health Guidance Manual for Hazardous Waste Site Activities.* NIOSH Publication N. 85–115. Cincinnati, OH: GPO, 1985.

National Institute for Occupational Safety and Health, Superintendent of Documents. *RTECS —Registry of Toxic Effects of Chemical Substances.* Washington, DC: Government Research Center, 1998. <grc.tis.gov/rtecs/htm>

Plog, B. A., Ed. *Fundamentals of Industrial Hygiene.* Itasca, IL: National Safety Council, 1996.

Schwope, A. D., P. P. Costas, J. O. Jackson, J. O. Stull, and E. J. Weitzman. *Guidelines for the Selection of Protective Clothing.* Cincinnati, OH: American Conference of Governmental Industrial Hygienists, 1987.

Scott, Ronald McLean. *Chemical Hazards in the Workplace.* Boca Raton, FL: Lewis Publishers, 1989.

Stellman, Jeanne M., Ed. *Encyclopedia of Occupational Health and Safety,* 4th ed. Volumes I and II. Geneva, Switzerland: International Labour Office, 1998.

Stern, A.C., Ed. *Air Pollution.* Volumes I–VIII. San Diego, CA: Academic Press, 1977–86.

Technical Resource, Inc. *NTP Annual Report on Carcinogens and Summary of the Annual Report on Carcinogens.* Springfield, VA: NTIS, 1991

Wadden, R. A. and P. A. Scheff. *Engineering Design for the Control of Workplace Hazards.* New York, NY: McGraw-Hill, 1987.

Wadden, Richard and Peter A. Scheff. *Indoor Air Pollution: Characterization, Prediction, and Control.* New York, NY: Lewis Publishers, 1982.

Walsh, P. J., C. S. Dudney, and E. L. Copenhaver. *Indoor Air Quality.* Boca Raton, FL: CRC Press, 1983.

Eldon L. Burkett has earned a Bachelor of Science Degree in Chemistry from Saginaw State College (University) and is a Registered Professional Safety Engineer (PE), Certified Safety Professional (CSP–Product Liability Aspects), as well as a Certified Hazardous Materials Manager at the Master's Level. He has prepared more than 700 Material Safety Data Sheets and continues to review MSDSs for safety, health (including Industrial Hygiene), environmental, chemistry, and engineering content for controlling risks.

Part IV

Land and Natural Resources

CHAPTER 17

The National Environmental Policy Act

James L. Oliver

Introduction

The National Environmental Policy Act (NEPA), (*42 USC §4321, et seq.*) was signed into law on January 1, 1970. The Act established a national environmental policy and set goals for the protection, maintenance, and enhancement of the environment as well as providing a process for implementing these goals throughout all Federal agencies. NEPA requires all Federal agencies to prepare a *detailed statement* on proposals for major Federal actions "significantly affecting the quality of the human environment." In addition, the Act also established the Council on Environmental Quality (CEQ).

In response to widespread and growing public concern regarding the quality of the nation's environment in the late 1960s, Congress passed the National Environmental Policy Act (NEPA) of 1969 which, President Nixon signed into law on

January 1, 1970. This Act was the first major environmental legislation passed by Congress and has profoundly influenced decision-making by Federal agencies for almost 30 years.

Compliance with NEPA is not discretionary with Federal agencies; rather, compliance is subject to review in Federal courts even though NEPA does not contain provisions for legal review. The legislative history of NEPA (discussed more fully in the next section of this chapter) does not indicate whether legal enforcement was discussed during development of the act. Regardless of whether some form of legal regulatory enforcement was discussed or not, a judicial review process has been established through the Federal courts and the resulting large amount of NEPA case law. Early court decisions held that Federal compliance with NEPA's environmental decision-making responsibilities was legally enforceable, thus making the Federal courts the chief enforcers of NEPA's environmental reporting mandates. The importance of NEPA resides in its role as a basic environmental responsibility of each Federal agency and their implementing regulations. Compliance with an agency's own NEPA implementing regulations does not excuse an agency from compliance with NEPA.

Passage of NEPA has been followed by the implementation of similar legislation by a number of states—commonly called State NEPA. Several of the states' legislative policies are identical to NEPA; however, some states (such as California) are considered more restrictive and require extensive additional requirements. Many of the State statutes require all State agencies and sometimes, local governments, to prepare appropriate environmental impact statements on their actions that significantly affect the environment. These State statutes generally do not apply to the private sector; however, some states require impact statements for specific land development projects.

NEPA (the Act) is a concise piece of legislation consisting of just over three pages of text and composed of two sections: Titles I and II. Title I contains a Declaration of National Environmental Policy, which requires the Federal government to use all practicable means to create and maintain conditions under which man and the environment can coexist in productive harmony and fulfill the social, economic, and other requirements of present and future generations of Americans. In order to carry out the policy set forth in the Act, Congress

identified in Section 101 the following six purposes for drafting NEPA:

1) Fulfill the responsibilities of each generation as trustees of the environment for succeeding generations.

2) Assure for all Americans safe, healthful, productive, and esthetically and culturally pleasing surroundings.

3) Attain the widest range of beneficial uses of the environment without degradation, risk to health or safety, or other undesirable and unintended consequences.

4) Preserve important historic, cultural, and natural aspects of our national heritage, and maintain, wherever possible, an environment which supports diversity, and variety of individual choice.

5) Achieve a balance between human population and resources which will permit a high standard of living while sharing these resources.

6) Enhance the quality of renewable resources and maximize the recycling of resources.

Section 102 of NEPA requires Federal agencies to

1) Use a systematic, interdisciplinary approach in their planning and decision-making

2) Identify and develop methods and procedures (in consultation with the Council on Environmental Quality—established by Title II of the Act)

3) Include in every report a detailed statement (commonly referred to as *environmental impact statements*, [EISs]) containing

   • The environmental impact of the proposed action

   • Any adverse environmental effects which cannot be avoided

   • Alternatives to the proposed action

   • The relationship between local short-term uses of man's environment and the maintenance and enhancement of long-term productivity

   • Any irreversible and irretrievable commitments of resources which would be involved in the proposed action should it be implemented

Title II of NEPA establishes the Council on Environmental Quality (CEQ) and requires the President to transmit to Congress, with the assistance of CEQ, an annual Environmental Quality Report on the state of the environment.

The Council on Environmental Quality, which is headed by a full-time chairperson, oversees NEPA with the assistance of a staff. The duties and functions of the Council are listed in Title II, Section 204 of NEPA, and include

- Gathering information on the conditions and trends in environmental quality

- Evaluating Federal programs in light of the goals established in Title I of the Act

- Developing and promoting national policies to improve environmental quality

- Conducting studies, surveys, research, and analyses relating to ecosystems and environmental quality

When Congress enacted NEPA, their intent was to foster better decision-making for all activities that could have a significant impact on the environment. As stated in the CEQ implementing regulations (40 CFR 1500.1[C]), "Ultimately, of course, it is not better documents but better decisions that count. NEPA's purpose is not to generate paperwork—even excellent paperwork—but to foster excellent action. The NEPA process is intended to help public officials make decisions that are based on understanding of environmental consequences, and take actions that protect, restore, and enhance the environment."

It should be remembered that a key component of NEPA is its role as an early planning document thus it does not prohibit development of the environ-ment, does not contain environmental regulatory requirements, and does not amend or pre-empt other Federal regulations.

> The legislative history of NEPA is limited; however, the birth of NEPA can be placed with the introduction of a bill by Senator James Murray in 1959 and with bills introduced by Senator Henry Jackson and Congressman John Dingell in 1969.

Legislative History of NEPA

The history of NEPA and the creation of the Council on Environmental Quality find their origin in a bill introduced by Senator James Murray in 1959. This bill was loosely based on an Act passed in 1946 in which a national employment policy was declared. This Act also established a Council of Environmental Advisors to the President. Senator Murray's bill declared a national policy on conservation and the use of natural resources, and likewise created a council of presidential environmental advisors. One of the functions of the council was to prepare an annual report for the President and/or Congress.

Senator Henry Jackson and Congressman John Dingell independently introduced legislative bills (Senate Bill 1075 and House Resolution 6750) in 1969 very similar to Senator Murray's bill. Both bills created a Council on Environmental Quality; however, only the Dingell bill provided a statement of environmental policy.

In April 1969, Senator Jackson's committee held hearings and afterwards, Senator Jackson introduced an amendment which incorporated a declaration of national policy. Another important amendment submitted by Senator Jackson required *findings* by Federal officials on the environmental impact of agency actions. This requirement was replaced in NEPA by the requirement that Federal agencies prepare *detailed statements*. This is the basis for today's environmental impact statement (EIS).

Significant among the many statements made during the various hearings was the statement by Professor Lynton Caldwell, an Indiana University professor, who believed the *findings* requirement to be an *action-forcing* component of NEPA, a phrase that is consistently used today in NEPA terminology. These *action-forcing* provisions require Federal decision-makers to take into account the potential consequences of their decisions on the quality of the human environment.

The Senate bill was unanimously adopted in July 1969 without debate. The House adopted its version of NEPA in September 1969, but added an amendment limiting the overall effect of the bill. However, that amendment was dropped in conference. Had this amendment remained, it would have limited NEPA's effect on agency decision-making.

Following passage of bills by both houses of Congress, it was referred back to the Senate, where additional amendments were added before the bill was sent to the Conference Committee. These amendments are known as the *Muskie-Jackson Compromise.* Senator Henry "Scoop" Jackson was the chief sponsor of NEPA in the Senate and Senator Edmund Muskie was the Chairman of the subcommittee responsible for pollution control and other environmental legislation. For some reason, the "compromise" was not well recorded in the legislative history, which may help explain the vague language presented in the compromise amendments. Nevertheless, both houses of Congress had agreed to the conference report by the middle of December 1969 and President Nixon signed NEPA into law on January 1, 1970. The Act set the course for a new direction of the entire Federal establishment, requiring all Federal agencies to consider and describe the environmental consequences of major decisions and alternative courses of action. Senator Jackson's description of NEPA at its passage almost 30 years ago as "the most important and far-reaching environmental and conservation measure ever enacted by Congress . . ." still rings true today. The following sections outline the process for implementation of NEPA, a simple act with far-reaching environmental implications.

Implementation of NEPA

Due to considerable uncertainty in what Congress intended by the enactment of NEPA, the Council on Environmental Quality (CEQ) in 1978 developed NEPA implementation regulations (40 CFR 1500–1508) which are binding on all Federal agencies. The regulations address the procedural provisions of NEPA and the administration of the NEPA process. Both NEPA and the CEQ implementing regulations contain *action-forcing* provisions to ensure that Federal agencies consider environmental information prior to making decisions on proposed actions. The NEPA process includes decision points at which the significance of environmental effects is assessed, project alternatives are evaluated, input from the public is obtained, and a decision prepared that publicly

states the alternatives considered and the decision reached by the Federal agency.

A part of the CEQ implementing regulations was a requirement that each Federal agency review their policies, procedures, and regulations accordingly and revise them as necessary to ensure full compliance with the purposes and provisions of NEPA. To this end, most Federal agencies have promulgated their own NEPA regulations and guidance which generally follow the CEQ procedures but are tailored for the specific mission and activities of the agency.

The NEPA Process

Before any major Federal action is approved, the Federal agency must assess the potential environmental impacts of the proposed action and reasonable alternatives. The NEPA process can range from one step to many sequential steps requiring extensive technical expertise. The NEPA process employs what has been identified as a *sliding scale* approach (see Figure 1.) This approach builds from the principles of CEQ guidance by recommending that agencies focus on significant environmental alternatives and by discussing impacts in proportion to their significance. Key principles utilized in this approach include the following:

1) Proposals fall on a continuum with respect to environmental impacts.

2) Greater potential for significant impacts requires more detailed analysis.

3) Likewise, small environmental impacts usually require less analysis.

4) Applies the process to identification of alternatives and to analysis of environmental impacts.

The NEPA process can be divided into three basic elements for compliance using the sliding scale approach. these consist of (1) identification of Federal action and NEPA documentation determination (*i.e.,* what type of document to prepare), (2) preparation of the appropriate NEPA documentation, and (3) decision-making and NEPA follow-up. This section provides details on each of these three basic elements.

Figure 1. The Sliding Scale as Used in the National Environmental Policy Act Process

The information presented in this chapter represents a general description of the NEPA process as it applies to Federal agencies. Many Federal agencies have developed detailed procedures and published these as regulations for their implementation of NEPA. The purpose of this chapter is to give our audience a general understanding of the NEPA process, and does not include detailed information that can be obtained from agency specific guidelines or the numerous books prepared on the subject.

The NEPA process consists of an evaluation of the environmental effects of a Federal undertaking including its alternatives. Therefore, before commencing any project, a Federal agency must determine whether the proposed project is a major Federal action and whether it may affect the quality of the human environment. The NEPA process can range from one step to many sequential developments. Additionally, NEPA evaluations should be initiated with other planning activities at the earliest possible time, to ensure that planning and decisions reflect environmental values, avoid delays later in the process, and head off potential conflicts.

Definition of the proposed action is very often the most difficult step in beginning the assessment process. Actions usually begin largely as concepts and in some cases as broad policies or programs whose associated biological, physical, and socioeconomic activities are minimal or unknown at that point. For example, a Federal agency is considering granting a lease of government-owned land to a private industry to provide a production facility which would serve both the Federal

government and industry or the general public. This action on the part of the government is administrative in nature but results in a range of potential biological, physical, and social impacts to the environment from construction and operation of the proposed facility by the industry. Therefore, to assess the environmental consequences of the proposed action, a firm description of the related physical activities must be prepared.

CEQ regulations require the use of *public scoping* (a process in which the agency involves the public and other State and Federal agencies in determining the range of alternatives and actions to be discussed in an environmental assessment [EA] or environmental impact statement [EIS]) to identify alternatives or other significant environmental issues that may have been overlooked by the agency in their development of the proposed action. Public involvement is one of the key components of the NEPA process, and scoping is just one method utilized to involve the public.

Identification of alternatives forms another key component in the preparation of an EA. An agency must identify and develop real alternatives as opposed to merely looking at options that speculation might suggest. On the other hand, NEPA does not require that all possible alternatives be evaluated, only the *reasonable* alternatives. In addition to these reasonable alternatives, NEPA also requires that agencies include the *no action* alternative. The **no action** alternative is defined as the action that would happen to the environment if the Federal agency's proposed action was not implemented.

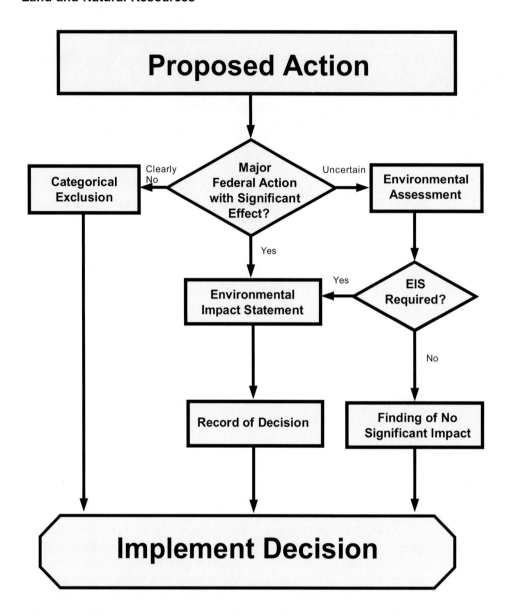

Figure 2. Flow Diagram for the General National Environmental Policy Act

There are three levels of analysis, depending on whether or not a proposed Federal action could significantly affect the environment. Under NEPA, the term *environment* encompasses the natural and physical environment (air, water, geography, and geology) as well as the relationship of people with that environment including health and safety, socioeconomics (jobs, housing, schools, transportation), cultural resources, noise, and aesthetics. These three levels include

1) Categorical exclusion determinations, and

2) Either preparation of an environmental assessment and its associated finding of no

significant impact (EA/FONSI)

3) Or, if there is a potential for significant impacts, preparation of an environmental impact statement (EIS)

See Figure 2.

Although guidance on the most appropriate level of NEPA documentation is generally provided by the respective Federal agency where they have identified classes of action that generally fall within one of the three levels for NEPA analysis, a modification of the sliding scale approach (Figure 3) illustrates the level of NEPA analysis required

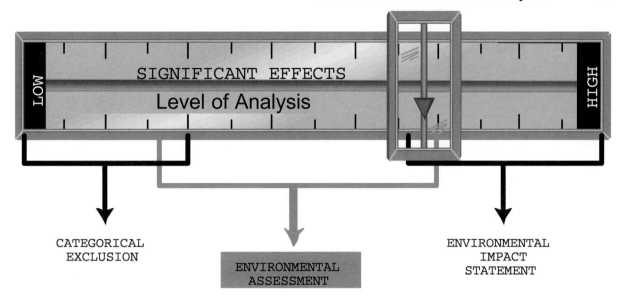

SIGNIFICANT EFFECTS

Level of Analysis

LOW

HIGH

CATEGORICAL
EXCLUSION

ENVIRONMENTAL
ASSESSMENT

ENVIRONMENTAL
IMPACT
STATEMENT

Figure 3. Use of the Sliding Scale in NEPA Documentation

and thus the resulting type of NEPA documentation necessary for a proposed action. If the proposed action will clearly have a significant effect, then the scale slides to the right and an environmental impact statement is required. On the other end of the scale lie insignificant effects that can be categorically excluded from further NEPA documentation. As depicted in the figure there are areas of overlap, and the effects of the proposed action are not clearly defined. In those cases, it is generally more conservative to prepare the next higher level of NEPA documenta tion.

Categorical Exclusions

In the first level, a Federal agency may *categorically exclude* an action from detailed environmental analysis if it meets certain criteria, which have been previously determined as having no significant environmental impact. CEQ regulations require Federal agencies to develop lists of actions, which normally are categorically excluded from environmental evaluation under their NEPA regulations. These are actions that do not individually or cumulatively have a significant effect on the quality of the human environment. For example, the routine maintenance (*e.g.*, oil changes) of a large number of vehicles for a Federal agency could impact the environment if not handled properly. In this case, other environmental precautions are in place to ensure that the used oil is disposed of correctly.

and this recurring action was previously discussed in a *Federal Register* notice available to the public. In this notice, the Federal agency outlined the environmental precautions that would be taken, and recommended that this recurring action be categorically excluded from NEPA documentation. Prior to the proposed action becoming a categorical exclusion, the agency must weigh the comments received by all commentors including individuals and State and Federal regulators.

Therefore, if an agency determines that the proposed action has no significant environmental impacts, is not controversial, and is on a list of agency actions categorically excluded from detailed analysis, no further action is needed. Most agencies require that some form of formal documentation be provided for this step. Although NEPA does not require documentation of a categorical exclusion, this usually consists of a written determination containing the proposed action and some type of checklist of environmental conditions that is generally reviewed by more than one qualified NEPA professional. The determination is filed and the proposed action can proceed.

Environmental Assessments

An *environmental assessment* (EA) is a concise public document prepared to determine whether to prepare an *environmental impact statement*

(EIS) or a *finding of no significant impact*. Federal agencies typically list classes of actions that normally require the preparation of an EA.

In the second level of NEPA analysis, a Federal agency has identified a major action but is not sure if there is a potential for significant environmental impacts (Figure 2). In this case the Federal agency prepares an environmental assessment (EA). An EA is a public document that briefly provides the evidence and analysis necessary to make a threshold determination of environmental significance. There is no defined methodology to be used in the preparation of an EA, but the selected method must be justifiable. An EA must include brief discussions of the need for the proposed action, alternatives to the proposed action, environmental impacts of the proposed action and alternatives, and a list of agencies and private parties consulted. The analysis presented in the EA is less rigorous than that performed for an EIS, and the analysis does not have to be as detailed. The EA should not form conclusions but should present sufficient information to reach a conclusion.

Specific page lengths for EAs are not presented in the CEQ regulations, but CEQ guidance recommends from 10 to 15 pages, with sufficient background data incorporated by reference. Over the past few years, the tendency for agencies is to prepare EAs with significantly greater number of pages. The courts have determined that larger EAs are not necessarily adequate and may require an agency to also prepare an EIS. An EA will not substitute for an EIS, especially if the agency has identified significant environmental effects for the proposed action and has not completed all the requirements for preparing an EIS, such as involving the public or other agencies in the process as defined by NEPA. In other words, if an agency believes there would be significant environmental effects, it should prepare an EIS rather than a lengthy EA.

Findings of No Significant Impact. Based on the EA, a Federal agency must either prepare an EIS if it has determined that its proposed action may have a significant impact on the environment or it must prepare a finding of no significant impact (FONSI) if it has determined that the action will not have significant environmental impacts.

Should it be determined that an EIS is required, the process outlined in the next section is followed. However, if a FONSI is prepared it must identify the reasons in some detail for the agency's determination. If certain factors used in the analysis are given greater weight than others, the agency must explain those reasons and present the information. Procedurally, the EA may be attached to the FONSI and incorporated by reference or it may be summarized in the FONSI. Other documents noted in the EA must be referenced completely.

Each EA and FONSI must be made available to the public. Each Federal agency may choose the most appropriate method of accomplishing this goal, as long as they ensure that all interested or affected parties are notified. Most Federal agencies maintain a list of interested individuals, agencies, organizations, *etc.*, that are interested in activities of that agency, and therefore the agency automatically sends each a copy of all EAs and FONSIs. In addition, CEQ recommends that each agency also place a notice of availability in the *Federal Register* and in national publications for proposed actions with a national scope. For regional or site-specific actions, a notice of availability may also be published in local newspapers.

Under limited circumstances, the CEQ recommends that FONSIs be published 30 days before an agency's final decision not to prepare an EIS. These circumstances are for two reasons: (1) the proposed action is, or is closely similar to, one which normally requires the preparation of an EIS or (2) the nature of the proposed action is one without precedent.

Environmental Impact Statements

The third level in compliance with the NEPA process involves preparation of an environmental impact statement (EIS) and constitutes the other end of the sliding scale or those actions that could significantly affect the environment. As the sliding scale implies, EISs are more formal than EAs and contain considerations of the environmental effects of the proposed action including: adverse effects that cannot be avoided; alternatives; the relationship between the short-term and long-term uses of the environment; and irreversible and

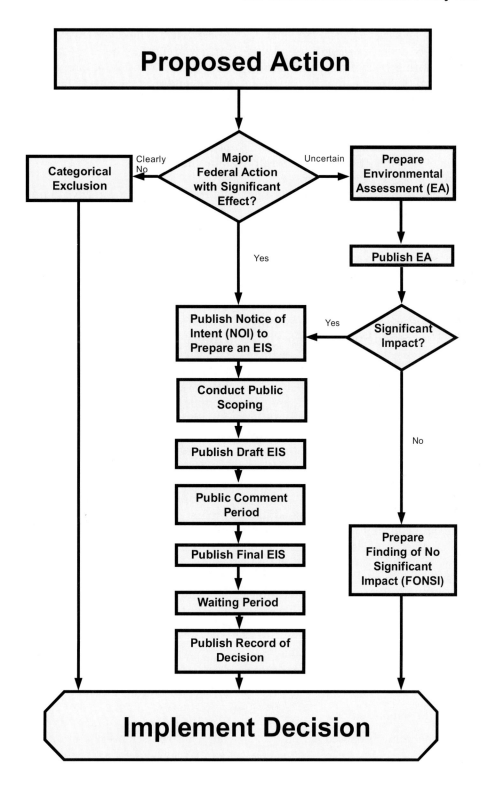

Figure 4. National Environmental Policy Act Process Flow Path

irretrievable commitments of resources as outlined in CEQ regulations.

The major steps in the NEPA process are outlined in Figure 4. This figure illustrates all three levels of NEPA analysis and provides additional detail on the steps in the preparation of an Environmental Impact Statement.

Once a decision has been made to prepare an EIS, the first step in the NEPA process is the public notification through publication of a *Notice of*

Intent (NOI) in the *Federal Register*. This notice describes the proposed action, possible alternatives the agency is considering, background information on issues and potential impacts, and the proposed scoping process including whether, when, and where any scoping meeting will be held. The NOI also contains a contact point within the agency.

After publication of the NOI, the agency generally begins the scoping process. As noted under EAs, scoping is the public process whereby an agency involves the public to determine the scope of issues to be addressed in an EIS and identifies the significant issues. Comments are solicited in a number of ways including in writing, voice mail, email, fax, and at public meetings and, possibly, public workshops.

The EIS is prepared in two stages: a draft EIS and a final EIS. The draft EIS describes, analyzes, and compares the potential environmental impacts of the alternatives that could be chosen to accomplish the purpose and need identified by the Federal agency. A draft EIS is issued to allow the public to review and comment on the proposed action and alternatives before any decisions are made. If the Federal agency has a preferred alternative at this stage, it will be identified in the draft EIS.

EIS Format. The CEQ regulations recommend that Federal agencies use a format for EISs which will encourage good analysis and clear presentation of the alternatives including the proposed action. The following standard format should be used in developing the EIS, unless the agency determines that there is a compelling reason to do otherwise:

- Cover sheet
- Summary
- Table of contents
- Purpose of and need for action
- Alternatives including proposed action
- Affected environment
- Environmental consequences
- List of preparers
- List of agencies, organizations, and persons to whom copies of the statement are sent
- Index
- Appendices (if any)

Cover Sheet. The cover sheet includes a list of the agency preparing the EIS; the title of the proposed action; the location of the proposed action; the name, address, telephone, fax, and email address (if available) of an individual from the agency who can supply additional information; whether the EIS is a draft or final; a one-paragraph abstract of the EIS; and the date by which all comments must be received.

Summary. The summary should normally not exceed 15 pages and should emphasize the major conclusions in the EIS and identify any controversial areas (including those identified during the public scoping period), as well as the issues to be resolved, such as choosing between alternatives.

Table of Contents. Consists of a listing of material presented in the EIS, including all tables and figures.

Purpose and Need for Action. This statement briefly specifies the underlying purpose and need to which the agency is responding in proposing the alternatives including the proposed action.

Alternatives. This section provides an objective evaluation of all reasonable alternatives as well as a discussion of all alternatives considered and eliminated from detailed study; identifies the preferred alternative if one exists; includes the alternative of no action; and presents a comparison of alternatives.

Affected Environment. The affected environment is the area affected or created by the alternatives presented in the EIS. The description of the affected environment must be sufficient for a reader to understand the alternative's environmental effects. The CEQ regulations state that this section be brief, with less important material being summarized, consolidated, or referenced.

Environmental Consequences. This section presents the scientific and analytic basis for the comparisons identified in the alternatives section. The discussion includes the environmental impacts of the alternatives including the proposed action; any adverse environmental effects which cannot be avoided should the proposal be implemented; and the relationship between short-term productivity and any irreversible or irretrievable commitments of resources which would be involved in the proposal should it be implemented. Duplication of information presented in the alternatives section should be avoided unless repetition is needed to increase clarification. In addition, direct and indirect effects and their significance should be presented as well as means to mitigate adverse environmental impacts.

List of preparers. The names and qualifications of people who were primarily responsible for preparing the EIS or other significant support documentation must be included in the EIS. Where possible, the particular person responsible for preparing a specific section or analysis should be identified.

Appendices (if any). The appendices should not contain analyses and information on environmental effects and alternatives; rather they should supplement information as appropriate to understanding the proposed action and alternatives. Appendices should also be used to contain responses to comments received by the agency on the draft EIS. If an appendix is prepared, it should be circulated with the EIS or be available upon request.

In addition to the items listed above, the EIS must also present a list of the agencies, organizations, and persons to whom the agency sent a copy of the EIS and an index. Many EISs frequently contain additional information to assist the reader in understanding the proposed action and may include maps, tables, listing of abbreviations, glossaries, *etc.*

Draft EIS. Once the Draft EIS is issued, a minimum of 45 days is provided for Federal agencies, State and local governments, Native American tribes, and stakeholders to comment on the Draft EIS. The public comment period begins upon publication of a Notice of Availability (NOA) for the Draft EIS in the *Federal Register.* Generally, at least one public meeting is held to solicit public input on the Draft EIS. Other methods for submission of comments for the EIS include written comments, fax, telephone, email, *etc.* All comments received by the Federal agency are considered to the extent practicable depending on whether they are received by the date indicated in the NOA and the EIS.

Final EIS. Following the public comment period, a Final EIS is published and distributed similar to the Draft EIS. The Final EIS reflects consideration of all comments received on the Draft EIS, contains the agency's responses to those comments, and provides revised EIS text. The Final EIS must also identify the preferred alternative or alternatives. The Final EIS is announced by again publishing a Notice of Availability (NOA) in the *Federal Register.*

Record of Decision. Once the Final EIS is published, a minimum 30-day waiting period (or 90 days after publication of the Draft EIS) is required by NEPA before a Federal agency can issue a ***Record of Decision*** (ROD), generally with an announcement in the *Federal Register.* The time periods are calculated from the date of publication of the Notices of Availability by the Environmental Protection Agency in the *Federal Register.* The ROD notifies the public of the agency's decision on the proposed action and the reasons for that decision. The ROD may also include consideration of other decision factors such as costs, technical feasibility, agency statutory mission, and/or national objectives.

The Federal agency must identify all alternatives considered by the agency in reaching their decision and must specify the alternative or alternatives which were considered to be environmentally preferable. The Federal agency is not required to select the environmentally preferable alternative; however, a discussion must be presented to support the selection of agency's preferred alternative. The agency's preferred alternative and the environmentally preferred alternative are many times the same.

Finally, the ROD must state whether all practicable means to avoid or minimize environmental harm from the alternative selected have been adopted, and if not, why. Where applicable for mitigation (refer to the "Mitigation" section for a definition relative to NEPA), a monitoring and enforcement program must be adopted and summarized.

Implement Action. Once the Federal agency has prepared a Record of Decision and met the time requirements, it is free to implement the action. Federal agencies may provide for monitoring to ensure that the decision(s) made are carried out in accordance with their Record of Decision. Mitigation and any other conditions established during the course of the EIS or during its subsequent review and committed as part of the ROD must be implemented by the agency.

Mitigation

Federal agencies shall mitigate actions to reduce otherwise significant environmental effects to a reduced level. NEPA requires that mitigation measures be subject to public comment. CEQ defines *mitigation* to include

- Avoiding the impact altogether by not taking a certain action or parts of an action

- Minimizing impact by limiting the degree or magnitude of the action and its implementation

- Rectifying the impact by repairing, rehabilitating, or restoring the affected environment

- Reducing or eliminating the impact over time through the use of appropriate preservation and maintenance operations during the life of the action

- Compensating for the impact by replacing or providing substitute resources or environments

The Benefits of Using NEPA in Decision-Making

The passage of NEPA has resulted in Federal agencies incorporating environmental values in their everyday decision-making process. For most agencies, conducting NEPA reviews is now an integral part of their program planning process. Numerous Federal agencies have developed multidisciplinary staffs to oversee and assist with NEPA compliance for their agency.

The primary benefit of implementing NEPA has been more protection for the environment in which Federal activities are undertaken. This protection has been enhanced because the NEPA review process resulted in positive environmental changes to the proposed projects before they were implemented. Specifically, these changes have been in alternatives to the project design, location of the proposed action or operation, review of a

greater range of alternatives presented by the Federal agency, implementation of mitigation measures, enhanced opportunity for public involvement in the decision-making process, and in some cases, reduced costs because of changes made in projects before implementation.

Finally, the NEPA process can assist other Federal agencies by allowing them to address regulatory compliance issues with other environmental laws as part of a single review process rather than as separate reviews under each law, thereby reducing the amount of paperwork, staff time and effort.

Bibliography

Bear, D. "NEPA at 19: A Primer on an 'Old' Law with Solutions to New Problems." *Environmental Law Reporter, News & Analysis*. 19: 10060; 1989.

"Council on Environmental Quality." *Code of Federal Regulations*. Title 40. Pts. 1500–1508.

Department of Energy, Office of NEPA Oversight. *Recommendations for the Preparation of Environmental Impact Statements*. Washington, DC: DOE, May 1993.

Environmental Protection Agency. Office of External Affairs, Office of Federal Activities. "Policies and Procedures for the Review of Federal Actions Impacting the Environment." Washington, DC: EPA, 3 October 1984.

"Filing System Guidance for Implementing 1506.9 and 1506.10 of the CEQ Regulations." *Federal Register* 54 (1989): 9592.

"Forty Most Asked Questions Concerning CEQ's National Environmental Policy Act Regulations." *Federal Register* 51 (25 April 1986): 15618.

"Scoping Guidance." *Federal Register* 46 (7 May 1981): 25461.

James L. Oliver is a Senior Project Manager with Tetra Tech NUS, Inc., in Aiken, South Carolina. Mr. Oliver has a BS in Biology from Murray State University. He has over 27 years professional experience in the management of a broad range of environmental and regulatory projects including: 13 years as a Fishery Research Biologist with the US Fish and Wildlife Service and over 14 years in the private environmental consulting field. Mr. Oliver has extensive experience with the National Environmental Policy Act process and, since 1994, has successfully managed or provided technical input and direction to nine major environmental impact statements. He has experience with the implementing requirements of the Council on Environmental Quality and the NEPA public involvement process. The projects he has participated in include environmental impacts associated with construction and operation of reservoirs and associated infrastructure (e.g., roads, bridges, and utilities), pumped storage hydroelectric operations for peaking electrical usage, small-scale hydroelectric redevelopment in rural and remote locations, environmental studies in large reservoirs, winter navigation in the Great Lakes and interconnecting river systems, environmental impacts of operating wastewater treatment plants as well as nuclear production and generation facilities. Mr. Oliver would like to thank Dr. Bruce H. Bradford for his technical review, comments, and guidance in the preparation of this chapter.

CHAPTER 18

Property Assessments

Keith Liner, MCE, CHMM

CERCLA Liability

The Comprehensive Environmental Response, Compensation, and Liability Act (CERCLA) was originally passed in 1980 to provide a response mechanism for uncontrolled release of hazardous substances to the environment. CERCLA provides authority for State and Federal governments to respond to release of hazardous substances into the environment. In order to provide funds for the cleanup actions a trust fund called *Superfund* was established by Congress. CERCLA includes the right for the State and Federal government to initiate cleanup and to seek reimbursement from the primarily responsible parties. The money collected from the responsible parties is intended to replenish the *Superfund*.

CERCLA includes two basic types of response actions: removal actions and remedial actions. A ***removal action*** is an immediate action taken over the short term to address a release or threatened release of a hazardous substance. A ***remedial action*** selects remedies that are protective of human health and the environment, that maintains protection over time and minimizes untreated wastes. A remedial action takes a longer time to implement than a removal action.

A ***release*** is defined in CERCLA as

> any spilling, leaking, pumping, pouring, emitting, emptying, discharging, injecting,

240   **Land and Natural Resources**

escaping, leaching, dumping, or disposing of hazardous substances to the environment.

CERCLA defines *hazardous substances* as:

such elements, compounds, mixtures, solutions, and substances which when released into the environment may present substantial danger to public health and welfare or the environment.

Hazardous substances under CERCLA include

* Any elements, compounds, mixtures, solutions, or substances designated by the United States Environmental Protection Agency (EPA) under Section 311 of the Clean Water Act (CWA), (40 CFR 116.4), or under Section 102 of the Comprehensive Environmental Response, Compensation, and Liability Act (CERCLA), (40 CFR 302.4)

* Any toxic pollutants listed under Section 307(a) of the CWA

* Any hazardous substances regulated under Section 311(b)(2)(A) of the CWA (40 CFR 110, 117, and 122)

* Any listed or characteristic Resource Conservation and Recovery Act (RCRA) hazardous wastes (40 CFR 262)

* Any hazardous air pollutants listed under Section 112(r) of the Clean Air Act (CAA) (40 CFR 68), or

* Any imminently hazardous chemical substances or mixtures regulated under Sections 7 of the Toxic Substances Control Act (TSCA) (40 CFR 761.120, *et seq.*).

CERCLA defines the environment as

(A) the navigable water, the waters of the contiguous zone, and the ocean waters of which the natural resources are under the exclusive management authority of the United States under the Magnuson Fishery Conservation and Management Act, and (B) any other surface water, ground water, drinking water supply, land surface or subsurface strator or ambient air within the United States or under the jurisdiction of the United States.

CERCLA imposes strict liability. The EPA and private parties can seek recovery of cleanup costs associated with releases of CERCLA hazardous substances from the primarily responsible party. The primarily responsible party is an

individual or company potentially responsible for, or contributing to the contamination problems at a Superfund site.

The EPA can use appropriate administrative and legal actions to require primarily responsible parties to clean up contaminated sites.

Under strict liability, any defense claiming that negligence did not occur because the activities were consistent with standard industry practice is no defense at all. CERCLA liability is joint and several. This means that if more than one party has disposed of hazardous substance in a common location, all are responsible. The EPA can divide the cleanup costs among all responsible parties regardless of their contribution to the problem. CERCLA liability is also retroactive. This means that parties may be found liable for actions taken long before passage of CERCLA, even if the disposal of the hazardous substance was lawful at the time of disposal.

CERCLA places liability on the present site owner and operator. Liability is a condition for land purchasers. One way for land purchasers to protect themselves is to invoke the *innocent landowner defense*. In order for the landowner to use the *innocent landowner defense* to CERCLA liability,

all appropriate inquiry must be made into the previous ownership and uses of property consistent with good commercial or customary practice.

Appropriate inquiry is made by conducting an environmental site assessment in accordance with the American Society for Testing and Materials (ASTM) Practice, E1527, "Environmental Site Assessment Transaction Screen" and E1528, "Environmental Site Assessment: Phase I Environmental Site Assessment Process." In court, the landowner can use the *innocent purchaser* defense. If the landowner uses the ASTM standards, he has exercised **due diligence**. *Due diligence* states that

at the time of land acquisition, all appropriate inquiry was made into the previous ownership and use of the property was consistent with good commercial or customary practice.

More extensive assessments called Phase II and Phase III are under development by ASTM and will not be covered in this chapter.

Transaction Screen Process

The Transaction Screen Process is intended to permit a user to satisfy one of the requirements to qualify for the *innocent landowner defense* to CERCLA liability. The Transaction Screen Process is intended to establish a customary practice for conducting an environmental site assessment of a parcel of land or property. The customary practice identifies recognized environmental conditions. ***Recognized environmental conditions,*** as defined by CERCLA, means

> the presence or likely presence of any hazardous substances or petroleum products on a property under conditions that indicate a release, a past release, or a material threat of any hazardous substances or petroleum products into structures on the property or into the ground, ground water, or surface water of the property

Recognized environmental conditions as defined under CERCLA do not include *de minimus* conditions that generally do not present a material risk or harm to public health or the environment and that generally would not be subject to an enforcement action if brought to the attention of appropriate governmental agencies.

The customary practice is designed to (1) synthesize and put in writing good commercial and customary practice for environmental site assessments for commercial real estate, (2) to facilitate high-quality standardized environmental site assessments, (3) to ensure that the standard appropriate practice is practical and reasonable, and (4) to clarify an industry standard for appropriate inquiry in an effort to guide legal interpretation of CERCLA's *innocent landowner defense.*

The customary practice assists the user in developing the information about the environmental condition of a property and has utility for a wide range of persons, including those who may have no actual or potential CERCLA liability and as such may not be seeking *the innocent landowner defense.*

The Transaction Screen Process is conducted by the user or by an environmental professional. A Phase I Assessment must be performed by an environmental professional.

Completion of the Transaction Screen Process in accordance with the ASTM standards may allow the user to conclude that no further inquiry is needed to assess the environmental conditions of the property. This will allow the user to invoke the *innocent landowner defense* without performing a Phase I Assessment. Completion of the Transaction Screen Process concludes that (1) no further inquiry into recognized environmental conditions of the property is needed for purposes of appropriate inquiry, or (2) further inquiry is needed to assess recognized environmental conditions appropriately for purposes of appropriate inquiry. If no further inquiry is needed, the user has completed appropriate inquiry into the property.

The process is not an exhaustive assessment of a clean property. A balance is established at a point at which the costs outweigh the benefits. The same level of inquiry will not be required for all properties. The type of property will guide the type and depth of inquiry.

The Transaction Screen Process consists of asking questions contained within the Transaction Screen questionnaire. This questionnaire is for owners and occupants of the property with direction provided by the Transaction Screen Process. The Questionnaire is then used to conduct a limited records review of government records and certain historical sources. The owners and occupants are asked the same questions.

The guide and process uses these steps: (1) inquiry into owners/occupants, (2) site visit, (3) inquiry into governmental records.

The questionnaire in the standard is divided into three sections:

1) Guide for Owner and Occupant Inquiry

2) Guide to Site Visit

3) Guide to Government Records/Historical Sources Inquiry

The questionnaire should be used to ask questions of the (1) current owner of the property (2) any major occupants of the property or if more than one occupant at least 10% of the occupants of the property, and (3) in addition to the current owner and occupants above, any that are likely to be using, treating, generating, storing, or disposing of hazardous substances or petroleum products on the property.

The evaluator uses the *site visit column* during the observation of the property including any buildings and other structures on the property. The questions on the questionnaire profile the property and ask about

- Current use and past use of the property
- Stored items on the property including drums, car batteries, paints
- The presence of ponds, pits, or lagoons
- Stained soil or foul-smelling drains
- The presence of hazardous substances, poly-chlorinated biphenyls (PCBs), among others

The respondents answer the questions *Yes, No,* or *Unknown* to the best of their knowledge.

The historical sources inquiry answers if the property is listed on the

- National Priorities List (NPL)
- RCRA Corrective Action Report System (CORRACT) facility list

or is near any of the following facilities:

- Hazardous waste sites being investigated for remediation
- Leaking underground storage tanks (USTs)
- Solid waste/landfill facilities

The guide and information obtained during the Transaction Screen Process should be used to conclude that no further inquiry is needed. If the user cannot conclude that no further inquiry is required based on their best business judgement and the questionnaire answers, then further inquiry may be required and the user should proceed to a full Phase I Environmental Site Assessment.

Phase I Assessment

The Phase I Assessment is a continuation of the Transaction Screen Process. The purpose of this practice is identical to the Transaction Screen. It is intended to define good commercial and customary practice in the United States for conducting an environmental assessment.

The Phase I Assessment purpose is to identify, to the extent feasible pursuant to the process prescribed herein, recognized environmental conditions in connection with the property.

A Phase I Environmental Site Assessment has four components:

- Records review
- Site reconnaissance
- Interviews
- Evaluation report

There is no sampling involved with the practice. The first three phases are intended to be complementary, in that if during a records review it is found that the property was used for a particular purpose, the site visit inspections should try to identify the previous use. An environmental professional must perform the interviews and the site visit. The environmental professional must also oversee the writing of the report and must review and interpret any information on which the report is based. The records review can be done by governmental agencies, through freedom of information requests, the user, occupants of the property, *etc.*, and should be done under the supervision of an environmental professional. The environmental professional is not required to independently verify the information provided by others unless the professional knows by actual knowledge that certain information is incorrect.

Records Review

The records review is intended to help identify recognized environmental conditions in connection with the property. The records should be obtained from reasonably ascertainable and standard sources. The records are reasonably ascertainable if they can be obtained with reasonable time and cost within 20 days of a written, verbal, or in-person request. The records should provide information that is relevant to the property. The records review should also include a review of property records for property adjacent to the property in question. A minimum reach distance is defined in the ASTM standard. This is intended to identify situations in which hazardous substances or petroleum products might migrate into the property in question. The approximate minimum reach distance may be reduced due to the (1) density of setting that the property is located, (2) distance hazardous or petroleum substances might migrate based on geological data, (3) other reasonable factors. The reduction should be documented in the final report. Typical minimum reach distances are:

Federal NPL Site List   1.0 mi

State Leaking USTs   0.5 mi

Federal RCRA CORRACTS TSD List   1.0 mi

Federal RCRA Non-CORRACTS TSD List   0.5 mi

The Federal NPL site list and Federal RCRA TSD list state that the minimum reach distance may not be reduced.

Typical local records include

- Lists of landfill/solid waste disposal sites
- Lists of underground storage tanks
- Records of contaminated public wells

A current United States Geological Survey (USGS) 7.5 Minute Topographic Map or equivalent showing the area that the property is located shall be reviewed. The map will provide information about geologic, hydrogeologic, or topographic characteristics of the site and is the only standard physical setting resource that is required to be evaluated. An equivalent map can be used if necessary.

In order to judge the past property use, historical resources should be used from the present back to the property's first obvious developed use, or back to 1940, whichever is earlier. Records should be used to develop a history of the property and surrounding properties. Some standard historical resources are

- Aerial photographs
- Fire insurance maps

- Property tax files
- Title records

Site Reconnaissance

The site reconnaissance is intended to obtain information and likelihood of identifying environmental conditions. An environmental professional must conduct the site reconnaissance. It is not expected that more than one site visit will be necessary. The environmental professional must visually and physically survey the property and any buildings on the property. The visit must be documented in the final report and include the method used for evaluation. The exterior of buildings as well as peripheral of the property must be inspected for any evidence of present or past contamination of hazardous substances or petroleum products. The interior of buildings must be visually and physically observed including commonly accessible areas. Any physical limitation of the inspection must be noted in the final report including physical obstructions, bodies of water, asphalt, snow, and rain. Occupants and past occupants must be questioned in accordance with the standard.

The current use of the property must be identified in the report. Any current uses likely to involve the use, treatment, storage, disposal, or generation of hazardous substances or petroleum products must be identified. The property must be inspected for past use in areas that the record review may have identified. A general description of buildings, roads, topography, water supply, sewage, and storage tanks must be included. Any odors, pools of liquid, or container storage must be identified and evaluated in the visit and documented in the report. Any equipment with PCBs must be identified in the report. Soil and pavement must be inspected for stains and evaluation completed. Vegetation shall be examined to see if it is stressed. Any drains, sumps, wells on the property must be documented in the report. The environmental professional must assess the entire property and surrounding area during the site visit.

Interviews with Owners and Occupants

Interviews must be conducted with owners and occupants in an attempt to identify recognized environmental conditions with the property.

Questions may be asked in person during the site visit, in writing, or by telephone. A key site manager must be interviewed. This individual should have a good knowledge of the uses and physical characteristics of the property. A reasonable number of occupants of the property must be interviewed. A representative number of individuals should be interviewed in order to assess any recognized environmental conditions. The report must document the individuals interviewed and the duration of their occupancy. Prior to the site visit the key site manager, property owner or user should be asked if the following documents exist and copies of them requested:

- Environmental audit reports

- Environmental site assessment reports

- Environmental permits

- Update on geological conditions

- Hazardous waste generator notices

- Any other applicable aspects of information

Interviews with Local Government Officials

At least one staff member from the local fire department, local health agency, or regional office of the State agency, and one from the local agency or regional office of State agency having jurisdiction over hazardous waste disposal or other environmental matters in the area in which the property is located, should be interviewed. The interviews with local officials are to obtain information on the property's environmental conditions. Questions may be asked by telephone or person to person. The content of the questions is at the discretion of the environmental professional and is aimed at gaining insight into current and past uses of the property.

Evaluation and Report Preparation

The final report should include documentation to support the analyses, opinions, and conclusions found in the report. All sources including those that revealed no findings should be documented to facilitate reconstruction at a later date. The report should follow the format contained in the

standard and should state if the user or property owner revealed that the property has an environmental lien encumbering the property. The report shall name the environmental professionals involved in conducting the Phase I Assessment. One of two options should be used to document qualifications of the persons involved: (1) the report shall include a qualification statement of the environmental professionals responsible for the assessment and preparation of the report and individual and correlate qualification or (2) a written qualification statement for the environmental professional responsible for the Phase I Assessment and preparation of the report shall be delivered to the user. All evidence of recognized environmental conditions shall be described in full. The report shall include the environmental professional's opinion of the impact of the recognized environmental conditions in connection with the property. The report shall include a findings and conclusions section that states one of the following:

> We have performed a Phase I Environmental Site Assessment in conformance with the scope and limitations of ASTM Practice E1527 of the property. Any exceptions to or deviations from this practice are described in Section [ ] of this report. This assessment revealed no evidence of recognized environmental conditions in connection with the property

or

> We have performed a Phase I Environmental Assessment in conformance with the scope and limitations of ASTM Practice E1527 of the property. Any exceptions to or deviations from this practice are described in Section [ ] of this report. The assessment revealed no evidence of recognized environmental conditions in connection with the property except for the following list . . . .

The environmental professional then signs the report and transmits it to the user.

Phase II standards are under development and will probably require sampling. Remediation techniques, liability/risk evaluation recommendations are outside the scope of the current standards.

Bibliography

American Society for Testing and Materials, "Standard Practice for Environmental Site Assessments: Phase I Environmental Site Assessment Process." ASTM Practice E 1527–97. Westconshohocken, PA: ASTM, 1997.

American Society for Testing and Materials, "Standard Practice for Environmental Site Assessments: Transaction Screen Process." ASTM Practice E 1528–96. Westconshohocken, PA: ASTM, 1996.

DOE/Westinghouse. "DOE/Westinghouse School for Environmental Excellence Manual, CERCLA Chapter." Hanford, WA: DOE/Westinghouse, 1993.

"Designation of Hazardous Substances." *Code of Federal Regulations*. Title 40, Part 302.4, 1997.

Keith Liner *is currently employed with the Westinghouse Savannah River Company, in Aiken, South Carolina. He is an Environmental Engineer and works at the Defense Waste Processing Facility within the High Level Waste Division where, for the past three years, he has been responsible for hazardous waste management. Mr. Liner is a Certified Hazardous Materials Manager and holds a BS in Chemical Engineering from the University of Pittsburgh and an MCE in Environmental Engineering from Auburn University. He would like to thank his wife, Chris, and his children, C.J. and Paul, for their love and support during the writing of this chapter.*

CHAPTER 19

Brownfields

John E. Milner, JD
Charles A. Waggoner, PhD, CHMM

Introduction

What Is a Brownfield?

Perhaps the most succinct general definition of **brownfields** can be found in *Superfund: Barriers to Brownfield Redevelopment*:

> abandoned, idled or under-utilized industrial and commercial facilities where expansion or redevelopment is complicated by real or perceived environmental contamination (GAO, 1996).

Historically, brownfields laws have typically targeted urban and heavily industrialized areas where there is significant contamination of extensive areas. However, brownfields are not limited to the classic *rust bucket* districts of major cities. They include small rural properties like gas stations where underground storage tanks may have leaked. Indeed, the brownfields concept has even expanded to include areas where there is only the perception of contamination due to a property's proximity to areas known to be contaminated.

Brownfields are often contrasted to **greenfields**, which are generally defined as uncontaminated properties outside of urban areas that have not

247

previously been sites of industrial or commercial uses. It is easy to understand why development of a greenfield property is a competing alternative to brownfield redevelopment. Greenfield properties do not require removal of old industrial or commercial structures and they demand little or no cleanup cost. Additionally, the time line for greenfield development does not include delays resulting from environmental investigation and remediation activities. Brownfield laws have been developed in an attempt to provide legal and economic incentives to encourage brownfield redevelopment as will be discussed below.

What Are the Problems Caused by Brownfields?

The General Accounting Office estimates indicate a range of 150,000 to 450,000 brownfield sites in the United States. This wide range of estimates is based on the varying definitions of brownfields and the properties that fit within those definitions. Left unremediated, brownfields pose significant potential problems. First, actual environmental contamination at brownfield sites obviously can create human health risks to industrial and commercial facility workers. More broadly, contamination can pose widespread risks through ground water to a broad scope of persons who live in the area of the contaminated properties, whether urban or rural. Second, the existence or even the perception of environmental contamination can depress property values of land in the area surrounding the brownfield. Third, local governments suffer on a variety of levels. The aesthetics can be a defining issue for the city and dramatically reduce the potential for bringing in new business interests. The loss of tax revenues due to reduced property values and the absence of ongoing economic activities compounds the problems associated with the increased overhead of providing infrastructure to nonproductive areas.

What Are the Sources of these Problems?

The primary sources of the problems creating the brownfield dilemma are Federal environmental laws imposing liability for the contamination that exists on these properties. Paramount among these laws is the Comprehensive Environmental Response, Compensation and Liability Act (CERCLA), also known as Superfund. It is appropriate to recall that CERCLA has a broad scope of liability for potentially responsible parties (PRPs), who are all persons having a contractual relationship with the contaminated site. This liability extending to PRPs under CERCLA is more authoritative and threatening than under any other Federal environmental statute. PRPs are exposed to liability which is strict, joint and several, and also retroactive to the time that the contamination occurred on the property.

The categories of PRPs under CERCLA who are potentially liable for cleanup costs are much broader than just the person(s) who caused the contamination. A current site owner can be compelled to clean up the property or completely underwrite the cost of site remediation, even if he/she has made no contribution to the contamination. Other categories of PRPs include site operators whether or not they have owned the property, generators of hazardous substances contaminating the site, and even individuals who arranged for transportation or disposal of the hazardous substances that ended up at the site.

In addition to CERCLA, sites that are not serious enough to be included on CERCLA's National Priorities List (NPL) may fall within the scope of the Resource Conservation and Recovery Act (RCRA). RCRA creates a *cradle-to-grave* regulatory scheme for management of hazardous wastes and also regulates underground storage tanks (USTs). The Environmental Protection Agency (EPA) has statutory authority to compel *corrective actions* to redress environmental contamination which has resulted from releases of hazardous wastes by RCRA regulated activities.

The Federal Clean Water Act and Clean Air Act can also impact brownfield properties. Both of these acts create stringent pollution control requirements applicable to new sources as well as modifications to existing sources. The reader should recognize that remediation of a brownfield property can be viewed as a new source of air emissions and regulated as such. Beyond these Federal laws, State environmental laws can also create obstacles for brownfield redevelopment. Many states have enacted their own *Superfund* laws that parallel the Federal CERCLA statute.

Voluntary cleanup of contaminated property has always been an option, but the degree of rigor required to obtain an official acknowledgment that effective cleanup has been accomplished has been the same for brownfields as for NPL sites. Clearly, the costs associated with obtaining such *clearance* can easily dwarf the cost of a small piece of property. Most regrettable is the fact that there is little incentive to clean up a site without the possibility of obtaining some sort of official recognition of the benefits. Consequently, these disincentives have led to abandonment or under-utilization of these sites because voluntary cleanup has not been economically feasible in many instances.

What Has EPA Done to Help Resolve the Brownfields Dilemma?

In the last several years, EPA has specifically targeted brownfield redevelopment through its *Brownfields Action Agenda*. The following components of the Agenda are significant.

The *kiss of death* for efforts to clean up a brownfield property has been the listing of it on EPA's NPL. Even listing of the site on EPA's much larger list of sites targeted for investigation and potential listing has been an obstacle to brownfields, although to a lesser extent. This larger investigatory list is called the Comprehensive Environmental Response, Compensation, and Liability Information System (CERCLIS). In an attempt to remove this stigma from undeserving sites, EPA has reviewed CERCLIS and deleted many of its listings. A Statement of Company Removal from the list is also provided to reinforce the fact that EPA has no further interest in the delisted sites.

EPA has also changed its policy so that it will not seek to impose cleanup obligations on persons having an interest in property adjoining a contaminated site where the contamination has migrated under the adjoining property. Therefore, many properties that previously would have been within the CERCLA liability scheme are now excluded.

Another important component of the Brownfields Action Agenda is EPA's effort to provide liability waivers called prospective purchaser agreements (PPAs) to a number of prospective purchasers of brownfield properties. Additionally, EPA has developed a guidance document regarding the issuance of *comfort letters* to sellers and buyers conducting voluntary cleanups. In these letters, EPA provides assurance that it does not intend to take enforcement action at the remediated site. EPA has also recently promulgated a directive allowing future land uses to be considered in selecting the appropriate remedial action at NPL sites.

Moreover, at least one EPA Region has entered into Memoranda of Understanding (MOU) with State environmental agencies under CERCLA. These memoranda provide assurances that EPA will not take enforcement action at sites where private parties have conducted cleanup under the state's direction or pursuant to State voluntary cleanup acts. At this point, MOUs do not appear to contain similar assurances under RCRA. EPA is continuing to refine and pursue MOUs with State environmental agencies.

Information concerning EPA's Brownfield initiatives is accessible over the Internet at

<www.epa.gov/brownfields>

What Have States Done to Resolve the Brownfield Problem?

Many states have developed and are continuing to develop legislation and regulations that narrow the scope of liability under State law, particularly regarding lenders, purchasers, and residential homeowners. Some states are attempting to provide greater certainty and flexibility in identification of cleanup standards, including allowing future land uses to be considered in selecting a remedy. These State laws consistently enact voluntary programs that provide brownfields cleanup incentives.

The two primary incentives that are the cornerstones of the State brownfields voluntary cleanup laws are

1) Utilization of risk-based cleanup criteria that are more specific to the risks encountered at a particular property

2) Liability protection against having to do further cleanup beyond the scope of the risk-based cleanup that is voluntarily agreed upon with the State agency

The utilization of risk-based cleanup criteria in most instances provides an opportunity for a less extensive and therefore, less expensive, cleanup. Integral to this risk-based process is an examination of the current and future land uses of the brownfield property. These land uses are often incorporated into the remediation plan since the land uses may eliminate or restrict the environmental risk associated with the contamination. These future land uses can then be legally incorporated into *institutional controls* such as deed restrictions and other legal agreements.

Deed restrictions provide public notification of use constraints and grant a variety of options for enforcing compliance with them. For example, if an owner of industrial property will agree through deed restrictions and other necessary legal devices that the property will only be used for industrial purposes in the future, he may select target cleanup criteria that are reflective of this intended use. Many states use less stringent concentration levels for industrial sites as compared to residential ones. By opting for these less stringent criteria, a less expensive cleanup may be designed which will not be less protective of human health and environment under the restricted use, so long as the use restrictions are honored.

The second lynch pin of these State brownfields laws is liability protection. Without legal limitation of liability, a brownfields cleanup would always be subject to agency action. The continuing ability to require additional cleanup in the future is an obvious detriment to redevelopment. In addition to future agency action, there is the potential liability to landowners and citizens for *third-party liability,* which includes personal bodily injury, diminution of property value, and destruction of natural resources. State brownfields laws, in most instances, strive to provide some measure of liability protection to the parties performing the agreed risk-based brownfield cleanup.

Although these State brownfields cleanup programs are important and laudable, it must be remembered that they are not binding upon EPA. Therefore, the existence of an MOU between the state and EPA that will bind EPA to the brownfields cleanup approved by the State agency is important. If a state does not have such an MOU, parties considering a voluntary cleanup under a State brownfields law must seek approval from EPA in conjunction with the State agency in order to have assurance that the agreed-upon risk-based cleanup will be approved without further additional activity in the future.

Analysis of State Brownfields and Voluntary Cleanup Laws

Due to the fact that Congress has not enacted Federal brownfields legislation, State governments have taken the initiative to enact laws to provide incentives for cleanup of brownfields properties. Without a Federal pattern for states to follow, every State law is different. We have attempted in this section to briefly analyze and compare these laws. The following categories are utilized for comparison:

1) Does the State law require fees and payment of agency oversight costs?

2) What is the extent of eligibility of properties and applicants for inclusion in the brownfields program?

3) Is risk-based analysis applied in the program?

4) Is liability protection provided by the State law?

5) Is a no further action letter or other similar documentation of acceptance of the cleanup by the State agency provided?

6) Does the program provide reopeners that could require additional cleanup under certain circumstances?

7) Does the program provide economic incentives to help pay for the cleanup of the brownfield property?

A table, located at the end of this chapter, identifies the states which currently have established a brownfields program. This table also contains attendant contact information, including Internet web sites, where available. A brief summary of the information found in each of the seven categories of analysis described above has been included. Finally, the table contains a synopsis of any unique features of the laws that did not fit into one of these seven categories. Please understand that this table has been developed for the purpose of providing a general *snapshot* of the current status of State brownfields programs.

The reader should note that State brownfields programs are currently undergoing a significant amount of change and it is impossible for this text to be completely accurate by the time that it becomes available. With that in mind, the authors have attempted to include in the summary table web page locations for State agency analogues to EPA, even if the State did not have a brownfield program at the time of submission for publication. Additionally, it is in the reader's best interest to download up-to-date information about a particular State program of interest.

The following subsections attempt to summarize trends that can be seen in each of the seven categories of analysis. The conclusions reached are very general in nature and therefore we encourage review of the tabular information for individual states. This is particularly true for states in which there is a vested interest. Web sites have been provided for nearly all states, but may be changed without notice. If interested in a particular state and the listed site is not found, a simple search can be initiated from the general State web page. The generic address of State home pages will follow the format:

<www.state.xx.us>   (*e.g.*, <www.state.ny.us> for New York State)

Most State home pages will have an *Agencies* icon. Click on it to get the listing or directory. EPA-related regulations are most commonly handled by the department of environment; however, they are administered by the department of natural resources or even the department of health in a few states.

Fees and Oversight Costs

The majority of the State brownfields voluntary cleanup programs require that the State environmental agency oversight costs be paid by the program applicant. This will normally consist of the agency's cost of processing and administration applicable to the particular brownfield project. Eight states require payment of agency oversight costs but have no specific application or other fees. Twelve states require both fees (ranging from $250 to $10,000) as well as payment of agency oversight costs. Twelve states do not require agency oversight costs, but do require some type of application or program fee ranging from $500 to $7,500. Only seven states require neither fees nor agency oversight costs.

Eligibility

The great majority (27) of the surveyed State programs allow anyone, including PRPs who caused or contributed to the contamination, to make application for inclusion into the program. Only six State programs do not allow PRPs to become an applicant while three states restrict eligibility to only potential purchasers of brownfield properties. The scope of most programs is broad with regard to the types of properties that may be included in the program, although most exclude NPL sites and sites that are under Federal orders requiring either RCRA corrective action or CERCLA cleanup. This is an important trend since it indicates a relatively common desire to include active industrial and commercial properties owned by persons that could be deemed to either have caused or contributed to the contamination under existing environmental laws. This indicates a trend toward encouraging as many properties to be cleaned up in the state as possible rather than restricting the brownfields incentives to prospective purchasers or innocent landowners.

Risk-Based Cleanup

Most of the states provide for risk-based cleanup of brownfields properties. There appears to be a trend towards allowing more emphasis on the cost effectiveness of the cleanup. Many states are utilizing institutional controls as part of the risk analysis in order to reduce the risk and therefore to reduce the cost of cleanup. This trend confirms that most states are utilizing risk-based cleanup in conjunction with institutional controls to provide significant incentives to clean up brownfield properties.

Liability Protection

Most states provide liability protection against further requirements by the State environmental agency for additional cleanup. Only three surveyed states do not provide for liability protection and only one state limited liability protection to prospective developers. Five states do not provide liability protection to PRPs. At the other end of the spectrum, six states not only provide that PRPs could obtain liability protection, but also extend liability protection to cover third-party liability.

No Further Action Letters

All of the surveyed State brownfields voluntary cleanup programs provide for documentation of approval of the risk-based cleanup through no further action letters (NFAs) or certificates of completion (CoCs). The programs were fairly evenly divided between these two types of approvals. This indicates the importance of NFAs and COCs for brownfields redevelopment to prospective purchasers and lenders. Documentation of the completion of the brownfield cleanup process and approval by the State agency provides a meaningful level of comfort for transactions involving these properties. It should also be pointed out that business owners who are not buying, selling, or lending also need this documentation in order to reactivate construction and use of these previously dormant properties.

Reopeners

Reopeners are legal requirements setting forth the bases on which additional cleanup of the brownfield property can be required in addition to what was originally agreed upon with the State agency. These reopeners vary from state to state, but in many states the following are included:

1) Fraudulent or misrepresented information has been provided to the agency.

2) An imminent threat to human health or the environment occurs on the property.

3) A previously unknown condition or new information comes to the agency's attention.

4) The risk on which the agreed-upon cleanup was based significantly changes so that additional cleanup to prevent endangerment to human health and the environment is required.

These reopeners are *chinks in the armor* of liability protection. However, it should be understood that reopeners can benefit the landowner, as well as call for increased levels of treatment. Although none of the stakeholders of a cleanup initiative wants to consider the possibility of reopening a completed project, inspection of factors that states have identified as triggering such action has obvious merit. State programs must provide necessary flexibility to meet both the land user's needs and also be protective of risks to human

health and the environment that were not contemplated by the initial agreed-upon risk-based cleanup.

Economic Incentives

A significant segment of the surveyed State programs, 13 states, do not provide economic incentives. Of the remaining states that do, 17 states provide tax exemptions, abatements or credits to encourage development of brownfields properties. 11 states provide monetary grants while the same number provide for low-interest loans to infuse capital into brownfields projects.

It is expected that in the future, more states will enact brownfields tax deductions due to the momentum provided by the passage by Congress in 1997 of legislation allowing eligibility for immediate tax deduction of environmental remediation expenditures for brownfields sites located in high poverty areas targeted for Federal empowerment zones or enterprise community efforts and announced as a brownfields pilot project by EPA. This tax treatment is scheduled to expire at sunset on December 31, 2000. It applies to neither rural brownfield sites nor to CERCLA NPL sites.

This summary analysis of State brownfields laws is only intended to provide a broad perspective on existing and developing trends in these State enactments. As stated earlier, we encourage the reader to review the attached table for clarification of any specific state's brownfields program and to further research the state's program by getting in touch with the State agency contact person and utilizing, if available, the State program's web site.

Implementation of State Brownfield Programs

State Brownfield Programs Just Extend CERCLA and RCRA to Less Contaminated Sites, Right?

Brownfield projects are voluntary efforts. The enabling statutes for State brownfields programs do not grant authority to mandate compliance unless the site is enrolled in the program. Individual states vary in the incentives offered property owners to increase the number of eligible sites and enrollment. To view brownfields programs as an attempt to extend the authority of RCRA and CERCLA is to lose sight of the incentive packages, a potentially costly mistake for many sites.

The differences between State programs preclude blanket statements being made, and that fact should have significant impact on how the prudent environmental professional evaluates potential brownfield programs. Be sure to establish as first priority the thorough review of the brownfield program in the state containing the project. This includes the eligibility requirements, the process and paperwork to be submitted, an estimate of the amount of time that regulators will need to review proposals and the results of sampling or modeling, the expected fees, the professional requirements called for by the agency, the liability protection offered and whether or not the state has signed a MOA with its EPA regional office, and, of course, the reopeners. For the purpose of discussing how Certified Hazardous Materials Managers (CHMMs) will likely interface with brownfield programs, we will assume that the state of interest has one of the more comprehensive statements of eligibility.

The first section of this chapter outlines the factors that have given rise to the numerous State brownfield programs that exist—predominantly the need to facilitate returning idle industrial and commercial property to productive use. This material has also indicated the individuality of the State programs. Even so, it is easy to note the similarity between elements of brownfields programs and the more established concepts of due diligence and CERCLA/RCRA-directed remediation efforts. Specifically, the determination of the extent of contamination that exists at a site is virtually synonymous with a Phase I and Phase II environmental property assessment or a rigorous site assessment under CERCLA/RCRA. The evaluation of technologies that may be used to remediate a brownfield site is very similar to what takes place in a CERCLA remedial investigation/feasibility study (RI/FS). It is, in fact, very desirable to conform to the requirements of these as much as possible. The more closely EPA, CERCLA, or RCRA protocols and methodology are followed in the characterization and remediation

of a brownfield site, the higher the confidence level for mitigation of long-term liability.

It is important to point out that there are at least three subtle differences between State brownfields programs and the older Federally mandated ones. These differences can have a significant impact on the decision to conduct the project under oversight of a State brownfield program. The first of these differences is so obvious that it is easy to overlook. Brownfields projects are voluntary cleanup activities. Keep in mind that there is a wide range of quality in voluntary cleanup efforts. There is a tendency to associate brownfields projects with redevelopment of derelict industrial sites, but voluntary cleanup efforts at active facilities can qualify for many of the State brownfields programs.

The absence of a regulatory mandate to remediate a site typically reduces the level of commitment to initiate a cleanup effort. This bias is one of the major factors that has motivated states to provide incentives for cleanup through brownfields programs. Consider the level of difficulty for convincing management of a marginally profitable industrial facility to undertake the expense of remedial activities that are not mandated. The reticence to initiate a voluntary cleanup project will be much stronger when there are homeowners near property boundaries that may become alarmed upon learning of the contamination.

The lack of a regulatory mandate removes the necessity to keep to a time line and it is easy for projects to bog down. Property redevelopment ventures are particularly susceptible to failure from this. The potential for all parties to simply walk away or defer action indefinitely kills many initiatives. It is prudent to go into any brownfield project with a heightened sensitivity to the need to provide compelling data that the proposed course of action is both economically and environmentally beneficial. Project initiation is facilitated by stressing the advantages of being proactive, clearly identifying economic benefits, emphasizing brownfield program incentives, providing concrete data regarding liability reduction, and presenting a realistic expectation for the reduction of excess risk to the general population.

The two remaining ways State brownfield restoration efforts differ from CERCLA/RCRA programs are best discussed together. These are: (1) the liability protection offered by State program oversight and (2) State program flexibility. Both of these result from the lack of a Federal statute requiring EPA to provide regulatory oversight of State programs. Absence of a requirement for equivalence provides states the flexibility to design a program with unique features to best serve the concerns of their constituents. But it also dramatically lessens the technical support they can receive from their EPA regional offices in these unique areas.

Federally mandated programs, although normally delegated to the same State agency overseeing the brownfield program, are still Federal programs and subject to oversight by EPA. It is currently impossible to determine the likelihood that a cleanup appropriately completed under a State brownfield program may be reopened under a Federal program (CERCLA or RCRA) because of the novelty of the State programs. This is especially true when there is not a memorandum of agreement between the state and EPA that includes those circumstances relevant to the site of interest.

One can reasonably expect more latitude for using innovative technologies in brownfield cleanups as compared to the more rigid RCRA or CERCLA allowances, but of course this will vary from state to state. Another way that some State programs differ from the Federal ones is by not requiring a public hearing to review the project before implementation.

Since brownfield initiatives are voluntary efforts, most States require the user to finance administrative costs via a fee structure. There are opportunities to take advantage of this and use State regulators as pseudoconsultants. Seek their input regarding reasonable measures that have proved effective in limiting the costs or time required for the completion of similar projects. As the site progresses through the process of characterization and assessment, significant time and cost reductions may be achieved by judiciously consulting the regulatory person reviewing the project.

What If the State Does Not Have a Brownfields Program?

In all likelihood, the state will still have a mechanism for handling voluntary cleanups in spite of the lack of a formal brownfields program. Contact the State environmental agency and provide a description of the site, its history, and the expected use upon completion of the remediation. The most likely places to start will be the Superfund or RCRA branches. Distill the facts regarding the site to a terse set of the most significant considerations. This should include whether the intended use of the remediated site will be residential, commercial, or industrial. Determine if there are sensitive receptors in the area such as drinking water wells, daycare centers, or hospitals. Have an inventory of the activities and chemicals that were used on the site. Review also the most typical cleanup targets that would be established by CERCLA or RCRA. If specific, insightful questions are asked, there is a much greater probability of obtaining the information necessary for decision making. A final recommendation: while discussing the brownfield project with regulators, ask questions, don't debate the issues. If information provided during the conversation does not appear to correlate with other data that have been collected, put that data in writing and submit it for review.

In the absence of brownfields guidance or process at the State level, follow CERCLA or RCRA cleanup policies and guidance as much as possible. The general rule of thumb for deciding which program will have jurisdiction over the site can be given as: (1) if the contamination originated from underground tank systems (either petroleum or chemical), it is probably a RCRA Subtitle I site; (2) if the contamination originated from industrial activity on a site which is currently regulated under RCRA Subtitle C (site has a hazardous waste generator identification number), then it will probably be a RCRA Corrective Action; otherwise, (3) the site is most likely a CERCLA site.

The best way to ensure long-term liability protection is to follow protocols equivalent to CERCLA/RCRA schemes. This is particularly true during the data-gathering stage. Even if the state has a voluntary cleanup program, all the site assessment data generated and the effectiveness of control measures implemented may ultimately be measured relative to CERCLA/RCRA yardsticks. Consider carefully the dangers associated with not doing an aggressive job of characterizing the site, designing the site sampling plan, implementing the sampling plan, using Quality Assurance /Quality Control (QA/QC) measures to ensure data quality, or using an analytical lab capable of producing Contract Laboratory Program (CLP) quality analytical data. A certain reopener for any project is to miss areas of contamination during the site investigation.

Data quality is a very important to EPA. Superfund program activities are particularly likely to require development and use of data quality objectives. If they are not employed from the outset of the project, State program approval of a site sampling plan is unlikely. The data generated by activities which were not developed consistent with data quality objectives may be deemed unusable by a regulatory agency or subjected to a very high degree of scrutiny.

What Are Data Quality Objectives and Why Are They Significant?

Over the past decade EPA has begun promoting and even requiring the use of Data Quality Objectives (DQOs) as a guiding principle for ensuring the defensibility of environmental data. In a document entitled *Data Quality Objectives Process for Superfund: Interim Final Guidance* (EPA, 1993), the agency sets forth a seven-step process for explicitly establishing acceptance and rejection criteria for information generated by each activity involved in the process of generating data. Details about DQOs and a number of useful tools can be obtained by searching the Office of Solid Waste and Emergency Response (OSWER) documents for DQOs from the EPA brownfield web site:

<www.epa.gov/swerosps/bf/>

The American Society for Testing and Materials (ASTM) has developed two standards related to the DQO process (the titles are included in the bibliography). The DQO process can be briefly summarized thus:

Step 1 State the problem that requires new environmental data.

Step 2 Identify the decision that must be resolved to address the problem.

Step 3 Identify the informational inputs (data) required to address the problem.

Step 4 Specify the spatial and temporal circumstances that are covered by the decision.

Step 5 Integrate the information developed in Steps 1–4 into a statement that describes the logical basis for choosing from among the alternate actions.

Step 6 Specify the acceptable limits on decision errors and the corresponding performance goals limiting uncertainty in the data.

Step 7 Design the most resource-effective sampling and analysis plan for generating the data expected to satisfy the DQOs.

DQOs are developed in the earliest stage of the project. They serve as the basis for judging both the appropriateness and the veracity of the data which will be generated.

What Tools Can I Use for the Site Evaluation Process?

The introduction of the concept of an *innocent landowner* in the 1986 Superfund Amendments and Reauthorization Act has given rise to the *due diligence investigations* of property prior to transfer of title. The environmental site assessment industry has matured to the point that there are now two ASTM standards describing the process. One describes the general Phase I Environmental Site Assessment process (E 1527–93) and the other is designed to address commercial real estate (E 1528–93). From a professional standpoint, one of the best protections against charges of negligence is to be proficient in the application of all applicable professional codes of practice and to follow them faithfully.

An environmental site assessment conducted in accordance with the two ASTM standards will provide a good basis from which to develop a site sampling plan. ASTM also has a set of standards on environmental sampling that can be purchased collectively (03–41–8097–38). These standards, in conjunction with guidelines set forth in SW–846 (EPA, 1997) and other EPA publications like the Region IV Environmental Investigation Standard Operating Procedures and Quality Assurance Manual (EISOPQAM), will be extremely helpful in planning for and conducting site sampling. This document can be easily downloaded from the web site:

<www.epa.gov/region04/sesd/eib/eisopqam.html>

All of these activities will be selected and conducted in a manner sufficient to achieve all DQOs. The list of activities will include

1) Identifying the analytes of interest

2) Targeting the size and types of samples to be taken

3) Developing a sampling scheme identifying the location of all samples to be collected

4) Establishing requirements for sampling tools, sample containers, and sample preservation for each type sample to be collected

5) Ensuring appropriate QA/QC measures are included in the sampling protocols

6) Providing forms for the necessary paperwork for documenting all samples, establishing chain of custody, and submitting the samples to an analytical laboratory for their appropriate analysis

States that have brownfields programs will have established forms and processes to be followed. It is highly likely that DQOs will be required as part of the process. It goes without saying that the development of a sampling plan followed by collection of environmental samples is an expensive process. No regulatory official wants to be put into a position of telling a property owner that the data they have paid for was collected incorrectly or is inappropriate for characterizing their site. By requiring DQOs to be included in the initial site description and Phase I characterization, an added measure of comfort can be provided to all parties that the approved scope of work detailed in subsequent stages will be adequate to achieve success.

Finally, there are a variety of software packages and guidance documents to assist in the development of a site safety and health plan. A commonly employed tool is the Health and Safety Plan (HASP) software available from EPA. HASP can be downloaded from the web site—

<http://204.46.140.12> (formerly <http://www.ert.org>)

and technical support can be obtained by calling (800) 999–6990. Keep in mind that the HASP will need to be updated or modified as the brownfield project moves from the initial screening stages through final remediation. It will also be necessary to ensure that all individuals involved in these activities have undergone the appropriate degree of Hazardous Waste Operations and Emergency Response (HAZWOPER) training.

What Kind of a Team Should Be Assembled for a Brownfield Project?

For purposes of discussion, it is probably best to divide the brownfield team into two groups, an administrative group to deal with financial and business considerations and a technical group to address issues related to site investigation and cleanup. The administrative team members would include the owner/site manager, capital source/chief financial officer, attorney, environmental professional, and a public relations person.

The technical team members will be equivalent to those that would be involved in a CERCLA cleanup. A CHMM needs to be involved to ensure that all applicable environmental regulations are properly taken into account. Many states require that a registered professional engineer submit the documentation associated with brownfield investigations and remediation. This engineer should be an expert in either the designing or construction of environmental remediation systems, or possibly both. A registered professional geologist may be necessary to assess soil samples and log the wells installed for ground water monitoring. The wells themselves should be drilled by a certified driller who has sufficient environmental experience to guarantee that wells and samples will not be contaminated by the drilling process. A certified industrial hygienist and/or a certified health physicist may be necessary to develop the HASP and assure worker safety. Site assessment may require the services of professionals experienced in the areas of ground water modeling, pollutant migration, or risk assessment. Other experts may be required to address endangered species, archeological considerations, Federal facility requirements, and wetlands.

A thorough discussion of selection of technologies applicable to a given site is beyond the scope of this chapter. The reader is referred to other sections of this text for guidance. It will also be helpful to review the EPA Superfund Office of Research and Development web site—

<http://www.epa.gov/ord/>

and Department of Energy (DOE) Environmental Management web site—

<http://www.em.doe.gov>

for information about innovative technologies.

Bibliography

"Amendment to the National Oil and Hazardous Substances Pollution Contingency Plan (NCP)." *Federal Register* 60 (29 March 1995): 16053. (Codified at 40 CFR 300).

American Society for Testing and Materials. *ASTM Standards on Environmental Sampling.* 2nd Ed. ASTM Publication Code Number 03–418097–38. West Conshoshoken, PA: ASTM, 1997.

American Society for Testing and Materials. *Standard Practice for Environmental Site Assessments: Phase I Environmental Assessment Process.* E 1527–93. West Conshoshoken, PA: ASTM, 1993.

American Society for Testing and Materials. *Standard Practice for Environmental Site Assessments: Transaction Screening Process.* E 1528–93. West Conshoshoken, PA: ASTM, 1993.

American Society for Testing and Materials. *Standard Practice for Generation of Environmental Data Related to Waste Management Activities Development of Data Quality Objectives.* D 5792–95. West Conshoshoken, PA: ASTM, 1995.

American Society for Testing and Materials. *Standard Practice for Generation of Environmental Data Related to Waste Management Activities Quality Assurance and Quality Control Planning and Implementation.* D 5283–92. West Conshoshoken, PA: ASTM, 1992.

Buonicore, A. J., Ed. *Cleanup Criteria for Contaminated Soil and Ground-Water.* DS 64. West Conshoshoken, PA: ASTM, 1996.

Davis, T. S., and K. D. Margolis. *Brownfields, A Comprehensive Guide to Redeveloping Contaminated Property.* Chicago, IL: American Bar Association, 1997.

Environmental Protection Agency. *Contract Laboratory Program National Functional Guidelines for Inorganic Data Review.* EPA/540/R–94/013. Springfield, VA: National Technical Information Service, 1994.

Environmental Protection Agency. *Contract Laboratory Program Statement of Work for Inorganic Analysis—Multi-Media, High Concentration ILM02.1.* EPA/540/R–94/095. Springfield VA: National Technical Information Service, 1991.

Environmental Protection Agency. *Contract Laboratory Program—Statement of Work for Organic Analysis, OLM03.1.* EPA/540/R–94/073. Springfield, VA: National Technical Information Service, 1994.

Environmental Protection Agency. *Contract Laboratory Program—Statement of Work for Turnaround Dioxin Analysis, Multi-Media.* EPA/540/R–94/091. Springfield, VA: National Technical Information Service, 1992.

Environmental Protection Agency. *Data Quality Objectives (DQO) Decision Error Feasibility Trials (DEFT) Version 4.0 (on diskette).* PB95–100418INC. Springfield, VA: National Technical Information Service, 1994.

Environmental Protection Agency. *Data Quality Objectives Process for Superfund.* EPA/540/R–93/071. Washington, DC: US Government Printing Office, 1993.

Environmental Protection Agency. *Environmental Investigations Standard Operating Procedures and Quality Assurance Manual.* Athens, GA: USEPA, Region IV, 1996.

Environmental Protection Agency. *Good Automated Laboratory Practices: Principles and Guidance to Regulations for Ensuring Data Integrity in Automated Laboratory Operations.* 2185. Research Triangle Park, NC: USEPA Office of Information, 1995.

Environmental Protection Agency. *Health and Safety Plan.* EPA/540/C–93/002. Springfield, VA: National Technical Information Service, 1990.

Environmental Protection Agency, *Organic Contract Compliance Screening System (OCCSS) Software (OLM01.8 Version 7).* PB94–504255INC. Springfield, VA: National Technical Information Service, 1994.

Environmental Protection Agency, *Quality Assurance/Quality Control Guidance for Removal Activities—Sampling QA/QC Plan and Data Validation Procedures.* EPA/540/G–90/004. Springfield, VA: National Technical Information Service, 1990.

Environmental Protection Agency. *Sampler's Guide to the Contract Laboratory Program.* EPA/540/R–96/032. Springfield, VA: National Technical Information Service, 1996.

Environmental Protection Agency. *Test Methods for Evaluating Solid Waste Physical/Chemical Methods, CD-ROM Version 2.* SW–846. Springfield, VA: National Technical Information Service, 1997.

Environmental Protection Agency. *User's Guide to Contract Laboratory Program.* EPA/540/8–89/012. Springfield,VA: National Technical Information Service, 1988.

"Final Policy Toward Owners of Property Containing Contaminated Aquifers." *Federal Register,* 60 (3 July 1995): 34790.

General Accounting Office. *Superfund: Barriers to Brownfield Redevelopment.* GAO/RED–96–195. Washington DC: Government Printing Office, US General Accounting Office, 1996.

Gerrard, M. B., Ed. *Brownfields Law and Practice.* New York, NY: Matthew Bender Publishing Company, 1998.

John E. Milner *is a partner in the Jackson, Mississippi law firm of Brunini, Grantham, Grower & Hewes, PLLC. He received his BA from the University of Mississippi in 1975 and his JD from the University of Mississippi Law School in 1978. His practice emphasizes environmental law, real estate due diligence, and contractual matters. Mr. Milner represents business and industrial clients in environmental litigation in State and Federal courts in the southeastern region. He has also had significant experience in environmental permitting and enforcement actions and regulation promulgation proceedings involving the US EPA and the Mississippi Department of Environmental Quality. Mr. Milner is currently serving as Environmental Counsel for the Mississippi Manufacturers Association, and has recently authored and lobbied to enact the Mississippi Brownfields Voluntary Cleanup and Redevelopment Act. He is the Mississippi consultant to Matthew Bender, Inc., for its multivolume reference publication* Brownfield Law and Practice. *He has authored numerous articles including "Overview of Major Federal Environmental Acts and Regulations for the General Practioner" (with Dr. Waggoner), for the* Mississippi Law Journal, *and "Environmental Justice," for the ABA's* Natural Resources & Environment.

Dr. Charles A. Waggoner *currently serves as the Manager of Safety, Excellence, and Environment for the Diagnostic Instrumentation and Analysis Laboratory at Mississippi State University (MSU). He holds a BS and MS in biochemistry and a PhD in physical chemistry. He has over 15 years experience in environmental management with particular emphasis on hazardous waste management and related issues. Dr. Waggoner's professional activities have included serving as the MSU Hazardous Waste Officer, Technical Director of Environmental Training for the MSU Division of Continuing Education, and Dean of Environmental Science and Technology at Chattanooga State Technical Community College. Dr. Waggoner has served as a member of the IHMM-ACHMM Advisory Committee, ACHMM Board of Directors, and as General Chairperson for the 1994 National Conference in Chattanooga, Tennessee. He has authored numerous articles, including, "Overview of Major Federal Environmental Acts and Regulations for the General Practioner" (with Mr. Milner), and the asbestos chapter in the* Handbook on Hazardous Materials Management, *5th edition, published by IHMM in 1995.*

The authors would like to express their appreciation to Gene Wasson and Richard Cirilli of the Brunini, Grantham, Grower, and Hewes law firm, for their assistance in the preparation of materials contained in this chapter.

Summary of State Brownfields Programs

State	Alabama	Alaska	Arizona	Arkansas
Program	Informal policy-based program only	Has discretion to allow alternative cleanup levels	Greenfields Pilot Program (GPP), 1997 Arizona Session Laws, Ch. 296 12. 4/30/97 and Voluntary Remediation Program (VRP), Arizona Revised Statutes Ann. 49-282.05, 4/29/97	Arkansas Code Ann. 8-7-110 et seq., 4/2/97
Contact Information	<www.state.al.us>	<www.state.ak.us/local/>	Arizona Dept. of Environmental Quality (ADEQ) 3033 N. Central Ave. Phoenix, AZ 85012 (602) 207-4109 <www.azleg.state.az.us>	Superfund Brownfields Branch Arkansas Dept. of Environmental Quality (ADEQ) PO Box 8913 Little Rock, AR 72219-8913 (501) 682-0744 <www.adeq.state.ar.us>
Fees			ADEQ's costs	
Eligibility			GPP: first 100 qualified sites – No groundwater impact – No UST sites – No site subject to current enforcement – No permitted hazardous waste site with a release to soil in volation of its permit VRP: anyone	Sites where no PRP can be found
Risk-Based Analysis			Predetermined risk-based levels based on future use of property or site specific standards based on risk assessment	Cleanup standards on case-by-case basis
Liability Protection				No
No Further Action (NFA)			Yes	CNS for past contamination
Reopeners			If audit shows site not cleaned in accordance with NFA letter	No
Economic Incentives			None	Loan fund
Unique Features				1997 statute provides for petition process to ease water quality standards for certain long-term improvement projects, including brownfields

State	California	Colorado	Connecticut	Delaware
Program	Dept. of Toxic Substances Control adopted a policy document in 1995	Voluntary Cleanup and Redevelopment Program (VCRP), 1994	Two programs: 95-183 and 95-190 Connecticut General Statutes 22a-133a et seq., 1995	1995
Contact Information	<www.state.ca.us/s/environ/>	Superfund/Voluntary Cleanup Unit Leader Colorado Dept. of Public Health and Environment (CDPHE) HazMat/Waste Management Div. 4300 Cherry Creek Dr. South Denver, CO 90222-1530 (303) 692-3300 <www.cdphe.state.co.us/hm/rp_gen.html>	Urban Sites Remedial Action Program Connecticut Dept. of Environmental Protection (CDEP) 79 Elm St. Hartford, CT 06106-5127 (860) 424-3800 <www.state.ct.us/agency.htm>	Manager Dept. of Natural Resources and Environmental Control (DNREC) Div. of Air and Waste Management SIRB 715 Grantham Lane New Castle, DE 19720 (302) 323-4540 <www.dnrec.state.de.us/>
Fees		$2000 filing fee plus costs	95-193: $2000 fee	$5000 oversight deposit
Eligibility		Owners of property that are not subject to any other government authority	95-183: 1) Sites on Connecticut's hazardous waste disposal site list 2) Groundwater class GA or GAA sites, and 3) "Establishments" under the Connecticut Property Transfer Act 95-190: Groundwater class GB and GC sites	Persons other than PRPs, numerous site exceptions
Risk-Based Analysis		Depends on future land use		State developed screening levels— varies with commercial and residential use
Liability Protection		No	No	Release from liability to DNREC for future release attributable to conditions existing before certification issued. Contribution protection under State CERCLA
No Further Action (NFA)		Yes	CNS if not a PRP	CoC or remedy
Reopeners		Failure to comply with cleanup plan	Cleanup not completed or land use restriction not filed or complied with	If work under plan is not complete
Economic Incentives		None	Loans	Matching grant to $25,000, tax credits, loans
Unique Features			CNS may in some cases cost 3% of the value of uncontaminated property	

See end of Table for list of abbreviations.

Summary of State Brownfields Programs (Continued)

State	District of Columbia	Florida	Georgia	Hawaii
Program	None	Brownfields Redevelopment Act Florida Statutes 288.107 and 376.77, *et seq.*, 5/30/97	1996 Amended 1998	Voluntary Cleanup Program (VCP) 7/7/97
Contact Information		Florida Dept. of Environmental Protection (DEP) 2600 Blair Stone Road MS 4505 Tallahassee, FL 32399-2400 (850) 488-0190 <www.dep.state.fl.us>	Georgia Dept. of Natural Resources (GDNR) Environmental Protection Div. (EPD) 205 Butler St., S.E., Suite 1462 Atlanta, GA 30334 (404) 657-8600 <www.ganet.org/dnr/environ/>	Hawaii Dept. of Health (HDH) Hazard Evaluation and Response Office, Rm 206 919 Ala Moana Blvd. Honolulu, HI 96814 <www.state.hi.us/doh/eh/ eiwmsg03.htm>
Fees				$1000 filing fee plus costs
Eligibility		Any person who has not contributed to contamination after 7/1/97; sites under active Federal enforcement and hazardous waste TSDs or post closure are excluded	Potential purchasers of sites on Hazardous Sites Inventory or State Superfund list – PRPs are excluded	Sites subject to current enforcement excluded
Risk-Based Analysis		Default cleanup standards with alternative of site-specific risk-based standards	Based on future use, then standard or site-specific risk assessment within the case classification	Yes
Liability Protection		From State enforcement and from third-party contribution while cleanup is ongoing	Limits third-party liability for contribution or damages; release of liability to the state	Liability protection from subsequent enforcement actions
No Further Action (NFA)		Yes	EPD will respond in writing as to whether it concurs with party's compliance status report	CoC
Reopeners		Fraud, new information of threat to environment or public health, cleanup not complete, increase in risk such as change in site use, new release occurs	No	No
Economic Incentives		Tax credit and loans	None	Tax exempt
Unique Features				

State	Idaho	Illinois	Indiana	Iowa
Program	VCP Enactment of statute, 3/14/95 Promulgation of regulations, 2/97	VCP 1989, amended 1995	Indiana Code 13.25-5-1 *et seq.*, 1992 Brownfields Financial Incentives, 1997	VCP Enactment of statute 1997 No regulations as of 1/1/99
Contact Information	Dept. of Health and Welfare (DHW) Division of Environmental Quality (DEQ) 1410 N. Hilton Boise, ID 83706 (208) 373-0276 <www.state.id.us/deq/>	Brownfields Coordinator Illinois Environmental Protection Agency (IEPA) Bureau of Land 1001 North Grand Ave. E. Springfield, IL 63702 (217) 785-3497 <www.epa.state.il.us/>	Brownfields Coordinator Indiana Dept. of Environmental Management (IDEM) 2525 N. Shadeland PO Box 6015 Indianapolis, IN 46202-6015 (317) 308-3058 <www.state.in.us/idem/oer>	Iowa Dept. of Natural Resources (IDNR) Environmental Protection Div. 900 E. Grand Ave. Des Moines, IA 50319 (515) 242-5817 <www.legis.state.ia.us>
Fees	Costs of review and oversight $250 filing fee	IEPA's oversight costs unless applicant retains and Illinois licensed professional engineer to oversee	$1000 application fee and pay costs of oversight	Review costs up to $7500
Eligibility	Sites subject to current enforcement excluded	Anyone – Specific site exceptions such as NPL sites	Excludes various sites such as imminent threat, current enforcement, *etc.*	Excludes NPL sites, USTs and animal feeding operations as well as sites subject to State or Federal enforcement
Risk-Based Analysis	Yes	Yes	Can either use standard or site-specific risk assessments	Standard or site-specific risk assumptions – Site-specific include future land use
Liability Protection	Liability protection from subsequent enforcement actions	Neither Illinois nor EPA will bring future enforcement actions under CERCLA or RCRA	CNS that protects against all public and private claims under State environmental laws	CNS that protects against further enforcement or liability for response actions – Exempt from liability under state environmental laws to the state or other parties
No Further Action (NFA)	CoC	Yes	CoC	Yes
Reopeners	No	Site is not managed in accordance with site plan	Conditions not known to IDEM	Imminent and substantial threat – Fraud or material misrepresentation – Corrective controls fail
Economic Incentives	Tax exempt	Grants, tax credits	Loans, grants, loan guarantees, *etc.* for local grants – Tax abatements for private entities	Fund for financial assistance – Local governments may provide tax relief
Unique Features				

See end of Table for list of abbreviations.

Summary of State Brownfields Programs (Continued)

State	Kansas	Kentucky	Louisiana	Maine
Program	VCP Enactment of statute, 1997 Promulgation of regulations, 1998	Brownfields Program 1996	VCP Enactment of statutes 7/1/96 No regulations as of 1/1/99	VCP 1993
Contact Information	Kansas Dept. of Health and Environment (KDHE) Bureau of Environmental Remediation Forbes Field, Bldg. 740 Topeka, KS 66620 (785) 296-1665 <www.kdhe.state.ks.us>	Superfund Branch Kentucky Natural Resources and Environmental Protection Cabinet (NREPC) Div. of Waste Management 14 Reilly Road Frankfort, KY 40601-1190 (502) 564-6716 <www.state.ky.us/>	Louisiana Dept. of Environmental Quality (LDEQ) Inactive and Abandoned Sites Div. PO Box 82282 Baton Rouge, LA 70884-2282 (504) 765-0487 <www.deq.state.la.us/oshw/ias/ias.htm>	Voluntary Response Action Program Maine Dept. of Environmental Protection (MDEP) Bureau of Remediation and Waste Management 12 State House Station Augusta, ME 04333-0017 (207) 287-2651 <www.state.me.us/dep>
Fees	$200 application fee and pay oversight costs		Agency costs	$500 application fee and oversight costs
Eligibility	Excludes various sites, such as NPL, imminent threat, current enforcement, *etc.*	Limited to public entities	PRPs are eligible but must remove all contaminants of concern Non-PRPs rely on easement restrictions, *etc.*	Anyone
Risk-Based Analysis	Site-specific, may include institutional controls, such as use restrictions	Yes	Standard and site-specific	Site-specific, including institutional controls
Liability Protection		Protection for public entity and any successor or assign from further remediation liability	Release from liability under State Superfund law – Does not extend to damage to third parties	Release of liability to state and protection against third party contribution
No Further Action (NFA)	Yes	Yes	CoC	CoC
Reopeners	Fraud or misrepresentation, failure to complete and maintain	No	Fraud or misrepresentation	No
Economic Incentives		None	None	None
Unique Features				

State	Maryland	Massachusetts	Michigan	Minnesota
Program	VCP Enactment of statute, 2/97	New law passed in 1998 No data available as of 1/1/99	Informal VCP that includes Brownfield financing—otherwise, no specific statute	Enactment of statute and promulgation of regulations in 1992
Contact Information	Maryland Dept. of the Environment Waste Management Administration 2500 Broening Hwy. Baltimore, MD 21224 (410) 631-3450 <www.mde.state.md.us/environment/>	<www.state.ma.us/>	<www.deq.state.mi.us/erd/>	Minnesota Pollution Control Agency (MPCA) Site Response Section 520 Lafayette Rd. St. Paul, MN 55155-4194 (612) 296-0892 <www.pca.state.mn.us/cleanup/index.html>
Fees	$6000 application fee and oversight costs			
Eligibility	Excludes sites such as NPL, current enforcement, and sites contaminated after 10/97, unless innocent purchaser or site-specific			Excluded sites include NPL sites and sites with drinking water contamination
Risk-Based Analysis	Standard or site-specific			Standard or site-specific
Liability Protection	Release from liability for further remediation under State law; protection from third party contribution			Release from state of responsibility for further remediation under State Superfund law
No Further Action (NFA)	CoC			CoC is available to non-PRPs
Reopeners	Imminent threat, fraud or misrepresentation, conditions previously unknown, *etc.*			New information
Economic Incentives	Loans, grants and tax abatement		Loans and tax credits	Loans, grants and tax credits
Unique Features				

See end of Table for list of abbreviations.

Summary of State Brownfields Programs (Continued)

State	Mississippi	Missouri	
Program	Mississippi Brownfields Voluntary Cleanup and Redevelopment Act (MBVCRA), 1988 Mississippi Laws, Chapter 528 Regulations are being promulgated as of 1/1/99	VCP 1993 Brownfields Redevelopment Program (BRP) 1995	
Contact Information	Brownfields Coordinator Mississippi Dept. of Environmental Quality (MDEQ) PO Box 10385 Jackson, MS 39289-0385 (601)961-5654 <http://www.welcome.to/brownfields>	VCP: Chief, Voluntary Cleanup Section Missouri Dept. of Natural Resources (MDNR) PO Box 176 Jefferson City, MO 65102 (573) 526-8913 <www.dnr.state.mo.us/deq/hwp/program.htm>	BRP: Manager, Finance Programs Missouri Dept. of Economic Development (MDED) PO Box 118 Jefferson City, MO 65102 (573) 751-0717
Fees	$2000 application fee and pay MDEQ's processing and administration costs	VCP: $200 application fee and initial deposit to cover oversight costs	
Eligibility	Anyone, including PRPs, except 1) Existing or proposed CERCLA NPL sites unless CoC has been issued 2) Sites under current CERCLA or RCRA orders and 3) RCRA sites undergoing RCRA corrective action unless it has been completed The commission on Environmental Quality has the discretion to determine qualifying projects	VCP: any person and most nonresidential property BRP: only certain types of limited projects	
Risk-Based Analysis	Risk-based cleanup criteria specific to the property mandated	BRP: allows for risk-based cleanups	
Liability Protection	Liability protection from further remediation – No third-party liability protection	BRP: liability protection consistent with the level of risk remaining at the site – Immunity from third-party civil action – Immunity from tort liability prior to issuance of NFA letter	
No Further Action (NFA)	Yes	Yes	
Reopeners	None, unless 1) Developer provides false information or fails to provide information 2) Previously unreported contamination 3) Changes in exposure 4) New information raises risk 5) Failure to file notice of Brownfields Party 6) Owner violates land use restrictions		
Economic Incentives	None, but MBVCRA requires MDEQ to propose incentives to the Mississippi legislature	BRP: Direct loans, guarantees, grants, and tax credits	
Unique Features			

State	Montana	Nebraska	Nevada	New Hampshire
Program	Voluntary Cleanup and Redevelopment Act (VCRA), 1995	Remediation Action Plan Moderating Act (RAPMA); 1/1/95	No VCP – Party may submit proposed remediation plan to the Nevada Dept. of Conservation and Natural Resources (NDCNR) for review and comment	New Hampshire Brownfields Program (NHBP), 7/1/96 No regulations as of 1/1/99
Contact Information	Montana Dept. of Environmental Quality (MDEQ) Remediation Division 2209 Phoenix PO Box 20091 Helena, MT 59620-0901 (406) 444-1420	RAPMA Program Coordinator Nebraska Dept. of Environmental Quality (NDEQ) The Atrium 1200 North Street, Suite 400 Lincoln, NE 68509 (404) 471-3387 <www.deq.state.ne.us/>	NDCNR, Div. of Environmental Protection Waste Management and Corrective Action 33 West Nye Lane Carson City, NV 89706 Superfund Branch Supervisor: (702) 687-4670 ext. 3022 Remediation Branch Supervisor: (702) 687-4670 ext. 3020	Supervisor of State Sites New Hampshire Dept. of Environmental Services (DES) PO Box 95 6 Hazen Drive Concord, NH 03302-0095 (603) 272-2456 or Brownfields Coordinator (603) 271-6771 <www.state.nh.us/des/hwrb>
Fees	Oversight costs	Applicant must submit payment plan and schedule for reimbursement along with application fees totaling $10,000		$500 application fee and $3000 initial program fee
Eligibility	Any person, but not sites subject to UST laws, agency orders, court actions or consent decrees; or sites proposed for the NPL	Any "entity"		Any person who did not cause or contribute to the contamination; NDES has discretion
Risk-Based Analysis	Yes	No		Yes
Liability Protection	Yes	No		Yes, but not responsible parties
No Further Action (NFA)	Yes	Yes		CoC
Reopeners	New information	If contamination is recurring, additional contamination is present which was not previously identified or remedial action was not taken		
Economic Incentives	None	No		Municipalities may grant tax abatements
Unique Features	Controlled Allocation of Liability Act (ALA) significantly modified State Superfund law. – Prospective purchaser agreements available			One of two states to receive $250,000 in Superfund program monies to undertake emergency removal work without direct EPA involvement

See end of Table for list of abbreviations.

Summary of State Brownfields Programs (Continued)

State	New Jersey	New Mexico
Program	VCP; Brownfields Act, 1/6/98	Voluntary Remediation Act (VRA), 3/21/97
Contact Information	New Jersey Dept. of Environmental Protection (DEP) Bureau of Field Operation Case Assignment PO Box 434 Trenton, NJ 08625-0434 (609) 292-2943 <www.state.nj.us/dep/srp/index.htm>	New Mexico Environment Dept. (NMED) Ground Water Quality Bureau Harold Runnels Bldg, Ste. N2300 1190 St. Francis Dr. Santa Fe, NM 87502 (505) 827-2918
Fees	Applicant must pay oversight costs	Application fee and oversight costs
Eligibility	Most contaminated properties and all parties	Current and prospective owners and operators – Excluding certain types of sites
Risk-Based Analysis	Yes	No
Liability Protection	CNS – However, liable parties under the Spill Compensation and Control Act are not eligible for a covenant not to sue	CNS releases purchaser from direct and future liability for claims to NMED, but does not protect against liability to Federal government, other State agencies, or third parties
No Further Action (NFA)	Yes	CoC
Reopeners	Contamination that has migrated – Negligent acts that aggravate or contribute to the contamination – Future noncompliance with environmental laws or – Noncompliance with the NFA	Unreasonable threat to human health or the environment, noncompliance with VRA or work plan, fraud or contamination not previously identified
Economic Incentives	Loans and loan guarantees – Real property tax exempts – Grants to municipalities and innocent parties – Corporate tax revenues allocated to site remediation – Reuse of sites for retail purposes is funded – Cost reimbursement development agreements	None
Unique Features		

State	New York	North Carolina	North Dakota	Ohio
Program	Articulated in various papers and speeches by officials of the New York State Dept. of Environmental Conservation (DEC) and subject to change without formal administrative process	Brownfields Property Reuse Act (BPRA), 10/1/97 VCP, State cleanup law was amended in 1994-5	No formal program, private party proposing a VCP would have to submit plan to North Dakota Dept of Health (NDDOH) for review and comment	Voluntary Action Program (VAP) Statute, 7/94 Regulations, 12/96
Contact Information	50 Wolf Road Albany NY 12233-7010 (518) 457-5861 <unix2.nysed.gov/ils/executive/encon/encon.htm>	Dept. of Environmental and Natural Resources (DENR) Div. of Waste Management Superfund Branch 401 Oberlin Rd. Raleigh, NC 27605 (919) 733-2801 <www.ehnr.state.nc.us/EHNR>	Coordinator of Hazardous Waste Program PO Box 5520 Bismark, ND <www.ehs.health.state.nd.us/ndhd/environ/wm/index.htm>	VAP Div. of Emergency and Remedial Response Ohio Environmental Protection Agency PO Box 1049 1800 Watermark Dr. Columbus, OH 43216-1049 (614) 644-2279 <www.epa.ohio.gov/derr/>
Fees		BPRA: $1000 initial fee plus $500 when developer submits final report		
Eligibility	All persons – Limited participation by PRPs	BPRA: only prospective developers or innocent landowners who did not cause or contribute to contamination VCP: any responsible party		All properties, unless specifically excluded
Risk-Based Analysis	No			Yes
Liability Protection	Yes	Prospective developers, but not responsible parties		Ohio EPA may issue CNS after receiving NFA from certified professional engineer
No Further Action (NFA)	Yes	BPRA: CNS VCP: a responsible party or developer my receive NFA from DENR for $500		Certified professional may issue the parties a NFA
Reopeners	Response plan proves insufficient to protect health and environment; site use changes to require more cleanup; fraud discovered; unknown environmental conditions discovered	Previously unreported contamination, changes in exposure, new information raises risk, failure to file Notice of Brownfields Party or owner violates land use restrictions		None
Economic Incentives	Funding is available to municipalities	None		Loans, grants, and tax abatements
Unique Features	1996 Environmental Bond Act – Municipalities can receive reimbursement up to 75% of costs – Municipality afforded liability protection – Urban renewal projects New York City has Brownfields Initiative			State oversight not required

See end of Table for list of abbreviations.

Summary of State Brownfields Programs (Continued)

State	Oklahoma	Oregon	Pennsylvania	Puerto Rico
Program	Brownfields Voluntary Redevelopment Act (BVRA), 1996	VCP, 1991 Oregon Recycled Lands Act (7/18/95)	VCP 7/18/95	None
Contact Information	Waste Management Div. Oklahoma Dept. of Environmental Quality (ODEQ) <www.deq.state.ok.us/brownfie.htm>	Oregon Dept. of Environmental Quality (ODEQ) 811 S.W. Sixth Ave. Portland, OR 97204 VCP: (503) 229-6834 or 1 (800) 452-4011 ODEQ: <www.deq.state.or.us> Portland Brownfields Initiative: <www.brownfield.org>	Director Land Recycling and Cleanup Program Pennsylvania Dept. of Environmental Protection (PDEP) PO Box 8471 Harrisburg, PA 17105-8471 (717) 783-7816 <www.dep.state.pa.us/dep/ deputate/airwaste/wm/landrecy/default.htm>	
Fees	Oversight costs	$5000 deposit to cover ODEQ's review and oversight costs, but if site will likely not need remediation, applicant can pay $2000 oversight fee to have ODEQ issue NFA		
Eligibility	Almost any individual or entity with some relationship to the property	Any person, including responsible parties	Responsible persons can participate	
Risk-Based Analysis	Yes	Yes	Parties may use site-specific standards	
Liability Protection	Applicant, lender, lessee, successor or assign are released from liablity; also protection from third party liability; responsible parties also eligible	CNS	Current or future owners; protection extends to citizen suits and contribution actions under state law	
No Further Action (NFA)	Either a CoC or a Certificate of No Action Necessary	Yes	If DEP approves the report, the developer and DEP will enter into an agreement that outlines the cleanup liability for the property	
Reopeners	Noncompliance with consent order or NFA		Fraud, new information, remediation fails to attain the selected standard, change in use of property creates unacceptable level of risk, remedy has becom technically and economically feasible	
Economic Incentives	Businesses which locate their principal operations on certain contaminated properties qualify for the incentive payments under the state Quality Jobs Act	Loans and grants are available	Two funds which are awarded on a competitive basis	
Unique Features	Sales tax exemption for purchases used in reducing volume or toxicity of hazardous waste		The American Legislative Exchange Council adopted the Pennsylvania program as the national model for industrial site recycling	

State	Rhode Island	South Carolina	South Dakota	Tennessee
Program	VCP, 1995	No statute but VCP for responsible parties was established in 1988 and expanded in 1995 to include nonresponsible party cleanups	South Dakota has a planned voluntary cleanup program (as of 1/1/99)	Voluntary Oversight and Assistance Program (VOAP), 1994
Contact Information	Rhode Island Dept. of Environmental Management (DEM) Div. of Site Remediation 291 Promenade Street Providence, RI 02908 (401) 277-3872 <www.state.ri.us/dem>	Bureau of Land Waste Management South Carolina DHEC 2600 Bull Street Columbia, SC 29201 (803) 896-4069 <www.state.sc.us/dhec/>	Dept. of Environment and Natural Resources Foxx Building 523 East Capitol Ave. Pierre, SD 57501-3182 (605) 773-5868 <www.state.sd.us>	Manager Voluntary Oversight and Assistance Program Tennessee Dept. of Environment and Conservation (TDEC) Div of Superfund 401 Church St. L&C Annex, 4th Floor Nashville, TN 37243 (615) 532-0912 <www.state.tn.us/environment>
Fees		Oversight costs		$5000 initial participation fee
Eligibility	Any person not a responsible party	NPL sites are not eligible		All parties are eligible, but only inactive hazardous substance sites are eligible to enter the VOAP
Risk-Based Analysis	Yes	Yes		Yes
Liability Protection	Person will not be considered a responsible party under the Industrial Property Remediation and Reuse Act; also, DEM may issue a CNS; additional protections provided for cleanups of CERCLIS sites	Responsible party: CNS, but no release for past contamination that is discovered at a later date. Nonresponsible party will receive full liability protection for all past contamination		VOAP parties protected from liability for further TDEC-mandated cleanups
No Further Action (NFA)	Letter of Compliance	CoC		Yes, in form of a letter of completion
Reopeners	No	No		If new or different information becomes known that questions the effectiveness of the selected remedial response
Economic Incentives	Investment tax credit, business tax credits and an interest income credit in terms of State income taxation and authorization for localities to grant property tax exempts to qualifying properties	None		No
Unique Features	Expedited environmental permit			

See end of Table for list of abbreviations.

Summary of State Brownfields Programs (Continued)

State	Texas	Utah	Vermont
Program	VCP, 9/1/95	VCP, 3/17/97	Redevelopment of Contaminated Properties Program (RCPP), 1995
Contact Information	Project Manager Voluntary Cleanup Section Texas Natural Resource Conservation Commission (TNRCC) PO Box 13087 Austin, TX 78711-3087 (512) 239-2252 or (512) 239-2498 <www.tnrcc.state.tx.us/>	Coordinator Voluntary Cleanup Program Div. of Emergency Response and Remediation Dept. of Environmental Quality 168 North 1950 West Salt Lake City, UT 84114-4810 (801) 536-4400	Chief Sites Management Section Dept of Environmental Conservation (DEC) Vermont Agency of Natural Resources (VANR) 103 S. Main Street Waterbury, VT 05671-0404 (802) 241-3491 <www.anr.state.vt.us>
Fees	$1000 application fee and costs	$2000 application fee and costs	Applicants must pay $500 fee and, following acceptance into the RCPP, must submit a $5000 fee
Eligibility	All persons and any contaminated property, subject to certain exceptions	Most sites and most parties	Only parties who are not responsible for the contamination; certain sites are excluded
Risk-Based Analysis	Yes	Yes	No
Liability Protection	Yes, but responsible parties will not be released from liability; CoC protects subsequent owners and lenders	Yes, for nonresponsible parties	No liability under the State hazardous waste management law
No Further Action (NFA)	CoC	CoC	CoC
Reopeners		Fraud, misrepresentation, or the knowing failure to disclose material information	When the eligible person or a successor engages in activities that are inconsistent with the approved correction action requirements or causes the release or contamination to become worse
Economic Incentives	Municipal tax abatements are available in reinvestment zones	None	None, but some are proposed
Unique Features	The TNRCC VCP has expanded its program to offer a limited number of federally funded brownfields site assessments at selected brownfields properties owned by qualifying governmental entities		

State	Virginia	Washington
Program	Voluntary Remediation Program (VRP), 1995	Person may undertake voluntary cleanup with formal, informal or no agency oversight under VCP of Department of Ecology (DOE)
Contact Information	Manager Voluntary Remediation Program Virginia Dept. of Environmental Quality (VDEQ) PO Box 10009 Richmond, VA 23240 (804) 698-4249 <www.deq.state.va.us/>	Department of Ecology PO Box 47775 Olympia, WA 98504-7775 (360) 407-6267 <www.wa.gov/ecology/tcp/>
Fees	$5000 fee or 1% of the cost of cleanup, whichever is less	For informal consultation, party must pay for agency service – $50 to $100 per hour and for more complex sites, the agency offers the option of entering into a prepayment agreement – Minimum deposit of 25% of estimated costs
Eligibility	Any persons who own, operate, have a security interest in or enter into a contract for the purchase or use of an eligible site; any property as long as remediation has not been mandated	Any person; however, where discussion or negotiations have already commenced, a potentially liable party cannot take remedial action unless 1) Such action does not foreclose negotiations or the selection of a cleanup action; and 2) The party has provided notice of the proposed action to DOE without objection
Risk-Based Analysis	Parties may select published, media specific cleanup levels or may develop site-specific cleanup levels through a risk assessment	No
Liability Protection	Immunity is limited to site conditions at the time of issuance of the certificate	Contribution protection and may contain CNS
No Further Action (NFA)	Certificate of Satisfactory Completion	Yes
Reopeners	If contamination posing an unacceptable risk to human health or the environment is rediscovered on the site, or if the parties supplied DEQ with false, inaccurate or misleading information	NFA does not preclude the DOE from requiring other further action based on reevaluation of the site or additional information
Economic Incentives	None	Limited "mixed funding" is available on a discretionary basis
Unique Features		Remedial Action Grants are available to municipalities

See end of Table for list of abbreviations.

Summary of State Brownfields Programs (Continued)

State	West Virginia	Wisconsin	Wyoming
Program	Voluntary Remediation Program (VRP), 7/1/96	Voluntary Party Exemption Program (VPEP)	Voluntary Corrective Action Order Program (VCAOP)
Contact Information	Chief West Virginia Div. of Environmental Protection (DEP) Office of Environmental Remediation Brownfields and LUST Remediation Programs 1356 Hansford Street Charleston, WV 25301 (304) 558-2508 <www.dep.state.wv.us/oer/index.html>	Wisconsin Dept. of Natural Resources (DNR) 101 S. Webster St. Box 7921 Madison, WI 53707-7921 (608) 261-6422 <www.dnr.state.wi.us>	Program Manager, VCAOP Wyoming Dept. of Environmental Quality (WDEQ) Solid and Hazardous Waste Div. Herschler Building 122 West 25th St. Cheyenne, WY 82202 (307) 777-7752 <www.state.wy.us>
Fees	Applicant must pay $1000-$5000 fee (based on a point system) upon filing	$250 application fee and must pay either $1000 or $3000 to cover administration and oversight costs	
Eligibility	Any person not causing or contributing to contamination; sites not subject to unilateral Federal or State enforcement actions or listed on NPL	Persons who are not deemed responsible for the contamination	Generators and transporters of hazardous waste and nonresponsible parties – Site conditions and hazardous waste release circumstances determine eligibility – TSD hazardous waste facilities are not eligible
Risk-Based Analysis	Yes	No	Yes
Liability Protection	Yes	Yes	No
No Further Action (NFA)	CoC which may be subject to a land use covenant	NFA at sites that don't need remediation – CoC affords protection against future State enforcement action	Yes
Reopeners	Fraud – New information – Level of risk has increased significantly or – Remediation method failed to meet standards	Failure to maintain the property as required by DNR; activities inconsistent with maintenance of the property	Subsequently discovered contamination
Economic Incentives	Loan fund	1) Agricultural Chemical Cleanup Program 2) Petroleum Environmental Cleanup Fund Act 3) Brownfields Environmental Assessment Program 4) State trust fund loan program 5) Tax credits 6) Business Improvement Districts 7) Tax-increment financing for municipalities	No

Abbreviations

CERCLA	Comprehensive Environmental Response, Compensation and Liability Act	**MOA**	Memorandum of Agreement	**TSD**	Treatment, Storage and Disposal Facility
CERCLIS	Comprehensive Environmental Response, Compensation and Liability Information System	**NFA**	No Further Action letter	**USEPA**	United States Environmental Protection Agency
CNS	Covenant Not to Sue	**NPL**	National Priorities List	**UST**	Underground Storage Tank
CoC	Certificate of Completion	**PRP**	Potentially Responsible Party	**VCP**	Voluntary Cleanup Program
IRRP	Industrial Property Remediation and Reuse Act	**RCRA**	Resource Conservation and Recovery Act	**VRP**	Voluntary Remediation Program

Part V

Management of
Hazardous Materials and
Hazardous Substances

CHAPTER **20**

Department of Transportation Hazardous Materials Regulations

Dorothy Bloomer, CHMM
Michael Eyer, CHMM

Introduction

The hazardous materials regulations (HMR) of the United States Department of Transportation (DOT) are promulgated with the goal of establishing shipping practices that would allow the movement of hazardous commodities without incident in the national and international transportation systems.

DOT hazardous materials regulatory authority is divided between several administrations: the Federal Highway Administration, Federal Railroad Administration, Coast Guard, Federal Aviation Administration, Research and Special Programs Administration (RSPA), and a pipeline safety agency. Although some hazardous materials regulations are written by the modal agencies, the largest part of the rules is generated by RSPA.

DOT also has a pipeline administration that regulates the movement of commodities via pipeline. Two of the Federal administrations, rail and highway, have also given some of their enforcement and penalty activities to the various states.

The hazardous materials/dangerous goods regulations that are used worldwide are based on suggestions made by the United Nations Committee of Experts and found in the "Orange Book," *Recommendations on the Transport of Dangerous Goods*, 10th Edition. Regulations for radioactive materials are developed separately by the International Atomic Energy Agency and are published in *Regulations for the Safe Transport of Radioactive Materials, Safety Series*. No. 6. DOT regulations apply to all modes of transport, but can be superseded by other nations' regulations and regulations issued by the International Maritime Organization and the International Civil Aviation Organization. To competently prepare a shipment of hazardous materials, you must be familiar with all of the possible regulatory agencies' rules, particularly if the movement will be intermodal or international.

Definitions and Regulatory Authority

When you see the words *hazardous materials*, you must understand that the definition applies to those materials in transportation, or in storage for the purpose of transport. By definition, all Environmental Protection Agency (EPA*) hazardous wastes* and *hazardous substances* are DOT hazardous materials. However, the Occupational Safety and Health Administration (OSHA), the Department of Agriculture, the Department of Health and Human Services, and the National Fire Protection Association (NFPA) define *hazardous materials* much more broadly than does the DOT. Unlike the hazardous waste regulations, under which states can adopt their own local waste codes, the DOT's definitions preempt all local, State, and regional definitions and codes. Both the EPA and OSHA are free to regulate to the full extent of their authority in areas where the DOT has no authority. However where DOT does regulate, the other agencies have only limited authority.

The principal exception to DOT's authority is the exception for noncommercial movement. This exception exempts Federal and State government agencies, school districts, *etc.* from most of these rules unless a state has specifically placed them under the authority of the regulation. This exemption applies only to the *Hazard Communication System* (HCS) parts of the regulations and does not extend to issues such as driver's licenses.

Regulations

The DOT regulations that address hazardous materials are found in the *Code of Federal Regulations*, Title 49, Parts 100–500 (49 CFR 100–500). The majority of the rules are in 49 CFR 170–180. In 49 CFR 171.8, the DOT defines a *hazardous material* as

> a substance or material, which has been determined by the Secretary of Transportation to be capable of posing an unreasonable risk to health, safety and property when transported in commerce, and which has been so designated. The term includes hazardous substances, hazardous wastes, marine pollutants, and elevated temperature materials as defined in this section, materials designated as hazardous under the provisions of [49 CFR] 172.101 of this subchapter, and materials that meet the defining criteria for hazard classes and divisions in [49 CFR] 173 of this subchapter.

The Secretary of Transportation can add, or delete, materials to the list at any time. For example, the Secretary added the entire category of *marine pollutants* following a major train incident that pollluted the Sacramento River in Dunsmuir, California. The products released were, at that time, DOT nonhazardous but were regulated by water shipment. When such changes occur, notices are provided through the *Federal Register*. If other agencies change their listings, (*i.e.*, the Comprehensive Environmental Response, Compensation, and Liability Act [CERCLA] designations in the EPA's hazardous substances table), the changes are automatically effective for the DOT, even though the change may not appear in Title 49 until the next printing. (Government titles are printed on an annual basis.) Title 49 has a cover date of October 1, and usually appears about mid-March. Without an update service of some kind, a shipper and/or carrier will always be at least five months out of date, even if he has the latest official edition. It is the responsibility of the shipper and the carrier to be aware of all changes.

Hundreds of other regulations and publications are *incorporated by reference* in the DOT regulations. These publications and regulations are indexed in 49 CFR 171.7. If a commercial company is shipping or carrying hazardous materials, the company is required to know the applicable regulations issued by the EPA, OSHA, the Department of Energy, the Nuclear Regulatory Commission, the Consumer Product Safety Commission, the Food and Drug Administration, *etc.* The shipper must also know the regulations promulgated by departments of the DOT, *i.e.*, National Highway Transportation Safety.

Historically, DOT regulations have been based on safety in transportation and have been developed independently from other nations' concerns. In the early 1980s, the DOT began to issue preliminary notices of its intent to harmonize with the United Nations (UN) rules used by the international community. Major changes were mandated beginning in 1992 and are continuing as of this writing. The entire Hazard Communication System (HCS) product classifications and ancillary information, labeling, marking and placarding, have been and are continuing to be substantively altered. Packaging moved from individualized safety-engineered packaging to performance oriented generic requirements. The DOT is one of the most fluid regulatory bodies. In order to be in full compliance with the rules, you will need frequent regulatory updates.

The DOT's hazardous materials regulations assume that all of the materials listed in the hazardous materials table of 49 CFR 172.101 and its two appendices are fully regulated in all quantities. The 49 CFR regulations give the shipper the option of removing specific commodities and/or groups of materials from the rules: *small quantities* are in 49 CFR 173.4, *limited quantities* are referenced in their specific hazard classes, *materials of trade* are discussed in 49 CFR 173.6. Specific commodities (*i.e.*, batteries 49 CFR 173.159) are also eligible for exception. RSPA, the principal rule-writing branch of the DOT, issues exemptions which release a commodity or package

from various regulatory requirements. As of late spring 1997, there were over 6,000 active exemptions. Exemptions are valid for two years and may be renewed on a continuing basis.

Shippers and carriers must know the DOT Hazard Communication System (HCS) in order to legally move a shipment of hazardous materials. The HCS requires each covered shipment to have a properly prepared shipping invoice, some form of package marking, some form of class symbols on the package (either labels or placards), and packaging that is properly designed to prevent the release of its contents during normal transportation.

The company must certify that all employees who handle hazardous materials meet the training requirements of 49 CFR 172.600, *et seq.*, which includes general awareness, safety, and function-specific areas. The training requirement is designed to increase safety awareness and to improve emergency preparedness for employees who must respond to transportation incidents and accidents. Testing and recordkeeping are mandatory. Retraining is required every three years. Anyone who directly affects hazardous materials transportation safety is a **hazmat employee**.

Before any hazmat employee performs a function specific to the HMR, the person must receive initial training in the performance of the function. If a new regulation is adopted, or an existing regulation is changed, affected employees must first be instructed in those new or revised requirements before carrying out their duties. An employee may perform a required function under the direct supervision of a properly trained and knowledgeable employee for 90 days, or until the training is provided.

The DOT requires a hazardous materials endorsement on a Commercial Driver's License for those drivers who haul certain quantities of hazardous materials. OSHA's hazard communication training may be used to meet the safety training portion of the DOT requirements.

Training is one of the most frequently violated sections of the HMR. The maximum DOT penalties go up to $27,500 per day/per violation, and if criminal sanctions are awarded in a criminal trial, up to 5 years in jail. These penalties can be levied in addition to any penalty from another agency; *i.e.*; a hazardous materials/waste transporter is subject to enforcement under both sets of regulations.

Shipping Papers

To prepare a legal shipping paper, the proper shipping information must be provided by the shipper, and, to the extent that common sense allows, checked by the carrier. The heart of the DOT hazardous materials regulations is found in 49 CFR 172.101, the hazardous materials table, its appendices and list of special provisions. A portion of the table is reproduced in Figure 1.

In Column 1 there are 5 possible symbols. The "+" fixes the proper shipping name, hazard class, and packing group, even though you know it might belong elsewhere. "A" and "W" indicate that the material is only regulated by *Air* or *Water*, unless the material is a hazardous substance or waste. "D" means the shipping name is allowable only for *Domestic* shipments. "I" is for *International* movements and the product is required to use this shipping name; the "I" is allowable for domestic use as well.

Column 2, **proper shipping name** (PSN), is the most important item in the DOT hazardous materials regulations. The PSN determines whether the material is legal to move, how to package, label, mark, and placard the shipment, whether special provisions apply, *etc.* It is the shipper's responsibility to determine the proper name; failing to properly identify a material can result in penalties, injuries, and deaths. A major contributing factor in the 1996 Valujet crash was failure to properly identify the oxygen generators.

There are about 2800 proper shipping names in the hazardous materials table. This list is insufficient to cover the tens of thousands of materials which are regulated (the current Chemical Abstract Society database lists nearly 2 million chemicals), so the DOT has also created approximately 170 generic names, like "solids containing corrosive liquid n.o.s." on the chart. The term "n.o.s." means **not otherwise specified** (*i.e.,* the material meets the definition of a DOT hazard class but is not listed by a specific name in the table).

If an n.o.s. entry is selected by the shipper as most accurately representing the hazard(s) of the commodity, the shipper must enter either the common or technical name in parentheses following the letters n.o.s.—*i.e.,* "flammable liquid, n.o.s. (xylene)"—unless the entry is "hazardous waste, liquid or solid, n.o.s.", Class 9, in which case the appropriate waste code may be used in lieu of the technical name in accordance with 49 CFR 172.203(k)(4)(i). Certain materials in Classes 6.1 and 2.3 (poisons and poison gases) also require additional technical information.

Only the names shown in **bold** type are legal shipping names The names in *italics* cannot be used as proper shipping names, but will refer you to a proper name; *e.g. "storage batteries, wet"* refers to "**batteries, wet.**" Material which is italicized following a proper shipping name may be used as supplemental information.

When deciding if a given shipment contains a hazardous material, remember that you cannot assume that simply because a material is not listed by specific name, or does not fit into a hazard class n.o.s. description, that the product is not a hazardous material. It may still be listed in either of the appendices. If the material exceeds the reportable quantity for a hazardous substance as found in Appendix A, or is to be placed in bulk (greater than 450 liters) packaging, or will move in part by water and is listed in Appendix B for marine pollutants, then the material is fully regulated.

Shipping papers must have the PSN and the following elements:

- Hazard class/division

- Identification number

- Packing group

- Total quantity by weight or volume

- Emergency response information

- Shipper's certification (if required)

Other information, like the Reportable Quantity (RQ) for EPA hazardous substances, may also be required. Additional shipping descriptions which are required by 49 CFR 172.203 can include

- The technical names of the chemicals for n.o.s. entries

- The wording "marine pollutants" for those materials in Appendix B to 49 CFR 172.101

- "Poison [or toxic]" for certain materials meeting Division 6.1 criteria and not otherwise identified

§ 172.101

§ 172.101 HAZARDOUS MATERIALS TABLE—Continued

Symbols (1)	Hazardous materials descriptions and proper shipping names (2)	Hazard class or Division (3)	Identification Numbers (4)	PG (5)	Label Codes (6)	Special provisions (7)	Packaging (§173.***)			Quantity limitations		Vessel stowage	
							Exceptions (8A)	Non-bulk (8B)	Bulk (8C)	Passenger aircraft/rail (9A)	Cargo aircraft only (9B)	Location (10A)	Other (10B)
	Sodium sulfide, anhydrous or Sodium sulfide with less than 30 percent water of crystallization.	4.2	UN1385	II	4.2	A19, A20, B106, N34	None	212	241	15 kg	50 kg	A	
	Sodium sulfide, hydrated with not less than 30 percent water.	8	UN1849	II	8	T8	154	212	240	15 kg	50 kg	A	26
	Sodium superoxide	5.1	UN2547	I	5.1	A20, N34	None	211	None	Forbidden	15 kg	E	13, 75, 106
	Sodium tetranitride	Forbidden											
	Solids containing corrosive liquid, n.o.s.	8	UN3244	II	8	49	154	212	240	15 kg	50 kg	B	40
	Solids containing flammable liquid, n.o.s.	4.1	UN3175	II	4.1	47	151	212	240	15 kg	50 kg	B	
	Solids containing toxic liquid, n.o.s.	6.1	UN3243	II	6.1	48	None	212	240	25 kg	100 kg	B	40
	Sounding devices, explosive	1.2F	UN0204	II	1.2F		None	62	None	Forbidden	Forbidden	E	
	Sounding devices, explosive	1.1F	UN0296	II	1.1F		None	62	None	Forbidden	Forbidden	E	
	Sounding devices, explosive	1.1D	UN0374	II	1.1D		None	62	None	Forbidden	Forbidden	B	
	Sounding devices, explosive	1.2D	UN0375	II	1.2D		None	62	None	Forbidden	Forbidden	B	
	Spirits of salt, see Hydrochloric acid												
	Squibs, see Igniters etc												
	Stannic chloride, anhydrous	8	UN1827	II	8	B2, T8, T26	154	202	242	1 L	30 L	C	
	Stannic chloride, pentahydrate	8	UN2440	III	8		154	213	240	25 kg	100 kg	A	
	Stannic phosphide	4.3	UN1433	I	4.3, 6.1	A19, B100, N40	None	211	242	Forbidden	15 kg	E	40, 85
	Steel swarf, see Ferrous metal borings, etc.												
	Stibine	2.3	UN2676		2.3, 2.1	1	None	304	None	Forbidden	Forbidden	D	40
	Storage batteries, wet, see Batteries, wet etc.												
	Strontium arsenite	6.1	UN1691	II	6.1		None	212	242	25 kg	100 kg	A	56, 58, 106
	Strontium chlorate	5.1	UN1506	II	5.1	A1, A9, N34	152	212	242	5 kg	25 kg	A	
	Strontium nitrate	5.1	UN1507	III	5.1	A1, A29	152	213	240	25 kg	100 kg	A	56, 58, 106
	Strontium perchlorate	5.1	UN1508	II	5.1		152	212	242	5 kg	25 kg	A	
	Strontium peroxide	5.1	UN1509	II	5.1		152	212	242	5 kg	25 kg	A	13, 75, 106
	Strontium phosphide	4.3	UN2013	I	4.3, 6.1	A19, N40	None	211	None	Forbidden	15 kg	E	40, 85

220

Figure 1. Excerpt from the DOT Hazardous Materials Table 172.101

- "Poison [or toxic] inhalation hazard" and the appropriate hazard zone for material meeting the poison or inhalation criteria

- "Hot" for elevated-temperature materials (*i.e.*, some asphalts), *etc.*

Shipping papers must be retained for one year, unless they are also an EPA hazardous waste manifest prepared in accordance with 40 CFR 262.20. In this case, there is a three-year retention rule. Under 49 CFR 172.205, an EPA form 8700–22/22A waste manifest can be used as a DOT shipping paper if it meets all of the DOT requirements.

Column 3, hazard class or division, refers to classes of materials developed under the aegis of the UN "Orange Book." The United Nations (UN) recognizes 9 hazard classes, and the United States (US) adds two additional classes: "combustible liquids" and "other regulated materials/ORM–D/consumer commodities." Some of the classes have subsets, called divisions. It is the shipper's responsibility to determine into which class their material falls.

All materials have at least *one* class/division; if they meet the technical requirements for a second/division, they may show the subsidiary class/division in parentheses following the PSN. The subsidiary class can usually be surmised if labels for two or more different classes are required. If a product meets the definition of more than one class, a hazard precedence table (found in 49 CFR 173.2[a]) determines which class is dominant. The essential order of precedence is: explosives, radioactive materials, poison gas, flammable gas, non-flammable gas, poison liquid, organic peroxide, and all others. Those materials which represent the greatest hazard in a possible transportation emergency are given precedence.

Note that although the DOT does sometimes show degrees of risk in transportation through sequential numbering (*i.e.*, the Packing Groups I, II, and III go from great to minor risk), the divisions do not represent any specific ranking. One exception to this is Class 1 explosive materials. For Class 1 materials, 1.1 is a much greater explosive risk than a 1.6.

Materials which are listed as "forbidden" in column 3, like *sodium tetranitride*, cannot be shipped. (Note: The name is italicized, therefore it is not a PSN.)

Figure 2 shows class numbers, division numbers, class or division names, and those sections of the subchapter which contain definitions for classify-

Class No.	Division No. (if any)	Name of class or division	49 CFR reference for definitions
None	....	Forbidden materials	173.21
None	....	Forbidden explosives	173.54
1	1.1	Explosives (with a mass explosion hazard)	173.50
1	1.2	Explosives (with a projection hazard)	173.50
1	1.3	Explosives (with predominately a fire hazard)	173.50
1	1.4	Explosives (with no significant blast hazard)	173.50
1	1.5	Very insensitive explosives; blasting agents	173.50
1	1.6	Extremely insensitive detonating substances	173.50
2	2.1	Flammable gas	173.115
2	2.2	Non-flammable compressed gas	173.115
2	2.3	Poisonous gas	173.115
3	....	Flammable and combustible liquid	173.120
4	4.1	Flammable solid	173.124
4	4.2	Spontaneously combustible material	173.124
4	4.3	Dangerous when wet material	173.124
5	5.1	Oxidizer	173.127
5	5.2	Organic peroxide	173.128
6	6.1	Poisonous materials	173.132
6	6.2	Infectious substance (Etiologic agent)	173.134
7	....	Radioactive material	173.403
8	....	Corrosive material	173.136
9	....	Miscellaneous hazardous material	173.140
None	....	Other regulated material: ORM–D	173.144

Figure 2. DOT Hazardous Materials Classes and Divisions

ing hazardous materials, including forbidden materials.

Prior to its harmonization with the UN classes used elsewhere throughout the world, the DOT used named hazard classes (*e.g.*, corrosive materials, now Class 8). The international classifications provide a numeric identification system based on the hazard class number and the class symbol that is theoretically recognizable worldwide.

Because the Orange Book and recommendations from the UN Committee of Experts are advisory, individual countries are allowed a degree of freedom within the UN system and may create their own additional classes and divisions as long as they do not supersede or differ substantially from the international standards. In the United States, the DOT changed the definition of Class 3, flammable liquid materials. If the shipper so chooses, any nonwaste hazardous substance, nonmarine pollutant Class 3 material with a flash-point between 38°C and 93°C can be reclassified as a combustible liquid in accordance with 49 CFR 173.120 and 173.150(f). The international standards define Class 3 materials as those with a flashpoint of 60.5°C and below, and have no regulation for materials with flashpoints above 60.5°C. This change allows (domestic only) shippers to remove themselves from some or all of the HCS requirements; *i.e.*, fuel oil haulers and paint manufacturers do not have to utilize specification packaging. Other countries also have changes: Anhydrous ammonia is a "2.3, poison gas" everywhere in the world except the US, where it is a "2.2, non-flammable gas," and Canada, where it is classified as a "2.4, corrosive gas." Canada and other countries divide Class 9, miscellaneous materials into three divisions. Many countries divide Class 8, corrosives into two divisions for acids and bases. The US has not yet adopted these latter two modifications of the basic system.

The US has also carried forward from pre-UN days "other regulated material-D/consumer commodities." Most hazard classes have sections identifying what requirements must be met to take advantage of the ORM-D exceptions, *e.g.*, 49 CFR 173.150 for Class 3 materials. Both are subsets of the nine classes and can be used optionally by shippers if all qualifications are met. Both allow the shipper and carrier to comply with fewer requirements. Consumer commodities are a subset of "limited quantities" and apply to those items intended for household use (*e.g.*, cosmetics which would normally be classified as flammables or corrosives could carry no hazard identifications of any kind if reclassified). Charcoal briquettes in packaging of up to 30 kilograms (66 pounds) are another example of a material which is frequently reclassified. Briquettes which would normally require shipping papers, labeling, and appropriate packaging are exempt from those portions of the HCS under 40 CFR 173.151(c), provided they are not a hazardous substance, hazardous waste, or marine pollutant.

Although other agencies use terminology similar to the DOT's, always remember that DOT terminology is preeminent in transportation. EPA waste manifests are required to list the DOT's proper shipping names. The DOT terminology sometimes differs from the EPA's. For example, EPA's "corrosivity" is pH-based, but DOT's Class 8, corrosives are based on several different criteria (found in 49 CFR 173.136), none of which involves pH. EPA's "ignitability" and DOT Class 3, "flammable liquids" are both determined by flashpoint, but use different flashpoints. In addition, the DOT's definition applies only to flammable liquids; other materials (gases and solids) covered in the EPA's definition are found in DOT Classes 2, 4, and 5.

Column 4, *identification number* (ID number), refers to a 4-digit number assigned to the commodity, or group of similar commodities, by the United Nations or the United States. If the material is identified in Column 1 with a "D," the PSN is allowed only for use in domestic transportation and in most cases the number will be prefaced "NA," North America. All other numbers are prefaced with a "UN," or United Nations. It is important not to confuse the 4-digit DOT hazardous material ID number with the 3-digit EPA waste code, since the waste code can appear as part of the PSN (49 CFR 172.203[c][1]). The ID number is not necessarily commodity-specific. See, for example the ID number 1993 in Figure 3. The ID number, in addition to being required on shipping papers, is also a requirement for package marking.

The ID number is most critical to emergency response, because it is used by responders to determine the chemicals' identities and what initial emergency steps to take. This information is found in the official DOT *North American Emergency Response Guidebook* which is a compilation of emergency response guides. The

ID No.	Guide No.	Name of Material
1993	128	Combustible liquid, n.o.s.
1993	128	Compound, tree or weed killing, liquid (flammable)
1993	128	Compounds, cleaning, liquid (flammable)
1993	128	Cosmetics, n.o.s.
1993	128	Diesel fuel
1993	128	Disinfectant, liquid, n.o.s.
1993	128	Drugs, n.o.s.
1993	128	Ethyl nitrate
1993	128	Flammable liquid, n.o.s.
1993	128	Fuel oil
1993	128	Heater for refrigerator car, liquid fuel type
1993	128	Medicines, flammable liquid, n.o.s.
1993	128	Refrigerating machine

GUIDE 128 — FLAMMABLE LIQUIDS (NON-POLAR/WATER-IMMISCIBLE)   NAERG 96

POTENTIAL HAZARDS

FIRE OR EXPLOSION
- HIGHLY FLAMMABLE: Will be easily ignited by heat, sparks or flames.
- Vapors may form explosive mixtures with air.
- Vapors may travel to source of ignition and flash back.
- Most vapors are heavier than air. They will spread along ground and collect in low or confined areas (sewers, basements, tanks).
- Vapor explosion hazard indoors, outdoors or in sewers.
- Some may polymerize (P) explosively when heated or involved in a fire.
- Runoff to sewer may create fire or explosion hazard.
- Containers may explode when heated.
- Many liquids are lighter than water.
- Substance may be transported hot.

HEALTH
- Inhalation or contact with material may irritate or burn skin and eyes.
- Fire may produce irritating, corrosive and/or toxic gases.
- Vapors may cause dizziness or suffocation.
- Runoff from fire control or dilution water may cause pollution.

PUBLIC SAFETY
- CALL Emergency Response Telephone Number on Shipping Paper first. If Shipping Paper not available or no answer, refer to appropriate telephone number listed on the inside back cover.
- Isolate spill or leak area immediately for at least 25 to 50 meters (80 to 160 feet) in all directions.
- Keep unauthorized personnel away.
- Stay upwind.
- Keep out of low areas.
- Ventilate closed spaces before entering.

PROTECTIVE CLOTHING
- Wear positive pressure self-contained breathing apparatus (SCBA).
- Structural firefighters' protective clothing will only provide limited protection.

EVACUATION
Large Spill
- Consider initial downwind evacuation for at least 300 meters (1000 feet).
Fire
- If tank, rail car or tank truck is involved in a fire, ISOLATE for 800 meters (1/2 mile) in all directions; also, consider initial evacuation for 800 meters (1/2 mile) in all directions.

Page 204

Figure 3.   DOT Hazardous Material ID No. 1993 and Its Emergency Response Guide

Guide for flammable liquids is shown in Figure 3. The official DOT *North American Emergency Response Guidebook*, (1996 edition), or an equivalent such as a Materials Safety Data Sheet (MSDS), must accompany every shipment in every mode. Information can be found in the *ERG* by looking up the ID number or the PSN. The *ERG* is issued every three years. The next edition is due in late summer, 1999. It is published in English, French, and Spanish, for use throughout North America.

Column 5, "PG," refers to packing groups, identified as I, II and III. The numbers I, II, and III represent great, medium and minor risks, respectively. Classes 2, gases; 7, radioactive materials; infectious substances and consumer commodities do not have packing groups. Gases and radioactive materials have specific packaging which is determined by other agencies. Consumer commodities theoretically represent a very low risk of hazard and are excepted from UN performance packaging. The other seven classes and combustible liquids are divided into three groups according to their degree of danger. Any one PSN may have up to three packing groups and the shipper is responsible for determining which packing group is correct for each shipment. *The packing group determines the proper packaging of a shipment.* Packing group criteria vary by hazard class, *i.e.*, for 6.1, poison liquids it is based on LD_{50} data; for Class 3, flammable liquids, the following criteria are used:

I   greatest danger, flashpoint (fp) and boiling point (bp) less than 35°C

II   medium danger, fp less than 23°C, bp greater than 35°C

III   minor danger, fp greater than 23°C, less than 60.5°C, bp greater than 35°C

Non-bulk packages are cross-indexed to the packing groups. They will show an "X" for those able to handle PG I (II and III), "Y" for PG II (and III), and "Z" for PG III only.

Non-bulk packages are defined as follows:

- Liquids–a maximum capacity of 450 liters or less

- Solids–a maximum net mass of 400 kg or less and capacity of 450 liters or less

- Gases–water capacity of 454 kg or less

Shipping papers must show the appropriate packing group Roman numerals (*e.g.*, II); however for Class 7, radioactive materials, the shipping papers may also be required to show other Roman numerals that mean the opposite of DOT's PGs (*e.g.*, where "I" is the lowest risk). Additional information on other commodities, such as pesticides, may contain potentially conflicting numbers and cannot be included in the basic description: PSN, hazard class/division, ID number, and packing group (if required). For example, pesticides use *signal words* like "Caution" which cannot be included in the basic description but are not precluded from appearing elsewhere on the shipping paper.

Column 6, labels, show by hazard class number the type(s) of labels that are required to be on each package in that class. Many products, such as stannic phosphide, require more than one label (in this case, 4.3 and 6.1). Label sizes and placement requirements are found in 49 CFR 172.407. Class 7, radioactive packages require two labels; all other classes require one label per package per class. In general, labels are applied to non-bulk packages and placards to bulk packages; however, numerous exceptions do apply.

Labels are not placards. Although the symbols and colors are the same, labels and placards differ in size and the number required to be placed on the package/vehicle. Unless excepted, labels are 100 mm x 100 mm, and placards are 273 mm x 273 mm. There are many exceptions from placarding; however, these exceptions do not apply to labels. In most cases, neither labels nor placards require the name, *e.g.*, corrosive, of a class. Usually, only the class icon and division/class/compatibility group letter are required. The exceptional cases that do require class names differ (*e.g.*, in keeping with the restrictive requirements noted above for shipping papers and labeling, Class 7, radioactive labels must have the class name shown in accordance with 49 CFR 172.405[a]).

If a package requires multiple labeling in Column 6, only the primary label may show the class number. The subsidiary labels must be blank in the bottom apex. The labels must appear near the marked PSN if the package dimensions allow. Labels may be on a tag attached by a wire or other means if needed. Cylinders may be labeled and marked in accordance with the Compressed Gas Association's *Pamphlet C-7* (1992). EPA regulations cross-reference the DOT requirements. A package may display the EPA and/or the OSHA Hazardous Materials Information System (HMIS) required labels while in transportation, but it must display the DOT label if not exempted.

Sample labels and placards with color and dimensional requirements can be found at the end of 49 CFR 172.400 and 172.500. DOT RSPA also supplies copies of the guide chart, currently chart 11, and the placards are shown (not the correct size) in the *ERG*.

Whenever inner containers of different classes or materials are placed in the same outside container, or overpack, the outside container must be labeled for each different class represented within.

Column 7 details any special provisions which may be applicable to the commodity. Such provisions may be found in 49 CFR 172.102, following the table's two appendices. These provisions can supplant, supplement, or change any of the table's or packaging section's requirements. If the special provision is a straight numeric designation, *e.g.*, 47 for "solids containing flammable liquid, n.o.s.," the provision applies to *all* modes of transportation. If the designation is a combined alphanumeric, (*e.g.*, B2, T8, or T26) it applies only in restricted circumstances:

- A   = air only
- B   = only bulk packages (except intermodal tanks)
- H   = highway
- N   = non-bulk packages
- R   = rail
- T   = intermodal tanks
- W   = water

The most common designator will be a straight numeric, 1–6 and 13 (which identifies the material as an inhalation hazard, requiring additional shipping paper entries, labeling, marking, possibly placarding, handling/stowage, driver's training, *etc.*). Be sure to read any listed special provisions before preparing hazardous materials for shipment. As an example: "cord detonating, flexible" is classified as a 1.1D where Division 1 represents the greatest risk of explosion. Special provision 102 allows the material to be reclassified as a 1.4D, a much lower category.

After the HCS requirements are met, the proper packaging is selected utilizing Column 8 of the Table and any special provisions found in Column 7. Column 8 is subdivided into three sections: 8A covers any exceptions (how the material could avoid the packaging requirements, *e.g.*, limited quantities or consumer commodities), 8B covers non-bulk packages, and 8C, bulk packages.

Shipping papers must show the following basic information: proper shipping name, hazard class/division, UN/NA number and packing group (if appropriate). An emergency response phone number, directly answerable by a knowledgeable person 24-hours a day (or as long as the material is in the transportation system); a listing of total quantity by weight or volume of hazardous materials; and, in most instances, a shipper's certification stating that the shipment is properly packaged, marked, labeled, and classified.

Labeling, Marking, and Packaging

Packages must remain marked and labeled until the container is cleaned and purged of all hazardous material, or unless it is filled with a nonhazardous material. This regulation is unlike the EPA's, which defines *empty* as less than 2% of the product remaining in the package.

Package marking is another way to communicate the nature of hazardous contents. Markings, like labels and placards, are integral parts of HCS. All of the HCS indications should indicate the same commodity/class.

Markings should convey the PSN, ID number, and hazard class. This information readily identifies a package that contains a hazardous material, and is used by carriers to ensure compliance with loading and stowage requirements. Markings help prevent the potentially dangerous situations which may occur when incompatible materials are loaded near each other; *e.g.*, poisons with foodstuffs, or certain Class 8's, corrosives, with Class 3's, flammables. The information aids in commodity identification for emergency responders, especially if the shipping papers have been destroyed. Markings must be in English, durable, and unobscured by other information.

Each person who offers a hazardous material for transportation must mark each package, freight container, and transport vehicle as required. Marking requirements vary depending upon package size, bulk/non-bulk packaging, commodity, and type of package.

Some examples of non-bulk markings which may be required are

- PSN
- ID number
- Package specification markings (including the X, Y, or Z that reflects the packing group)
- Shipper or consignee name and address
- International Organization for Standardization orientation arrows, for liquids packed in a container inside an outer package
- Hazardous substance RQ notation
- Inhalation hazard
- The embossed word "poison"

Marking on bulk containers might display some or all of the following:

- PSN
- ID number

- Package specification numbers (without the "X, Y, Z")

- Elevated temperature (HOT)

- Commodity/class name (inhalation hazard)

- Marine pollutant ("dead fish"/"fish and chips")

On bulk packages, the ID number is normally displayed on the placards. If a shipment contains more than 4000 kg of any one non-bulk commodity, the container's exterior must show that ID number; if a bulk package is inside an enclosed outer container, the outer container must show the ID number. Bulk containers may display ID numbers on orange panels or white-square-on-point display configurations that are the same size as placards. The display configurations may also be used for any HOT or inhalation hazard markings. Specific bulk packages such as portable tanks, cargo tanks, and tank cars may also require the commodity name.

Placards provide a readily visible warning that hazardous materials are present. Placards are normally what emergency responders first see when approaching an incident. Both placards and labels use the color, symbol, and hazard class to indicate the contained danger.

Placards convey information about the hazardous material in several ways:

- Color (red for flammable liquids/solids/gases/combustibles)

- Symbols (skull/crossbones for poisons, skull/crossbones with a black background for inhalation hazards)

Additional information may include

- Name text (if required; *e.g.*, radioactive)

- ID number (1830 on a Class 8 for sulfuric acid)

- Hazard class/division/compatibility group letter (1.1A for an explosive)

Cargo tanks, rail cars, and certain portable tanks must be placarded both on the sides and the ends, and must remain placarded until cleaned and purged.

The shipper must provide the carrier/driver with the appropriate placards if they are not already displayed. In certain instances, subsidiary placarding may be required per 49 CFR 172.505, for inhalation hazards, and dangerous when wet materials. *Permissive* placarding is allowed as long as the placard correctly represents the hazard(s).

All bulk shipments must be placarded for all classes of materials. The hazard class/division of non-bulk shipments determines whether or not they must be placarded. Materials listed in Table 1 of 49 CFR 172.504 must be placarded for any quantity. These materials include 1.1, 1.2, 1.3, and certain types of commodities in 2.3, 4.3, 5.2, 6.1, and 7. Any materials not meeting the 49 CFR 172.504, Table 1 definitions fall into Table 2. Table 2 allows up to 454 kg *aggregate gross weight* (material and its packaging) to be shipped without placards. This means most freight forwarders, *e.g.*, UPS or FedEx, can carry hazardous materials without external identification.

If a shipment contains two or more 49 CFR 172.504 Table 2 classified materials, and no single class weighs more than 1000 kg, the vehicle may be required to display a "Dangerous" placard. If any single class exceeds 1000 kg, placarding for the specific class is mandated.

Placarding is the final segment of the HCS: shipping papers, labeling, marking, and placarding. All of the elements should be identical, so that all personnel and any emergency responders in the transport loop are fully aware of the hazards that they are facing.

Proper packaging is the first line of defense in ensuring that hazardous materials will not be released during transportation. A poorly packaged commodity may not be offered for transportation, accepted, or transported. The HMR specify various performance levels for the packing used with hazardous materials based on the nature and levels of risk the hazards pose. All packing must be designed to ensure that there will be no releases under the normal conditions of transportation. Except for packaging designed for Class 2,7, and some 6.1 materials, all DOT UN non-bulk performance packaging has to pass a series of performance oriented tests: stacking, bumping, dropping, *etc*. The tests are designed to be progressively more challenging for each packing group. If the packaging passes the tests, the DOT allows the shipper some flexibility in actual package design.

Training

Subpart H, 49 CFR 172.200, *et seq.*, of the regulations details the requirements of the mandatory US DOT training. Every person who in any way handles, moves, bills, loads, transports, or is responsible for any of those actions, in association with hazardous materials, must be trained at least every three years. The training is divided into three sections: general awareness/familiarization, function-specific, and safety. Employees are not allowed to perform their job functions unless they are trained or under the direct supervision of someone who is trained. However, in all cases, the training must be completed within 90 days. Any time an employee changes job functions, the employee must be trained again.

Each employee must also be tested in the three areas. Employers are required to keep copies of tests, training materials, and employee training records for as long as the employee is with the company plus 90 days. The section does not specify who will do the training, what the trainers' qualifications should be, what kind of test is required, or what is a passing score. This section of the regulations is one of the most frequently cited by regulators; any exception to other sections can also be construed as a failure to properly train.

Reporting and Penalties

The DOT regulations also contain ancillary requirements for shippers and carriers. These ancillary provisions deal mainly with incident reporting, national registration, and training.

Companies that are responsible for the movement of certain classes/quantities of materials (defined in 49 CFR 107.601) are required to pay a $300 annual Federal registration fee. A copy of the registration certificate must be carried at all times by affected highway carriers. This program is administered by RSPA, and the majority of the money ($250) is sent to the states for the purpose of training emergency responders. States and local jurisdictions frequently have their own registration/permitting programs. It is the responsibility of the shipper and/or carrier to be aware of these programs. RSPA has formulated a national Uniform Program for registration and permitting; if enacted it will probably begin in 2000. The current program returns about $7,000,000 a year to the states for training. It is up for review in 1999, and RSPA is actively looking for ways to increase the revenues collected to meet target goals originally promised the states.

If an incident occurs involving hazardous materials, a report must be filed with the Federal government. This report is required if hazardous materials are within the same unit, even if the hazardous material itself is not directly involved. Certain circumstances require immediate telephone reporting and all require written follow-up according to the requirements outlined in 49 CFR 171.15 and 49 CFR 171.16. Failure to report, even though all corrective action was successfully accomplished, has resulted in severe financial penalties. If a hazardous material is involved in a spill and the commodity is also a hazardous substance and an amount greater than the reportable quantity is spilled, EPA must also be immediately called. The DOT reporting call will not satisfy the EPA and State/local requirements. Separate calls must be made.

DOT penalties are established by RSPA and are periodically adjusted for cost of living. The penalties are found in Appendix A to 49 CFR 107. By way of example, for "using a shipping name and hazard class that is incorrect such that a material is mis*described*," the baseline penalty assessment is $3,700, and if, using the above description, the "material is mis*classified*," the baseline is $6,200. The penalties are per shipment/per day. If willful intent to avoid the law is determined, criminal penalties (including up to 5 years in prison) are assessable.

Summary

Shippers and carriers have the following responsibilities:

- Shipping documents must include the PSN, hazard class, ID number, packing group (except classes 2 and 7), shipper certification, proper count and weight, as well as emergency response phone number and information.

Additional information may be required depending upon the PSN, class, and special provisions.

- Proper packaging must be selected.

- Packages must be properly closed and show no evidence of leakage.

- Packages and/or vehicles must be properly labeled, marked, and placarded.

- Incompatible materials must not be packaged together in the same packaging.

- All required training must be accomplished and documented for everyone involved at each step.

- Proper emergency response information must accompany the shipment.

- Personnel must be aware of any other agencies' rules which may affect transportation; *e.g.*, EPA spill reporting requirements.

- Highway shippers may need to ensure that drivers have a Commercial Driver's License with appropriate endorsements.

- Both shippers and carriers need to be registered under the DOT registration programs, depending on the type and quantity of material shipped and/or carried.

- Some materials may require route plans, copies of exceptions or exemptions.

The DOT regulations are dynamic. They have undergone constant change since they were first promulgated 90 years ago. If all affected parties comply in full, the nation will continue to experience the excellent safety record and benefits the commodities bring.

Resource Information

Companies providing 24-hour phone response (note these are subscription services; companies can provide their own response, if qualified):

CHEMTREC: (800) 424–9300

PERS: (800) 728–2482

National Response Center for reporting a spill of hazardous substances greater than the RQ:

(800) 424–8802 (24 hours)

Internet Resources

<http://hazmat.dot.gov>(US DOT RSPA hazardous materials Web site)

<http://www.hmac.org>(Hazmat Advisory Council)

Dorothy Bloomer has over 24 years of experience in the chemical industry, with 10 years as Shipping Supervisor for Elf Atochem N.A. Elf Atochem N.A. manufactures and ships approximately 1000 tons of chemicals a day. She is also President of her own company, Pacific Training, a consulting firm for the transportation of hazardous materials. Ms. Bloomer has an AA in Chemistry, BS in Business Management, and has been a Certified Hazardous Materials Manager since 1990. She is Past President of the Willamette Columbia Chapter in Portland, Oregon, and Portland's Transportation Person of the Year in 1997.

Michael Eyer is currently a Hazardous Materials Specialist, Rail Division for the State of Oregon. He has over 23 years of experience in the hazardous materials management field, including 12 years as a District Inspector with the Bureau of Explosives, three years with Oregon's Department of Environmental Quality managing the spill response and illegal drug lab cleanup programs, as well as eight years with the Oregon PUC/DOT as a Transportation Specialist. Mr. Eyer is a Certified Hazardous Materials Manager, USDOT-certified Highway and Rail Vehicle Inspector/Trainer, USDOT Hazardous Materials Inspector/Trainer and USDOE Radioactive Materials Inspector/Trainer. He has also served as Regional Chair of the Cooperative Hazardous Materials Enforcement and Development (COHMED) program (1993-9) and is the 1999 National chair. Mr. Eyer has been a guest instructor at UCLA, OSU, and University of Michigan.

CHAPTER 21

Underground Storage Tanks

Alan A. Eckmyre, CHMM

Introduction

Underground tanks that contain either petroleum or hazardous substances are subject to the Federal Underground Storage Tank (UST) regulations. These regulations, issued by the Environmental Protection Agency (EPA) under authority of Subtitle I of the Resource Conservation and Recovery Act (RCRA) (Section 9003 of the Hazardous and Solid Waste Amendments [HWSA] of 1984), established standards for installation, operation, release detection, corrective action, repair, and closure of USTs and attached piping. EPA also authorized individual states to implement their own UST regulatory programs in place of the Federal requirements. To receive this authorization, the State program must be *no less stringent* than the Federal requirements. In authorized states, the UST requirements may be more stringent than the Federal requirements. When a State program receives authorization, it becomes the *implementing authority* for the UST regulations. This means that the State agency has primary enforcement responsibility for its UST program. Parties in these states must comply with State requirements that may be more stringent than Federal regulations.

Federal regulations govern roughly 1.1 million active USTs. About 96 percent of these contain petroleum products, including used oil. Fewer than 1 percent contain hazardous materials, and about 2 percent are empty. Petroleum or hazardous substance releases from USTs can occur during tank filling. They can also occur from leaks in tanks and piping that result from corrosion, structural failure, or faulty installation. As of September 1996, EPA has reported nearly 318,000 confirmed releases at Federally regulated USTs. More are expected. These releases can contaminate soil and groundwater and cause fires or explosions.

Identification and Classification

The UST regulations differentiate between existing USTs and new USTs and between petroleum USTs and hazardous substance USTs. The regulations also exclude certain USTs and defer requirements for other USTs. Therefore, correct application of the UST regulations requires accurate identification and classification of regulated tanks and piping.

An *underground storage tank* is defined by Subtitle I as any tank or combination of tanks (including connected underground pipes) that is used to contain an accumulation of regulated substances, the volume of which (including the volume of connected underground pipes) is 10 percent or more beneath the surface of the ground (40 CFR 280.12).

Connected piping means all buried piping, including valves, elbows, joints, and flexible connectors, attached to tank systems through which regulated substances flow. For the purpose of determining how much piping is connected to any individual UST system, the piping that joins two UST systems should be allocated equally between them (40 CFR 280.12).

The regulations hold owners and operators responsible for compliance. An UST owner is

- Any person who owns an UST system on November 8, 1984, or who brought the system into use after that date

- In the case of any UST system in use before November 8, 1984, but no longer in use on that date, any person who owned an UST immediately before the discontinuation of its use

An *operator* is any person in control of or having responsibility for the daily operation of the UST system (40 CFR 280.12).

RCRA Subtitle I gave EPA the authority to distinguish types (storing petroleum or hazardous substances), classes (excluded, deferred, or fully regulated), and age.

Categories of USTs

There are two broad categories (types) of USTs: petroleum USTs and hazardous substance USTs. *Petroleum USTs* contain petroleum or a mixture of petroleum with *de minimis* quantities of other regulated substances. These systems include tanks containing motor fuel, jet fuel, distillate fuel oil, residual fuel oil, lubricants, petroleum solvents, or used oil. *Hazardous substance USTs* contain either

1) Hazardous substances as defined in Section 101(14) of the Comprehensive Environmental Response, Compensation, and Liability Act (CERCLA) (40 CFR 302.4), but not including any of the substances regulated as a hazardous waste under RCRA Subtitle C

2) Any mixture of such CERCLA-listed substances and petroleum that is not a petroleum UST system (40 CFR 280.12)

In order to determine whether a tank system contains a regulated substance under Subtitle I, the following steps should be taken:

- First, determine if the substance belongs to one of the seven general categories of regulated petroleum substances (*i.e.*, motor fuel, jet fuel, distillate fuel oil, residual fuel oil, lubricants, petroleum solvents, and used oil).

- Second, determine whether the stored material is included within the *production process and physical properties* description for petroleum products (*i.e.*, petroleum-based substances comprised of a complex blend of hydrocarbons derived from crude oil through the processes of separation, conversion, upgrading, and finishing, or any fraction of crude oil that is liquid at 60°F and 14.7 lbs/in² absolute, standard temperature and pressure).

• Third, determine whether the substance is listed as a hazardous substance under Section 101(14) of CERCLA and is not a hazardous waste under Subtitle C of RCRA.

If the substance meets any of these three criteria, then it is a regulated substance.

Hazardous substance USTs have more stringent release detection requirements (including secondary containment and interstitial monitoring) than petroleum USTs. All other technical requirements are the same for UST systems storing regulated substances.

UST Exclusions and Exemptions

In identifying UST classes, there are two ways a tank can be excluded from the UST regulations. One way is for the tank to be excluded from the statutory definition of an UST (RCRA Section 9001 [1][A – I]). The other way is for the tank to be excluded from the applicability section of the regulations (40 CFR 280.10[b]). Changes in the statutory exclusions must be made by Congress, while changes in the regulatory exclusions may be made by EPA. For example, USTs used for storing heating oil for consumptive use on the premises were exempt by the statute, as were septic tanks and stormwater collection systems. Tanks excluded from the regulations by EPA include hydraulic lift tanks and tanks that have a capacity of less than 110 gallons.

Some tanks were excluded because they pose a negligible risk to human health and the environment. These USTs were excluded to allow Federal, State, and local agencies as well as UST owners and operators to focus their limited resources on those USTs that pose a greater risk. Other tanks were excluded because they were sufficiently regulated under other statutes.

Excluded USTs are not regulated by the Federal UST regulations, but they may be subject to State laws, other Federal laws, or industry codes or practices. States may write regulations for tanks that have been excluded from the Federal UST regulations. This has been the case in some states that have regulated tanks containing heating oil. Federal laws, other than RCRA Subtitle I, may apply to excluded tanks depending on their function. For example, any UST system holding hazardous wastes listed or identified under Subtitle C of the Solid Waste Disposal Act is exempt from Subtitle I but regulated by RCRA Subtitle C.

The following categories of tanks have been excluded from the definition of an UST:

• Farm or residential tanks of 1,100 gallons or less capacity used for storing motor fuel for noncommercial purposes

• Tanks used for storing heating oil for consumptive use on the premises where stored

• Septic tanks

• Pipeline facilities regulated under the Natural Gas Pipeline Safety Act of 1968 or the Hazardous Liquid Pipeline Safety Act of 1979, or that are intrastate pipeline facilities regulated under State laws comparable to the provisions of these acts

• Surface impoundments, pits, ponds, or lagoons

• Storm water or wastewater collection systems

• Flow-through process tanks

• Liquid traps or associated gathering lines directly related to oil or gas production and gathering operations

• Storage tanks situated in an underground area (such as a basement, cellar, mineworking, drift, shaft, or tunnel) if such tanks are situated upon or above the surface of the floor

The term *underground storage tank* does not include any pipes connected to any tank that is described above.

In addition to the statutory exclusions, EPA excluded the following tanks from the RCRA Subtitle I regulations:

• Any UST system holding hazardous wastes listed or identified under Subtitle C of the Solid Waste Disposal Act or a mixture of such hazardous waste and other regulated substances

• Any wastewater treatment tank system that is part of a wastewater treatment facility regulated under Section 402 or 307(b) of the Clean Water Act

- Equipment or machinery that contains regulated substances for operational purposes such as hydraulic lift tanks and electrical equipment tanks

- Any UST system whose capacity is 110 gallons or less

- Any UST system that contains a *de minimis* concentration of regulated substances

- Any emergency spill or overflow containment UST system that is expeditiously emptied after use

The types of tanks excluded because they are considered equipment or machinery includes any tank that is part of a piece of equipment or machinery (*e.g.*, hydraulic lifts and electrical equipment) and that meets two criteria:

1) The equipment or machinery contains small amounts of regulated substances solely for operational purposes.

2) Faulty operation of the equipment or machinery will cause a loss of regulated substance.

In addition, EPA does not specify a maximum time a tank may hold an emergency spill or overflow material, but intends this exclusion to apply to many types of sumps and secondary barrier tanks that are rarely used and are emptied shortly after use. The purpose of this exemption is to allow immediate response to emergency situations. It is analogous to the exclusion for emergency response treatment and containment under RCRA Subtitle C (40 CFR 264.1 and 265.1). The exclusion does not include sumps designed to store petroleum or hazardous substances during periodic cleaning or maintenance of machinery or equipment (*e.g.*, turbine oil sumps that are used during maintenance of electric power generation turbines) (40 CFR 280.10[b][6]).

Any UST system holding a mixture of hazardous waste identified under Subtitle C of the Solid Waste Disposal Act and petroleum or nonpetroleum substances regulated by Subtitle I is excluded from the Federal UST regulations. EPA excluded these tanks from Subtitle I regulation because USTs containing this kind of mixture would be subject to dual jurisdiction from Subtitle C and Subtitle I.

Very small, *de minimis*, concentrations can occur in an UST accidentally (through contamination)

or by design (for example, underground tanks storing potable water that has been treated with chlorine). EPA has not defined a specific percentage as the *de minimis* cutoff because of the many difficulties with measuring tank contents for low concentrations of regulated substances. Instead, implementing agencies determine on a case-by-case basis if tanks that hold very low concentrations of regulated substances are excluded via *de minimis* amounts (40 CFR 280.10 [b][5]).

Radioactive mixed waste (RMW) contains both a hazardous component regulated under RCRA and radioactive material regulated under the Atomic Energy Act. The applicability of Subtitle I to RMW depends on whether those USTs are regulated under RCRA Subtitle C. If an UST containing RMW is regulated under Subtitle C, it is exempt from Subtitle I in accordance with 40 CFR 280.10 (b)(1).

To determine if an UST is subject to Subtitle C regulation, it is necessary to know the authorization status of a particular state's hazardous waste program. In most states, RMW is regulated as hazardous waste under RCRA Subtitle C either because the state has an authorized RCRA program and has received a modification of their authorization to cover RMW, or because the state does not have RCRA authorization and therefore the hazardous waste component of the mixed waste is subject to the Federal RCRA Subtitle C program. If an UST located in one of these states contains RMW, it is excluded from the Subtitle I regulations.

On the other hand, if the UST is located in a state that has an authorized RCRA program but has not yet received a modified authorization to cover RMW, then the RMW is not recognized or regulated as a hazardous waste under Subtitle C. In these circumstances, the UST is regulated under Subtitle I.

Also, states without mixed waste authorization may regulate underground storage tanks containing RMW pursuant to their State regulations. In these cases the wastes may be regulated by both EPA and the state.

Tanks containing radioactive and other materials (not hazardous wastes) fall under the deferred tank classification.

Deferred USTs

A *deferred UST* refers to an UST for which EPA has deferred the Subpart B (design, construction, installation, and notification), Subpart C (general operations requirements), Subpart D (release detection), Subpart E (release reporting, investigation, and confirmation), and Subpart G (closure) regulations. Until EPA decides how to regulate these USTs fully, the only regulations that apply are Subpart A (interim prohibition) and Subpart F (release response and corrective action). Examples of deferred tanks include underground, field-constructed, bulk storage tanks, and UST systems that contain radioactive wastes.

UST systems that store fuel solely for use by emergency power generators are deferred from Subpart D (release detection) requirements for these tanks, because those requirements currently mandate frequent monitoring that may be unworkable. These tanks often are used to store diesel fuel as a source of backup power in unmanned, remote locations (*e.g.*, telephone switching locations). The other Subtitle I requirements apply fully to these USTs.

Tanks containing radioactive wastes and other radioactive materials at commercially licensed nuclear facilities are regulated under the Atomic Energy Act of 1954 (AEA) and by the Nuclear Regulatory Commission (NRC). EPA deferred these systems because they are already subject to stringent regulation.

New and Existing USTs

EPA classifies tanks for which installation commenced on or before December 22, 1988, to be *existing USTs*. Any UST for which installation commenced after this date is a *new tank system*. All new tank systems must meet the requirements of Subparts B, C, and D before they can commence operation (40 CFR 280.12).

Installation is considered to have commenced if

- The owner or operator has obtained all Federal, State, and local approvals or permits necessary to begin physical construction or installation

- Either a continuous on-site physical construction or installation program has begun

- The owner/operator has entered into contractual obligations (which cannot be canceled or modified without substantial loss) for physical construction at the site or installation of the tank system

New USTs have stringent installation requirements that must be met before the UST may be used. Existing USTs must be upgraded to meet the requirements of Subparts B and C (for corrosion protection, overfill controls, and operating requirements) by December 22, 1998. To do this, they must either be replaced with tanks that meet these standards or upgraded in place. Upgrades must include overfill controls and corrosion control (an interior lining, cathodic protection, or another acceptable method of protecting the tank system from corrosion and structural failure). The release detection requirements for existing USTs were phased in on a schedule based on the age of the tank and took effect no later than December 22, 1993.

New USTs must prevent releases due to structural failure, corrosion, or spills and overfills. Spill and overfill controls must be installed with each new UST, and the installation must be properly conducted and certified by at least one of several methods described below. By December 22, 1998, all existing USTs must meet the new UST standards by being replaced or upgraded with lining and/or cathodic protection and must comply with the new UST spill and overfill control requirements (40 CFR 280.20 and 280.21).

The new UST system must be designed and constructed to provide protection for buried components from corrosion by using either noncorrodible materials or a cathodic protection system. EPA requires that the installation be conducted in accordance with a code of practice developed by a nationally recognized association or independent testing laboratory and in accordance with the manufacturer's instructions. In addition, the installation must be certified, tested, or inspected by one of six procedures provided in the regulation. These are:

1) The installer has been certified by the tank and piping manufacturers.

2) The installer has been certified or licensed by the implementing agency.

3) The installation has been inspected and certified by a registered professional engineer with training and experience in tank installation.

4) The installation has been inspected and approved by the implementing agency.

5) All work listed in the manufacturer's installation checklist has been completed.

6) The owner/operator may use another method if approved by the implementing agency.

Anyone bringing an UST system into use after May 8, 1986, must within 30 days of bringing such tank into use, submit the "Notification for Underground Storage Tanks" form (EPA form 7530–1) or the corresponding State form to provide notice of the existence of the tank to the administering agency. This notification also includes a certification of compliance with the installation certification requirement (40 CFR 280.20[a], [d], and [e], and 280.22[a] and [b]).

Existing tanks may meet the new tank performance standards or they may be retrofitted with an internal lining, cathodic protection, or an internal lining combined with cathodic protection. In addition, EPA has established inspection requirements for these upgraded tanks. If an existing tank does not meet the upgrading requirements by December 22, 1998, it must be replaced or closed.

UST Procedures and Operating Requirements

Spill and overfill procedures must be followed, and corrosion protection must be operated and maintained, USTs must be compatible with their contents, reports must be made and records must be kept, and any repairs must be performed in accordance with the regulations in order to met the day-to-day operating requirements of UST systems (40 CFR 280.30–34).

In addition, before a transfer is made, owner/operators must ensure that the volume in the tank is greater than the volume of product to be transferred. The transfer operation also must be monitored continuously to prevent overfilling and spilling.

The following UST reporting is mandatory:

- Notification for all UST systems, which includes certification of installation for new UST systems

- Reports of all releases including suspected releases, spills, and overfills, and confirmed releases

- Notification of corrective actions planned or taken including initial abatement measures, initial site characterization, free product removal, investigation of soil, groundwater cleanup, and corrective action plans

- Notification before permanent closure or change in service (40 CFR 280.34[a])

All of the following records must be kept, if applicable:

- A corrosion expert's analysis of site corrosion potential if corrosion protection equipment is not used

- Documentation of corrosion protection equipment operation

- Documentation of UST system repairs

- Recent compliance with release detection requirements

- Results of the site investigation conducted at permanent closure (40 CFR 280.34[b])

Leak detection methods used by the owner/operator must be able to detect a release from any portion of the tank and piping system that routinely contains product. Allowable methods of release detection include

- Monitoring for vapors in the excavation area

- Monitoring for petroleum floating on the water table

- Secondary containment with interstitial monitoring

- Continuous product level gauging

- Tank system integrity (or tightness) testing

- Inventory control

Monitoring for vapors or liquids in the environment around the tank or in the interstitial space of secondary containment can be done manually or automatically. If done manually, the sampling activity must be done once every 30 days. Tank tightness testing, however, is required only once every five years if the tank meets the corrosion protection and overfill requirements. If the tank does not meet these requirements, testing must be done once each year. In both cases, tank inventories must be recorded daily and reconciled monthly.

Whichever method is selected, the release detection system must be installed, calibrated, operated, and maintained in accordance with the manufacturer's instructions. In addition, the method(s) used must meet EPA performance standards specified for each type. Performance standards have been promulgated for each type of release detection method (40 CFR 280.43 and 44).

Owners/operators are required to maintain all written performance claims for release detection systems for five years. In addition, the following release detection records must be maintained. The results of any sampling, testing, or monitoring must be maintained for at least one year except that the results of tank tightness testing must be retained until the next test is conducted. Written documentation of all calibration, maintenance, and repair of release detection equipment permanently located onsite must be maintained for at least one year after the servicing work is completed. Any schedules of required calibration and maintenance provided by the release detection equipment manufacturer must be retained for five years from the date of installation. The implementing agency is authorized to modify recordkeeping time limits. However, prudent recordkeeping that entails record retention for the life of the UST system is recommended to assist with compliance enforcement mitigation and/or penalty avoidance.

Release Reporting

When should a release be suspected and reported? The following would be considered causes to suspect releases:

- Discovery of released regulated substances at the UST site or in the surrounding area (such as the presence of vapors in soils, basements, or sewer and utility lines, or emanating from nearby surface water).

- Unusual operating conditions (such as the erratic behavior of product-dispensing equipment, the sudden loss of product from the UST system, or an unexplained presence of water in the tank).

- Monitoring results from a release detection method indicating that a release may have occurred. Suspected releases must be reported to the implementing agency within 24 hours.

The following types of spills and overfills must be reported and are subject to corrective action:

- Spills or overfills of petroleum that result in the release of 25 gallons or that cause a sheen on nearby surface water

- Spills and overfills below 25 gallons if cleanup of the spill or overfill cannot be accomplished within 24 hours

The implementing agency has the authority to establish reasonable quantities or time periods different from those listed above. If they do, then an owner/operator should follow the limits established by the implementing agency.

When a release from an UST has been confirmed, the UST owner/operator must perform

- Initial response actions

- Initial abatement measures

- A site check

- A site characterization

- Free product removal if necessary

- Investigation for soil and ground water cleanup if certain conditions exist

In addition, upon request of the implementing agency or by the decision of the UST owner or operator, a corrective action plan may need to be developed. For each confirmed release that requires a corrective action plan, the implementing agency provides for public participation.

The following initial response actions must be taken within 24 hours of a release or within another reasonable period of time determined by the implementing agency:

- The release must be reported to the implementing agency (*e.g.*, by telephone or electronic mail).

- Immediate action must be taken to prevent any further release of the regulated substance into the environment.

- Fire, explosion, and vapor hazards must be identified and mitigated (40 CFR 280.61).

After a confirmed release, the owner/operator must

- Remove as much of the regulated substance from the UST system as is necessary to prevent further release to the environment.

- Visually inspect any above-ground or exposed below-ground releases and prevent further migration of the released substance into surrounding ground water and soils.

- Continue to monitor and mitigate fire and safety hazards.

- Remedy hazards posed by soils and, if necessary, treat and dispose of soils according to applicable State and local requirements.

- Measure for the presence of a release where contamination is most likely to be present at the UST site, unless the presence and source of the release has already been confirmed.

- Investigate to determine the possible presence of free product and begin free product removal as soon as practicable. Within 20 days after release confirmation, submit a report to the implementing agency summarizing the initial abatement steps taken.

Owners and operators are encouraged, in the interest of minimizing environmental contamination and promoting effective cleanup, to begin cleanup of soil and ground water before the corrective action plan is approved by the implementing agency, provided that they

- Notify the implementing agency of their intention to begin cleanup

- Comply with any conditions imposed by the implementing agency, including halting cleanup or mitigating adverse consequences from cleanup activities

- Incorporate these self-initiated cleanup measures in the corrective action plan that is submitted to the implementing agency for approval (40 CFR 280.66)

If an UST is located at a RCRA-permitted hazardous waste treatment, storage, or disposal (TSD) facility and the UST contained hazardous waste, then the corrective action must be performed in accordance with the hazardous waste regulations under Subtitle C of RCRA found in 40 CFR 264.100 and 264.101.

UST Change in Service and Closure

If a facility stops using an UST, the UST must undergo temporary closure, a change in service, or permanent closure. A change in service occurs when an UST system is no longer used to contain a regulated substance. If an UST is closed for longer than 12 months, it must be permanently closed (40 CFR 280.70 and 71).

For an UST to remain temporarily closed, the owner or operator must continue operation and maintenance of corrosion protection and release detection. However, if the UST system has been emptied, then release detection is not required (the UST system is empty when all materials have been removed using commonly employed practices so that no more than 2.5 cm [1 inch] of residue or 0.3% of the weight of the total capacity of the UST system remain). When an UST is temporarily closed for three months or more, the vent lines must be left open and functioning, and all other lines, pumps, manways, and ancillary equipment must be capped and secured. Any UST system temporarily closed for more than 12 months must be permanently closed unless it meets the new UST performance standards (40 CFR 280.20 and 280.21). Under this upgrading provision, an owner/operator of an operating existing UST has until December 22, 1998, to meet the new tank requirements. Therefore, the owner/operator may postpone upgrading his or her temporarily closed UST until December 22, 1998. After December 22, 1998, any tank that is temporarily closed for more than 12 months must be permanently closed unless it meets the new UST standards of 40 CFR 280.20 or the technical upgrading requirements under 40 CFR 280.21 (See 40 CFR 280.70).

Continued use of an UST system to store a nonregulated substance is considered a change in service. Before a change in service, the UST must be emptied and cleaned by removing all liquid and accumulated sludge, and a closure site assessment must be performed. At least 30 days before beginning a change in service, or within another reasonable time period determined by the implementing agency, the implementing agency must be notified of the intent to perform a change in service. Closure records also must be maintained.

To permanently close a tank, it must be emptied and cleaned by removing all liquids and accumulated sludges. All tanks taken out of service permanently must either be removed from the ground or filled with an inert solid material. (Note: Many State regulations require that closed USTs must be removed unless the removal would endanger a building, road, or other structural foundation.) At least 30 days before beginning permanent closure, or within another reasonable time period determined by the implementing agency, the implementing agency must be notified of the intent to close the UST. Before permanent closure is granted by the implementing agency, owners and operators must perform a closure site assessment. If during the assessment a release is confirmed, then corrective action would need to be commenced.

Records demonstrating compliance with the closure requirements must be maintained, in addition to the results of the excavation zone assessment, for the closure or change in service site assessment for at least three years after completion of permanent closure or change in service. Records may be maintained by the owner or operator who took the UST system out of service or by the current owner or operator of the UST system. The records may be mailed to the implementing agency if they cannot be maintained at the closed facility (40 CFR 280.74).

UST Upgrade Requirements

By December 22, 1998, all existing petroleum and hazardous substance USTs must be equipped with spill protection, overfill protection, and corrosion protection devices. Owners and operators have three choices for complying with these requirements. These are:

- Add spill, overfill and corrosion protection. (USTs that never receive more than 25 gallons at a time are not required to meet the spill and overfill protection requirements.)

- Close the existing UST.

- Replace the closed existing UST with a new UST. (When new USTs are installed, they must have spill, overfill, and corrosion protection, and leak detection devices.)

Briefly, the basic upgrading requirements for existing USTs are presented below.

Spill Protection. Spills cause releases at many UST sites. Generally, spills occur at the fill pipe when a delivery truck hose is disconnected. Such spills usually are small, but repeated small releases can lead to significant environmental problems. The regulations require that by December 22, 1998, existing tanks must have catchment basins to contain spills from delivery hoses.

Overfill Protection. Overfilling a tank can lead to large releases at the fill pipe and through loose fittings on the top of the tank. Existing USTs must have overfill protection devices by December 22, 1998, that will do one of the following:

- Automatically shut off flow into the tank when the tank is no more than 95 percent full

- Alert the operator when the tank is no more than 90 percent full

- Restrict for 30 minutes prior to overfilling, alert the operator 1 minute prior to overfilling, or automatically shut off flow so that none of the fittings on the top of the tank is exposed to product

Corrosion Protection. Corrosion occurs when bare metal, soil, and moisture combine to produce an underground electric current that destroys hard metal. Because unprotected steel USTs can corrode and release product through corrosion holes, Federal regulations require owners or operators to install corrosion protection in existing tanks by December 22, 1998. Existing tanks may already meet the corrosion protection requirements if one of the following performance standards is satisfied:

- The tank and piping are made entirely of noncorrodible material, such as fiberglass.

- The tank and piping are made of steel having corrosion-resistant coating and having cathodic protection.

- The tank is made of steel clad with a thick layer of noncorrodible material.

Because it is impractical to coat or clad unprotected steel USTs, owners or operators of such tanks must choose one of the three following methods to provide corrosion protection:

- Add cathodic protection.

- Add interior lining to the tank.

- Combine cathodic protection interior lining.

The regulations also require that by December 22, 1998, existing piping have one of the following characteristics:

- Uncoated steel piping has cathodic protection.

- Steel piping has corrosion-resistant coating and cathodic protection.

- Piping is made of, or enclosed in, noncorrodible material (*e.g.*, fiberglass).

Besides the spill, overfill, and corrosion protection upgrades required of all existing petroleum and hazardous substance USTs, hazardous substance USTs must meet additional leak detection requirements by December 22, 1998.

The UST regulations are intended in part to ensure that releases or leaks from USTs are discovered before contamination can spread. EPA has found that hazardous substances that have leaked into the soil are more difficult to detect and to clean up than petroleum leaks. Consequently, leak detection requirements for new hazardous substance USTs are more stringent than those for new petroleum USTs. Thus, while new petroleum USTs can meet leak detection requirements by selecting one of several specified leak detection methods (these methods include secondary containment and interstitial monitoring, automatic tank gauging systems, vapor monitoring, groundwater monitoring, statistical inventory reconciliation, manual tank gauging, and tank tightness testing, and inventory control), new hazardous substance USTs must be equipped with secondary containment systems with monitoring devices. By December 22, 1998, all existing hazardous substance tanks must meet the more stringent leak detection requirements for new hazardous substance USTs.

If an existing petroleum or hazardous substance UST system is not upgraded by December 22, 1998, it must be properly closed by that date. After the existing system has been closed, it may be replaced by installing a new UST that meets the new tank performance standards. As December 1998 approached, increased demand to upgrade existing USTs led to higher charges for contractors and supplies. Upgrades can take several months. However, citations and fines can result if the 1998 deadline has been missed.

Bibliography

Environmental Protection Agency. *Must for USTs*, EPA/530/UST–88/008. Washington, DC: EPA, 1991.

Grace H. Weaver. *Strategic Environmental Management*. New York, NY: John Wiley and Sons, Inc., 1996.

"Technical Standards and Corrective Action Requirements for Owners and Operators of Underground Storage Tanks." *Code of Federal Regulations*. Title 40, Pt. 280.

The Superfund Amendments and Reauthorization Act of 1986. PL 99–499, 100 Stat.1613 (codified in various sections of 42 USC §§9601–9675, 1988).

Thomas F. P. Sullivan, Ed. *Environmental Law Handbook*. 13th ed. Rockville, MD: Government Institutes, 1995.

Alan A. Eckmyre has a BS in Business Administration from the New York Institute of Technology, a BE in Nuclear Engineering from the State University of New York, an MBE from Emporia State University, and an MS in Environmental Engineering from the University of Kansas. Mr. Eckmyre has been a Certified Hazardous Materials Manager since 1991 and has served as the Chair of the National Academy Publications Committee, Secretary and Director of the Heartland Chapter in Kansas, and President of the Magnolia Chapter in South Carolina. He is also a member of the American Society of Testing and Materials' Environmental Assessment Committee. Mr. Eckmyre has over 20 years of experience in nuclear and environmental engineering, consulting, business development and program management. His project management experience includes: UST system closures, State UST ground water/soil remediation investigation and design projects, Phase I and II site assessments, Federal facility-specific and programmatic environmental audits, RCRA permitting and corrective measures studies, CERCLA preliminary assessments and Remedial Investigation/Feasibility Studies, NPDES permitting, construction remediation, expert witness testimony, and managment of technical staff. He currently provides environmental engineering support to DOE–HQ EM–40 programs. Mr. Eckmyre has published a number of reports relating to field evaluation, planning, and analysis of the West Valley Demonstration Project, Tar Creek Superfund Site, Paducah Gaseous Diffusion Plant, Savannah River Federal River Federal Facility, Culvert Cliffs Nuclear Power Plant, and the Department of Energy's spent nuclear fuel and waste management programs.

CHAPTER 22

Toxic Substances Control Act Basics

Valentino P. De Rocili, PhD, CHMM

History and Overview of the Toxic Substances Control Act

The Toxic Substances Control Act (TSCA) of 1976 was enacted by Congress in order to test, regulate, and screen all chemicals produced or imported into the United States. It was designed as a regulatory framework to deal with the risks posed by the manufacture and use of chemical substances. Many thousands of chemicals and chemical compounds are developed each year with unknown toxic or other dangerous characteristics. To prevent tragic consequences, TSCA requires any chemical that reaches the consumer marketplace to be tested for possible toxic effects before commercial manufacturing begins. TSCA requires corrective action in cases of cleanup of toxic materials contamination. TSCA was also designed to supplement other Federal statutes, including the Clean Air Act and the Toxic Release Inventory under the Emergency Planning and Community Right-to-Know Act (EPCRA).

The enactment of TSCA brought about the following basic changes to the environmental regulatory requirements:

- Created a screening mechanism for the production of new chemicals and new uses of existing chemicals

- Prescribed Environmental Protection Agency (EPA) control actions in the manufacture, use, and disposal of chemicals

- Conferred upon EPA extraordinary authority to address imminent hazards

- Banned the manufacture and use of polychlorinated biphenyls (PCBs)

- Gave EPA the power to require the testing of chemical substances that present a risk of injury to health and the environment

- Regulated the production and distribution of new chemicals and governed the manufacture and use of existing chemicals

- Regulated the cleanup, disposal, and record-keeping of asbestos and PCBs (Lee, 1996)

Any existing chemical that poses health and environmental hazards is tracked and reported under TSCA.

Even before the EPA was created in 1970, political pressure to develop this regulation focused on several issues, including the perceived need to regulate chemicals. In 1962, *Silent Spring* by Rachel Carson publicized the misuse of chemicals. According to Carson, this misuse of chemicals threatened to upset the balance of nature and enter the food chain. The book was so powerful that some feel it marked the beginning of the environmental movement in the United States.

Congress' intent in creating TSCA was to provide a comprehensive *catchall* law to close the loopholes that existed in previous chemical laws, such as the Federal Food, Drug, and Cosmetic Act (FFDCA) of 1938, and the Federal Insecticide, Fungicide, and Rodenticide Act (FIFRA) of 1947.

Today, the United States is involved in a major drive toward international cooperation in the control of chemicals. Whenever international agreement is reached on this issue, the EPA has indicated that it would be prepared to modify its policy, if necessary, in order to conform to international standards. The agency believes that the international harmonization of efforts in the control of chemicals will protect health and the environment, while fulfilling its obligations under the Trade Agreements Act of 1979 (40 CFR 707.20[b][3]).

TSCA Authority

The principal authority of TSCA was provided by the Congress in 1976 as stated in the *United States Code* (USC, Title 15 [Commerce and Trade], Chapter 53 [Toxic Substances Control], Subchapter I [Control of Toxic Substances]). In this statute, Congress stated three important findings:

- Human beings and the environment are exposed each year to many chemical substances and mixtures.

- There are some chemical substances and mixtures that may present an unreasonable risk of injury to health or the environment.

- Regulation of both interstate and intrastate commerce is required.

From these findings, Congress established the policy of the United States as follows:

- Develop and collect adequate data regarding the effect of chemical substances and mixtures on health and the environment from those who manufacture and process chemical substances and mixtures.

- Provide regulatory authority regarding chemical substances and mixtures that present an unreasonable risk of injury to health or the environment.

- Require that regulatory authority not impede unduly or create unnecessary economic barriers to technological innovation.

The definition of *chemical substances* does not include pesticides (as defined by the Federal Insecticide, Fungicide, and Rodenticide Act); any source material, special nuclear material, or by-product material (as defined in the Atomic Energy Act of 1954); any article the sale of which is subject to the tax imposed by the Internal Revenue Code of 1986; and any food, food additive, drug, cosmetic, or device (as defined by the Federal Food, Drug, and Cosmetic Act). In addition, TSCA does not regulate tobacco, any tobacco product, or firearms and ammunition, because other laws regulate these articles.

When it created TSCA, Congress' intent was that the Administrator should carry out his/her responsibilities in a reasonable and prudent manner, considering the environmental, economic, and social impact of any action taken (15 USC §2601). In 1978, under Executive Order No. 12088, provisions for Federal compliance with pollution control standards were introduced, making the head of each executive agency responsible for that agency's compliance with any applicable standards of pollution control (43 FR 47707).

The current TSCA regulations are found in Title 40, *Code of Federal Regulations*, Parts 700–799 (40 CFR 700–799). The statutes behind the regulation can be found in 15 USC §§2601–2629. Notable changes to TSCA requirements have been made from 1980 through 1998. In 1986, the Asbestos Hazard Emergency Response Act was enacted.

Regarding enforcement, in 40 CFR 710, Inventory Reporting Regulations, EPA says that it "does not intend" to concentrate its enforcement efforts on insignificant clerical errors in reporting, but rather on the overall goal of TSCA. However, the potential penalty for incomplete or missing PCB Annual Reports as required by 40 CFR 761 is a fine of $10,000 per year.

How TSCA Works

Under TSCA, EPA has the authority to approve all chemical substances used for manufacture, importation, or processing in commercial applications before they are introduced into commerce in the United States.

Manufacturers and importers must notify the EPA at least 90 days in advance of manufacturing a new chemical or importing a new chemical in bulk for commercial purposes. The EPA has 90 days to complete its review, and either approves the production of the chemical substance or acts to ban manufacture. EPA may extend this review period for an additional 90 days. If additional time is needed, the review may be interrupted to develop additional data.

The Toxic Substances Control Act is intended to be comprehensive, and ensure the protection of health and the environment from unreasonable risks associated with the use of chemicals. For this reason, TSCA includes importation in the Act's definition of the term *manufacture*: "manufacture means to import, produce, or manufacture" (15 USC §2602[7]). Thus, importers are responsible for making sure that their methods of importing chemicals comply with TSCA just as domestic manufacturers are responsible for ensuring that their methods for manufacturing chemicals comply with TSCA (40 CFR 707.20[b]).

The Toxic Substances Control Act supplements other Federal statutes, including the Clean Air Act and the Toxic Release Inventory, under the Emergency Planning Community Right-To-Know Act (EPCRA) (EPA, 1997). The disposal of PCBs is regulated by TSCA; however, hazardous wastes mixed with PCBs are regulated under the Resource Conservation and Recovery Act (RCRA).

Under Section 6 of the Act, four chemical substances are regulated because they pose unreasonable risk. These chemicals are asbestos, tetrachlorodibenzo-*p*-dioxin (TCDD), chlorofluorocarbons (CFCs), and PCBs.

Section 5(a)(2) of the Act defines the uses of chemical substances that EPA considers to be significant new uses. EPA imposes the Significant New Use Rule (SNUR) after the chemical is approved for manufacture. Under the SNUR, the use or production volume of the chemical substance may be restricted. In September 1984, the first SNUR was issued. It required manufacturers to notify the EPA if the use of the specified chemical substance exceeded 5 percent of a consumer product. The SNUR also specifies the procedures that manufacturers, importers, and processors must use in order to report on any significant new uses.

Under TSCA, a Premanufacture Notification (PMN) must be filed with the EPA for chemical substances that are not already listed on the inventory. The PMN must include chemical identification, information on use, method of disposal, production levels, work exposure, potential by-products or impurities, and health and environmental effects. The inventory of chemical substances is contained in 40 CFR 721. Exemptions from the PMN included in 40 CFR 723, Subpart B are:

- Chemical substances manufactured in quantities of 10,000 kilograms or less per year.

- Chemical substances with low environmental releases and human exposures.

- Some high molecular weight polymers.

- Chemical substances used for the manufacture or processing of instant photographic and peel-apart film articles have special requirements.

TSCA Regulatory Requirements

The Toxic Substances Control Act was designed by the Federal government to provide the following 21 basic functions:

1) **Collection of Fees and Cost Reimbursement.** Fees are required from manufacturers, importers, and processors who submit notices and applications to EPA under Section 5 of the Toxic Substances Control Act (15 USC §2604) in order to defray part of EPA's cost of administering the Act (40 CFR 700.40[a], Subpart C). 40 CFR 791, Data Reimbursement, establishes procedures and criteria to be used in determining fair amounts of reimbursement for the costs of testing incurred under Section 4(a) TSCA (15 USC §2603[a]).

2) **Mechanism for Civil Actions.** Section 20 (a) (1) and (2) of the TSCA authorizes any person to begin a civil action to compel the EPA to perform TSCA nondiscretionary acts or duties; to restrain any violation of TSCA, or of any rule promulgated under Sections 4, 5, or 6, or of any order issued under Section 5 of TSCA. The purpose of this regulation is to outline the procedures for submitting a notice of intent to file suit as required by Section 20(b) of TSCA before beginning civil actions (40 CFR 702.60, Subpart C).

3) **Reporting and Recordkeeping Procedures.** 40 CFR 704 specifies reporting and recordkeeping procedures under Section 8(a) of TSCA for manufacturers, importers, and processors of chemical substances and mixtures. The reporting and recordkeeping provisions are chemical-specific (7 USC §§136, et seq.).

4) **Important Definitions.** Title 15, Section 2602 of the USC includes the following important definitions:

Administrator means Administrator of the Environmental Protection Agency.

Chemical substance means any organic or inorganic substance of a particular molecular identity, including

- Any combination of substances occurring in whole or in part as a result of a chemical reaction or occurring in nature
- Any element or uncombined radical

Chemical substance *does not* include

- Any mixture
- Any pesticide (as defined by the Federal Insecticide, Fungicide, and Rodenticide Act) when manufactured, processed, or distributed in commerce for use as a pesticide (7 USC §§136, *et seq.*)
- Tobacco or any tobacco product
- Any source material, special nuclear material, or by-product material (as such terms are defined in the 1954 Atomic Energy Act)
- Any article the sale of which is subject to the tax imposed by the 1986 Internal Revenue Code
- Any food, food additive, drug, cosmetic, or device as defined by the Federal Food, Drug, and Cosmetic Act

Mixture means any combination of two or more chemical substances if the combination does not occur in nature and is not, in whole or in part, the result of a chemical reaction. There are two exceptions to this definition: (1) the combination is *not* a mixture if it occurs as the result of a chemical reaction in which there are no new chemical substances, and (2) the combination is *not* a mixture if it could have been manufactured for commercial purposes without a chemical reaction at the time the combination's chemical substances were combined.

New chemical substance means any chemical substance that is not included in the chemical substance list compiled and published under Title 15, Section 2607(b) of the *United States Code*.

5) **General Import Requirements and Restrictions.** TSCA addresses aspects of the regulation promulgated by the United States Customs Service (Customs), Department of the Treasury (19 CFR 12.118–12.127, 127.28

[amended], 15 USC §2612). Section 13 requires the Secretary of the Treasury to refuse entry into the Customs territory of the United States of any chemical substance, mixture, or article if it does not comply with the rules that are in effect under TSCA (40 CFR 707.20, Subpart B).

6) **EPA Enforcement Procedures.** The EPA and Customs monitor chemical imports to determine if imported shipments comply with the certification requirements and the substantive mandates of TSCA. Customs is authorized to refuse entry to any shipment until the certification is properly submitted, and is allowed to detain a shipment if reasonable grounds exist to believe that such shipment or its import violates TSCA's regulations or orders. A shipment that is in violation must either be brought into compliance, exported, destroyed, or voluntarily abandoned within the time periods prescribed in 19 CFR 12.124 of the Section 13 rule. When EPA determines that a shipment should be detained, EPA will identify the reasons for the detention and the actions required of the importer in order to bring the shipment into compliance with TSCA (40 CFR 707.20[c], Subpart B).

7) **Inventory Reporting Procedures.** Those who manufacture, import, or process chemical substances for commercial purposes under Section 8(a) of the Toxic Substances Control Act must report according to regulations (15 USC §2607[a]). Section 8(a) authorizes the Administrator to require reporting of information necessary for administration of the Act, and requires EPA to issue regulations that allow it to compile an inventory of the chemical substances manufactured or processed for a commercial purpose. Following the initial reporting period, EPA published an initial inventory of chemical substances manufactured, processed, or imported for commercial purposes. The EPA periodically amends the inventory to include the new chemical substances reported under Section 5(a)(1) of the Act. The agency also revises the categories of chemical substances and makes other amendments as appropriate. Under Section 15(3) of TSCA, it is unlawful for any person to fail or refuse to submit the information that is required under these reporting regulations. In addition, Section 15(3) states that it is

unlawful for any person to fail to keep, and allow access to, the records that are required by these regulations. Section 16 states that any person who violates a provision of Section 15 is liable to the United States for a civil penalty and may be criminally prosecuted. Pursuant to Section 17, the Government may seek a court order to compel submission of Section 8(a) information and to otherwise restrain any violation of Section 15 (40 CFR 710.1, [a], [b]).

8) **Chemical Information Rules.** Procedures are outlined by which chemical manufacturers and processors must report production-, use-, and exposure-related information on listed chemical substances. The rules stipulate the requirements that apply to all reporting (40 CFR 712.1, Subpart A). 40 CFR 750 establishes procedures for all rulemakings under authority of Section 6 of the Toxic Substances Control Act (TSCA), 15 USC §2605.

9) **Requirements for Health and Safety Data Reporting.** TSCA requires manufacturers, importers, and processors to submit lists and copies of health and safety studies on the chemical substances and mixtures selected under Section 4(a), and on other chemical substances and mixtures for which EPA requires health and safety information.

10) **Records and Reports for Adverse Chemical Substance Reactions.** Section 8(c) of TSCA requires manufacturers, processors, and distributors of chemical substances and mixtures to

- Keep "records of significant adverse reactions to health or the environment, as determined by the Administrator by rule, alleged to have been caused by the substance or mixture"

- "Allow inspection and submit copies of such records," upon request of any designated representative of the Administrator

This rule implements Section 8(c) of TSCA. It describes the records to be kept and prescribes the conditions under which certain firms must submit or make the records available to a duly designated representative of the Administrator (40 CFR 717.1 [a] and [b], Subpart A)

11) Rules for Reporting New Chemical Substances. 40 CFR 720 establishes the procedures by which new chemical substances must be reported by manufacturers and importers under Section 5 of the Toxic Substances Control Act (15 USC §2604). The rule defines the persons who are subject to the reporting requirements, and the chemical substances that must be reported. It prescribes the contents of Section 5 notices, and establishes procedures for submitting notices. The rule also outlines the EPA's policy regarding claims of confidentiality for, and public disclosure of, various categories of information that are submitted about Section 5 notices. A complete list of chemicals is located in 40 CFR 721.

12) Specific Use Requirements for Certain Chemical Substances. 40 CFR 747 provides specific requirements for metal-working fluids including mixed mono- and diamides of an organic acid (40 CFR 747.115), triethanolamine salt of a substituted organic acid (40 CFR 747.195), and triethanolamine salt of tricarboxylic acid (40 CFR 747.200). 40 CFR 749.68 furnishes the requirements for water treatment chemicals in air-conditioning and cooling systems, such as hexavalent chromium-based water treatment.

13) Prohibitions of, and Requirements for, the Manufacture, Processing, Distribution in Commerce, Use, Disposal, Storage, and Marking of PCBs and PCB Items. 40 CFR 761 applies to all persons who manufacture, process, distribute in commerce, use, or dispose of PCBs or PCB Items. Substances regulated by this rule include, but are not limited to, dielectric fluids, contaminated solvents, oils, waste oils, heat-transfer fluids, hydraulic fluids, paints, sludges, slurries, dredge spoils, soils, materials contaminated because of spills, and other chemical substances or combinations of substances, including impurities and by-products and any by-product, intermediate, or impurity manufactured at any point in a process. Most of the provisions of this part apply to PCBs only if PCBs are present in concentrations above a specified level. For example, Subpart D applies generally to materials at concentrations of 50 parts per million (ppm) and above. Also, certain provisions of Subpart B apply to the PCBs that are inadvertently generated in manufacturing processes at concentrations specified in the definition of PCB under 40 CFR 761.3. No provision specifying a PCB concentration may be avoided because of dilution, unless otherwise specifically provided (40 CFR 761.1, Subpart A).

14) PCB Spill Cleanup Policy. In 1987, EPA promulgated a national PCB spill cleanup policy. The policy requires notification and recordkeeping of PCB spills into sensitive areas such as open water, near wells, vegetable gardens, and grazing lands. It requires a specific cleanup and cleanup verification function for any spills of PCB in concentrations of 50 ppm or above. Additional requirements are prescribed for spills of oil containing PCBs at levels greater than 1 pound and 10 pounds (40 CFR 761.120–135).

15) Asbestos-Containing Materials. 40 CFR 763 regulates asbestos abatement projects and Asbestos-Containing Materials (ACM) in schools. It also prohibits the manufacture, importation, processing, and distribution in commerce of certain asbestos-containing products and establishes the labeling requirements for these products.

16) Requirements for Testing for Dibenzo-*p*-Dioxins / Dibenzofurans. 40 CFR 766 outlines the requirements for testing to ascertain whether certain specified chemical substances may be contaminated with halogenated dibenzodioxins (HDDs) and/or dibenzofurans (HDFs) as defined in 40 CFR 766.3, and requirements for reporting under Section 8 of TSCA (15 USC §2607).

17) Rules that Govern Testing Consent Agreements and Testing. TSCA establishes procedures for gathering information, conducting negotiations, and for developing and implementing test rules or consent agreements on chemical substances and mixtures under Section 4 of TSCA. Section 4 of the Act authorizes the EPA to require the manufacturers and processors of chemical substances and mixtures to test these chemicals in order to determine whether they have adverse health or environmental effects. Section 4(a) empowers the Agency to require such testing. In addition, the EPA has the implied authority to enter into enforceable consent agreements that require testing where the Agency provides procedural safeguards equivalent to those that apply where the testing is conducted by rule rather than by agreement. The EPA uses enforceable consent agreements to carry out testing where a consensus exists between the EPA, the affected manufacturers and/or processors, and interested members of the public concerning the need for testing. If such a consensus does not exist and the Agency believes that it can produce the findings specified in Section 4(a), then the EPA will initiate proceedings to declare test rules that will be codified in 40 CFR 799.

18) **Good Laboratory Practice Standards.** Good laboratory practices for conducting studies relating to health effects, environmental effects, and chemical fate testing are established. The standards are intended to ensure the quality and integrity of the data submitted pursuant to testing consent agreements and test rules issued under Section 4 of the TSCA (PL 94–94–469, 90 Stat. 2006, 15 USC §§2603, *et seq.*).

19) **Provisional Test Guidelines.** TSCA provides provisional chemical fate guidelines, environmental effects guidelines, health effects guidelines, and exposure and toxicity guidelines (40 CFR 795–798).

20) **Identifies Specific Chemical Substance and Mixture Testing Requirements.** 40 CFR 799 identifies the chemical substances, mixtures, and categories of substances and mixtures for which data will be developed, specifies the persons who are required to test (manufacturers, including importers, and/or processors), specifies the test substance(s) in each case, prescribes the tests that are required and their standards, and provides deadlines for the submission of reports and data to EPA. This part requires manufacturers and/or processors of the chemical substances or mixtures that are identified in Subpart B to submit letters of intent to test, exemption applications, and study plans according to the EPA test rule. This part also requires the manufacturers and/or processors of the chemicals identified in Subpart B to conduct tests and to submit data following the test standards contained in this part in order to enable the EPA to develop data on the health and environmental effects and other characteristics of these chemicals. This

data is used to assess the risk of injury to human health or the environment that is presented by these chemicals (40 CFR 799.1 [a], [b], and [c]). This includes hazardous waste constituents subject to testing, (40 CFR 799.5055) and drinking water contaminants subject to testing (40 CFR 799.5075).

21) **Lead-Based Paint Poisoning Prevention in Certain Residential Structures.** TSCA requires the disclosure of known lead-based paint and/or hazards upon sale or lease of residential property, and regulates lead-based paint activities (40 CFR 745). On June 1, 1998, the EPA promulgated a final rule which changes the requirements for hazard education before renovation of target housing by requiring that a lead hazard informational pamphlet be provided to owners and occupants of such housing prior to commencing of a renovation. In addition, the rule requires notification on the nature of renovation activities in certain circumstances involved with multifamily housing (63 FR 29908).

Highlights of Regulatory Changes for the Disposal of PCBs

On June 29, 1998 in the *Federal Register*, the Environmental Protection Agency promulgated its final rule under TSCA for the disposal of PCBs (63 FR 35383–35474). Changes made to the existing regulations became effective on August 28, 1998. The changes affected requirements under both 40 CFR 750 and 40 CFR 761.

The EPA amended the portion of its rules under TSCA which addresses the manufacture, processing, distribution in commerce, use, cleanup, storage, and disposal of PCBs. The rule provides some flexibility regarding selection of disposal technologies for PCB wastes and expands the allowable procedures for decontamination of PCB on surfaces and water. The rule also modifes the requirements regarding the use and disposal of PCB equipment. It also codifies policies that EPA has developed and implemented over the 19 years prior to this ruling.

Regulatory changes that affect PCB concentration assumptions for electrical equipment are as follows:

- The PCB concentration assumption rule was changed, allowing any person to assume that electrical transformers, circuit breakers, reclosers, oil-filled cable, and rectifiers containing less than 1.36 kg (3 lbs) of fluid whose PCB concentration has not been established contain less than 50 ppm of PCBs, and therefore non-PCB rated equipment (63 FR 35436).

- Any person may assume that mineral oil-filled electrical equipment, including transformers, that was manufactured before July 2, 1979, and whose PCB concentration is not established, is PCB Contaminated Electrical Equipment. It will be assumed that the equipment will contain 50 ppm or greater PCBs, but less than 500 ppm PCBs (63 FR 35436).

- Pole-top and pad-mounted distribution electrical transformers manufactured before July 2, 1979, are assumed to be mineral oil–filled and PCB-Contaminated Electrical Equipment while in use, unless the concentration has been established (63 FR 35436).

- If the date of manufacture of mineral oil–filled electrical equipment is unknown, any person must assume it to be PCB Contaminated Electrical Equipment (63 FR 35437).

- Any person may assume that oil-filled electrical equipment, including transformers, that was manufactured after July 2, 1979, and whose PCB concentration is not established, is Non-PCB Rated Equipment, containing less than 50 ppm of PCBs (63 FR 35437).

- Any person must assume that an electrical transformer manufactured before July 2, 1979, that contains 1.36 kg (3 lbs) or more of fluid other than mineral oil, and whose PCB concentration has not been established, is a PCB Transformer, containing 500 ppm or greater PCBs (63 FR 35437).

- Any person may assume that a capacitor manufactured after July 2, 1979, is Non-PCB Rated Electrical Equipment, containing less than 50 ppm PCBs (63 FR 35437).

- Any person may assume that a capacitor manufactured before July 2, 1979, whose PCB concentration is not established, is a PCB Capacitor, containing 500 ppm or greater of PCBs (63 FR 35437).

- If the date of manufacture is unknown, any person must assume that a capacitor contains 500 ppm or greater of PCBs (63 FR 35437).

- Any person may assume that a capacitor marked at the time of manufacture with the statement "No PCBs" in accordance with the regulations is Non-PCB Rated Equipment (63 FR 35437).

Assumption policies do not apply when electrical equipment is being disposed of. At the time of disposal, the owner or operator of PCB equipment must know its actual PCB concentration and use the proper disposal method (63 FR 35389).

In addition to changes in assumption policies for electrical equipment, the EPA required that registration of PCB Transformers be completed no later than December 28, 1998. This new registration requirement includes PCB Transformers in use or in storage for reuse, even if a specific PCB Transformer was registered under the previous regulatory requirements (63 FR 35440).

If an electrical transformer owner who assumes that a transformer is PCB Contaminated, and discovers after December 28, 1998 that it is in fact a PCB Transformer, the owner of the transformer must register the newly identified PCB Transformer, in writing, with the EPA no later than 30 calendar days after it is identified as such. This requirement does not apply to transformer owners who have previously registered with the EPA PCB Transformers located at the same location as the newly identified PCB Transformer (63 FR 35440).

The new rule also addresses persons who take possession of a PCB Transformer after December 29, 1998. For this condition, the owner is not required to register or re-register the transformer with the EPA (63 FR 35440).

Owners of PCB Transformers are required to maintain, with an annual log, proof of registration with the EPA, local fire response, and/or building owners (63 FR 35440).

PCB cleanup levels for soil have changed in this new ruling. The soil cleanup level for high occupancy (*e.g.*, residential areas) is less than 1 ppm, or less than 10 ppm if the contaminated soil is capped (63 FR 35390). For PCB-contaminated soil in low occupancy (*e.g.*, electrical substations), the soil cleanup level is less than 25 ppm to less than 100 ppm, depending upon site conditions (63 FR 35390).

Decontamination standards for PCB-contaminated surfaces are as follows:

- Nonporous surfaces in contact with liquid PCBs destined for reuse: less than 10 micrograms of PCBs per 100 square centimeters (63 FR 35390).

- Nonporous surfaces in contact with nonliquid PCBs destined for reuse: as required by the National Association of Corrosion Engineers (NACE) Visual Standard No. 2, Near-White Blast Cleaned Surface Finish (63 FR 35390).

- Concrete surfaces containing fresh PCB spills: less than 10 micrograms of PCBs per 100 square centimeters (63 FR 35390).

- PCB spills to water: less than 5 micrograms of PCBs per liter for unrestricted use (63 FR Page 35390).

The handling of PCB wastes was addressed in this new rule as follows:

- If the PCB component of a waste is approved for disposal at a facility, the approval for the disposal of the other regulated waste components must be addressed by all other statutes or regulatory authorities (63 FR 35390).

- Unless otherwise noted in the regulations, references to weights or volumes in 40 CFR 761 apply to the total weight or volumes of the PCB-containing material (*e.g.*, oil, soil, *etc.*), not to the calculated weight or volume of only the PCB fraction within that substance (63 FR 35390).

- Restrictions regarding manufacturing, processing, distribution in commerce, and use of PCBs were addressed in the new ruling. The EPA also revised the definition of PCB remediation waste and addressed sewage sludge containing PCBs (63 FR 35392).

New marking requirements were provided in the ruling requiring the M_L mark for storage units, PCB Large Low Voltage Capacitors, and PCB Equipment (including equipment in use) that contain PCB Transformers, PCB Large High or PCB Large Low Capacitors (63 FR 35426).

For the first time, natural gas pipeline systems are included under the PCB regulation. The EPA has initiated a compliance monitoring program for companies with greater than 50 ppm PCBs in their pipelines, to address contamination issues in natural gas pipelines and associated equipment (63 FR 35395).

Where to Get More Information

In support of Title 15, Section 2625, the EPA Office of Prevention, Pesticides, and Toxic Substances (OPPTS) provides a TSCA Assistance Information Service. The service provides information and assistance to chemical manufacturers, processors, users, storers, disposers, importers, and exporters, concerning the regulations under TSCA.

As of June 1998, OPPTS could be reached at (202) 554–1404 or at the following Internet address:

tsca-hotline@epamail.epa.gov.

Also, check for TSCA information on the EPA homepage:

<http://www.epa.gov>

Bibliography

"Agriculture." *US Code.* Title 7, Secs. 136, *et seq.*

"Chemical Fate Testing Guidelines." *Code of Federal Regulations.* Title 40, Pt. 796.

"Chemical Import and Exports." *Code of Federal Regulations.* Title 40, Pt. 707.

"Chemical Information Rules, General Provisions." *Code of Federal Regulations.* Title 40, Pt. 712, Subpart A.

"Citizen Suit." *Code of Federal Regulations.* Title 40, Pt. 702, Subpart C.

"Control of Toxic Substances, Definitions." *US Code.* Title 15, Sec. 2602.

"Control of Toxic Substances, Findings, Policy, and Intent." *US Code.* Title 15, Sec. 2601.

"Disposal of Polychlorinated Biphenyls (PCBs), Final Rule." *Federal Register* 63:124. (29 June 1998) pp. 35383–35474.

"Entry into Customs Territory of the United States." *US Code.* Title 15, Sec. 2612.

"Environmental Effects Testing Guidelines." *Code of Federal Regulations.* Title 40, Pt. 797.

Environmental Protection Agency, Office of Public Affairs. "Toxic Substances Control Act (TSCA) Assistance Information Service." Washington, DC: EPA, 13 March 1997.

"Fees." *Code of Federal Regulations.* Title 40, Pt. 700, Subpart C.

"General Import and Exports." *Code of Federal Regulations.* Title 40, Pt. 707

"General Import Requirements and Restrictions." *Code of Federal Regulations.* Title 40, Pt. 707, Subpart B.

"General Reporting and Recordkeeping Provisions for Section 8(a) Information-Gathering Rules." *Code of Federal Regulations.* Title 40, Pt. 704, Subpart A.

"Health and Safety Data Reporting, General Provisions." *Code of Federal Regulations.* Title 40, Pt. 716, Subpart A.

"Health Effects Testing Guidelines." *Code of Federal Regulations.* Title 40, Pt. 798.

"Identification of Specific Chemical Substance and Mixture Testing Requirements." *Code of Federal Regulations.* Title 40, Pt. 799.

"Inventory Reporting Regulations." *Code of Federal Regulations.* Title 40, Pt. 710.

Lee, C. C. *Dictionary of Environmental Legal Terms.* New York, NY: McGraw-Hill, 1996.

"Manufacturing and Processing Notices." *US Code.* Title 15, Sec. 2604.

"Metal Working Fluids." *Code of Federal Regulations.* Title 40, Pt. 747.

"Polychlorinated Biphenyls (PCBs) Manufacturing, Processing, Distribution in Commerce, and Use Prohibitions." *Code of Federal Regulations.* Title 40, Pt. 761.

"Premanufacture Notification." *Code of Federal Regulations.* Title 40, Pt. 720.

"Premanufacture Notification Exemptions." *Code of Federal Regulations.* Title 40, Pt. 723.

"Provisional Test Guidelines." *Code of Federal Regulations.* Title 40, Pt. 795.

"Public Health and Welfare." *US Code.* Title 42, Secs. 2011, *et seq.*

"Records and Reports of Allegations That Chemical Substances Cause Significant Adverse Reactions to Health or the Environment, General Pro-visions." *Code of Federal Regulations.* Title 40, Pt. 717, Subpart A.

"Reporting and Retention of Information." *US Code.* Title 15, Sec. 2607.

"Testing of Chemical Substances and Mixtures." *US Code.* Title 15, Sec. 2603, *et seq.*

"Water Treatment Chemicals." *Code of Federal Regulations.* Title 40, Pt. 749.

Valentino P. De Rocili is the President of Compliance Environmental, Inc., an environmental health and safety consulting firm located in Dover, Delaware. He has managed over $55 million of environmental and safety-related projects in more than 20 years of hands-on work experience in areas including hazard and risk assessment, PCBs, lead-based paint inspection and abatement, waste management, underground storage tanks, indoor air quality, wastewater treatment plants, solid waste landfills, and compliance requirements. He is currently an Assistant Professor at the University of Delaware, and the co-founder as well as Vice President of the Delaware Chapter of Hazardous Materials Managers. He serves on several technical committees including the Environmental Technical Review Committee for the United States Air Force at Dover Air Force Base, and the Lead Advisory Group for the State of Delaware Department of Health and Social Services. In 1996, he received the Champion of Excellence Award from the National Academy of Certified Hazardous Materials Managers.

CHAPTER 23

Federal Insecticide, Fungicide, and Rodenticide Act

Keith Trombley, CHMM

Introduction

The Federal Insecticide, Fungicide, and Rodent-icide Act (FIFRA) was first passed in 1947. Its primary purpose was to require the registration of pesticides to protect consumers from mis-branding, adulteration, and ineffective pesticides.

Original jurisdiction for this law was given to the United States Department of Agriculture (USDA), but this was transferred to the United States Environmental Protection Agency (EPA) when it was created in 1970. FIFRA has been amended many times by Congress over the years, most recently through the Food Quality Protection Act (FQPA) in 1996 (Public Law [PL] 104–170).

One important distinction between FIFRA and most of the other environmental protection laws is that it was originally a *registration* rather than *reporting* standard (however, more and more reporting requirements have been added to the act). The registration concept tends to limit the amount of liability incurred by the industry once their products are legally on the market. This benefit is often offset because the registration process has become expensive and difficult. Most products that make any pesticidal claims on their label must go through a far more extensive testing

305

and approval process than most of the other chemicals we may use in our daily lives. This effects not only a user's cost, but also the variety of chemicals that are available to consumers.

The text of FIFRA can be found under the reference 7 *United States Code* (USC) §§136, *et seq*. The regulations promulgated from this law are found in the *Code of Federal Regulations* at 40 CFR 150–189 (Subchapter E, Pesticide Programs). However, knowing the law and regulations is only the beginning step to compliance and getting a pesticide registered and on the market. At the Federal level there are guidance manuals, interpretation documents, and many specific bureaucratic procedures that must be followed to successfully negotiate the registration process. Nearly every state has its own level of regulations, some of which are just as detailed as those at the Federal level. The following text is an overview of some of the more notable issues of the pesticide regulation process.

The Law

FIFRA can trace its legislative lineage as far back as 1906 to the Pure Food Act. It is still closely linked to the present-day Federal Food, Drug, and Cosmetic Act (FFDCA). The text of the act is broken down into several sections, the most important being: Sections 3 and 4 on registration and reregistration (7 USC §136[a]), Section 7 on registrations of establishments (7 USC §136[e]), Section 10 on protection of trade secrets and other information (7 USC §136[h]) and Section 11 on use of restricted-use pesticides and applicators (7 USC §136[i]).

As with some of the other environmental laws presently being enforced, it is worthwhile to be familiar with both the law and the regulations. References to the act, the law (*i.e.*, *United States Code*), and the regulations are interchanged frequently.

Definitions and Exceptions

As with any law and subsequent regulations, certain terms must be specifically defined so everybody understands and accepts what everybody else is taking about. Listed below are some of the more key definitions and notable exceptions to the rules.

Pest—An organism deleterious to man or the environment if it is: any vertebrate animal other than man; any invertebrate animal, including but not limited to any insect, other arthropod, nematode, or mollusk such as a slug and snail, but excluding any internal parasite on living man or other animals; any plant growing where not wanted, including any mosses, alga, liverwort, or other plant of any higher order, and any plant part such as a root; or any fungus, bacterium, virus, or other microorganism, except those on or in living man or other living animals and those on processed food or processed animal feed, beverages, drugs, and cosmetics (as defined under FFDCA Sections 201[g][1] and 201[i]).

Pesticide—Any substance or mixture of substances intended for preventing, destroying, repelling, or mitigating any pest, or intended for use as a plant regulator, defoliant, or desiccant, other than any article that is a new animal drug under the FFDCA Section 201(w); an animal drug that has been determined by regulation of the Secretary of Health and Human Services not to be a new animal drug; or an animal feed under the FFDCA Section 201(x) that bears or contains any substances described above.

Pesticide product—A pesticide in a particular form (including composition, packaging, and labeling) in which the pesticide is, or is intended to be, distributed or sold. The term includes any physical apparatus used to deliver or apply the pesticide if distributed or sold with the pesticide.

End use product—A pesticide product whose labeling: includes directions for use of the product (as distributed or sold, or after combination by the user with other substances) for controlling pests or defoliating, desiccating, or regulating the growth of plants; and does not state that the product may be used to manufacture or formulate other pesticide products.

Manufacturing use product—Any pesticide product that is not an end use product.

Label—The written, printed, or graphic matter on, or attached to, the pesticide or device or any of its containers or wrappers.

Labeling—All labels and other written, printed, or graphic matter which: accompanies the pesticide or device at any time; or, reference is made on the label or in literature accompanying the pesticide or device, except to the current official publications of the EPA, the USDA, Department of the Interior (DOI), Department of Health and Human Services (DHHS), state experiment stations, state agricultural colleges, and other Federal or State institutions or agencies authorized by law to conduct research in the field of pesticides.

Some products that are not considered pesticides, even though the above definitions seem to include them, are: drugs; fertilizers; deodorizers, bleaches and cleaning agents (if they make no pesticide label claims); barriers that contain no toxicants (such as screens or sheet metal); treated articles or substances (such as treated lumber); natural cedar; and a listing of 31 specific *minimum-risk pesticides* (*e.g.*, cinnamon, garlic, peppermint, zinc metal strips, *etc.*).

Many things that are not directly part of the product's label are still considered labeling. For example, Material Safety Data Sheets (MSDSs), while not directly regulated by the EPA, are considered product labeling under FIFRA and should include the same directions and precautions as found on the product label.

The Regulated Community

The main groups of people or businesses directly regulated by FIFRA fall into two basic categories: manufacturers and formulators. A manufacturer is a company that actually produces the chemical that is the active ingredient in a pesticide. These companies typically are very large chemical firms, such as Dow, Dupont, or Monsanto. There are probably less than 30 companies in the United States that fall into this category. Conversely, there are probably 100 times this number of pesticide formulators in the United States. A formulator is a firm that mixes active ingredient(s) with various inert ingredients, such as diluents, carriers, propellants, *etc.*, to create a product for an end user. Consequently, pesticides are categorized for manufacturing or end use. While most pesticide manufacturers make both manufacturing use and end use products, formulators focus almost entirely on end use products.

The Registration Process

When registering a pesticide, there are three main items that will be reviewed by the EPA before the product is allowed to be sold: the label, the formula, and the supporting data. All of the required documents and forms must be grouped together in an application that is submitted to the EPA for review and, hopefully, approval.

The Paperwork

There are numerous applications, forms, and documents that must be submitted in the correct amount, format, and order. Failure to meet these administrative requirements will result in immediate rejection of the application. Once the application is determined to be administratively correct, then all submittals, especially the label and formula, will be reviewed in detail. Depending on the type of product and the length of the label, this process can be quite time-consuming, and there is always the chance that something that was accepted during a first review will not be accepted on a subsequent review. The review process for a registered pesticide is never really over. Sometimes products that have been in use for years will be required to have their label changed because something that was accepted in the past is now "unacceptable."

Some of the documents that will have to be submitted with a registration application are

- Application for Pesticide Registration (EPA Form 8570–1)
- Confidential Statement of Formula (CSF) (EPA Form 8570–4)
- Formulator's Exemption Statement (EPA Form 8570–27)*
- Certification with Respect to Citations of Data (EPA Form 8570–34)
- Data Matrix (EPA Form 8570–35)
- Labels
- Supporting data

* Only needed when using a registered pesticide as an ingredient in the formula.

These forms can be found on EPA's Office of Pesticide Programs Internet site at

<http://www.epa.gov/opprd001/forms/>

The EPA publishes a very helpful guidance document (called the Blue Book) which is indispensable to even the most experienced applicant. (Note: At the time of this writing, the Blue Book was being revised). It gives detailed guidance for correctly submitting applications, amendments, notifications, and other various documents.

The Label

The EPA must approve all pesticide labels. There are very specific requirements on what and where required language must appear on a pesticide's label. The product label is usually the only source of instructions and precautions given to a user. While most of the FIFRA requirements for labels are straightforward, some requirements are less clear and sometimes interpreted differently by the registration applicant than by the EPA. The root of this "conflict" is usually due to the registrant's desire to provide user-friendly instructions that do not meet the EPA's specific statements for directions and precautions. Both parties have an interest in developing a *good* label, but finding a happy medium can result in a lengthy review process and a delay in getting a product approved and on the market.

As anyone who has ever read the label of a pesticide will profess, it is not something that is read unless necessary. Pesticide labels are typically very long, often complicated, and not easy to understand. While both the EPA and the regulated community are trying to make labels less unpleasant to read, this will not be corrected overnight. Pesticide registration applicants should keep in close contact with the regulators that will be providing the approval. Early communication can eliminate later misunderstandings.

All pesticide labels must include the following items:

- Ingredient statement
- Restricted use pesticide statement*

---

* Required only when applicable to that pesticide.

- "Keep Out of Reach of Children" statement
- Precautionary statements with signal word
- Skull and crossbones symbol and the word "Poison"*
- First aid statements ("statement of practical treatment")
- Environmental hazard statements
- Physical or chemical hazard statements
- Directions for use
- Re-entry statement
- Product name
- Storage and disposal statements
- Registrant name and address
- Net contents statement
- EPA registration and establishment numbers

There are also requirements regarding where on the label most of these items must be placed, the order in which the statements must appear on the label, and the size of the text. While this may seem rather restrictive, it does allow for a somewhat standardized format and easy retrieval of information once the consumer is familiar with it. This can be very handy in an emergency.

The label section that is usually the hardest to get approved is the **directions for use** statements. This is the part of the label where the registrant describes how much of the product to use, for what applications it can be used (*e.g.*, which crops, pests, locations, *etc.*), and how to apply it. This is a key portion of the product label and will always start with the statement "It is a violation of Federal law to use this product in a manner inconsistent with its labeling." This is the EPA's method of putting all users on notice that they must follow the label directions.

The directions for use statements are also where most companies' marketing departments get involved and try to make performance claims. One person's acceptable claim is another person's outrageous claim, and this difference of opinion often occurs between the registrant and the regulator. Even the proposed product name can become a bone of contention, because of perceived or implied claims made by the name. Again, communication with the regulators is the key to avoiding roadblocks and delays.

Formula

All pesticides must have a specific written formula with certifiable concentration limits of active ingredients. A Confidential Statement of Formula (CSF) must list the names of all the ingredients, the suppliers of the ingredients, the amount (by weight and percentage of total formula) of each ingredient in the final formula, and the reason for adding each ingredient to the formula. In addition, any registered pesticide used in the formula must include the EPA registration number. Including registered pesticides in formulae is a cost-efficient way of making an end use product. This is the method used by most pesticide formulators to avoid thousands of dollars in costs for supporting data. The subject of supporting data will be discussed in the next section.

While the EPA does allow for substituting different ingredients in a formula, all of these substitutions must be spelled out in the statement of formula. The statement of formula goes back to one of the original reasons behind FIFRA, *i.e.*, the protection of consumers from adulterated products. A pesticide that does not conform to its specified statement of formula is considered adulterated and illegal to sell.

Product formulae are considered Confidential Business Information (CBI) and very tightly controlled by businesses as well as the EPA. Product formulae are exempt from the provisions of the Freedom of Information Act (FOIA).

Supporting Data

Supporting data are all of the scientific and clinical data required before a product is considered safe enough to be registered. Areas where a registrant could be required to submit supporting data to the EPA are:

- Product properties
- Product performance
- Fate, transport, and transformation
- Health effects
- Occupational and residential exposure
- Spray drift
- Ecological effects
- Residue chemistry

There are also specific testing procedures for biochemical and microbial pesticides.

Determining which supporting data must be submitted is dependent on the use pattern of the pesticide. Typically, a formulator will need product property data and the *6-pack* of acute toxicity studies:

1) Acute oral toxicity

2) Acute dermal toxicity

3) Acute inhalation toxicity

4) Primary eye irritation

5) Primary dermal irritation

6) Dermal sensitization

Products that make claims to protect human health (*e.g.*, sanitizers and rodenticides) will usually have to submit product performance data also.

Supporting data can be either submitted or *cited*. If a product is identical or similar to an existing registered product, permission to reference these data that are already on file with the EPA can be given. There is also a provision in the law for an applicant to cite another registrant's data, or even "cite-all" data on file, without the other registrants' immediate permission, by making an *offer to pay* for these data references. Using this method opens an applicant to potential arbitration procedures to determine the actual amount to be paid to the data holders.

Most manufacturers of active ingredients have developed numerous end use formulae for their manufacturing use products. The manufacturer will usually supply a formulator with the formulae and permission to cite the supporting data already on file with the EPA, as long as the ingredient is purchased from them. In fact, the majority of all data are collected and submitted by pesticide manufacturers in support of their widely used active ingredients. Supporting data are where the bulk of the pesticide industry's financial investment is found. The cost and time involved for a chemical manufacturer to discover, develop, and test a new active ingredient can be staggering. Consequently, additional protections have been built into FIFRA to protect a manufacturer's investment. There is a 15-year window where the first submitter of supporting data for an active ingredient is given *exclusive use* protection and other applicants cannot cite their data without specific permission.

As with the unending label revisions, all registrants can look forward to eventually being tasked with a Data Call-In (DCI) notice. The EPA has the authority to require registrants to submit additional supporting data for their products as new issues arise. When this happens, registrants are faced with deciding to generate the new data (either alone or with other registrants), or canceling their product registration.

The Food Quality Protection Act

As everyone should recognize, a pesticide is a poison. In fact, early legislation actually referred to them as *economic poisons*. When supporting data first started being submitted for review, analysis was done on a cost-benefit basis. Applying a poison to our environment has a certain cost to the environment. The benefit of applying this poison might be increased crop yields, or decreased health hazards. Up until 1996, all pesticides were subjected to this cost-benefit analysis when they were registered.

However, when Congress passed the FQPA in 1996, analysis shifted from a cost-benefit to a health-based analysis. This changed the *tone* of the law, even though the EPA was already using many of the procedures mandated by the new law. It also fixed many discrepancies in the law, which could have eventually caused some major problems for the agriculture industry.

Under FQPA, the benefits of a pesticide could only be used to sway a registration decision in specific situations or when certain conditions were met. Specifically, if the health effects of using a pesticide are less than the health effects of not using it, or if not using it would jeopardize the domestic food supply, then a pesticide's benefits can be used to influence registration decisions. An example of this could be if using a certain pesticide was the only presently available way to combat a certain pest on a certain crop (*e.g.*, blueberries), then the environmental cost of using that pesticide would be weighed against the possibility of having no blueberries.

One of the major *discrepancies* corrected by FQPA was with the **Delaney clause**, which was part of the FFDCA. This clause was added to the law in 1958 and basically stated that carcinogens could not be purposely added to food. Prior to passage of the FQPA, there was a division of responsibilities for setting acceptable pesticide tolerances in food. FIFRA handled tolerances on raw agricultural commodities and the FFDCA handled tolerances on processed foods. Delaney prohibited the setting of tolerances for carcinogens on processed foods, even though a tolerance may already have existed for the raw commodity from which the processed food was derived. This system functioned appropriately at first, but as time went on and scientific detection methods improved and more toxicological data was collected, certain beneficial pesticides were in danger of being restricted in their use because of their low-level carcinogenic effects. In order to handle this situation, over the years an *interpretation* of the Delaney clause was developed that allowed for tolerances on processed foods for carcinogens when they posed negligible risk. However, in the early 1990s environmental groups were successful in demanding strict enforcement of Delaney's *zero* carcinogen level, by suing the EPA in Federal court. Congress then passed the FQPA (after several years of debate), which created a food tolerance procedure more compatible with the advances of science and the realities of agriculture.

Adverse Effects Reporting

One of the reporting requirements of FIFRA is in the area of unreasonable adverse effects, found under Section 6(a)(2) of the law (40 CFR 159). This requirement is very similar to the reporting required under the Toxic Substances Control Act (TSCA), Section 8(c). Registrants must report to the EPA when they have knowledge of unreasonable adverse effects to humans or the environment. There are specific reporting procedures and deadlines, depending on the actual adverse effect. Failure to follow this reporting requirement could result in very costly enforcement actions.

Reregistration

FIFRA went through a major revision in 1988, which started a process called **reregistration.** The EPA systematically categorized and grouped all active ingredients in the formulae of all registered pesticides. Active ingredients that had been initially registered before 1984 are put through a process of complete reevaluation. This evaluation covers labels, formulae, and especially supporting data. A determination is made whether or not an active ingredient should be eligible for re-registration, and what additional supporting data, if any, is needed to complete the cost-benefit picture. The Reregistration Eligibility Document (RED) for that active ingredient is published quantifying the results of the evaluation. DCIs are issued for any missing or inadequate supporting data. If registrants do not meet the requirements of the RED within the set time periods, then that product's registration is canceled.

One of the outgrowths of this program has been the registration maintenance fee. Each registrant must pay an annual fee to the EPA based on the number of registrations they have. This money is used to fund the reregistration program. The EPA's original goal was to complete the reregistration process for all active ingredients by 1997, at which time their authority to collect the maintenance fee would expire. However, the reregistration program is still in progress, and the authority to collect fees has been extended (as part of the FQPA).

The FQPA also requires the EPA to evaluate the pesticide tolerances for *all* pesticides using the new health-based risk analysis. All active ingredients that had already gone through the reregistration process will have to have their tolerances reevaluated using the new risk analysis scheme. There is also a requirement under FQPA to periodically review all registrations on a 15-year cycle, to ensure that accurate and complete data are maintained.

The reregistration program has been a massive task for the EPA, but once completed, all active ingredients will be essentially *up-to-date* in terms of labels, formulae, and supporting data.

Applicators and Worker Protection

A *restricted use pesticide* is a pesticide whose toxicity (typically, acute toxicity) is high enough that the EPA restricts its use. Applicators of restricted use pesticides must be certified. This certification process is handled at the State level, usually by the State departments of agriculture. Applicators are tested and certified in the techniques and hazards of pesticide application. Applicators fall into two categories: commercial and private. Commercial applicators are those which apply pesticides for a living (*e.g.*, exterminators). Private applicators apply pesticides only to their own property (*e.g.*, farmers). Most states have specialized certification tests for different applications, such as structural pest control, pools, turfgrass, *etc.*

The EPA has also developed the Worker Protection Standard (WPS) for pesticide application. While the EPA does not usually regulate occupational hazards, they took the lead when it came to occupational exposures to pesticides in the agriculture industry. This standard requires employers in agriculture (farms), silvaculture (forests), and horticulture (nurseries and greenhouses) that use pesticides to follow some basic procedures in hazard communication, first aid and emergency medical treatment, and personal protective equipment. It also requires employers to restrict workers from entering treatment areas during a pesticide's Restricted Entry Interval (REI).

The States

Each state has its own pesticide registration requirements. While this usually only means submitting copies of the EPA approvals, labels, formula, *etc.*, along with a fee, some states, particularly California, have a registration process that nearly duplicates the EPA's. In fact California's supporting data requirements are usually more involved and more difficult to maneuver than EPA's. It is standard to require product performance data with all pesticide registration applications in California (something the EPA requires only for certain types of products). Some states may not accept certain label claims, or may prohibit the sale of a product's active ingredient. This can be very important to know before marketing a product nationally.

Also regulated at the State level are products that often are associated with pesticides: fertilizers, soil amendments, seeds, *etc.* The labeling and formula requirements for these types of products are just as stringent as those for pesticides. Some states may even require supporting data for some label claims made on these products.

Finally, there are the additional costs for registering products with the states. While the annual registration maintenance fees required by the EPA under the reregistration program are very high (easily in the five-digit range for even a small formulator), paying a fee to each state (remember, there are 50) for each product can easily add up to a significant amount of money.

Conclusion

Pesticides are a reality of our modern world. There probably isn't a household in the United States that doesn't have at least one pesticide in the garage, bathroom, or under the kitchen sink. Pesticides are used for sanitation and vector control, as well as insect and weed control in agricultural and residential locales. We can debate the value, the costs, and the benefits of different pesticide uses. Deciding what is safe, acceptable, or beneficial will continue to be debated in the scientific, regulatory and political arenas. FIFRA provides an adjustable process for evaluating and controlling these useful but potentially hazardous chemicals, while still protecting our environment.

Internet Resources

<http://ww.epa.gov/pesticides> (Environmental Protection Agency Office of Pesticide Programs Homepage; links to forms and other sites)

<http://www.epa.gov/opprd001/forms/> (Environmental Protection Agency, list of and links to pesticide registration forms)

Keith Trombley is an Industrial Hygienist on staff with the University of Michigan in Ann Arbor. He works in the University's Department of Occupational Safety and Environmental Health (OSEH). He is primarily responsible for handling the safety and health concerns of the University's Plant Operations Department, which includes approximately 1200 employees in the construction trades, grounds keeping, custodial services, utilities, and transportation. Mr. Trombley has been working in the hazardous materials management, occupational safety, and environmental health fields for over 12 years, in consulting, manufacturing, and the service industries.

CHAPTER 24

Incident
Response

Marilyn L. Hau, MS, RN-C, COHN-S,
EMT-P, OHST, ASP
Daryl W. Dierwechter, REA, CHMM

Hazardous materials incidents are among the most dangerous that responders may face. The combination of chemical and physical threats compounded by the uncontrolled and often unknown nature of emergency events mandate that special precautions be taken to ensure the safety of response personnel.

Types of Emergencies

Minor emergencies (minor event alert)–handled on a regular, day-to-day basis by routine procedure, such as an employee with a laceration requiring first aid.

Limited emergencies (standby notification alert)–require a limited staff for the Command Post, which may be a temporary setup in an office or near the scene, such as a break in a water pipe in an office area requiring immediate maintenance attention. No public danger.

Potential disaster (response alert)–is one step beyond a limited emergency. Under these conditions, additional staff are required at the Command Post, and the Incident Command System in its more formal structure is implemented. A hazardous materials release being handled by

First Responder Operations–trained employees would be a possible example. Potential public impact.

Full emergency (major event alert)–necessitates full mobilization of facility and community resources, such as in a major plant explosion. Impacts the public.

Stages of Response

There are 5 basic stages of response to an emergency or disaster:

1) Recognition
2) Notification/warning
3) Immediate employee safety
4) Community/public safety
5) Property protection
6) Environmental protection

The length of each stage depends upon the emergency situation. For example, the notification/warning stage for a hurricane may be several hours, whereas the notification stage for an explosion may be minutes or only seconds. Warning categories are Alert, Minor, Major, and Catastrophic. Each stage depends on effort in earlier stages. The extent to which evacuation ensures immediate employee safety greatly influences later tasks in providing environmental protection. Property protection deals with property at the emergency scene as well as protecting property on which the event may impinge. Environmental protection involves reduction and elimination of emergency incidents affecting air, waterways and ground water, soil and wildlife. *The priority in emergency response is: Persons, then Property, then Environment.* Although environmental damage is usually considered the lowest exposure priority, it is still a very important concern. If possible, all released products should be contained and held until their impact on the environment can be determined.

Response Decision-Making

A strategic decision that must be made in hazardous substance emergencies is whether to respond in a defensive or an offensive mode. Offensive mode response requires special chemical and Self-Contained Breathing Apparatus (SCBA) personal protective equipment (PPE) with personnel who have been trained to this level of response. Defensive mode is no intentional contact with the product, wearing routine PPE only.

Tactical decisions that must be made as soon as possible during the emergency are:

• Isolation of the site
• Rescue of people inside the isolation area
• Protection of exposures (people, property, and environment)
• Fire extinguishment
• Confinement of the substance
• Recovery

Not all of these goals may be accomplished. Be sure you know your capabilities and do not exceed them. For example, rescue may not be possible due to risk to the rescue personnel. A fire may be beyond your level of equipment and training and should be left in the hands of the community fire department.

Risk analysis is an important tool to use in emergency response decision-making. What is the hazard? How vulnerable are the exposures? What is the risk in the proposed action or inaction? This must be a continuous decision-making process because emergency incidents seldom are stable. A hazardous materials release may escalate. An incipient fire may progress to a higher level. A medical emergency may develop at any time.

A decision-making process used to train Hazardous Materials Technicians under 29 CFR 1910.120(q) for hazardous materials incidents is the *DECIDE Process*.

Personal Protective Equipment

The need for proper protective clothing and equipment during a hazardous material incident response is obvious. Unfortunately, no single type of PPE offers total chemical protection under all conditions. Before a technician responds to a hazardous material incident, he/she must be familiar with the types and levels of protective clothing available. The Occupational Safety and Health Administration has developed a classification scheme for the various levels of chemical protective clothing (29 CFR 1910.120, Appendix B). The levels are defined as follows.

Level A–protection that should be worn when the highest level of respiratory, skin, eye, and mucous membrane protection is needed (for example—SCBA and gas-tight, totally encapsulating chemical suit).

Level B–protection that should be selected when the highest level of respiratory protection is needed, but a lesser level of skin and eye protection (for example—SCBA plus nongas-tight encapsulated or hooded suit, gloves, and boots). Level B protection is the minimum level recommended on initial site entries until the hazards have been further identified and defined by monitoring, sampling, and other reliable methods of analysis and personal protective equipment corresponding with those findings utilized.

Level C–protection that should be selected when the type of airborne substance is known, concentration is measured, criteria for using air-purifying respirators are met, and skin and eye exposure is unlikely. Periodic monitoring of the air must be performed.

Level D–primarily, a work uniform. It should not be worn on any site where respiratory or skin hazards exist.

The DECIDE Process

Detect hazardous materials presence (**D**etermine the nature and extent of the emergency.)

Estimate likely harm without intervention.

Choose response objective.

Identify action options.

Do best option.

Evaluate progress.

Incident Management

Management of an emergency response is handled through the *incident command system (ICS)*. The ICS is based upon basic business management practices. In a business, leaders perform the basic tasks of planning, directing, organizing, co-ordinating, communicating, delegating, and evaluating. The same is true in incident command. These functional areas are under the overall direction of the Incident Commander. The system can be implemented on a company, community, state and/or national basis, and under a variety of conditions. It does not require the disruption of existing reporting channels. The Superfund Amendments and Reauthorization Act requires that organizations which deal with hazardous material incidents respond under an ICS. Most fire and Emergency Medical Services (EMS) departments implement the ICS at fire scenes and in Mass Casualty Incidents. Certain insurance companies and local regulations require implementation of an ICS.

External Resources

An important resource in a hazardous materials emergency is the Chemical Transportation Emergency Center (CHEMTREC)—access via telephone, (800) 424–9300 or (202) 483–7616 in Washington, DC. CHEMTREC is a clearinghouse that provides a 24-hour telephone number for chemical transportation emergencies. It covers more than 3,600 chemicals that have been submitted by manufacturers as the primary materials they ship. CHEMTREC is sponsored by the Chemical Manufacturers Association, although nonmembers also are served. The emergency telephone number is widely distributed to emergency service personnel, carriers, and the chemical industry. The number usually appears on the bill of lading. When an emergency call is received by CHEMTREC, the person on duty writes down the essential information. As much information as possible is obtained by phone. The person on duty gives the caller information on hazards of spills, fire, or exposure that the manufacturers of the chemicals involved had furnished. The person on duty then notifies the shipper of the chemical by phone. At this point, responsibility for further guidance passes to the shipper. CHEMTREC's function is basically to serve as the liaison between the person with the problem and the chemical shipper and/or the manufacturer, the people who know the most about the product and its properties.

Another important resource is the local poison control center. Many chemical manufacturers in the United States will list the phone number of a poison control center on their Material Safety Data Sheets (MSDSs).

Recordkeeping

Detailed recordkeeping is an integral part of any incident response. Complete and accurate incident information allows you to evaluate/critique the response, revise response plans based upon lessons learned, provide data for governmental compliance/reporting needs, provide data for management/corporate needs, help facilitate future financial planning and allocations, and be of possible beneficial use in case of litigation. The following can be used as a guide for the types of information you may want to collect during an incident.

Incident Phase.

1) Document all action—telephone/radio traffic included.

2) Take photographs/videotapes.

3) Retain all documents, *i.e.*, shipping papers, *etc.*

4) Work within defined responsibilities and levels of authority.

5) Document reasoned actions, decisions, and responses.

Post-Incident Phase.

1) File records and reports promptly.

2) Consult with legal counsel.

3) Keep any notes you may have made.

4) Correct noted response deficiencies.

5) Mitigate identified hazards.

Media Relations

The media have an important role to play in our society. However, emergency responders may forget the importance of the media's role in covering an emergency when they are forced to deal with what they view as hordes of reporters at the site of a crisis. On the other hand the media also may view emergency responders as difficult to deal with, because of the lack information that is given to the press. The media represents the public's First Amendment right-to-know at a news incident. Media representatives feel they have an ethical responsibility to report news to the public, using the means provided by their specific media branch. As such, it is the responsibility of the press to find out what has taken place and to provide objective and responsible reporting to the public based on the best available information that can be obtained. *Most reporters take their roles seriously and make sincere efforts to responsibly communicate news events.*

Different media outlets have different needs. A single statement issued by a public information officer may not meet the needs of everyone. The television reporter is looking for a good shot. Radio reporters want good sound bites. Newspaper reporters want details and background information. And specialty services may want to cover "the story behind the story."

The incident site is the first place the press will go. It provides the best source of information for the first-in journalist. The second location where the press will show in force are at command-and-control locations. Command Posts, Emergency Operations Centers (EOCs), police and fire stations, and the offices of public officials are just a few of the locations where reporters may appear. Your system must be prepared to monitor and, if necessary, counter misinformation. A good defense during an initial encounter with the media is to have a preselected site for media operations and interviews and to distribute a packaged press kit. This kit, which can be prepared with generic information months in advance, may buy you time to gain control of the situation and what is taking place. Sample information contained in package press kits can include

- Telephone numbers for press lines at the emergency operations center

- Background information on emergency service units
- Background information on emergency response teams
- Glossary
- Diagrams of specialized equipment
- Training photographs
- Explanation of procedures; *e.g.*, use of the Chlorine B Kit
- Safety information
- Interview procedures and policies
- Information on past incidents/disasters

Encounters with the press should be planned. Designate persons with media training to provide interface with them. Instruct all others to refer the media to this persons or persons. They should avoid saying "No Comment!," as this only fuels media curiosity and doubt.

Post-Incident Follow-up

1) Ensure safety of responders and community.
2) Prepare an accurate record of the incident.
3) Revise response plans from lessons learned.
4) Share experiences with other response organizations.
5) Organize data for governmental compliance needs.
6) Organize data for management/corporate needs.
7) Establish contact persons for further activity needs.
8) Stimulate financial planning and allocations.
9) Prevent long-term emotional aftereffects.

Inaccurate information after the incident can result in responder illness; improper cleanup; unsafe disposal; inaccurate public, management, and news media perceptions; and failure to obtain maximum benefit from the incident experience. A Comprehensive Event Analysis should be conducted, with a reconstruction of the incident and the response activities to create a clear picture of everything that occurred. This is completely independent of any investigations for probable cause for administrative, civil, or criminal proceedings or litigation. A completed report is disseminated to management, department heads, training personnel, and team leaders/coordinators responsible for development of standard operating procedures.

Critical Incident Stress Debriefing

Emotional support and psychological relief should be considered through critical incident stress debriefing (CISD). Intervention by a CISD team can help prevent post traumatic stress disorder for which emergency responders may be at risk. Most public safety service response agencies, such as fire departments, have CISD teams available for rapid activation. These can be a resource to you and your industrial responders. CISD is conducted in one session. It should be done within 24 to 48 hours and should be considered mandatory for all responders, to avoid the *macho* declination of participation that later results in emotional disability.

Bibliography

Carlson, Gene P. *Hazardous Materials for First Responders*. Stillwater, OK: Fire Protection Publications, 1988.

Hau, Marilyn L. *While Help Is On The Way*. Chicago, IL: Health Products Marketing, 1994.

National Fire Academy. *The Incident Command System*. Emmitsburg, MD: National Emergency Training Center, 1989.

Marilyn L. Hau *has a graduate degree in Occupational Safety and Health, is a certified community health registered nurse, a certified occupational health nurse specialist, and a certified firefighter and paramedic. Ms. Hau holds safety certifications as an OSHT and ASP from the Board of Certified Safety Professionals and the American Industrial Hygiene Association. Ms. Hau also is trained in Incident Command by the National Fire Academy and has 31 years experience in health, safety, and emergency response in communities and industry. She is a HAZWOPER Technician Instructor and On-Site Incident Command Instructor. In addition, she is a frequent speaker on hazardous materials and emergency response, including several annual sessions of the National Safety Congress and American Occupational Heath conference. Ms Hau is listed in the 1988* National Distinguished Service in Nursing Registry *and the 1993-4,* Who's Who in American Nursing. *She is employed as Corporate Manager, Health and Safety, at McWhorter Technologies, Inc. Her publications related to HAZWOPER and emergency response include:* While Help Is on the Way, *"Emergency Care in Acute Chemical Exposures* (AAOHN Journal*), "Emergency Action Plans: Is Yours Just an Illusion?"* (Safety and Health), *and a chapter on recognizing and preventing occupational injury and illness in* Core Curriculum for Occupational Health Nurses.

Daryl W. Dierwechter *is Vice President, Environment, Health and Safety for McWhorter Technologies, Inc., and has over 20 years of experience in the environmental health and safety field. He has been a CHMM since 1986 and is a Fellow of the Institute of Hazardous Materials Management. Mr. Dierwechter currently is on the Board of Directors for ACHMM.*

CHAPTER 25

Natural Disasters and Induced Hazardous Materials Releases

Brett A. Burdick, PG, CHMM

Introduction

During the weekend of October 15, 1994, more than 20 inches of rain fell on Southeast Texas. The San Jacinto River in Houston, which normally flows at a depth of two and one-half feet, crested to a record flood level of 23.5 feet on October 21st and forced the evacuation of more than 13,000 residents (Fedarko, 1994; National Response Team, 1997). Some rivers in Southeastern Texas spread nine miles outside of their levees, impacting homes where flooding had never before been seen. At least 6,000 residences were affected and some 18 persons died as a result of the flooding, prompting the Governor to declare Southeast Texas a disaster area and to seek a Federal Disaster Declaration. The President declared 26 counties in Southeast Texas and the City of Houston a Disaster Area on October 18th (National Response Team, 1997). Eventually, 36 counties were included in the Federal Disaster Declaration.

As the storms eased and the San Jacinto River began to subside, two petroleum pipelines, originally buried to a depth of three feet beneath the riverbed, lay exposed. The pipelines may have been damaged by debris or may have simply collapsed, but around breakfast time that October morning they ruptured, releasing about 200,000 barrels of gasoline and diesel fuel into the still swollen stream. Seconds later the fuels ignited, pushing flames more than 100 feet into the air and sending a burning torrent flowing downstream at nearly 80 miles per hour. The inferno engulfed trees, vessels, and more than one house as it sped onward.

The two pipelines carried a huge quantity of petroleum—about one-sixth of the nation's daily fuel consumption—from Texas to New Jersey. The spill occurred within the refining and petrochemical district in Houston, severely impacting fuel availability and fuel prices throughout the east. The Houston Ship Channel was closed for several days. Several other petroleum pipelines were also leaking and discharging oil into Galveston Bay (Fedarko, 1994). The fuel spill, in addition to the fire and environmental consequences, resulted in severe financial disruption beyond the devastation already caused by the flooding.

Induced Hazardous Materials Releases

Any natural or man-caused disaster may be the trigger of an induced hazardous materials release (IHR). The IHR may be extremely localized—such as the spillage of household cleaners from a shelf at a local grocery store during an earthquake—or, as illustrated above, a relatively widespread problem that spreads destruction throughout a community and severely impacts communities and economies well outside of the disaster area. It is possible that the hazardous materials component of the incident may become the major factor in terms of actual harm resulting from the disaster. In some emergency situations, as seen in the 1984 release of methyl isocyanate in Bhopal, India, and the 1986 nuclear reactor accident in Chernobyl, the hazardous materials incident is itself the disaster. In other cases, the harm done by the release of hazardous materials is minimal except to those who happen to be in the immediate vicinity of the release. The full spectrum of hazardous materials incidents has been ex-perienced as IHRs.

As a component of the larger emergency situation, the IHR must be managed to a successful and desirable outcome as part of the overall response and recovery operations. The goal of these operations is to alleviate suffering and to protect human health and the environment. These activities may need to be performed under extremely difficult circumstances, such as when electrical power is lacking due to a damaged supply grid, with ongoing adverse weather situations, with the occurrence of additional aftereffects as the disaster continues, and with the potential of public panic and hysteria. There are also the challenges of competing priorities, resources, and personnel availability in the wake of a major disaster response operation. Any IHR associated with a community's response to a disaster situation is, to one degree or another, a critical incident. While the largest IHRs will undoubtedly be a focus of government response activities, a given IHR at a given facility may not rank high enough in priority to warrant government response in comparison to all of the other issues at hand. Yet, the protection of human health and the environment demands that the release be abated.

The response to hazardous materials incidents must be a labor-intensive exercise if it is to be performed safely and effectively. Hazardous materials response is different from other emergency management functions in that it is governed by Federal law and regulation, most notably under the Occupational Safety and Health Act and its regulations at 29 CFR 1910.120. Even the exigencies of a disaster situation do not preclude the need to operate in a safe manner compliant with the governing regulations. These two factors—being labor-intensive and requiring compliance with Federal rules—induce a third factor into the equation of hazardous materials incident response—that of time. The issue of the time required to abate any hazardous materials incident makes dealing with IHRs especially challenging. Just when the community is involved in a full-scale response to the disaster, the response to IHRs requires the community to devote a significant amount of the available resources for a considerable length of time to doing hazardous materials (HAZMAT) management. The need to address IHRs under disaster conditions and to have the mechanism in place to do this should be understood by all Certified Hazardous Materials Managers (CHMMs).

There is a linkage and a synergy between disasters and hazardous materials incidents. This linkage has only recently begun to be studied, to define the degree of association and the true magnitude of the problem. Recent investigations, however, have shown that almost all natural and man-caused disasters involve some hazardous materials component. Understanding this linkage is vital to the ability of private and public safety officials to respond to and recover from natural and man-caused disasters.

The purpose of this chapter is twofold—to introduce CHMMs to the field of Emergency Management as it is practiced in the United States and to examine the IHR problem. Emergency Management has changed significantly over the past few years and is now recognized as a separate and distinct profession that unites and coordinates the traditional public safety areas of firefighting and hazardous materials response, law enforcement, emergency medical services, and mass care. Specifically, we will investigate how hazardous materials management fits into the larger emergency management picture and how hazardous materials incidents might be addressed within this larger context.

The rest of this chapter will introduce the general concepts of managing emergency events through the concept of the emergency management cycle. It will then review the major types of IHRs as the kinds of hazardous materials response challenges IHRs pose. Finally, the Chapter will review the Federal response structure and the roles of State and local government and of industry in dealing with IHRs.

The Emergency Management Cycle

The Federal Emergency Management Agency (FEMA) was established as a Federal agency in 1979 and charged with coordinating the national efforts in emergency management. FEMA defines the term **emergency** as *any event which threatens to, or actually does, inflict damage to property or people*, and the term **management** as *the coordination of an organized effort to attain specific goals or objectives*. Within this conceptual framework, it is clear that hazardous materials

incidents fall under the umbrella of emergency management (FEMA, 1993).

Implicit within this concept is the assumption that the natural or man-caused disaster interacts with people or their property to the detriment of people. This important concept is sometimes forgotten. A strong coastal storm or hurricane near a large city can be a disaster. Along an open coastline in an unpopulated area it is just another natural event. It is the interaction of the event with humans and their works that makes it a disaster.

The field of Emergency Management addresses the preparation for, response to, and recovery from any natural or man-caused disaster. This *All Hazards* approach has allowed Emergency Management to evolve from an exercise of "throwing assets at problems as they arise" to a legitimate field of study and endeavor where scientific and managerial principles are applied to the four phases of Emergency Management. These four phases are conceptually laid out as a cycle following an event and are, in order of occurrence, Response, Recovery, Mitigation, and Preparedness. Each of these four phases is briefly discussed below (FEMA, 1993).

The Event

Within the context of All Hazards Emergency Management, an event can be anything that precipitates a community-wide response. It could be a natural disaster (flood, hurricane, tornado, or earthquake), a man-caused disaster (hazardous materials release, nuclear power plant accident, act of terrorism), or even a planned event (political rally, concert, State fair, sporting event). For our purposes, the event is viewed as the beginning of the active phase of Emergency Management where significant community resources are or will be committed to intervene in the natural course of events resulting in successful and more desirable outcomes than would otherwise be expected.

It is usually assumed that the event is of such importance that both governmental and private resources must be involved in the resolution of the event to avoid undesirable outcomes. This may be a result of the scale of the event being outside of those normally encountered or a result of some particular complexity that requires unusual resources to be involved. The event is not defined by normal, day-to-day activities within the routine capabilities of public safety agencies and the private sector. The scale of the event, therefore, is such that its resolution may involve the coordination of public and private assets and capabilities, multiple legal authorities, and possibly multiple levels of government. The coordinated response to and recovery from this type of event is the mainstay of Emergency Management.

Response

Response activities occur during and immediately after the event and are designed to provide emergency assistance to victims of the event and to reduce the likelihood of secondary damage. In the case of response to a disaster, the traditional life-saving activities, law enforcement actions, and environmental protection measures come into play during this phase.

Response involves the effective and efficient application of assets and activities to resolve the immediate impacts of an event. In the case of a planned event, response activities include the application of sufficient resources to ensure that the event occurs without undue or unexpected undesirable outcomes.

Recovery

Recovery activities occur until all community systems return to normal or nearly normal conditions. This includes both short-term and long-term recovery actions. *Short-term recovery* is aimed at returning vital life-support systems to minimum operating standards, while *long-term recovery* continues until the entire disaster area is completely redeveloped, either as it was in the past or for entirely new purposes that are less disaster-prone. This may and frequently does take many years to accomplish. Recovery involves those assets and activities designed to return the community to normalcy following an event.

It is important to note that government recovery activities will not restore a community or individuals within that community to 100 percent of their preevent status. It is more involved with ensuring that a community and its citizens have the opportunity to recover through their own efforts. Recovery is a community undertaking and

may involve such diverse fields as housing, economic development, and community investment.

Mitigation

Mitigation refers to those actions and activities taken to reduce or eliminate the chance of occurrence or the effects of a disaster. These include restricting development in flood plains, requiring building codes that will increase the likelihood of surviving a given wind load, and providing secondary containment for the storage of oil and hazardous materials. FEMA (1993) professes that much can be done to either prevent a major emergency or disaster from ever happening or, at least, to reduce the impact of such an event.

Preparedness

Preparedness is planning how to respond in case an emergency or disaster occurs and working to increase the resources that are available to respond effectively. The concept of preparedness is intended to address the capabilities of a community to respond to the next disaster in a safer, more efficient, more effective manner. Preparedness activities are designed to help save lives and minimize damage by preparing people to respond appropriately when an emergency is imminent. This involves planning, exercising of the plans, identifying hazards and risks, and public education through outreach and awareness. Preparedness is the process of preparing for the next event, whether the event is planned or unexpected.

The four phases of the Emergency Management Cycle often overlap and co-occur. Mitigation and Preparedness, for instance, go hand in hand before the event occurs. The transition from Response to Recovery is seldom clean, and in fact this transition may occur at different times in different geographical areas impacted by the same event (such as when a flood event works its way downstream into new areas). The utility of the Emergency Management Cycle model, however, lies in its ability to help us conceptualize the major aspects of dealing with any emergency or disaster.

It is appropriate for Certified Hazardous Materials Managers to understand the Emergency Management System as it exists in the United States today and how that system is designed to respond to hazardous materials releases associated with disasters. Disasters of all kinds will continue to involve hazardous materials releases. CHMMs are uniquely positioned to provide the linkage between disaster activities and protection of the public from the release of hazardous materials.

Disasters and IHRs—Some Examples

The model of the Emergency Management Cycle can be applied with great success to hazardous materials incidents, and specifically to those incidents precipitated by natural or man-caused disasters. In this section we will examine how natural disasters can give rise, often unexpectedly and improbably, to hazardous materials incidents. Later, we will examine how the field of Emergency Management and the programs supporting the Emergency Management Cycle can assist in resolving these incidents as part of the emergency management system.

The Federal Emergency Management Agency (FEMA) has identified at least 15 kinds of natural and man-caused hazards that may impact local communities. These include severe thunderstorms, floods and flash floods, landslides and mudflows, tornadoes, hurricanes, winter storms, drought and extreme heat, wildfire, earthquake, tsunami, volcanic eruption, dam failure, structural fires and explosions, radiological accidents, and hazardous materials releases (FEMA, 1994). Many other hazards exist that are within the domain of emergency management. These include transportation incidents, mass casualty incidents, and public health incidents, for example.

Several of these 15 natural and man-caused hazards can and have resulted in a secondary release of hazardous materials. These secondary or induced hazardous materials releases (IHRs) compound the problems facing emergency managers and, in fact, sometimes become the most significant hazard encountered during the disaster.

One review of the occurrence of IHRs caused by natural disasters in the United States over the period of 1980 through 1989 was accomplished by surveying the states and asking them to respond to a questionnaire (Showalter and Myers, 1994). The survey revealed that of 311 incidents identified in 20 states, the vast majority of IHRs were the result of earthquakes (73 percent). This might be expected due to the large area of impact an earthquake can have and the fragility of hazardous materials storage and transportation systems when exposed to significant ground shaking and displacement. Earthquake-related IHRs were followed in descending order of frequency by hurricanes (8.3 percent) and floods (5.1percent). The remaining 14 percent of IHRs were attributed to lightning, winds, storms, dam breaks, fire, fog, ice storms, tornadoes, and landslides. It follows from this that any natural event could result in the release of hazardous materials, but that the most common causes of IHR are earthquakes, hurricanes, and floods.

The authors do not claim that this analysis is all-inclusive. It is evident that their results may be biased by over-reporting by states prone to earthquakes and by poor recordkeeping among the states in general. It is also difficult to understand what exactly the states meant, for instance, by a "flood-induced hazardous materials incident." If a flood resulted in the dislocation and subsequent recovery of 100 hazardous materials containers throughout the flood plain, is this one incident or 100? If an earthquake causes 100 separate IHRs in the impacted area, how are these tallied? (For example, the Loma Prieta Earthquake in 1989 resulted in 300 separate IHRs; see below.) Nevertheless, the Showalter and Myers (1994) study is instructive in that it is an initial quantification of the number and geographic extent of IHRs. At least 311 incidents in 20 states and 10 years indicate that the occurrence of IHRs is not a trivial problem.

Results from this study suggest a correlation between the number of IHRs experienced during an event and the magnitude of the causative event, although this general correlation must be tempered to consider the degree of industrialization in the affected area. In a large event in an urban setting, the cumulative problem of IHR would predictably be quite serious. The authors also conclude that when communities and commercial facilities are simultaneously impacted by a natural disaster, response agencies will first become involved with assisting the citizenry directly, at the expense of responding to damaged facilities. This means that hazardous materials facility response plans that rely on the full cooperation of public fire, police, and other emergency response departments are inherently unworkable during a natural disaster. The ability of any individual community to respond to a HAZMAT incident varies. CHMMs involved in IHR planning must consider the local capabilities and those of industry within the community when formulating realistic response plans.

The above discussion suggests that three phenomena—earthquakes, hurricanes, and floods—result in the majority of IHRs. Each of these phenomena results in a different pattern of IHR that should be understood by all emergency and hazardous materials managers. The rest of this section looks at each of these natural events and the kinds of IHR they may generate.

Earthquakes

The release of hazardous materials caused by earthquakes is the most studied of all IHR events. Many authors conclude that this event is the most likely kind of natural disaster to result in an IHR, due in part to the widespread impacts of ground shaking over a large geographical area and the proximity of some major metropolitan and industrialized areas to active faults (Los Angeles, San Francisco, Mexico City, and Tokyo, as examples). The occurrence of strong aftershocks tending to follow major earthquakes may also result in IHRs among already damaged structures or in otherwise undamaged communities miles away from the original epicenter and for months following the original earthquake (Hough and Jones, 1997).

The risk posed by earthquake-induced hazardous materials releases (EIHRs) can be explored from an historical standpoint (Lindell and Perry, 1996 and 1997). The EIHR experience in recent earthquakes in this country includes the following:

1) The 1971 San Fernando earthquake (magnitude 6.6) resulted in 100 fires requiring fire department response and 18 documented hazardous materials releases.

2) The 1983 Coalinga earthquake (magnitude 6.7) resulted in many natural gas leaks and oil pipeline ruptures and 9 documented hazardous materials releases. In addition, chlorine tanks at the local sewage treatment plant slid 10 inches and threatened to release.

3) The 1987 Whittier Narrows earthquake (magnitude 6.1) resulted in 1,411 natural gas line ruptures and 30 hazardous materials releases, one of which involved the release of chlorine (see below). Another event involved the toppling of a container containing metallic sodium under kerosene at a college science laboratory. A water leak from a safety shower reacted with the sodium to produce sufficient heat and hydrogen gas to cause a fire. The fire resulted in the vaporization of mercury elsewhere in the laboratory, and firefighting activities exposed asbestos incorporated into the building construction. The cleanup of this laboratory cost nearly one-quarter of a million dollars.

4) The 1989 Loma Prieta earthquake (magnitude 6.7) resulted in more than 300 natural gas leaks and over 300 hazardous materials releases. Fifty of these involved hazardous gases other than natural gas. One incident involved the rupture of a 2-inch anhydrous ammonia line at a food-processing plant, resulting in the release of between 5,000 and 20,000 pounds of ammonia.

5) The 1994 Northridge earthquake (magnitude 6.8) resulted in more than 15,000 natural gas leaks linked to 3 street fires, 51 structure fires, and 172 mobile home fires. The earthquake induced at least 130 other hazardous materials releases. One of these was an oil pipeline rupture which released 173,000 gallons of product and ultimately cost $30 million to remedy. Another was a railcar release of 2,000 gallons of sulfuric acid. The authors indicate that hazardous materials releases occurred at 10 percent of the industrial facilities and 5 percent of the commercial facilities in the zone of greatest impact.

What is most striking about this listing of EIHRs is the number of hazardous materials incidents that are reported to be associated with earthquakes. Addressing the 130 IHRs reported during the 1994 Northridge Earthquake could take weeks or months to accomplish. The effort required to resolve the 300 nearly simultaneous IHRs documented during the 1989 Loma Prieta Earthquake is mind-boggling.

In contrast to Showalter and Myers (1994), Lindell and Perry (1996) find little direct correlation between earthquake magnitude and the number of EIHRs, due to other effects including soil conditions, mitigation measures employed in the area, and the specific location and concentration of hazardous materials sources. They conclude, however, that a thorough postevent investigation will reveal the presence of EIHRs in almost all cases of damaging earthquakes. They also indicate that the actual numbers of EIHRs probably are significantly under-reported (Lindell and Perry, 1997) due to the general confusion that reigns following a significant earthquake, the independent actions of facility personnel who clean up releases without reporting them, and those, perhaps, ignored and not reported by the responsible parties.

Some of the sequences of events that can be documented associated with EIHRs are quite improbable (Lindell and Perry, 1996). One example comes from an unidentified locality during the 1964 Alaskan earthquake. The earthquake generated a tsunami that resulted in the rupture of some oil storage tanks and in a break in the hose connecting a tank vessel loading diesel fuel in the harbor. The spilled fuel ignited. A second tsunami drove the burning fuels inland, igniting the oil in several railroad tank cars. The tank cars exploded sequentially in a chain reaction which ultimately resulted in the ignition of a bulk fuel storage yard many blocks away. If anything can go wrong, it will.

The pattern of the impacts of EIHRs is unique and has relevance to the planning process. An earthquake may induce widespread and nearly simultaneous releases of hazardous materials. It is common for multiple hazardous materials incidents to require response at the same time. Most hazardous materials response planning assumes that hazardous materials events are nearly random and statistically independent events (Noll, et al., 1988). The capabilities of any community's hazardous materials response programs will be seriously taxed in the event of a widespread disaster of this type.

Beyond the pure numbers that indicate the magnitude of the EIHR problem, it is instructive to examine the specific hazardous materials issues that may be involved in the management of the response to and recovery from such a disaster. The two examples that follow demonstrate the types of demands that may be placed on emergency managers confronted with an EIHR event.

In 1987, the Whittier Narrows earthquake in southern California resulted in the rupture of a chlorine storage tank in the city of Santa Fe Springs (Seligson, et al., 1996). The ruptured tank released approximately 240 gallons of chlorine gas which drifted as a plume toward the city of Whittier prompting the evacuation of some areas. People living in southern California experience magnitude 5.9 earthquakes on a relatively frequent basis. Hough and Jones (1997) report that there have been 74 magnitude 5.5 and larger events in Southern California since 1932, so it is probable that a magnitude 5.9 occurs at least once every few years. It was the co-location of the earthquake epicenter near the chlorine storage facility in Santa Fe Springs, not just the magnitude of the earthquake, that resulted in this release.

Just how much of a risk does chlorine storage pose to residents in the Los Angeles area? One study (Seligson, et al., 1996) has modeled the potential impacts of a large earthquake and concluded that nearly 7,000 people could have been exposed to this release. While this earthquake would not be strong enough to cause major damage, it clearly could have a significant impact on the population through secondary release of hazardous materials. Thus, the hazardous materials issue becomes the dominant source of harm in this scenario.

The simulations analyzed by Seligson and coworkers also included a far worse case event, that of a magnitude 7.0 earthquake along the Newport-Inglewood Fault which cuts through the Los Angeles Basin. After identifying 22 sources of chlorine and ammonia in the area and assuming that the storage vessels were all compromised, the model suggests that as many as 133,000 people—or 2 percent of the population of Los Angeles County—would be exposed to harmful concentrations of these two chemicals released during the earthquake. This scenario would produce a public health emergency that probably could not be handled effectively by a medical infrastructure already damaged by an event of this magnitude. Protective actions involving 130,000 people would stretch the capabilities of any emergency services organization under more routine circumstances, let alone following a destructive earthquake.

A second example of the impacts of EIHR comes from the earthquake that impacted Japan in the vicinity of the City of Kobe in January, 1995. In addition to the hazardous materials releases caused directly by the Kobe earthquake, Japan's environmental agency reported that the collapsed buildings resulted in a significant increase in the concentration of asbestos fibers in the ambient air

(Hadfield, 1995). Japanese construction commonly utilized asbestos until 1965 for fireproofing and soundproofing, even in residential buildings. Many of the structures destroyed in the event had been constructed prior to 1965. For days following the earthquake it was reported that a "haze of asbestos dust hung over the city . . ." and that the areas where demolition crews were working had especially high levels of airborne asbestos (Hadfield, 1995).

The 1993 national average concentration for ambient asbestos samples in Japan was reported to be 0.43 fibers per liter of air. Measured concentrations in Kobe and a nearby town in February following the earthquake averaged nearly twice this (0.8 fibers per liter of air), and in the area where building demolition was ongoing the average level was 11.2 fibers per liter. This value exceeds the maximum concentration of 10 fibers per liter allowed under Japan's Air Pollution Control Law that governs asbestos removal actions (Hadfield, 1995). Buildings undergoing demolition in Japan are usually sprayed with water to avoid raising clouds of asbestos dust. Following the earthquake, however, water pipes and roof tanks were damaged to the point that water was not available during demolition.

In this example, the actions of cleanup crews involved in the recovery from the earthquake were coupled with the direct impacts of the disaster to compound an already bad situation. The risks to public health from the airborne asbestos could, theoretically, result in many more deaths than the 5,000 resulting directly from the earthquake.

It seems to be prudent to assume that any building involved in a major earthquake is likely to be a source of IHRs until proven otherwise. For the case of collapsed buildings, the probability that hazardous materials are involved is near certainty.

As can be seen, the IHR problems associated with earthquakes are impressive. Generally speaking, the most severe public safety impacts would be expected to come from releases of toxic or poisonous gases and long-term impacts from the exposure to asbestos. Other localized impacts may result from spills and releases of other hazardous materials classes, especially flammable liquids. Earthquakes result in the most widespread and immediate IHR impacts of the natural phenomena studied to date.

Hurricanes

The second most important sources of IHRs are hurricanes. Impacting wide areas of coastal communities where they strike, hurricanes result in wind damage, flooding, and the destruction of buildings that may release hazardous materials. Direct impacts could include the destruction of storage tanks, the flooding of warehouses and docks, and the destruction of storage areas containing hazardous materials. The overriding concern, however, is the accumulation, staging, and disposal of tremendous volumes of debris, some of it hazardous.

One example that illustrates some of the risks is from southern Florida in 1992. Hurricane Andrew has been described as one of the costliest natural disasters in United States' history. Several examples of localized hazardous materials impacts were identified among the devastation. As in Kobe, asbestos was again a prominent hazard.

The sheer volume of the hurricane-derived debris was staggering. In all of south Florida the hurricane resulted in 40 million cubic yards of debris. This is equivalent to one football field stacked *five miles* high (National Park Service, 1994)! On September 21, 1992, at the peak of the cleanup activity, the effort involved more than 4,000 personnel, 2,158 trucks, and 720 loaders. Contractors moved 290,000 cubic yards of debris on that day alone (Emergency Management Institute, undated).

Approximately 23 percent of the storm-generated debris were composed of building materials, including a significant amount of asbestos (National Park Service, 1994). Other hazardous materials were also released by the destruction of storage sheds and buildings. Household hazardous and toxic wastes (HHW) are a major concern in residential areas. HHW can include common household chemicals, propane tanks, oxygen bottles, batteries, and other chemicals such as pesticides and paints (Emergency Management Institute, undated). HHW and other hazardous wastes must be separated from the waste stream and staged for appropriate removal by trained contractors.

One significant obstacle to the cleanup following a hurricane is that there is often no way to tell where the hazardous materials are present until debris removal begins. Crews involved in debris clearance must be aware of the potential for IHRs and must be able to notify appropriate authorities if IHRs are encountered.

Most of the asbestos encountered by the National Park Service within the National Parks of south Florida resulted from roofing shingles and siding stripped from buildings by the hurricane's winds. Some damage was caused by the direct impact of storm surge and waves on asbestos-containing structures. Much of the asbestos was blown and washed into adjacent water bodies. The National Park Service addressed the asbestos problem by contracting for the removal and disposal of asbestos debris, as well as the other hazardous materials released during the storm (National Park Service, 1994).

The woody and other organic debris generated by hurricanes and staged during the cleanup will, in time, decompose and emit toxic or volatile vapors (Emergency Management Institute, undated). Debris-burning activities will also produce some adverse air quality impacts. All debris storage and disposal areas require active air monitoring to ensure that workers and the general public are not exposed to hazardous air pollutants.

A similar set of problems was encountered by the United States Army Corps of Engineers in the cleanup following Hurricane Marilyn in the Virgin Islands (Anonymous, 1995). The hurricane generated more than 300,000 cubic yards of debris in September, 1995 that required disposal. The landfills on the islands of St. Croix and St. Thomas were full, so the Corps opted to burn or bury the waste. Hazardous waste was reported to be barreled and transported to the United States mainland. The lack of a disposal facility on the islands resulted in the suggestion that the debris be moved to the mainland. The Corps estimated that the shipping would cost $31 million.

Catallo, Theberge, and Bender (1989) discuss the problems associated with waste disposal in coastal areas. They suggest that a direct result of human preferences for living in areas subject to coastal storms and hurricanes is the disposal of wastes, including hazardous wastes, in these same areas. They estimate that half of the nation's Superfund sites occur in or near estuarine, riparian, or coastal lowland settings. This fact, coupled with the demonstrable eustatic rise in sea level observed in long-term tidal records, suggests that future impacts of coastal storms and hurricanes on waste disposal areas could result in an increased likelihood of IHRs through time.

Dealing with the debris resulting from a major hurricane shows the interplay of hazardous materials management and solid waste disposal. While not a large proportion of the debris and waste disposal stream, the sheer magnitude of the volume of debris involved raises the hazardous and asbestos-laden waste to the level of more than merely a trivial problem.

Floods

Floods represent the most common natural hazard in the United States (FEMA, 1994). The power of flowing water in storm-swollen streams is immense. Structures containing hazardous materials, disposal sites, hazardous materials storage areas in the flood plains, and—as we saw in the example of the San Jacinto River—utility crossings of flooding streams are significant potential sources of secondary releases to the environment. Despite zoning and siting criteria, there are large quantities of hazardous materials in areas inundated by flood waters.

Denning (1992) has investigated the types of hazardous materials problems that may be associated with a flood. Based upon the experience of an unnamed southern state in the spring of 1991, Denning's instructive study classified the specific hazardous materials–related problems experienced during this flood as follows:

1) *The displacement of underground storage tanks and their contents by flood waters.* Some tanks floated out of the ground while others filled with water and discharged their inventories, becoming both an environmental problem and a fire hazard.

2) *The flooding of reserve pits, tanks, and wells within oil fields.* We could add the flooding and displacement of all pits and lagoons located in the flood plain whether they were oil field–related or not, as well as drum and container storage areas inundated by the flood waters.

3) *The interruption of ongoing hazardous materials site cleanup.* Flooding can redistribute contaminants and impact previously uncontaminated areas, and certainly can interfere with critical construction timetables.

4) *Wind-induced damages, such as at an airport where a hangar collapsed on aircraft and*

resulted in a release of aviation fuel. Aircraft (and vehicles) are a source of both toxic and flammable fluids.

5) *The inundation of landfills by the flooding.* Landfills can be damaged, submerged, or washed out by periodic flooding. Floods can remove daily cover from putrefying waste and, under some conditions, result in compromising the structural integrity of the landfill. The release of municipal waste to the environment can result in a widespread public health emergency. Wright, Inyang, and Myers (1993) discuss ways to reduce the risk of landfill releases due to floods and other natural hazards through the planning process.

6) *The inundation of sewage systems and wastewater treatment plants.* In addition to the interruption of sewage service, the plants themselves handle hazardous materials and can be a source of hazardous materials releases.

7) *The contamination of wells by flood waters.* This is an acute problem in rural areas, but also in urban areas where the municipal well fields are impacted. Floodwaters contaminate aquifers by directly entering the wellbore and recharging the ground water with bacteria and chemical-laden flood waters.

8) *The inundation of automobile junkyards.* Junkyards, even the best managed of them, are sources of petroleum, coolant, and debris. Those located in the flood plain are clear sources of IHR.

9) *The release of transformer fluids into flood waters.*

10) *Barge accidents on waterways releasing cargoes and fuels.* Flooded streams are always hazardous to navigate, and moored barges and vessels, exposed to great stress from the flow, may be cast adrift during the flood.

11) *Trucking accidents in bad weather and the release of diesel fuel and hazardous materials cargoes.*

12) *And, as a last example from this study, a fire and explosion at a chemical plant manufacturing anhydrous ammonia and other hazardous chemicals.* The explosion killed 8 and injured 100 employees, required the evacuation of 1,500 residents including the local hospital and its 20 patients. (It was noted by Denning that the chemical plant explosion was not caused by the floods; but its timing coincident with the flooding demonstrates the types of incidents that can happen when communities are already responding to emergency situations. Again, if anything can go wrong, it will. The explosion and chemical release would be difficult enough to deal with by the community. During the flooding, the additional requirements of responding to the explosion nearly stretched local responders ". . . to their limits" (Denning, 1992),

Denning's (1992) list is long. It points out several specific types of impacts any jurisdiction might expect in a disastrous flood, but even it is not all-inclusive. To it we could undoubtedly add many more potential hazardous materials impacts, including, perhaps, the flooding of facilities at nuclear weapons plants (Anonymous, 1996). All of the impacts listed above are secondary to the primary disaster of the flood. Local resources, already committed to life-saving missions along compromised transportation systems, using overloaded communications systems, and without electrical power, are hard pressed to deal effectively with public safety issues caused by hazardous materials releases. Most of the hazardous materials releases involved in this flood were small and easily dealt with during the recovery period (Denning, 1992), but there is no reason to believe that this must be the case.

Some of the released materials will, with luck, be diluted or dispersed by the flood waters, and in that way, public safety impacts are minimized. Alternatively, some flood processes may act to concentrate the occurrence of hazardous materials, such as the trapping of drums and containers on the upstream side of bridges and culverts. Some of these hazardous materials incidents will require significant resources to resolve. Secondary hazards caused by hazardous materials releases were real concerns to responders in this study.

Another example of flooding causing IHR incidents comes from the experience of Virginia following two flooding disasters in 1996 (Virginia Department of Emergency Services, 1996). In January, snow melt and a warm rain resulted in significant flooding, primarily to the west of the Blue Ridge in the Shenandoah Valley but also involving several communities on the eastern side of the Appalachians. That September, widespread flooding impacted the same areas as Tropical Storm Fran moved into Virginia after coming ashore as Hurricane Fran in North Carolina.

These flooding events resulted in a significant number of containers being swept downstream throughout much of the Commonwealth. The drainages involved included the Dan, James, Maury, Rapidan, Roanoke, and Shenandoah Rivers—approximately one-third of the land area of the state. The location, recovery, and identification of the contents of these containers involved a significant effort. The field activities following the January flooding had not been completed before Tropical Storm Fran hit, adding to the number of containers involved and requiring that all areas investigated during the January floods be surveyed again. In all, 2,838 containers of varying sizes ranging from gasoline cans to propane and petroleum storage tanks were recovered from the streams and the flood plains, characterized, and disposed of properly. The cost of the effort was approximately one quarter of a million dollars. Understandably, most of the containers were empty at the time of the flooding, but a significant number of them—perhaps 20 percent—contained hazardous materials or residues. Notwithstanding the actual number of containers that carried hazardous materials downstream, a significant effort was undertaken to find those that were actually a hazard. Public safety considerations demand that all of the containers be found and investigated.

These examples serve to demonstrate the magnitude of IHRs in flooding events and the difficulty of dealing with these hazardous materials incidents. The costs of these incidents fall primarily on government, as the releases are the result of an *Act of God* and exempted from much regulatory authority, including cost recovery.

The Federal Government Response Organization

There is a high probability that some IHRs will accompany any significant disaster. The systematic approach taken by the Federal government to respond to all disasters is detailed in the *Federal Response Plan* (FRP). The FRP, prepared by FEMA and adopted by 27 Federal departments and agencies, is designed to facilitate the delivery of all types of Federal response assistance to states to help them deal with the consequences of large-scale disasters. It should be remembered that the FRP details the support that the Federal government gives *to the states*. The states, in turn,

provide both their own support and the Federal support to the localities. The system is designed so that localities remain in charge of disaster response in their own communities and that State and Federal assistance is provided to augment local response capabilities as the conditions warrant. While support is given primarily to public sector organizations and agencies (fire, police, emergency medical services, *etc.*) the Federal-State-local partnership also assists private organizations and, indeed, integrates public support to the community with support from private organizations.

The FRP is organized around 12 Emergency Support Functions (ESFs). Each of these EFSs has a single Federal agency or department which leads the Federal response under that ESF. The ESFs and the lead Federal agencies are identified in Table 1. The Corps of Engineer's operations in the Virgin Islands following Hurricane Marilyn, for example, were the result of the Corps being the lead Federal agency for ESF #3–Public Works and Engineering, and responsible for debris clearance and disposal in this disaster. The FRP allows these Federal agencies and departments to plan their own response strategies, and these individual agency plans are incorporated into the FRP by reference. It is FEMA's responsibility to ensure that all of the plans mesh.

Recognition of the reality of hazardous materials disasters is underlined by the creation of ESF #10 –Hazardous Materials. While developing the plan, FEMA understood that there is a need to prepare for the direct response to man-caused disasters (such as the Bhopal and the Chernobyl Incidents) as well as a secondary hazardous materials response in any disaster situation. The Lead Federal Agency under ESF #10 is the Environmental Protection Agency (EPA).

The EPA's concept of operations under ESF #10 is described in the FRP. The purpose of ESF #10 is to provide Federal support to State and local governments in response to an actual or potential discharge of hazardous materials following a catastrophic earthquake or other catastrophic disaster. As such it is intended primarily to address IHRs. EPA has provided for a coordinated response by placing the response mechanisms of the National Contingency Plan (NCP) within the FRP coordination structure. This means that the response to all oil and hazardous substance releases associated with a natural or man-caused

Table 1. Emergency Support Functions and the Lead Federal Agencies under the Federal Response Plan

Emergency Support Function	Lead Federal Agency
1. Transportation	Department of Transportation
2. Communications	National Communications System
3. Public Works and Engineering	U.S. Army Corps of Engineers
4. Firefighting	U.S. Forest Service
5. Information and Planning	Federal Emergency Management Agency
6. Mass Care	American Red Cross
7. Resource Support	General Services Administration
8. Health and Medical Services	U.S. Public Health Service
9. Urban Search and Rescue	Department of Defense
10. Hazardous Materials	Environmental Protection Agency
11. Food	Department of Agriculture
12. Energy	Department of Energy

disaster will be in accordance with the NCP. EPA, the United States Coast Guard, Department of Energy, or Department of Defense On-Scene Coordinators (OSCs) will be assigned to each release or to the multiple releases, and the OSC will be responsible for implementing the appropriate remedy.

The FRP provides that the EPA Co-Chair of the Regional Response Team (RRT) will be assigned to the FEMA Disaster Field Office (DFO) working directly under the FEMA Federal Coordinating Officer (FCO) to coordinate all activities of the OSCs. All DFO personnel working in support of ESF #10 are, to the extent possible, members of the RRT. Among the planning assumptions under ESF #10 are that emergency exemptions for the disposal of contaminated material will be required. At the national level, the support structure is implemented by the EPA Director of the Chemical Emergency Preparedness and Prevention Office (CEPPO) in the Office of Solid Waste and Emergency Response (OSWER).

Presence of ESF #10 in the DFO allows incoming information regarding hazardous materials releases to be acted upon in a coordinated fashion. Response priorities are established at the regional level with input from EPA Headquarters and State, and local representatives.

When EPA becomes involved in a disaster response, it has the authority to involve other Federal agencies in support of ESF #10. For example, EPA can enlist the assistance of the Occupational Safety and Health Administration (OSHA) to ensure that site-specific health and safety plans are reviewed in a timely manner, and the Department of Agriculture to assist in the development of natural resources damage assessments. Disposal of wastes and discharge of treated wastewater can go forward under existing EPA authority without the need for obtaining permits, thus streamlining the response effort.

Much of this coordination will be transparent to the local government and to private industry response organizations. What is significant is that a process for coordinating the response to hazardous materials releases exists within the overall disaster response structure. The upshot of this is that allocation of resources and priority of response assets will take place in a coordinated fashion at the DFO.

Federal and State Support to Local Government

As we saw with the San Jacinto River incident, there is a formal process by which the governors of the states can request assistance from the Federal government to respond to and recover from major disasters. The flooding in Southeast Texas was of sufficient magnitude that the Governor declared a State of Emergency and requested a Federal Disaster Declaration. In declaring a State of Emergency, the Governor is invoking a range of state laws that allow for assistance to be given local governments (among other provisions) in their response to a disaster. The range and scope of these laws vary from state to state; however, they generally allow for augmentation of local capabilities using State assets such as the National Guard. They also, in general, provide for an alternative funding mechanism and procurement system so that assistance can be provided in a timely manner.

State law also may provide for the ability of local government to declare a State of Emergency. In Virginia, for example, a local declaration allows the local Director of Emergency Services to expedite contracts and to control the sale and distribution of food (including alcohol), fuel, and other commodities. If the local jurisdiction requests additional assistance, the Governor may declare a State of Emergency. This declaration would then implement several additional powers at the State level, including the ability of the Governor to control access and egress from areas of the Commonwealth.

FEMA's support to localities is always channeled through the states. FEMA support, through the Stafford Act, is initiated when the Governor requests the President to provide Federal assistance. The request requires considerable documentation of the severity of the disaster to support a Federal declaration, and requires a clear statement on the part of the Governor that State capabilities are or will be overwhelmed. The most common Federal assistance will be in the form of recovery assistance, although FEMA does have formidable response assets as well.

Other Federal agencies have the ability to support local governments and the states directly under existing authorities and do not require a presidential declaration before implementation. Two of the most important are the assistance provided by EPA and the Military Aid to Civil Authorities provided by the Department of Defense.

EPA can respond to the release of oil and hazardous substances at any time under its existing Comprehensive Environmental Response, Compensation, and Liability Act (CERCLA) and the Oil Pollution Act (OPA) authorities, including during disasters. An EPA or Coast Guard OSC

has the capability to address hazardous materials releases without a presidential declaration. In addition, any Department of Defense base commander can respond to a request for assistance from local government in order to protect property or to alleviate human suffering under his or her own authority. These capabilities are frequently exercised when a quick response involving a large number of personnel is needed; such as during floods or when an impending levee break is imminent. Some Defense Department bases have a hazardous materials response capability that could be utilized.

The overall system of Emergency Management revolves around local government and private response capabilities. As this brief discussion implies, the problem of responding to hazardous materials emergencies, even in conjunction with a disaster, is fundamentally a local problem, and includes a mechanism by which additional assets can be obtained in a graduated and measured way if the local capabilities are overwhelmed.

Local governments and industries are responsible for doing all they can to address any IHR issues. Capabilities vary from locality to locality across the nation. If they are overwhelmed, local government can request mutual aid from adjacent jurisdictions, assistance from any Department of Defense installations nearby, or from EPA, but the most probable next step is to contact the State Emergency Management agency. The state's job is to coordinate the mobilization of State-level assets (including the National Guard) and to prioritize the requests for assistance they receive from local government. In large-scale disasters, the state will also coordinate the request for Federal assistance and coordinate the application of Federal assistance to localities, to businesses, and to private citizens impacted by the disaster, including when IHRs are involved.

Local Government and Private Sector Roles and Responsibilities

The bottom line with IHRs, and with all emergency management for that matter, is that local government and industry are ultimately responsible for effective hazardous materials response during a disaster. This may occur under extremely adverse conditions of weather and infrastructure damage, and will need to be prioritized along with all other demands of the moment. Some issues can be addressed by effective planning; for instance, by industry not relying exclusively on government capabilities in responding to hazardous materials releases.

The protection of public safety is an ethical issue for both government and industry. While no activity is risk-free and while many necessary industrial activities involve hazardous materials that may be released, the goal of all response and hazardous materials professionals must be to minimize the potential for releases and to optimize our capabilities in responding to those releases that do occur. As we have seen, IHRs may occur at the most difficult times and under the most bizarre of circumstances. It is incumbent upon local government and industry to operate safely and in a manner protective of public safety.

Industry has a vested interest in minimizing the impacts of IHRs. All hazardous materials releases must eventually be remediated and, as we have seen through many years of regulatory action, it is usually cheaper in the long run to prevent or minimize releases facility-wide than to remediate even one. It is foolish and unethical to view IHRs as a cost of doing business. In addition, local impacts can be severe, and the proximity of industry to any release suggests that the facilities themselves and their personnel may be the recipients of the greatest impacts. Ultimately, of course, industry owns the hazardous materials involved and is ultimately responsible for the safe and effective custody of those materials. Industry has the greatest knowledge of the hazards associated with any release of these materials and the greatest expertise in the response to and cleanup of spills and releases. Industries in hazardous areas—especially those prone to earthquakes, hurricanes, and floods—must prepare for the potential for IHRs.

Government also has a vested interest in ensuring that the impacts of IHRs are minimized. Response to hazardous materials releases anywhere outside of facility operational areas usually falls to public sector responders. IHRs occur at times when other issues may be more pressing, or the IHR itself may be overwhelming. In either case, local capabilities are limited, so it makes sense to attempt to minimize the number and severity of IHRs through effective planning. Government may also be impacted by IHRs through the ownership of

infrastructure. Transportation routes are owned by the public and are impacted during releases. Publicly owned water utilities and sewage treatment plants can be severely impacted by IHRs. Let us not forget, also, that government owns a lot of hazardous materials as well.

The studies of IHR suggest that the planning process can be used to predict vulnerabilities to IHR and that mitigation efforts can be put in place to minimize their impacts. CHMMs and other managers and planners should serve as those individuals responsible for considering these impacts as part of local government and industry *all hazards* planning.

Summary

IHRs usually are secondary events that occur coincident with and as a direct result of a larger natural or man-caused disaster. In many instances the IHRs add to and complicate an already difficult situation. In other instances the IHR may predominate over the initial disaster event in terms of potential harm and resources required to respond effectively. The IHR may result in more local damage than the original event and, due to the number and severity of the IHRs that occur, may be a significant challenge to public safety in the long term. The co-location of a natural or man-caused disaster and hazardous materials is required for an IHR to occur, and the severity of the IHR is related to the proximity to population centers as well as the properties of the materials involved. The response to IHRs is labor- and time-intensive, but these resources must be applied to the remediation of IHRs if the public safety is to be protected. The ability of local governments to respond to IHRs varies across the nation.

A small but significant number of IHRs occur each year. Current research concludes that IHRs are associated with every type of disaster that has been studied, and some studies suggest that, all other things equal, larger events give rise to more numerous or more severe IHRs. Three phenomena—earthquakes, hurricanes, and floods—give rise to approximately 90 percent of all reported IHRs, with earthquakes accounting for nearly three-fourths of the documented incidents. It follows that industries, companies, and government agencies (and the surrounding communities) that use hazardous materials located in earth-quake-, hurricane-, or flood-prone areas are at risk for the occurrence of IHRs.

The emergency management system in place in the United States is adaptable to the response to hazardous materials emergencies as well as other types of natural or man-caused disasters. The Federal response structure includes the potential occurrence of IHRs in the *Federal Response Plan* under ESF #10–Hazardous Materials, with the EPA as the lead Federal agency. The Federal response structure will follow the National Contingency Plan in addressing IHRs. The *Federal Response Plan* is also sufficiently robust to allow for a coordinated Federal response even if the disaster is itself a hazardous materials release, such as occurred at Bhopal and Chernobyl. In addition, many Federal agencies and de-partments can assist local governments through existing authorities in responding to IHRs. These include the EPA and the Department of Defense.

State response and recovery structures vary from state to state, but in general all allow for the coordination of State-level assets (including the National Guard) to support the efforts of local government and industry at the local level. Legal options exist in each state whereby the Governor can invoke a Declaration of Emergency in support of local efforts.

Ultimately, however, the response to and recovery from any disaster—whether it involves IHR or not—is a local responsibility. Government has well-defined roles in the protection of public safety, the providing of human services, and the pro-tecting of infrastructure. Industry has important roles that are both appropriate and that positively impact on the company's profitability. Local response to disasters is a true public–private partnership. After all, both government and industry are part of the impacted community.

Proper planning and mitigation efforts will help reduce the occurrence and the severity of the impacts of IHRs. Strengthening and hardening hazardous materials against disasters that are likely to occur within a community is a prudent mitigation step. In many cases, hazardous materials have been released when there has been little or no structural damage to the building, suggesting that hazardous materials storage systems are less robust than they should be.

Realistic hazardous materials response systems should be adopted, such that industries do not rely exclusively on local government response teams in the event of an IHR. While specific IHR incidents may become a high priority in any disaster response, governmental resources will not be available to deal with every release associated with a disaster.

Finally, the hazardous materials planning performed under Title III of the Superfund Amendments and Reauthorization Act (SARA) should be used to include IHRs in a possible response scenario. For industries located in those communities prone to IHRs, a reasonable and expected worst-case release should include loss of power, unavailable governmental response assets, and a damaged transportation and communi-cations infrastructure as part of their planning as-sumptions. Anything that can go wrong really will.

Disclaimer

The opinions expressed in this chapter are solely those of the author and may not be representative of the position of the Virginia Department of Emergency Services or the Commonwealth of Virginia.

Bibliography

Anonymous. "Corps of Engineers Looking for New Ways to Dispose of Hurricane Marilyn Debris, Waste." *HAZMAT News*, Vol. (6): p. 7, 20 November 1995.

Anonymous. "Report Says Most Nuclear Weapons Plants at Risk from Natural Disasters." *HAZMAT News*. Vol. (7): pp. 1-2, 29 July 1996.

Catallo, W. J., N. B. Theberge and M. E. Bender. *Sea Level Rise and Hazardous Wastes in the Coastal Zone: An Ecological Perspective*, in Coastal Zone '89, Proceedings of the Sixth Symposium on Coastal and Ocean Management. O. T. Magoon, Ed. Charleston, SC: ASCE, 1989.

Denning, E. J. *Hazardous Materials as Secondary Results of Flooding: A Case Study of Planning and Response*. Quick Response Report #49. Boulder, Co: University of Colorado Natural Hazards Center, 1992.

Emergency Management Institute. *Debris Management Course Participant Manual.* Emmitsburg, MD: Emergency Management Institute.

Fedarko, K. "Flood, Flames, and Fear." *Time.* Vol (148): 31 October 1994.

FEMA. "The Emergency Program Manager." Emergency Management Institute Home Study Course IS-1. Emmitsburg, MD: FEMA, 1993.

FEMA. "Emergency Preparedness U.S.A." Emergency Management Institute Home Study Course HS-2, Emmitsburg, MD: FEMA, 1994.

Hadfield, P. "Asbestos Shock in Kobe Cleanup." *New Scientist.* p. 6, 4 March 1995.

Hough, S. E., and L. M. Jones. "Aftershocks: Are They Earthquakes or Afterthoughts?" *EOS*, Vol (78), No (45). pp. 505–508, 11 November 1997.

Lindell, M. K., and R. W. Perry. "Addressing Gaps in Environmental Emergency Planning: Hazardous Materials Releases during Earthquakes." *Journal of Environmental Planning and Management.* Vol (39), No (4), pp. 529–543, 1996.

Lindell, M. K., and R. W. Perry. "Hazardous Materials Releases in the Northridge Earthquake: Implication for Seismic Risk Analysis." *Risk Analysis.* Vol (17), No (2), pp. 147–156, 1997.

National Park Service. *Hurricane Andrew, 1992: The National Park Service Response in South Florida.* Washington, DC: US Department of the Interior, 1994.

National Response Team. *Information Exchange— Lessons Learned from Exercises and Incidents.* Vol (3), No (1), 1997.

Noll, G., M. Hildebrand, and J. Yborra. *Hazardous Materials: Managing the Incident.* Stillwater, OK: Oklahoma State University Fire Protection Publications, 1988.

Seligson, H. A., R. T. Eguchi, K. J. Tierney, and K. Richmond. *Chemical Hazards, Mitigation and Preparedness in Areas of High Seismic Risk: A Methodology for Estimating the Risk of Post-Earthquake Hazardous Materials Release.* Technical Report NCEER–96–0013. Buffalo, NY: National Center for Earthquake Engineering Research, 1996.

Showalter, P. S. and M. F. Myers. "Natural Disasters in the United States as Release Agents of Oil, Chemicals, or Radiological Materials between 1980–1989: Analysis and Recommendations." *Risk Analysis.* Vol (14): pp. 169–182, 1994.

Virginia Department of Emergency Services. *Annual Report.* Richmond, VA: Technological Hazards Division, Virginia Department of Emergency Services, 1996.

Wright, F. G., H. I. Inyang, and V. B. Myers. "Risk Reduction through Regulatory Control of Waste Disposal Facility Siting." *Journal of Environmental Systems.* Vol (22), No (1): pp. 27–35, 1993.

Brett A. Burdick *is the Environmental Programs Manager with the Virginia Department of Emergency Services (VDES) in Richmond, Virginia, where his primary responsibility is directing the remediation and regulatory compliance of several State-owned properties contaminated with oil and hazardous materials. He also serves as the Program Manager overseeing the Commonwealth's Terrorism Consequence Management Preparedness Program. Prior to coming to VDES, Mr. Burdick served as Director of the Office of Environmental Response and Remediation in the Waste Division of the Virginia Department of Environmental Quality, and as the Ground Water Services Manager overseeing the Undergound Storage Tank Program in the Northern Regional Office of the State Water Control Board. He holds a BA in Geology and has completed graduate studies in Hydrogeology and Marine Science as a Doctoral Candidate. He is a Registered Professional Geologist in two states and is a Certified Hazardous Materials Manager at the Master's Level. Over the last 20 years, Mr. Burdick has published several technical papers and presented research findings in the fields of geology, hydrogeology, oceanography, hazardous materials management, and terrorism consequence management.*

CHAPTER **26**

Radiation Safety Principles

George D. Mosho, CHMM

Introduction

Radioactive material is a hazardous material. Although obvious, that had to be stated. With that in mind, radioactive material can be understood and safely managed as other hazardous materials are managed.

This chapter presents some basic information in the field of radiation safety, which is commonly known as *health physics*. As basic information, it is given as a primer in radiation safety. It is not the intent to belabor the reader with details concerning some topics such as nuclear theory, biological effects of radiation, radiation-producing devices, regulations, *etc*. However, for the sake of completeness, some of these topics shall be briefly covered in this chapter. There are many books and documents that have been written that treat those and other topics with the attention they deserve. A bibliography and list of professional organizations are supplied at the end of the chapter for additional information. Some common terms in the field of radiation safety are defined in the Glossary at the end of the book. Glossary words found in the text are ***bold italicized***. Specific questions concerning radiation safety, radioactive materials, and radiation-producing devices should be addressed to the governmental agency that has regulatory jurisdiction for such matters in your state.

ALARA

Radiation is the emission and propagation of energy through space. Radiation can be classified as either ionizing radiation (alpha particles, beta particles, gamma rays, x-rays, neutrons, high-speed electrons, high-speed protons, and other particles capable of producing ions) or nonionizing radiation (sound, radio, microwaves, or visible, infrared, and ultraviolet light).

When radiation is to be used in a process or as an end product, a basic question should be asked: "Does the *benefit* outweigh the *risk*?" If it does not, the *exposure* (of people and the environment) is unnecessary and should not occur. If, however, the benefit does outweigh the risk, then a review should follow to investigate any potential change in the operation or product that could further minimize that risk while considering the impact such changes may have on the total cost, or the efficacy of the operation, or the quality of the product. This process is a cornerstone of radiation safety programs. It is commonly known as *ALARA*.

ALARA is an acronym for *As Low As Reasonably Achievable*. The goal of radiation protection is to reduce and maintain all necessary exposures (internal and external) to radiation to ALARA levels. This requires actions as well as reviews to implement successfully. The design of workplaces, equipment, process systems, and procedures needs to be done by proactively integrating ALARA. This proactive approach is also evident in training programs emphasizing ALARA techniques as daily work habits. These habits are better described in the following section on radiation protection principles. Finally, on-the-job surveillance and monitoring for radiological exposure and contamination sources are imperative to adequately define actual workplace conditions and the effectiveness of prejob ALARA engineering.

Obviously, the ALARA process has a cost. The cost may be time, equipment, training, process redesign and reengineering, new materials, or some other parameter. The principal determination of the reasonable and appropriate cost for ALARA is based upon the level of accepted risk. There may also be regulatory drivers stating or inferring these acceptable risk levels. Whatever the case, these elements define ALARA at a workplace or facility.

Radiation Protection Principles

As stated earlier in the introduction, radioactive material is a hazardous material. To better express that statement, Table 1 was extracted from a recent *Health Physics* journal article. In reviewing these 10 principles or commandments, it is obvious to anyone who works with hazardous materials and risk management that this list is very applicable to almost all hazardous materials.

"Time, distance, and shielding" had long been the mantra for radiation protection. It is easy to acknowledge that the reduction of time (#1) spent at or near a radiation source would effect a reduction in exposure from that source. It is also easy to understand that the reduction of the concentration (#3) or of the size (#4) of the radiation source or the substitution of a non-radioactive material for the source or adoption of an entirely different process (#9) would consequently reduce or eliminate the exposure. Keeping the material contained (#5) and off your body (#6), as well as quickly and effectively decontaminating personnel who have been contaminated (#7), is also important.

Another time-related exposure-reduction method may be available when dealing with short-lived radionuclides. All radiation decays exponentially. *Radioactive decay* describes the process where an energetically unstable atom transforms itself to a more energetically favorable, or stable, state. The unstable atom can emit ionizing radiation in order to become more stable. This atom is said to be *radioactive*, and the process of change is called *radioactive decay*.

Each radionuclide decays at a different rate. The term *half-life* describes the time required to deplete the radioactive source to ½ of its original activity. Radionuclide half-lives range from extremely small fractions of a second to hundreds

Table 1. Principles and Commandments of Radiation Protection

Principle	Commandment (familiar)	Commandment (technical)
1. Time	Hurry (but don't be hasty).	Minimize exposure/intake time.
2. Distance	Stay away from it.	Maximize distance; stay upwind.
3. Dispersal	Disperse it and dilute it.	Minimize concentration, maximize dilution.
4. Source Reduction	Use as little as possible.	Minimize production and use of radiation and radioactive material.
5. Source Barrier	Keep it in.	Maximize absorption (shield); minimize release (contain and confine it).
6. Personal Barrier	Keep it out.	Minimize entry into the body of radiation and radioactive materials.
7. Decorporation (Internal and Surface Irradiation only)	Get it out of you and off of you.	Maximize removal or blocking of materials from the body (after intake or skin contamination).
8. Effect Mitigation	Limit the damage.	Optimize exposure over time and among persons; scavenge free radicals; induce repair.
9. Optimal Technology	Choose the best technology.	Optimize risk-benefit-cost figure.
10. Limitation of Other Exposures	Don't compound risks (don't smoke).	Minimize exposures to other agents that may work in concert with radiation (e.g., genotoxic agents or those that may cause initiation, promotion, or progression of tumors).

Reproduced from the journal *Health Physics* with permission from the Health Physics Society.

(Strom, 1996)

of thousands of years. It may be possible, if the material is one that quickly decays, to wait some length of time before doing the needed task. This would also decrease the overall potential exposure. An example is given below that depicts the relationship of radioactivity and time.

This relationship is described by the following equation:

$$A = A_o e^{-\ln 2 \, (t/T)} \tag{1}$$

where:

A = Activity remaining at time = t
A_o = Activity at time = 0
t = Elapsed time
T = Half-life of the radioisotope

For example, the activity of a 15 mCi source with a 30-year half-life would decrease to 13.4 mCi after 5 years:

A_o = 15 mCi
t = 5 y
T = 30.0 y [^{137}Cs]

Therefore:

A = 13.4 mCi

Two other principles of radiation protection involve biological mechanisms:

- Permit the body to heal itself between exposures, and medically assist in that repair if necessary (#8).

- Limit or remove other causal factors (#10) that may work as catalysts or enhancers for the undesired effect such as cancer.

Probably one of the most effective ways to minimize exposure is to maintain a distance from the source. For illustration purposes, assume the source is a *point*. The radiation field about a point source is spherical in shape, similar to light from a light bulb, and therefore follows what is called the *inverse square law*. This physical law is expressed by the formula:

$$I_2 = I_1 (d_1^2/d_2^2) \tag{2}$$

where:

I_1 = Intensity of radiation (exposure rate) at position #1

I_2 = Intensity of radiation (exposure rate) at position #2

d_1 = Distance position #1 is from source

d_2 = Distance position #2 is from source

As an example, the exposure at 2 meters from a 120 mR/h source, decreases to 30 mR/h when the distance is increased to 4 meters:

I_1 = 120 mR/h
d_1 = 2 m from the source
d_2 = 4 m from the source

Therefore:

I_2 = 30 mR/h

The importance of this and the other radiation protection principles cannot be overemphasized. They are simple, straightforward, generally easy to employ, and very effective.

Operational Radiation Safety Program

The cornerstone of radiation safety is an effective program. Essential elements in such a program are presented in the following list.

Operational Radiation Safety Program Elements. (In no particular order)

- Establish a radiation safety organization.

- Obtain and maintain a valid radioactive material license or other such documentation of authorization.

- Construct and maintain an inventory of radioactive material sources, or radiation-producing devices.

- Identify responsible users.

- Establish an exposure (external and internal, if needed) tracking program, including external dosimetry and bioassay services.

- Establish an ALARA program as well as a job safety or hazard analysis program to review proposed, current, and completed jobs or tasks with respect to personnel exposure to hazards.

- Develop, institute, and maintain procedures for the purchase, use, transfer, and disposal of radioactive material sources or radiation-producing devices, and emergency response.

- Institute a surveillance and monitoring program using qualified individuals with appropriate instrumentation.

- Maintain pertinent records.

- Perform self-assessments of the radiation safety program elements for quality assurance.

There is much that can be said about each program element listed. As a matter of regulatory compliance, usually these elements are specified further by the governmental agency that has jurisdiction over radiological issues.

The size of a radiation safety organization usually is not critical. It can be as small as one (*i.e.*, a dentist) or many people (*i.e.*, Argonne National Lab-East has 13 health physicists, 2 chief technicians, and 32 technicians in the Health Physics Section alone). What is important is the qualifications (both knowledge and experience) of the individuals, and the direct support of management in executing their respective duties.

Radiation Safety Training. As implied previously, these radiation safety personnel and the responsible users require training appropriate to their needs. Pertinent topics for such training are listed below (*Note: The depth and detail of these topics are dependent on the specific responsibilities of the individual as well as the scope of work* *performed with radioactive materials and radiation-producing devices)*:

- Review of basic mathematics and science
- Theory of radioactivity
- Sources of radiation
- Interaction of radiation with matter
- Biological effects of ionizing radiation
- Radiation units
- Radiation protection standards and regulations
- Dose assessment
- Radiation detector theory
- Contamination control
- Exposure (external and internal) control
- Waste management
- Emergency response actions

Radiation Theory Overview

This section is a very simple overview of some of the information that would be presented in a radiation safety training program.

Types of Radiation. *Nuclear radiation* is all around us and in us. Radiation sources can be either natural (terrestrial and extraterrestrial) or man-made. The radiation emitted by these sources can be any one or a combination of forms of electromagnetic energy: gamma, and x-rays, and particles: alpha, beta, and neutron.

Gamma (γ) rays, x-rays, and *neutrons* are highly energetic and have the capability to cause ionization by penetrating the human body and other materials. Many diagnostic instruments such as moisture density gauges, industrial radiography cameras, and medical x-ray units utilize this feature to provide *in situ* analyses of various materials. Since these types of radiation can penetrate the human body, they are considered as external hazards. That is, they have the potential to harm an exposed individual from outside the body. Shielding with high-Z materials (*i.e.*, lead, steel, *etc.*) for both γ and x-rays can effectively reduce exposure rates. However,

similar shielding for neutrons would not be as effective. Fairly large quantities of hydrogenous material (*i.e.*, water, concrete, plastic, *etc.*) are more appropriate to achieve significant attenuation of neutrons.

Alpha (α) particles are rather large on the atomic scale. Each α particle consists of two protons and two neutrons. The mass and electric charge of such a particle precludes it passing through the dead layer of skin on the outside of the human body. These particles can travel about 3 cm in air and <0.003 cm in human skin tissue. When the particle strikes the dead layer of human skin, its energy is transferred to the tissue, causing ionization in the dead skin. Therefore, α particles are not an external hazard. However, if they are inhaled, ingested, or they enter the body in some other manner, they can be deposited directly against sensitive body tissues. These particles have a significant potential for harm within the body and are listed as an internal hazard. A single sheet of paper can effectively stop α radiation. Efforts should be taken to minimize the inhalation and ingestion pathways when working with α particulate radiation.

Beta (β) particles are essentially electrons. This type of particulate radiation can travel to a maximum distance of roughly 5 m in air and 1.5 cm in body tissue. It is considered to be both an internal and external hazard to the body. Although β radiation is more penetrating than α radiation, adequate shielding can generally be accomplished by using plastic, cardboard, wood, or heavy clothing. High-Z materials should not be used as shielding due to the potential to produce secondary radiations from the interaction of the β radiation with those types of materials.

Interaction of Radiation with Matter. What all these types of radiation have in common is that they can interact with atoms causing a loss of electrons and resulting in the creation of ions. In short, they are *ionizing radiation*. Nonionizing radiation cannot do this. Examples include microwave, laser, and radio wave. Obviously, when the material that ionizing radiation interacts with is human tissue, damage to the tissue can occur and there is reason to be concerned.

Many years ago, as a conservative approach to radiation exposure, an assumption was formed from a dose-response model. That model (the

Linear No Threshold Hypothesis) depicted that any exposure to radiation incurred a deleterious effect. This assumption did not take into account the duration of the exposure or time between exposures, the repair mechanisms of the human body, and many other parameters. Today, this model is highly debated by radiation safety professionals but is still used as the standard.

The effect that radiation has on any material is determined by the **dose**. Radiation dose is simply the quantity of radiation energy deposited in a material. There are several additional descriptive terms used in radiation protection which provide precise information on how the dose was deposited, the method used to calculate the dose, and how the radiation energy deposited in tissue will affect humans.

Absorbed dose is the amount of energy deposited in any material by ionizing radiation. The unit of absorbed dose, the rad, is a measure of energy absorbed per gram of material. The unit used in countries other than the US is the *gray*. One gray equals 100 rad.

The concept of **equivalent dose** involves the impact that different types of radiation have on humans. Not all types of radiation produce the same effect in humans. The equivalent dose takes into account the type of radiation and the absorbed dose. For example when considering beta, x-ray, and gamma ray radiation, the equivalent dose (expressed in rems) is equal to the absorbed dose (expressed in rads). For alpha radiation, the equivalent dose is assumed to be 20 times the absorbed dose.

With that in mind, dose can be received from chronic and/or acute **exposures** to radiation. An *acute exposure* is one where a large exposure (high exposure rate) is received over a very short period of time, usually minutes or seconds. A *chronic exposure* is a long term or possibly constant exposure to low-level radiation. Cell repair mechanisms have demonstrated that cumulative doses to radiation resulting from chronic exposures have less impact than similar doses from acute exposures.

The effect of radiation on the human body can be somatic and/or genetic. The term *somatic* relates to the person exposed. These somatic effects may be early or late depending upon when the effect is evident. *Early effects* occur within hours or days of an acute exposure. Examples of early effects are erythema (reddening of the skin), loss of hair, and, in extreme cases, death. *Late effects* appear many years, possibly 20 or 30, after an acute exposure. Cancer induction is an example of a late effect. Another form of somatic effect is known as teratogenic effects. Children *in utero* that are exposed to radiation may develop and exhibit certain birth defects. Radiation-damaged DNA may cause genetic effects. Such an effect may become evident in future progeny.

Radiation Units. Throughout the world, most countries have adopted radiation units defined by the International System of Units (SI). However, despite efforts to change to SI in the United States, these units seem to be used routinely only in transportation documentation, as prescribed by the United States (US) Department of Transportation. A comparison of the SI and US systems of units is presented in Table 2.

Type	US	SI	US → SI Multiply by
Activity	curie (Ci)	becquerel (Bq)	3.7 E+10
Absorbed Dose	rad (r)	gray (Gy)	1.0 E-02
Effective Dose Equivalent	rem (rem)	sievert (Sv)	1.0 E-02
Exposure	roentgen (R)	coulomb per kilogram (C/kg)	2.58 E-04

Table 2. Radiation Unit Comparison

In discussing data with radiation control technicians (RCTs) or health physics technicians (HP Techs), often the term **dose** is incorrectly used interchangeably for **exposure**. Most often, radiation data are expressed as *rates*. Portable survey meters display values in terms such as *counts per minute* (cpm), *disintegrations per minute* (dpm), and *$\mu R/h$* or *mR/h*. Contamination measurements are denoted for the type of radiation and for a specific area (*i.e.,* 320 dpm α/100 cm^2) which is usually the active area of the survey probe or, if a smear, the area smeared for counting.

Radiation Detection. The whole issue of radiation safety would be irrelevant if there was no practical way to detect radiation. Fortunately, that is not the case. In fact, detecting low-level radiation is usually easier than detecting many hazardous materials. Detectors used routinely by RCTs or HP Techs are either portable or nonportable equipment. Portable detectors or survey meters are designed for a specific purpose such as contamination or exposure monitoring, but not both. They are also specific to the type of radiation that they can detect. For example, a *rem ball* is a neutron-only detector; a ZnS scintillator detects exclusively α contamination; whereas a *pancake* Geiger-Mueller (G-M) is good for β-γ contamination, and so on. Sorry, no magic box yet that can detect, discriminate, and quantify all types of ionizing radiation.

To measure the dose an individual may actually receive from penetrating radiation, there are dosimeters. One of the first discoveries about the effect of radiation was that certain radiations (γ and x-rays) caused fogging on photographic film. This was later advantageously transformed into a film badge with various attenuators to determine the energies of the radiation affecting the film. This type of badge is still used today; however, many facilities have upgraded to a thermo-luminescent dosimeter (TLD). This badge is composed of several crystal chips with attenuators. These dosimeters are much better than the film and they are reusable after readout and annealing and therefore more cost-effective in the long run.

Whether TLD or film, the badge is worn on the chest to measure the quantity of dose received by the *whole body*. Extremity badges for hands, feet, head, *etc.* may be important in operations where exposure fields are not uniform. In addition, self-reading dosimeters (SRDs) may be used to supplement the *whole body* badge by giving the worker a method to check on his/her exposure during the course of a job.

There are also several types of portable detectors that can be readily used in the field for radiation detection and quantification. There are detectors for contamination that detect either α, β, or β-γ, and some that can do both α and β. Exposure rate meters are designed specifically for x-rays, γ rays, and fast and thermal neutrons. Special detectors exist to perform tasks such as floor monitoring, *in situ* γ spectroscopy, and ^{3}H vapor detection. Some of the more common detectors are listed in Tables 3 and 4.

Finally, there are some counting systems that, due to the method of detection used or the precision needed in the resultant data, are nonportable. A few of the typical types of counters are the gross α - β counting system, spectroscopy (α, β, or γ) system, personnel contamination monitors, and continuous air monitors (CAMs).

Table 3. Portable Gamma Exposure Detectors

Instrument (Example types)	Detector Description	Measurement Units
"µR meters" Eberline PRM-3; Ludlum 19	Scintillator (internal) NaI 1" x 1"	µR/h
Bicron Microrem	Scintillator (internal) Loaded plastic 1" x 1"	µrem/h
Eberline RO-20, RO-2, RO-2A, PIC-6; Victoreen 450, 450P	Ionization Chamber (internal)	mR/h; R/h
Teletectors, Ludlum 77-3, Eberline 6112B; Xetec 302-B	Geiger-Mueller (G-M) energy compensated	mR/h; R/h

Table 4. Portable Surface Contamination Detectors

Instrument (Example types)	Detector Description	Measurement Units	Radiation Type
NE Technology, Ltd. Electra	Dual scintillator, 100 cm^2 sampling area	cpm; dpm	α and β
Eberline PAC4G-3	Gas proportional, 61 cm^2 sampling area	cpm	α
Ratemeter with Eberline HP-260, HP-210 (T, L; AL); Ludlum Model 44-9	Geiger-Mueller (G-M) "pancake probe" 4.4 cm diameter window	cpm	β-γ
Ratemeter with Eberline AC-3; Ludlum Model 43-5	ZnS 59 cm^2 sampling area	cpm	α
Ludlum Model 3	Air proportional, 50 cm^2 sampling area	cpm	α

Radiation Protection Planning

The optimal way to ensure that ALARA, radiation protection principles, and good radiation safety practices, are a routine way of business is by getting in at the planning stage. Many facilities use a radiation work permit (RWP) process to do just that. The permit process predicates having all interested parties sit down and think the job through before actually doing it. Previous radiation survey data and discussion of the task to be performed help the radiation safety personnel (probably a health physicist) to visualize the job and estimate potential dose received for doing that job. A workplace hazard mitigation plan is developed using engineering techniques and personnel protective equipment if necessary. Considerable thought is placed upon this topic. Often it is found that protecting an individual from one hazard causes him/her to be at greater risk from another hazard (*i.e.*, poly-coated coveralls *vs.* heat stress). Once again, does the benefit outweigh the risk? The plan is noted formally in the RWP and the permit is read and signed off by all job participants. After the work has been completed, a review of the lessons learned and the actual dose (depicted from SRDs) accumulated during the job aid in future training, and in job planning.

Summary

Despite what reservations you may have initially had concerning radiation, radioactive material and radiation is really no different than other hazardous materials. The basic philosophies used to minimize exposure (time, concentration, size, optimal technology, containment, personal protection, and decontamination) are the same. Both radioactive and other hazardous materials need to be treated with the appropriate respect, understanding, and care.

Bibliography

American Society for Testing and Materials. "Standard for the Use of the International System of Units (SI): The Modern Metric System." IEEE-ASTM–SI–10; IEEE/ASTM SI–10. West Conshocken, PA: ASTM, 1999.

Cember, H. *Introduction to Health Physics*, 3rd ed. Elmsford, NY: Pergamon Press, 1996.

Eisenbud, M. and T. F. Gesell. *Environmental Radiation from Natural, Industrial, and*

Military Sources, 4th ed. San Diego, CA: Academic Press, 1997.

Gollnick, D. A. *Basic Radiation Protection Technology*, 3rd ed. Altadena, CA: Pacific Radiation Corporation, 1994.

Reynolds, A. B. *Bluebells and Nuclear Energy*. Madison, Wisconsin: Cogito Books, 1996.

Shapiro, J. *Radiation Protection: A Guide for Scientists and Physicians*, 3rd ed. Cambridge, MA: Harvard University Press, 1990.

Strom, D. J. "Ten Principles and Ten Commandments of Radiation Protection." Rev. 8/6/96. *Health Physics*. Vol. (70) No. (3): pp. 388-393, 1996.

Turner, J. E. *Atoms, Radiation, And Radiation Protection*, 2nd ed. New York, NY: John Wiley and Sons, Inc., 1995.

Internet Resources

<http://www.ans.org/> (American Nuclear Society)

<http://www.hpinfo.org/> (Borders' Dictionary of Health Physics)

<http://www.hps.org/> (Health Physics Society)

<http://www.nrrpt.org/> (National Registry of Radiation Protection Technologists)

<http://www.umich.edu/~radinfo/> (Radiation and Health Physics Homepage)

George D. Mosho is a health physicist at Argonne National Laboratory-East, near Chicago. He received his BS in Physics at The Citadel: The Military College of South Carolina (1977), and a MS in Radiological Health at the University of Pittsburgh (1985). Currently, his duties include the role of staff health physicist for the Intense Pulsed Neutron Source (IPNS) Accelerator Facility, team commander for the US Department of Energy (DOE) Radiological Assistance Program (RAP) for DOE Region 5, and radiological instructor for the US National Domestic Preparedness Program (Weapons of Mass Destruction) NBC Technician and Awareness Courses. He has spent over 15 years in radiological characterization, decontamination, and decommissioning, and emergency response operations.

CHAPTER 27

Lead

Chris Gunther, CHMM

Background

Lead was considered to be one of the seven metals in antiquity, partially because of an inherent property to bond well with many different types of substrates. Archeologists have found lead-based pigments on buildings that were constructed 5,000 years ago. It is therefore not difficult to understand why lead-based pigments have been used as an additive in paint for centuries.

In the 18th century, however, harmful health effects from exposure to lead paint were noticed. In 1786, Benjamin Franklin wrote a letter to a friend named Benjamin Vaughn. In the letter, Mr. Franklin discussed in some detail his observations regarding "the mischievous effects of lead." The effects outlined in the letter included poisonings, partial paralysis, and severe gastric disorders. The observations included secondhand information that Mr. Franklin collected from conversations with knowledgeable people, as well as from direct personal experience. Mr. Franklin ended his letter by writing a statement that has unfortunately been validated in recent times:

> This, my dear friend, is all I am at present to recollect on the Subject. You will see by it, that the Opinion of this mischievous Effect from Lead, is at least above Sixty Years old; and you will observe with concern how long a useful Truth may be known, and exist, before it is generally receiv'd and practis'd on.

In the twentieth century, lead compounds, such as lead chromate and lead carbonate, have been used widely as pigments in paint and, to a lesser extent, in varnishes and primers. Although the use of lead, particularly on interior surfaces, has declined over the years, most housing units constructed before 1980 have some lead-based paint. Although the paint can be found anywhere in a building, it was used primarily on areas of high impact, friction, or humidity.

Presently, the principal use of lead is in the manufacturing of electrical storage batteries, and as solder in electronics. Other current uses include ammunition, chemicals, and sinkers for fishing. The use of lead as a paint additive and gasoline additive has been greatly reduced (if not eliminated) in the United States, along with lead solder and piping. Unfortunately, the residual effects from past uses are still present.

Routes of Exposure and Health Effects

Although lead occurs naturally in small quantities in the earth's crust, the greatest exposure to lead stems from manmade processes and products. Lead serves no useful purpose in the body; rather, it is a poison that alters or hinders vital biological reactions. Some of the areas in the body that are impacted by the presence of lead include the brain and central nervous system, kidneys, heart, bones, and liver.

The major sources of lead poisoning in children are surface dust and soil that have been contaminated by deteriorated lead paint. The primary routes of exposure among children are through ingestion (from hand-to-mouth activity), inhalation of the dust, and by drinking water that has passed through plumbing that contains lead solder or brass valves.

In construction and general industry, the primary routes of exposure are still inhalation and ingestion. However, inhalation of lead dust and fumes is more common among workers in these areas than is ingestion. For example, many welders have been lead-poisoned by welding steel containing lead as a primer coat.

Lead is absorbed through the stomach in the same manner that it absorbs the vital metals (calcium, zinc, iron, chromium, boron, *etc.*). The absorption of lead into the blood stream appears to be diet-related. The higher the fat content in a person's diet, the higher the rate of lead absorption appears to be. A healthy diet (rich in protein and metals such as calcium, zinc, iron, *etc.*) tends to decrease the absorption rate of lead.

Exposure to high concentrations of lead can cause retardation, convulsions, coma, and death. Children (especially those under the age of six) are especially vulnerable to lead because of their metabolism, developing nervous systems, and their tendency to absorb nutrients into developing bones and muscles. Even low levels of lead consistently absorbed into a child's body during childhood have been known to slow the child's development. This retarded development can also lead to learning and behavioral problems.

Symptoms of lead poisoning are not readily observed because they are not specific. The most common short-term symptoms are abdominal pain, headaches, constipation, and aches in the joints. Unfortunately, all of these symptoms can be attributed to causes other than lead poisoning, such as the common cold or influenza. The main difference however, is that these *flu-like* symptoms don't subside in the same manner as a real flu. Instead, the symptoms linger or even become worse if the poisoning continues.

If the poisoning continues over time, then the lead will accumulate in the body, particularly in the bone structure. Even if the exposure to lead is suddenly halted, the detrimental effects of the metal can linger for months afterwards. In some conditions, such as pregnancy or aging, the effects of lead poisoning can recur as bone tissue breaks down and releases the absorbed lead into the bloodstream.

Over the past 20 years, the Centers for Disease Control and Prevention (CDC) has responded to the increasing information about the effects of lead poisoning in children by progressively lowering the blood lead level that requires medical intervention. The current goal is for all children to have lead levels below 10 micrograms per deciliters of blood (µg/dl).

The present level of concern (10 µg/ dl) is far above the *natural background* of the blood level of people living in remote areas of the world. The fatal dose for a young child is 100–150 μg/dl, or roughly 10–15 times the level of concern.

Regulations

Regulatory Overview

Today, lead typically is present in small detectable quantities in paints that are for sale both commercially and publicly. In 1976, the Lead-Based Paint Poisoning Prevention Act (LBPPPA) established an upper limit of lead in paint at 600 ppm (.06%). In 1977, the Consumer Product Safety Commission (CPSC) adopted this standard for most residential and commercially available paints, with the intent of reducing lead exposures in facilities where children may be present. However, CPSC has no jurisdiction over worker exposure, and OSHA does not recognize this arbitrary limit. Instead, OSHA considers any paint with detectable levels of lead to pose a potential exposure to employees who are involved in *trigger tasks*. The Occupational Safety and Health Administration (OSHA) interprets a **trigger task** as one that may result in occupational exposures above the permissible exposure limit for lead (50 μg/m^3). These tasks include

- Sanding
- Scraping
- Manual demolition
- Heat gun removal
- Spray painting
- Abrasive blasting
- Torch burning
- Welding
- Cutting
- Tool cleaning

The OSHA permissible exposure limit is based upon airborne concentrations of lead dust, not the lead content of the painted surface impacted by an employee. In situations where paint with detectable lead levels is expected to be disturbed, all OSHA worker protection measures and work practices apply during trigger tasks. Since air concentrations of disturbed lead dust or fumes are usually unknown, then the employer is required to protect affected workers with respiratory and body protection, along with engineering controls to control the migration of the lead aerosols.

The presence of lead must be established by recognized sampling protocols and methods of analysis. The most common analytical methods include atomic absorption spectrometry (AAS), inductively coupled plasma-atomic emission spectroscopy (ICP-AES), or x-ray fluorescence analysis (XRF). The use of over-the-counter colorimetric tests is not a valid method for determining the presence of lead in paint.

In addition, the Department of Housing and Urban Development (HUD) has adopted several Environmental Protection Agency (EPA) guidelines for abating lead hazards, and has defined inspection protocols and risk assessment methodologies.

On the Federal level, the following documents regulate lead exposure:

- 29 CFR 1910.1025 (OSHA)
- 29 CFR 1926.62 (OSHA)
- 40 CFR 745 (EPA)
- 40 CFR 50 (EPA)
- 40 CFR 122 (EPA)
- 40 CFR 261 (EPA)
- *Guidelines for the Evaluation and Control of Lead-Based Paint Hazards in Housing* (HUD)

Lead Standard for General Industry

29 CFR 1910.1025. The general industry standard does not apply to the construction industry, even though this standard shares many aspects of the construction industry standard. For example:

- The regulation allows an airborne action level (AL) for lead concentration of 30 μg/m^3 as an 8–hour time-weighted average (TWA).
- The regulation allows an airborne permissible exposure limit (PEL) for lead concentration of 50 μg/m^3 as an 8–hour TWA.
- The regulation requires biological monitoring at all times and medical removal if blood lead becomes too elevated (>50 μg/dl).

- The regulation requires the employer to institute a medical monitoring program for all employees exposed above the AL for more than 30 days per year. The employer must also make available the results of the blood testing to the employees.

Lead Standard in the Construction Industry

29 CFR 1926.62. The construction industry standard applies to construction-related activities including demolition, removal, encapsulation, alteration, transportation, storage, and disposal of lead (metallic lead, inorganic lead compounds, and organic lead soaps). The standard also establishes maximum limits of airborne exposure to lead for all covered workers, including a PEL of 50 μg/m^3 (TWA) and an AL of 30 μg/m^3 (TWA).

If the employee's initial exposure to lead is at or above the action level, the employer must implement a site-specific lead compliance manual, and collect samples that represent the *regular daily exposure* of a full work shift, one sample for each shift. In addition, a full medical examination with extensive testing must be available for each employee who has been subject to exposure at or above the action level more than 30 days per year.

The employer must notify all affected employees of the examination and test results within five days of receiving the medical results. If the employee has a final medical determination or detected medical condition that places him/her at increased risk of material impairment from exposure to lead, the employee must be temporarily removed from exposure, with medical removal benefits still in effect. An employee whose blood lead level exceeds 50 μg/dl is subject to removal if the employer has no other work under the AL to offer. The employee(s) can return to the job when: (1) two consecutive blood sampling tests indicate a blood level of lead below 40 μg/dl, and (2) a subsequent medical examination indicates that there is no longer a detectable medical condition that increases risks to health from lead exposure. All medical removals and returns are subject to physician review.

Lead; Requirements for Lead-Based Paint Activities in Target Housing and Child-Occupied Facilities; Final Rule

40 CFR 745. This document became effective on August 29, 1996, and was written to ensure that

- Individuals who conduct activities involving lead-based paint in target housing and child-occupied facilities are properly trained and certified

- Accredited training programs provide instruction in these activities

- These activities are conducted according to reliable, effective, and safe work practices, as presented in *The HUD Guidelines for the Evaluation and Control of Lead-Based Paint Hazards in Housing*

The regulation seeks to ensure that a trained and qualified workforce is available to identify and address the hazards associated with lead-based paint and to protect the public from exposure to these hazards.

Air Programs

40 CFR 50. The portions of the *Code of Federal Regulations* that address the Clean Air Act can be found in 40 CFR Subchapter C, "Air Programs," encompassing Parts 50 through 99. Restrictions on lead emissions can be found in 40 CFR 50.12, "National Primary and Secondary Ambient Air Quality Standards for Lead." This section establishes a limit of 1.5 μg/m^3 in the ambient air; this is the maximum arithmetic mean averaged over a calendar quarter.

Water Programs

40 CFR 122. A National Pollutant Discharge Elimination System (NPDES) permit must be obtained for the discharge of a pollutant into the waters of the United States. This part of the regulation was written to address pollutants that are piped or channeled into the water from facilities. Construction projects require permits if more than five acres will be disturbed (or less than five acres if the parcel is part of a larger project).

Resource Conservation and Recovery Act

40 CFR 261. The Resource Conservation and Recovery Act of 1976 (RCRA) lists lead as a toxic hazardous waste. The analytical test used to determine toxicity under RCRA is the Toxicity Characteristic Leaching Procedure (TCLP), which is also referenced as EPA Method 1311. Wastes containing lead concentrations greater than 5 ppm by this test are regulated as hazardous wastes. Paint chips and other nonrecyclable painted components are typically regulated as potential hazardous wastes under RCRA.

Solid waste landfills generally do not accept non-recyclable lead-painted construction debris, if the debris has failed the TCLP test (>5 ppm). This situation is unusual, however, because other non-lead construction debris in the mix tends to dilute the amount of leachable lead in the test. Construction debris that does fail the TCLP test should be sent to a Treatment, Storage, and Disposal Facility (TSDF) for ultimate treatment and disposal. Metallic lead flashing, lead batteries, and lead-painted metal building components that are recycled are exempt from RCRA.

In addition to paint, lead wastes from other sources and operations may fall under RCRA. Lead dust in battery-manufacturing facilities, or lead dust and solder from an electronics plant may be subject to RCRA when these wastes are generated during renovation or demolition. Likewise, the lead-contaminated debris, equipment, clothing, and wastewater resulting from a lead abatement project are subject to RCRA.

The amount of lead waste emanating from a renovation or demolition project can have important ramifications for the hazardous materials manager. For example, these projects have the potential to temporarily alter a facility's generator permit status from a conditionally exempt generator to a small quantity generator. Some states may allow a special one-time exemption for such an event, so it is important to plan ahead.

The regulation and disposal of lead-contaminated wastes is continuing to evolve. The EPA has stated that a decision regarding the exemption of lead-painted architectural components from RCRA will be made sometime during the spring of 1999. In addition, compliance with RCRA will continue to vary from state to state, and even from landfill to landfill. It is therefore imperative that the hazardous waste manager stay abreast of the regulations.

The HUD Guidelines

Formally known as *The HUD Guidelines for the Evaluation and Control of Lead-Based Paint Hazards in Housing*, this document (the *Guidelines*) provides detailed comprehensive, technical information on how to identify lead-based paint hazards and how to control such hazards safely and effectively. The goal is to assist property owners, private contractors, and government agencies in reducing children's exposure to lead without unnecessarily increasing the cost of housing. The *Guidelines* provide more complete guidance than do the regulations on *how* lead abatement activities should be implemented, and *why* certain measures are recommended. Lead exposure that comes from air emissions, Superfund sites, drinking water, ceramics, home (folk) remedies, cosmetics, and foods are *not* the focus of the *Guidelines*.

Additional Lead Regulations

Many states and other jurisdictions have recently enforced regulations dealing with the presence of lead-based paint, particularly in residential properties. For example, Maryland has recently adopted a regulation that allows a limited liability cap for rental property owners. One of the conditions under which the limited liability can occur is if the property can be certified as *lead-free*. Under the *Code of Maryland Regulations* (COMAR 26.16.02.02[B][5]), *lead-free* means

- Containing no lead paint (<0.7 mg/cm^2 by XRF or <0.5% by paint chip analysis)

- All interior surfaces of the affected property contain no lead-based paint

- All exterior surfaces of the affected property coated with lead-based paint that were chipping, peeling, or flaking have been restored without lead-based paint

- No exterior surfaces coated with lead-based paint are chipping, peeling, or flaking

In most instances, the properties in question have been certified as lead-free by means of inspections with an x-ray fluorescence analyzer.

Related Protocols

The American Society for Testing Materials (ASTM) has issued protocols for collecting samples of dust, soil, paint and air for subsequent analysis for lead. These protocols are summarized below.

E1727–95. *Standard Practice for Field Collection of Soil Samples for Lead Determination by Atomic Spectrometry Techniques.* This protocol deals with the collection of soil samples using coring and scooping methods. The protocol is not suitable for areas that are paved, nor does it address the development of sampling plans such as grid patterns, *etc.* Lead concentration is measured in ppm.

E1728–95. *Standard Practice for Field Collection of Settled Dust Samples Using Wipe Sampling Methods for Lead Determination by Atomic Spectrometry Techniques.* This protocol covers the collection of settled dust on hard surfaces using the wipe sampling method. The standard does not address criteria for sampling design. Lead concentrations are presented in $\mu g/ft^2$.

E1729–95. *Standard Practice for Field Collection of Dried Paint Samples Using Wipe Sampling Methods for Lead Determination by Atomic Spectrometry Techniques.* This protocol deals with the collection of dried paint samples or other coatings from buildings and related structures. The procedure is used to collect samples for subsequent analysis on an area basis (milligrams of lead/area sampled) or on a concentration basis (milligrams of lead/gram of dried paint solid). The protocol does not include sampling plan criteria that are used for risk assessments or other purposes.

E 1975–98. *Standard Practice for Collection of Surface Dust by Air Sampling Pump Vacuum Technique for Subsequent Lead Determination.*

This procedure describes the vacuum collection of surface dusts onto filters using portable, battery-powered air-sampling pumps. The protocol is designed for the subsequent analysis for lead on a loading basis (micrograms of lead/area sampled) unless preweighed filters are used. The amount of lead can also be determined on a concentration basis (micrograms of lead/gram of dust collected) if preweighed filters or filter cassettes are used.

Bibliography

American Society for Testing and Materials. *ASTM E1727–95 Standard Practice for Field Collection of Soil Samples for Lead Determination by Atomic Spectrometry Techniques.* West Conshohocken, PA: ASTM, 1995.

American Society for Testing and Materials. *ASTM E1728–95 Standard Practice for Field Collection of Settled Dust Samples Using Wipe Sampling Methods for Lead Determination by Atomic Spectrometry Techniques.* West Conshohocken, PA: ASTM, 1995.

American Society for Testing and Materials. *ASTM E1729–95 Standard Practice for Field Collection of Dried Paint Samples Using Wipe Sampling Methods for Lead Determination by Atomic Spectrometry Techniques.* West Conshohocken, PA: ASTM, 1995.

American Society for Testing and Materials. *ASTM E1975–98 Standard Practice for Collection of Surface Dust by Air Sampling Pump Vacuum Technique for Subsequent Lead Determination.* West Conshohocken, PA: ASTM, 1998.

Department of Housing and Urban Development. *The HUD Guidelines for the Evaluation and Control of Lead-Based Paint Hazards in Housing.* Washington, DC: GPO, 1995.

"Lead in Construction." *Code of Federal Regulations.* Title 29, Pt. 1910.26.

Chris Gunther *is the Environmental Manager of Aerosol Monitoring & Analysis, Inc. (AMA) in Hanover, Maryland. He has been employed in the field of hazardous materials management for over 12 years, primarily in the disciplines of lead, asbestos, environmental risk assessment, and indoor air quality. In addition to performing lead inspections and risk assessments, Mr. Gunther also teaches courses in AMA's Training Division. Mr. Gunther thanks Mr. Greg Baker, CIH, CSP, CHMM, and Mr. E. Rush Barnett, CSP, for their review of this chapter.*

CHAPTER 28

Asbestos

Margaret V. Naugle, EdD, CHMM
Charles A. Waggoner, PhD, CHMM

Asbestos, what is it? Where is it found? How is it used? What are the health hazards associated with exposure? These are questions with clear, concise answers. But when it comes to asbestos and the regulations, duties, and responsibilities of property owners and employers, the answers are more difficult. The information that follows offers a response to the basic background information and addresses the current regulations affecting employers and property owners.

Background Information

Asbestos is a group of naturally occurring fibrous minerals that have been used since the time of the ancient Greeks. Most of the world's asbestos comes from Canada and South Africa. It has also been mined in Russia, Australia, Finland, and the United States.

There are six recognized asbestiform minerals composed predominantly of magnesium silicate. The precise chemical formulation of each species varies with the location from which it was mined. They are classified based upon their crystal structure and the presence of contaminating metals (iron, calcium, and sodium) within the mineral matrix. Asbestos includes: chrysotile, crocidolite, amosite, tremolite asbestos, actinolite asbestos, anthophyllite asbestos, and any of these minerals that have been chemically treated and/or altered.

Asbestos minerals are classified into one of two groups: (1) serpentine or (2) amphibole. The serpentine group consists of only one asbestos mineral, chrysotile or white asbestos. Chrysotile accounts for approximately 95% of all asbestos found in buildings in the United States. Serpentine is identified by its layered or sheet-type crystal structure. Its fibers are strong, flexible, and straw-like fibers. These characteristics make it an excellent composite material allowing it to be spun and woven into fabrics. The hollow straw-like structure of the fibers contributes to the material's ability to absorb water. This important physical property allows *wet methods* to be used as an engineering control to prevent fiber release during abatement activities.

The amphibole group is characterized by a chain-like crystal structure and the presence of metals other than magnesium. Asbestos minerals included in this group include: amosite (brown asbestos), crocidolite (blue asbestos), and the asbestiform minerals: tremolite, actinolite, and anthophyllite. Amphiboles are more resistant to heat and chemical degradation than is chrysotile. Therefore, amosite and crocidolite are frequently found where high temperature insulation applications are found. The unique crystalline structure of amphiboles makes it more difficult to wet, and requires the addition of surfactant to water when wet methods are used to control fiber releases.

The physical properties of asbestos include: high tensile strength, chemically inert, noncombustible, and heat-resistant. Asbestos has a high electrical resistance and good sound-absorption properties. It can be woven into cables, fabrics, or other textiles, and can be matted into asbestos papers, felts, or mats.

The physical and chemical properties of asbestos coupled with the economics of production led to the manufacture of more than 3,000 products in the United States. Asbestos has been used in the manufacture of heat-resistant clothing, automotive brake and clutch linings, as well as thermal, fireproof, and acoustical insulation. Other building materials which are asbestos-containing include floor tiles, roofing felts, ceiling tiles, asbestos cement pipe and sheets, and fire-resistant drywall. It has been used as a strengthening agent in concrete, mortar, grout, and drywall spackling compounds. It may present in pipe and boiler insulation materials and in sprayed-on decorative materials. Asbestos-containing materials (ACM) are found on beams, in crawl spaces, and within the wall cavities.

Asbestos is a versatile material, ideal for many uses. It is found everywhere. It has been widely used in construction, it occurs naturally in our environment, and, because of its unique properties, it is found in many manufactured products. Why then is it so highly regulated? The answer is related to the potential adverse health effects and the associated liability.

Health Effects

Asbestos is a toxic substance. The toxic effects of asbestos seem to stem from particle size and shape more than from chemical composition. The primary routes of exposure are inhalation and ingestion. Inhalation is considered the most dangerous mode of exposure. Most asbestos-related illness is the result of inhaling microscopic asbestos fibers. Exposure via ingestion is believed to be linked to an increased incidence of cancer in the alimentary canal. Dermal exposure is known to cause asbestos warts on the hands of abatement workers. These warts tend to disappear after exposure is ended. Dermal exposure to asbestos is not believed to produce dangerous or irreversible effects.

Asbestos is a known carcinogen. It causes disabling respiratory disease and various types of cancers. Six common disorders associated with exposure to airborne asbestos include: lung cancer, asbestosis, mesothelioma, pleural effusion, pleural plaque, and pleural thickening. These diseases are progressive and the latency period of each varies.

Lung Cancer. Inhalation of asbestos fibers has been directly linked to the development of all types of lung cancer, a malignancy of the covering of the bronchial tubes. As with most asbestos-related diseases, lung cancer follows a dose-response relationship. Risk to the individual increases as the length of exposure time increases and as levels of exposure increase. A latency period of 20 to 30 years can be expected between initial exposure to airborne asbestos fibers and declaration of the disease. There is a synergistic effect between smoking tobacco and exposure to airborne asbestos fibers. Exposure to industrial concentrations of asbestos fibers increases the likelihood of lung cancer approximately five times. Smoking alone increases the risk of cancer ten times. Individuals who both smoke and are exposed to asbestos increase the likelihood of contracting cancer by as much as 80 to 100 times.

Asbestosis. *Asbestosis*, also known as *white lung*, is the scarring of lung tissue that results from an accumulation of asbestos fibers in the lung. Asbestosis is the most common asbestos-related disease. The symptoms of asbestosis are similar to those of emphysema. All forms of asbestos minerals have demonstrated the ability to cause asbestosis. The latency period for asbestosis ranges from 15 to 30 years and follows a dose-response relationship. In most cases, individuals will not develop asbestosis unless exposed to high concentrations of airborne fibers for an extended period of time (years). Risk of contracting asbestosis can be reduced by lowering either exposure concentration or exposure time, and therefore represents one of the most preventable of the asbestos-related diseases.

Mesothelioma. *Mesothelioma* is a malignancy of the lining of the chest or abdominal cavity. Pleural mesothelioma is a malignant growth of the pleura, the exterior lining of the lungs. Peritoneal mesothelioma is a malignancy of the peritoneum of the abdominal cavity. Either form of the disease spreads quickly and is always fatal. Pleural mesothelioma gained public notoriety when Steve McQueen was diagnosed and later died of it. The disease has a latency period of greater than thirty years and represents a severe management problem because it does not follow a dose-response relationship. Lack of a dose-response relationship between exposure and incidence of this disease implies there is no known safe level of exposure.

Pleural Effusion and Thickening. *Pleural effusion* is a collection of fluid around the lung and is the most common effect of inhalation of asbestos dust. It is probably the only effect which occurs during the first ten years of exposure. Diffuse *pleural thickening* is often associated with the occurrence of pleural effusion. It is a thickening of the visceral (lung) and/or parietal (chest wall) pleura. The thickening can vary from 0.5 to 2 cm in thickness and results in increased difficulty in breathing.

Pleural Plaque. *Pleural plaque* is a thickening of tissue under the parietal pleura, which can become calcified. An associated latency period of 20 years exists and the condition is typically asymptomatic. Pleural plaques are the most important x-ray abnormalities found as a result of asbestos exposure and are used as diagnostic signposts.

The best noninvasive method for monitoring exposure is by radiographic examination. Pleural plaques are more likely to be identified using x-rays. Only fifteen percent of the pleural plaques found in asbestos workers during autopsy were capable of being diagnosed by preautopsy x-ray examination. This underscores the fact that in addition to being a known human carcinogen, asbestos exposure leads to physiological changes in a majority of individuals who are significantly exposed.

Asbestos-related diseases are well known for their long latency periods which vary from 10 to 40 years. All asbestos-related diseases are progressive; therefore, treatment is intended to retard the rate at which a patient's condition degrades. It is important to note that most of the asbestos-related diseases are asymptomaticis; that is, the patient never demonstrates physical symptoms. This fact coupled with the long latency period cause many workers to become cynical to the danger of exposure. Although not an immediate health risk, asbestos does cause adverse health effects. The best weapon is prevention; therefore, worker education is essential to minimize unwarranted exposure.

The Management Process

The questions of what asbestos is, how it is used, and where it can be found have been addressed. The wide use of asbestos-containing materials (ACM) and their associated health effects explain why it is so highly regulated. So, how does one ensure the health and safety of people, protect the environment, and manage asbestos-containing material? The first step is to identify the presence of asbestos and assess the risks.

One might expect that a person with experience and knowledge about where ACM is located and how it is used could identify materials that contain asbestos. A building owner might turn to the maintenance man for help. Beware! Both the building owner and the maintenance man incur liability if errors are made. The only way to be certain that a suspect material contains asbestos is by laboratory analysis. So, think again when you undertake to make your facility a safe workplace, for by law

- Only accredited inspectors may identify and assess asbestos-containing materials

- Only certain laboratories may analyze samples of suspect materials

- Only certain methods of analysis can verify if the material contains asbestos in sufficient quantity to meet the legal definition of ACM

- Only certain response actions are available to the building owner

- Only certain persons may design the appropriate response action

- Only certain contractors may perform the response action

- Only certain supervisors may oversee the management of the response action

- Only certain workers may perform the work

- Only clearance air monitoring can determine if the job is done

The total process of managing asbestos, whether maintaining it in place or removing it, can best be found in the Asbestos Hazard Emergency Response Act (AHERA) regulations, Title II of the Toxic Substances Control Act (TSCA). AHERA, as revised, continues to represent the state of the art in the asbestos abatement industry. The Model Accreditation Plan found in Appendix C of 40 CFR 763 defines the qualifications of the various accredited individuals who are required by law to perform the many tasks which are a part of any asbestos-related project. The first rule: *No one does anything unless accredited.*

The process begins with identification of ACM and presumed asbestos-containing material (PACM) as required by the Occupational Safety and Health Administration (OSHA). This is done by an AHERA-accredited inspector who makes a physical assessment to determine the quantity, the location, and condition of ACM.

A statistically random and scientifically valid sampling scheme must be developed. The sampling scheme describes where and how many samples must be taken from each homogeneous area. The inspector must physically touch each homogeneous area to determine *friability*. The inspector collects the appropriate number of bulk samples. The entire process requires specific documentation, *i.e.*, sample locations on schematic diagrams, and a chain-of-custody form. The samples are properly packaged and shipped to an accredited laboratory for analysis. EPA requires bulk sample analysis by Polarized Light Microscopy (PLM) methods. The inspector signs and dates a written inspection report certifying to the building owner that the entire process was conducted pursuant to the regulations. The inspection report includes all supporting documentation developed during the inspection process, and the analytical report which is signed and dated by the technician performing the analysis. This document becomes part of the management plan.

Note: The building owner may avoid inspection by assuming that all suspect materials are asbestos-containing and treat them accordingly. However, this management strategy extends potential liability and places a real financial burden on the owner; i.e., quantifying and labeling materials, purchasing and maintaining special equipment, training and monitoring of personnel, and recordkeeping.

Based on this inspection report, an AHERA-accredited management planner makes a hazard assessment. The hazard assessment is based on the type and condition of the ACM and the potential for disturbance from noise, vibration, or contact. Using the hazard assessment the management planner outlines the appropriate response actions to be considered by the building owner. The management plan includes an operations and maintenance program to manage asbestos-containing material remaining in the facility. The plan becomes a *living document*; it remains with the facility being updated as periodic surveillance is performed, the 3-year reinspections are conducted, emergency response actions occur, and abatement activities are completed. The management plan is transferred to new owners.

Note: With the exception of the accredited management planner, the Asbestos School Hazard

Abatement Reauthorization Act extended AHERA accreditation requirements to public and commercial building owners who undertake abatement activities. However, OSHA requires building owners to have a written plan to manage the ACM/PACM at their facilities. With so many regulations impacting the decision-making process, it would be prudent to find experienced professionals to assist in the development of the written plan.

In any building, it is the building owner who selects the response actions. Response actions include: enclosure, encapsulation, repair and maintenance, or removal. Removal is required only before demolition or renovation pursuant to the National Emissions Standards for Hazardous Air Pollutants (NESHAP). The building owner must choose appropriate response actions which *protect human health and the environment.* The building owner may choose that action which is *the least burdensome method* (40 CFR 763.90).

Once the response action has been chosen, an AHERA-accredited project designer prepares the design to accomplish the appropriate response.

An accredited contractor develops an abatement plan. The Asbestos NESHAP requires that a 10-day advance written notification be given to the agency with jurisdiction before abatement work begins. If the project does not start on that date, a new 10-day notice is given. AHERA-accredited supervisors who are *OSHA-competent* manage the abatement project and AHERA-accredited workers perform the abatement work. All OSHA health and safety standards are applicable. Each project is unique; *methods of compliance* are specified in the regulations.

Pursuant to AHERA regulations (40 CFR 763.90 (i)(1–6) response actions require clearance by air monitoring. Clearance testing methods are determined by the size and type project. Transmission Electron Microscopy (TEM) clearance test methods are required in schools for projects greater than 260 linear or 160 square feet. Phase Contrast Microscopy (PCM) clearance test methods are required in most public and commercial building projects.

All waste must be properly packaged and labeled. It must be shipped with a Waste Shipment Record (WSR) or waste generator's manifest. DOT regulations apply. Waste is stored at an EPA-approved landfill. Any discrepancies in the waste shipment must be reported to EPA. The landfill covers the waste with six inches of dirt within 24 hours. In keeping with the *cradle-to-grave* concept, waste continues to be the property of the building owner.

Note: (1) The building owner should be aware that a conflict of interest exists between the air monitoring function and the abatement contractor. The building owner should consider a third party to verify clearance of the response action. (2) Additional State and local regulations may apply. (3) Most states require asbestos professionals to be licensed in addition to maintaining accreditation training.

Regulatory Overview

The Federal, State, and local regulations that address the manner in which asbestos can be used or handled are many and complex. There are numerous references and differing definitions that open the door to various interpretations. In addition, specific regulations often target definite businesses, as is the case with schools, both public and private. Business and industry were caught off-guard by not having monitored the promulgation of regulations under the Asbestos School Hazard Abatement Reauthorization Act which successfully extended AHERA training requirements to public and commercial buildings.

The people impacted by the wide range of regulations are far-reaching. Building owners and occupants, engineers, construction contractors, as well as other service crews like custodial and maintenance workers, electricians, cable and telephone service personnel, and plumbers are all touched by asbestos regulations. The list goes on to include lawyers, insurance professionals, analytical laboratories, training providers, health care providers, and health and safety equipment suppliers.

The Federal agencies that impact the use of asbestos include the Environmental Protection Agency, the Occupational Health and Safety Administration, the Department of Transportation and the Consumer Product Safety Commission (CPSC). Regulated activities include mining and milling of asbestos ores, manufacturing of asbestos-containing products, construction, removal of ACM, disturbance of ACM while performing custodial and maintenance activities, repair of automotive brakes and clutches, as well as transportation and disposal of asbestos-containing waste. Anyone who *may* come into contact with ACM is afforded protection by law.

The following information is a broad overview of current Federal regulations. A chronology of the regulations appears in Table 1, and a cross-reference guide to asbestos regulations is shown in Table 2.

The regulations can be explained by examining them according to the agency with oversight. Three Federal agencies have primary regulatory jurisdiction over asbestos: the Department of Transportation (DOT), the Occupational Safety and Health Administration (OSHA) and the the Environmental Protection Agency (EPA). DOT regulations represent a good starting point since they are the most brief.

Department of Transportation

The Hazardous Materials Transportation Uniform Safety Act (HMTUSA) falls under the auspices of DOT. This act specifies packaging, marking, labeling and placarding of materials shipped by rail, aircraft, vessel, and public highway. Under the HMUSA, asbestos is categorized as Class 9, miscellaneous. If the material is only moving domestically, it may be billed as: Asbestos, 9, UN2212, III, RQ.

Note: The RQ is only required if > 1 lb of pure, friable asbestos is being shipped.

If the type of asbestos is known, then the following proper shipping names may be used:

* Blue Asbestos, 9, UN2212, III [RQ optional], if crocidolite

* Brown Asbestos, 9, UN2212, III [RQ optional], if amosit/mysonite

* White Asbestos, 9, UN2212, III [RQ optional], if chrysolite/actinolite/anthophylite/tumolite

One of the latter three names must be used if the shipment is international or includes an air or water leg. Other requirements may also apply if the shipment is going by air or water.

Containers appropriate for shipment of asbestos are found in 49 CFR 173.216 for non-bulk packaging and 49 CFR 173.240 for bulk packaging. Containers can be rigid or nonrigid but must be air-tight. Asbestos or ACM which are fixed or immersed in a natural or artificial binder (cement, plastic, asphalt, *etc.*) are exempt (removed) from the DOT regulations.

Note: See relevant EPA and OSHA regulations. In some circumstances, there may be a conflict between the DOT, EPA, and OSHA requirements.

Occupational Safety and Health Administration

The first OSHA asbestos standard appeared in 1971. Over the years it has been revised and updated many times. A totally revised Asbestos Standard (in effect, three standards) was published August 10, 1994. The Asbestos Standards focus on protecting workers from harmful exposure to asbestos fibers. A brief description follows.

General Industry. In the past, the General Industry Standard (29 CFR 1910.1001) had been considered to be the *generic* asbestos standard. It covers all activities (except agriculture) that are not covered by the construction or shipyard standards. The two largest segments covered by 29 CFR 1910.1001 are *brake and clutch repair* and custodial workers who are not involved with construction activities. General industry maintenance activities and custodial personnel who may, in the performance of their duties, *disturb* ACM and be exposed to asbestos fibers, have been specifically assigned to and included in provisions of the construction and shipyard standards.

Construction Industry. The Construction Industry Standard at 29 CFR 1926.1101 covers the activities that may involve asbestos such as demolition, removal, alteration, repair, maintenance, installation, cleanup, emergency response to spills

Table 1. Chronology of Asbestos Regulations

Year	EPA	OSHA	Other
1971	Asbestos listed as a hazardous air pollutant	PEL set at 5 f/cc	
1973	Spray application of friable ACM prohibited under NESHAP		
1973	Standard for milling, manufacturing, and building demolition under NESHAP		
1975	NESHAP extended to waste collection and processing industries not previously covered		
1976		PEL lowered to 2 f/cc	
1977			Consumer Product Safety Commission prohibits asbestos in patching compounds
1978	All friable spray on ACM prohibited—demolition and renovation covered by NESHAP		
1979	Technical assistance program to schools to identify and control friable ACM.		DOT regulates ACM
1982	Reporting required for production of products containing asbestos (TSCA)		
1982	Asbestos in School Rule–identification and notification of friable ACM (TSCA)		
1984	Asbestos School Hazard Abatement Act–loan and grant program to help schools eliminate hazard		
1987	Worker Protection Rule–extended OSHA to government employees		
1987	Asbestos Hazard Emergency Response Act–established model accreditation program, extended responsibilities for schools to identify, notify, and respond, added criminal and civil liabilities (TSCA)	General Industry Standard Revised–29 CFR 1910.1001	
1989	Ban and Phase Out Rule	Construction Standard Implemented 29 CFR 1926.58	
1990	NESHAP Revised		
1991	NESHAP Revised –Training Requirements		
1992	Asbestos School Hazard Abatement Reauthorization Act (ASHARA)–extended AHERA training requirements to public and commercial buildings		
1994	ASHARA Revised Model Accreditation Plan	Asbestos Standards Revised –Construction changed from 29 CFR 1926.58 to 29 CFR 1926.1101, Maritime added 29 CFR 1915.1001, General Industry Amended 29 CFR 1910.1001	
1995		Asbestos Standards Compliance Dates Extended 2/21 Revised–6/28, Corrected–9/29	
1998		Respiratory Protection Revised 29 CFR 1910.134	

Table 2. Cross-Reference Guide to Asbestos Regulations

Agency	Reference	Standard
EPA	40 CFR 61, Subpart M	Asbestos NESHAP
	40 CFR 763, Subpart E	Asbestos in Schools (AHERA/ASHARA) Appendix C Model Accreditation Plan
	40 CFR 763, Subpart F	Friable Asbestos-Containing Materials in Schools
	40 CFR 427	Effluent Standards (Asbestos Manufacturing)
	40 CFR 763, Subpart G	EPA Worker Protection Rule
	40 CFR 763, Subpart I	Asbestos Ban and Phase Out Rule
OSHA	29 CFR 1910.1001	General Industry Asbestos Standard
	29 CFR 1915.1001	Shipyard Asbestos Standard
	29 CFR 1926.1101	Construction Industry Asbestos Standard
	29 CFR 1926	General Health and Safety
	29 CFR 1926.59	Hazard Communication Standard
	29 CFR 1910.134	Respiratory Protection Standard
MSHA	30 CFR 57, Subpart D	Surface Mining Asbestos Standard
	30 CFR 57, Subpart D	Underground Mining Asbestos Standard
DOT	49 CFR 171 and 172	Hazardous Materials Transportation Act
CPSC	16 CFR 1304	Consumer Products Bans

containing asbestos, transportation, storage, and disposal of contaminated debris. Previously identified as 29 CFR 1926.58, this standard was redesignated as 29 CFR 1926.1101 to reflect the reorganization of health standards covering construction made on June 30, 1993 (58 FR 35076).

Shipyards. OSHA received many comments complaining that placing shipyards under the construction standard caused confusion because it was never clear when 29 CFR 1910 applied or when 29 CFR 1926 was correct. Therefore, OSHA produced a *vertical* standard in 29 CFR 1915.1001 for shipyards that is different in some ways but, overall, is meant to be neither more nor less rigorous than the general industry or construction standards.

The 1994 regulatory overhaul was in response to issues raised by the DC Circuit Court of Appeals when it remanded the 1986 Asbestos Standard back to OSHA for reconsideration. The issues raised by the court were: (1) establishment of operation-specific exposure limits, (2) small scale, short duration definition, and (3) extension of

reporting and information-transfer requirements. These *new* rules were challenged in the courts and subsequently stayed and amended. At issue were management and work practices involving asbestos-containing roofing and flooring materials.

The final amended rule resulted in many significant changes—some that were not at issue in the initial court's remand. These changes included but are not limited to: a change in permissible exposure levels; a change in the definition of the term *asbestos* to include *presumed asbestos-containing material* (PACM); an expanded responsibility and liability of building owners; and the addition of a classification system for work activities (Class I, II, III and IV) involving the potential contact, possible disturbance, and removal of ACM/PACM.

Exposure Levels. The permissible exposure limit (PEL) for airborne asbestos fibers was lowered to 0.1 fibers per cubic centimeter (cc) of air over a time-weighted average of 8 hours. The action level was eliminated. The short-term exposure of workers to airborne asbestos fibers or excursion limit (EL) is 1.0 fiber/cc on a 30-minute time-weighted average.

Asbestos Defined.

Asbestos includes chrysotile, amosite, crocidolite, tremolite asbestos, anthophyllite asbestos, actinolite asbestos, and any of these minerals that has been chemically treated and/or altered. For purposes of this standard, *asbestos* includes PACM, as defined below.

PACM means *Presumed Asbestos-Containing Material*. Presumed Asbestos-Containing Material means thermal insulation and surfacing material found in buildings constructed no later than 1980. The designation of material as *PACM* may be rebutted pursuant to paragraph (k)(4) of this section.

This definition of asbestos, coupled with the economics of treating all thermal system insulation and all surfacing material as ACM, forces property owners to pursue the *rebuttal criteria* as specified in the Standard. Unless a property owner can prove otherwise through (1) documentation obtained from an inspection performed by an Inspector accredited under AHERA with (2) sample analysis by an accredited laboratory, then any surfacing material and any thermal system insulation that is in a facility built not later than 1980 shall be considered to be asbestos.

Note: See relevant EPA regulations. EPA regulations under ASHARA require that inspections, response actions, and design undertaken by commercial and industrial facilities must be performed by persons who are accredited under the Model Accreditation Plan as described by AHERA (40 CFR 763). This is more stringent than OSHA, which allows a Certified Industrial Hygienist (CIH) to take samples.

Owner Responsibility and Liability. Greater regulatory liability is placed upon the building owner. New and expanded requirements with *specific information-conveying and retention duties* are assigned to the building owner. Notification *shall be in writing or shall consist of a personal communication between the owner and the person to whom notification must be given or their authorized representative.* Owner responsibilities include but are not limited to the following:

- Communication about the presence, location, quantity, and condition of all ACM and PACM to employees, to prospective employees, to all employers of employees, to tenants, and to future building owners.

- Communication through signs, labels, training, and site-specific written operations and maintenance plans.

- Confirmation through an *initial exposure assessment* that employees are not exposed to ACM/PACM in the performance of their duties.

- Maintenance of records for the life of the facility. The records will remain with the facility and be transferred to future owners.

- Notification to OSHA if *methods of compliance* involve the use of of work practices other than those in the Standard.

In addition to the changes described above, the Standard specifically addresses *General Contractors* of construction projects. General contractors are held liable over work covered by the standard even though the general contractor may not have specific knowledge of asbestos abatement practices. The contractor *shall ascertain whether the asbestos contractor is in compliance with the standard and shall require such con-tractor to come into compliance with this standard when necessary.*

Classification of Asbestos Work I-IV. OSHA added four classes of activities to the Construction and Shipyard Standards. These classes trigger different provisions in the standard. OSHA deleted *small scale, short duration* from its final standard.

Note: See related EPA regulations.

The Agency, through its four classes of work, distinguishes high-risk from lower-risk operations. Class I activities represent the greatest risk, with decreasing risk potential associated with each successive class. Work that in the 1986 standard was considered *small scale, short duration* falls into Class II and III in the new standard. The Construction and Shipyard Standards regulate Classes I–III, and all three standards regulate Class IV.

Note: "Small scale, short duration projects" and "major" and "minor" fiber release episodes are defined by EPA. See 40 CFR 763.

Classes I–III are intended to cover the kinds of work which, under the 1986 Construction Standard, were designated *asbestos removal, demolition and renovation operations, including small scale, short duration operations such as pipe repair, valve replacement, installing electrical conduits, installing or removing drywall, roofing and other general building maintenance or renovation.*

By establishing Class IV, OSHA rejected the idea that some activities potentially involving asbestos disturbance would result in a *de minimis* risk, and as such should not be regulated.

Class I–means activities involving the removal of Thermal System Insulation (TSI) and surfacing ACM and PACM.

Class II–means activities involving the removal of ACM that is not TSI or surfacing material (miscellaneous materials). Examples are removal of floor or ceiling tiles, siding, roofing, and transite panels.

Class III–means repair and maintenance operations where ACM, including TSI and surfacing materials, are likely to be disturbed. It includes repair and maintenance activities involving intentional disturbance of ACM or PACM. It is limited to ACM/PACM.

Class IV–means maintenance and custodial work, including cleanup, during which employees contact ACM or PACM.

Note: Class I, II, and III work shall be conducted in a " regulated area". See 29 CFR 1926(b) Definition of "Regulated Area" and (e).

Note: Workers who perform tasks in regulated areas are required by EPA through AHERA/ ASHARA to be accredited under the Model Accreditation Program.

The OSHA standards require that regulated work areas be established where airborne concentrations of asbestos fibers might be expected to equal or exceed the PEL. The establishment and supervision of a regulated area is to be under the direction of a **competent person**. The qualifications of a competent person vary depending upon the classification of work being supervised; *i.e.*, the greater the risk of the work being overseen, the more experience and training the supervisor needs. The responsibilities of the competent person are to

- Establish negative pressure enclosures where necessary

- Supervise exposure monitoring

- Designate appropriate personal protective equipment

- Ensure training of workers with respect to proper use of said equipment

- Ensure the establishment and use of hygiene facilities

- Ensure that proper engineering controls are used throughout the project

Respiratory Protection. OSHA significantly overhauled the Respiratory Protection Standard (29 CFR 1910.134) on January 8, 1998. Employers are required to provide respirators to all workers who may be exposed to greater than the PEL of airborne asbestos fibers during their work activities. The employer must ensure that the provided equipment is capable of giving adequate protection to the exposed individuals. Respirator selection for asbestos exposure is to be accomplished in accordance with the OSHA guideline listed in 29 CFR 1926.1101(h). A *site-specific* written respiratory protection program shall be established. Selection of proper respiratory protective equipment is based on area air sampling and personnel monitoring, which is required prior to and throughout the project. Sampling and analysis conducted for this purpose follows the National Institute for Occupational Safety and Health (NIOSH) Method 7400 (PCM). All employees covered by this respiratory protection program are to be notified of the results of personal exposure monitoring either by personal contact or the posting of the test results.

Medical Surveillance. Workers who (1) are exposed to concentrations of asbestos fibers at or above the PEL for 30 calendar days or more in a year, or (2) are required to wear a negative pressure respirator, must be included in a medical surveillance program. The medical examinations include: collection of a medical history; completion of a standardized questionnaire contained in Appendix D of 29 CFR 1910.1001, 1915.1001, and 1926.1101; a chest x-ray; examination of the pulmonary and gastrointestinal system; as well as a pulmonary function test by forced vital capacity or forced expiratory volume. Examinations are to be completed yearly, with the cost being borne by employers.

Other OSHA Standards. In general, OSHA standards are directed at protection of workers and ensuring a safe work environment. All general health and safety standards apply. Due to its carcinogenic nature, OSHA considers asbestos a hazardous chemical and therefore covered under the Hazard Communication Standard (29 CFR 1910.1200). Other OSHA standards which should be reviewed when undertaking action involving asbestos include: Lock Out/Tag Out, Confined Space Entry, Fall Protection, Ventilation, and General Health and Safety.

Environmental Protection Agency

EPA has promulgated a number of rules intended to limit environmental release of asbestos fibers. The jurisdiction of EPA, as it relates to asbestos, involves the enforcement of various legislative acts.

National Emmissions Standard for Hazardous Air Pollutants

The regulations promulgated pursuant to the Clean Air Act Section on the National Emissions Standard for Hazardous Air Pollutants, better known as the *Asbestos NESHAP*, are found in 40 CFR 61, Subpart M. These regulations are some of the most frequently cited EPA asbestos regulations.

The currently enforced asbestos NESHAP was published in final form on November 20, 1990 (55 FR 48406); some significant revisions were made at that time. A number of definitions were either revised or added (see 40 CFR 61.141), including those for adequately wet, nonfriable, regulated ACM, and waste shipment record. In addition, the regulations uniformly established a 10-day written notice prior to the commencement of defined abatement activities.

The asbestos NESHAP addresses: (1) fugitive emissions from work sites, (2) required removal before renovation or demolition, (3) notification requirements before disturbing ACM, and (4) disposal of asbestos-containing waste materials in landfills.

No Visible Emissions. The *no visible emissions* statement for releases from mining, milling, and manufacturing processes is probably the most commonly quoted part of the asbestos NESHAP. This restriction extends to renovation and building demolition projects. It is important to distinguish between the OSHA PEL and the EPA *no visible emissions level*. OSHA establishes exposure standards for inside the work space while EPA establishes limits for nonpoint source emissions to the environment. The EPA no visible emissions limit is not intended to serve as a worker protection standard.

Required Removal. A second requirement of the asbestos NESHAP is that all regulated asbestos-containing materials (RACM) must be removed from a building prior to demolition. This is the only place in Federal regulations in which asbestos removal is required.

Notification. Before any renovation, remodeling, or removal activities involving disturbance of RACM can be undertaken, the State agency responsible for Clean Air Act enforcement must be provided written notification 10 days in advance of beginning work. This advance notification is required in the event that 260 linear feet or 160 square feet of RACM will be affected, or if 35 cubic feet of asbestos-containing waste material will be generated. Should the starting date of the removal project change prior to initiation of work, a revised notification form must be submitted and 10 working days must separate submission of this form from the start of work. Finally, the delegated State agency must be provided with project updates in the event the amount of asbestos affected changes by 20 percent.

Waste Disposal. Asbestos-containing waste materials are hazardous *materials,* not hazardous *waste.* As hazardous materials, asbestos-contaminated waste can be disposed of in EPA-approved landfills. The asbestos NESHAP establishes guidelines for proper disposal of such asbestos wastes. These guidelines include provisions for restricting emissions, adequate wetting of waste materials, use of hazard warning labels, indelible marking of each container with information about the generator, and use of waste shipment records in a fashion analogous to the uniform manifest for hazardous waste disposal. This documentation of waste shipments includes the requirement to submit an *exception report* if written verification of disposal of the asbestos waste is not received from the designated disposal facility within 45 days.

Note: See other relevant EPA regulations and relevant DOT and OSHA regulations.

Toxic Substances Control Act

Asbestos is explicitly addressed in the Toxic Substances Control Act (TSCA) as one of five chemicals specifically regulated under TSCA and found in 40 CFR 763.

Ban and Phase out Rule. The EPA published the Ban and Phase out rule on July 12, 1989, and it became effective August 25, 1990. The purpose of the rule was to *prohibit at staged intervals, the future manufacture, importation, processing, and distribution in commerce of asbestos in almost all products.* The first stages of the rule concentrated on nonfriable forms of ACM such as floor tile, textiles, and asbestos cement products. Implementation was scheduled in three phases. If the restrictive ruling had not been challenged in the courts it eventually would have eliminated 94 percent of all asbestos-containing materials from the United States marketplace by 1997. However, on October 21, 1991, the Federal Appeals Court sent the rule back to EPA and effectively suspended the rule except for those products which were out of production prior to July 12, 1989.

Notifications and reports which must be submitted by miners, primary processors, secondary processors, manufacturers, and importers, are outlined in 40 CFR 763, Subpart D. In certain cases, information must also be submitted to the Consumer Product Safety Commission as well as the EPA.

Worker Protection Rule. When OSHA was first created, its purview was limited to the private sector. Federal agencies were incorporated into the OSHA sphere of authority in 1979 via executive order of the President. Approximately half the states in our nation have statutes placing their State and local government agencies subject to OSHA standards. EPA addressed this small group of OSHA-exempt workers in something commonly referred to as the *Worker Protection Rule* (40 CFR 763, Subpart G).

The purpose of the EPA Worker Protection Rule is to extend the OSHA worker safety guidelines to employees of State and local governments, especially those who perform asbestos abatement work. The regulations became effective March 27, 1987.

AHERA and ASHARA. The Asbestos Hazard Emergency Response Act (AHERA) of 1986 and the Asbestos School Hazard Abatement Reauthorization Act (ASHARA) of 1990 go hand-in-hand. They are Title II of TSCA and found in 40 CFR 763, Subpart E.

AHERA established state-of-the-art practices for the asbestos abatement industry. It included the *Model Accreditation Plan* (MAP). This was the first time in our nation's history that government set forth specific training requirements for the work force. Only EPA-approved training providers may conduct accreditation training programs. AHERA initially applied to schools, public and private (see 40 CFR 763, Appendix C).

Definitions. As with any discussion involving environmental regulations, one must begin with *definitions* (40 CFR 763.83). An **asbestos-containing material** (ACM) is considered to be any material or product containing greater than 1 percent asbestos. An **asbestos-containing building material** (ACBM) is defined to be ACM found in or on interior structural members or other parts of a building. EPA further defines asbestos-containing materials by grouping them into one of three categories: surfacing materials, thermal system insulation, and miscellaneous ACM. Each of these categories can be further described as being either *friable* or *nonfriable*. **Friable ACM** is defined as material which, when dry, may be crumbled, pulverized, or reduced to powder by hand pressure, and includes any damaged nonfriable material. There are also key definitions for removal, enclosure, encapsulation, and repair which define *abatement* activity.

Other important asbestos definitions include:

Surfacing Material—ACM that has been sprayed or troweled on surfaces (walls, ceilings, structural members) for acoustical, decorative, or fireproofing purposes. This includes plaster and fireproofing insulation.

Thermal system insulation—Insulation used to inhibit heat transfer or prevent condensation on pipes, boilers, tanks, ducts, and other components of hot and cold water systems and heating ventilation and air condition (HVAC) systems. This includes pipe lagging and pipe wrap; block, batt, and blanket insulation; cement and *mud*, and other products such as gaskets and ropes.

Miscellaneous materials—Other, mostly nonfriable and friable products and materials such as floor tile, ceiling tile, roofing felt, concrete pipe, outdoor siding, and fabrics.

Model Accreditation Plan—The Model Accreditation Plan (MAP) established five areas of accreditation as the appropriate training requirements for individuals who would: (1) perform inspections, (2) prepare management plans for buildings, (3) supervise abatement projects, (4) serve as workers on abatement projects, or (5) design abatement projects. Accreditation in each of these areas is achieved by attending an EPA-approved course and passing a written examination. To maintain accreditation, individuals must attend annual refresher classes and may be required to pass written examinations. In addition, many states now require asbestos professionals to have a registration or license.

Asbestos School Hazard Abatement Reauthorization Act. This law (Public Law 101–637) was passed on November 20, 1990. The highlights of this act includethe following:

- As of April 4, 1994, the Revised Model Accreditation Plan was implemented on an interim basis, to be followed by full implementation (under ASHARA and the Revised MAP) by October 4, 1994.

- Any person who conducts asbestos work in schools as an Inspector, Management Planer, Worker, Contractor/Supervisor, and/or Project Designer must still be accredited.

- Any person who conducts asbestos work in public or commercial buildings as an Inspector, Worker, Contractor/Supervisor, and/or Project Designer must be accredited under the provisions of ASHARA and the Revised MAP.

- The length of training programs for accreditation as a Worker has increased from 3 days to 4 days, with a minimum of 14 hours of hands-on training.

- The length of training programs for accreditation as a Contractor/Supervisor has increased from 4 days to 5 days, with a minimum of 14 hours of hands-on training.

- The extended training requirements of the Contractor/Supervisor course also satisfy the training requirements for Workers.

- The Contractor/Supervisor course no longer meets the accreditation requirements for Project Designers.

- A person must take the refresher course specific to the area in which he or she seeks to maintain accreditation; e.g., Design Update, Inspector Update.

- There are some grandfather provisions in the Revised MAP:

  - anyone holding valid accreditation in any of the five disciplines as of April 3, 1994, will be accredited (grandfathered) under the Revised MAP.

  - the 12–month grace period for obtaining renewal accreditation is in effect *if* a state allows it.

  - for nonaccredited persons as of April 3, 1994, a person must either take a course which meets the upgraded training requirements noted above, or take a course meeting the old MAP requirements and then take an upgraded course before October 4, 1994. *In effect, if* a person took a course between April 4 and October 4, 1994, that had not been upgraded, that person would need to retake the initial training again before October 4, 1994.

AHERA and ASHARA carry the full enforcement authority of TSCA, with criminal and civil penalties for failure to comply. States were required to adopt the AHERA guidelines or to develop their own that were *at least as stringent as* the Federal guidelines. Because of the far-reaching impact of the expanded regulations under ASHARA, states once again had to take legislative action to adopt guidelines at least as stringent as the new Federal guidelines. These regulations are required reading for anyone who is dealing with asbestos manage-ment or abatement projects.

Inspection and Reinspection. 40 CFR 763.85 requires looking at the inspection procedure and floor plans along with record drawings and specifications of construction projects in order to identify locations of suspected ACBM. Homogeneous areas defined as an area of surfacing material, thermal system insulation material, or miscellaneous material which is uniform in color and texture are then established for the purpose of sampling.

Sampling. 40 CFR 763.86 requires that bulk samples be taken from each homogeneous area in a statistically random fashion, with the number of samples collected based upon the square or linear footage present in the homogeneous area. Each sample must be collected by an accredited inspector. Documentation of the entire process is specific and includes unique numbering system, diagrams of sample location, date, signature, and accreditation number of the inspector, a description of the homogeneous area, and a chain of custody form.

Analysis. 40 CFR 763.87 requires sample analysis using the PLM point count method in a laboratory which has been accredited by the National Institutes of Standards and Technology (NIST). The analytical report must be signed and dated by the individual analyzing the sample. If one sample from a homogeneous area is determined to contain greater than one percent asbestos, the homogeneous area must be classified as ACBM. The results of the analysis must be submitted to the designated asbestos program manager within 30 days of analysis. In lieu of sampling and analyzing suspected ACM, the building owner is allowed to assume the material is ACM and manage it accordingly.

Assessment. 40 CFR 763.88 requires that the final stages of inspecting a building include an assessment of the physical condition of all materials identified or assumed to be ACBM. The inspector indicates the type ACM and the degree to which damage is present (poor, fair, or good). The inspector indicates the potential for disturbance (high, moderate, or low) as probability of human contact, noise, and/or sound vibration. The inspection report submitted to the building owner will include the sampling scheme, sample locations, names of inspectors collecting samples, analysis reports for all samples, classification of homogeneous areas as ACBM or non-ACBM, and physical assessment and description of all homogeneous areas.

Response Actions. According to AHERA (40 CFR 763.90), an accredited management planner will review information contained in the inspection report and conduct a hazard assessment. The hazard assessment serves as an indicator of risk for building occupants. Based upon this hazard assessment, the management planner will recommend a response option for each homogeneous area: removal, or enclosure, encapsulation, repair, and maintenance. In conjunction with these response option recommendations, the management planner will also prescribe an operation and maintenance plan for all ACBM left in the building. The management planner must sign, date, and record their accreditation number on recommendations made to the building owner.

Any response to a *major fiber release* must be completed only by accredited personnel: project designer, project supervisor, and project workers. A ***major fiber release*** is defined to have occurred when *greater than three square or linear feet of ACBM becomes dislodged from its substrate.*

In the event a major fiber release occurs, the facility environmental manager should take immediate action. The area where the release occurred should be isolated by modifying or shutting off the heating, ventilation, and air conditioning (HVAC) systems. Entry to the affected area should be restricted and signs posted warning of the hazard. An air sampling firm should be contacted to have nonisolated areas tested for airborne asbestos fibers. Finally, an accredited project designer should be contacted to coordinate the response action.

Operations and Maintenance. 40 CFR 763.91 contains specific work practices for operations and maintenance activities: worker protection, initial cleaning, major and minor fiber-release episodes. Maintenance activities involving the disturbance of quantities of ACM greater than three linear feet or three square feet must be designed and conducted by persons accredited under AHERA. More specific work procedures are identified in the regulations.

Training and Periodic Surveillance. AHERA also establishes training requirements for those individuals who carry out the operations and maintenance plan (40 CFR 763.92). Any custodial or maintenance personnel working in a building containing ACBM must be given two hours of awareness training, whether they work with ACBM or not. Those individuals who may *disturb* ACBM during their duties must receive an additional 14 hours of training. Employee training must be site-specific, and include locations of ACBM in each building. Operations and maintenance personnel routinely document periodic surveillance every six months to ensure that no change in condition or use of the ACM has occurred.

Finally, AHERA does not supersede those requirements established by OSHA standards, the Asbestos NESHAP, or the EPA Worker Protection Rule. EPA has begun integrating concepts of AHERA into other asbestos regulations, as is reflected by the updated NESHAP. AHERA has become the *state of the art* in asbestos management and abatement.

Summary

Asbestos is a naturally occurring mineral that has been widely used. It is a known carcinogen that also causes disabling respiratory disease. There is no known safe level of exposure, because all of the illnesses are not dose-response related. There are many regulations that impact the manner with which asbestos is used and encountered in the workplace. The regulations are complex and intertwined; they are cross-referenced one to the other. There are differing definitions that open the door to various interpretations. The regulations often are targeted at specific businesses, as is the case with schools. All public and commercial buildings are regulated. Only accredited individuals may *disturb* ACM. Waste generated that contains asbestos is believed to be the property of the building owner. Property owners and employers should read the regulations and seek the assistance of asbestos professionals.

Asbestos Laws and Regulations

A list of useful asbestos laws and regulations is provided below.

Clean Air Act (CAA): 42 USC §§7401, 7412, 7414, 7416, and 7601.

Asbestos listed as a hazardous air pollutant: CAA, 42 USC §7412(b)(1).

Toxic Substances Control Act (TSCA), Title II: 15 USC §2601.

National Emissions Standards for Hazardous Air Pollutants (Asbestos NESHAP): 40 CFR 61.140–61.157 (Subpart M); *see also* Asbestos NESHAP Clarification, 58 *Federal Register* (FR) 51784 (10/5/93); 59 FR 542 (1/5/94); 60 FR 38725 (7128195); and 60 FR 65243 (12/19/95).

Asbestos Hazard Emergency Response Act of 1986 (AHERA): 15 USC §§2641–2656 (dealing with asbestos in schools), and amended by The Asbestos School Hazard Abatement Reauthorization Act (ASHARA), 15 USC §§2641–2654; 40 CFR 763, Subpart E.

OSHA General Industry Standard: 29 CFR 1910.1001, 1910.1200.

OSHA Construction Industry Standard: 29 CFR 1926.1101.

OSHA Shipyard Employment Standard: 29 CFR 1915.1001.

State and Local Employees Worker Protection Rule: 40 CFR 763.

Effluent standards for asbestos manufacturing source categories: 40 CFR 427.

Safe Drinking Water Act: 42 USC §300(f) (municipal annual water testing for asbestos).

Asbestos labeling: 29 CFR 1910.1001(g)(2)(ii) and 40 CFR 61.149.

Hazardous Materials Transportation Uniform Safety Act of 1990: 49 USC §§5101–5127, Asbestos transportation: 49 CFR 171–172.

Asbestos manufacture, importation, processing, and distribution prohibitions, and labeling requirements—Asbestos mining activities: 30 CFR 56, Subpart D; 30 CFR 57, Subpart D.

OSHA and EPA Guidance Documents

Environmental Protection Agency. *Asbestos: Waste Management Guidance,* EPA 530SW85007. Washington, DC: EPA.

Environmental Protection Agency. *A Guide to Performing Re-Inspections under the Asbestos Hazard Emergency Resource Act (AHERA).* ("Yellow Book") EPA 700/B–92/001. Washington, DC: EPA, Office of Pollution Prevention and Toxic Substances, February 1992.

Environmental Protection Agency. *A Guide to Normal Demolition Practices under the Asbestos NESHAP.* EPA 340/1–92–013. Washington, DC: EPA, Office of Air Quality Planning and Standards, September 1992.

Environmental Protection Agency. *A Guide to Respiratory Protection for the Asbestos Abatement Industry.* ("White Book") EPA 560–OPTS–86–001. Washington, DC: EPA, 1986.

Environmental Protection Agency. *Abatement of Asbestos-Containing Pipe Insulation, Asbestos-in-Buildings,* Technical Bulletin 1986–2. Washington, DC: EPA, 1986.

Environmental Protection Agency. *Advisory to the Public: On Asbestos in Buildings.* EPA 745K93014. Washington, DC: EPA.

Environmental Protection Agency. *EPA Guidance for Service and Maintenance Personnel.* EPA 560/5–85–018. Washington, DC: EPA, 1985.

Environmental Protection Agency. *Guidelines for Asbestos NESHAP Demolition and Renovation Inspection.* EPA 340190007. Washington, DC: EPA.

Environmental Protection Agency. *Guidance for Controlling Asbestos-Containing Materials in Buildings.* ("Purple Book") EPA 560/5–85–024. Washington, DC: EPA, 1985.

Environmental Protection Agency. *EPA Model Curriculum for Training Asbestos Abatement Contractors and Supervisors.* CX–820760–01–0. Atlanta, GA: Safety, Health, and Ergonomics Branch, Electro-Optics, Environment, and Materials Laboratory, Georgia Tech Research Institute, October 1995.

Environmental Protection Agency. *EPA Model Training Course Materials for Accrediting Asbestos Building Inspectors and Management Planners in Accordance with AHERA.* Washington, DC: EPA, 1995.

Environmental Protection Agency, *EPA Model Curriculum for Training Asbestos Abatement Project Designers Instructor's Manual,* CX–816386–0. Marietta, GA: Georgia Environmental Institute, 1995.

Environmental Protection Agency. *Fact Sheet: Asbestos (HTML),* EPA 745F93007. Washington, DC: EPA.

Environmental Protection Agency. *Managing Asbestos in Place.* ("Green Book") EPA 20T–2003. Washington, DC: EPA, 1990.

Environmental Protection Agency. *Managing Asbestos In-Place: A Building Owner's Guide to Operations and Maintenance for Asbestos-Containing Material.* EPA 745K93013. Washington, DC: EPA.

Environmental Protection Agency. *Simplified Sampling Scheme for Surfacing Materials.* ("Pink Book") EPA 560/5–85–030a. Washington, DC: EPA, 1986.

Occupational Safety and Health Administration. *Asbestos Standard for Construction Industry.* OSHA 3096. Washington, DC: OSHA, 1995.

Occupational Safety and Health Administration. *Asbestos Standard for Shipyards.* OSHA 3145. Washington, DC: OSHA.

Occupational Safety and Health Administration. *Chemical Hazard Communication.* OSHA 3084. Washington, DC: OSHA, 1998.

Occupational Safety and Health Administration. *Personal Protective Equipment.* OSHA 3077. Washington, DC: OSHA.

Occupational Safety and Health Administration. *Respiratory Protection.* OSHA 3079. Washington, DC: OSHA, 1998.

Dr. Margaret V. Naugle is President of Environmental Training Corporation of Birmingham, Alabama and teaches graduate courses in environmental management at Birmingham Southern College. She has been a consultant to business and industry in the area of environmental training and regulatory compliance for over 20 years. Her experience includes planning, designing and implementing the State of Mississippi's Model Accreditation Plan under the Asbestos Hazard Emergency Response Act; serving on the State of Misissippi Joint Legislative Committee for Environmental Matters; giving presentations at local, state, and national conferences; and publishing articles and training materials about various hazardous materials. Dr. Naugle currently serves on the National Advisory Board of Business and Legal Reports. She is Past-President of the Academy of Certified Hazardous Materials Managers and the Alabama Society of Hazardous Materials Managers. The Academy recognized Dr. Naugle for her leadership as President with their first honorary lifetime membership. Dr. Naugle is active in the National Environmental Training Association and the American Industrial Hygiene Association.

Dr. Charles A. Waggoner currently serves as the Manager of Safety, Excellence, and Environment for the Diagnostic Instrumentation and Analysis Laboratory at Mississippi State University (MSU). He holds a BS and MS in biochemistry and a PhD in physical chemistry. He has over 15 years experience in environmental management, with particular emphasis on hazardous waste management and related issues. Dr. Waggoner's professional activities have included serving as the MSU Hazardous Waste Officer, Technical Director of Environmental Training for the MSU Division of Continuing Education, and Dean of Environmental Science and Technology at Chattanooga State Technical Community College. Dr. Waggoner has served as a member of the IHMM-ACHMM Advisory Committee, ACHMM Board of Directors, and as General Chairperson for the 1994 National Conference in Chattanooga, Tennessee. He has authored numerous articles, including "Overview of Major Federal Environmental Acts and Regulations for the General Practioner," and the asbestos chapter in the Handbook on Hazardous Materials Management, 5th *edition, published by IHMM in 1995.*

Part VI

Air Quality

CHAPTER 29

The Clean Air Act

Adriane P. Borgias, MSEM, CHMM

The purpose of this chapter is to provide the hazardous materials manager with an overview of the Clean Air Act. The Clean Air Act governs a broad range of activities from manufacturing and processing, to transportation and management of hazardous chemicals. This overview covers the aspects of the Clean Air Act that will be of particular use to the hazardous materials manager. These include: important terms and concepts used in the Act, a brief summary of the provisions of each Title, and a brief guide to the implementing regulations. The more subtle nuances of the Clean Air Act, such as special requirements relating to research programs, energy emergencies, Federal facilities, sewage treatment, and other special-interest measures, will not be covered in depth this chapter. For more information on these topics, the reader is urged to refer to the Act and implementing regulations. Reference tables to the Act and implementing regulations can be found at the end of the chapter.

A Brief History of the Clean Air Act

The Clean Air Act was originally promulgated in 1955. Prior to 1970, only minor changes were made to the original law. In 1970, the basic principles that form the backbone of the Act as it exists today were developed. The 1970 Clean Air Act was written because of the air pollution resulting from increasing urban growth and industrialization. Major amendments were again made to the Act in 1990. The most significant aspects of these amendments were the inclusion of an operating permit program (a program *intended* to simplify the complex permitting requirements that were currently in place) and a major revision of the way in which hazardous air pollutants (HAPs) are regulated. The new amendments gave EPA specific direction and a schedule for implementation of HAP regulations.

Air quality in the United States has improved significantly since the 1970 version of the Clean Air Act. According to the Environmental Protection Agency (EPA), the ambient concentrations of the six major air pollutants have dropped by about a third in the past 25 years. The 1990 amendments and recent regulations focus on continued improvement in air quality. In particular, this includes the control of hazardous air pollutants, ozone, particulates, and visibility.

The Clean Air Act, As Amended in 1990

Today's Clean Air Act, as amended in 1990, consists of 11 Titles that regulate a wide variety of activities associated with the control of air pollutants. The first six Titles are of the most importance to the hazardous materials manager. Only minor changes were made to the first three Titles. The next three Titles were either new to the Clean Air Act or were existing laws which were strengthened and folded into the Act.

The Act was written because of our growing reliance on industries that produce a complex set of pollutants, particularly in urban areas. These pollutants threaten not only our health but our environment and economic infrastructure as well (agriculture, property, air, and ground transportation). Consequently, the primary stated goal of the Clean Air Act is to promote pollution prevention.

The intent of the Act is to *protect and enhance the quality of the nation's air resources*. This goal is to be accomplished through a national air pollution prevention and control research program, technical and financial assistance to State and local governments, and the development of regional programs where needed. This important goal is often lost in the day-to-day worries concerning compliance with the complex requirments of the Act.

Of the 11 Titles, the first six are summarized below.

Title I: Air Pollution Prevention and Control. Title I focuses on the control of pollution in urban areas. This Title describes air quality criteria, control regions, and implementation plans for the improvement and maintenance of healthy air. A significant amendment to this Title is the direction provided to EPA to set emissions standards for what is now 188 Hazardous Air Pollutants by the year 2000.

Title II: Emissions Standards for Moving Sources. Title II regulates motor vehicle and aircraft emissions. The 1990 amendments refined the requirements of this Title in response to the substantial increase over the past 20 years in pollutants from these sources. The regulation of fuels, including the clean fuels program, fleets, and clean-fuel vehicles is included here.

Title III: General. Title III contains general provisions and procedures relating to the act. Included in this Title are sections on citizen suits, employment effects, and employee (whistleblower) protection.

Title IV: Acid Deposition Control. This Title regulates the acid emissions from electric generation sources. The control of acid emissions from power plants was first legislated in 1980 as the Acid Precipitation Act. Actual controls of acid rain emissions, however, were delayed until 1990 when expanded requirements were folded in. These new requirements consist of a program that provides market-based incentives for sulfur dioxide and nitrogen oxides reduction.

Title V: Permits. This Title, added with the 1990 amendments, requires operators of major sources of air pollution to obtain operating permits prior to construction and operation. This title outlines the basic requirements of the operating permit program.

The Title V program has a number of significant impacts. First, it is intended to strengthen the ability of the urban areas to come into compliance with the air quality standards by including previously excluded major sources in the permitting arena. Second, the program is intended to *standardize* the permitting process by implementing a nationwide program similar to the existing National Pollutant Discharge Elimination System (NPDES) under the Clean Water Act. Third, it creates a self-funded permit program by assessing the emitting sources fees on a per-ton-of emissions basis.

Title VI: Stratospheric Ozone Protection. Originally included in Part B of Title I of the Act, the 1990 amendments created a new Title VI that regulates the use and phaseout of ozone-depleting substances. It provides for proper handling, recycling, and disposal of these substances as well as a mechanism for introducing ozone-safe alternatives.

The new Title VI strengthens the requirements of the original law by creating a structured program to identify, evaluate, manage, and phase out these substances. National recycling and emissions reduction programs were established. Economic incentives are provided via an allowance trading program.

Titles VII–XI. The remaining five Titles of the Act are not covered in detail in this chapter. Their topics include

- *Title VII.* Provisions Relating to Enforcement
- *Title VIII.* Miscellaneous Provisions
- *Title IX.* Clean Air Research
- *Title X.* Disadvantaged Business Concerns
- *Title XI.* Clean Air Employment Transition Assistance

Important Concepts and Terms Used throughout the Act

There are several key concepts and terms used throughout the Act that are important to know. Most of these concepts are introduced in Title I

and are used in the other Titles as well. These concepts are described below.

Responsibility for Air Pollution Prevention and Control

Congress determined that the states and local governments are primarily responsible for air pollution prevention and control. As a result, EPA was given the authority to delegate to the states many provisions of the Act. The most significant of these are the development of State plans for achieving and maintaining healthy air; the issuance of construction and operating permits; as well as motor vehicle inspection and maintenance programs.

Congress also believed that Federal leadership and financial assistance is crucial to the success of the clean air program. Cooperative State and local programs are encouraged through research and development programs, training grants, program development funds, and, when approved by Congress, the partial funding of interstate commissions.

Not all provisions of the Act can be delegated to the states. Programs such as the marketing and trading of emissions allowances under the acid deposition program, new motor vehicle emissions standards, and stratospheric ozone protection have EPA oversight.

National Ambient Air Quality Standards

A key component of the Act is the development and use of the National Ambient Air Quality Standards (NAAQS). NAAQS are established for air pollutants that threaten public health and welfare, whether from diverse mobile or stationary sources. The criteria for setting NAAQS include variable factors such as atmospheric conditions, types of pollutants present and their interactions, as well as the potential adverse effects on public health and welfare.

Primary NAAQS are developed for a particular pollutant to protect public health with an *adequate* margin of safety. Sensitive populations, such as children and the elderly, are used as the basis for calculation of the safety margin. Secondary NAAQS are intended to protect public welfare. This includes the effects of air pollution on vegetation, materials, structures, and visibility.

The EPA develops the NAAQS using all available peer-reviewed scientific data. The standards are reviewed extensively by the scientific community, public interest groups, and the general public. The group which has the most credibility in the review process is the Clean Air Scientific Advisory Committee (CASAC), a Congressionally mandated group of independent scientists and technical experts. The standards require that the NAAQS be reviewed/revised once every five years.

Note that the cost-benefit of controlling a criteria pollutant is not one of the factors considered when a NAAQS is promulgated. However, simultaneous to issuing a NAAQS, EPA is required to issue data relating to the cost of controlling the pollutant, the benefits and environmental impact of control, as well as alternative methods for its control (such as alternative fuels, processes, and operating methods).

To date, NAAQS have been issued for six pollutants: ozone, carbon monoxide (CO), particulate matter (PM_{10} and $PM_{2.5}$), lead, nitrogen dioxide (generally measured as nitrogen oxides or NO_x), and sulfur oxides (SO_2). These pollutants are commonly known as *criteria pollutants* and their standards are summarized in Table 1.

These criteria pollutants are generally associated with urban areas because of both the number of people impacted by their presence (an estimated 130 million in 1990) and the concentration of sources in these areas. Most large urban areas in the United States and some regions exceed the NAAQS for ozone. CO and PM also exceed the NAAQS in many urban areas. The remaining three pollutants (lead, SO_2, and NO_2) rarely exceed the NAAQS.

Attainment, Nonattainment, and Unclassified Areas

When EPA issues a NAAQS, states are required to monitor and evaluate the quality of the ambient air in their jurisdiction. Areas in which the air quality is cleaner than the NAAQS (*i.e.*, the concentrations of criteria pollutants in the ambient air are below the standard) are considered by EPA to be in *attainment*. Areas in which the air quality is poorer than the NAAQS are considered by EPA to be in *nonattainment* and, depending on the pollutant, are further classified as marginal, moderate, serious, severe, or extreme.

Table I. National Ambient Air Quality Standards for Criteria Pollutants

Criteria Pollutant	Primary Standard	Secondary Standard
Carbon Monoxide	9 ppm [Note 1]	None
	35 ppm [Note 2]	
Lead	1.5 $\mu g/m^3$ [Note 3]	1.5 $\mu g/m^3$ [Note 3]
Nitrogen dioxide (NO_x)	0.053 ppm	0.053 ppm
	(100 $\mu g/m^3$) [Note 4]	(100 $\mu g/m^3$) [Note 4]
Ozone	0.12 ppm	0.12 ppm
	(235 $\mu g/m^3$) [Note 5]	(235 $\mu g/m^3$) [Note 5]
	0.08 ppm [Note 6]	0.08 ppm [Note 6]
PM_{10}	50 $\mu g/m^3$ [Note 4]	50 $\mu g/m^3$ [Note 4]
	150 $\mu g/m^3$ [Note 7]	150 $\mu g/m^3$ [Note 7]
$PM_{2.5}$	15 $\mu g/m^3$ [Note 4]	15 $\mu g/m^3$ [Note 4]
	65 $\mu g/m^3$ [Note 7]	65 $\mu g/m^3$ [Note 7]
Sulfur oxides (SO_2)	0.030 ppm [Note 4]	0.05 ppm [Note 8]
	0.14 ppm [Note 7]	

Notes

1) Maximum 8-hour average; not to be exceeded more than once per year

2) Maximum 1-hour average; not to be exceeded more than once per year

3) Maximum arithmetic mean averaged over a calendar quarter

4) Annual arithmetic mean concentration

5) Maximum 1-hour average; not to be exceeded more than 1 day per calendar year

6) Daily maximum 8-hour average

7) 24-hour average

8) 3-hour average

Some parts of the country, primarily rural areas, are considered to be **unclassified** because there is insufficient available air quality data to make a determination.

These designations are important because they are used throughout the Clean Air Act to prescribe the requirements for states as well as operators of air pollution sources. Note that the concept of attainment is pollutant-specific. An area can be in *attainment* for one criteria pollutant but in *non-attainment* for another; thus the area is in both attainment and nonattainment at the same time.

Class I, II, and III Areas

In order to protect public health and welfare, as well as existing air quality in attainment and unclassified areas, the Act further designates areas on the basis of land use. Attainment areas of special national or regional natural recreation, scenic, or historic value are further designated as Class I, II, or III. The intent of these classifications is threefold:

1)   To ensure economic growth

2)   To prevent significant deterioration (PSD) of the existing air quality

3)   To ensure public participation in the decision-making process (*i.e.*, air permitting)

The class designations affect operators of new or modified major sources wishing to obtain a permit in an attainment area. As discussed later, the PSD procedures require explicit consideration of the impacts a source may have on these areas. The resulting permit limitations are based on the class of the impacted areas.

Class I areas allow for very little deterioration of air quality and include all international parks; national wilderness areas and memorial parks larger than 5000 acres; and national parks exceeding 6000 acres in size. States have the ability to designate similar areas if the size of the area exceeds 10,000 acres. In addition, tribes can designate lands within the boundaries of their reservations as Class I areas.

All other areas in the United States are ***Class II areas*** but can be redesignated as Class I or Class III. Class II areas allow for moderate deterioration of air quality. Some Class II Federal lands can only be redesignated as Class I. These include the following lands if they exceed 10,000 acres: national monuments, primitive areas, recreational areas, wildlife refuges, wild and scenic areas, and newly established national parks. The Act also contains provisions for redesignating Class II areas to ***Class III***, the least restrictive class. No areas in the United States have been reclassified as Class III.

State and Tribal Implementation Plans

The states and tribes create detailed descriptions regarding how they will fulfill their responsibilities towards achieving and maintaining clean air. The ***State Implementation Plan***, also known as a ***SIP***, and the ***Tribal Implementation Plan***, or ***TIP***, are essentially collections of the regulations the state or tribe uses to control pollutants.

The Clean Air Act lays out the requirements for these plans. An implementation plan is a publicly available document that describes in detail the state's or tribe's requirements for implementation, maintenance, and enforcement of the primary and secondary standards. The plan also identifies, or designates, areas of attainment and nonattainment. EPA subsequently promulgates these area designations.

In order to obtain EPA approval, an implementation plan must include certain requirements:

1)   Enforceable emissions limitations and other control measures with schedules for compliance. These control measures can include economic incentives such as fees, marketable permits, and auctions of emissions rights.

2)   A system to monitor, compile, and analyze data on ambient air quality for any pollutant with a NAAQS.

3)   An enforcement program.

4)   Assurances by the state that adequate personnel and resources are available to enforce the plan provisions. Note that funding for the State programs is to come from the air pollution sources.

5)   A contingency plan which describes the specific measures the state or tribe will take in order to assure *reasonable further progress* toward attainment of NAAQS.

6)   A periodic review and revision of the plan.

The plan must also

1)   Be developed in consultation with the local political subdivisions affected by the plan

2)   Prohibit a source from any activity which contributes significantly to nonattainment

3)   Prohibit a source from any activity which interferes with measures to protect air quality or visibility

The last two provisions are related to the concepts of *New Source Review* and *Prevention of Significant Deterioration* and will be discussed more thoroughly in a later section.

An implementation plan can have a direct effect on existing stationary (nonvehicular) sources of air pollution. For example the plan can require, if necessary, the installation, maintenance and replacement of equipment in order to bring an area into compliance with the NAAQS, as well as the monitoring and reporting of emissions. An indirect source review program can also be included in the plan. **Indirect sources** are sources that do not produce pollution but have the ability to attract vehicles, or other mobile sources of pollution. Examples of indirect sources include buildings, parking lots and garages, roads, and highways.

Enforcement of an Implementation Plan

Federal enforcement of an implementation plan is the responsibility of the EPA. The EPA can enforce on a person (owner/operator of a source) or a state that has violated or failed to enforce the plan.

Enforcement on an individual can result in the issuance of a compliance order, an administrative penalty, or a civil action. Civil penalties of up to $25,000 per day for each violation (to a maximum of $200,000) are allowed. Higher penalties can be assessed if the EPA and Attorney General determine that it is appropriate. This determination is not subject to judicial review.

A Federal enforcement action against a state can be initiated if the state is not adequately managing its implementation plan. This can occur when the EPA notices trends in its enforcement of individuals. The EPA then notifies the state of the trends. Once this happens, EPA retains its authority to enforce against individuals until it is satisfied that the state is adequately managing its plan.

The EPA can impose sanctions against states that fail to:

1)   Submit an implementation plan

2)   Meet plan requirements

3)   Submit required information to EPA

4)   Gain EPA approval of the required submittals

5)   Implement the plan

The sanctions that EPA can impose on states with nonattainment areas include the prohibition of new transportation projects and stricter emissions-offset requirements for new sources.

Exempted from the sanctions are transportation projects which can be demonstrated to result in a safety improvement, public transit projects, high-occupancy vehicle roads and parking, work trip reduction programs, traffic flow improvements which achieve net emissions reductions, as well as accident and information programs to reduce congestion.

Sources: Mobile, Stationary, Major, Synthetic Minor, Area, New, Modified, and Existing

Sources are emitters of air pollution. This can include the criteria pollutants, for which NAAQS have been issued, as well as the hazardous air pollutants (HAPs). (HAPs are discussed in more detail in the next section.)

The Clean Air Act broadly regulates sources as mobile or stationary. **Mobile sources** include vehicles, airplanes, and other forms of transportation. On an individual basis, these sources are relatively small emitters of criteria pollutants and HAPs. Therefore, regional air pollution caused by mobile sources can generally be thought of as being caused by an aggregation of individual activities. This creates the need for a different regulatory framework than is in place for stationary sources.

Stationary sources, are sources of air pollution which don't move, such as a building, structure, facility, or other installation. A stationary source is considered to be a *major source* if it emits or has the potential to emit a significant quantity of one or more air pollutants. The significant quantity varies with the type of pollutant and is illustrated in Table 2.

A *synthetic minor source* has the capacity to operate as a major source but the owner/operator has voluntarily accepted an enforceable limitation to keep its emissions below the levels listed in Table 2.

Area sources are stationary or nonroad sources that are too small and/or too numerous to be included in a stationary source inventory. Examples of area sources include: water heaters, gas furnaces, fireplaces, and wood stoves.

As will be discussed in more detail in a later section, air pollution from stationary sources is generally controlled through the State or local permitting programs. These programs distinguish between new, modified, and existing sources. A *new source* is one that has not started construction. A source has been *modified* when it has been physically changed or is operated in a manner that increases the emissions of existing or new pollutants. A major modification is any physical change in the method of operation of a stationary source that would relate to a significant net emissions increase of any regulated pollutant. The definition of *significant* is specifically provided in the form of various pollutant emissions rates. An *existing source* is any source other than a new source.

Prevention of Significant Deterioration, New Source Review, and New Source Performance Standards

Prevention of Significant Deterioration. The concept of the prevention of significant deterioration (PSD) applies to permitting of stationary sources of criteria pollutants in attainment areas. Note that the PSD concept does not apply to the regulation of HAPs. In attainment areas, sources are required to evaluate their impact on the ambient air quality prior to starting construction.

The purpose of the PSD program is to further the goal of the Clean Air Act through the implementation of the following objectives:

1) Protect public health and welfare.

2) Preserve, protect, and enhance air quality in our scenic and wild areas.

3) Prevent the significant deterioration of air quality.

4) Have informed public participation in the decision-making process.

The Prevention of Significant Deterioration program is required to be part of the SIP, although EPA has the authority to regulate for PSD if a SIP is not in place. EPA can also stop construction of a source at any time, even if the state has an authorized air permitting program.

An owner wishing to construct a new or modify an existing source in an attainment area must first provide the permitting agency with enough

Table 2. Categories of Major Sources Based on Potential to Emit

Type of Pollutant	Type of Source	Single Air Pollutant	Combination of Air Pollutants
Criteria Pollutant	Specifically listed in §169 (1) of the CAA	$\geq$ 100 T/yr	Not applicable
Criteria Pollutant	Any other source not listed in §169 (1) of the CAA	$\geq$ 250 T/yr	Not applicable
Hazardous Air Pollutant	Any stationary source or group of stationary sources within a contiguous area and under common control	$\geq$ 10 T/yr	$\geq$ 25 T/yr

information for the agency to set the permit conditions. With the information provided, the agency must ensure that the new source will not violate the state's control strategy or interfere with the attainment of a NAAQS in the local area and region. Therefore, a permit application is submitted to the permitting agency that provides specific details about the source (such as nature and amounts of emissions as well as location, design, construction, and operation of the source). In some cases, air quality modeling may be required.

Obtaining a permit under the PSD rules is a public process. Any information submitted by the owner as well as any determinations made by the agency are public information. In addition, public participation is invited via a notice and comment period. In some cases a public hearing is also required.

In an attainment area, the emissions from a new source are limited to levels below the **Maximum Allowable Increase** (MAI) as well as any other applicable standard. Maximum Allowable Increases are calculated as a function of location or class of the source. They can also be included as a SIP requirement. In the regulations, MAI is referred to as an *ambient air increment* and the permitted emissions levels result in *consumption* of the increment.

The resulting PSD permit provides the owner or operator with specific allowable emissions. **Allowable emissions**, under the rules, are the more stringent of the applicable New Source Performance Standards (see discussion at the end of this section), SIP limitations, or Federally enforceable permit conditions.

New Source Review. In nonattainment areas, a similar permitting process is followed. The process, called **new source review** (NSR), entails a scoping of the proposed project including emissions limitations, and public involvement similar to the PSD process. The primary difference between the PSD and NSR programs is that the PSD program is intended to *preserve* the air quality of the attainment area whereas the NSR program is intended to *make reasonable further progress* toward achieving attainment. Therefore, sources being permitted through the NSR process are required to obtain **emissions offsets**, or surplus emissions reductions that compensate for the sources' impact on air quality. The offsets concept provides a source with an opportunity to find the most cost-effective way to reduce emissions (offsets, for example, may be obtained by changes in the plant process or may be purchased and sold to other facilities).

New Source Performance Standards. Another mechanism the Clean Air Act uses to ensure reasonable progress toward the attainment of NAAQS, is the development and implementation of *new source performance standards.*

New source performance standards (NSPS) are source-specific emissions-control limitations and requirements intended to promote the *best* technological system of continuous emission reduction. The NSPS take into account factors such as cost, as well as non-air health and environmental impacts, and energy requirements. The Clean Air Act requires that NSPS be reviewed once every eight years. The emissions control requirements for new or modified sources must be stricter than the NSPS.

National Emissions Standards for Hazardous Air Pollutants

Sources that emit hazardous air pollutants must conform to the National Emissions Standards for Hazardous Air Pollutants (NESHAPS). EPA develops NESHAPS for each source category and subcategory. The NESHAPS require the maximum degree of reduction in HAP emissions. If feasible, this maximum degree of reduction may be accomplished by prohibiting the emission of a particular substance.

The factors EPA must use when setting NESHAPS include cost, non-air quality environmental and health impacts, and energy requirements. The resulting standards can require a variety of methods to achieve emissions reduction including process changes; material substitutions; closed systems; collection, capture, or treatment of pollutants; as well as design, equipment, work practice, or operational standards (including operator certification and training).

The Alphabet Soup of Control Technologies: BACT, LAER, MACT, RACT, BART, GACT, RACM, and BACM

The Clean Air Act specifies the use of a variety of control technologies to minimize the emissions of air pollutants. This alphabet soup of control technologies is used throughout the different Titles of the Clean Air Act. The specific terms are defined below.

BACT or *Best Available Control Technology* is required for new or modified sources being permitted in an attainment area under the Prevention of Significant Deterioration rules, and may also be used in nonattainment areas under the New Source Review rules. Sources implementing BACT are using the most effective control technology that is available. The statutory definition of BACT is

> An emission limitation based on the maximum degree of reduction of each pollutant subject to regulation . . . taking into account energy, environmental, and economic impacts and other costs . . . is achievable for the facility through the application of production processes, and available methods, systems, and techniques. . .

BACT determinations are made by the permitting agency on a case-by-case basis and must ensure that the resulting source emissions level is in compliance with the NSPS for that source.

LAER, or *Lowest Achievable Emissions Rate*, is used by sources in nonattainment areas. The statutory definition of LAER is the more stringent of the following conditions:

A) the most stringent emission limitation [for that class or category of source] which is contained in the implementation plan of any state . . . unless the owner or operator . . . demonstrates that such limitations are not achievable.

B) the most stringent emission limitation which is achieved by practice [for that class or category of source]

Like BACT, LAER determinations are made on a case-by-case basis and are limited by the emissions levels set in the NSPS.

LAER does not take into direct consideration economic factors. In actual practice, and after subsequent EPA interpretation, there is little difference between BACT and LAER.

The EPA, in cooperation with the states, maintains a BACT/LAER Clearinghouse database of sources. The clearinghouse can be accessed through the Clean Air Technology web site on the Internet–

<http://www.epa.gov//ttn/CATC/>

and the query page can be found at

<http//mapsweb.rtpnc.epa.gov/RBLC/Web/rbqry.html>

MACT, or, *Maximum Achievable Control Technology*, is a concept used in the permitting of hazardous air pollutants. The control technology for new and existing sources is determined by establishing the *MACT floor* or baseline. For existing sources, the MACT floor must equal

> the average emission limitation achieved by the best performing 12% of the existing sources . . .

Or, if there are fewer than 30 sources in the category:

> the average emission limitation achieved by the best performing 5 sources.

The MACT floor for new sources is

> emission control that is achieved in practice by the best controlled similar source.

The intent of the MACT standards is to provide a performance-based method for reducing toxic emissions. Facilities with good emissions controls have an economic advantage when the standards are developed. The real challenge in determining MACT for a particular source is the definition and refinement of the source category and the subsequent *best performers*.

The 1990 amendments mandated EPA to develop MACT standards for an initial list of 190 HAPs by the year 2000. Subsequent revisions to the law have resulted in a current list of 188 HAPs. HAP requirements are discussed more thoroughly in a later section.

RACT, or *Reasonably Achievable Emissions Control Technology*, is incorporated by a state into its SIP in order to achieve compliance with the NAAQS in nonattainment areas. RACT is defined as

> Devices, systems, process modifications, or other apparatus or techniques that are reasonably available and take into account:
>
> 1) the necessity of imposing such controls, and
>
> 2) the social, environmental, and economic impact of such controls

BART, or *Best Available Retrofit Technology*, can potentially be required on existing major sources in areas affected by the regional haze rule. EPA has only recently addressed visibility protection nationwide. BART, which would be potentially applicable to 26 industrial source categories in 34 states, would take into account availability of the control technology, cost of compliance, current pollution control equipment used by the source, the remaining useful life of the source, and the expected improvement in visibility.

GACT, or *Generally Achievable Control Technology*, is an alternative standard which, at the EPA's discretion, may be applied to HAP area sources in order to enable control the emissions of the most significant HAPs.

BACM and RACM are what EPA considers the *Best Available* and *Reasonably Available Control Measures* for the management of particulate matter less than 10 microns (PM_{10}). EPA has issued guidance on BACM and RACM for emis-sions from fugitive dust, residential wood combustion, and prescribed burning.

Title I—Air Pollution Prevention and Control

Title I of the Clean Air Act, as amended in 1990, consists of three parts: air quality and emissions limitations, prevention of significant deterioration of air quality, and plan requirements for non-attainment areas. Some of the significant features of Title I not discussed in the previous section are covered below.

Air Quality and Emissions Limitations

Title I of the Act is an overview of Congress's philosophy and intent. In Title I, air pollution is identified as being primarily an urban problem associated with increasing industrialization and urban growth. In fact, at the time of the Clean Air Act Amendments, over 100 cities were in non-attainment for ozone. Pollutants threaten not only public health but also the public infrastructure such as agriculture, property, and transportation.

One of the fundamental tenets of the Clean Air Act is that states and local governments are primarily responsible for air pollution prevention and control. The Federal government is responsible for providing cooperative oversight, financial assistance, and leadership. This philosophy can be seen throughout the Act in the State Implementation Plan requirements, permitting programs, research programs, and grants.

Because air pollution doesn't necessarily stop at the State line, cooperation between states, and formation of regional pacts and commissions, are encouraged. Regional pacts, however, require congressional approval.

Title I also gives the states the primary responsibility and requirements for designating attainment and nonattainment areas as well as developing and implementing the State Implementation Plans.

EPA promulgates the air quality designations, and will allow redesignation of an area if it meets the following conditions:

1) The air quality meets the NAAQS.

2) The SIP has been approved.

3) The improvement in air quality has been due to permanent and enforceable reductions in emissions resulting from implementation of the SIP, Federal rules, and other permanent and enforceable reductions.

4) A maintenance plan has been approved.

5) The state meets all the requirements for implementation plans and plan requirements for nonattainment areas.

Title I establishes the development by EPA of primary and secondary NAAQS as well as a transportation air quality planning process.

Title I also introduces the concept of New Source Performance Standards and specifies that any permit issued to a new or modified stationary source must comply with the standards of performance applicable to that source. This requires the source to employ a *technological system of continuous emissions reduction*. In addition, the construction and operation of a new or modified source must be in compliance with all other requirements of the Act.

Hazardous Air Pollutants

Hazardous Air Pollutants (HAPs) have been part of the Clean Air Act since 1970. However, because of the way the 1970 law was written, over the next 20 years EPA was only able to set regulatory standards for seven HAPs. The 1990 amendments increased the number of regulated HAPs to 189 and established a 10-year schedule for developing performance-based compliance standards. Congress required that EPA promulgate certain specific standards within two years; 25 percent of the standards within four years; an additional 25 percent within seven years; and the remaining 50 percent within 10 years, or by the year 2000.

Hazardous Air Pollutants (HAPs) are pollutants that are having or may have an adverse effect on human health. HAPs may be carcinogenic, mutagenic, teratogenic, neurotoxic, chronically toxic, bioaccummulative, or cause reproductive dysfunction. An initial list of HAPs is included in the Act. The list is periodically reviewed and revised, and minor changes have been made since its initial promulgation.

In addition to defining HAPs, Title I also regulates HAP sources. EPA is required to maintain a list of HAP sources, both stationary and area. MACT and GACT are used to control emissions of HAPs from these sources. EPA has completed promulgating MACT standards for its mandated 2-, 4- and 7-year schedules.

Incorporated into the Act is the incentive for sources to reduce HAPs prior to a regulatory mandate. If a source can achieve a 90 percent reduction in HAPs (or, for particulate HAPs, a 95 percent reduction) before a MACT standard has been issued, then the source can continue to operate at the reduced level for up to six years from the compliance date of the applicable standard. This six-year window allows the source to continue operation under the Title V Operating Permit program until the next renewal period. Operating Permits, as will be discussed in the next section, are on a 5-year renewal cycle.

An increase in awareness to community needs resulted in the inclusion in the 1990 amendments of the accidental release provisions for 77 HAPs. Stationary sources that produce, process, handle, or store certain substances are required to implement an accidental release prevention program. This program is commonly known as the Section 112(r) requirements, referring to that section of the Clean Air Act Amendments. It is also known as the Risk Management Plan (RM Plan), or Risk Management Program (RMP).

An RMP is required if the 77 regulated substances (which have the potential to cause death, injury, or serious adverse effects to human health or the environment) are stored in quantities exceeding a threshold amount. The accidental release prevention program includes hazard assessment techniques, safe design and maintenance of the facility, as well as the means to minimize the consequences of a release.

Owners and operators affected by the RMP rules are required to comply with the requirements by June 21, 1999. Refer to the Risk Management Progam chapters in this book for more information.

Federal Enforcement of the Clean Air Act

Like many of the other environmental statutes, Federal enforcement of the Clean Air Act is particularly onerous. Title I outlines the enforcement procedures for states as well as corporations and individuals that fail to meet the provisions of the Act.

EPA has the responsibility for Federal enforcement that can result in a number of actions, including

- Issuance of an administrative penalty order

- Issuance of a compliance order

- A civil action

- A criminal action

The type of enforcement action taken depends on the type of violation, its seriousness, and the good-faith actions taken by the violator to comply with the requirement.

Civil penalties can be assessed up to $25,000/day with a $200,000 maximum. Criminal penalties can apply to knowing violations of the act. Conviction of a criminal violation can result in a fine and/or imprisonment. In determining the amount of penalty to be assessed, the following factors are taken into account:

- The size of the business

- The economic impact of the penalty on the business

- The compliance history

- Good-faith efforts towards compliance

- Duration of the violation

- Previous penalties for the same violation

- The economic benefit of noncompliance

The Clean Air Act Amendments of 1990 also introduced a field citation program and bounty hunter provisions for citizens reporting. The purpose of the field citation program is to allow an EPA inspector to address minor violations. Civil penalties of less than $5000 can be issued to a facility immediately by an EPA inspector.

Under the bounty hunter provisions, EPA can pay an award of up to $10,000 to anyone who provides information that ultimately leads to a criminal conviction or civil penalty.

Visibility Protection

Title I includes in the PSD program a national goal to protect the visual resources of the United States. The Act protects visibility in Class I Federal areas (in particular, wilderness areas and national parks) by remedying existing and preventing any future impairment of visibility in these areas.

This program requires EPA to evaluate sources in Class I areas as well as other regions with clean or visibility-impaired air in order to issue regulations and ensure reasonable progress towards this goal. Existing sources that are contributing to visibility impairment can be required to install BART under these rules.

Visibility impairment is caused by a number of factors. Particulates such as dust and wood smoke have a direct impact on visibility. Gaseous pollutants, such as SO_2 and NO_x, are precursors to sulfates and nitrates—particulates that contribute to regional haze. Because the pollutants that can contribute to visibility impairment are capable of migrating long distances from their sources, Congress authorized the establishment of visibility transport regions and commissions. The commissions recommend measures to improve visibility, develop clean air corridors, alternative siting measures, and long-range strategies. The Grand Canyon Visibility Transport Commission, for example, was formed under the authority of the CAA to improve visibility in the 16 national parks and wilderness areas located on the Colorado Plateau. This transport commission is networked with other government and collaborative efforts, such as the National Park Service Visibility Program and the Western Regional Air Partnership.

General Requirements for Nonattainment Areas

In order to achieve improvements in air quality, Title I contains some general provisions for nonattainment areas. The intent of these provisions is for states to achieve *reasonable progress* toward attainment of NAAQS.

Once a nonattainment area is identified, EPA publishes a notice in the *Federal Register*. Subsequent to identification, a state must ensure that its SIP contains a plan to implement the following:

1) The implementation of reasonably available control technology (RACT) on existing sources

2) Reasonable further progress (RFP) toward attaining NAAQS

3) An inventory that identifies and quantifies current emissions levels

4)   Construction and operating permits for new and modified sources

5)   Emissions limits and other control measures (including economic incentives such as fees, marketable permits, and auctions of emissions limits)

6)   Measures to be taken if an area fails to make RFP

A state's schedule for attainment of NAAQS is set out in the SIP. In general, for a primary NAAQS, the attainment date must occur within five years of its promulgation, with extensions possible of no more than 10 years. A secondary NAAQS does not have a statutory deadline for attainment but must be achieved as expeditiously as possible.

Any source in a nonattainment area is potentially under more stringent permitting and operating conditions than if it were located in an attainment area. States and local agencies may impose additional requirements in these areas that exceed the Federal standards. An operator wishing to construct or modify a source must demonstrate that the source will be beneficial when weighed against the environmental and social costs. The operator of the source must obtain a permit from the regulating agency prior to starting construction or modification. The source is under strict emissions limitations. The source emissions characteristics must meet NSPS, LAER must be installed, and the source must find offsetting emissions reductions from existing sources in the region.

As discussed earlier, the use of offsets is a market-based approach toward reducing air pollution and achieving attainment. The offsetting emissions reduction obtained from the existing sources must exceed the proposed increase. Offsets can be obtained from a different stack on the same source, the same facility, or other facilities within the nonattainment area. Trading of offsets is allowed.

One of the requirements for new projects in a nonattainment area is that these projects (which are generally transportation related) must conform to the SIP. SIP conformity means that the project promotes the SIP's overall purpose in eliminating or reducing the severity and number of NAAQS exceedances.

Once an area has passed its scheduled attainment date, EPA evaluates the area's air quality and publishes the results. If an area fails to attain NAAQS, the state must submit a revised SIP to EPA for approval. The attainment clock for the failed area starts over from the date of the published notice. The state has a statutory deadline of five to 10 years to come into attainment with the primary NAAQS.

Should a state fail to attain NAAQS, EPA has the ability to impose sanctions. As discussed previously, the EPA can take sanctions against a state for failure to prepare an adequate plan as well as failure to implement a plan. The sanctions include limiting funding for transportation projects, as well as increasing the offset ratio of emissions reductions to emissions increases.

These general provisions for nonattainment areas apply to all of the criteria pollutants, including sulfur oxides, nitrogen dioxide, and lead. Some of the criteria pollutants—ozone, carbon monoxide, and particulate matter—have additional requirements. These are summarized in the following sections.

Ozone Nonattainment Areas

Ground-level ozone is the primary constituent of smog. Ozone at ground level causes a number of deleterious health effects including respiratory infection, lung inflammation, and an aggravation of preexisting respiratory diseases such as asthma.

Like carbon monoxide, ambient ozone concentrations have decreased substantially over the years. However, recent changes in the ozone NAAQS will impact the regulation of sources that emit VOCs and NO_x.

In 1997, EPA revised the NAAQS for ozone, changing the standard from 0.12 ppm, averaged over a 1-hour period, to 0.08 ppm, averaged over an 8-hour period. This change in the NAAQS created a significant increase in number of new nonattainment areas.

Smog forms when ozone, unburned hydrocarbons (VOCs), and NO_x combine in the atmosphere. The smog-forming chemical reaction requires heat, sunlight, and time. Thus, depending on the meterological conditions, smog can form many miles away from the emitting sources. It is because of the fact that ozone is formed in the atmosphere rather than emitted directly from a source, that control of ozone pollution is achieved primarily through the regulation of VOCs and NO_x.

Ozone nonattainment areas are classified as marginal, moderate, serious, severe, and extreme. Each of the classifications has prescribed actions and schedules for attainment. It is possible for areas that have been previously unclassified to subsequently become designated as nonattainment. An area can become reclassified to a higher (more restrictive) classification should it fail to attain NAAQS within six months of its attainment date. In addition, states can voluntarily reclassify to a more restrictive standard.

Rural transport areas are rural nonattainment areas associated with a larger metropolitan area. These areas are handled as marginal for ozone nonattainment.

Single ozone nonattainment areas that cover more than one state are called *multistate ozone nonattainment areas*. Multistate nonattainment areas are required to coordinate their implementation plans. A state in such an area may petition EPA, and, through modeling or other methods, demonstrate its ability to be in attainment if it weren't for the failure of other states to achieve attainment.

In order to manage ozone, the CAA specifies both Federal and State actions. At the Federal level, the Clean Air Act requires that EPA issue guidelines for cost-effective control of Volatile Organic Compounds (VOCs) and NO_x.

The EPA also has the ability to set up interstate regions and commissions in order to control interstate ozone. The findings of these commissions can result in recommendations for SIP revisions and other actions, which need to be taken in order to bring the area under control.

The Ozone Transport Assessment Group (OTAG), for example, is a partnership between the EPA, states, industry, and environmental groups that was formed to address ozone transport over the Eastern United States. OTAG is a workgroup whose mission is to assess the ozone transport problem and develop a regional strategy for achieving attainment.

For severe and extreme ozone nonattainment, failure to attain NAAQS can result in fees to major sources of $5K/ton of VOC.

Carbon Monoxide Nonattainment Areas

Carbon monoxide (CO) is a colorless, odorless gas that is formed during incomplete combustion of fuels. Carbon monoxide is poisonous at high concentrations, replacing oxygen in the blood and consequently reducing the amount of oxygen available to the body's organs and tissues. Ambient carbon monoxide concentrations in the United States have decreased by 37 percent between 1987 and 1996, despite a 26 percent increase in the number of vehicle-miles traveled during this same period.

Carbon monoxide nonattainment areas are classified as moderate or serious. Like the provisions for ozone nonattainment, an unclassified carbon monoxide area can be designated as nonattainment. Moderate areas that fail to achieve attainment in a timely manner can be reclassified as serious. Also, states which are part of multistate nonattainment areas must coordinate actions taken under their SIPs.

For carbon monoxide, both the moderate and serious classifications have prescribed actions and schedule for attainment. The types of actions required to control carbon monoxide become more stringent with poorer air quality. Depending on the attainment classification, measures such as a vehicle-miles inventory, use of oxygenated gasoline, and increased regulation of stationary sources may be required.

If a serious area fails to achieve attainment, then the state must revise its SIP, committing to an annual 5% decrease until NAAQS is achieved.

Particulate Matter Nonattainment Areas

Particulate matter (PM) is a general term for a mixture of solid and liquid particles in air. The particles originate from manmade sources (such as power plants, motor vehicles, and wood stoves) and natural sources (such as wind-blown dust).

PM contributes to a series of health effects including increased respiratory disease, decreased lung function, and death. Particulate matter also contributes to regional haze, and therefore is a component of the visibility requirements under the Act.

The ambient concentrations of *particulate matter less than 10 microns* (PM_{10}) decreased by approximately 25% between 1988 and 1996. In 1997, EPA revised the particulate matter NAAQS to include particles less than 2.5 microns in diameter ($PM_{2.5}$). As a result of this revision, it is anticipated that the number of counties in the United States that are in nonattainment for particulate matter will be triple the current number. The exact impact of the revised NAAQS will not be known until a monitoring network has been established to determine ambient $PM_{2.5}$ levels.

PM_{10} nonattainment areas are classified as moderate and serious, depending on the air quality of the area. Again, each of the classifications has increasingly stringent prescribed actions and schedules for attainment. Measures that may be used to control PM_{10} include the use of RACT, BACT, as well as the control of PM_{10} and PM_{10} precursors from stationary sources.

Title II—Emission Standards for Moving Sources

Moving sources, such as automobiles and airplanes, represent a special regulatory challenge. Emissions from moving sources originate from a multitude of individual sources, and it is the aggregate emissions from these sources that creates regional air quality problems. For example, motor vehicles produce up to half of the smog forming VOCs and nitrogen oxides, release greater than 50% of the HAPs and 90% of the CO in urban air. Although today's cars produce 60–80% fewer pollutants that the cars of the 1960s, and although people are using more mass transit today than they did then, the levels of pollutants from mobile sources have not substantially decreased. The reasons for this are that

- Americans drive more (1 trillion-vehicle miles in 1970 *vs.* an estimated 4 trillion in 2000).

- People live far from work and/or may not have access to mass transit.

- Reformulated unleaded gasoline can release more smog-forming gases (VOCs) to the atmosphere.

Motor Vehicle Emissions Standards

Congress decided to approach the regulation of moving sources from several directions. In ozone and CO nonattainment areas, where air quality problems are significant, vehicle maintenance and inspection programs are required.

Congress also mandated the control of air pollution from new motor vehicles such as automobiles, heavy-duty trucks, and motorcycles. Not only are emissions standards required, but fueling and onboard vapor recovery standards are prescribed, as well as special requirements for high-altitude vehicles.

Manufacturers fund the motor vehicle program. New vehicles must have a certificate of conformity with these standards as well as an emissions system warrantee. Manufacturers are required to maintain records of the performance tests.

As an example, the purchase of a new car today includes under-the-hood systems and dashboard warning lights for the emissions-control system. These devices must be operable for 100,000 miles.

Unlike other portions of the CAA, the regulation of vehicle emissions standards cannot be delegated to the states. Other than California, states cannot set independent standards for new motor vehicles. Only California can receive EPA authorization to adopt and enforce other standards. Once California standards have been adopted by EPA, other states can follow suit.

Fuel Standards

Another approach used to control air pollution from moving sources is the regulation of fuel and fuel additives. For example, the use of tetraethyl lead in gasoline has been phased out and sulfur, which contributes to acid rain and smog, is required to be removed from diesel fuel.

EPA can register new fuels and fuel additives (including lead substitutes) only after the manufacturer demonstrates that the product and its emissions will not endanger public health and welfare. Lead substitutes must also be shown to be effective in reducing valve seat wear while avoiding other engine side effects. Like the new vehicle standards, the introduction of new fuels and fuel additives cannot be delegated to the states.

Certain additives are expressly forbidden from inclusion and/or use in fuels. The prohibitions are summarized in Table 3.

In the United States, nine of the major metropolitan ozone nonattainment areas are required to use reformulated gasoline as a means of controlling emissions of volatile organics. The purpose of the reformulation is to decrease benzene, a carcinogen, and other smog-forming chemicals. In order to recertify reformulated gasoline, it must meet certain criteria, summarized in Table 4.

The use of reformulated gasoline in nonattainment areas is controlled by a series of options, economic incentives, and other requirements. For example, states can petition EPA to voluntarily opt in to the reformulated fuels program. Manufacturers of reformulated gasoline can generate credits by creating formulas that exceed the minimum standards, and those credits can be transferred for use within the same nonattainment area. Also, manufacturers of reformulated gasoline are required to ensure that their average per-gallon emissions of VOCs, oxides of nitrogen, carbon monoxide, and toxic air pollutants are less than the 1990 baseline year.

In carbon monoxide nonattainment areas (39 urban areas in the United States), oxygenated fuels of not less than 2.7% O_2 by weight are required. The requirement is typically in effect during the winter months only. The added oxygenated compounds make the fuel burn more efficiently, resulting in a reduction of CO emissions.

Retailers of oxygenated gasoline are required to label the fuel dispensers, indicating that the fuel is oxygenated and will reduce CO emissions from the vehicle. Marketable credits for oxygenated gasoline are available for use within the nonattainment area, similar to the reformulated gasoline credits.

The control of emissions from urban buses built since 1994 is affected through the use of low-polluting fuels. The goal is to obtain a 50% reduction in PM emissions. Urban buses in metropolitan areas that had a 1980 population exceeding 750,000 are required to retrofit engines replaced or rebuilt after January 1, 1995.

In order to qualify as a *clean alternative fuel*, the fuel must meet certain standards and be

Table 3. Fuel Standards

Additive or Characteristic	Condition or Limitation
Any fuel or additive	Must be substantially similar to the fuel or additive used to certify vehicles after 1974.
	Must not cause or contribute to the failure of an emissions control device or fuel system.
Manganese	0.0625 g/gallon fuel.
Leaded gasoline	Prohibited from use entirely after 12/31/95. Prior to 12/31/95, prohibited in a vehicle which is labeled "Unleaded Gasoline Only"
Sulfur	0.05% by weight in diesel.
	0.10% by weight for heavy-duty diesel vehicles and engines.

Table 4. Reformulated Gasoline Criteria

Criteria	Specification
NO$_x$ emissions	Less than the levels of NO$_x$ emitted by baseline vehicles using baseline gasoline.
Oxygen content	Greater than or equal to 2% by weight (could be waived if this condition prevents or interferes with attainment of a NAAQS).
Benzene content	Less than 1% by volume.
Heavy metals (lead and manganese)	None allowed.
Aromatics	Less than 25% by volume.
Detergents	Sufficient to prevent the accumulation of deposits in engines and/or fuel supply systems.
VOC emissions	15% less than vehicles using baseline gasoline before 2000. 25% less after 2000, measured on a mass basis.
Toxic air pollutants	15% less than vehicles using baseline gasoline before 2000, 25% less after 2000, measured on a mass basis.

useable by a certified clean-fuel vehicle. Examples of clean alternate fuels include methanol, ethanol or other alcohols, gasohol mixtures greater than 85% alcohol by volume, reformulated gasoline or diesel, natural gas, liquefied petroleum gas, hydrogen, or an alternate power source such as electricity.

Nonroad Engines

Nonroad engines include a wide variety of internal combustion engines from small lawn mower type engines to heavy-duty construction equipment. Locomotives, marine engines, and aviation equipment are also in this category. Historically, emissions control from these engines has not been considered in their design. Therefore they emit relatively higher levels of hydrocarbons, particulate matter, NO$_x$, CO, and CO$_2$ than other types of moving sources.

Emissions standards have been set for many of these engines. Depending on the engine, these standards are being phased in through the year 2008. Engine manufacturers are required to ensure that their engines meet the standards. The International Civil Aviation Organization stan-

dards were adopted for aircraft emissions. These standards are enforced by the Federal Aviation Administration.

Clean Fuel Vehicles

The Clean Fuel Vehicle program, which is mandated in ozone nonattainment areas, is a special program intended to encourage the use of both low-emission vehicles and clean alternative fuels.

The Clean Fuel program was initiated and tested in California as a means to control air pollution in ozone nonattainment areas. Included as part of the State Implementation Plan are requirements that clean fuels be made available, with sufficient facilities for refueling. The state has the ability to provide economic incentives toward the use of clean fuel vehicles by issuing emissions credits.

Clean fuel vehicle requirements apply only to vehicles of up to 8500 pounds gross vehicle weight rating. The requirements include specifications for on-board diagnostics, evaporative emissions, as well as emissions standards for Non Methane Organic Gas (NMOG), CO, and NO$_x$.   The emissions standards for different types of vehicles are on phase-in schedules through the year 2000.

Fleets of greater than 10 vehicles owned by a single entity (including the government) and operated in populated ozone and CO nonattainment areas are required to phase in the purchase of clean fuel vehicles. After the year 1999, 50% of the new fleet vehicles are required to have clean fuel capabilities.

The phase-in of the Clean Fuel program is required to be written into the SIP. Other requirements to be included in the SIP are

- The availability of clean alternative fuels at central fueling areas.

- The issuance of credits to the fleet operator for exceeding the minimum requirements of the program. The credits may be held, traded, sold, or banked for later use.

States can voluntarily opt in early to the clean fuel program for any ozone nonattainment areas.

Inspection and Maintenance Programs

The inspection and maintenance (I/M) programs are a third mechanism used to control emissions from moving sources. I/M programs, run at State and local levels, ensure that vehicles and their emissions systems are adequately maintained. Approximately 110 cities and states have or are required to have programs in place to check motor vehicle tailpipe emissions.

Title III—General

Title III of the Clean Air Act contains general permitting requirements. This section describes the procedures the EPA must use to delegate air programs to the state. The procedures are intended to

- Ensure regional fairness and uniformity in criteria, procedures, and policy

- Ensure auditability of each states' performance, compliance with, and enforcement of the act

- Provide a mechanism for the identification and standardization of inconsistent or varying criteria, procedures, and policies

With the exception of appropriations under the CAA, tribes are treated the same as states. Although a tribe is not eligible for the same amount of money as a state (up to 0.5% of the annual appropriations for the program), a tribe is eligible for grant and contract assistance to carry out the Act. Furthermore, tribes with governing bodies capable of administering the Act, can be delegated certain responsibilities and programs. For example, tribes can prepare their own Implementation Plans and permitting programs.

EPA's Duties and Authorities

Title III outlines EPA's duties and authorities as well as other procedures and standards. For example, EPA has the right to immediately restrain the operation of any source that is presenting an *imminent and substantial endangerment to public health and welfare.* EPA has the ability to issue other orders necessary for the protection of public health, welfare, or the environment. EPA can also issue subpoenas for the purpose of obtaining information about any investigation, monitoring, reporting requirement, entry, compliance inspection, or administrative enforcement proceeding under the Act.

Title III contains the procedural requirements for petitioning EPA for review of a NAAQS, the rulemaking process, and judicial review, as well as congressional appropriations.

Another important aspect of Title III is that it gives EPA the authority to review legislation, regulations, and projects for their potential environmental impact.

Civil Suits

The provisions for civil suits are found in Title III. Any citizen can take civil action against a source or the EPA for

1)   An alleged violation of an emission standard, limitation, or order

2)   EPA's failure to comply with the act

3)   Failure to obtain a permit prior to modification or construction, or an alleged violation of a permit condition

For the purposes of a civil action, the term *emission standard or limitation* refers to State or Federal requirements relating to the quantity or rate of emissions. Note that the term is, in fact, defined quite broadly, and includes

1)   A schedule or timetable of compliance, and emission limitation, standard of performance, or emissions standard

2)   A control or prohibition of a motor vehicle fuel or fuel additive

3)   A permit condition or requirement relating to attainment and nonattainment areas

4)   Any state implementation plan requirement relating to transportation control measures, air quality maintenance plans, vehicle inspection and maintenance programs, vapor recovery, visibility, and ozone protection

5)   A standard, limitation, or schedule issued under the Title V program

Civil penalties under the citizen's suits provisions can, at the discretion of the court, be deposited in a penalty fund or used to fund beneficial mitigation projects. The maximum penalty in a citizen's action is $100,000.

Economic Impact

The Clean Air Act Amendments monitor the cost effectiveness of air pollution control. Under the Act, EPA is required to submit various reports to Congress regarding the economic impact of its air quality programs. These reports include:

- A biennial economic impact analysis for the criteria air pollutants as well as hazardous air pollutants, mobile sources, in addition to SO_2 and NO_x substitutes

- The costs and effects for the New Source Review, New Source Performance Standards, and the regulations controlling ozone, Prevention of Significant Deterioration, emissions standards, fuels, and aircraft

One important aspect of the economic picture is the potential change in employment resulting from the Act. This change includes threatened and actual plant closures, layoffs, and other adverse employment affects. EPA has the ability to conduct full investigations of these matters upon request. The request can come from an employee or employees' representatives.

Employees who believe they may have been discharged because of participation against an employer in an enforcement proceeding are provided with *whistleblower* protection under the Act. Employees in this situation may file a written complaint with the Secretary of Labor and an investigation of the incident will be conducted. If the employee prevails in the complaint, then the employer is liable for all costs and expenses, as well as compensatory damages to the employee.

Air Quality Monitoring

Title III authorizes the EPA to establish an air quality monitoring network. The primary purpose of the monitoring network is to track air quality throughout the United States using uniform monitoring methodology and criteria.

The states and local agencies implement the program, which has been set up by EPA. It consists of four categories of monitoring stations.

SLAMS, or *State and Local Air Monitoring Stations*, is a network of 4000 monitoring stations as determined by the needs of the state or local agency. SLAMS are generally located in urban areas.

NAMS, or *National Air Monitoring Stations*, are 1080 key sites within the SLAMS network. They are also located in urban areas, focusing on areas of maximum pollutant concentrations and high population density.

SPMS, or *Special Purpose Monitoring Stations*, are used to supplement the SLAMS. They are not permanent stations but are installed by the State and local agencies as needed in order to support their air programs.

PAMS is a network of about 90 *Photochemical Assessment Monitoring Stations*. PAMS is required in any ozone area that is designated as serious, severe, or extreme.

Title IV—Acid Deposition Control

Acid deposition or *acid rain* is believed to be caused by sources that emit two of the by-products of combustion: SO_2 and NO_x. Acid rain affects forest ecosystems, damages structures, impairs visibility, and impacts public health. Fossil fuel–fired electrical generators have been identified as significant sources (in particular, the coal-burning power plants of the Midwest). SO_2 and NO_x emitted from these facilities travels toward the east coast of the United States and Canada where it falls to the earth as rain, gases, or dust. As a result, fossil fuel–fired power plants are regulated as a means to control acid rain.

In the Clean Air Act amendments, Congress set as a goal the control of acid rain through the reduction of SO_2 by 10 million tons and NO_x by 2 million tons from the 1980 levels. Congress dictated that energy conservation and renewable, clean alternatives are to be taken into consideration when calculating emissions reductions.

The most significant aspect of the acid rain law was the control of acid rain emissions through *economic incentives* as opposed to *command and control* emissions limits.

Sulfur Dioxide

The economic incentives used by the acid rain program consist of annual SO_2 emissions allowances which are allocated to the operator of the source. An **allowance** is defined as authorization for an affected unit to emit one ton of SO_2 during a calendar year. Every January 1, an affected utility unit is allocated allowances based on historical fuel consumption and a unit-specific emissions rate. The allowances have economic value and can be sold and traded by the sources. After January 1, 2000, the number of new allowances will be limited. These allowances are assignable by the owner/operator to other units under the owner/operator's control. The owner/operator must request reassignment in the permit application.

In simple terms, for each ton of SO_2 emitted, one allowance is consumed (*i.e.,* turned over to EPA). If there are allowances left over at the end of the year, they can be transferred within the utility system or sold to another facility for future use. If, on the other hand, the source does not hold enough allowances to cover the emissions, it must procure them from someone (buy or trade), or be in noncompliance.

The reduction in SO_2 emissions from fossil fuel–fired electric generation plants is phased in over a five-year period. The first phase took place on January 1, 1995. On that day, the initial emissions allowances for approximately 300 large acid rain sources were set. Also on that day, reserve allowances were calculated based on the units' capacities and actual usage in 1985, adjusted to 1995 usage. Allowances and reserves for 1995 were subsequently calculated and allocated through the year 1999, on a steadily decreasing scale. By 1995, SO_2 emissions from Phase I units were reduced by 40% below their required level.

In the acid rain program, assigning bonus allowances for avoided emissions encourages energy conservation and renewable energy. Avoided emissions can result from such activities as the installation of clean coal technology, use of solar or wind power, and implementation of customer energy conservation programs. Utilities are eligible for these allowances if they do all of the following:

1)   Directly fund the qualified energy conservation and renewable energy measures.

2)   Quantify the avoided SO_2 emissions.

3)   Adopt and implement a State-approved plan for least cost energy conservation and power which evaluates new power supplies, energy conservation, and renewable energy resources.

4)   Certify their rates and charges by the Department of Energy, and ensure that they do not negatively impact the utility's net income.

Allowances for energy conservation and renewable energy are allocated from the Conservation and Renewable Emissions Fund. A total of 300,000 allowances are in the fund, which is reduced on an annual basis from the year 2000 to 2009. Allowances remaining after 2010 are allocated as part of the Phase II program.

Phase II implementation of the acid rain program begins in the year 2000. Phase II will affect approximately 2000 electric generation units. Included in Phase II are units with an output capacity exceeding 25 megawatts and all new units. In addition to tightening the restrictions on Phase I plants, Phase II will set new restrictions on smaller units.

Nitrogen Oxides

Nitrogen oxides (NO_x) are also regulated under the acid rain program. Unlike the SO_2 program, NO_x emissions are not tradable or capped. NO_x emission rates are established for utility boilers. The rates are based on low NO_x burn technology and the degree of reduction achievable through retrofit technology. Source operators have the option of complying with an individual emission rate for the boiler or averaging emissions over two or more boilers.

The control of NO_x began with Phase I in 1996. Phase I included approximately 170 dry-bottom wall-fired and tangentially fired boilers. These boilers were required to achieve a NO_x limitation of 0.45 lbs. NO_x/mmBtu per year.

Phase II begins in the year 2000. Phase II lowers the emissions rates for the Phase I boilers and establishes emissions limitations for boilers using cell-burner technology, cyclone, wet bottom, and other types of coal-fired boilers.

Permitting

Acid rain sources are regulated through the issuance of permits. The permit program, which is enforced as part of the Title V operating program, can be delegated to the states. If a source has already received a Title IV permit, then that permit is in effect until the Title V permit program is implemented in that state. Refer to the next section for a discussion of Title V permits.

The acid rain provisions prohibit annual emissions of sulfur in excess of the number of allowances, exceedance of applicable emissions rates, and the use of any allowance prior to its allocated year. Certain sources not normally covered by the acid rain provisions can voluntarily opt into the program and be eligible for allowances.

Each initial permit is accompanied by a compliance plan that describes how the source will meet the proscribed SO_2 and NO_x emissions limitations. Each acid rain source must have a permit and operate in accordance with its terms and the terms of the compliance plan. The CAA specifies that acid rain sources install and operate continuous emissions monitoring systems (CEMS). The purpose of the CEMS is to ensure that data is available for SO_2, NO_x, opacity, and the volumetric flow rate of each unit.

Although operations of an electric utility cannot be terminated for not having a permit or compliance plan, the strict enforcement provisions and penalties described below still apply.

The penalties for exceeding emissions limitations are $2000/ton excess NO_x emitted or $2000/$SO_2$ allowance held by the source. The SO_2 owner/operator is also liable for offsetting the excess emissions through additional emissions reductions the following year. For enforcement purposes, each ton of SO_2 emitted over the allowance amount is considered to be a separate violation.

Title V—Permits

The Title V Operating Permit program was added with the 1990 amendments. The Title V program is modeled after the Federal National Pollutant Discharge Elimination System (NPDES), a permitting program that is currently used to control point source water pollution.

Prior to 1990, the permitting of sources in the United States varied greatly depending on the location and type of industry. Under the Federal rules, major sources were only required to obtain a Permit to Construct (PTC). Although the PTC usually carried with it operating restrictions, Operating Permits were generally not required. Facilities that had multiple activities occurring at the same location could be regulated under multiple complicated and sometimes conflicting pemits.

The 1990 amendments established a Federal operating permit program. The purpose of the program, which is administered by EPA, is to *simplify* permitting of a source by bringing all requirements into a single document.

A source that is covered under the operating permit program is required to submit a timely application to the EPA or regulating agency. The agency is under an obligation to review and issue a permit in a timely manner. The 1990 Amendments provide specific guidelines to achieve this.

The Operating Permit program can be delegated by EPA to the states. State permit programs can be more stringent than the EPA's. Once a state has an operating permit program in place, then covered sources are required to pursue a permit with that state.

The permit process begins with the submission of an application. The application summarizes the processes occurring at the facility, the terms and conditions of existing permits, and the compliance status of the facility, not only with permits but with all the sections of the Clean Air Act. If the facility is out of compliance, then a compliance plan must be submitted with the application. If the facility is in compliance, then a *responsible official* (generally an officer of the company) certifies compliance. The application, which is a public document, must be timely, and complete.

Once submitted, the applicant may receive a **permit shield.** When granted a permit shield, the applicant is deemed by the permitting agency to be in compliance with all other aspects of the Clean Air Act. It does not extend to non-compliance with a permit condition, applicable requirements that have not been covered in the application, or new requirements promulgated after the issuance of a Title V permit. Prior to issuance, EPA must review the permit. If other states may be impacted by the source's operations, then EPA ensures that these states are notified.

The agency, in issuing a permit, must include enforceable emissions limitations and standards, as well as schedules for compliance and compliance monitoring.

EPA has the option of developing a general permit for numerous similar sources. Title V also contains provisions for temporary sources.

Title VI—Stratospheric Ozone Protection

Scientists have been monitoring the stratospheric ozone layer since the late 1970s. The stratospheric ozone layer, which is 9–31 miles above the earth, shields the earth from harmful ultraviolet rays. Ultraviolet radiation can cause eye cataracts and skin cancer as well as possible harm to agriculture (plants) and plant life in the Antarctic seas. Scientific studies have led to the discovery that certain manmade chemicals, (in particular, certain organic chemicals containing halogens), were responsible for drastic deterioration of the earth's ozone layer. These chemicals have the effect of catalyzing and thus enhancing a chemical reaction that destroys stratospheric ozone.

The Montreal protocol, originally signed by nations around the world in 1987, is an international agreement to limit the worldwide production and use of certain halogenated hydrocarbons. Because of the long atmospheric lifetime of these substances, stratospheric ozone depletion is expected to continue well into the next century. Even with the controls on these substances that are stipulated in the Montreal Protocol, it will not be until the year 2045 that the ozone losses which have already occurred are predicted to recover.

Title VI of the CAA prescribes the process the United States will use to meet the terms of the protocol. Ozone-depleting substances are divided into two classes based on their ozone-depleting potential. Class I substances have the highest ozone-depleting potential and contribute significantly to harmful effect on the stratospheric ozone layer. Class I substances include certain chlorofluorocarbons, halons, carbon tetrachloride, and methyl chloroform. Class II substances are chemicals which are *known to* or *may reasonably be anticipated to* have harmful effects on the ozone layer. Class II substances include certain hydrochlorofluorocarbons.

The Clean Air Act requires EPA to prepare a list of Class I and Class II substances. New substances can be added to the list and Class II substances can be recategorized as Class I substances. Class I substances, however, cannot be removed. In addition, EPA is required to publish the chlorine and bromine loading potential, the atmospheric lifetime, and global warming potential of each listed substance.

Producers, importers, and exporters of Class I and II substances are required to report to EPA. The EPA uses this information to prepare its report to Congress on the production, use, and consumption of these substances. This report is prepared every three years. Every six years, the EPA reports to Congress on the environmental and economic effects of any atmospheric ozone depletion.

Note that phaseout of both Class I and II substances can be accelerated for a particular substance if

1)   There is scientific evidence that a more stringent schedule is necessary in order to protect the environment

2)   Substitutes and/or other technology is available that makes it possible to achieve a more stringent schedule

3)   The Montreal Protocol phaseout schedule is modified

In fact, EPA's current regulations on stratospheric ozone protection, require an accelerated phaseout of Class I and Class II substances. The production and consumption of most of the Class I substances were completely phased out by 1995. Methyl chloroform will be phased out by 2000. (There are, however, limited exceptions for essential uses of certain substances for the purposes of aviation safety, medical devices, export to developing countries that are parties to the Montreal Protocol, national security, and fire suppression.)

Class II substances are scheduled for phaseout beginning in the year 2003. At that time, certain Class II substances introduced into commerce will have production, import and use restrictions. The production phaseout of all Class II substances must be completed by the year 2030. There are exceptions to this schedule for medical devices and developing countries.

Nonessential products, such as plastic party streamers, cleaning fluids, and other consumer products that contain chlorofluorocarbons, were phased out from 1992 through 1994.

As part of the phaseout program, a trading scheme has been devised which encourages greater reductions in the use of Class I and Class II substances. Transfers between pollutants based on their ozone-depleting potential are allowed, as well as trade of consumption allowances between persons under certain circumstances. Production allowances may also be transferred between other parties to the Montreal Protocol under certain conditions.

The EPA regulates the use, labeling, and consumption of Class I and II substances by requiring recycling of these substances whenever possible. Specific regulations regarding the use and disposal of these substances during service, repair, and disposal of appliances have been issued. For example, anyone servicing a motor vehicle air conditioner must be properly trained and certified to do so. In addition, the equipment used to service these air conditioners must also be certified. Release of Class I and Class II substances to the atmosphere is specifically prohibited.

In order to encourage the replacement of Class I and Class II chemicals with environmentally acceptable alternatives, the EPA has created the Significant New Alternatives Policy Program (SNAP). The purpose of this program is to identify alternative refrigerants, solvents, and fire retardants that decrease the environmental risk. EPA maintains a clearinghouse of alternative chemicals, product substitutes, and alternative manufacturing processes. EPA also publishes by specific use a list of the prohibited substitutes as well as safe alternatives. Substances can be added to the list by petition. Health and safety studies are required from any person producing a Class I chemical substitute, prior to introducing that chemical substitute into commerce.

Implementing Regulations

The Clean Air Act, as amended in 1990, is administered and enforced by EPA. The Act has become, upon implementation, complex and difficult to interpret. The application of its requirements to a particular facility requires a good understanding of the law, the Federal, State, and, in some cases, local regulations. As a place to start, the Federal implementing regulations for the Clean Air Act are found in 40 CFR 50–88. Table 5 is a guide to locating in the regulations specific topics covered by the Clean Air Act.

Table 5. The Implementing Regulations of the Clean Air Act

Clean Air Act Topic	CFR Reference
National Ambient Air Quality Standards	40 CFR 50
State Implementation Plans and Prevention of Significant Deterioration	40 CFR 51–52
Ambient Air Quality and Monitoring, and Planning	40 CFR 53, 58, 81
Citizen Suits	40 CFR 54
New Source Performance Standards	40 CFR 60
National Emission Standards for Hazardous Air Pollutants	40 CFR 61, 63
Enforcement	40 CFR 65-67
Risk Management Plans	40 CFR 68
State Operating Permits	40 CFR 70
Acid Rain	40 CFR 72–78
Fuel Additives	40 CFR 79–80
Stratospheric Ozone Protection	40 CFR 82
Motor Vehicle Controls	40 CFR 85–88

Internet Resources

<http://www.epa.gov/airs/aewin/> (Environmental Protection Agency. *Aerometric Information Retrieval System [AIRS]*)

<http://www.epa.gov/> (Environmental Protection Agency. *Acid Rain Program Overview*)

<http://www.epa.gov/ttn/catc/> (Environmental Protection Agency. *Clean Air Technology Center*)

<http://mapsweb.rtpnc.epa.gov/RBLC/rbqry.html> (Environmental Protection Agency. *Clean Air Technology Center Query Page*)

<http://www/epa.gov.ozone/title6/snap/lists/index.html> (Environmental Protection Agency. *Lists of Substitutes: Ozone Depleting Substances*)

<http://www.epa.gov/ttn/uatw> (Environmental Protection Agency. *MACT Implementation Strategy*. EPA–456/R–97–003, 9/97)

<http://www.epa.gov/oar/emtrnd> (Environmental Protection Agency. *National Air Pollution Trends Report: 1900–1966*)

<http://www.epa.gov/oar/aqtrnd96> (Environmental Protection Agency. *National Air Quality and Emissions Trends Report*, 1996)

<http://www.epa.gov/oar/oaqps/> (Environmental Protection Agency. *Office of Air Quality Planning and Standards Homepage*)

<http://www.epa.gov/oms/omshome2.htm> (Environmental Protection Agency. *Office of Mobile Sources Homepage*)

<http://www.epa.gov/oar/oaqps/peg-caa/pegcaain.html> (Environmental Protection Agency. *The Plain English Guide to the Clean Air Act*. EPA 400–K–93–101, 4/93)

<http://www.epa.gov/ozone> (Environmental Protection Agency. *Stratospheric Ozone Protection*)

<http://www.epa.gov/oar/oaqps/takingtoxics/> (Environmental Protection Agency. *Taking Toxics Out of the Air: Progress in Setting 'Maximum Achievable Control Technology Standards under the CAA'*. EPA/451/K–98–001, 2/98)

<http://www.epa.gov/ttn/otag> (Environmental Protection Agency. *TTN Web OTAG*)

<http://www.al.noaa/wwwhd/publdocs/wmounepa4.html> (National Oceanic and Atmospheric Administration. *WMO/UNEP Scientific Assessment of Ozone Depletion*, 1994)

Adriane P. Borgias is the owner of fusion environment&energy, LLC, an environmental consulting firm that specializes in meeting the environmental needs of the energy industry. Ms. Borgias has a BS in Chemistry from the University of California, Berkeley, and a MS in Environmental Management from the University of San Francisco. She has over 20 years of experience in the energy industry in research, corporate, and operating departments, as well as an independent consultant. She has over15 years of experience in environmental and hazardous materials management, with particular experience in agency collaboration, regulatory review and development; environmental management systems, auditing and continuous improvement; hazardous materials and waste compliance activities; permitting; pipeline construction and inspection; as well as site assessment, investigation, and remediation. Ms. Borgias is a former licensed nuclear reactor operator and has been a Certified Hazardous Materials Manager since 1986. She has served the Academy at the local and national level. In addition to being co-founder of the Northern California Chapter of ACHMM, Ms. Borgias has served as a National Committee Chair, Board Member, Treasurer, and President. In 1990, Ms. Borgias received the President's Award for Outstanding Service to the Academy. She was also the 1996 recipient of the Academy's Founder's Award. Ms. Borgias has contributed chapters to Women in Chemistry and Physics, a Biobibliographic Sourcebook *and is profiled in* Northwest Women in Science: Women Making a Difference. *She would like to acknowledge in particular the invaluable assistance of her peer reviewers: Daniel L. Todd, QEP, CHMM, and Beth S. Fifield, PE, as well as her husband, Brandan A. Borgias, PhD.*

CHAPTER 30

Prevention of Accidental Releases of Chemicals: Risk Management Programs

Ken B. Baier, CSP, CEA
Alan R. Hohl, CSP, CEA, CHMM
John S. Kirar, CSP, CEA, CHMM

Introduction

The Environmental Protection Agency's (EPA) most recent Risk Management Program (RMP) requirements are found in the *Code of Federal Regulations* (40 CFR 68) and Section 112 (r), of the Clean Air Act (CAA) Accidental Release Provisions. The Rule and regulations, promulgated in 1996, can be seen as an important element of an integrated approach to chemical safety. The RMP requirements complement related industry standards and practices and build upon the chemical safety requirements established by the Occupational Safety and Health Administration's (OSHA) 29 CFR 1910.119 Process Safety Management (PSM) Standard promulgated in 1992.

Purpose

The purpose of the RMP Rule, as it is integrated with other environmental and safety regulations, is to ensure that the public will be properly informed about chemical risks in their communities, and that Federal, State and local regulators will have more effective tools to assist with lowering chemical accident risk. The RMP Rule provides a systematic protocol for facilities to identify, assess, and document chemical hazards, and describes prevention and emergency response programs for covered processes. After full disclosure of the hazard and risk information to the public and other stakeholders, industry, government, and the public can work together toward reducing the identified chemical risks to a community's health and environment.

Application

While it is EPA public policy that industry, state and Federal governments, and the public have joint responsibility for the prevention of accidental releases of hazardous chemicals, the CAA RMP Rule clearly states that facilities that handle hazardous chemicals bear the primary responsibility for ensuring their safe use. Approximately 60,000 to 70,000 facilities that store, produce, distribute, handle, or process certain chemicals are affected by the RMP requirements. Such facilities include manufacturers in the chemical, petrochemical, and refining industries and other manufacturers in various sectors such as pulp and paper, organic and inorganic chemicals, plastics and resins, nitrogen fertilizers, and agricultural chemicals. Also included are facilities that use ammonia as a refrigerant, such as food processors, distributors, and refrigerated warehouses. Public water treatment systems, chemical retailers, Federal facilities, and some service industries are also affected.

Owners or operators of stationary sources that are affected by the RMP Rule must comply with the requirements of 40 CFR 68 no later than the latest of the following dates: (1) June 21, 1999, (2) three years after the date on which a regulated substance is first listed, or (3) the date on which a regulated substance is first present above a threshold quantity (TQ) in a process.

Important Terms Defined

The CAA-mandated RMP Rule addresses the prevention of *accidental/unintended releases* of *regulated substances* from *stationary sources* to the air in order to prevent death, injury, or adverse effects to human health and the environment. The Rule applies to all facility *processes* using *regulated substances* in excess of a *threshold quantity (TQ)*. The heart of the Rule concerns the development of an RMP that includes a *process hazard analysis, worst-case and alternative release scenario hazard assessments, off-site public and environmental consequence analyses*, a *prevention program*, and *emergency response plans*. The RMP is summarized in a Risk Management Plan (RM Plan) which is to be submitted to the EPA. RMP terms are briefly described below in the context of the RMP Rule; the terminology is expanded in subsequent sections that explain RM Plan requirements.

Accidental release—An unanticipated emission of a regulated substance or other extremely hazardous substances (EHS) into the atmosphere from a stationary source.

Alternative release scenario (ARS) analysis—The owner or operator of a stationary source must analyze one ARS to represent all flammable substances in a covered process and one ARS for each toxic substance in a covered process. The owner or operator must choose a scenario that is more likely to occur than the Worst-Case Scenario (WCS) and that will reach an endpoint off-site. Even if no ARS reaches an off-site endpoint, the source is still required to perform an ARS that is more likely to occur than the WCS. When selecting an ARS, the source must consider releases that have been documented in the five-year accident history, or failure scenarios identified through the process hazard analysis, or a hazard review. Sources should consider several types of events such as process piping releases, vessel overfills, and shipping container mishandling.

Environmental receptor—Natural areas such as national or State parks, forests, or monuments; officially designated wildlife sanctuaries, preserves, refuges, or areas; and Federal wilderness areas that could be exposed to an accidental release. A stationary source owner or operator may rely on information provided on local US Geological Survey (USGS) maps or on any data source containing USGS data to identify these environmental receptors.

Extremely hazardous substance (EHS)—Any substance that, if released to the atmosphere, could cause death or serious injury because of its acute toxic effect, or as a result of an explosion or fire, or which causes substantial property damage by blast, fire, toxicity, reactivity, flammability, volatility, or corrosivity.

General duty clause—The statutory principle that facilities are responsible for designing and maintaining a safe plant, identifying the hazards, and minimizing the consequences of accidental releases of chemicals. Owners and operators of stationary sources at which regulated or any other EHSs are present in a process, regardless of the amount of the substance, are subject to the RMP general duty clause.

Hazard assessment (HA)—The purpose of the HA is to evaluate the off-site consequences of accidental releases to public health and the environment. HAs must be conducted for each regulated substance and must include the following:

- An analysis of a worst-case release scenario for Program 1, 2, 3 processes
- An analysis of an alternative release scenario for Program 2, 3 processes
- A five-year history of accidental releases for Program 1, 2, 3 processes

Mitigation system—Mitigation means specific activities, technologies, or equipment designed or deployed to capture or control substances upon loss of containment to minimize exposure to the public or the environment. Passive mitigation means equipment, devices, or technologies that function without human, mechanical, or energy input; common examples include dikes, berms, and flame arresters. Active mitigation systems are equipment, devices, or technologies that require human, mechanical, or energy input to function. Examples include relief and isolation valves, sprinkler systems, and fire brigades.

Offsite consequence analysis (OCA)—Areas beyond the property boundary of the stationary source, or areas within the property boundary to which the public has routine and unrestricted access, both during or outside business hours, are considered *offsite*. An OCA must include an analysis of at least one worst-case release scenario and one alternative release scenario and address the impacts to offsite populations and the environment. The EPA has developed and summarized offsite consequence modeling via an OCA guidance document. The document contains simplified *lookup* tables of dispersion and explosion models to minimize a source's modeling efforts.

Owner/operator—Any person who owns, operates, leases, controls, or supervises a stationary source and its processes. For RM Plan registration purposes, the owner or operator is normally the highest-ranking company executive onsite.

Process and covered process—*Process* means any activity involving a regulated substance, including any use, storage, manufacturing, handling, or onsite movement of such substances, or combination of these activities. A *covered process* is a process that contains a regulated substance in excess of a TQ. Any group of vessels that are interconnected, or separate vessels that are located such that a regulated substance could be involved in a potential release, are considered a single process.

Process hazard analysis (PHA)—An analysis that identifies, evaluates, and controls the hazards involved with the design, operation, and maintenance of processes containing regulated substances. The PHA level of detail is commensurate with the complexity of the process and must consider the extent of the process hazards, the number of potentially affected employees, age of the process, and the operating history of the process. PHAs developed to meet the OSHA PSM Standard are acceptable to meet the requirements of the RMP Rule.

Program 1, 2, or 3 eligibility—EPA has established three program levels. All stationary sources must assign each covered process a program level. The requirements for Programs 1, 2, and 3 are based on the size and complexity of a source's processes. Program 1 applies to sources that have not had a significant accidental release within the previous five years.

Program 1 sources require only registration and a certification that the process's worst-case release would not affect off-site populations or the environment. Program 2 requires sources to register, conduct an off-site consequence analysis, document a five-year accident history, and summarize prevention and emergency response steps taken. Prevention requirements are modeled after the OSHA 29 CFR 1910.119, PSM Standard. Program 3 sources require full compliance with the RMP Final Rule.

Public receptor—Receptors that include off-site residences, institutions (schools, hospitals), industrial, commercial, and office buildings, parks, or recreational areas occupied by the public at any time without restriction where members of the public could be exposed to toxic concentrations, radiant heat, or overpressure, as a result of an accidental release of a regulated substance. Roads are not included as public receptors but should be considered when coordinating with emergency planning organizations.

Regulated substance—The EPA List Rule (40 CFR 68.130) presents the list of regulated substances that includes 77 volatile toxic substances and 63 flammable substances. Toxic and flammable substances each constitute a separate hazard class. Toxic substances are included on the list based on their toxicity, physical state, vapor pressure, production volume, and accident history. Toxicity criteria used to identify chemicals as EHSs under the Emergency Planning and Community Right-To-Know Act (EPCRA) were used as criteria for the list. TQs are established for toxic substances ranging from 500 to 20,000 pounds.

Flammable gases and volatile flammable liquids were included on the list based on the flashpoint and boiling point criteria used by the National Fire Protection Association (NFPA) for its highest flammability hazard ranking (flashpoint below 73°F and boiling point below 100°F). All listed flammables have a TQ of 10,000 pounds. At this time, the EPA does not expect to add any chemicals to the list of regulated substances. The statute, however, requires EPA to review the list at least every five years.

Several important exemptions are also defined in the List Rule. Exemptions include pipelines, transfer stations, associated storage, Department of Transportation (DOT) shipping containers

under active shipping orders, and other related activities covered by DOT under 49 CFR 192, "Transportation of Natural and Other Gas by Pipeline," 49 CRF 193, "Liquefied Natural Gas Facilities," or 49 CFR 195, "Transportation of Hazardous Liquids by Pipeline." Additional exemptions include manufactured items (*e.g.,* batteries), products used for routine janitorial maintenance, employee use of foods, drugs, cosmetics, or other personal items, laboratory activities, and ammonia used as an agricultural nutrient.

Stationary source—All buildings, structures, equipment, or substance-emitting stationary activities which belong to the same industrial group, located on one or more contiguous properties, under the control of the same person(s), from which accidental release may occur. Examples include chemical manufacturing plants, oil refineries, and refrigerated (ammonia) warehouses.

Threshold quantity (TQ)—The maximum amount of a substance used in a single process that triggers the requirement for the substance to be reported and included in the emergency response plans for chemical spills and releases. The TQ is determined by the maximum amount of a substance used in a single process, not the maximum quantity on-site. The entire weight of a mixture containing a regulated flammable substance must be counted for TQ determination if the mixture itself meets the NFPA "4" criteria. Some TQ examples for toxic chemical and flammable substances include:

Toxic Chemical	RMP TQ	Toxic Endpoint
Ammonia (≥ 20%)	10,000 lb	0.011 mg/l
Hydrogen Sulfide	10,000 lb	0.042 mg/l
Nitric Acid (≥ 80%)	15,000 lb	0.026 mg/l

Flammable Substance	RM TQ
Butane	10,000 lb
Propane	10,000 lb
Vinyl Chloride	10,000 lb

Worst case release scenario (WCS) analysis—The WCS is the scenario in which the release reaches the greatest distance beyond the stationary source's boundary. In order to accurately determine which WCS should be specifically analyzed and documented in the RM Plan, the owner or operator will need to evaluate the WCS (*i.e.,* determine the scenario with the greatest distance to the endpoint) for every covered process. If any particular covered process contains more than one regulated substance in excess of a TQ, WCSs involving each of those substances will need to be analyzed.

RM Plan Requirements

The RM Plan is envisioned by the EPA as a multi-purpose document that will be useful to the public, the facility, and regulators. The EPA intends each RM Plan to attain the following major goals:

- The stationary sources will develop RM Plans with the requisite chemical accident hazard assessment, prevention, and emergency response program planning.

- The public will be provided with sufficient information in an understandable form to encourage improvement in communication between the public and the sources on prevention and preparedness issues.

- The regulators will be provided with information of sufficient detail to enable audit and inspection of the plans against RMP requirements.

The EPA plans to reduce the cost of RM Plan development for small businesses by developing *Model RM Plans*. The "Model Risk Management Program and Plan for Propane Users and Retailers," is a good example of an available model and can be located on the Internet at

<http://www.epa.gov/swercepp/acc-pre.html>

<http://www.epa.gov/ceppo/pubs/plan/material. html>

Note: Model Plans for chemical distributors, warehouses, ammonia refrigeration, *etc.*, are also found at the Universal Resource Locator (URL) addresses listed above.

The RMP Rule requires the RM Plan to indicate compliance with the regulations, include a hazard assessment, and include prevention and emergency response programs. Figure 1 depicts the major elements of an RM Plan.

Figure 1. RM Plan Major Elements

Registration

RM Plans must be registered with the EPA by June 21, 1999, or three years after the date on which a regulated substance is first listed, or the date on which a regulated substance is first present above a TQ in a process, whichever is later. RM Plans must also be submitted to the Chemical Safety and Hazard Investigation Board, the implementing agency (the state or EPA), the State Emergency Response Commissions (SERCs), and Local Emergency Planning Committees (LEPCs). The plans will also be made available to the public. Each regulated substance handled in a covered process requires its own individual registration.

Process Safety Information

Current written process safety information must be compiled for process chemicals, process technology, and source equipment before a PHA can be performed. Such information provides the owner or operator and operations personnel with an understanding of the hazards involved in a particular process.

For process chemicals, toxicity, physical, reactivity, corrosivity, chemical, and thermal stability data must be obtained. Information on chemical exposure limits (*e.g.*, permissible exposure limits [PEL], threshold limit values [TLVs], Emergency Response Planning Guidelines [ERPGs], *etc.*) and the effects of inadvertent mixing of incompatible materials must be included, also. Material Safety Data Sheets (MSDS) that are compliant with 29 CFR 1910.1200 (OSHA Hazard Communication Standard) may be used to satisfy these information requirements.

Equipment and technology data and hazards should include the materials of construction, a simplified process flow or block diagram, process chemistry, maximum intended inventory, safe upper and lower temperature, pressure, and compositions. The written information must list the consequences of deviations from standard operation, Piping and Instrumentation Diagrams (P&IDs), electrical classification, relief system design and basis, ventilation system design, design codes and standards used, material and energy balances, and safety systems such as interlocks, detection, and protection equipment. The owner or operator must document that the process equipment and the processes comply with generally accepted good engineering practices. For existing equipment that was designed and constructed in accordance with codes and practices no longer in general use, the owner or operator must determine and document that the equipment has been maintained, tested, and is operating safely.

Process Hazard Analysis

The owner or operator must perform a Process Hazard Analysis (PHA) on processes covered by the RMP Rule. The PHA must identify the potential causes of a mishap (such as a failed valve, malfunctioning gauge, human error), evaluate the possible consequences of a mishap, and control the hazards involved with the design, operation, and maintenance of processes containing regulated substances. The PHA's level of detail should match the complexity of the process and must take into account the number of potentially affected employees, age of the process, and the operating history of the process. The analysis must describe and analyze both engineering and administrative controls and their interrelationships. A hazard evaluation team with engineering and process operations expertise must perform the PHA, and at least one team member must be knowledgeable of the PHA methodology being used. Program 3 processes require the completion of a PHA, while Program 2 processes must contain a hazard review as part of its prevention program. The hazard review is similar to a PHA, and must identify the hazards associated with the process and regulated substances. A hazard review should also address opportunities for equipment malfunctions and human errors, safeguards needed to control the hazards, and any steps used or needed to detect or monitor releases. Program 1 processes do not require a PHA.

The PHA must be updated and revalidated by a hazard evaluation team at least every five years. The owner or operator must retain PHAs, updates, and revalidations for the life of the process. PHAs developed to meet OSHA's PSM Standard are acceptable to meet the requirements of the RMP Rule.

Hazard analysis methodologies that can be used to complete a PHA are What-If Analysis, Checklist, Hazard and Operability Study (HAZOP), Failure Mode and Effects Analysis (FMEA), Fault Tree

Analysis (FTA), or an appropriate equivalent methodology. The techniques of each method are briefly described below.

What-If. A What-If analysis considers the consequences of events that occur as a result of failures involving equipment, design, or procedures. The What-If analysis allows the hazard evaluation team to examine hypothetical failures of the system in terms of *what-if* questions. All possible system failures are collected in checklist form and then evaluated. The people compiling the list of failures must have a basic understanding of what is intended in the analysis, and the ability to combine or synthesize possible deviations and reject incredible scenarios.

Checklist. A Checklist is an analytical technique that involves developing a checklist of failure areas and reviewing each area to determine the possible effects of a failure.

What-If/Checklist. This methodology combines the What-If and Checklist analysis techniques to identify and evaluate process hazards.

Hazard and Operability Study. Hazard and Operability (HAZOP) analyses are conducted by brainstorming teams to systematically identify hazards or operability problems throughout a process via the use of certain guidewords such as *no flow* and *no cooling*. The consequences of the deviation associated with the guidewords are assessed, and credible deviations are identified and addressed.

Failure Mode and Effects Analysis. The Failure Mode and Effects Analysis (FMEA)is a methodology that tabulates a process's equipment, failure modes, and the effect of each failure mode on the process, and then ranks each failure mode.

Fault Tree Analysis (FTA). Fault Tree Analysis (FTA) is a deductive technique that focuses on a top-level accident event and provides a method for determining causes of the event. The fault tree is a graphic display that shows all of the various combinations of equipment faults and failures that can result in a release.

Hazard Assessment

An RMP Hazard Assessment (HA) evaluates the impact of significant accidental releases on public health and the environment, and develops a history of such releases. The RMP Rule does not specify what the HA results are to be used for; however, the results are to made available to the public. The results should be used by the LEPC for emergency response planning.

Owners and operators must review and update the HA at least every five years. A change in the covered process, management, facility, or community requires amendment of the HA within 60 days. An HA must be performed for each regulated substance and include the procedures shown below.

Off-Site Consequence Analysis

The Off-site Consequence Analysis (OCA) must establish the rate, duration, and quantity of substance released into the air, the distance in all directions in which exposure could occur in a WCS, and the most common meteorological conditions for the area. An OCA must also consider off-site populations within three-mile distances that could be exposed to a vapor cloud, and the environmental damage that could occur. An OCA must include an analysis of at least one WCS and one ARS. To reduce the cost burden on facilities and to allow consistent and streamlined assessments, the EPA has developed off-site consequence modeling guides. The RMP OCA Guidance document–

<http://www.epa.gov/swercepp/acc-re.html#OCA Guidance>

contains simplified dispersion and explosion impact distance *lookup* tables. The use of such tables can provide considerable savings to smaller sources or those sources without the technical expertise to purchase and use models that are available commercially. Sources using the guidance document assume a generic 3 meters per second wind speed, "D" atmospheric stability class, 25°C and 50% humidity.

EPA also provides RMP*Comp, an electronic tool used to perform the OCA under the Risk Management Program Rule. Previously, EPA has referred to this tool as "RMP Calculator" or "RMP Assistant." RMP*Comp is provided to help with your initial risk management program planning and evaluation. Macintosh and PC-compatible formats are available for download at

<http://response.restoration.noaa.gov/chemaids/ rmp/rmp.html>

The costs of commercially available transport and dispersion models range from approximately $100 to $17,000 depending on the uncertainties in the model, types of calculations performed, and the level of refinement. Some models are better for use with a specific gas release scenario and some are accepted more widely by public stakeholders. The lookup tables, while inexpensive, could result in overly conservative consequence analyses compared to some commercially available models. Available dispersion models and databases (such as ARIP, ALOHA, CAMEO, and Landview™) are provided by the EPA at

<http://www.epa.gov/swercepp/tools.html>

If the owner or operator is able to demonstrate that the applicable local meteorological data indicate a higher minimum wind speed or less stable atmospheric conditions over the course of the previous three years than what is shown in the *lookup* tables, then the local data may be used for the OCA.

Information on the locations of chemical accident receptors (public and environment) is readily available from the United States Census, the USGS, and the EPA. These agencies have made available software such as LandView™ II, which is an innovative community right-to-know DOS-based software tool. LandView is published on CD-ROM in the format of an electronic atlas, and can also be downloaded from the Internet at

<http://www.rtk.net/landview/>

LandView™ runs on standard personal computers. It displays county-by-county maps and tables that combine EPA databases with geographic features and statistics on demographics and economics from the 1990 Census. Landview™ can be used to help local communities evaluate environmental risks and identify areas of concern for environmental justice. Commercial sources of RMP software are also available to assist with RM Plan development, analysis, and implementation.

Since the EPA focuses on exposures that result in serious, irreversible health effects, the agency has chosen to use Emergency Response Planning Guideline (ERPG)–2 values to determine the toxic endpoints for OCAs. The ERPG–2 values indicate the maximum airborne concentrations of a substance to which nearly all individuals could be exposed for up to one hour without experiencing or developing irreversible or other serious health effects or symptoms that could impair their abilities to take protective action. For toxic substances that have no ERPG–2 values, determine endpoints by using the level of concern (LOC), as identified in the Technical Guidance for Hazards Analysis published jointly by the Federal Emergency Management Agency, the EPA, and the DOT. LOCs are intended for the protection of the general public for exposure periods of up to one hour.

Single Worst-Case Release Scenario

A Worst-Case Release Scenario (WCS) is the event that would result in a release that traveled the greatest distance to an end point beyond the plant or stationary source boundary. In order to accurately determine which WCSs should be specifically documented in the RM Plan, the owner or operator may need to analyze the WCS for every covered process. Each WCS assumes either an instantaneous release (in the case of toxic substances that normally are liquids) or a 10-minute release duration (in the case of toxic substances that normally are gases, and refrigerated liquids) that encountered only passive mitigation. If any particular covered process contains more than one regulated substance in excess of a TQ, WCSs involving each of those substances may need to be analyzed.

A scenario involving a smaller quantity of regulated substance handled at a higher process temperature or pressure, as well as a scenario involving a smaller quantity located closer to the boundary of the stationary source, may result in the WCS.

One WCS must be analyzed and reported in the plan for each Program 1 process. In order for any process to be eligible for Program 1 requirements, the owner/operator must demonstrate that a WCS from that process would not affect any public receptor. For Program 2 and 3 processes, a single WCS analysis may be reported to represent all regulated toxic substances, and a single WCS analysis will be acceptable to represent all regulated flammable substances. If WCSs from other covered processes potentially affect public receptors other than those outlined in the first WCS, additional WCSs must be analyzed and reported for Program 2 and 3 processes.

Owners and operators are only required to analyze a regulated substance for the hazard for which it is listed. For example, ammonia is listed as a regulated toxic substance, so ammonia's WCS must be modeled as a toxic release. In order to make the public and first responders aware of additional hazards, an owner or operator may also want to analyze the effects of an explosion or fire involving the listed toxic substance. In the Program 2 hazard review and Program 3 PHA, all hazards of a regulated substance should be considered.

If the WCSs for a regulated toxic substance and for a regulated flammable substance in Program 2 and 3 processes come from the same process, then the two WCSs must be analyzed separately.

An Example of a WCS. Failure of a large storage tank when filled to the greatest amount allowed would release 222,000 pounds of propane. Company policy limits the maximum filling capacity of this tank to 88% at 60°F. Assuming that the entire contents of the tank are released as vapor that finds an ignition source, 10% of the released quantity can be assumed to participate in the resulting explosion. Using the OCA *lookup* tables, a one-pound-per-square-inch (psi) over-pressure for this WCS is a distance of 0.47 miles.

Alternative Release Scenarios

The owner/operator of a stationary source must analyze one Alternative Release Scenario (ARS) to represent all flammable substances in a covered process and one ARS for each toxic substance in a covered process. The owner/operator must choose a scenario that is more likely to occur than the WCS and that will reach an end point off-site. Even if no ARS reaches an off-site end point, the source is still required to perform an ARS that is more likely to occur than the worst-case scenario. When selecting an ARS, the evaluators must take into account releases that have been documented in the five-year accident history, or failure scenarios that were identified through the PHA or hazard review. As many types of events as are applicable to the process should be considered for the ARS.

The owner or operator should use typical meteorological conditions for the facility (*e.g.*, wind speed, atmospheric stability, temperature, and humidity) when compiling an ARS. Typical meteorological

data can be gathered at the stationary source or at a local meteorological station.

An Example of an ARS. A vehicle pull-away causes the failure of a 25-foot length of 4-inch hose during the loading of propane. The excess flow valves engage to stop the flow of propane. The full content of the hose is released and the resulting unconfined vapor travels to the lower flammability limit. The distance to the end point for the lower flammability limit for the alternative scenario is less than 317 feet. This release has the possibility of extending beyond the facility boundary.

Five-Year History of Accidental Releases

The owner or operator of each process must keep a separate record of each accidental release from a covered process that has occurred within the last five years and that resulted in deaths, injuries, or significant property damage on-site. The accident history must also record any known off-site deaths, injuries, evacuations, sheltering in-place, property damage, or environmental damage. No process is eligible for Program 1 if it has had an accidental release of a regulated substance which resulted in off-site death, injury, or response or restoration activities at an environmental receptor in the five years prior to the submission date of the RM Plan.

The five-year history must include

- Quantity of regulated substance released
- Date and time of release
- Duration
- Type of release (such as a gas release, fire, spill, or explosion)
- Concentration of regulated substance that was released
- Description of off-site consequences

Prevention Program Requirements

The elements of an RMP prevention program adopt and expand OSHA's PSM standard. At this time four states, New Jersey, California, Nevada, and Delaware, have accidental release prevention regulations. The Chemical Manufacturers Association (CMA), the American Institute of Chemical Engineers (AIChE), the American Petroleum Institute (API), and the European Community (EC) all have programs of chemical PSM.

The prevention program is designed to evaluate the hazards present at the facility, and to find the best ways to control the hazards and detect and prevent accidental releases. The plant's owner/operator must implement an integrated management system suitable to the hazard level of the facility.

Management System

To ensure the effective integration of all accident prevention elements, the owner/operator is required to implement a management system for facilities with Programs 2 and 3 processes. The requirements for Program 1 systems are found in the RMP Rule general requirements section. Owners or operators are required to designate a single person or position that has overall responsibility for the development, implementation, and integration of RMP requirements.

The management system is essentially a system defined by facility managers for integrating the implementation of the RMP elements and assigning responsibility. The extent of the management system will depend on the size and complexity of the source. At many small sources, the appointment of a person or position that has the overall responsibility for the development, implementation, and integration of the risk management program elements may satisfy the management system requirement. For larger sources, separate divisions may be responsible for overseeing different elements of the RMP.

The management system should document the integration of a source's operations. For example if process equipment is changed, process safety information must also be changed, training may need to be revised, and both operators and maintenance staff must be informed of the change. The management system is a means to ensure that each employee involved in a process knows his or her responsibilities, the lines of authority, and the contacts when changes or issues arise.

There is no standard format for the management system. An organization chart may document the system or other methods may also be used. The rule provides the flexibility for the source to design, implement, and maintain its management system as it sees fit. Sources are not required to submit the management system to the EPA.

Standard Operating Procedures

Facilities subject to RMP must develop and implement written operating procedures that provide clear instructions for safely conducting the activities involved in each covered process. Standard Operating Procedures (SOPs) must be consistent with the process safety information. SOPs must include initial startup, normal, temporary, and emergency operations, normal and emergency shutdown, operating limits, safety and health considerations, chemical hazards, and the consequences of deviation. SOPs must be reviewed and updated as often as necessary to ensure they reflect current operating practice, including the changes that result from process chemical, technology, and equipment modifications, and changes to the facility.

Training

The training program must provide every employee who operates a process with initial training that emphasizes specific safety and health hazards, emergency operations (including shutdown) and safe work practices. Refresher training must be provided at least every three years. Proof of training must show that each employee involved in operating a process has received and understood the required training. The documentation must also show the date of the most recent training, the type of training provided (classroom, on-the-job, or a combination), and the type of competency testing used.

Maintenance (Mechanical Integrity)

A maintenance program must ensure the mechanical integrity of process equipment. The maintenance program's procedures must be written, and training must be provided for all employees who are involved in maintenance activities. The maintenance personnel must inspect and test process equipment in accordance with industry standards, codes, or good engineering practice.

Pre-Startup Review

A pre-startup review must be performed for new facilities and for modified stationary sources when the modification is significant enough to require a change in the process safety information and HA.

Management of Change

The owner/operator must establish and implement written procedures to manage changes (not including *replacements in kind*) and to address process chemicals, technology, equipment, and procedures, as well as changes in the stationary sources that affect a covered process. The procedures must address the technical basis for the proposed change, impacts to safety and health, modifications to operating procedures, and the authorization requirements for the change. Employees must be informed of and trained on the change.

Compliance Audits

Owners/operators must certify that they have evaluated compliance with the applicable prevention program provisions at least once every three years to verify that established procedures and practices are adequate and are being followed. The audit must be conducted by at least one person knowledgeable in the process. A report of the findings must be developed, and the owner or operator must promptly document responses to each finding.

Accident Investigation

The owner or operator must have procedures to investigate each incident that resulted in, or could reasonably have resulted in, a catastrophic release of a regulated substance within 48 hours of the incident. A summary report must be prepared that includes the date, description, and contributing factors to the incident, and recommendations, resolutions, and corrective actions implemented. For a good EPA description of an accident

investigation with RMP/PSM findings and recommendations listed, see

<http://www.epa.gov/swercepp/acc-is.html#terra press release>

(Terra International, Inc., Port Neal Complex accident.)

Employee Participation

The owner/operator must develop a written plan of action to implement an employee participation program. The owner or operator must consult with employees on the conduct and development of PHAs as well as elements of the PSM. All employees must have access to this process safety information.

Emergency Response Program

EPCRA assists local communities in preparing for and responding to chemical incidents/accidents. EPCRA requires communities to develop emergency response plans based on information from industry concerning hazardous chemicals. As with EPCRA, the RMP emergency response program's purpose is to prepare to respond to, and mitigate, accidental releases. Preplanning can limit the severity of accidental releases and their subsequent impact to public health and the environment. RMP emergency response planning considers the steps to be taken by the owners or operators of stationary sources, their workers, LEPCs, and the public.

To prevent unnecessary duplication of such emergency response plans, the EPA has agreed to implement an interagency *One-Plan* guidance for release of oil and hazardous substances. *One-Plan* guidance was developed under the auspices of the National Response Team with participation by industry and environmental groups, State agencies, and many Federal agencies, including the EPA. The guidance provides facilities with a common-sense option for meeting multiple emergency planning requirements under the numerous Federal regulations. The guidance includes development of a core facility response plan for releases of oil and hazardous substances. The core plan will be supplemented with annexes

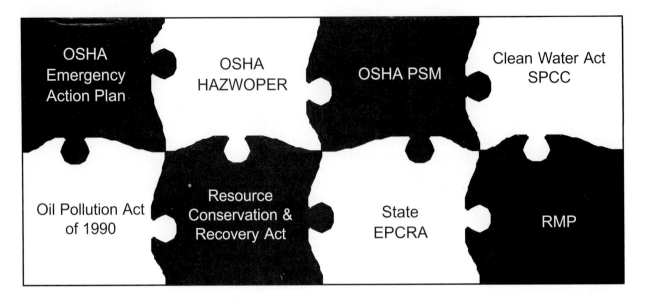

Figure 2. The Emergency Response Plan Puzzle

that contain specific information such as a facility's incident command system and specific facility and process hazards required to meet the eight sets of different federal regulatory requirements. Facility plans prepared in accordance with *One-Plan* guidance will satisfy facility emergency response planning requirements and will minimize duplication of effort.

Figure 2 depicts the interlaced nature of Federal and State regulations that may impact a stationary source's emergency response plans and that may be integrated by the EPA's *One Plan*.

The owner or operator must develop and implement an emergency response program for the purpose of protecting public health and the environment. Emergency response plan elements include

- Procedures for informing the public and local emergency response agencies about accidental releases, documentation of proper first aid and emergency medical treatment necessary to treat accidental human exposures, and procedures for emergency response after an accidental release of a regulated substance. (Sources are not required to have a plan if public responders or other nonemployees handle all response activities.)

- Procedures for the use of emergency response equipment and for its inspection, testing, and maintenance.

- Training for all employees in relevant procedures.

- Procedures to review and update, as appropriate, the emergency response plan to reflect changes at the source and ensure that employees are informed of these changes.

Summary Risk Management Plan

The owner or operator of a source must submit a single Risk Management (RM) Plan for all covered processes. The RM Plan must be submitted using a method and format specified by the EPA prior to June 21, 1999. Sources will be required to submit the plan electronically to a central point specified by the EPA, as discussed below. The RM Plan will be made available to the EPA Chemical Safety and Hazard Investigation Board, State and local governments, SERCs, LEPCs, and the public. RM Plan Data Element Instructions, specific industry guidance and industry model plans, as well as submission and access systems information can be found at

<http://www.epa.gov/ceppo/acc-pre.html>

<http://www.epa.gov/ceppo/rmpsubmt.html>

EPA provides a free, official EPA, personal computer software program, RMP*Submit, for facilities to use in submitting Risk Management Plans (RMP). Portions of the RMP submitted to EPA will be publicly available on the Internet via RMP*Info. RMP*Submit downloads are available at

<http://www.epa.gov/ceppo/rmpsubmt.html>

RMP*Submit is also available free of charge through the National Center for Environmental Publications and Information (NCEPI) at

    Phone: (800) 490–9198
    Fax: (513) 489–8695
    Email: necpi.mail@epamail.epa.gov

RMP*Submit may also be ordered through the on-line NCEPI Publication Request Form at

<http://www.epa.gov/ncepihom/nepishom>

RMP*Submit

- Has all the data elements identified in the RMP regulations
- Verifies completion of all required data

- Allows correction of errors before submitting your RMP
- Accepts limited graphics
- Provides on-line help

With the exception of the Executive Summary, most of the data elements are submitted in a checklist format consisting of check-off boxes, numerical entries, and yes/no answers. Major elements, depicted in Figure 3, are described below.

Executive Summary

The executive summary section must include brief descriptions of

- The accident release prevention and emergency response policies at the stationary source
- The stationary source itself and the regulated substances handled
- The WCS(s) and the ARS(s), including administrative and mitigation measures to limit the endpoint distances for each scenario

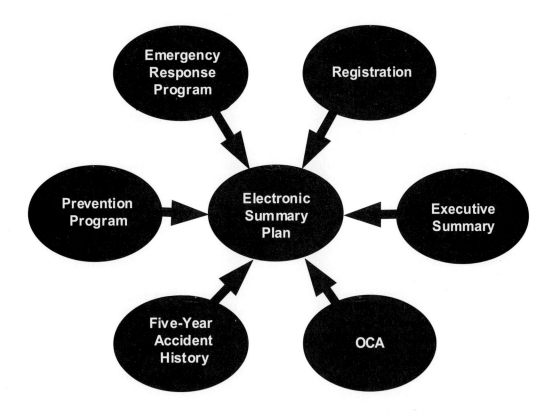

Figure 3. Risk Management Plan

- The general accident release prevention program and chemical-specific prevention steps
- The five-year accident history
- The emergency response program
- Planned changes to improve safety

Registration

Registration consists of a single registration form. The form covers all regulated substances handled in a covered process. The required information includes

- Source identification
- Dun and Bradstreet Number, if applicable
- Owner or operator names and mail addresses
- Business phone number
- Name of person responsible for RMP implementation
- Emergency contact name, title, and phone number
- Chemical names, Chemical Abstracts Service (CAS) registration numbers, quantity, North American Industry Classification Code (NAICS), which replaces the Standard Industrial Classification (SIC)
- RMP program level
- EPA Identification Number
- Number of full-time employees or equivalents
- OSHA PSM, EPCRA, and CAA Title V operating permit applicability
- The date of the last safety inspection

Facilities must amend their registration within 60 days after changing a chemical, the use of a chemical, or the amount of a chemical used in a process.

Off-Site Consequence Analysis

Depending upon each process's program level, include the WCS(s) for toxic and flammable substances above a TQ, and ARS(s) for toxic and

flammable substances above a TQ. The data to be included is basically a synopsis of analyses performed, as previously described in this chapter's OCA, WCS, and ARS sections. The off-site consequence analysis (OCA) data will not be posted on the Internet.

Five-Year Accident History

A separate record for each accidental release from covered processes that has occurred within the last five years resulting in deaths, injuries, or significant property damage on-site, or known off-site deaths, injuries, evacuations, sheltering-in-place, property damage, or environmental damage must be submitted. Data required to be submitted includes dates, times, release durations, chemical quantities released, type of release, weather conditions, initiating events, and contributing factors.

Prevention Program

Depending upon each process's program level, sources will supply WCS(s) for toxic and flammable substances above a TQ, and ARS(s) for toxic and flammable substances above a TQ for each process. Data to be included is basically a synopsis of the PHA, including technique used (such as What-If, FMEA, or FTA), changes resulting from the PHA, major hazards identified, mitigation systems, monitoring/detection systems, and any changes in inventory or processes since the last PHA. Management of change, SOP, and incident investigation information are detailed in this section.

Emergency Response Program

In this data element, the source is queried regarding written emergency response plans, procedures, training, LEPC contact coordination data, and other regulations the facility is subject to such as EPCRA, OSHA, and RCRA.

Certification (Program 1)

The owner or operator of a Program 1 process must certify the following to the EPA:

Based on the criteria in 40 CFR 68.10, the distance to the specified end point for the worst-case accidental release scenario for the following process(es) is less than the distance to the nearest public receptor: [list process (es)]. Within the past five years, the process (es) has (have) had no accidental release that caused offsite impacts provided in the risk management program rule (40 CFR 68.10[b][1]). No additional measures are necessary to prevent off-site impacts from accidental releases. In the event of fire, explosion, or a release of a regulated substance from the process(es), entry within the distance to the specified end points may pose a danger to public emergency responders. Therefore, public emergency responders should not enter this area except as arranged with the emergency contact indicated in the RMP. The undersigned certifies that, to the best of my knowledge, information, and belief, formed after reasonable inquiry, the information submitted is true, accurate, and complete. [Signature, title, date signed].

RM Plan Developments

On April 17, 1998, EPA amended the RMP rule to add four mandatory and five optional RMP data elements, establish specific procedures for protecting confidential business information (CBI) when submitting RMPs, replace the use of Standard Industrial Classification (SIC) codes with the North American Industry Classification System (NAICS) codes, and make technical corrections and clarifications to 40 CFR 68.

The four mandatory elements added are: latitude/longitude method and description, CAA Title V permit number, percentage weight of a toxic substance in a liquid mixture, and NAICS code for each process that had an accidental release reported in the five-year accident history. EPA also added five optional data elements: Local Emergency Planning Committee (LEPC) name, source or parent company email address, source homepage address, phone number at the source for public inquiries, and status under OSHA's Voluntary Protection Program (VPP).

EPA clarified how confidential business information (CBI) submitted in the RMP will be handled. EPA determined that the information required by certain RMP data elements does not meet the criteria for CBI and, therefore, may not be claimed. EPA also requires submission of substantiation at the time a CBI claim is filed.

Finally, EPA promulgated several of the technical corrections and clarifications, as proposed in the *Federal Register*, April 17, 1998 (63 FR 19216).

On November 5, 1998, the Director of the EPA Chemical Emergency Preparedness and Prevention Office issued a memorandum stating that, due to security and potential terrorist threat concerns, the off-site consequence analysis (OCA) data will not be posted on the Internet, as previously planned.

As shown above, it is important that the reader stay abreast of the latest EPA RM Plan electronic submission requirements by periodically accessing the Internet address

<http://www.epa.gov/ceppo/acc-pre.html>

Recordkeeping

The owner or operator must retain detailed records supporting the implementation of Program 1 and 2 process RM Plan for five years. Program 3 PHAs, updates, and revalidations must be kept for the life of the process. Accurate written records must support the prevention program requirements as previously presented. The records can be a focus of regulator inspections and audits as presented in the Regulatory Audits and Inspections Section, below.

Public Participation and Risk Communication

Public Participation

The EPA has not adopted specific public participation requirements. As described previously however, the EPA hopes that making RM Plans immediately available to the public, other stakeholders, and LEPCs will present new opportunities for all parties to better understand a community's chemical industries, the chemicals used, prevention practices, and community concerns. RM Plans available for public review will not include classified information; such data will be included in a classified annex that is reviewable only by State and Federal representatives with appropriate security clearances.

Risk Communication

The EPA will make RM Plans available to the public, other stakeholders, LEPCs, as well as the regulatory agencies. Consequently, early communication of the existence, nature, form, severity, and acceptability of the hazards and risks involved with the source's processes can be extremely important. The process of risk communication between those responsible for risk management, the risk bearers, and the wider community as a whole can be a difficult one, especially if the facility has little or no experience or capability of doing so. It may be problematic for some sources that are required to develop RM Plans, as they often begin the process from a position of distrust with some community groups due to the types of chemicals used.

The risk communication process involves four major components: (1) the sender of the message or, in the RMP case, the *source*, (2) the risk message itself, (3) the receiver of the message or the *audience*, and (4) the feedback loop. If there is no feedback component to the process, the sender is not listening to what the audience, the public, other stakeholders, and the regulators are saying. There exists a body of risk communication experience and lessons learned that is available to facilities and that includes tactful and truthful methods of persuasion, as well as program design and implementation processes. To help ensure a successful RMP risk communication program, historical environmental risk communication lessons learned should be reviewed and include the following points:

- Accept and involve all stakeholders as partners in the RM Plan risk communication process.

- Know who all your audience is: local public, professional environmentalists, LEPC members, regulators, *etc.*

- Listen to their concerns and respond positively with truthful answers.

- Speak *their language*; speak clearly and with compassion; be honest, open, and frank.

- Disclose all RM Plan information.

- Meet the needs of the media.

- Establish consistent RM Plan assumptions and content in OCA, WCS, and ARS scenarios as well as presentation materials.

- Present specific examples of how the facility prevents accidents, spills, and releases to the environment; openly present a commitment to safe production, storage, handling, and processing of chemicals.

- Use risk communication programs early in the RM Plan development stage to begin building credibility and relationships with the local media, LEPCs, the public, and other important stakeholder groups.

- Cooperate and collaborate with neighboring facilities, industries, and trade groups; sources may find strength in numbers.

- Participate in LEPC planning and exercises.

- Plan to continuously improve documentation and communication processes.

- Use professional, trained, experienced contractors to assist with RM Plan development and risk communication if internal company capabilities do not exist.

Regulatory Audits and Inspections

The EPA and other implementing agencies have general audit, inspection, and enforcement authority under the RMP to compel source owners and operators to correct deficiencies in their RM Plan. The EPA intends to use the audit process as a way to verify and modify the quality of the RM Plan elements submitted. Audits may be detailed paper reviews or may be performed at the source to confirm that on-site documentation is consistent with reported information.

Additionally, the air permitting authority, whether State or Federal, may also perform audits of a source's RM Plan for completeness rather than the quality of the RM Plan content. Inspections, on the other hand, are generally more thorough than audits and may review all 40 CFR Part 68 requirements, as well as RM Plan elements, including operating practices.

The EPA has not included public petitions as a mechanism to trigger audits and inspections of a source's RMP. However, it should be noted that states that have implemented RMP and therefore have regulatory authority may adopt more stringent requirements than exist in the EPA RMP Rule.

The implementing agency will select stationary sources for audits based on any of the following criteria: (1) accident history of the stationary source, (2) accident history of other stationary sources in the same industry, (3) quantity of regulated substances present at the stationary source, (4) location of the stationary source and its proximity to the public and environmental receptors, (5) the presence of specific regulated substances, (6) the hazards identified in the RM Plan, and (7) a plan providing for neutral, random oversight. Audit findings will be submitted in writing to owners or operators of the source.

It is important to note that those sources with a Star or Merit ranking under OSHA's Voluntary Protection Program (VPP) are exempt from RMP audits.

Implementation and Integration with State Programs

The EPA intends that RMP planning requirements be eventually implemented and enforced at the State or local level, and encourages states to seek delegation of the program under the CAA Section 112(l) rules. If a state chooses not to implement the RMP, the EPA will, by default, be the implementing agency.

The EPA estimates that about 10 to 15 percent of the sources subject to the RMP Rule must also obtain operating permits under CAA Title V. The RMP Rule is an applicable requirement for facilities subject to Title V requirements. The EPA believes, however, that the RMP planning requirements should not be in the permit, and has proposed a set of Title V permit conditions to meet the Section 112(r) requirements. The State or local

air permitting agency would be required to determine whether the permit conditions have been met and the RM Plan is complete. A decision whether the plan is complete could, however, be made by the RMP implementing agency under a cooperative agreement with the CAA Title V permitting agency.

Enforcement

The EPA has the authority to bring administrative and judicial actions against violators. Judicial actions may be either civil and/or criminal in nature. The CAA Section 113(a)(3) authorizes the Agency to order violators to comply with the RMP requirements. Under Section 113(b), the Agency may initiate civil judicial enforcement for violations of the RMP to assess penalties up to $25,000 per day for each violation. Under Section 113(c), the Agency may seek criminal penalties for knowing violations of the RMP. Under Section 113(d) the Agency may assess administrative civil penalties of up to $25,000 per day for each violation. Administrative actions initiated under Section 113(d) cannot exceed $200,000 unless approved by the Department of Justice. In addition to the authority to bring administrative and judicial actions against violators, the Agency may issue orders under CAA Sections 112(r)(9) and CAA Section 303 when there is an imminent and substantial threat of an actual or potential release.

Summary

The EPA requires sources with processes having regulated substances above a TQ to carry out risk management planning and to submit RM Plans for existing covered processes by June 21, 1999. The EPA will collect RM Plans electronically to allow wide dissemination of RM Plan data to emergency planning organizations and the public.

In order to implement RMP requirements and ensure compliance with the RMP Rule, owners or operators of processes involving toxic and flammable substances should: (1) determine the extent of coverage, (2) determine status of OSHA PSM implementation activities in order to minimize duplicate effort, (3) determine the status of existing RMP activities, (4) determine existing RMP capabilities, (5) develop a corporate strategy to implement RMP, (6) develop RMP implementation plans for covered processes and facilities, (7) execute RMP implementation plans, and (8) monitor RMP implementation.

Small and medium-sized businesses may receive information about the RMP Final Rule through each state's Small Business Assistance Program, the Federal Small Business Assistance Program, Small Business Development Centers, and the EPCRA Hotline: (800) 424–9346 or (703) 412–9810.

The EPA plans to publish general technical guidance, guidance to states regarding implementation, guidance to LEPCs on ways to disseminate RMP data to their communities, and additional *model plans* for certain industry sectors and regulated substances. The EPA plans to produce training packages and, in cooperation with industry and engineering societies, convene workshops around the country to introduce RMP requirements to interested stakeholders.

Bibliography

The following publications and Internet addresses can be reviewed or accessed for further information and the latest information regarding EPA RMP Amendments, Notices, Agreements, Guidance Documents, and training.

ABS Groups, Inc. (formerly JBF Associates, Inc.) RMPlanner ™, (Windows program for RM Plan development and maintenance). Knoxville, TN: ABS Groups, Inc.

"Accidental Release Prevention Requirements." *Code of Federal Regulations*. Title 40, Pt. 68.

"Clean Water Act, Oil Spill Prevention Control and Countermeasures (SPCC) Plan." *Code of Federal Regulations*. Title 40, Pt. 112.

"Emergency Planning and Community Right To Know, Appendices A and B." *Code of Federal Regulations*. Title 40, Pt. 355.

Environmental Protection Agency. *EPA Guideline for Air Quality Models*. Washington DC: GPO, 1986.

Environmental Protection Agency. *EPA PCRAM-MET User's Guide*. Washington DC: GPO, 1995 (<http://www.epa.gov/ncepihom/nepishom/>).

Environmental Protection Agency. *EPA SCREEN2 Model User's Guide*. EPA–450/4–92–006, Washington DC: GPO, 1992 (<http://www.epa.gov/cepihom/nepishom/>).

Environmental Protection Agency. *EPA User's Guide to TSCREEN. Model for Screening Toxic Air Pollutant Concentrations*. EPA–454/B–94–023. Washington DC: GPO, 1994 (<http://ww.epa.gov/ncepihom/nepishom/>).

Environmental Protection Agency. *EPA Workbook of Screening Techniques for Assessing Impacts of Toxic Air Pollutants*. EPA–454/R–92–024. Washington DC: GPO, 1992 (<http://www.epa.gov/ncepihom/nepishom/>).

Environmental Protection Agency. *RMP Offsite Consequence Analysis Guidance*. Washington, DC: GPO (<http://www.epa.gov/ncepihom/nepishom/>).

Environmental Protection Agency. *The Seven Cardinal Rules of Risk Communication*. EPA 230K92001. Washington DC: GPO, 1992 (<http://www.epa.gov/ncepihom/nepishom/>).

Environmental Protection Agency, Federal Emergency Management Agency. *Handbook of Chemical Hazard Analysis Procedures*. PB–93158756. Washington,DC: GPO, 1995 (<http://www.epa.gov/ncepihom/nepishom/>).

Environmental Protection Agency, Federal Emergency Management Agency and Department of Transportation. *Technical Guidance for Hazards Analysis*. Washington, DC: GPO, 1987 (<http://www.epa.gov/ncepihom/nepishom/>).

Handmer, John and Edmund Penning-Rowsell, Eds. *Hazards and the Communication of Risks*. Brookfield, VT : Gower Publishing Company, 1990.

"Hazard Communication." *Code of Federal Regulations*. Title 29, Pt. 1910.1200.

"Hazardous Materials Transportation." *Code of Federal Regulations*. Title 49 Pts. 172.101 and 172.194.

"Hazardous Waste Operations." *Code of Federal Regulations*. Title 29, Pt. 1910.120.

Meyer, Eugene. *Chemistry of Hazardous Materials*, 2nd ed. Englewood Cliffs, NJ: Brady Prentice Hall, Career and Technology, 1990.

National Fire Protection Association. *Fire Protection Guide on Hazardous Materials*, 8th ed. Quincy, MA: NFPA, 1984.

National Safety Council. *Accident Prevention Manual for Business and Industry, Environmental Management*. Itasca, IL: National Safety Council, 1995.

National Safety Council. *Chemicals, the Press, and the Public: A Journalist's Guide to Reporting on Chemicals in the Community*. Itasca, IL: National Safety Council, 1992.

"OSHA Emergency Action Plan." *Code of Federal Regulations*. Title 29, Pt. 1910.38.

Process Safety Institute. *Course # 101, Hazard Analysis for DOE SAR's and QRA's*. Knoxville, TN: ABS Groups, Inc. (formerly JBF Associates, Inc.).

Process Safety Institute. *Course #302, Performing Hazard Assessments to Comply With EPA's RMP Rule*. Knoxville, TN: ABS Groups, Inc. (formerly JBF Associates, Inc.).

"Resource Conservation and Recovery Act (RCRA)." *Code of Federal Regulations*. Title 40, Pts. 264, 265, 279.52.

Internet Resources

<http://198.6.4.175/docs/educate/crsindex.asp> (AIChE RMP, PSM Education Services Courses)

<http://www.jbfa.com> (ABS Group, Inc. Homepage; complete references, RMP and PSM courses, links)

<http://www.beeline-software.com> (Air modeling software and links, RMP Modeling course)

<http://www.cmahq.com/cmawebsite.nsf/pages/compliance> (Chemical Manufacturer's Associ-

ation, Compliance Assistance Center; regulatory monitoring service, guidance documents)

<http://www.nsc.org/ehc/cameo.htm> (Computer-Aided Management of Emergency Operations)

<http://www.riskworld.com/EEI/Courses/process.html> (EEI Corp. RMP, PSM, and Risk Communication Course information)

<http://www.epa.gov/docs/ngispgm3/enviro/index.html> (Envirofacts Data Warehouse)

<http://www.epa.gov/ceppo/pubs/112rlist.htm> (EPA CEPPO Regional Contacts for RMP)

<http://www.epa.gov/swercepp/> (EPA Chemical Emergency Preparedness and Prevention Office [CEPPO], Chemical Accident Prevention and Risk Management Plan, Draft RM Plan Data Elements and Instructions, Off-site Consequence Analysis; Model Ammonia and Propane Plans, Terra International, Inc., Port Neal accident)

<http:/www.epa.gov/ncepihom/nepishom/> (EPA Environmental Publications Ordering site; order over 6,000 environmental publications online; snail-mail address provided also)

<http://www.epa.gov/swercepp/pubs/oshaepa.html> (EPA/OSHA Implementation of Chemical Accident Prevention Provisions)

<http://www.access.gpo.gov/su_docs/aces/aces140.html> (Government Printing Office. Sources for Code of Federal Regulations on the Internet)

<http://www.osha-slc.gov/Preamble/toc_preambles.html> (OSHA Hazard Communication, Hazardous Waste Operations and Emergency Response [HAZWOPER,], PSM)

<http:www.rtk.net> (RTK site for Landview™ Risk Communication, Community Right-to-Know, CAA Amendments background information)

<http://www.sba.gov> (Small Business Administration; important links to help for businesses)

<http://www.thompson.com> (Thompson Publishing Group Homepage; RMP and PSM references, links, publications)

Ken B. Baier, Alan R. Hohl, and John S. Kirar are Vice Presidents of Engineered Safety Source Company, providing safety engineering and environmental management services to government, environmental, and commercial clients. The authors work as nuclear safety engineers at the Department of Energy's Rocky Flats Environmental Technology Site in Golden, Colorado.

Mr. Baier has 20 years experience in safety engineering and risk management, systems engineering, project management, product assurance, and environmental regulatory management. He has worked on multiple Department of Defense, National Aeronautics and Space Administration, and Department of Energy projects, managing the risk to workers, the public, and the environment.

Mr. Hohl has 25 years experience in safety engineering and risk management, project management, waste management, and environmental regulatory compliance. He has worked on a wide variety of Department of Defense, National Aeronautics and Space Administration, and Department of Energy projects, managing the risk to workers, the public and the environment, as well as the clients' facilties and hardware.

Mr. Kirar has 20 years experience in safety engineering and risk management, systems engineering, project management, environmental health and safety management, and regulatry compliance. He has worked in the hazardous matrerials management field for seven years, evaluating public, worker, and environmental risks associated with the release of chemicals and radiological contaminants.

CHAPTER 31

A Step-by-Step Guide to Risk Management Planning

Dale M. Petroff, NRRPT, CHMM

Introduction

On June 20, 1996, The Environmental Protection Agency (EPA) published the final rule for Section 112(r) of the Clean Air Act (CAA), otherwise called the Risk Management Program (RMP) Rule (40 CFR 68).

The purpose of this paper is to provide a step-by-step guide for the Certified Hazardous Materials Manager (CHMM) assigned to develop and implement a Risk Management Plan (RM Plan). Additionally, the use of the *One Plan* concept is discussed for the integration of the RM Plan with the other spill control, emergency response, and hazardous materials contingency plans required under other regulations.

The intent of this chapter is to guide the user through the RM Plan development process. The EPA's general guidance document, *General Guidance for Risk Management Programs* (40 CFR 68) (*General Guidance*), forms the basis for this paper and is the best source of information for the RM Plan process. The sections of this paper follow those of the guidance document with a few exceptions.

Background

The RM Plan requirement was established in response to the increased threat from chemical releases to the public. After the Bhopal incident in 1984, the general population in the United States became aware of the potential risks posed by various industrial sites in their communities.

An estimated 64,000 facilities are subject to the RMP Rule based on the quantity of regulated substances they have on-site. These facilities are required to implement a Risk Management Program and submit a summary of the program (called the Risk Management Plan) to a central location specified by EPA by June 20, 1999.

Public awareness of the potential danger from accidental releases of hazardous chemicals has increased over the years as serious chemical accidents have occurred around the world (*e.g.*, the 1974 explosion in Flixborough, England, and the 1976 release of dioxin in Seveso, Italy). Public concern intensified following the 1984 release of methyl isocyanate in Bhopal, India, that killed more than 2,000 people living near the facility. A subsequent release from a chemical facility in Institute, West Virginia sent more than 100 people to the hospital and made Americans aware that such incidents can and do happen in the United States. The Risk Management Program is the EPA's response to that concern.

As a Certified Hazardous Materials Manager (CHMM), you may be called to develop and implement an RM Plan for your facility. In preparing for that task, there are some key elements to remember.

First, if you are required to develop an RM Plan for your facility, then you are probably required to implement the Occupational Safety and Health Administration's (OSHA) Process Safety Management (PSM) Standard (29 CFR 1910.119) since the regulated chemicals and quantities are similar. The Clean Air Act Amendments (CAAA) of 1990 places responsibility for the prevention of accidental chemical releases on both OSHA and EPA. OSHA has responsibility for the protection of workers from accidental chemical releases under Section 304 of the Amendments and has promulgated the PSM Standard (29 CFR 1910.119) to satisfy this requirement. EPA has responsibility for protection of the general public and the environment from accidental chemical releases under Section 112(r) of the Amendments. EPA promulgated the Risk Management Program rule (40 CFR 68) to satisfy this requirement.

The EPA believes that one chemical accident prevention program can serve to protect workers, the public, and the environment, and has incorporated the OSHA PSM Standard as the chemical accident prevention program for certain facilities subject to both rules.

Therefore, if required, a PSM program should be developed prior to the developing an RM Plan. The PSM analysis will provide key information on the development of accident information used to determine off-site impact.

The second key element to remember is that the results of the Off-Site Consequence Analysis (OCA) may become an issue with the local emergency management agency and/or members of the public. All Consequence Analysis results should be kept confidential until they have been thoroughly reviewed and validated.

The next key element is the development of an emergency response plan that addresses both on-site worker safety and potential off-site impacts. The public and local authorities will want notification even if the Worst-Case Release Scenario has no off-site impact. Companies need to demonstrate that they are good neighbors, or they may find their facilities forced out of the area.

The RM Plan must be made available to the public, and a key factor in gaining the public's acceptance is to ensure that concerns of stakeholders are addressed. The involvement of the local emergency management agency and local government through the Local Emergency Planning Committee (LEPC) or the Chemical Accident Emergency Response (CAER) group is critical in gaining that acceptance.

The RMP is seen as a continuation of the chemical safety work begun under the Emergency Planning and Community Right-to-Know Act (EPCRA) (also known as the Superfund Amendments and Reauthorization Act [SARA] Title III), and the PSM regulations of OSHA.

EPCRA is intended to help local communities prepare for chemical accidents. It requires communities to develop emergency response plans, based on information from industry concerning hazardous chemicals.

The PSM program, as discussed in this chapter, requires the analysis of the processes, procedures, and operations at a facility that may cause a chemical accident. The objective of this hazard analysis is to determine how the accident scenarios may be avoided. The PSM rules are based on the assumption that the listed chemicals, if released, will result in significant chemical concentrations within the boundaries of the facility. Again, the focus of this regulation is worker safety.

Risk management planning, as part of an integrated safety management program, will relate to local emergency preparedness and response, to pollution prevention, and to worker safety. The RMP is an element of an integrated approach to safety and complements existing industry codes and standards.

Preventing accidental releases of hazardous chemicals is the shared responsibility of industry, government, and the public. The first steps toward accident prevention are identifying the hazards and assessing the risks. Once information about chemicals is shared, industry, government, and the community can work together toward reducing the risk to workers, public health, and the environment.

Preparation

The first step in any project is preparation. The EPA has published an excellent guide for the RMP process entitled *General Guidance for Risk Management Programs (40 CFR Part 68)*. This guide is the basis for discussion in this chapter. The guide, which is approximately 100 pages in length, is a good place for anyone contemplating the development of an RM Plan.

The guide can be found at EPA's Chemical Emergency Preparedness and Prevention Office website. The reference is:

<http://www.epa.gov/ceppo/>(Environmental Protection Agency. *General Guidance for Risk Management Programs (40 CFR Part 68)*. RMP Series. EPA 550–B–98–003. July, 1998.)

The document is maintained as an Adobe-formatted document and the EPA provides a link where a free copy of Adobe reader program may be obtained. The document is broken into several parts and 1.25 megabytes of disk space is required to download the entire document. The guide is organized into 11 chapters with 6 appendices.

Table of Contents for:
General Guidance for Risk Management Progams

Chapter 1: General Applicability

Chapter 2: Applicability of Program Levels

Chapter 3: Five-Year Accident History

Chapter 4: Off-Site Consequence Analysis

Chapter 5: Management System

Chapter 6: Prevention Program (Program 2)

Chapter 7: Prevention Program (Program 3)

Chapter 8: Emergency Response Program

Chapter 9: Risk Management Plan

Chapter 10: Implementation

Chapter 11: Communication with the Public

Appendices

Appendix A: 40 CFR 68 (1997 edition)

Appendix B: Selected NAICS Codes

Appendix C: EPA Regional Contacts

Appendix D: OSHA Contacts

Appendix E: Technical Assistance

Appendix F: OSHA Guidance on PSM

Careful study of the guide should answer most questions about the development and implementation of the RM Plan. Once you are familiar with the contents of the guide, the process of determining applicability and data collection can begin.

Applicability

One of the main factors contributing to the number of people killed or injured as a result of the Bhopal accident is that a large amount of the material released had accumulated waiting for the final process line to be repaired. The intent of this regulation is the prevention of accidents and/or their mitigation.

The first step in determining applicability is to determine if your facility is required by the regulation to develop an RM Plan. Note that, while there is a great deal of overlap with the PSM Program, the PSM requirements do not completely cover the RMP requirements.

In 40 CFR 68.10, *applicability* is defined as follows:

> An owner or operator of a stationary source that has more than a threshold quantity of a regulated substance in a process, as determined under 40 CFR 68.115, shall comply with the requirements of this part no later than the latest of the following dates:
>
> (1) June 21, 1999;
>
> (2) Three years after the date on which a regulated substance is first listed under 40 CFR 68.130; or
>
> (3) The date on which a regulated substance is first present above a threshold quantity in a process.

The table in 40 CFR 68.130 that lists regulated toxic substances and their threshold quantities is not reproduced here, as it is subject to change. Check in the *Federal Register* for the latest revision prior to making an initial determination. After the initial determination has been made, it should be reviewed whenever a change to the process is being evaluated and on a regular basis. If your facility does not subscribe to a regulations update service, then the review should be conducted in April and October when the Semi-Annual Regulatory Agenda is printed in the *Federal Register*.

In the determination process, two definitions in the regulation are critical. These are *stationary source* and *threshold quantity*.

Stationary source means

> any buildings, structures, equipment, installations, or substance emitting stationary activities which belong to the same industrial group, which are located on one or more contiguous properties, which are under the control of the same person (or persons under common control), and from which an accidental release may occur.

A stationary source includes transportation containers that are no longer under active shipping papers and transportation containers that are connected to equipment at the stationary source for the purposes of temporary storage, loading, or unloading. The term *stationary source* does not apply to transportation, including the storage incident to transportation, of any regulated substance or any other extremely hazardous substance under the provisions of this part, provided that such transportation is regulated under 49 CFR 192, 193, or 195. Properties shall not be considered contiguous solely because of a railroad or gas pipeline right-of-way.

Threshold quantity means:

> the quantity specified for regulated substances pursuant to Section 112(r)(5) of the Clean Air Act as amended listed in 40 CFR 68.130 and determined to be present at a stationary source as specified in 40 CFR 68.115 of this part.

Again, if the process or the products at the facility change, then the facility must prepare an RM Plan at the time a regulated substance is first present in a process and is above a threshold quantity.

If it is determined that storage of feed materials, intermediate products, or finished products requires the development of an RM Plan, then a review of the necessary quantities should be made. If the facility can operate with amounts below the threshold quantities, then a change to the amounts stored should be considered. Also, the replacement of the material with a less toxic product should be considered.

Program Eligibility

The EPA has established three tiers of requirements to reduce the level of effort for facilities with lower risk of off-site impacts. These tiers are referred to as Programs 1, 2, and 3. Program 1, for *no impact* facilities, has the fewest requirements, while Program 3 has the most. The program tier that the facility falls under defines the level of effort needed to establish and maintain the RM Plan; it also defines the information that must be certified in the final report. It may not be possible to determine final eligibility until a Consequence Analysis has been completed.

In order to determine which program is applicable, several Process Hazard Assessment (PHA) tasks should be completed for the facility. The various program eligibility requirements are listed below, but verify these in a current revision of the regulation before proceeding.

Program 1 Eligibility Requirements

A covered process is eligible for Program 1 requirements as provided in 40 CFR 68.12(b) if it meets all of the following requirements:

1) For the five years prior to the submission of an RM Plan, the process has not had an accidental release of a regulated substance where exposure to the substance, its reaction products, overpressure generated by an explosion involving the substance, or radiant heat generated by a fire involving the substance led to any of the following off-site:

    a) Death

    b) Injury, or

    c) Response or restoration activities for an exposure of an environmental receptor.

2) The distance to a toxic or flammable end point for a worst-case release assessment conducted under Subpart B and 40 CFR 68.25 is less than the distance to any public receptor, as defined in 40 CFR 68.30.

3) Emergency response procedures have been coordinated between the stationary source and local emergency planning and response organizations.

Program 2 Eligibility Requirements

A covered process is subject to Program 2 requirements if it does not meet the eligibility requirements of either Program 1 or Program 3.

Program 3 Eligibility Requirements

A covered process is subject to Program 3 if the process does not meet the requirements of Program 1 of this section, and if either of the following conditions is met:

1) The process is in Standard Industrial Code (SIC) numbers 2611, 2812, 2819, 2821, 2865, 2869, 2873, 2879, or 2911.

2) The process is subject to the OSHA Process Safety Management standard, 29 CFR 1910. 119.

If at any time a covered process no longer meets the eligibility criteria of its program level, the owner or operator shall comply with the requirements of the new program level that applies to the process and update the RM Plan as provided in 40 CFR 68.190.

Once applicability and a program determination have been made, then the Consequence Analysis should be conducted. It may not be possible to make a final determination of program eligibility until the Consequence Analysis has been completed.

Consequence Analysis

The Consequence Analysis is the keystone to the RMP. The results of the Consequence Analysis define which program the facility falls under and determine the level of emergency planning that must be done.

You can decide to conduct the Consequence Analysis at your facility or hire a contractor to accomplish this. The EPA has published guidance documents that provide detailed information on conducting the Consequence Analysis by facility personnel without previous experience.

The guidance documents are intended for facilities that plan to do their own air dispersion modeling. The EPA has prepared a guidance document, *RMP Off-Site Consequence Analysis Guidance,* that is available at the same website as the *General Guide* and can be downloaded as a WordPerfect file. This file will require 1.25 megabytes of space to download.

The *RMP Off-Site Consequence Analysis Guidance* provides simple methods and reference tables for determining distance to an end point for Worst-Case and Alternative Release Scenarios. This method will tend to be very conservative and overestimate off-site impacts. If by using this method no off-site impacts are identified, then the analysis can be concluded and documented.

If off-site impacts are shown by the use of the reference table methodology, then a more advanced Consequence Analysis should be performed using one of the many computer software programs available.

In conjunction with the National Oceanic and Atmospheric Administration (NOAA), the EPA has developed a software program, RMP* Comp™, that performs the calculations described in the *RMP Off-Site Consequence Analysis Guidance*. This software is available for free from the NOAA Internet website at

<http://www.noaa.gov>

In addition, EPA is preparing industry guidance for several industries covered by 40 CFR 68. In these documents, EPA provides chemical-specific modeling for the covered industries. All of the information provided in Chapter 4 of the *General Guidance* is also included in EPA's *RMP Off-Site Consequence Analysis Guidance* and the industry-specific guidance documents available from EPA.

If you intend to use the cited guidance documents to carry out the Off-Site Consequence Analysis, Chapter 4 of the *General Guide* may be skipped. If you plan to do your own modeling, Chapter 4 will provide you with the information you need to comply with the rule requirements; it does not, however, provide methodologies.

Whether the analysis at the facility is conducted by facility personnel or contractors, the analysis must be acceptable to the EPA. As discussed earlier, you may use EPA's *RMP Off-Site Consequence Analysis Guidance*. Results obtained using the methods in EPA's guidance are expected to be conservative. The conservative assumptions were introduced to compensate for high levels of uncertainty. The use of the EPA guide is optional,

and you are free to use other air dispersion models, fire or explosion models, or computation methods, provided that

- They are publicly or commercially available or are proprietary models that you are willing to share with the implementing agency

- They are recognized by industry as applicable to current practices

- They are appropriate for the chemicals and conditions being modeled

- You use the applicable definitions of Worst-Case Release Scenarios

- You use the applicable parameters specified in the rule

Figure 1, "Considerations for Choosing a Modeling Method" has been copied from the *General Guide*. This chart has been designed to provide additional suggestions on making the decision of whether to use a complex model or the simpler methodology in the EPA guidance document.

When conducting an Off-Site Consequence Analysis, it is a good practice to have all the analyses independently developed and reviewed by two different individuals. This method is more costly and time-consuming, but the impact caused by the results analysis may be even more costly in terms of resources devoted to emergency preparedness, pubic relations, source reduction, and additional equipment.

The Off-Site Consequence Analysis consists of two elements: a Worst-Case Release Scenario and an Alternative Release Scenario. The development of these scenarios is discussed in detail in the *General Guide*. An overview follows.

A *Worst-Case Release Scenario Analysis* is applicable to all covered processes, regardless of program level. In order to determine whether a process is eligible for Program 1, you must evaluate the Worst-Case Release Scenarios for each toxic and flammable substance in the process that is above the threshold quantity. The process is eligible for Program 1 if there are no public receptors within the distance to an end point for all of the worst-case scenarios analyzed for the process. (The other Program 1 criteria must also be met—see Chapter 2 of the *General Guide*). For every Program 1 process, you must report on the

Approach	Examples	Advantages	Disadvantages
Simple guidance	EPA's *Off-Site Consequence Analysis Guidance*	• Free • No computer requirements • Simple to use • Provides all data needed • Provides tables of distances • Ensures compliance with rule	• Conservative results • Few site-specific factors considered • Little flexibility in scenario development
Simple computer models	EPA models such as RMP*Comp™	• No/low cost • May be simple to use • Can consider some site-specific factors	• Some may not be simple to use • Likely to give conservative results • May not accept all of EPA's required assumptions • May not include chemical-specific data • May not address all consequences
Complex computer models	Commercially available models	• May address a variety of scenarios • May consider many site-specific factors	• May be costly • May require high level of expertise
Calculation methods	"Yellow Book" (Netherlands TNO)	• Low cost • No computer requirements	• May require expertise to apply methods • May required development of a variety of data

From: Environmental Protection Agency, *General Guidance for Risk Management Program.* July, 1998.

Figure 1. Considerations for Choosing a Modeling Method

Worst-Case Release Scenario with the greatest distance to an end point.

If your site has Program 2 or Program 3 processes (processes that are not eligible for Program 1— see Chapter 2 of the *General Guide*), you must analyze and report on

- One worst-case analysis representing all toxic regulated substances present above the threshold quantity

- One worst-case analysis representing all flammable regulated substances present above the threshold quantity

You may need to submit an additional worst-case analysis if a worst-case release from elsewhere at the source would potentially affect public receptors different from those affected by the initial worst-case scenario(s).

An *Alternative Release Scenario Analysis* is applicable to all Program 2 and Program 3 processes. Alternative Release Scenarios should be those that may result in concentrations, overpressures, or radiant heat levels that reach the end points specified for these effects beyond the fenceline of your facility.

You must present information on one Alternative Release Scenario Analysis

- For each regulated toxic substance held above the threshold quantity, including the substance considered in the worst-case analysis

- To represent all flammable substances held above the threshold quantity

If the distance to the end point for your worst-case release just reaches your fenceline, you may not have an alternative release scenario with a distance to an end point that goes beyond the fenceline. However, you still must report an alternative release scenario. You may want to explain in the RM Plan Executive Summary why the distance does not extend beyond the fenceline.

In developing the Consequence Analysis, the complex models that can account for many site-specific factors may give less conservative estimates of off-site consequences than the simplified methods in EPA's guidance. This is particularly true for alternative scenarios, for which EPA has not specified many assumptions.

However, complex models may be expensive and require considerable expertise to use. These may be beyond the ability of a facility to utilize. The EPA's optional guidance is designed to be simple and straightforward. You will need to consider the trade-off in deciding how to carry out your required consequence analyses.

Whether you use EPA's guidance or another modeling method, you should bear in mind that the results you obtain from modeling your Worst-Case or Alternative Release Scenarios should not be considered to predict the likely results of an accidental release. The worst-case assumptions are very conservative and, regardless of the model used you can expect very conservative results. Although the results from modeling alternative scenarios will be less conservative, you still must use conservative end points.

In addition, results of an actual release will depend on many site-specific conditions (*e.g.*, wind speed and other weather conditions) and factors related to the release (*e.g.*, when and how the release occurs, how long it takes to stop it). You should make reasonable assumptions regarding such factors in developing your alternative scenarios, but the circumstances surrounding an actual release may be different.

Different models will likely provide different results, even with the same assumptions, and most models have not been verified with experimental data. Therefore, results of even sophisticated modeling have a high degree of uncertainty and should be viewed as providing a basis for discussion, rather than as predictions. Modeling results should be considered particularly uncertain over long distances (*i.e.*, 10 kilometers or more).

The *General Guide* does identify possible sources of assistance for the conduct of modeling the various scenarios required. Chapter 4 of the *General Guide* discusses the parameters of developing the scenarios and conducting the analysis of Worst-Case and Alternative Release Scenarios. Prior to choosing an approach, this chapter should be studied carefully.

As part of the Consequence Analysis, receptor end points must be determined. The rule requires that you estimate in the RM Plan residential populations within the circle defined by the end point for your Worst-Case and Alternative Release Scenarios (*i.e.*, the center of the circle is the point of release and the radius is the distance to the end point).

In addition, you must report in the RM Plan whether certain types of public receptors and environmental receptors are within the circles. These end points determine the potential impact of your facility and which program is applicable to the facility. Chapter 4 in the *General Guide* provides excellent information on how to determine the receptors and the sources of information needed to make the determination.

When the Consequence Analysis has been completed, then the facility must determine the applicable program. At this point, the facility processes should be reviewed to determine if the reduction of source terms is more cost-effective than implementing the applicable program.

Administrative controls will probably not be acceptable to the EPA or the State agency regulating this program, as they are too easily overcome by human error. Positive controls, such as a lockout or alarm system to control levels, with a double contingency to prevent the accumulation of the larger source terms, may be acceptable.

Once the program applicability has been determined, the prevention program and Risk Management Plan itself must be developed.

Management System

If you have at least one Program 2 or Program 3 process (see Chapter 2 of the *General Guide* for information on determining the program levels of your processes), the management system provision in 40 CFR 68.15 requires the facility to

- Develop a management system to oversee the implementation of the risk management program elements

- Designate a qualified person or position with the overall responsibility for the development, implementation, and integration of the risk management program elements

- Document the names of people or positions and define the lines of authority through an organizational chart or other similar document (if the responsibility for implementing individual requirements of the risk management program is assigned to people or positions other than the person or position with overall responsibility for the risk management program).

The Management System Provision

Management commitment to process safety is a critical element of your facility's risk management program. Management commitment should not end when the last word of the risk management plan is composed. For process safety to be a constant priority, the facility must remain committed to every element of the risk management program.

The rule takes an integrated approach to managing risks. Each element must be implemented on an ongoing, daily basis, becoming an integral part of the facility operations. Therefore, your commitment and oversight must be continuous. By satisfying the requirements of the management system provision, you are ensuring that

- The risk management program elements are integrated and implemented on an ongoing basis

- All groups within a facility and/or process understand the lines of responsibility and communication

How to Meet the Management System Requirements

The sources covered by the rule are diverse. The EPA has recognized the need to maximize flexibility in complying with this program. Facility personnel are usually the best resource to use in deciding the appropriate methods for implementation and incorporation of the risk management program elements.

A key element in the performance of any program is that responsibility is assigned and the authority is granted to carry out that responsibility. A small facility may name an individual in the risk management plan as being responsible for the program. A medium or large facility may have more managerial turnover than smaller sites. For this reason, it is recommended in Chapter 5 of the *General Guide* that the facility identify a position, rather than the name of the specific person, with overall responsibility for the risk management program elements. Remember, the only element of the management system that must be reported in the RM Plan is the name of the qualified person or position with overall re-sponsibility. *Note that changes to this data element in your RM Plan do not require you to update your RM Plan.*

All of the positions identified in the RMP documentation must report their progress to the person/position with overall responsibility for the program. However, nothing in the risk management program rule prohibits you from satisfying the management provision by assigning process safety committees with management responsibility—provided that an organizational chart or similar document identifies the names or positions and lines of authority.

Defining the lines of authority and roles and responsibilities of staff that oversee the risk management program elements will help to

- Ensure effective communication about process changes between divisions

- Clarify the roles and responsibilities related to process safety issues at the facility

- Avoid problems or conflicts among the various people responsible for implementing elements of the risk management program

- Avoid confusion, and allow those responsible for implementation to work together as a team

- Ensure that the program elements are integrated into an ongoing approach to identifying hazards and managing risks

Management's commitment to the Risk Management Program is crucial to its success. Only with the commitment of the management will the RM Plan be prevented from being just another stack of paper written up to meet a regulatory requirement.

The RMP with the PSM are part of an integrated safety management system that can prevent severe losses due to accident or the failure to mitigate that accident when it happens. In most cases if there is a loss of life, medium to small companies will not be able to withstand the losses. Therefore remember you generally only get one mistake; the *3-Strike Rule* is not in effect.

In addition to financial losses, the loss of the goodwill of the community may result in the facility becoming a focal point for State and local regulators. For example, during a recent incident in the southeast, a processing facility had two releases in two days and some off-site injury which resulted in a stop-work order from the EPA as well as the regulatory focus of two states and several counties. So—learn from the mistakes of others and use the ounce of prevention, as a pound of cure may not be enough.

The CHMM standard of ethics requires you to be committed to the safety of your facility personnel, the general public, and the environment.

Prevention Program

The best accident to have is the one that has been prevented. In addition to determining off-site impacts, a major goal of the RMP is the development of an accident prevention program. The required prevention programs are keyed to the Program levels 2 or 3 determined earlier in the process.

Most Program 2 processes are likely to be relatively simple and may be located at small businesses. The EPA has developed the Program 2 prevention program by identifying the basic elements that are the foundation of sound prevention practices—safety information, hazard review, operating procedures, training, maintenance, compliance audits, and accident investigation.

By meeting other Federal regulations, State laws, industry codes and standards, as well as good engineering practices, a Program 2 facility probably has already met most of the Program 2 prevention element requirements.

Many Program 3 facilities will need to do little that is new to comply with the Program 3 prevention program, because they should already have the OSHA PSM program in place.

Keep in mind if you have a Program 3 facility, whether you're building on the PSM standard or creating a new program, that EPA and OSHA have different legal authority:

* EPA for off-site consequences

* OSHA for on-site consequences

If you are already complying with the PSM standard, your Process Hazard Analysis (PHA) team may have to assess new hazards that could affect the public or the environment offsite. Protection measures that are suitable for workers (*e.g.*, venting releases to the outdoors) may be the very kind of thing that imperils the public.

To integrate the elements of your prevention program, you must ensure that a change in any single element of your program leads to a review of other elements to identify any effect caused by the change.

The details for developing and implementing a Program 2 prevention program are contained in Chapter 6 of the *General Guide*. The requirements for a Program 3 prevention program are detailed in Chapter 7 of the *General Guide*.

Most importantly, make accident prevention an institution at your site. Like the entire risk management program, a prevention program is more than a collection of written documents. It is a way to make safe operations and accident prevention the way you do business everyday.

Emergency Response

One of the most effective methods used to mitigate an accident is to develop a comprehensive emergency response program. If you have at least one Program 2 or Program 3 process at your facility, then 40 CFR 68 may require you to implement an emergency response program, consisting of an emergency response plan, emergency response equipment procedures, employee training, and procedures to ensure the program is up-to-date. This requirement applies if facility employees will respond to some releases involving regulated substances.

Nonresponding Facilities

The EPA recognizes that, in some cases (particularly for retailers and other small operations with few employees), it may not be appropriate for employees to conduct response operations for releases of regulated substances. Chapter 8 of the *General Guide* covers how to develop and implement an emergency response program as well as how to determine if a facility can opt out of an employee response program.

For example it would be inappropriate, and probably unsafe, for an ammonia retailer with only one full-time employee to expect that a tank fire could be handled without the help of the local fire department or other emergency responder. The EPA does not intend to force such facilities to develop emergency response capabilities. At the same time, facility personnel are responsible for ensuring effective emergency response to any releases at the facility.

If your local public responders are not capable of providing such response, you must take steps to ensure that an effective response is available (*e.g.*, by hiring response contractors or providing support to the local government to enable them to respond).

The EPA has issued guidance for responding and nonresponding facilities. EPA has also adopted a policy for nonresponding facilities similar to that adopted by OSHA in its Hazardous Waste Operations and Emergency Response (HAZ-WOPER) Standard (29 CFR 1910.120), which allows certain facilities to develop an emergency action plan to ensure employee safety, rather than a full-fledged emergency response plan. If your employees will not respond to accidental releases of regulated substances, then you need not comply with the emergency response plan and program requirements. However, you must ensure that some sort of response will be provided.

If employees are not required to respond to regulated substance releases at the facility, then the facility is required to coordinate with local response agencies to ensure that they will be prepared to respond to an emergency at your facility. This will help to ensure that your community has a strategy for responding to and mitigating the threat posed by a release of a regulated substance from your facility. To do so, you must ensure that you have set up a way to notify emergency responders when there is need for a response. Coordination with local responders also entails the following steps:

- If you have a covered process with a regulated toxic, then work with the local emergency planning entity to ensure that the facility is included in the community emergency response plan prepared under EPCRA regarding a response to a potential release.

- If you have a covered process with a regulated flammable, then work with the local fire department regarding a response to a potential release.

Although you do not need to describe these activities in your risk management plan, document your efforts by keeping a record of

- The emergency contact (*i.e.*, name or organization and number) that you will call for a toxic or flammable release

- The organization that you worked with on response procedures

Again, a facility located in an area such Houston, Texas, or Mobile, Alabama, where there are a large number of facilities having regulated substances and local emergency response organizations that are trained and equipped to respond, the non-response option is normally very acceptable.

If you are in a rural county without a hazardous material response capability, then the non-response option would be acceptable only with limitations in the quantity of hazardous material on-site, if you have a small staff on-site and a very low population in the receptor zones.

Responding Facilities

If your facility employees will respond to releases of regulated substances from the facility, your emergency response program must contain an emergency response plan (maintained at the facility) that includes

- Procedures for informing the public and emergency response agencies about releases

- Documentation of proper first aid and emergency medical treatment necessary to treat human exposures

- Procedures and measures for emergency response

- Procedures for using, inspecting, testing, and maintaining your emergency response equipment

- Training for all employees in relevant procedures (personnel must be qualified in accordance with the HAZWOPER standard)

- Procedures to review and update, as appropriate, the emergency response plan to reflect changes at the facility and ensure that employees are informed of changes

Finally, your plan must be coordinated with the community plan developed under the Emergency

Planning and Community Right-to-Know Act (EPCRA). In addition, at the request of local emergency planning or response officials, you must provide any information necessary for developing and implementing the community plan.

In Chapter 8 of the *General Guide,* the *concept* of a Local Emergency Planning Committee (LEPC) is explained. This has been reproduced in Figure 2. If your county or city does not have an LEPC you may want to join with other regulated companies to organize one or to develop a Chemical Accident Emergency Response (CAER) group using the guidance developed by the Chemical Manufacturers Association.

Although EPA's required elements are essential to any emergency response program, they are not comprehensive guidelines for creating an adequate response capability. Rather than establish another set of Federal requirements for an emergency response program, EPA has accepted the concept

of an Integrated Contingency Plan (ICP), which is explained in detail in a later section of this chapter.

If your facility has a regulated substance on site, you are already subject to at least one emergency response rule: OSHA's emergency action plan requirements (29 CFR 1910.38). Under OSHA HAZWOPER, any facility that handles *hazardous substances* (a broad term that includes all of the CAA regulated substances and thus applies to all facilities with covered processes) must comply with either 29 CFR 1910.38(a) or 1910.119(q). If you have a HAZMAT team, you are subject to the 29 CFR 1910.119(q) requirements.

Various Federal regulations requiring emergency response plans have been reproduced from Chapter 8 of the *General Guide* and are shown in Figure 3. If you determine that the emergency response programs you have developed to comply with these other rules satisfy the elements listed at the beginning of this section, you will not have

---

What is a Local Emergency Planning Committee?

Local emergency planning committees (LEPCs) were formed under the Emergency Planning and Community Right-to-Know Act (EPCRA) of 1986. The committees are designed to serve as a community forum for issues relating to preparedness for emergencies involving releases of hazardous substances in their jurisdictions. They consist of representatives from local government (including law enforcement and fire fighting), local industry, transportation groups, health and medical organizations, community groups, and the media. LEPCs

- Collect information from facilities on hazardous substances that pose a risk to the community

- Develop a contingency plan for the community based on this information

- Make information on hazardous substances available to the general public

Contact the mayor's office or the county emergency management office for more information on your LEPC.

From: Environmental Protection Agency, *General Guidance for Risk Management Program.* July, 1998.

Figure 2. The Concept of a Local Emergency Planning Committee

to do anything additional to comply with these elements. Additional guidance on making this decision is provided in Chapter 8 of the *General Guide*.

Be careful not to confuse writing a set of *emergency response procedures* in a plan with developing an *emergency response program*. An emergency response plan is only one element of the integrated effort that makes an emergency response program. Although the plan outlines the actions and equipment necessary to respond effectively, other aspects of the program such as training, program evaluation, equipment maintenance, and co-ordination with local agencies must occur regularly if your plan is to be useful in an emergency.

Again, it will take management commitment to ensure that an emergency response is established and capable of responding to an emergency. An emergency response program that integrates emergency operating procedures and alarm response can become the most effective means of preventing damage, injury, or loss of life. The goal of the program is to enable you to respond quickly and effectively to any emergency.

Figure 4 contains useful references listed in the *General Guide*. They will be helpful in developing specific elements of your emergency response program.

Risk Management Plan

As stated previously, if you have a facility that falls under the provisions of the Risk Management Program, you must submit a Risk Management Plan to EPA.

Elements of the Risk Management Plan

The length and content of your RM Plan will vary depending on the number and program level of the covered processes at your facility. Any facility with one or more covered processes must include in its RM Plan

- An executive summary (40 CFR 68.155)

- The registration for the facility (40 CFR 68.160)

- The certification statement (40 CFR 68.185)

- A worst-case scenario for each Program 1 process; at least one worst-case scenario to

---

Hazardous Materials Emergency Planning Guide (NRT–1), National Response Team, March 1987. Although designed to assist communities in planning for HAZMAT incidents, this guide provides useful information on developing a response plan, including planning teams, plan review, and ongoing planning efforts.

Criteria for Review of Hazardous Materials Emergency Plans (NRT–1A), National Response Team, May 1988. This guide provides criteria for evaluating response plans.

North American Emergency Response Guidebook (NAERG96), US Department of Transportation, 1996. This guidebook lists over 1,000 hazardous materials and provides information on their general hazards and recommended isolation distances.

Response Information Data Sheets (RIDS), US EPA and National Oceanic and Atmospheric Administration. Developed for use with the Computer-Aided Management of Emergency Operations (CAMEO) software, these documents outline the properties, hazards, and basic safety and response practices for thousands of hazardous chemicals.

From: Environmental Protection Agency, *General Guidance for Risk Management Program.* July, 1998.

Figure 3. Federal Guidance on Emergency Planning and Response

The following is a list of some of the Federal emergency planning regulations:

- EPA's Oil Pollution Prevention Regulation (SPCC and Facility Response Plan Requirements): 40 CFR 112.7(d) and 112.20–112.21

- MMS's Facility Response Plan Regulation: 30 CFR 254

- RSPA's Pipeline Response Plan Regulation: 49 CFR 194

- USCG's Facility Response Plan Regulation: 33 CFR 154, Subpart F

- EPA's Risk Management Programs Regulation: 40 CFR 68

- OSHA's Emergency Action Plan Regulation: 29 CFR 1910.38(a)

- OSHA's Process Safety Standard: 29 CFR 1910.119

- OSHA's HAZWOPER Regulation: 29 CFR 1910.120

- OSHA's Fire Brigade Regulation: 29 CFR 1910.156

- EPA's Resource Conservation and Recovery Act Contingency Planning Requirements: 40 CFR 264, Subpart D; 40 CFR 265, Subpart D; and 40 CFR 279.52

- EPA's Emergency Planning and Community Right-to-Know Act Requirements: 40 CFR 355. (These planning requirements apply to communities, rather than facilities, but will be relevant when facilities are coordinating with local planning and response entities.)

- EPA's Storm Water Regulations: 40 CFR 122.26

From: Environmental Protection Agency, *General Guidance for Risk Management Program.* July, 1998.

Figure 4. Federal Emergency Planning Regulations

cover all Program 2 and 3 processes involving regulated toxic substances; at least one worst-case scenario to cover all Program 2 and 3 processes involving regulated flammables (40 CFR 68.165[a])

- The five-year accident history for each process (40 CFR 68.168)

- A summary of the emergency response program for the facility (40 CFR 68.180)

Any facility with at least one covered process in Program 2 or 3 must also include in its RM Plan

- At least one Alternative Release Scenario for each regulated toxic substance in Program 2 or 3 processes, and at least one Alternative Release scenario to cover all regulated

flammables in Program 2 or 3 processes (40 CFR 68.165[b])

- A summary of the prevention program for each Program 2 process (40 CFR 68.170) and RM Plan

- A summary of the prevention program for each Program 3 process (40 CFR 68.175)

Subpart G of 40 CFR 68 (see Appendix A of the *General Guide*) provides more detail on the data required for each of the elements. The actual RM Plan form, however, will contain more detailed guidance to make it possible to limit the number of text entries. For example, the rule requires you to report on the major hazards identified during a Process Hazard Assessment (PHA) or hazard review and on public receptors affected by Worst-Case and Alternative Release Scenarios.

By January 1999, EPA will make RMP*Submit™ available for completion and filing of your Risk Management Plan. RMP*Submit™ allows the electronic submission of the Risk Management Plan. The instructions for using the software will be made available with the software and will be posted on EPA's web site. The instructions cover each of the data elements to be reported in the RM Plan with RMP*Submit™. The instructions will explain each data element and help you understand what acceptable data are for each. Prior to the availability of the RMP*Submit™ software there are instructions and electronic version of the forms to be used that can be downloaded from the EPA website.

The RM Plan will provide a list of options for you to check for these elements. Except for the executive summary, the RM Plan will consist primarily of yes/no answers, numerical information (e.g., dates, quantities, distances), and a few text answers (e.g., names, addresses, chemical identity).

Where possible, RMP*Submit™ will provide *pick lists* to help you complete the form. For example, RMP*Submit™ will provide a list of regulated substances and automatically fill in the CAS numbers when you select a substance.

Information and instructions on trade secrets and confidentiality are covered in the *General Guide*, Chapter 9.

When Does the Off-Site Consequence Analysis (OCA) Need to Be Revised? You'll need to revise your OCA when a change at your facility results in the distance to an end point from a worst-case release rising or falling by at least a factor of two. For example, if you increase your inventory substantially or install passive mitigation to limit the potential release rate, you should reestimate the distance at an end point. If the distance is at least doubled or halved, you must revise the RM Plan. For most substances, the quantity that would be released would have to increase by more than a factor of five to double the distance to an end point.

Can a Facility File Predictively? Predictive filing is an option that allows a facility to submit an RM Plan that includes regulated substances that may not be held at the facility at the time of submission. This option is intended to assist facilities such as chemical warehouses, chemical distributors, and batch processors whose operations involve highly variable types and quantities of regulated substances, but who are able to forecast their inventory with some degree of accuracy. Under 40 CFR 68.190, you are required to update and re-submit your RM Plan no later than the date on which a new regulated substance is first present in a covered process above a threshold quantity.

By using predictive filing, you will not be required to update and resubmit your RM Plan when you receive a new regulated substance if that substance was included in your latest RM Plan submission (as long as you receive it in a quantity that does not trigger a revised Off-Site Consequence Analysis as provided in 40 CFR 68.36).

If you use predictive filing, you must implement your Risk Management Program and prepare your RM Plan exactly as you would if you actually held all of the substances included in the RM Plan. This means that you must meet all rule requirements for each regulated substance for which you file, whether or not that substance is actually held on site at the time that you submit your RM Plan. Depending on the substances for which you file, this may require you to perform additional Worst-Case and Alternative Release Scenarios and to implement additional prevention program elements. If you use this option, you must still update and resubmit your RM Plan if you receive a regulated substance that was not included in your latest RM Plan. You must also continue to comply with the other update requirements stated in 40 CFR 68.190.

How Do You De-Register a Facility? If your facility is no longer covered by this rule, you must submit a letter to the RMP Record Center within six months indicating that your stationary source is no longer covered.

Implementation

The implementing agency is the Federal, State, or local agency that is taking the lead for implementation and enforcement of 40 CFR 68. The implementing agency will review RM Plans, select some plans for audits, and conduct on-site inspections. The implementing agency should be your primary contact for information and assistance.

Under the CAA, EPA will serve as the implementing agency until a State or local agency seeks and is granted delegation under CAA Section 112(l) and 40 CFR 63, Subpart E. You should check with the EPA Regional Office to determine if your state has been granted delegation or is in the process of seeking delegation. The Regional Office will be able to provide contact names at the State or local level. Appendix C of the *General Guide* has addresses and contact information for EPA Regions and State implementing agencies.

Delegated Programs

If the program is delegated, what does that mean to the facility? To gain delegation, a State or local agency must demonstrate that it has the authority and resources to implement and enforce 40 CFR 68 for all covered processes in the State or local area. Some states may, however, elect to seek delegation to implement and enforce the rule only for sources covered by an operating permit program under Title V of the CAA.

When EPA determines that a State or local agency has the required authority and resources, EPA may delegate the program. If the state's rules differ from 40 CFR 68 (a state's rules are allowed to differ in certain specified respects, as discussed below), EPA will adopt, through rulemaking, the State program as a substitute for 40 CFR 68 in the state, making the State program Federally enforceable.

In most cases the state will take the lead in implementation and enforcement, but EPA maintains the ability to enforce 40 CFR 68 in delegated states. Should EPA decide that it is necessary to take an enforcement action in the state, the action would be based on the State rule that EPA has adopted as a substitute for 40 CFR 68. Similarly, citizen actions under the CAA would be based on the State rules that EPA has adopted.

Under 40 CFR 63.90, EPA will not delegate the authority to add or delete substances from 40 CFR 68.130. EPA also plans to propose, in revisions to 40 CFR 63, that authority to revise Subpart G (relating to RM Plans) will not be delegated. With respect to RM Plans, you would continue to be required to file your 40 CFR 68 RM Plan, in the form and manner specified by EPA, to the central location EPA designates.

You should check with your state to determine whether you need to file additional data for state use or submit amended copies of the plan with the state to cover State elements or substances. If your state has been granted delegation, it is important that you contact them to determine if the state has requirements in addition to those in 40 CFR 68. State rules may be more stringent than 40 CFR 68. The EPA guidance documents do not cover State requirements.

Reviews/Audits/Inspections

Reviews. The implementing agency is required under 40 CFR 68 to review and conduct audits of RM Plans. *Reviews* are relatively quick checks of the plans in order to determine whether they are complete and whether they contain any information that is clearly problematic. For example, if an RM Plan for a process containing flammables fails to list fire and explosion as a hazard in the prevention program, then the implementing agency may flag that as a problem. The RMP data system will perform some of the reviews automatically by flagging RM Plans that have been submitted without the necessary data elements completed.

Facilities may be selected for audits based on any of the following criteria, set out in 40 CFR 68.220:

* Accident history of the facility
* Accident history of other facilities in the same industry
* Quantity of regulated substances handled at the site
* Location of the facility and its proximity to public and environmental receptors
* The presence of specific regulated substances
* The hazards identified in the RM Plan
* A plan providing for random, neutral oversight

Audits and Their Conduct. Under the CAA and 40 CFR 68, audits are conducted on the RMP. *Audits* will generally be reviews of the RM Plan to review its adequacy and require revisions when necessary to ensure compliance with 40 CFR 68.

Audits are used to identify whether the underlying risk management program is being implemented properly. The implementing agency will look for any inconsistencies in the dates reported for compliance with prevention program elements. For example, if you report that the date of your last revision of operating procedures was in June, 1998 but your training program was last reviewed or revised in December, 1994, the implementing agency will ask why the training program was not reviewed to reflect new operating procedures.

The agency will also look at other items that may indicate problems with implementation. For example, if you are reporting on a distillation column at a refinery, but used a checklist as your PHA technique, or you fail to list an appropriate set of process hazards for the process chemicals, the agency may seek further explanations as to why you reported in the way you did. The implementing agency may compare your data with that of other facilities in the same industrial sector using the same chemicals to identify differences that may indicate compliance problems. If audits indicate potential problems, they may lead to requests for more information or to on-site inspections.

The number of audits conducted will vary from state to state and from year to year. Implementing agencies will set their own goals, based on their resources and particular concerns.

Inspections. *Inspections* are site visits to check on the accuracy of the RM Plan data and on the implementation of all 40 CFR 68 elements. During inspections, the implementing agency will probably review the documentation for rule elements, such as the PHA reports, operating procedures, maintenance schedules, process safety information, and training.

Unlike audits, which focus on the RM Plan but may lead to determinations concerning needed improvements to the risk management program, inspections will focus on the underlying risk management program itself. Implementing agencies will determine how many inspections they need to conduct.

Audits may lead to inspections or inspections may be done separately. Depending on the focus of the inspection (all covered processes, a single process, or particular part of the risk management program) and the size of the facility, inspections may take several hours to several weeks.

Relationship with the Title V Permit Programs

40 CFR 68 is an applicable requirement under the CAA Title V permit program and must be listed in a Title V air permit. You do not need a Title V air permit solely because you are subject to 40 CFR 68. If you are required to apply for a Title V permit because you are subject to requirements under some other part of the CAA, you must

- List 40 CFR 68 as an applicable requirement in your permit
- Include conditions that require you to either submit a compliance schedule for meeting the requirements of 40 CFR 68 by the applicable deadlines or
- Include compliance with 40 CFR 68 as part of your certification statement

You must also provide the permitting agency with any other relevant information it requests.

The RM Plan and supporting documentation are not part of the permit and should not be submitted to the permitting authority. The permitting authority is only required to ensure that you have submitted the RM Plan to the EPA and that it is complete. The permitting authority may delegate the review of the plan to other agencies.

If you have a Title V permit and it does not address the 40 CFR 68 requirement, you should contact your permitting authority and determine whether your permit needs to be amended.

Penalties for Noncompliance

Penalties for violating the requirements or prohibitions of 40 CFR 68 are set forth in CAA Section 113. This section provides for both civil and criminal penalties. EPA may assess civil penalties of not more than $27,500 per day per violation.

Anyone convicted of knowingly violating 40 CFR 68 may also be punished by a fine pursuant to Title 18 of the *United States Code* or by imprisonment for no more than five years, or both. Anyone convicted of knowingly filing false information may be punished by a fine pursuant to Title 18 or by imprisonment for no more than two years.

One Plan

A facility subject to the Risk Management Program of the Clean Air Act is usually subject to a number of other regulations in regard to hazardous materials. In the past, this has meant separate plans and documents for each of these regulations. The One Plan (also known as the Integrated Compliance Plan, or ICP) concept has been developed to allow the combination of these requirements.

The National Response Team (NRT) announced the *One Plan* guidance (*ICP Guidance)* for integrated contingency planning in June 1996. This guidance provides the method to consolidate multiple plans that a facility may have prepared in order to comply with various regulations into one functional emergency response plan.

The *ICP Guidance* was developed to

- Provide a mechanism for consolidating multiple facility response plans into one plan that can be used during an emergency

- Improve coordination of planning and response activities within the facility and with public and commercial responders

- Minimize duplication and simplify planning

The *ICP Guidance* resulted from recommendations in the December 1993 NRT Report to Congress: *A Review of Federal Authorities for Hazardous Materials Accident Safety*. The NRT received input from representatives from State and local agencies, industry, and environmental groups prior to developing the guidance. Five agencies signed the one-plan guidance: the Environmental Protection Agency (EPA), the Coast Guard, the Occupational Safety and Health Administration (OSHA), the Office of Pipeline Safety of the Department of Transportation (DOT), and the Minerals Management Service (MMS) in the Department of the Interior. The NRT and the agencies responsible for reviewing and approving compliance with hazardous materials regulations agree that integrated response plans prepared in accordance with this guidance will be acceptable and will be the Federally preferred method of response planning.

The *ICP Guidance* gives facilities a common-sense option for meeting multiple emergency requirements under nine different regulations. The *ICP Guidance* is the outgrowth of the 1994 presidential review of Federal authorities related to hazardous materials accident prevention, mitigation, and response. That review identified multiple and overlapping facility emergency response plans as a problem area. Within the guidance document is a core facility response plan for releases of oil and hazardous substances. Plans prepared by facilities in accordance with the guidance will satisfy requirements of the five participating agencies and will be the Federal preferred method of such planning.

Regulations Covered by the ICP Guidance

Rather than a regulatory initiative, the ICP document is guidance. It presents a sample contingency plan outline that addresses requirements of the following Federal regulations:

- The Clean Water Act (CWA) (as amended by the Oil Pollution Act [OPA]) Facility Response Plan Regulations (EPA, Coast Guard, DOT, MMS)

- EPA's Risk Management Program Regulation, Oil Pollution Prevention Regulation, and the Resource Conservation and Recovery Act (RCRA) Contingency Planning Requirements

- OSHA's Emergency Action Plan Regulation, Process Safety Management Standards, and the Hazardous Waste Operations and Emergency Response (HAZWOPER) Regulation

Format of the ICP

A facility may use the ICP sample format or use an alternate format. The ICP sample format includes the following three sections:

1) Plan introduction

2) A core plan that serves as the primary response tool

3) A series of annexes that provide more detailed supporting information and regulatory compliance documentation

The ICP sample format is based on the Incident Command System (ICS). By organizing an integrated contingency plan according to the structure of the ICS, a facility will allow the plan to dovetail with established response management practices. This should promote its usefulness in an emergency.

Cross-References

The *ICP Guidance* supports the use of linkages (*i.e.*, references) to facilitate coordination with other facility plans and with external plans such as Local Emergency Planning Committee (LEPC) plans and Area Contingency Plans. When a facility submits a plan for Federal agency review, it must provide a table indicating where the regulatory required elements can be found in the one-plan format. The *ICP Guidance* includes tables that cross-reference the requirements of individual regulations with the ICP sample format.

The NRT intends to continue promoting the use of the *ICP Guidance* by regulated industries, and encourages Federal and State agencies to rely on the *ICP Guidance* when developing future regulations. The *ICP Guidance* was published in the *Federal Register* on June 5, 1996 (61 FR 28642).

For copies and more information, call the

RCRA, RCRA Contingency Planning Requirement, Superfund, and EPCRA HOTLINE: (800) 424–9346 (TDD: [800] 553–7672)

In the Washington, DC area, call the HOTLINE at

(703) 412–9810

The ICP Guidance is also available on EPA's Chemical Emergency Preparedness and Prevention Office website at

<http://www.epa.gov/swercepp/pub/one-plan.html>

For questions on the interface of the *ICP Guidance* with specific regulations, call the contact listed for that particular regulation:

Coast Guard Facility Response Plan Regulation: (202) 267–1983

DOT/Research and Special Programs Administration (RSPA) Pipeline Response Plan Regulation: (202) 366–8860

EPA Oil Pollution Prevention Regulation: (703) 603–8735

Spill Prevention, Control, and Countermeasures (SPCC) Information Line: (202) 260–2342

MMS Facility Response Plan Regulation: (703) 787–1567

EPA Risk Management Planning Regulation: (202) 260–0030

For OSHA regulations and standards, contact either the regional or area OHSA office.

The one-plan approach will minimize duplication of effort and unnecessary paperwork burdens. The consolidation of the various plans will allow for the development of an integrated emergency plan and training of facility personnel.

Communication with the Public

Possibly the most important part of the RMP process is communicating results of the RM Plan to the public. The *Not In My Backyard* response is what you do not want when communicating your results to the public. Once you have prepared and submitted your RM Plan, EPA will make it available to the public. Public availability of the RM Plan is a requirement under Section 114(c) of the Clean Air Act (the Act provides for protection of trade secrets, and EPA will accordingly protect any portion of the RM Plan that contains Confidential Business Information).

Therefore, you can expect that your community will discuss the hazards and risks associated with your facility as indicated in your RM Plan. You will necessarily be part of such discussions. The public and the press are likely to ask you questions, because only you can provide specific answers about your facility and your accident prevention program.

This dialogue is a most important step in preventing chemical accidents and should be encouraged. You should respond to these questions honestly and candidly. Refusing to answer, reacting defensively, or attacking the regulation as unnecessary are likely to make people suspicious and willing to assume the worst.

If the people and business residents in the vicinity of your facility believe that your operations threaten their health and safety, they will take action. They will contact their elected officials and demand that steps be taken. Local governments can regulate your facility out of business if a threat is perceived and there is no confidence that your facility is committed to prevention of and effective response to incidents.

A basic fact of risk communication is that trust, once lost, is very hard to regain. This has been a problem for the Nuclear Power Industry since the Three Mile Island incident. The industry did not inform the public of risks prior to the accident, and during the event failed to effectively communicate what was happening and was perceived as not telling the truth. One example of good risk communication is found in the airline industry. When fatal crashes have occurred, those airlines with an effective public communications/relations plan have fared better than those without.

You should prepare as early as possible to begin talking about these issues with the community, Local Emergency Planning Committees (LEPCs), State Emergency Response Commissions (SERCs), other local and State officials, and other interested parties. Communication with the public can be an opportunity to develop your relationship with the community and build a level of trust among you, your neighbors, and the community at large.

It is important to ensure that the public understands that by complying with the RMP Rule, you are taking a number of steps to prevent accidents and protect the community. These steps are the individual elements of your risk management program. A well-designed and properly implemented risk management program will set the stage for informative and productive dialogue between you and your community.

Some industries have developed guidance and other materials to assist in this process—contact your trade association for more information.

Risk communication means establishing and maintaining a dialogue with the public about the hazards at your operation, and discussing the steps that have been or can be taken to reduce the risk posed by these hazards. Of particular concern under this rule are the hazards related to the chemicals you use and what would happen if you had an accidental release. Chapter 11 of the *General Guide* covers the subject in depth and offers excellent suggestions for developing your risk communications program. This area is one where expert assistance should be seriously considered.

Conclusion

The RMP is a demanding task. It requires expertise in a number of specialties, and is one of the few programs that require off-site impacts of your facility to be analyzed and provided to the public.

The key points in the implementation of the RM Plan are:

- Identify the regulating agency.

- Understand that audits and inspections will occur on some basis that may not be the result of actions or incidents at your facility.

- The RMP must be included in Title V Permit documents.

- There are penalties for noncompliance.

If you are the person most likely to bear the responsibility for the RMP at your facility, always keep in mind that the implementation and maintenance of the RMP are as critical as any other aspect. Having an analysis on a bookshelf or in a computer file does not mean you are done and ready for the next job.

The Worst-Case Release Scenarios are usually well beyond the probabilities considered in the design and construction of your facility. However the potential is there, and one accident resulting in serious injury or death to members of the general public or workers can mean the termination of your facility and possibly the financial ruin of your company.

Bhopal was the result of inattention to the consequences of accumulating an extremely hazardous material that was an intermediate product. Controls in process, inventory, can result in substantial long-term savings. The RMP is a challenge, but when added to the rest of the regulations it fills in the last piece of the puzzle. The completed picture is an integrated safety management program that safeguards facility workers and the general public.

Dale M. Petroff is currently the Technical Coordinator for Systematic Management Service, Inc. in Aiken, South Carolina. Mr. Petroff has 23 years of experience in hazardous materials management at military bases, DOE sites, waste-processing facilities, and nuclear and fossil fuel power plants, as well as industrial sites and facilties. His areas of expertise include radiation protection/industrial safety, environmental compliance, waste management, security/intelligence, hazards and safety analysis, training (development/instruction), and emergency management/ preparedness. His contributions to the field of hazardous materials management include the development of a job task analysis for response by local responders to radiological accidents, a methodology for the development of emergency preparedness hazard analysis, and a methodology to calculate the release of radioactive materials resulting from decontamination and dismantlement operations. Mr. Petroff has published "Emergency Response Planning Guidelines: A Step by Step Approach," in Environmental Solutions, *as well as technical reports on decontamination of buildings containing radioactive materials.*

Part VII

Water Quality

CHAPTER 32

Clean Water Act

John M. Higgins, PhD, PE, CHMM

The Clean Water Act is the principal law governing pollution control and the quality of the nation's waters. The basis for the current law was first promulgated in 1972 when many of the nation's waters were seriously degraded. Many rivers received raw or poorly treated sewage. Some beaches were closed due to contamination, and aquatic life in many lakes and estuaries was endangered. Congress responded by enacting the Federal Water Pollution Control Act (PL 92–500), commonly referred to as the Clean Water Act. The legislation totally revised the conceptual foundation of policy for the control of water pollution and called for a new approach to water quality standards, facility planning, regulatory programs, discharge permitting, enforcement, and Federal funding.

Today, the Clean Water Act is one of the most highly regarded environmental statutes in existence and has brought about a significant improvement in the quality of the nation's waters. The latest national assessment indicates that 60% of the waters that were evaluated met the Federal standards for water quality, up from only 36% in 1972; the number of people served by secondary sewage treatment facilities has more than doubled to over 62%.

The first section of this chapter will look at the historical development of the Act. The remaining sections will summarize the provisions that are most relevant to hazardous material managers.

Clean Water Act

Introduction

Federal involvement in the management of water resources dates back to 1824 when the Corps of Engineers was authorized to clear a navigable channel on the Ohio and Mississippi Rivers. The Rivers and Harbors Act of 1899 prohibited unauthorized obstruction or alteration of the navigable waters of the United States. Section 13 of the Act, known as the Refuse Act of 1899, authorized the Secretary of the Army to promulgate regulations and prohibit the discharge of refuse (or any other activity) that would affect the course, location, condition, or physical capacity of navigable waters. This law served as the primary basis for the Federal regulation of water pollution until 1972, when authority over the discharge of refuse was transferred to the Environmental Protection Agency (EPA) and the states under Sections 402 and 405 of the Clean Water Act.

Other legislation included the Public Health Service Act of 1912, which contained a section on waterborne diseases and established a Stream Investigation Station in Cincinnati to conduct water pollution research. The Oil Pollution Act of 1924 was passed to prevent the discharge of oil into coastal waters. The first comprehensive national legislation came in 1948 with the passage of the Water Pollution Control Act (PL 80–845). The Act required the United States Surgeon General to develop comprehensive programs to reduce the pollution in interstate waters. Enforcement of the Act was largely left to the states and some funds were provided for the treatment of municipal wastewater.

Subsequent statutes facilitated the development of the information, methodologies, and institutions for water quality management. These statutes included the Federal Water Pollution Control Act of 1956 (PL 84–660), the 1961 amendments to that Act (PL 87–88), the Water Quality Act of 1965 (PL 89–234); and the Clean Water Restoration Act of 1966 (PL 89–753). These statutes provided increased funding for municipal wastewater treatment and for scientific studies of pollution problems. Greater emphasis was placed on standards for water quality and the criteria for assessing pollution problems and priorities. Enforcement, however, remained primarily a State responsibility and usually required significant evidence of damage that could be attributed to a specific facility. The enforcing agency was essentially required to prove that the facility was violating a State standard or endangering public health and welfare, and the cost of abatement was less than the corresponding benefits.

By the late 1960s, water quality continued to decline while the public's interest in environmental quality increased. On October 18, 1972, Congress overrode a presidential veto and enacted the Federal Water Pollution Act Amendments of 1972 (PL 92–500). The needs of water quality took precedence over economic considerations as minimum levels of treatment and a *zero discharge* goal were established. The Act shifted the burden of proof for pollution control from the regulatory agency to the polluter. Before the Act was passed, the government had to prove harm by a specific facility in order to justify an action. After the Act passed, the facility had to explain why the discharge should not be eliminated.

The objective of the 1972 Act, as stated in Section 101, was "to restore and maintain the chemical, physical, and biological integrity of the Nation's waters." The following goals and policies were established:

1) The discharge of pollutants into the navigable waters would be eliminated by 1985.

2) Wherever attainable, an interim goal of water quality that provides for the protection and propagation of fish, shellfish, and wildlife, and provides for recreation in and on the water, would be achieved by July 1, 1983.

3) The discharge of toxic pollutants in toxic amounts would be prohibited

4) Federal financial assistance would be provided to construct publicly owned waste treatment works

5) Planning processes for the management of area-wide waste treatment would be developed and implemented to ensure adequate control of the sources of pollution in each state.

6) A major research and demonstration effort would be undertaken to develop the technology required to eliminate the discharge of pollutants into the navigable waters, waters of the contiguous zone, and the oceans.

7) Programs for the control of nonpoint sources of pollution would be developed and implemented as quickly as possible in order to enable the goals of the Act to be met through

the control of both point and nonpoint sources of pollution.

Enforcement authority was vested in the administrator of the EPA. Effluent limitations and a national discharge permitting program (NPDES) were established. Every *point source* facility was required to obtain a permit from the EPA or an authorized State agency by the last day of 1974. Technology-based effluent limits and minimum treatment requirements were also established. By July 1, 1977, all facilities had to achieve a wastewater treatment level that represented the *best practicable control technology* currently available, (BPT) and by July 1, 1983, had to achieve the *best available technology* economically achievable (BAT). EPA was required to publish industrial pretreatment standards, a list of toxic pollutants, and effluent standards for each substance.

By 1977, the new approach to water pollution control was established, and the water quality was improving in many areas. Some elements of the Act were too ambitious, however. The Clean Water Act of 1977 (PL 95–217) fine-tuned the 1972 Act, while confirming the basic programs and directions. The 1983 *fishable and swimmable* and the 1985 *zero discharge* goals remained. The 1983 BAT requirement now applied only to toxic substances, rather than to all discharges. A new term, *best conventional pollutant control technology* (BC-PCT), was introduced for conventional pollutants like biochemical oxygen demand and suspended solids. The states were to lead in managing the NPDES permit and construction grants programs.

The 1977 Act expanded EPA's mandate to control the release of toxic pollutants. The EPA had focused on conventional pollutants and made little progress in dealing with the more complex toxic pollutants. Environmental groups sued the EPA in 1976 for its failure to meet statutory deadlines. The resulting *consent decree* laid out a detailed strategy for the control of toxic waste that became the foundation of the toxic pollutant provisions of the 1977 Act. Under the decree, the EPA was required to promulgate effluent guidelines, performance standards for new sources, and pretreatment standards for 65 *priority pollutants* in 21 major industrial categories (later expanded to 129 chemical substances and 34 industrial categories).

The Water Quality Act of 1987 was passed despite a presidential veto. The Act focused on the most significant remaining water quality problems: *nonpoint* (diffuse) *sources* and toxic pollutants. The EPA was directed to supplement technology-based standards with a water quality–based approach. Waters that did not meet the water quality standards due to nonpoint and toxic pollutants were to be identified and restored. In addition, the discharge permit program was expanded to require stormwater discharge permits for industrial facilities and large cities.

The 1987 amendments established a schedule and guidelines for replacing the construction grants program, which had helped to establish municipal wastewater treatment plants, with revolving loans from the states. Federal assistance was to be phased out in 1994 in favor of a continuing source of funds that was to be provided by these State loans. The 1987 amendments also funded demonstration programs that would encourage watershed planning to clean up the Great Lakes and the Chesapeake Bay. Other provisions of the Act limited certain pollution control exemptions for industrial facilities; prohibited *backsliding* (lowering of requirements) when permits were renewed; and required regulations to control the toxic pollutants in sewage sludge.

In summary, the Clean Water Act provides a foundation for water quality management. Zero discharge, fishable and swimmable waters, and no toxic discharges in toxic amounts are the Act's goals. Key components of the Act include designated water uses, water quality standards that support each use, effluent limits and minimum treatment requirements, a permit program that regulates discharges, and antidegradation and antibacksliding provisions to prevent future regression. Key administrative elements of the Act include

- Primary responsibility for enforcement belonging to the states
- National effluent limits and water quality guidelines
- Minimum requirements
- Federal funding for publicly owned facilities
- Setting of priorities among pollution sources (*e.g.*, toxic pollutants, nonpoint sources, accidental spills, watershed approach)

Discharge of Pollutants

The Clean Water Act is founded upon the prohibition of the *discharge* of pollutants. The Act, regulations, and court decisions have defined how this prohibition is implemented. A discharge is "any addition of any pollutant to navigable waters from any point source." The EPA and the courts have interpreted the term *addition* to mean any introduction of a pollutant into a body of water. Two exceptions have been allowed, however. The first excludes pollutants that are present in the discharge only because they were present in the intake water. This scenario applies if the discharge is released back into the same body of water and the pollutant is not removed by normal operations. The second exception excludes the discharge of water from dams.

The term *pollutant* includes dredge spoil, solid waste, incinerator residue, sewage, garbage, sewage sludge, munitions, chemical wastes, biological materials, radioactive materials, heat, wrecked or discarded equipment, rock, sand, cellar dirt, and industrial, municipal, and agricultural waste discharged into water. The courts have broadened the definition to include virtually any material or characteristic (*e.g.*, toxicity or acidity).

A *point source* is "any discernible, confined and discrete conveyance, including, but not limited to, any pipe, ditch, channel, tunnel, conduit, well, discrete fissure, container, rolling stock, concentrated animal feeding operation, or vessel or other floating craft from which pollutants are or may be discharged." Point sources can include vehicles, natural conveyances, and intermittent sources, such as stormwater outfalls.

Navigable waters are all "waters of the United States," including waters used in interstate commerce, waters subject to tides, interstate waters, and intrastate lakes, rivers, streams, and wetlands. Waters used by interstate travelers and migratory birds are also included. Ground water is not included unless there is a hydrological connection between the ground water and surface water or unless ground water is specifically included under a State program.

NPDES Permits

Point source discharges are regulated through the NPDES permit program. The states administer this program in accordance with the EPA's guidelines and oversight. Section 303 of the Act allows states to adopt water quality standards that are based on the designated use of each body of water. An antidegradation provision is required to ensure that the use classifications and the existing water quality will not be downgraded. The states must submit proposals for receiving water standards and implementation plans for enforcement to the EPA, along with the effluent limitations that are needed to maintain or attain the water quality necessary for designated uses. If a state's standards are deemed inadequate, the EPA must adopt Federal standards for the state. If Federal effluent limitations are too low to meet the state's water quality standard, the EPA must specify more stringent effluent limitations. The states must also establish a maximum allowable pollution load (Total Maximum Daily Load–TMDL) for those waters that will not meet water quality standards by following the Federal effluent limitations. Public hearings to review the standards for water quality must be conducted at least every three years.

The effluent limitations for industrial operations are technology based (*i.e.*, dependent not only on what is needed to maintain water quality, but also on the currently available treatment technologies). The EPA can, however, establish water quality–based effluent limitations. A public hearing is required to determine the social and economic costs of achieving the proposed standard. If an industrial point source can demonstrate that there is no reasonable relationship between the benefits and costs, then the EPA must relax the proposed effluent limitations that are based on water quality.

Section 402 of the Act requires all point source dischargers to obtain an NPDES permit. The permit is the mechanism for implementing and enforcing the water quality standards and effluent limitations. The permit identifies all wastewater discharges (by location, flow rate, and character and volume of pollutants), effluent limits, monitoring requirements, administrative procedures, and any special provisions. An industry that discharges wastewater into a publicly owned treatment works (POTW) may be required to pretreat the waste if the discharge is not susceptible to or would interfere with the treatment provided by the POTW. Permit renewals are subject to an antibacksliding provision that prevents any lowering of the effluent limits that were in the previous permit.

An NPDES permit is required for stormwater discharge if it contributes to a violation of a water quality standard or is a significant contributor of pollutants. Stormwater permits apply to industries and to municipalities that serve more than 100,000 people. Section 401 of the Act requires any applicant for a Federal license or permit to obtain a State water quality certification if the activity may result in a discharge to navigable waters. The certification seeks to ensure that the activity or facility will comply with applicable water quality standards of the state.

Toxic Pollutants and Spills

Section 307 requires the EPA to maintain a list of toxic pollutants that are subject to effluent limitations based on the best available technology that is economically achievable. The list must be reviewed and revised at least every three years. The factors to be considered include

- The toxicity of the pollutant
- Its persistence
- The presence of the affected organism in any waters
- The importance of the affected organisms
- The nature and extent of the effect of the toxic pollutant on the organisms

Today, the list consists of 65 priority pollutants and their compounds (129 total chemical substances) that must be considered in developing NPDES permits for 34 industrial categories (40 CFR 122 Appendix A).

The Act requires that an "ample margin of safety" be provided in setting effluent standards for toxic

pollutants. The EPA must publish information on their methods for establishing criteria and measuring toxicity, including biological monitoring and assessment of synergistic effects. New permits may require toxicity testing of effluent streams. Where toxicity is found, a toxicity reduction evaluation may be required to identify and control the sources of toxicity.

Section 311 prohibits the discharge of oil or hazardous substances into navigable waters, adjoining shorelines, or waters of the Outer Continental Shelf. The EPA has identified almost 300 substances as hazardous when spilled. The National Response Center must be notified of spills that exceed the minimum reportable quantity.

The Act establishes a National Contingency Plan and a funding mechanism for cleaning up oil and hazardous substance spills. Federal and State governments are authorized to react to spills as necessary, to recover the costs of pollution control and environmental damage, and to assess civil and criminal penalties. Cleanup funding is consolidated under the Comprehensive Environmental Response, Compensation, and Liability Act (CERCLA), which covers spills both on land and in the water. A Spill Prevention, Control and Countermeasure (SPCC) Plan is required if petroleum products exceed the designated minimum quantities (i.e., 1320 gallons above ground, 42,000 gallons below ground, or 660 gallons in a single above-ground tank). The plan must identify the facility components that are specifically designed to control spills, on-site control measures, and off-site cleanup measures. The plan must be updated every three years and approved by a registered engineer. Companies can be fined up to $5000 per day for failure to have an SPCC plan.

Dredged and Fill Material

Section 404 establishes a separate permit program for the disposal of dredged and fill material in the waters of the United States, including wetlands. The program is administered by the Secretary of the Army, acting through the Chief of Engineers. The Corps of Engineers issues permits for specific sites after the public has been notified and an opportunity for public hearings has been given.

Disposal sites must comply with EPA guidelines that are developed in conjunction with the Secretary of the Army. Ocean disposal must comply with the dumping criteria set forth in the Marine Protection, Research, and Sanctuaries Act. The EPA may forbid or restrict the use of a proposed or existing area if it determines that there could be unacceptable effects on municipal water supplies, shellfish beds, fishery areas, wildlife, or recreational activities. Such a determination must be made in consultation with the Corps of Engineers and the permit applicant.

Dredged material and sediments beneath navigable waters must be tested and evaluated for their suitability for disposal. The EPA and the Corps of Engineers have developed specific guidelines for testing, dredging, and disposal. Any proposed actions that meet these guidelines may be permitted in one of several ways. General Permits may be issued on a State, regional, or national basis for activities that are similar in nature and have minimal individual and cumulative adverse effects. General Permits are valid for up to five years. Minor activities within certain categories are excluded or have been preauthorized for approval under a Nationwide Permit. Actions that may cause significant environmental effects must be reviewed separately and obtain an Individual Permit.

Wetlands are defined as "areas that are inundated or saturated by surface or ground water at a frequency and duration sufficient to support a prevalence of vegetation typically adapted for life in saturated soil conditions." In order for a site to be classified as a wetland, the appropriate vegetation, soils, and hydrology must be present.

Executive Order 11990 directs Federal agencies to avoid the unnecessary alteration or destruction of wetlands. Agencies must not undertake or assist in new construction in wetlands unless no practical alternative is available and all practical mitigation measures must be included. These requirements apply to projects that receive Federal permits and funding. Public notice and the opportunity for public hearings must be provided, even for projects that are not considered significant enough to require an environmental impact statement.

Reauthorization

Reauthorization of the Clean Water Act has been considered by the 103rd and 104th Congresses, but no legislation has been enacted. The House passed a comprehensive reauthorization bill (HR 961) during the 104th Congress, but the Senate did not act on it. There has been no legislative activity in the 105th Congress, and none is expected. Key issues in the reauthorization include whether regulatory requirements should be made less burdensome, implementation of the phaseout of Federal assistance to states and communities, control of nonpoint sources (*e.g.*, agricultural runoff, urban stormwater, and sewer overflows), and the regulation of wetlands as an intrusion on private land rights. Continued disagreement on the type of reforms needed is likely to delay reauthorization until a clear consensus emerges.

Bibliography

92nd Congress. *Federal Water Pollution Act Amendments of 1972.* PL 92–500, as amended, 1972.

Arbuckle, J. C., *et al. Environmental Law Handbook,* 14th ed. Rockville, MD: Government Institutes, Inc., 1997.

Copeland, Claudia. *Clean Water Act Reauthorization in the 105th Congress.* Washington, DC: Congressional Research Service, Environment and Natural Resources Division, 1998.

Gallagher, L. M., and Miller, L. A. *Clean Water Handbook.* Rockville, MD: Government Institutes, Inc., 1996.

Kovalic, Joan M. *The Clean Water Act of 1987.* Water Pollution Control Federation, No. P0070JR. Alexandria, VA: Water Pollution Control Federation, 1987.

Wolf, Sidney M. *Pollution Law Handbook: A Guide to Federal Environmental Law.* New York, NY: Quorum Books, 1988.

John M. Higgins *is a Senior Water Quality Engineer for the Tennessee Valley Authority. He has over 30 years experience in project management, water supply planning, pollution control, hazardous materials, and water resource management. Dr. Higgins' projects and responsibilities have included international projects, operation of aeration systems at TVA dams, and reservoir and watershed management, as well as environmental permitting. Dr. Higgins has written articles on reservoir release improvements (published in the* Energy Division Journal, *American Society of Civil Engineers), integrated watershed management (Proceedings, CONSERV93, American Water Works Association), and phosphorous retention models for Tennessee Valley Authority reservoirs (*Water Resources Research*).*

CHAPTER **33**

Oil Pollution Act

Mark L. Bricker, PE, CHMM

Introduction

The Oil Pollution Act (OPA) was signed into law in August, 1990. The OPA was the initial congressional response to the *Exxon Valdez* oil spill of March, 1989. The *Valdez* spill, which has been called the "Pearl Harbor" of the United States environmental movement, galvanized public support behind legislation to ensure that future oil spills are minimized, that effective responses are made to those that do occur, and that those responsible pay for the damages and are subject to severe penalties.

It would be a mistake to view the OPA as a response only to the *Valdez* spill and other tanker incidents that occurred in 1989 and 1990 (*e.g.*, the *Mega Borg* fire and explosion in the Gulf of Mexico in 1990, the *American Trader* oil spill near the southern California coast in 1990, and a rash of incidents in later June, 1989). Congress had been working for almost 15 years to consolidate and rationalize oil spill response mechanisms under various Federal laws, including Section 311 of the Federal Water Pollution Control Act, (Clean Water Act [CWA]), the Deepwater Port Act of 1974, the Trans-Alaska Pipeline Authorization Act of 1973 (TAPAA), and the Outer Continental Shelf Lands Act Amendments of 1978 (OCSLA). Congress was also trying to harmonize these various oil spill laws with State laws, international conventions, and other Federal environmental laws, especially the Comprehensive Environmental Response, Compensation, and Liability Act of 1980 (CERCLA or Superfund).

The main elements of the OPA are as follows:

1)   A comprehensive Federal liability scheme, addressing all discharges of oil to navigable waters, the exclusive economic zone (a zone contiguous to the territorial sea extending 200 miles from shore), and shorelines.

2)   A single, unified Federal fund, called the Oil Spill Liability Trust Fund, to pay for the cleanup and other costs of federal oil spill response. The Fund was authorized at $1 billion per spill incident.

3)   Stronger Federal authority to order removal action or to conduct the removal action itself.

4)   Drastically revised Spill Prevention, Control, and Countermeasure Plan requirements for onshore facilities, offshore facilities, and vessels.

5)   Tougher criminal penalties.

6)   Higher civil penalties for spills of oil and for spills of hazardous substances.

7)   Tighter standards and reviews for licensing tank vessel personnel, and for equipment and operations of tank vessels, including requiring the use of double hulls.

8)   No pre-emption of State laws, and an endorsement of the United States' participation in an international oil spill liability and compensation scheme.

9)   Several provisions pertinent to Prince William Sound, to Alaska, and to other provisions of the United States.

The OPA improved the nation's ability to prevent and respond to oil spills by: (1) establishing provisions that expand the Federal government's ability and (2) providing the money and people necessary to respond to oil spills. Under the OPA, the United States (US) Coast Guard is the designated Federal response authority. Rule-making authority for onshore facilities was delegated to the Coast Guard and the Environmental Protection Agency (EPA).

The Oil Pollution Act Law

Far more comprehensive and stringent than any previous US or international oil pollution liability and prevention law, the OPA comprises nine titles.

Title I–Oil Pollution Liability and Compensation.
Title I contains the definitions used in the OPA, establishes the liability scheme for oil spills, provides the mechanisms for recovery from the Oil Spill Liability Trust Fund and from responsible parties, and establishes financial responsibility requirements.

Title II–Conforming Amendments.
Title II makes conforming changes in the Intervention on the High Seas Act, the CWA, the Deepwater Port Act, and the OCSLA.

Title III–International Oil Pollution Prevention and Removal.
Title III outlines participation in international oil spill prevention and removal regimes and directs the Secretary of State to review international agreements and treaties and to negotiate agreements with Canada regarding oil spills on the Great Lakes, Lake Champlain, and Puget Sound.

Title IV–Prevention and Removal.
Title IV has three subtitles. *Subtitle A–Prevention* provides for the review of information contained in the National Driver Register for issuing licenses, certificates of registry, and merchant marines' documents; provides for the suspension and revocation of those documents for alcohol and drug abuse incidents; and establishes prevention measures. *Subtitle B–Removal* provides Federal removal authority and requirements for the national planning and response system. *Subtitle C–Penalties and Miscellaneous* strengthens the civil and criminal penalties available to the government.

Title V–Prince William Sound Provisions.
Title V contains several provisions designed specifically to avoid future spills in Prince William Sound.

Title VI.
Contains miscellaneous provisions.

Title VII.
Provides for an oil pollution research and development program.

Title VIII. Contains provisions dealing with the Trans-Alaska Pipeline System and oil spills in the Arctic Ocean.

Title IX. Transfers funds from the several pre-existing Federal oil spill funds into the Oil Spill Liability Trust Fund.

Summary of Key Terms

Section 1001 of the OPA contains 37 definitions used throughout the Act. The OPA restates verbatim many of the definitions of the CWA. Definitions of a few of the key terms in the OPA are described below.

Navigable waters—Navigable waters are broadly defined to include all waters that are used in interstate or foreign commerce, all interstate waters including wetlands, and all intrastate waters, such as lakes, rivers, streams, wetlands, sloughs, prairie potholes, wet meadows, playa lakes, or natural ponds.

Vessel—Vessel is defined to include every description of watercraft or other artificial contrivance used, or capable of being used, as a means of transportation on water, other than a public vessel. Public vessels are noncommercial government vessels.

Tank vessels—These are vessels constructed, adapted to carry, or that carry oil or hazardous materials in bulk as cargo or cargo residue and that are US documented vessels, operate in US waters, or transfer oil or hazardous material in a place subject to the jurisdiction of the US.

Facility—Facility is any structure, group of structures, equipment, or device (other than a vessel) which is used for exploring, drilling, producing, storing, handling, transferring, processing, or transporting oil. The term also includes any motor vehicle, rolling stock, or pipeline used for these purposes. Facilities are further subdivided into onshore and offshore facilities. The boundaries of a facility may depend on several site-specific factors, such as the ownership or operation of buildings, structures, and equipment on the same site and the types of activity at the site.

Nontransportation-related facilities—Nontransportation-related facilities refer to all fixed facilities, including support equipment (but excluding certain pipelines) railroad tank cars en route, transport trucks en route, and equipment associated with the transfer of bulk oil to or from water transportation vessels. The term also includes mobile or portable facilities such as drilling or workover rigs, production facilities, and portable fueling facilities while in a fixed, operating mode.

Summary of Key Provisions

A summary of key provisions in the OPA is provided below.

§1002(a). Provides that the responsible party for a vessel or facility from which oil is discharged, or which poses a substantial threat of a discharge, is liable for: (1) certain specified damages resulting from the discharged oil and (2) removal costs incurred in a manner consistent with the National Contingency Plan (NCP).

§1002(c). Exceptions to the CWA liability provisions include: (1) discharges of oil authorized by a permit under Federal, State, or local law, (2) discharges of oil from a public vessel, or (3) discharges of oil from onshore facilities covered by the liability provisions of the Trans-Alaska Pipeline Authorization Act.

§1002(d). Provides that if a responsible party can establish that the removal costs and damages resulting from an incident were caused solely by an act or omission by a third party, the third party will be held liable for such costs and damages.

§1004. The liability for tank vessels larger than 3,000 gross tons is increased to $1,200 per gross ton or $10 million, whichever is greater. Responsible parties at onshore facilities and deepwater ports are liable for up to $350 million per spill; holders of leases or permits for offshore facilities, except deepwater ports, are liable for up to $75 million per spill, plus removal costs. The federal government has the authority to adjust, by regulation, the $350 million liability limit established for onshore facilities.

§1016. Offshore facilities are required to maintain evidence of financial responsibility of $150 million, and vessels and deepwater ports must provide evidence of financial responsibility up to the maximum applicable liability amount. Claims for removal costs and damages may be asserted directly against the guarantor providing evidence of financial responsibility.

§1018(a). The Clean Water Act does not pre-empt state law. States may impose additional liability (including unlimited liability), funding mechanisms, requirements for removal actions, and fines and penalties for responsible parties.

§1019. States have the authority to enforce, on the navigable waters of the State, the OPA requirements for evidence of financial responsibility. States are also given access to Federal funds (up to $250,000 per incident) for immediate removal, mitigation, or prevention of a discharge, and may be reimbursed by the Trust Fund for removal and monitoring costs incurred during oil spill response and cleanup efforts that are consistent with the NCP.

§4202. Strengthens planning and prevention activities by: (1) providing for the establishment of spill contingency plans for all areas of the US, (2) mandating the development of response plans for individual tank vessels and certain facilities for responding to a worst-case discharge or a substantial threat of such a discharge, and (3) providing requirements for spill removal equipment and periodic inspections.

§9001(a). Amends the Internal Revenue Act of 1986 to consolidate funds established under other statutes and to increase permitted levels of expenditures. Penalties and funds established under several laws are consolidated, and the Trust Fund borrowing limit is increased from $500 million to $1 billion.

Planning and Response System

The OPA addresses development of a National Planning and Response System. Under our current national emergency response infrastructure for oil spills, planning occurs on four basic levels:

1) National

2) Area

3) Local

4) Facility

In the event of an oil spill, the facility response plan is immediately activated. Depending on the nature of the spill, local, area, or Federal plans may also be put into motion.

National Response

The National Oil and Hazardous Substances Pollution Contingency Plan, more commonly called the National Contingency Plan (NCP), is the Federal government's blueprint for responding to both oil spills and hazardous substance releases. The NCP is the result of our country's efforts to develop a national response capability and promote overall coordination among the hierarchy of responders and contingency plans. The NCP differs somewhat from the other types of contingency plans in that it provides the framework for our National Response System, and it serves as a guide for the way in which the different levels of responding organizations coordinate their efforts. The first NCP was developed and published in 1968, in response to a massive oil spill from an oil tanker off the coast of England the year before. The NCP was revised in 1994 to reflect the oil spill provisions of the OPA.

Area Response

Area response plans are often brought into action when facilities are unable to handle spills on their own. Under the OPA, EPA initially established 13 Areas covering the US and convened Area Committees composed of Federal, State, and local government agencies to prepare contingency plans for the designated Areas. The Area Contingency Plans include detailed information about resources (such as equipment and trained response personnel) available from the government agencies in the Area. They also describe the roles and responsibilities of each responding agency during a spill incident, and how the agencies will respond if they are called upon in an emergency. These plans also describe how two or more Areas might interact, such as when a spill occurs in a river that flows between Areas, to ensure that a spill is controlled and cleaned up in a timely and safe manner.

Local and Facility Response

When a release or spill occurs, the party responsible for the release, its response contractors, the local fire and police departments, and the local emergency response personnel provide the first line of defense. If needed, a variety of State agencies stand ready to support, assist, or take over response operations if an incident is beyond local capabilities. Individual facility requirements are described in the section below.

Facility Oil Pollution Prevention Regulations

EPA first promulgated oil pollution prevention regulations in 1973 in the *Code of Federal Regulations* (40 CFR 112). The oil pollution prevention regulations establish requirements for facilities to prevent oil spills from reaching the navigable waters of the US or adjoining shorelines. The regulations apply to owners or operators of certain facilities that drill, produce, gather, store, process, refine, transfer, distribute, or consume oil.

These regulations form the basis of EPA's Spill Prevention, Control, and Countermeasure (SPCC) program, which seeks to prevent oil spills from certain above-ground storage tanks (ASTs) and underground storage tanks (USTs). An owner or operator of a nontransportation-related facility is required to prepare an SPCC Plan if

- The facility has an above-ground storage capacity of more than 660 gallons in a single tank, an aggregate above-ground storage capacity of more than 1,320 gallons, or a total underground storage of 42,000 gallons

- The facility could be expected to discharge oil in harmful quantities into navigable waters of the United States

In 1990, Congress enacted the OPA which, among other things, required certain oil storage facilities to prepare facility response plans. In response, EPA proposed revisions to the oil pollution prevention regulations in two phases. EPA proposed its first set of revisions to the regulation on October 22, 1991. These proposed revisions, in addition to strengthening and clarifying previous regulatory language, outline the additional requirements for regulated oil storage and handling facilities. On July 1, 1994, EPA issued the Phase II SPCC revisions, which increase the preparedness and response capabilities of onshore facilities through an expanded regulatory framework. The revisions incorporate the new requirements added by the OPA that direct facility owners or operators to prepare, and in some cases submit to the Federal government, plans for responding to a worst-case discharge of oil.

Facility owners or operators that are SPCC-regulated facilities must determine whether they could *substantially* harm the environment in the event of an oil spill. Owners or operators of *substantial harm facilities* are required to prepare and submit a Facility Response Plan to EPA. A Facility Response Plan is a separate document from a SPCC Plan. A facility is defined as a **substantial harm facility** if it meets the following criteria (see also Figure 1):

- The facility transfers oil over water to or from vessels, and has a total oil storage capacity of at least 42,000 gallons; or

- The total oil storage capacity at the facility is at least 1 million gallons and at least one of the following criteria is met:

  1) The facility's secondary containment for each AST area will not hold the volume of the largest single AST plus sufficient freeboard for precipitation.

  2) A discharge could injure fish, wildlife, or sensitive environments.

  3) A discharge would shut down operations at a public drinking water intake.

  4) The facility has had a reportable spill of at least 10,000 gallons within the last five years.

The sections below describe a SPCC Plan and a Facility Response Plan.

SPCC Plans

The oil pollution prevention regulations require each owner or operator of a regulated facility to have a fully prepared and implemented SPCC Plan. A SPCC Plan is a detailed, facility-specific, written

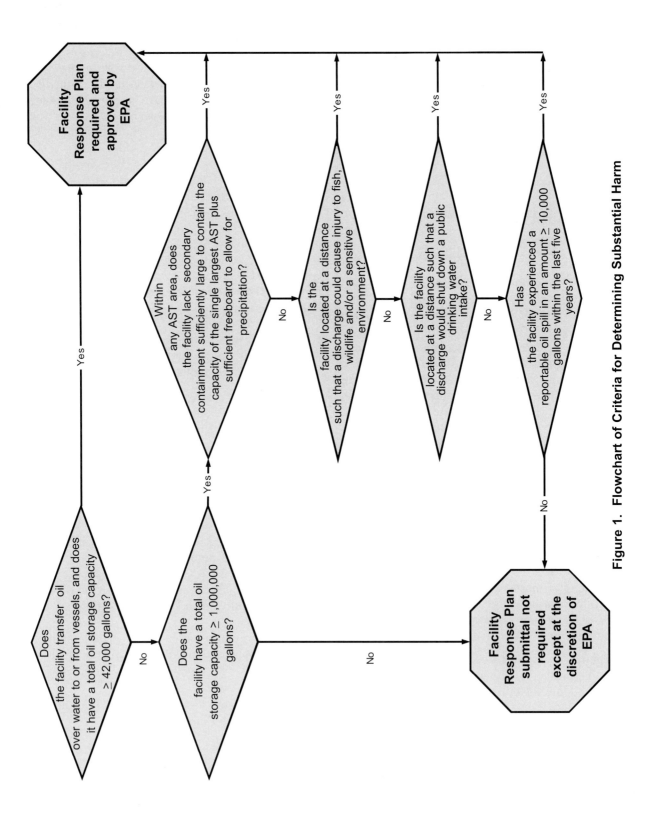

Figure 1. Flowchart of Criteria for Determining Substantial Harm

description of how a facility's operations comply with the prevention guidelines in the oil pollution prevention regulation. These guidelines include such measures as secondary containment, facility drainage, dikes or barriers, sump and collection systems, retention ponds, curbing, tank corrosion protection systems, and liquid level devices.

There is no rigid format for a SPCC Plan. Each SPCC Plan is unique. The regulations in 40 CFR 112.7 require certain elements to be included in the plan. These elements are identified and described below.

Name of Facility. The name may or may not be the business name.

Type of Facility. A brief description of the business activity.

Date of Initial Operations. The date the facility began operations.

Location of Facility. The location of the facility must be described in words or using a physical street address. The location of the facility must be supported by area maps.

Name and Address of Owner. The name of the owner of the facility must be identified. A physical street address for the owner must be given if the owner conducts business at an address other than the facility.

Designated Person Responsible for Oil Spill Prevention. Each facility needs to delegate an individual with overall oil spill responsibility. This individual should be thoroughly familiar with the regulations and the SPCC Plan.

Oil Spill History. This section of the plan can be either a reactive declaration or a detailed history of significant spill events that occurred in the previous three years. Typical information for spills would include: the type and amount of oil spilled; the location, date, and time of the spill; any surface water affected; a description of any physical damage; the cause of the spill; and any action taken to prevent recurrence.

Management Approval. A signed statement of an individual with the authority to commit management and the financial resources necessary to implement the plan. This individual is typically the facility manager.

Certification. A signed and certified statement of a registered professional engineer. The certifying engineer does not have to be registered in the state in which the facility is located. The registered professional engineer must be familiar with the provisions of the oil pollution prevention regulations, and must have examined the facility. The engineer's name, registration number, and state of registration must be included as part of the SPCC Plan. In addition, the engineer's seal must be affixed to the SPCC Plan as part of the certification.

Periodic Review. The periodic review section is a signed and dated statement by the owners or operators of the facility that a review and evaluation of the SPCC Plan has been performed. A review of the plan must be completed at least once every three years. As a result of the review evaluation, the regulations require the owner or operator of a facility to amend the SPCC Plan within six months to incorporate more effective control and prevention technology if: (1) the technology will significantly reduce the likelihood of a release and (2) if the technology has been field-proven at the time of the review.

Facility Analysis. A portion of the plan should include a description of the facility operation, which generally indicates the magnitude of spill potential. For example, the amount and type of storage, normal increments of transfer of patterns of usage, distribution, processes, *etc.* In this analysis, the direction of flow of spilled oil must be indicated along with any factors which are pertinent to or influence spill potential (*e.g.*, secondary containment, isolation valves, catch basins, *etc.*). This type of information can be supported with charts, tables, plot plans, or maps as necessary, to aid clarity or improve brevity.

Location of Facility. The geographical location of the facility should be described. A general location and topographic map should be included.

Amendment of SPCC Plans. Owners or operators of facilities must amend the SPCC Plan whenever there is a change in facility design, construction, operation, or maintenance which materially affects the facility's potential for the discharge of oil into or upon the waters of the United States or adjoining shorelines. Amendments must be fully implemented as soon as possible, but no later than six months after the change occurs. In addition, EPA may require amendments to the SPCC Plan following a single discharge at the facility in excess of 1,000 gallons, or following two discharges within any 12-month period. Amendments must be certified by a registered professional engineer.

Facilities must implement the plan, including carrying out the spill prevention and control measures established for the type of facility or operations, such as measures for containing a spill (*e.g.*, berms). In the event that a facility cannot implement containment measures, the facility must develop and incorporate a strong spill contingency plan into the SPCC Plan. In addition, facility owners or operators must conduct employee training on the contents of the SPCC Plan. Facilities must prepare a SPCC Plan within six months of the date they commence operations; they must implement the Plan within one year of the date operations begin. A certified copy of the SPCC Plan must be available at the facility for review by EPA or a State regulatory agency if requested. If the facility is not attended, then the SPCC Plan must be kept at the nearest company office.

Facility Response Plans

A *Facility Response Plan* addresses how a facility will respond to a worst-case oil discharge and a substantial threat of such a discharge. A worst-case oil discharge is the largest foreseeable discharge in adverse weather conditions, as determined using worksheets supplied in the regulations (40 CFR 112, Appendix E). The facilities that must prepare and submit a Facility Response Plan are a small, high-risk subset of the SPCC regulated community. A Facility Response Plan includes

- An emergency response action plan

- Facility information

- Emergency response information such as a list of authorities to notify, response equipment list, schedule of response equipment testing and deployment, list of personnel, evacuation plans, and description of each individual's duties

- A hazard evaluation, including hazard identification, vulnerability analysis, analysis of oil spill potential, and reportable oil spill history

- A small, medium, and worst-case discharge scenario

- A description of discharge detection systems

- A description of plan implementation

- A schedule and description of self-inspection, drills, exercises, and response training

- Diagrams of the facility, including storage tanks, transfer areas, containment systems, and hazardous material storage areas

- A description of security measures, including automatic and manual measures

An *emergency response action plan* (ERAP) is the heart of a Facility Response Plan. The ERAP contains all information that is needed to combat a spill, organized for easy reference. The following information should be included in the ERAP:

- Information on the individual who certified the plan

- Emergency notification phone list

- Spill response notification form

- List of response equipment and its location

- Schedule of response equipment testing and deployment drills

- List of response team members

- Evacuation plan

- Description of the response resources for small, medium, and worst-case discharges

- Facility diagram

The worst-case discharge volume is an essential element of a Facility Response Plan. It is calculated on a facility's oil storage capacity, not a facility's inventory. The calculation method typically uses the oil storage capacity of the largest tank or a permanently manifolded group of tanks within a secondary containment area. Facilities without secondary containment must use the capacity of all storage tanks.

Besides planning for a worst-case scenario, facility owners and operators must plan for small and medium spills. A small spill is one whose volume is 2,100 gallons or less, provided the amount is less than the worst-case discharge. A medium spill is greater than 2,100 gallons and less than or equal to 36,000 gallons or 10 percent of the largest tank's capacity, whichever is less. Addressing the medium-spill scenario is necessary only if the medium-spill volume is less than the worst-case discharge.

In certain cases, information required in a Facility Response Plan is similar to that contained in a SPCC Plan. In such cases, owners or operators may photocopy the information for the Facility Response Plan. The Facility Response Plan does not need to be certified by a registered professional engineer, but the facility owner or operator must certify that the information in the plan is accurate.

Facility owners or operators must demonstrate access to the required response personnel and equipment to respond effectively to identified spill scenarios. If the Facility Response Plan relies on facility-owned equipment, an inventory must be provided. Facilities that rely on other arrangements must include evidence of contracts.

Vessel and Offshore Oil Pollution Prevention Regulations

Owners and operators of tank vessels and offshore facilities are required to prepare and submit for approval response plans similar in content to a Facility Response Plan. The tank vessel and offshore facility response plans must be consistent with the NCP and Area Contingency Plans. Tank vessel and offshore facility response plans must identify the qualified personnel having full authority to implement removal actions, and the plans must require immediate communications between Federal officials and private removal contractors.

The tank vessel and offshore facility must identify and ensure by contract or other approved means the availability of private personnel and equipment necessary to remove to the maximum extent practical a worst-case discharge and to mitigate or prevent a substantial threat of such a discharge. A worst-case discharge for a vessel is a discharge of its entire cargo in adverse weather conditions. For an offshore facility, the worst-case discharge is the largest foreseeable discharge in adverse weather conditions.

The response plans must also describe training, equipment testing, periodic unannounced drills, and response actions of vessel and offshore facility personnel to mitigate or prevent the discharge. The plans must be updated periodically and be resubmitted for approval for each significant change.

The OPA requires licensing and drug and alcohol testing for new applicants, renewal applicants, and current holders of merchant marine pilots licenses. Holders of licenses are tested for drugs and alcohol on a random, periodic, reasonable cause, and post-accident basis. The terms of merchant marine pilot licenses were changed to five years. The OPA also added additional grounds for suspensions and revocations of licenses. For instance, the individual in charge of a vessel may be relieved of command if found to be under the influence of alcohol or drugs and incapable of commanding the vessel.

Tank vessels are required to have a double hull. This requirement for new vessels went into effect immediately after OPA was adopted. A phaseout schedule for existing vessels began in 1995 and runs until 2015. Older and larger vessels are retired first.

Oil Spill Liability Trust Fund

Under the Oil Pollution Act of 1990, the owner or operator of a facility from which oil is discharged (also known as the *responsible party*) is liable for the costs associated with the cleanup of the spill and any damages resulting from the spill. The EPA's first priority is that the responsible parties clean up their own oil releases. However, when the responsible party is unknown or refuses to pay, funds from the Oil Spill Liability Trust Fund (Fund) can be used to cover removal costs and/or damages resulting from discharges of oil. The Fund is administered by the US Coast Guard's National Pollution Funds Center (NPFC).

The Fund can provide up to $1 billion for any one oil pollution incident, including up to $500 million for the initiation of natural resource damage assessments and claims in connection with any single incident. The main uses of Fund expenditures are:

- State access for removal actions
- Payments to Federal, State, and Indian tribe trustees to carry out natural resource damage assessments and restorations
- Payment of claims for uncompensated removal costs and damages
- Research and development and other specific appropriations

Penalties

Under the Clean Water Act, as amended by the OPA, EPA has greater authority to pursue administrative, judicial, and criminal penalties for violations of the regulations and for discharges of oil and hazardous substances. Under the new penalty system, three different courses of action are available to EPA in the event of a spill: (1) an administrative penalty may be assessed against the facility, (2) a judicial penalty may be assessed against the facility in the Federal court system, or (3) a criminal action may be sought against the facility in the Federal court system.

EPA may assess administrative penalties against oil or hazardous substance dischargers as well as facility owners or operators who fail to comply with oil pollution prevention regulation. The administrative penalty amounts that violators must pay have increased under the OPA, and a new system of administrative penalties was created based on two classes of violations. Class I violations may be assessed an administrative penalty up to $10,000 per violation, but no more than $25,000 total. The more serious Class II violations may be assessed up to $10,000 per day, but no more than $125,000 total. However, a facility that has been assessed a Class II administrative penalty cannot be subject to a civil judicial action for the same violation.

Judicial penalties may be assessed against facility owners or operators who discharge oil or hazardous substances, who fail to properly carry out a cleanup ordered by EPA, or who fail to comply with the oil pollution prevention regulation. Courts may assess judicial penalties for discharges as high as $25,000 per day or up to $1,000 per barrel of oil spilled (or $1,000 per reportable quantity of hazardous substance discharged). For those discharges that result from gross negligence or willful misconduct, the penalties increase to no less than $100,000 and up to $3,000 per barrel of oil spilled (or per unit of reportable quantity of hazardous substance discharged). Owners and operators of facilities that fail to comply with an EPA removal order may be subject to civil judicial penalties up to $25,000 per day, or three times the cost incurred by the Oil Spill Liability Trust Fund, as a result of their failure to comply. Finally, if the facility fails to comply with its EPA-approved SPCC plan, the civil judicial penalty may reach $25,000 per day of violation.

EPA may pursue criminal penalties against facility owners or operators who fail to notify the appropriate Federal agency of a discharge of oil. Specifically, under the Clean Water Act, the Federal government can impose a penalty up to a maximum of $250,000 for an individual or $500,000 for a corporation, and a maximum prison sentence of five years.

Conclusion

The OPA sets out general requirements for the progressive oil pollution prevention, and a liability and compensation regime. The OPA took well over a decade to develop and enact, and will take years to implement fully. With the OPA's emphasis on planning, preparedness, and prevention, the need to respond to worst-case incidents and those posing substantial threats to public health and the environment should be reduced. The OPA's true effectiveness will be measured over time by the absence of oil spill catastrophes and near-misses, and by the reduction in the thousands of smaller spills that occur each year.

Bibliography

Gokare, Manjunath A. and J. Ronald Lawson. "Preparing for Disaster." *Industrial Wastewater*. p. 21, November/December 1994.

Olney, Austin. *Environmental Law Handbook,* 14th ed. Rockville, MD: Government Institutes, Inc., 1997.

Openchowski, Charles. "Federal Implementation of the Oil Pollution Act of 1990." *Environmental Law Reporter*. Vol. (21): p. 10605, October 1991.

Rogers, William H., Jr. *Environmental Law*, 2nd ed. St. Paul, MN: West Publishing Co., 1994.

Russell, Randle V. "The Oil Pollution Act of 1990: Its Provisions, Intent, and Effects." *Environmental Law Reporter*. Vol. (21): p. 10119, March 1991.

Internet Resource

<http://www.epa.gov/superfund/programs/er/index.htm> (EPA's Emergency Response Homepage)

Mark L. Bricker is a Senior Environmental Engineer at CH2M Hill's Portland, Oregon office where he helps private industry clients meet regulatory requirements while achieving their operational needs and objectives. Mr. Bricker has over 20 years of environmental and engineering experience with a former Fortune 500 company and as a consultant. Mr. Bricker has been involved with hazardous materials management for 10 years. He directed corporate oversight of an environmental management program for operations in five states, and has worked closely with State and Federal regulatory agencies to negotiate permit and regulatory compliance issues. Mr. Bricker has conducted several environmental due diligence investigations for major merger and acquisition transactions. He also has extensive negotiation experience in a regulatory and business setting. He has conducted environmental assessment and compliance audits at client-owned sites and third-party disposal facilities nationwide.

CHAPTER 34

Ground Water

Michael R. Matthews, PE, PG, CHMM

Ground Water

In the field of environmental management, ground water is arguably the medium of most interest and concern. This resource is used as a source of drinking water for humans and livestock, for irrigation, and for numerous industrial applications throughout the country. Therefore, the degradation of ground water has the potential of impacting the health and well-being of people, but plants and animals can also be affected.

The degradation of ground water can last for centuries, and contaminants can move many miles from the original source of release. The potential impact of ground water contamination can be devastating. It behooves any environmental manager to understand the basic concepts of ground water occurrences, movement, and quality.

Source of Ground Water

It is difficult to determine the beginning or the end of the *hydrologic cycle*, the constant movement of water by precipitation, overland runoff, and evaporation. For our discussion on ground water we must focus on precipitation (rain, snow, or hail). Rainwater that percolates into the soil is the ultimate primary source of ground water. Rainfall varies with the seasons, and in the United States most ground water percolation occurs

461

during the winter months. Any particular rainfall droplet may suffer a variety of fates. It may: 1) run off of the soil surface and coalesce with other runoff to form streams or rivers, 2) run directly into lakes or reservoirs, 3) evaporate directly into the atmosphere, 4) infiltrate into the soil and become part of soil moisture, 5) be absorbed by plants, or 6) percolate into the ground, through soil, and become part of a ground water aquifer. Higher evaporation rates, water storage, and plant uptake and transpiration processes use a major portion of spring, summer, and fall rainfall.

Rainfall that has entered the earth is stored in aquifers. An *aquifer* is any geologic formation capable of holding water in sufficient quantities to produce *free water*. **Free water** is water not incorporated into the soil as moisture. Aquifers can be unconsolidated soil or rock formations. The area where rainfall infiltrates through the soil and percolates down into the aquifer is known as a *recharge area*. Recharge areas are very difficult to map, even though the concept is relatively simple. Man can contribute to and influence the recharge process by influencing recharge conditions (*e.g.*, land use activities) and by contributing to ground water contamination.

Just as water can enter the ground and become stored in aquifers, it can also leave the subsurface and discharge at points or areas where the aquifer contacts the surface terrain. In the eastern United States, nearly all streams and rivers normally are discharge points for ground water. This discharge is what keeps rivers flowing during summer drought conditions. Springs and artesian wells are also discharges of ground water to the surface. In some regions streams can be recharge as well as discharge areas (*e.g.*, carbonate bedrock areas).

Aquifer Characteristics

An aquifer is any geologic formation that contains water. An aquifer is not the water itself! Aquifers can be earth, gravel, or porous stone. Beneath any one land surface location there can exist multiple aquifers, each with different characteristics. Impervious layers called *confining beds* separate these aquifers. The first aquifer that is usually encountered is unconsolidated materials (*e.g.*, soil). Beneath the soil is bedrock. Going

deeper into the earth one finds changes in geology. Each geologic change constitutes a different aquifer with different characteristics.

The United States Geological Survey classifies aquifers as excellent (water exists in quantities for large users), good (water is sufficient for domestic use), or poor (the water that is present is not capable of being pumped or yields less than 5 gallons per minute, dries up, *etc.*). Figure 1 illustrates a multiaquifer system.

Saturated and Unsaturated Zones

Directly beneath the soil surface lies an area in which the soil particles are surrounded in varying degrees by air and water. This area is called the **unsaturated zone** (or sometimes the **vadose zone**). Beneath this unsaturated zone is an area where soil particles are completely surrounded by water and no air is present. This is called the **saturated zone** and makes up the aquifer itself. The boundary between the unsaturated and saturated zones, called the **potentiometric surface,** is the area where the pressure exerted by the surface of the water table is equal to the atmospheric pressure at that depth.

Unconsolidated Aquifers

Unconsolidated material (soil) can be stream-deposited material (alluvium), wind deposits (sand dunes), glacial outwash, or deeply weathered bedrock. Water is present in and around the particles of soil and rock in the aquifer.

Consolidated Aquifers

Unconsolidated materials become rocks through a process called *lithification* (compaction and cementation). Sedimentary aquifers are divided into carbonate (limestone or dolomite) and non-carbonate (*e.g.*, sandstone or shale) aquifers.

In carbonate aquifers, water is present in the enlarged solution cavities. These cavities vary greatly in size and length. Different types of carbonates dissolve differently. Some are very large and form huge cave systems.

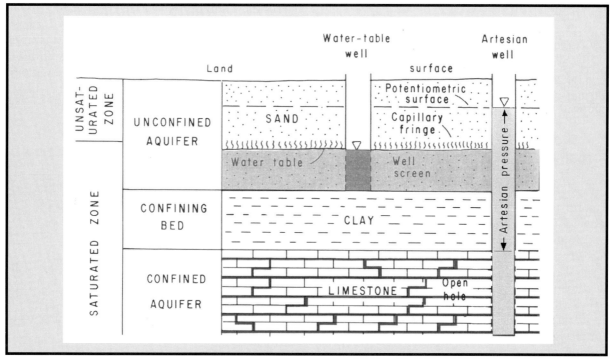

Source: Heath, 1998

Figure 1. Aquifers and Confining Beds

Consolidated noncarbonate aquifers hold water in the spaces around the soil and rock particles. How much water the aquifer can hold depends on the degree of cementation and fracturing that is present in the aquifer and the size of the particles in the rock.

Sedimentary rocks such as sandstone hold water around the sand grains. Igneous and metamorphic rocks generally do not contain much water unless they are fractured from folding or faulting activities. When folding or faulting occurs, rocks fracture. Water can accumulate in the fractures and flow along the fracture zone. Lava can have coarse-grained material deposited between flows and these flows can be interconnected by cooling fractures into a very complex water storage pattern.

Unconfined Aquifers

Unconfined aquifers are also called **water table aquifers**. Water table aquifers are usually in unconsolidated material. The water table rises and falls seasonally. Water tables tend to rise in winter when more rainfall infiltrates and tend to fall in summer due to less rainfall infiltration.

Movement of water is by gravity. Water is going to flow downhill. The water table surface can respond to changes in atmospheric pressure, since the void spaces above the water table are directly exposed to the atmosphere through interconnected spaces and pores in the soil. Water table aquifers are usually nearest the land surface and are most often monitored for contamination.

Confined (Artesian) Aquifers

The defining feature of these aquifers is a confining bed that overlies the water-bearing formation. The water beneath this confining layer is under pressure and wants to get into balance with atmospheric pressure and recharge potential. The imaginary point where this balance occurs is called the *potentiometric* or *piezometric surface*. This is where the term *piezometer* comes from. A piezometer is installed primarily for the measurement of water table elevations, whereas a monitoring well is intended for sampling. Both are essentially the same with minor construction differences.

Perched Aquifers

In a perched aquifer, percolating water is held up locally by some type of confining bed (hard pan, clay layer, *etc.*). This can be a permanent or seasonal situation and can be confused with the true water table.

Porosity

Any soil, sand, gravel, or fractured rock consists of solid particles and empty spaces (voids) in between the particles. The ratio of voids to the total volume of a soil or rock is called its **porosity**. Porosity is expressed as a decimal fraction or a percentage. Porosity tells us the maximum amount of water a rock or soil can hold if it is saturated. Related characteristics of aquifers are the terms specific yield and specific retention. **Specific yield** is the volume of water that will drain by gravity from a specific volume of rock or soil. **Specific retention** is the volume of water that is retained as a coating on soil or rock particles after gravity draining. Both specific yield and specific retention are expressed as percentages. Porosity can be described by the following equation.

Porosity = Specific Yield + Specific Retention

Typical values of porosity are presented in the following table:

Material	Porosity (percent)
Soil	55
Sand	50
Clay	75
Gravel	20
Limestone	10

Water Movement

Unconsolidated Aquifers

Water flows by gravity downhill. In unconsolidated aquifers this movement is in and around the rock and soil particles. This movement can be estimated using Darcy's Law. Darcy's Law is usually valid in unconsolidated materials (water table aquifers), but is not valid where aquifer characteristics are not uniform (carbonate aquifers or water movement in rock fractures).

Darcy's Law

The formula is usually expressed as

$$Q = KA \, (H_1 - H_2) \, / \, L \qquad (1)$$

Where:

Q = the total flow or volume (gallons or cubic feet/ unit of time)

K = the hydraulic conductivity (potential velocity of movement)

A = the area the water moves through (horizontally)

$(H_1 - H_2)/L$ = a dimensionless friction or head pressure loss term. $(H_1 - H_2)$ is the difference in head (elevation), usually expressed in feet. L is the distance between the H_1 and H_2 locations.

Another way the change in elevation over some distance can be viewed is as a slope. The steeper the slope, the faster the water moves and the more water that moves with time. **Hydraulic conductivity** (K) is the length (distance traveled)/ time (potential velocity) value.

Typical K Values (in cm/sec)	
Gravel	100–0.1
Sand	1–10^{-5}
Clay	10^{-5}–10^{-10}

The most important K value to remember is 10^{-7} cm/sec (clay). This value has typically been selected for protection of the ground water resource in many regulations. The finer the material, the smaller the K value.

Vertical permeability is usually less than horizontal permeability, and it is important to recognize this when assessing ground water movement.

Example

Ground surface elevation is 1500' above mean sea level (MSL). Water level in a well 400' (L) from stream is 1480' MSL (H_1). Water level in the stream is 1475' (H_2). Soil in the area is sand (K = 100 ft/day). Sand aquifer is 50' thick and lies above a shale layer.

Flow rate =
$$Q/A = K(H_1 - H_2)/L$$
$$Q/A = 100 \text{ ft/day}(1480 - 1475)/400$$
$$= 1.25 \text{ ft/day}$$

Transmissivity

The *transmissivity* of an aquifer is a measure of its capacity to deliver water. Transmissivity is the product of the hydraulic conductivity and the aquifer thickness.

$$T = Kb \qquad (2)$$

Where:

T = Transmissivity

K = Hydraulic Conductivity

b = Aquifer Thickness (variable thickness)

Transmissivity in a particular aquifer varies in time and space. In an unconfined aquifer the transmissivity is almost always seasonally variable throughout the aquifer. In a confined aquifer transmissivity is constant at a particular point in the aquifer as long as both layers are confining, but varies from place to place within the aquifer.

Consolidated Aquifers

Consolidated aquifers are much more difficult to measure and estimate flow. Movement in certain formations that may have relative uniform thickness can sometimes be assessed in a manner similar to unconsolidated aquifers (*e.g.*, sandstone aquifers). However, ground water movement is more often in the fractures or solution cavities of these rocks. The fractures are not uniform. The fractures vary in width and length. Fractures and solution cavities are not necessarily continuous and can somewhat randomly change directions.

In monitoring ground water movement in consolidated aquifers fracture locations, the size of openings, length of the fracture or cavity, and the pattern of the fractures or cavities are of vital importance in determining where ground water will move and how fast it will move.

Water movement in fractures and solution cavities is also by gravity. Water will move fairly quickly as it flows through the fracture or void, much like water flows in a pipe or river. Water movement in fractures and solution cavities can move great distances in relatively short periods of time.

Sinkholes

The formation of sinkholes is a process that takes a long time. Carbonate rocks with cracks or fissures provide openings for rainfall to enter. In general, rainwater is naturally slightly acidic. The acidic water dissolves the calcium carbonate and forms a *solution cavity*. With time this cavity becomes bigger (similar to cave formations). Once the opening is big enough it can no longer support the soil above it and the soil collapses (sometimes catastrophically, like in Florida).

Water enters from openings at the surface and is directly recharged (injected) into the aquifer without the benefit of filtration. People with wells connected to these water sources can see nearly instant changes to water quality during storm events as sediment is washed into their drinking water. Pollutants and other debris can enter just as easily.

Measuring Aquifer Ground Water Flow

Hydraulic conductivity measurements in the laboratory (usually taken from field samples) do not always equal actual field conditions. Natural conditions are subject to varying conditions that are not easy to define.

In addition there are many different types of flow. One-directional flow obeys Darcy's Law. This flow can be steady but more typically varies with time and is considered unsteady (*e.g.*, water tables rise and fall seasonally). Water movement can also be radial. This is best understood by imagining movement toward a large pumping ground water well. This movement can also be steady or unsteady depending on how much water is pumped and when it is pumped. Large ground water withdrawals from industry or agriculture can influence ground water movement and must be considered when assessing ground water flow patterns.

The boundaries between different types of aquifers impact ground water movement in different ways. All aquifers have boundaries and are interconnected. Water percolates down through unconsolidated aquifers until it reachs bedrock. What it does next depends on the type of bedrock it encounters.

If shale or slate is encountered, the water may not infiltrate and may move along the soil and bedrock interface until it discharges somewhere. If limestone is encountered, the water will move along the rock surface until a solution cavity is encountered where it will enter and follow the cavities until discharged.

In consolidated rocks that have been folded or faulted, the infiltrating water could flow along the bedrock surface until a subsurface structure (*e.g.*, fault) causes the water to follow it preferentially down a path of less resistance.

Boundary conditions between aquifers can and do modify flow. Sometimes these boundaries act as recharge points to lower aquifers and discharge points to aquifers located above the lower aquifers.

When geology changes occur, so do aquifer characteristics. Geologic features also are frequently hydraulically connected with streams. Streams that act as discharge points for ground water are called *gaining streams*. Sometimes (particularly in carbonate areas) streams can actually help recharge the ground water. These are called *losing streams*.

Water Table

Depth to ground water is measured from the ground surface and is correlated to a set elevation scheme (mean sea level). Management of the water table is useful in controlling ground water contamination. When ground water is removed from a pumping well (*e.g.*, irrigation), the water withdrawal depresses the water table around the well. This water table depression is referred to as a *cone of depression* (Figure 2) for the geometric shape it assumes around the well.

Since water flows downhill due to gravity, a cone of depression can be used to artificially control ground water movement toward a pumping well. This is particularly useful in the management of contaminants that tend to float on the surface of the water table due to being less dense than water. If contaminants move toward the pumping well they can be easily recovered and removed from the aquifer.

The cone of depression of large well fields can be very large, and impact the movement of ground water for miles in all directions from the wellhead if the removal of water is steady and of great quantity. This condition exists around the city of Memphis, Tennessee, which utilizes ground water for its drinking water. The area of ground water in an aquifer that is influenced by the removal of ground water is referred to as the *zone of influence* for a specific well.

Visual images of the top of water tables can be depicted in *potentiometric maps* (Figure 3). These maps are a visual image of the water table. Flow direction is determined by first mapping the elevation of the water at numerous locations. Contour lines called *isopleths* are drawn between points of equal ground water elevations. The flow direction is determined by a perpendicular line drawn from the isopleth with the highest elevation, so that the line intersects the next highest isopleth perpendicularly. This process is repeated from isopleth to isopleth until the flow direction is determined.

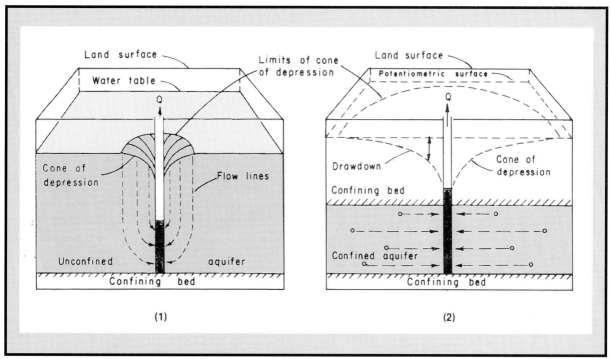

Source: Heath, 1998

Figure 2. Cone of Depression

Ground Water Quality

The primary origin of ground water is rainfall. Therefore, the original quality of ground water is essentially equal to the quality of the rain as it enters an aquifer. Characteristics of rainfall generally are that: 1) it will act as a solvent because it is slightly acidic; 2) its density is about 1.0; 3) it has a quality that is determined primarily by the dissolved gases (CO_2, SO_2, N_2, O_2) and particulates (dust, minerals, dichlorodiphenyltrichloroethane [DDT], *etc.*) picked up as the rain falls through the atmosphere.

Rainfall is naturally slightly acidic. Acid rainfall has a pH much less than natural rainfall (often 4.0–5.0). Because of its acidic nature, rainfall can dissolve rock (primarily carbonate rocks). As the rock dissolves away, the minerals in the rock are released to the ground water and impact its quality. The release of carbonates into ground water acts to buffer ground water and raise its pH to above 7.0. Ground water that has not been in the subsurface for a long time will have a density of around 1.0 (fresh water). The longer ground water is underground the more chance it has to pick up dissolved minerals. The more minerals in the ground water, the more dense it becomes.

Denser liquids tend to sink below lighter liquids. Therefore, fresh water that has few minerals will float on salt water (full of minerals). As a general rule, ground water becomes more saline with depth.

The nature of geologic formations can impart quality changes to ground water. Carbonate formations can release many minerals such as calcium and magnesium to the water. The concentration of calcium and magnesium will determine whether or not ground water is hard or soft. Inert sand and gravel does not tend to dissolve in ground water so these kinds of formations do not have as much of an impact on water quality.

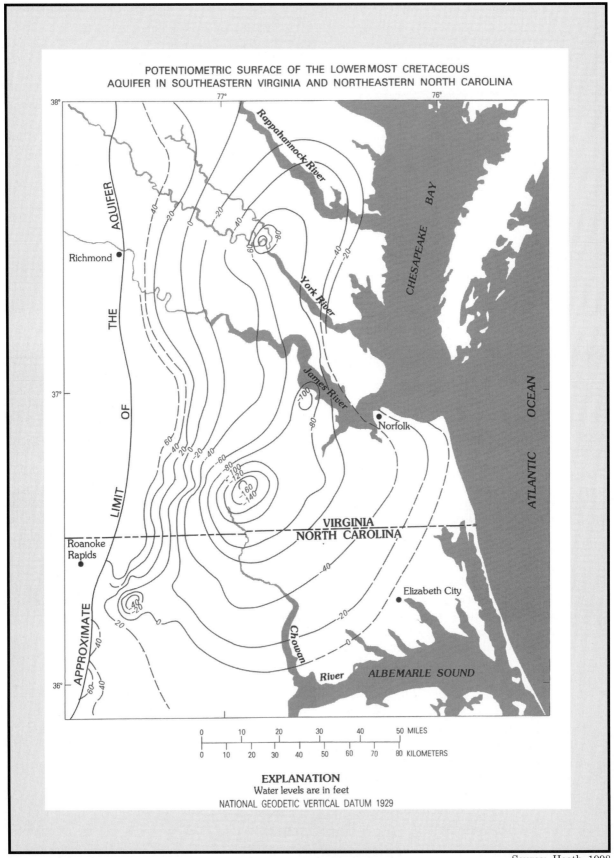

POTENTIOMETRIC SURFACE OF THE LOWERMOST CRETACEOUS
AQUIFER IN SOUTHEASTERN VIRGINIA AND NORTHEASTERN NORTH CAROLINA

EXPLANATION
Water levels are in feet
NATIONAL GEODETIC VERTICAL DATUM 1929

Source: Heath, 1998

Figure 3. Potentiometric Map

Contaminants in the Subsurface

Vadose (Unsaturated Soil) Zone. In the vadose zone, inorganics are removed from ground water by the cation exchange capacity of the soil. Generally soils have a net negative charge, and this charge attracts and holds metals but it is not necessarily a permanent removal mechanism. Once the charges are matched, further removal ceases.

The net negative charge of soil repels negatively charged inorganics like chlorides, sulfides, cyanides, arsenates, *etc.* These inorganics move quickly and easily through soil. Also, plants take up nutrients in the soil and effectively remove them from the vadose zone.

Biodegradation is the primary means of attenuation for organics in the vadose zone. Movement depends on solubility of the organic in question. Organics tend to leave residues in soil that leach into ground water over long periods of time.

Ground Water. Inorganics are not easily removed from ground water. Cations are attenuated sooner than anions, and dilution occurs with dispersion. Hence, the concentration of contaminants typically is reduced with distance from the source. Ground water is a preservation medium (usually acidic, little oxygen, and cool). These are the very conditions that are used to preserve metals for laboratory analysis.

Processes/Reactions Affecting Pollutants. Dilution is usually slow unless ground water is in large solution cavity situations where rapid mixing can occur particularly after storm event. It has been observed that organic phases can flow at rates greater than or slower than water flow.

Nature is dynamic, and numerous factors can change contaminants once introduced into the environment. Geochemical reactions can occur such as: precipitation, acid-base reactions (*e.g.*, limestone-dissolving and -releasing minerals), adsorption/desorption, and complexation. There are biochemical processes than can also affect pollutants: organic decomposition, cell synthesis, transpiration (plant use), and respiration (bacteria use). Physical processes move and remove pollutants by transporting contaminants to other locations, mixing contaminants and water together, dispersing (scattering) pollutants in the subsurface, filtering out solids, evaporating liquids, and removing pollutants by gas movement. Pathogens can be transported, filtered, and reproduced in certain situations in a series of biophysical processes.

Plume Migration. Contaminant plume shape and migration can be affected by additional factors that must be considered. Density factors of the pollutants will determine whether the contaminant will settle to the bottom of the aquifer, like DNAPL—dense non-aqueous phase liquids (*e.g.*, polychlorinated biphenyls [PCBs]). LNAPL—light non-aqueous phase liquids (*e.g.*, many organics), tend to float on the water table surface, or mix throughout the aquifer.

How the contaminant is released (continuous, spill, or intermittent) will determine the shape of the contaminant plume. Continuous releases over long periods of time become long, tear-shaped plumes as dispersion and mixing spreads the contaminants out with distance. Plumes from spills will move with the ground water in a mass, but leaving a smear of contaminants on soil and rock that it moves through. With time and distance the plume may effectively disappear, due to the various processes noted above and since additional material is not added. Intermittent releases results in plumes where the concentrations of contaminants rise and fall as the mass of the various releases moves past a measuring location.

Regulations

Clean Water Act

The Clean Water Act primarily addresses ground water through the underground injection control (UIC) program (40 CFR 144). Injection activities that could allow for contaminants to enter underground sources of drinking water are strictly regulated. Injection wells fall into five primary categories. Classes I–IV are wells used for injection of hazardous wastes, solution mining, and oil or gas production and storage. Class V wells include those wells not specifically addressed in Classes

I–IV: sinkholes, leach pits, cesspools, and septic systems. However, individual and single-family septic systems (serves fewer than 20 persons per day) used solely for sanitary wastes are exempt. This act also provides for protection of sole-source aquifers and wellhead protection.

Resource Conservation and Recovery Act

The Resource Conservation and Recovery Act (RCRA) is extremely comprehensive and includes solid wastes, hazardous wastes, and underground storage tanks. Ground water is addressed in this act primarily through ground water protection requirements (how facilities must be constructed or operated to protect ground water), resource assessment and characterization (gather ground water movement and quality data), and monitoring (to assess compliance with or the effects of waste management activities on ground water) of the ground water resources.

Comprehensive Environmental Response, Compensation, and Liability Act

The Comprehensive Environmental Response, Compensation, and Liability Act (CERCLA) addresses accidental spills and abandoned hazardous waste sites. CERCLA has many steps that must be performed: 1) Preliminary Assessment/Site Investigation, 2) Remedial Investigation/Feasibility Study, 3) Remedial Design, and 4) Operation and Maintenance.

This act manages ground water primarily through resource assessment/characterization (gather ground water movement and quality data in the early assessment phases) and monitoring (to assess compliance with Record of Decision or Consent Agreement) of the ground water resources.

Bibliography

Heath, Ralph C. *Basic Ground Water Hydrology*. US Geological Survey Water-Supply Paper 2220. Washington, DC: USGS, 1998.

Michael R. Matthews has spent nearly 25 years working in the environmental field, primarily in solid waste, hazardous waste, and ground water management. His early career (eight years) focused on the design and management of solid waste transportation and disposal systems. For four years he was responsible for Tennessee Valley Authority's ground water quality program (in seven states), and implemented ground water protection programs at the local level. Mr. Matthews has worked in the field of hazardous materials management since 1985. He has been responsible for the management and remediation of numerous hazardous waste sites. He has also performed ground water assessments, remedial investigations/feasibility studies, and risk assessment. Mr. Matthews has experience in Phase I and II site assessments, UST removal and corrective actions, air pollution permitting, RCRA Part B permitting, and other similar work.

Part VIII

Management of Hazardous Wastes

CHAPTER 35

RCRA Overview: A Generator Perspective

Gregory C. DeCamp, MS, CHMM

Introduction

Purpose, Scope, and Approach

Compliance with the Resource Conservation and Recovery Act (RCRA) hazardous waste management requirements is a serious challenge faced by most hazardous material managers at one time or another. This chapter is intended to help the hazardous material manager successfully meet this challenge. An overview is provided of most major elements of the RCRA regulations. However, emphasis is placed on hazardous waste determination, accumulation, storage, and related waste generator issues that often must be solved by managers themselves with little specialized assistance. Key regulatory requirements are presented, practical approaches to compliance are noted, and references to sources of additional information are listed.

Regulatory compliance is addressed in this chapter in terms of the Federal RCRA hazardous waste management program in Subtitle C of the RCRA statute and implementing regulations codified by the Environmental Protection Agency (EPA) at 40 CFR 260–266, 268, 270, 271, and 273. (See 40

CFR 250–299 for the full spectrum of RCRA rules, including those related to nonhazardous waste and the underground storage tank program.) Although the RCRA program is designed to be implemented by the states, no attempt is made to address State or local hazardous waste require-ments. It is helpful to know that State hazardous waste programs must be consistent with and at least as stringent as the Federal RCRA program (RCRA §3006, 40 CFR 271) and that the major provisions of State regulations closely approximate the Federal regulations. Requirements applicable in a specific locale nonetheless typically differ from those of the Federal program, particularly in cases where State or local authorities have adopted more stringent rules or where an authorized state has not yet *caught up* with the Federal program by formally adopting recently promulgated Federal rules. Therefore, it is essential for hazardous materials managers to become familiar with applicable State and local rules.

In a similar vein, some hazardous wastes are also directly regulated under Federal laws and regulations other than RCRA. Examples include polychlorinated biphenyl (PCB) contaminated waste regulated under the Toxic Substances Control Act (TSCA) and radioactive waste

Table 1. The Federal Hazardous Waste Management Regulations

40 CFR Part	Title	Description
260	Hazardous Waste Management System–General	Generally applicable information, including RCRA overview, procedures for change petitions, generally applicable definitions
261	Identification and Listing of Hazardous Waste	Solid and hazardous waste definitions/criteria
262	Standards Applicable to Generators of Hazardous Waste	Rules for Hazardous waste determination, on-site accumulation and treatment, preparation for transport, import/export
263	Standards Applicable to Transporters of Hazardous Waste	Rules for transporters of hazardous waste
264	Standards for Owners and Operators of Hazardous Waste Treatment, Storage, and Disposal Facilities (TSDFs)	Rules governing virtually all aspects of permitted TSDFs
265	Interim Status Standards for Owners and Operators of Hazardous Waste TSDFs	Rules for TSDFs conditionally allowed to operate under *interim status authorization* until a RCRA permit is obtained. Similar to, but less stringent than, 40 CFR 264 rules.
266	Standards for the Management of Specific Hazardous Waste and Specific Types of Hazardous Waste Management Facilities	Rules for management of certain hazardous wastes that are recycled, boilers and industrial furnaces (BIFs), and military munitions
268	Land Disposal Restrictions (LDRs)	LDR provisions, including treatment standards
270	The Hazardous Waste Permit Program	RCRA permit requirements, including permit application contents and procedures
271	Requirements for Authorization of State Hazardous Waste Programs	Protocols for states to obtain authority to implement parts of the Federal RCRA program[1]
272	Approved State Hazardous Waste Management Programs	List useful for determining the extent to which a state is authorized to implement the Federal RCRA program[1]
273	Standards for Universal Waste Management	Standards that may be used for specified common hazardous wastes as a more lenient alternative to the standard RCRA hazardous waste rules[1]

[1] This topic not substantially addressed in this chapter.

regulated under the Atomic Energy Act (AEA). These *mixed wastes* present special management problems that are generally too detailed to be dealt with in this chapter. However, it is important to understand these additional regulations and how they are accommodated while also complying with requirements imposed by RCRA.

Hazardous waste generation and storage topics are addressed in this chapter roughly in the order they are addressed in EPA's hazardous waste regulations. This approach is generally logical and promotes familiarity with the regulations, a necessity for practicing hazardous materials managers. Table 1 lists those parts of the RCRA regulations normally associated with the RCRA Subtitle C program.

Historical Background

This chapter is not intended to provide a history of RCRA. However, the following historical information provides some insight into the origins and intent of the regulations that may be useful in efforts to develop regulatory interpretations, *etc*.

RCRA (42 USC §6901, *et seq*.) had its beginnings in the Solid Waste Disposal Act (SWDA) of 1965, which generally addressed solid waste problems and gave states the responsibility for solid waste management plans. The Resource Recovery Act of 1970 amended SWDA to include EPA funding for resource recovery provisions, but had little practical effect on waste management practices in the United States. However, the modern-day RCRA came into being with passage of the Resource Conservation and Recovery Act of 1976, which amended and replaced SWDA entirely. This comprehensive legislation was the first to meaningfully address hazardous waste (in Subtitle C), and also addressed nonhazardous solid waste (in Subtitle D) as well as resource recovery. EPA's implementing regulations for this legislation, which initiated the *basic RCRA program*, were issued in May, 1980 and were effective on November 19, 1980. The basic law has been amended several times. However, the Hazardous and Solid Waste Amendments (HSWA) of 1984 are by far the most notable amendments to date. HSWA significantly expanded RCRA to include highly prescriptive requirements such as the topics listed below.

Land Disposal Restrictions (LDRs). Standards hazardous waste must meet to ensure threats to human health and the environment are minimized prior to placement in or on the land.

Minimum Technological Requirements. Requirements for double liners, leachate collection systems, as well as ground water monitoring for hazardous waste surface impoundments, landfills, and waste piles.

Corrective Action Program. Requirements to remediate past releases of hazardous constituents from solid waste management units at facilities that have or are seeking a RCRA permit.

Underground Storage Tank (UST) Program. Comprehensive requirements in RCRA Subtitle I for underground tanks which store hazardous substances as defined under the Comprehensive Environmental Response, Compensation, and Liability Act (CERCLA), except RCRA hazardous waste (hazardous waste tanks are addressed under RCRA Subtitle C).

EPA continues to promulgate regulations to fully implement HSWA, particularly in the areas of LDR and corrective action. Where these regulations impose requirements that are more stringent than those in effect in an authorized state, they become applicable upon promulgation and are enforced by EPA until the state adopts the rule, so it is especially important to be aware of new regulations even in authorized states.

Compliance Strategy

The ideal strategy for compliant management of hazardous waste is developed and implemented in the context of a well-organized environmental management system. A system that adheres to the basic principles of International Organization for Standardization's ISO 14001 or an equivalent system helps ensure that activities are planned and carried out safely with minimal potential for noncompliance. The hazardous waste management strategy should also reflect consistency with EPA's Pollution Prevention Hierarchy (Pollution Prevention Act, §6602[b]), which calls for environmentally responsible source reduction and recycling as the first and second priorities in dealing with pollution, including hazardous waste. Avoiding or minimizing generation of hazardous waste in the first place and recycling of waste unavoidably generated should always be investigated as potentially cost-effective initial steps in any hazardous waste management strategy. Stringent RCRA regulations can also be partly or entirely avoided by legitimately recycling the waste. Where waste generation is unavoidable, the Pollution Prevention Hierarchy calls for environmentally safe treatment whenever feasible, and disposal only as a last resort. The following sequence of activities is suggested as part of an orderly approach to RCRA compliance:

- Become familiar with RCRA hazardous waste regulations and determine requirements potentially applicable to your operation.

- Evaluate pollution prevention/waste minimization opportunities and incorporate feasible and cost-effective options into projects beginning in the early planning and design phases.

- Plan for management of any waste well before it is generated, if possible, particularly if special management requirements are likely to apply (*e.g.*, RCRA requirements for hazardous waste).

- Determine if the waste meets the RCRA definition of solid waste and, if so, whether it is considered RCRA hazardous waste.

- Determine hazards to human health and the environment posed by the waste, and measures necessary to address these hazards, including measures specifically required by RCRA.

- Implement measures determined to be necessary to address hazards and achieve or maintain RCRA compliance.

- Evaluate effectiveness and make improvements where necessary.

The RCRA Hazardous Waste Regulations—General Considerations

There seems to be no easy way to develop a working knowledge of the maze of RCRA regulations, even if one's interest is limited to that of the waste generator, as is the case for most hazardous materials managers. Even persons who have specialized in hazardous waste management for many years can be easily stumped. The moral is: Overviews of the sort presented in this chapter or in typical introductory or review courses are a good place to start, but are no substitute for careful study of current regulations and guidance documents and consultation with experts when one is attempting to address real compliance issues. In this context, it is helpful to know that RCRA regulations are codified in the *Code of Federal Regulations* (CFR) only once per year (each July 1), so full knowledge of current regulations requires familiarity with both the CFR and the *Federal Register* (FR) issued since the last codification. Review of the preambles to proposed and final rules in the FR is also very often helpful for regulatory interpretation. Finally, direct consultation with EPA or the applicable State RCRA regulatory authority is usually very helpful, and may be essential in cases where compliance determinations remain unclear or where regulatory discretion can be exercised. See the Bibliography for useful sources of guidance.

Waste Management Planning

Good planning is the most effective way to ensure that hazardous wastes are managed safely and in accordance with RCRA requirements. RCRA provides no *grace period* for compliance; regulatory requirements apply at the time and place the waste is generated (*i.e.*, at the *point of generation of the waste*). This implies that adequate resources necessary to comply with RCRA requirements must be in place at the time of generation or soon thereafter. Depending on the specific RCRA requirements that apply, needed resources may include the following:

- Containment facilities and supplies (*e.g.*, containers, tank systems, containment buildings) adequate to safely accumulate and store the waste when generated and facilitate subsequent management (*e.g.*, analysis, storage, treatment, disposal)

- Security to prevent contact of the waste by unauthorized persons (*e.g.*, locks, fences, signs)

- Release prevention and response provisions (*e.g.*, communication system, fire and spill control equipment, contingency plan)

- Waste sampling and analysis provisions (*e.g.,* facilities, equipment, contract with qualified laboratory)

- Provisions for waste inventory, inspections, and monitoring

- Provisions for recordkeeping and reporting

- Personnel training

- Provisions for subsequent management (*e.g.,* transportation, treatment, disposal facilities or vendor contracts)

Planning for generation and storage of hazardous waste should address all hazardous wastes that could reasonably be expected to occur at a facility, not only those which are normal or routine. Examples of potential nonroutine wastes for which contingent hazardous waste management plans should be developed include

- Commercial chemical products upon discard

- Residue from spills of commercial chemical products, process intermediates

- Abandoned or *orphaned* waste containers

- Waste from nonroutine activities (*e.g.,* remodeling, maintenance, changes in process)

Upon generation of a waste, a generator must notify EPA of that activity and obtain an EPA Identification (ID) Number or amend their previous notification to reflect the new activity, as discussed later in this chapter.

RCRA Introductory Regulations

The introductory part of the RCRA regulations (40 CFR 260) includes information which is useful to persons engaged in virtually any aspect of hazardous waste management.

Confidentiality of Information

Normally, any information provided to EPA under the RCRA regulations is available to the public under the Freedom of Information Act. However, some protection from disclosure of proprietary information can be obtained by asserting a claim of business confidentiality as described in this section (40 CFR 260.2).

Definitions

Many of the terms used in the RCRA regulations, even seemingly common words, have special meanings which must be understood to interpret the regulations correctly. This section (40 CFR 260.10) defines terms used throughout the regulations. Additional definitions are provided in individual parts of the regulations.

Rulemaking Petitions

Virtually any provision in the RCRA regulations can be changed using processes described in 40 CFR 260, Subpart C. Therefore, if requirements are clearly inappropriate and potential advantages outweigh the substantial time and money typically required, petitioning for a regulatory change may be a good idea. A firm technical and legal basis for any petition is essential. Petitions most likely to be considered by generators or storers of hazardous waste are *delisting petitions* to remove a waste at a particular facility from the hazardous wastes listed by EPA in 40 CFR 261, Subpart D, and petitions to exclude certain wastes that are recycled.

Overview of Hazardous Waste Management Regulations

EPA provides an overview of the RCRA regulations in about two pages of text and four flowcharts in 40 CFR 260, Appendix I. This information merits a quick read, in that it is useful as a general indication of how the basic RCRA program is structured. However, it was last updated April 1, 1983, and much of the detail is no longer correct. As a result, it no longer entirely fulfills EPA's intent to help those unfamiliar with the program to determine RCRA regulations that apply to them.

Waste Determinations

Determining whether a material is a hazardous waste is the generator's first hurdle in complying with RCRA's hazardous waste (Subtitle C) rules. Under RCRA, hazardous waste is a subset of solid waste, and each term has a very specialized meaning. Therefore, the waste determination process first involves the application of criteria, including specific attributes and exceptions, to determine if a material qualifies as a RCRA solid waste. If the material is a RCRA solid waste, additional criteria must then be applied to determine if the solid waste qualifies as hazardous waste.

The determination process must be undertaken carefully in view of the risk of noncompliance resulting from an incorrect determination. Each waste determination should involve a careful, stepwise consideration of each element of the solid and hazardous waste definitions in 40 CFR 261 with close attention to the definitions in 40 CFR 260.10. Computer-based and hard copy guidance specifically designed to assist in negotiating this maze is available; the RCRA Hotline (800) 424–9346) and documented regulatory interpretations involving a wide variety of wastes from EPA's Office of Solid Waste are particularly helpful.

Highlights of the solid and hazardous waste definitions and determination process are described below.

Solid Waste Determination

RCRA basically defines **solid waste** as any garbage, refuse, sludge, and other discarded material, including solids, semisolids, liquids, and contained gases. Note that solid waste is not restricted to solid phase material. Note also that uncontained gases, which generally are regulated under the Clean Air Act, are not RCRA solid wastes. This statutory definition includes specific exceptions for several wastes, chiefly on the basis that they are regulated under other statutes. In practice, one must refer to 40 CFR 261.

In determining whether a material is a solid waste, first determine whether it is specifically excluded. A material may be excluded by either (1) a variance granted in response to a petition under 40 CFR 260.30 or .31 or, (2) by being a material specifically described in 40 CFR 261.4(a). The former case is rare and readily determined by simple inquiry to the owner of the waste or the regulator. 40 CFR 262.4(a) lists over a dozen materials that are excluded from being solid waste, as summarized in Table 2. If a material is excluded from being a solid waste, it cannot be a hazardous waste.

If a material is not specifically excluded in 40 CFR 262.4(a), then the next step in waste determination is deciding if the material is **discarded**. As defined in 40 CFR 261.2, a material is discarded and thus a solid waste if it is

1) **Abandoned**—Disposed of, burned or incinerated, or accumulated, stored, or treated before or in lieu of being disposed of, burned or incinerated.

2) **Recycled**, or accumulated, stored, or treated before recycling—by being

   a) **Used in a "manner constituting disposal"**—placed on the land directly or as an ingredient in a product (except commercial chemical products ordinarily applied to the land).

   b) **Burned for energy recovery or used to produce a fuel product** (except commercial chemical products that are already fuels).

   c) **Reclaimed**—Regenerated or processed to recover a usable product (except sludges and by-products that exhibit a characteristic of hazardous waste but which are not listed as a hazardous waste, and commercial chemical products).

   d) **Accumulated speculatively**—(Except commercial chemical products). This provision is specifically intended to discourage *sham recycling*. If a person can demonstrate that a material is potentially recyclable, a feasible means of recycling is available, and at least 75 percent of the material present on January 1 of a calendar year is recycled or transferred to another site for recycling within that year, then the material is not considered to be speculatively accumulated per 40 CFR 261.1(c)(8).

3) **Inherently waste-like**—(*i.e.*, Specific dioxin and furan-containing wastes and other materials that are ordinarily managed as waste or which pose a substantial hazard when recycled).

4) **A military munition specifically identified as a solid waste** in 40 CFR 266, Subpart M.

Materials that do not otherwise qualify as discarded as listed above are not RCRA solid wastes. In particular, materials that can be shown

Table 2. Major RCRA Solid Waste Exclusions
(See 40 CFR 262.4[a] for a complete list.)

- Domestic sewage and any mixture of domestic sewage and other waste that passes through a sewer system to a publicly owned treatment works for treatment

- Point source industrial wastewater discharges regulated under Clean Water Act, Section 402 (*i.e.*, National Pollutant Discharge Elimination System [NPDES]), not including wastewater prior to discharge or sludge from treatment

- Irrigation return flows

- Source, special nuclear, or by-product material as defined under the Atomic Energy Act

- Materials subjected to *in situ* mining techniques that are not removed from the ground during extraction process

- *Excluded scrap metal* being recycled (processed scrap metal from all sources; unprocessed, as-generated scrap metal from steel mills, foundries, refineries, and metal working/fabrication industries)

- Various conditionally exempted materials generated in industrial processes that are being recycled, often by being used/reused in the process, including the following: pulping liquors, spent sulfuric acid, secondary materials maintained in a closed-loop system, wood-preserving solutions and wastewater, recovered oil, certain materials generated in the coking and steelmaking industries

to be recycled in the following ways are not RCRA solid wastes:

- Materials used or reused as ingredients in an industrial process to make a product if not first reclaimed or land-disposed (*i.e.*, placed virtually anywhere except in a tank, container, containment building, or drip pad as defined in RCRA)

- Materials used or reused as effective substitutes for commercial products

Hazardous Waste Determination

The RCRA statute broadly defines **hazardous waste** as solid waste that, because of quantity, concentration, or physical, chemical, or infectious characteristics: (a) causes or significantly increases mortality or serious irreversible or incapacitating reversible illness, or (b) poses a substantial present or potential hazard to human health or the environment when improperly managed. EPA thus has Subtitle C jurisdiction over a correspondingly broad range of wastes. In attempting to establish a practical way to implement this definition, EPA has focused on potentially harmful physical or chemical characteristics of wastes and on the toxicity of specific chemicals in the waste (*e.g.*, poisons, carcinogens, teratogens, mutagens), and has established two ways of designating wastes as hazardous: (1) listings, and (2) characteristics. Each listed waste and each hazardous waste characteristic is denoted by a specific Hazardous Waste Number or *Code* assigned by EPA. Some states define more wastes as hazardous than are acknowledged in the Federal regulations; examples include used oil, PCB-contaminated waste, and other toxics.

In 40 CFR 262.11, EPA directs that a generator of a RCRA solid waste must accurately determine if that waste is hazardous by

1) Determining if the waste is excluded from regulation in 40 CFR 261.4.

2) Determining if the waste is listed as a hazardous waste in 40 CFR 261, Subpart D.

3) Determining if the waste exhibits a characteristic of hazardous waste as defined in 40 CFR 261, Subpart C using either analytical methods or by applying knowledge of the hazardous characteristic in light of the materials or process used. EPA's approved analytical methods are documented in *Test Methods for Evaluating Solid Waste, Physical/ Chemical Methods*, EPA Publication SW–846.

4) If the waste is determined to be hazardous, the generator must refer to 40 CFR 261, 264, 265, 266, 268, and 273 to determine any further exclusions or restrictions pertaining to management of a specific waste.

The key steps in this process are discussed below.

Exclusions from the Hazardous Waste Definition. In a similar manner to that described previously for solid waste determinations, a solid waste may be excluded from the hazardous waste definition by petition (granted under 40 CFR 260.20 and 260.22) or by specific exclusion. The former exclusions are listed by EPA in 40 CFR 261; Appendix IX. The latter, listed in 40 CFR 261.4(b), are solid wastes that are excluded from being hazardous wastes due to practicalities of enforcement, economics, relatively low hazard potential, resource recovery potential, or other reasons. Table 3 is a useful summary of these exclusions; however, one must consult the regulations for an actual determination.

Listed Hazardous Waste. EPA lists a waste on the basis that it may exhibit hazardous characteristic(s) (*i.e.*, ignitability [I], corrosivity [C], reactivity [R], and toxicity [E] as described below for characteristic waste) or otherwise may be toxic (T) or acutely hazardous (H). These ***listed hazardous wastes*** and their cause(s) for listing are tabulated in 40 CFR 261, Subpart D under four series of hazardous waste codes (F-, K-, P-, and U- series), as described below.

F-Listed Wastes. These are generic wastes from nonspecific sources that are produced by a variety of industries. Common examples include F006 sludges from treatment of wastewater from electroplating operations, F039 leachate resulting from land disposal of two or more listed hazardous wastes, and the F001–F005 spent solvents summarized below with the basis for listing in parentheses:

- F001–specified spent halogenated solvent/ solvent mixtures from degreasing operations (T)

- F002–specified spent halogenated solvent/ solvent mixtures *not* from degreasing (T)

- F003–specified spent nonhalogenated solvent/ solvent mixtures (I)

- F004–specified spent nonhalogenated solvent/ solvent mixtures (T)

- F005–specified spent nonhalogenated solvent/ solvent mixtures (I, T)

Hazardous waste determinations for spent solvents require very close attention to the listing descriptions in 40 CFR 261.31. The F001–F005 listings apply only to chemicals/mixtures which contain, before use, constituents and constituent amounts specified in the listing description, are used for their solvent properties (*e.g.*, degreaser, extractant), and are spent (*i.e.*, no longer usable without being reclaimed). A given spent solvent waste may qualify for more than one of these five spent solvent codes.

Of nearly thirty F-listed wastes designated to date, only six are listed on the basis of acute toxicity (H): F020–F023, F026, and F027, pesticide manufacturing wastes containing dioxins. The hazardous constituents prompting the assigned basis codes for each F-listed waste are provided in 40 CFR 261, Appendix VII.

K-Listed Wastes. These are hazardous wastes from specific industrial processes. K-listed wastes are grouped by industry category, including the following: wood preservation, inorganic pigment, organic chemical, inorganic chemical, pesticides, explosives, petroleum refining, metal production (various), veterinary pharmaceuticals, ink formulation, and coking. Typical wastes include wastewater treatment sludges and distillation bottoms from specified manufacturing operations. Some examples are

- K011–bottom stream from the wastewater stripper in the production of acrylonitrile (R,T)

- K088–spent potliners from primary aluminum production (T)

Table 3. RCRA Hazardous Waste Exclusions[1]

- Household waste, including any solid waste derived from residences, hotels, campgrounds, *etc.*

- Agricultural crop and animal waste returned to the soil as fertilizers

- Mining overburden returned to the mine site

- Fly ash, bottom ash, slag, and flue gas emission control wastes derived primarily from fossil fuel combustion (except as specified in 40 CFR 266.112 from hazardous waste combustion)

- Drilling fluids and other wastes from exploration, development, or production of crude oil, natural gas, or geothermal energy

- Certain chromium-containing waste, mostly from the leather tanning industry, where it can be assured the chromium is not or would not be converted to a toxic (hexavalent) form

- Solid waste from extraction, beneficiation, and specifically listed processing of ores and minerals

- Cement kiln dust waste

- Discarded arsenical treated wood or wood products that fail the Toxicity Characteristic Leaching Procedure (TCLP) test (discussed later in this chapter) for Hazardous Waste Codes D004 through D017 but which are not otherwise hazardous waste, if discarded by persons who use the treated wood and wood product for their intended use

- Petroleum-contaminated media and debris that fail the TCLP for D018 through D043 only and are subject to RCRA Underground Storage Tank (UST) corrective action provisions

- Used chlorofluorocarbon refrigerants from totally enclosed heat transfer equipment that are reclaimed for further use

- Non-terne plated used oil filters that have been gravity hot-drained

- Used oil re-refining distillation bottoms used as feedstock for asphalt products

[1] See 40 CFR 261.4[b] for a complete description.

- K099–untreated wastewater from production of 2, 4–D (T)

None of the K-listed wastes to date has been designated on the basis of acute toxicity (H). The hazardous constituents prompting the assigned basis codes for each K-listed waste are provided in 40 CFR 261, Appendix VII.

P-Listed Wastes (Acutely Hazardous) and U-Listed Wastes. Both P-listed and U-listed wastes consist of the following:

- Specified commercial chemical products or manufacturing chemical intermediates (in-

cluding those which are off-specification) that are discarded unused

- Residues thereof in containers or container liners (unless *RCRA empty*)

- Residue or contaminated media or debris resulting from cleanup of spills of such chemical products or intermediates

These listings include commercially pure and technical grades of these chemicals, and products in which the listed chemical is the sole active ingredient. Products in which a listed chemical is not the sole active ingredient do not qualify as P- or U-listed wastes.

The difference between the P- and U-listed chemicals is that the P-list are those considered to be acutely hazardous (coded *H*) by EPA. Acutely hazardous wastes are subject to special controls as discussed later in this chapter.

Characteristic Hazardous Waste. The determination of hazardous waste listings as described above most often requires knowledge of the waste's origin, and a given listed waste code generally applies to a narrow category of wastes. In contrast, determining if a waste is a characteristic hazardous waste requires evaluation of specific physical and chemical properties of the waste and applies to a broad spectrum of solid wastes

Table 4. Characteristic Hazardous Waste Summary: Ignitable-Corrosive-Reactive Wastes

EPA Hazardous Waste Number	Characteristic	Description and Specified Analytical Methods
D001	Ignitability	• Liquids (other than <24 vol% aqueous alcohol solution) with flash point <140°F (60°C), using specified ASTM standard closed cup methods • Nonliquids capable, under standard temperature and pressure, of causing fire through friction, absorption of moisture, or spontaneous chemical changes and, when ignited, burn vigorously enough to create a hazard • Ignitable compressed gas (as defined by DOT in 49 CFR 173.300) (Note: The ignitability characteristic is different than the various classifications of flammable material under the International Classification System for hazardous materials used by DOT.) • Oxidizers (as defined in by DOT in 49 CFR 173.151) *Examples*: nonhalogenated organic solvents, pressurized acetylene cylinders, propane tank, concentrated nitric acid
D002	Corrosivity	Liquids that • Are aqueous and exhibit a pH $\leq$ 2 or pH $\geq$ 12.5, or • Corrode steel (SAE 1020) at a rate of 6.35 mm/yr (0.25 in/yr) at 130°F (55°C) as determined using specified methods from SW–846 *Examples*: strong acids and bases, including nitric acid, sulfuric acid, hydrochloric acid, sodium hydroxide (caustic), potassium hydroxide, *etc.*
D003	Reactivity	Solid waste with any of the following properties: • Normally unstable and undergoes violent change without detonating • Reacts violently with water • Forms potentially explosive mixtures with water • Generates toxic gases, vapors, or fumes in sufficient quantity to pose a threat to human health or the environment when mixed with water or (cyanide- or sulfide-bearing waste only) when exposed to pH conditions between 2 and 12.5[1] • Capable of detonation or explosive reaction if subjected to strong initiating source or if heated under confinement • Readily capable of detonation or explosive decomposition at standard temperature and pressure • Forbidden, Class A, or Class B explosive as defined in 49 CFR 173 *Examples*: peroxides formed from outdated ether, elemental sodium, elemental phosphorus, magnesium powder, hydrazine, picric acid, and possibly pressurized aerosol cans.

[1] Toxic quantities are not specified, but 250 mg HCN or 500 mg H_2S per kg of solid waste is noted as a common reference threshold in Department of Energy RCRA Information Brief EH–231–007/1291 available online; see Bibliography.

regardless of origin. EPA has established characteristic hazardous waste on the basis of four properties and has assigned D–series waste codes, as follows: ignitability, D001; corrosivity, D002; reactivity, D003; and toxicity (Code *E*, not to be confused with toxicity Code *T* as a basis for listing), D004–D043.

Table 4 contains the specific criteria for the first three types of characteristic waste (the so-called *ICR wastes*), analytical methods prescribed for their determination, and common examples of such wastes. Because the potential ill effects of these wastes is immediate (*e.g.*, fire, explosion, chemical burns), they are given special consideration in the RCRA regulations (*e.g.*, see minimum setback from property lines for storage at 40 CFR 264/265.17).

The characteristic of **toxicity** is based on the toxic properties of (to date) 8 metals and 32 organic compounds, as listed in Table 5. The hazard of greatest concern with regard to such toxins is contamination of ground water and exposure via ingestion of contaminated ground water, so EPA

Table 5. Characteristic Hazardous Waste Summary: Toxicity Characteristic Wastes

EPA HW No.	Constituent Name	Regulatory Level (mg/l)[1]	EPA HW No.	Constituent Name	Regulatory Level (mg/l)[1]
Metals (8)					
D004	Arsenic	5.0	D008	Lead	5.0
D005	Barium	100.0	D009	Mercury	0.2
D006	Cadmium	1.0	D010	Selenium	1.0
D007	Chromium	5.0	D011	Silver	5.0
Organics (32)					
D012	Endrin	0.02	D028	1,2–Dichloroethane	0.5
D013	Lindane	0.4	D029	1,1–Dichloroethylene	0.7
D014	Methoxychlor	10.0	D030	2,4–Dinitrotoluene[2]	0.13
D015	Toxaphene	0.5	D031	Heptachlor (and its epoxide)	0.008
D016	2,4–D	10.0	D032	Hexachlorobenzene[2]	0.13
D017	2,4,5–TP (Silvex)	1.0	D033	Hexachlorobutadiene	0.5
D018	Benzene	0.5	D034	Hexachloroethane	3.0
D019	Carbon tetrachloride	0.5	D035	Methyl ethyl ketone	200.0
D020	Chlordane	0.03	D036	Nitrobenzene	2.0
D021	Chlorobenzene	100.0	D037	Pentachlorophenol	100.0
D022	Chloroform	6.0	D038	Pyridine[2]	5.0
D023	*o*–Cresol[3]	200.0	D039	Tetrachloroethylene	0.7
D024	*m*–Cresol[3]	200.0	D040	Trichloroethylene	0.5
D025	*p*–Cresol[3]	200.0	D041	2,4,5–Trichlorophenol	400.0
D026	Cresol[3]	200.0	D042	2,4,6–Trichlorophenol	2.0
D027	1,4–Dichlorobenzene	7.5	D043	Vinyl chloride	0.2

[1] Based on Toxicity Characteristic Leaching Procedure (TCLP), SW–846 Method 1311.
[2] Quantitation limit is > calculated regulatory level and therefore becomes the regulatory level.
[3] If *o*–, *m*–, and *p*–cresol concentrations cannot be differentiated, total cresol (D026) regulatory level is used.

requires use of a standard test protocol, the Toxicity Characteristic Leaching Procedure (TCLP), to simulate leaching of these constituents from waste that has been land-disposed (*e.g.* in a landfill) from percolation of water through the waste. The regulatory level of each constituent, also shown in Table 5, is expressed as a risk-based concentration in the TCLP extract, a simulated leachate. The TCLP test procedure, detailed in SW–846, is different for liquid and solid phase waste. For liquid wastes (those containing < 5% dry solids), the waste after filtration is defined as the TCLP extract. For solid phase wastes, the procedure involves prescribed preparation of a representative sample of the waste (including size reduction, if necessary) followed by simulated leaching of the waste sample in mildly acidic aqueous buffer solution for 16–20 hours, followed by analysis of potential toxic constituents in the extract. A minimum sample size of 100 grams is recommended for both liquid and solid phase samples. For the latter, an amount of extraction fluid equal to 20 times the weight of the sample is specified. This aspect of the test procedure is occasionally useful in process knowledge waste determinations, in that if the total concentration of a constituent in a solid phase waste is less than 20 times the TCLP regulatory standard (the dilution factor in the TCLP), then the waste cannot exhibit the toxicity characteristic for that constituent.

The wastes shown in Tables 4 and 5 are commonly referred to as **characteristic hazardous waste** and bear all D–series codes that apply. Waste that exhibits any one or more of the first three of these characteristics, also known as **Ignitable, Corrosive, Reactive (ICR) wastes** (Table 4), have chemical properties that can be immediately dangerous if improperly managed and thus merit special protective measures in the regulations.

The Mixture and Derived-from Rules and Related Criteria. EPA has expanded the hazardous waste universe beyond that established by the criteria described above, to include two additional classes of listed waste. The final step in the hazardous waste determination process is to ascertain whether the solid waste meets the criteria for either or both of these classes of waste. This involves evaluation with respect to EPA's so-called *mixture rule* and *derived-from rule*.

The **mixture rule** (40 CFR 261.3[a][2][iii–iv]) applies to listed hazardous waste mixed with solid waste, and can be generally stated as follows:

> A mixture of listed hazardous waste (in any amount) and a nonhazardous solid waste is also a listed waste and bears the same listed waste codes as the listed portion of the mixture, unless that portion was listed on the basis of hazardous characteristic(s) only (*i.e., basis code I, C, R, E as described above*) and the mixture exhibits none of the characteristics.

A major exception to this rule is made for mixtures of wastewater regulated under the Clean Water Act with certain listed hazardous wastes (*e.g.*, toxic chemicals from laboratory operations and certain listed solvents in amounts resulting in very small concentrations at the headworks of the wastewater treatment facility, including *de minimis* losses of U- and P-listed wastes from manufacturing operations, such as rinsate from empty containers or containers rendered empty by that rinsing).

There are additional considerations regarding mixtures that are pertinent to mention at this point.

Rebuttable Presumption for Used Oil. With certain limited exceptions, used oil that contains more than 1000 ppm of total halogens is presumed to be a hazardous waste due to mixing with listed halogenated hazardous waste (*e.g.*, spent solvents). This presumption is rebuttable by demonstrating that such hazardous waste is not present (40 CFR 261.3[a][2][iv]).

Impermissible Dilution. Mixtures of characteristically hazardous (but not listed) waste and nonhazardous solid waste are hazardous only if the mixture exhibits one or more characteristics and is not otherwise excepted. This is not meant to suggest that mixing nonhazardous solid waste with a characteristic hazardous waste is in all cases allowable; such mixing for purposes of avoiding legitimate treatment would constitute impermissible dilution (40 CFR 268.3).

Contained-In Policy. The mixture rule only addresses mixtures of hazardous waste and solid waste. In cases where the material mixed with the hazardous waste is not a solid waste, EPA's so-called **contained-in policy** governs. Under this policy, a mixture of hazardous waste and material other than solid waste, commonly environmental media (*e.g.*, soil, ground water, and debris), must be managed *as if the entire mixture was hazardous waste* unless or until the hazardous waste is determined to no longer be present (*e.g.*, via a *contained-in determination* from the regulator), at which time the nonsolid waste material escapes Subtitle C regulation.

The ***derived-from rule*** (40 CFR 261.3[c][2]) can be practically stated as follows:

> Any solid waste generated from the treatment, storage, or disposal of a listed hazardous waste is also a listed waste and bears the same listed waste codes regardless of whether or not the hazardous waste was listed on the basis of characteristic(s) only. Any solid waste generated from the treatment, storage, or disposal of a waste that is hazardous by characteristic only is hazardous only if it exhibits one or more hazardous characteristic.

Sludge derived from treatment of a listed hazardous waste at a wastewater treatment plant, and ash derived from incineration of listed waste, are examples of so-called *derived-from* waste.

Determining Regulations Applicable to a Hazardous Waste

Once it has been established that a waste is a hazardous waste and all hazardous waste codes applicable to that waste have been identified, the next challenge is to determine the specific regulations that apply. As with the waste determination process, determining the extent to which a hazardous waste is regulated requires methodical review of numerous conditions, the most commonly encountered of which are summarized in this section. As a general rule, a hazardous waste is subject to *full Subtitle C regulation* (*i.e.*, 40 CFR 262–266, 268, 270), unless it is entirely or partly exempted or subject to alternative management standards as briefly described below.

Nonrecycled but Exempted Hazardous Waste

The following hazardous wastes are conditionally exempt from virtually all RCRA Subtitle C regulations (see 40 CFR 261.4–7). Further, except as noted for treatability study samples, persons whose only hazardous waste activity is generation of these wastes are not required to notify EPA or obtain an EPA Identification Number.

***In-Process* Wastes.** Hazardous wastes generated in product or raw material tanks, transport vehicles or vessels, pipelines, manufacturing process units (except surface impoundments), until removed or until 90 days after the process (*i.e.*, manufacturing, product or raw material storage or transport) ceases. (40 CFR 261.4[c])

Characterization Samples. Samples collected to determine characteristics or composition during collection, transport to the laboratory and back to the generator, analysis, and associated legitimate storage. Conditions include compliance with Department of Transportation (DOT), United States Postal Service (USPS), or other applicable packaging and transport requirements. (40 CFR 261.4[d])

Treatability Study Samples. Samples collected for purposes of conducting a treatability study during collection and transport and associated preparation and storage. Conditions include amount restrictions (*e.g.*, 1,000 kg nonacute hazardous waste, 1 kg acute hazardous waste; larger amounts for debris waste) and compliance with DOT, USPS, or other applicable packaging and transport requirements. Samples undergoing treatability studies and the associated laboratory or testing facility are similarly exempt. Conditions include notification, obtaining EPA Identification Number, timing and amount restrictions, recordkeeping, and reporting. (40 CFR 261.4[e], [f])

Conditionally Exempt Small Quantity Generator (CESQG, or "sea-squeegee") Waste. Hazardous waste generated by a CESQG, up to specified quantities. (40 CFR 261.5)

Residues of Hazardous Waste in Empty Containers. Hazardous waste that remains in an empty container or an inner liner removed from an empty container. The term *empty container* is highly specific as described below. (40 CFR 261.7)

Recyclable Materials

Hazardous wastes that are legitimately recycled, termed *recyclable materials*, are generally subject to less than full Subtitle C regulation in view of RCRA's mandate to encourage recycling. In general, they fall into one of the categories specified in 40 CFR 261.6.

Totally Exempted Recyclable Materials. Certain hazardous wastes that are being legitimately recycled, including industrial ethyl alcohol and scrap metal not qualifying as excluded scrap metal under 40 CFR 261.4(a), and several petroleum industry wastes .(40 CFR 261.6[a][3])

Recyclable Materials Subject to Special Standards of 40 CFR 266. Recyclable materials used in a manner constituting disposal, hazardous waste burned in boilers and industrial furnaces (BIFs), recyclable materials utilized for precious metal recovery, and spent lead acid batteries being reclaimed (40 CFR 261.6[a][2]). The 40 CFR 266 regulations are discussed in more detail in a separate section below.

Other Recyclable Materials. For most other recyclable materials, the recycling operation itself is subject to little or no RCRA regulation, while associated storage and transport are fully regulated (40 CFR 261.6[b], [c]). For example, generators and transporters of such recyclable materials are required to submit notification to EPA of their hazardous waste activities and comply with applicable standards for generators (40 CFR 262) or transporters (40 CFR 263). Owners and operators of facilities that store recyclable materials before they are recycled are likewise required to submit notification to EPA and comply with 40 CFR 264, 265, 268, and other applicable standards. In contrast, RCRA regulation of the recycling operation itself is limited to EPA notification, pertinent hazardous waste manifest requirements for receipt or return of such materials and reporting of discrepancies, and, if the recycling operation occurs at a facility that is otherwise subject to RCRA permitting requirements, the RCRA air emission standards at 40 CFR 264, 265; Subparts AA and BB.

Alternate Management Standards for Universal Waste and Used Oil

As a means of encouraging responsible management, EPA has developed management standards for the following two classes of hazardous waste that may be used in place of management under 40 CFR 260–270 (40 CFR 273, 279). Both of these sets of standards are less stringent than the predecessor Federal programs, so authorized states are not obligated to adopt them.

Universal Wastes. Certain hazardous wastes that are generated in large quantity by a variety of generators, so-called *universal wastes*, may be managed in accordance with EPA's Standards for Management of Universal Waste at 40 CFR 273 to the extent they are adopted by authorized states. EPA has so far designated only three types of hazardous waste as universal wastes: batteries, pesticide stocks that have been recalled or which are being collected and managed under a waste pesticide collection program, and mercury-containing thermostats. These regulations provide alternative standards for universal waste management from generation through accumulation and transport to destination facilities for treatment, recycle, or disposal. Treatment, recycling, and disposal of universal wastes remain subject to RCRA rules that would otherwise apply. The Universal Waste Regulations also include procedures for petitioning EPA to add other wastes to the universal waste management system and criteria for inclusion. Common wastes that have been suggested as universal wastes are fluorescent light bulbs, solvent-contaminated wipers, and aerosol cans. Some states have issued management standards for these types of waste independent of the universal waste rules.

Recycled Used Oil. Recycled used oil which exhibits a hazardous characteristic but which is not a listed hazardous waste (*e.g.*, by mixing with listed waste) may be managed in accordance with EPA's Used Oil Management Standards at 40 CFR 279 to the extent adopted by authorized states. Certain recycled mixtures of used oil and other materials (*e.g.*, fuel) may also be managed as used oil under these regulations.

Empty Containers and Tanks

Under RCRA rules, ***containers*** are portable devices ranging up to and including transport vehicles, and ***tanks*** are stationary devices made primarily of nonearthen materials (40 CFR 260.10; 59 FR 62917). Since hazardous waste remaining in an empty container or an inner liner removed from an empty container is not subject to hazardous waste rules, it is important to know EPA's definition of empty container at 40 CFR 261.7, essential elements of which are as described below.

Compressed Gas Containers (Including Aerosol Cans).
These containers are considered empty when pressure in the container approaches atmospheric pressure.

Containers/Inner Liners That Contained Acute Hazardous Waste.
These containers are considered empty when triple-rinsed with solvent capable of removing the waste, or cleaned by another method demonstrated in the scientific literature or generator testing to achieve equivalent removal, or (containers with liners only) when the intact liner is removed.

Other Containers.
All other containers are considered empty when all waste is removed that can be removed using practices commonly employed for that container type, *and*

- ≤ 2.5 cm (1 in.) of residue remains in the bottom of the container or inner liner, *or*

- For containers ≤ 110 gallons, $\leq 3\%$ (by weight) of the total capacity remains, *or*

- For containers > 110 gallons, $< 0.3\%$ (by weight) of the total capacity remains

EPA has not included in the regulations a specific definition of empty with respect to hazardous waste tanks, so consultation with the regulator may be advisable.

Determining Generator Category

As a practical matter, EPA imposes less stringent rules on generators of relatively small quantities of hazardous waste. Three generator categories are recognized on the basis of type and amount of hazardous waste generated and accumulated onsite in any given calendar month: conditionally exempt small quantity generator (CESQG), small quantity generator (SQG), and large quantity generator (LQG). The generator categories, category criteria, and extent of regulation for each are summarized in Table 6. A more detailed comparison of regulatory differences is presented later.

Not all hazardous wastes are counted in the determination of generator category. Wastes that are excluded are specifically listed in 40 CFR 261.5(c) and (d). Examples include certain wastes exempted from RCRA (*e.g.*, some recyclable materials), universal wastes managed under 40 CFR 273, and wastes managed immediately on-

Table 6. Hazardous Waste Generator Categories and Summary of Key Regulatory Differences

Regulatory Aspect	CESQG 40 CFR 261.5	SQG 40 CFR 62.34 (d) and (f)	LQG 40 CFR 262.34(a–c)
Generation Limits (kg/month):			
Acute Hazardous Waste	≤ 1	≤ 1	> 1
Acute Hazardous Waste Spill Residue	≤ 100	–	–
Total Hazardous Waste	≤ 100	100–1000	> 1000
Maximum Storage Limit (kg):			
Acute Hazardous Waste	1	1	> 1
Acute Hazardous Waste Spill Residue	100	100	> 100
Total Hazardous Waste	1000	600	> 6000
General Extent of Regulation	Exempt from Subtitle C except 40 CFR 262.11 (hazardous waste determination)	Reduced generator accumulation requirements	Full Subtitle C regulation

site in elementary neutralization units or wastewater treatment plants. The status determination can be quite complicated for generators whose waste generation is subject to large fluctuations, because EPA has taken the position that a generator can be subject to different standards at different times, depending on generator status with respect to a waste in a give calendar month. Similarly, the Biennial Report (described later) of a generator whose status fluctuates in a year would cover only activities for those months the generator has LQG status. (51 FR 10146)

Standards for Generators, Importers, and Exporters of Hazardous Waste

RCRA regulations at 40 CFR 262, the so-called *generator standards*, initiate RCRA's *cradle-to-grave* management scheme for most hazardous wastes. In this respect, the rules are designed to ensure that waste generators recognize hazardous waste as such and manage it under RCRA Subtitle C from the moment it is generated through its initial accumulation, storage, and preparation (*e.g.*, packaging) for subsequent transport, storage, treatment, or disposal as required. These standards also set forth rules for importers and exporters of hazardous waste.

The Cradle: Hazardous Waste Determination, Notification of Hazardous Waste Activity, and Obtaining an EPA ID Number

As described earlier in this chapter, Subpart A, 40 CFR 262.11 mandates that generators of solid waste determine if their waste is hazardous waste by methodically determining if it is excluded, listed as a hazardous waste, or (by testing or process knowledge) exhibits any of the characteristics of hazardous waste. Upon positive determination, 40 CFR 262.11 directs that the generator review the remainder of the regulations to determine the extent to which that waste is regulated. These requirements apply to all generators, regardless of category.

EPA further ensures through standards in 40 CFR 262.13 that hazardous waste is promptly brought under RCRA management standards by requiring virtually all hazardous waste generators, except those qualifying as CESQGs per criteria in 40 CFR 261.5 (see Table 6), to notify EPA or the authorized State regulator of their hazardous waste generation activities. The notification is made on EPA Form 8700–12 to EPA (or on a corresponding State-specific form to State authorities) and EPA (or the authorized state) issues a unique identification number (*i.e.*, the EPA ID Number) which is included on virtually all subsequent regulatory submittals. Persons engaged in any other hazardous waste activity (including transport, storage, treatment, recycling, and disposal) must also provide regulatory notification of that activity and obtain an EPA ID Number.

Land Disposal Restriction Determination

Upon determining that their waste qualifies as hazardous waste, generators (except CESQGs) must also make a Land Disposal Restriction (LDR) determination. A waste that is an ***LDR restricted waste*** has a promulgated treatment standard that must be met before it can be land-disposed. If the waste is restricted, it is then also important to know if the waste is a ***prohibited waste***, or actually prohibited from land disposal. Generally, a restricted waste is a prohibited waste upon the effective date of the treatment standard if the waste does not meet the applicable treatment standard and no other variances are applicable (see detailed discussion of LDR, 40 CFR 268, later).

The LDR standards apply at the point of generation of the waste, and are generally designated for each applicable Hazardous Waste Number (*i.e.*, Hazardous Waste Code) by type of waste (*e.g.* wastewater, nonwastewater) and treatment subcategory. Therefore, an accurate determination of applicable treatment standard(s) requires knowledge of the waste when it is generated. This information is needed to plan and carry out any required treatment, whether on-site or off-site. The determination as to whether a waste meets a treatment standard can be made by either testing or knowledge of the waste.

Assuming that a waste is a hazardous waste and not subject to any variances from the LDR treatment standards (*e.g.*, treatability variance, national capacity variance, no migration variance), the following basic items of information are needed at the point of generation of a hazardous waste to determine what, if any, treatment standards apply:

- All Hazardous Waste Codes that apply to the waste

- Type of waste (*i.e.*, debris, wastewater, non-wastewater as specifically defined in 40 CFR 268.2)

- Treatment/regulatory subcategory (as listed in Table 268.40) for each Hazardous Waste Code

- All physical/chemical parameters for which the waste may require analysis to confirm that treatment is necessary and/or demonstrate that it meets applicable treatment standards, including

  - parameters/constituents contributing to its Hazardous Waste characteristic(s) or listing(s) (see 40 CFR 261, Subparts C and D, and Appendix VII)

  - any underlying hazardous constituents (as defined in 40 CFR 268.2) reasonably expected to be present

  - any specified technologies required for treatment of the waste, from Table 268.40

This information can then be used to look up all treatment standards applicable to the waste and to plan for any analyses that may be necessary to determine what treatment is required and/or demonstrate compliance with applicable treatment standards. The standards are variously expressed as total constituent concentrations or constituent concentrations in TCLP waste extract (using methods specified in SW–846), or as a requirement to apply a specified technology (see Tables 268.40, 268.42[a]). Alternative standards are available for certain waste forms including hazardous debris (certain solid phase wastes with predominant particle size > 60 mm), contaminated soil, and lab packs. Additional information is provided below in the discussion of LDR regulations.

Considerations for Sampling, Analysis, and Characterization

As previously indicated, generators must characterize their waste sufficiently to enable them to manage that waste in full compliance with RCRA. Required information includes that needed to determine if the waste is hazardous waste, all hazardous waste codes that apply, and compliance status of the waste with respect to applicable LDR treatment standards. Of primary importance, however, is the need to be thoroughly familiar with all potential hazards posed by the waste and measures that must be taken to address these hazards, regardless of whether or not the hazard is RCRA-regulated. For example, the predominant risk posed by **mixed waste** (hazardous waste that is also radioactive) may be posed by radiation, which becomes a controlling factor in most aspects of its management.

A thorough knowledge of the chemical compatibility of the waste with containment provisions of associated management units (*e.g.*, tanks, containers, secondary containments) and other wastes or materials with which it may be mixed or otherwise come in contact with (intentionally or unintentionally) is important for the safe management of virtually all hazardous wastes. Once the physical and chemical properties of a waste are known, compatibility charts are available to identify potentially incompatible materials (See 40 CFR 264/265, Appendix V). Examples of commonly encountered incompatibility problems are use of incompatible replacement fittings in tank systems containing corrosives and container storage areas with incompatibles in common containment (*e.g.*, organics such as used oil with oxidizers like concentrated nitric acid, or strong acids with strong bases).

Waste characterization must be conducted using a representative sample of the waste. Approved sampling and analysis procedures are described in SW–846. Other EPA-approved methods (*e.g.*, American Society for Testing and Materials [ASTM] methods) are also acceptable. Methods for obtaining representative samples of various waste forms are described in 40 CFR 261, Appendix I and other documents referenced therein, including SW–846, and several ASTM standards.

As is the case for all hazardous waste management activities, the safety of personnel performing sampling and analysis is a first-order concern. Such activities should be undertaken only after thorough identification and analysis of the hazards involved (*e.g.*, physical, chemical, radiological, biological), development of a plan to address those hazards, and training on that plan. Depending on the situation, pre-eminent hazards may be associated with the hazardous properties of the waste itself (particularly for ICR, incompatible, and acutely toxic wastes) or other factors (*e.g.*, drowning hazard at impoundments, injury from operating machinery). Use of appropriate personal protective equipment (PPE) and other safety-related facilities and equipment and safety-conscious procedures are key.

Hazardous Waste Accumulation, Storage, and Treatment by Generators

EPA allows conditional generation and storage of hazardous waste on-site without a RCRA permit or interim status authorization. The applicable conditions, detailed at 40 CFR 262.34, provide for accumulating small amounts of hazardous waste close to points of waste generation, and for accumulating or consolidating larger amounts of hazardous waste (*e.g.*, for shipment off-site) for limited time periods (90 days for LQGs, 180 days for SQGs). Generators may also treat their wastes without a permit or interim status provided that the 40 CFR 262.34 provisions and certain other requirements are met, as discussed below.

Satellite Accumulation. EPA's initial regulations promulgated on November 19, 1980, were substantially modified five years later to provide practical rules for generators to accumulate small quantities of hazardous waste at so-called *satellite accumulation* areas (49 FR 49568). Table 7 lists key requirements for satellite accumulation areas, which are applicable to both SQGs and LQGs. CESQGs are exempt from these and other Subtitle C requirements (40 CFR 292.34[c]).

90-Day and 180/270-Day Accumulation Areas. Many hazardous waste generators ship their wastes off-site for treatment or disposal. Therefore, EPA conditionally allows generators to temporarily accumulate or store hazardous wastes moved from satellite accumulation areas or accumulated directly from generator processes on-site without a RCRA permit or interim status

authorization. The amount of hazardous waste that can be temporarily accumulated or stored in this manner is not specifically limited. However, as noted above for satellite areas, any storage of hazardous waste that is restricted from land disposal is prohibited except to accumulate sufficient quantity to facilitate proper recovery, treatment, or disposal (40 CFR 268.50). These accumulation areas are often established by generators to consolidate and package waste for shipment to an off-site Treatment, Storage, and Disposal Facility (TSDF). Table 8 lists the main conditions under which this temporary accumulation/storage is allowed (40 CFR 262.34).

Treatment without a RCRA Permit. EPA allows generators to treat their hazardous wastes on-site without a RCRA permit or interim status authorization, provided that this treatment take place in accumulation containers, tanks, or containment buildings (but not drip pads) managed in accordance with 40 CFR 262.34 and other management standards invoked therein as summarized in Tables 7 and 8. (51 FR 10168, 57 FR 37194). If this treatment is done for the purposes of meeting LDR treatment standards other than the alternate standard for debris at 40 CFR 268.45, then the generator must also develop and follow a written waste analysis plan that describes procedures to be used to comply with the treatment standards (40 CFR 268.7[a][4]). Some states (*e.g.*, Maryland) do not allow treatment in accumulation areas.

Preparing and Manifesting Hazardous Waste for Off-Site Transport

RCRA rules seek to ensure safeguarding and continuous accountability for hazardous waste being shipped off-site by requiring that generators

- Package, label, mark, and placard waste shipments in accordance with United States Department of Transportation regulations.

- Use only transporters and destination treatment, storage, and disposal facilities (TSDFs) that have an EPA ID Number (40 CFR 262.12).

- Track all such shipments using the prescribed hazardous waste manifest.

In addition, the designated TSDF is required to send written notification to the generator that the

Table 7. Key Requirements for Satellite Accumulation Areas

Regulatory Topic	Key Requirements
Maximum Allowable Accumulation Volumes	Each satellite area is limited to the following cumulative volume, regardless of the number of containers of waste accumulated at that satellite area: $\leq$ 55 gallons of nonacutely hazardous waste or $\leq$ 1 quart of acutely hazardous waste[1].
Allowable Type, Number, and Condition of Accumulation Units	• Only containers are allowed (*i.e.*, portable devices per 40 CFR 260.10). Accumulation in any other allowable accumulation device (*e.g.*, tanks) must meet 40 CFR 262.34 requirements for 90–day (LQGs) or 180–day (SQGs) accumulation areas[2]. • There is no limit on the number of containers at a satellite area. • Containers must be in good condition and compatible with wastes they contain per 40 CFR 265.171, 172.
Location Restriction and Control	The satellite area must be located at or near the point of generation of the waste[3] and must be under the control of the operator of the process generating the waste[4].
Labeling	All containers must be marked with the words *Hazardous Waste* or other words that identify the content of the containers.
Container Management	All containers must be kept closed except when adding or removing waste (40 CFR 265.173[a])[5]. If a container begins to leak, the waste must be transferred to a compliant container or otherwise managed in compliance with RCRA (40 CFR 265.171).
Limit on Number of Satellite Areas, Total Volume Accumulated, and Duration of Accumulation	• EPA has not limited the number of satellite areas at a generator's site or the cumulative total volume of waste that may be accumulated in satellite areas, but considers establishment of multiple satellite areas for purposes of circumventing regulatory responsibilities to be noncompliance (49 FR 49569, 12/20/84). • EPA has not set specific limits on the length of time hazardous waste is left in a satellite accumulation area • The basic definition of satellite accumulation area at 40 CFR 262.34(c) requires the existence of a process generating the accumulated waste. Therefore, when a process is permanently discontinued, waste being accumulated at its satellite area(s) must be placed in 90-day (180-day if SQG) storage or a permitted or interim status hazardous waste management unit (HWMU). It may be advisable to consult with the regulator in the case of extended temporary process shutdowns. • Storage of hazardous wastes restricted from land disposal, including that in satellite areas, is prohibited except to accumulate sufficient quantity to facilitate proper recovery, treatment, or disposal (40 CFR 268.50).
Training Requirements	EPA has specified no particular training requirements for personnel working in or near satellite accumulation areas. However, as a condition for 90-day (LQGs) or 180-day (SQG) storage of hazardous waste without a permit or interim status, LQG employees must be trained in accordance with 40 CFR 264.15 or 265.15, and SQG employees must be familiar with waste-handling and emergency procedures relevant to their responsibilities (40 CFR 262.34[d][5]).
Transfer of Excess Hazardous Waste	In the event the volume accumulated in a satellite area exceeds 55 gallons of nonacutely hazardous waste or 1 quart of acutely hazardous waste, that excess amount must be moved within 3 days to temporary storage allowable under 40 CFR 262.34 or a permitted or interim status hazardous waste management facility, and the container(s) holding that excess amount must be marked with the date the excess accumulation began (40 CFR 262.34[c][2]), thereby providing evidence of compliance with the transfer and/or storage time limits.
Other Transfer Restrictions	Hazardous waste from a satellite accumulation area may not be transferred to another satellite accumulation area. It must be moved to either a 90-day (LQG) or 180-day (SQG) accumulation area or to a permitted or interim status hazardous waste management facility.

[1] Individual states may interpret these limitations differently.

[2] See Table 8.

[3] The meaning of *at or near* is subject to interpretation, so it may be advisable to consult with the regulator in cases where the desired location is not proximate to the point of generation. Achieving substantial risk reduction (*e.g.*, reduced fire hazard by placing a satellite area at somewhat greater distance) without incurring substantial loss of control by the operator can be a factor in obtaining regulatory concurrence on the location.

[4] What constitutes adequate control is subject to interpretation, so it may be advisable to consult with the regulator in cases where adequate control is questionable. The extent of security provisions in place at the process location can be a factor in obtaining regulatory concurrence. Locking satellite area containers, cabinets, or enclosures in cases where the operator is not always present in the immediate area is a control measure that can be a factor in obtaining regulatory concurrence.

[5] The term *closed* is subject to interpretation. However, a good conservative rule of thumb in most circumstances is to ensure that little or no container contents could escape even if the container were upset. Therefore, firm connection and closure devices should be provided on filler funnel devices, all screw-type lids and bungs should be at least hand tight, drum locking rings should be firmly secured, *etc.*

Table 8. Key Requirements for Temporary On-site Accumulation/Storage without a Permit or Interim Status Outside of Satellite Accumulation Areas

Regulatory Aspect	Key Requirements	
	SQG 40 CFR 262.34(d) and (f)	LQG 40 CFR 262.34(a–c)
Accumulation Time Limit (from time of introduction of first waste to the unit)	*180 days (270 days if transporting to TSDF > 200 miles away),* extendable by 30 more days for unforeseen, temporary, and uncontrollable circumstances[1].	*90 days,* extendable by 30 more days for unforeseen, temporary, and uncontrollable circumstances.
Allowable Waste Accumulation/ Storage Units and Management Standards	*Containers* managed under Subpart I (except minimum requirements for setback from property line for ignitable and reactive waste). *Tanks* managed in accordance with relaxed SQG tank standards at 40 CFR 265.201). Tanks must be emptied completely every 90 days (records required).	*Containers* managed under Subpart I *Tanks* managed under Subpart J (with minor exceptions). Tanks must be emptied completely every 90 days. *Drip Pads* managed under Subpart W (plus maintain procedures and records to demonstrate that wastes are removed at least every 90 days). *Containment Buildings* managed under Subpart DD (plus obtain PE certification that building complies with Subpart DD design standards, and maintain procedures and records to demonstrate that wastes are removed at least every 90 days).
Air Emission Standards (40 CFR 265; Subparts AA, BB, CC) Compliance	Not required (40 CFR 262.34[d]).	Required[2]
Labeling and Marking of Tanks and Containers (40 CFR 262.34[a][2,3])	Same as LQG.	*Containers*–accumulation start date must be clearly marked and visible. *Containers and Tanks*–clearly labeled with all EPA Hazardous Waste Numbers applicable to contents and the following words: *Hazardous Waste – Federal law prohibits improper disposal.*
Compliance with 40 CFR 265.16, Training Requirements	Not required, but generator must ensure all employees are familiar with waste-handling and emergency procedures (40 CFR 262.34[d][5]).	Required
Compliance with 40 CFR 265, Subpart C; Preparedness and Prevention	Same as LQG.	Required[2]
Compliance with 40 CFR 265, Subpart D; Contingency Plan and Emergency Procedures	Not required, but emergency response information must be posted. Emergency Coordinator must be on premises or on call, and must ensure proper response (*e.g.*, fire, spill response, notification) (40 CFR 262.34[d][5]).	Required[2]
Compliance with LDRs, 40 CFR 268	Same as LQG.	Required[3]
Allowable Disposition of Waste	RCRA Subtitle C, TSDF; legitimate reclaimer/recycler, or (for universal waste) a 40 CFR 273 regulated facility.	RCRA Subtitle C TSDF; legitimate reclaimer/ recycler, or (for universal waste) a 40 CFR 273 regulated facility.
Compliance with 40 CFR 265, Subpart G; Closure and Postclosure Care	Not required, but all hazardous waste must nonetheless be removed and properly disposed of at closure (40 CFR 265.201[d]).	Only 40 CFR 265.111 and 114 apply, requiring that applicable closure standard be met and that contaminants be removed and disposed of properly. Failure to *clean close* incurs postclosure care.

[1] Some states do not allow the extra time for remoteness from a TSDF.
[2] See discussion on key provisions under 40 CFR 264/265.
[3] See discussion on key provisions above and under 40 CFR 268.

facility is permitted to receive the waste and will accept it (40 CFR 264/265.12), so it is important to make these advance arrangements. As a final caution, generators intending to ship their waste to another state should become familiar with and observe not only regulations applicable in the originating state, but also applicable rules of the destination state and all states through which the shipment passes.

172. Markings consist of detailed information (*e.g.*, shipping names, identification numbers) designed to facilitate proper handling. All markings must be durable, in English, affixed to the surface of the container or on a tag or sign, displayed on a background of contrasting color, and be unobscured. In addition, containers ≤ 110 gallons must be marked with the words shown in Figure 1.

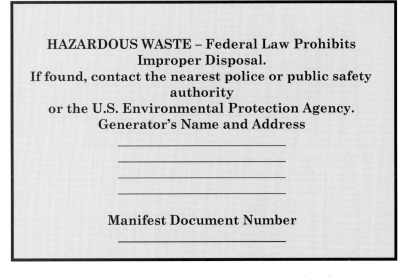

**HAZARDOUS WASTE – Federal Law Prohibits
Improper Disposal.
If found, contact the nearest police or public safety
authority
or the U.S. Environmental Protection Agency.
Generator's Name and Address**

_____

_____

_____

_____

Manifest Document Number

_____

Figure 1. Hazardous Waste Label Information for Containers

Packaging, Labeling, Marking, and Placarding. RCRA basically defers to the Department of Transportation requirements for packaging, labeling, marking, and placarding hazardous waste shipments. The following points are pertinent to these activities:

• Packaging must be in accordance with DOT packaging regulations (49 CFR 173, 178, 179), which are designed to ensure the waste will be safely contained during transport.

• Each package must be labeled in accordance with DOT regulations at 49 CFR 172 (especially 172.101, 172.102, 400–406) to ensure that persons handling them can quickly identify the waste and the potential hazards it poses. With certain exceptions, labels generally must be affixed to the surface of the container near the proper shipping name. If a container has a volume ≥ 64 cubic feet (approximately 480 gallons), labels must be placed on at least two sides.

• Containers must also be marked in accordance with DOT marking requirements at 49 CFR

One special form of packaging problem commonly encountered by generators is that associated with numerous small containers of hazardous waste (*e.g.*, discarded chemicals). These wastes may be specially packaged as so-called *lab packs* in accordance with DOT standards (49 CFR 173, 178, 179) by overpacking in containers of no more than 416-liter (110 gallon) capacity, typically in fiber drums, and surrounded by compatible absorbent material sufficient to absorb all liquids. Incompatibles must be segregated and reactives must generally be rendered nonreactive (40 CFR 264/ 265.316). Vendors are often used to ensure compliance.

It is the generator's responsibility—not the transporters—to either placard the transport vehicle or freight container, or offer the initial transporter appropriate placards in accordance with DOT regulations at 49 CFR 172, Subpart F (40 CFR 262.33). Placarding is not required in a few instances; *e.g.*, where the vehicle or freight container holds <1001 pounds of hazardous materials (unless considered a poison inhalation hazard) or rail cars loaded with transport vehicles or freight containers (49 CFR 172.504, 172.505). In general, placards are specific to the waste being transported and are placed on the front, back, and both sides of the vehicle as specified, such that they will be readily visible.

Manifesting. Hazardous waste shipments generally must be tracked using the manifesting system prescribed in 40 CFR 262, Subpart B and DOT regulations at 49 CFR 172.205. The Federal Uniform Hazardous Waste Manifest (EPA Form 8700–22) or a corresponding State manifest must be used except for certain cases in which SQG waste is being reclaimed under a contractual agreement that specifies the type of waste and frequency of shipments, or cases where the off-site transport is essentially confined to public or private roadways through or bordering the site (40 CFR 262.20).

The following points are useful to consider when manifesting interstate shipments of hazardous waste:

- Use the Federal manifest (EPA Form 8700–22) in cases where neither the state of origin nor the destination state has its own hazardous waste manifest.

- Use the State manifest in cases where either the state of origin or the destination state, but not both, has its own hazardous waste manifest.

- In cases where both the state of origin and the destination state has its own hazardous waste manifest, the manifest of the originating state is generally used; however, check with both states to be sure.

The following key items of information must be included on the manifest:

- Document number

- Generator name, mailing address, telephone number, and EPA ID Number

- Name and EPA ID Number of each transporter to be used for the shipment

- Name, address, and EPA ID Number of the intended destination facility and, at the generator's option, an alternate facility

- Description of the waste as required by DOT regulations at 49 CFR 172.101, 202, and 203 (*e.g.*, proper shipping name)

- Total quantity of each hazardous waste by units of weight and type and number of containers

- Special handling instructions, if any

- Generator's certification (signature and date) that the information supplied on the manifest is complete and accurate, that packaging is in accordance with applicable national (DOT) and international regulations, and that the generator, if a LQG, has a waste minimization program in place or has made a good-faith effort to minimize waste generation, if a SQG, and that an appropriate waste management method has been selected

EPA Form 8700–22, its continuation sheet, and detailed instructions for completing it, are provided in an Appendix to 40 CFR 262.

The *hazardous waste manifest* basically provides a chain of custody for the waste during shipment. With certain exceptions for domestic United States shipments entirely by water or originating by rail, each party taking custody of the waste signs and dates the manifest and provides a signed copy to the party from whom the waste was received. In addition, upon acceptance of the waste the consignee (or TSDF) must send a signed and dated copy of the manifest back to the generator to confirm that the waste has arrived at its destination. Therefore, sufficient copies of the manifest must be prepared to allow one copy for the generator and each transporter and two copies for the designated TSDF, one of which must be returned to the generator upon acceptance of the waste (40 CFR 262.22, 262.23).

Exception Reporting. If a LQG does not receive a signed and dated copy of the manifest from the designated facility within 35 days of the date the waste was accepted by the initial transporter, the LQG must contact the transporter and/or TSDF to determine the status of the shipment. Further, LQGs must send to EPA a formal Exception Report, which includes a copy of the pertinent manifest and a cover letter explaining efforts to determine status of the shipment, if the signed manifest is not received within 45 days of initial transporter acceptance. Exception Reports are also required of SQGs, but are more lenient in that they need not be initiated until 60 days have elapsed and may merely consist of a copy of the pertinent manifest with notation that status of the shipment is unconfirmed.

LDR Notification. Shipments of hazardous waste must not only be manifested, as detailed above, but at a minimum, the initial shipment of a waste must also be accompanied by a notification prepared by the shipper (*e.g.*, generator) that serves to inform the consignee and/or otherwise document the LDR treatment standards applicable to the waste and the status of the waste with respect to these standards. (Individual states may require notification with each shipment.) The notification requirements for generators, detailed in 40 CFR 268.7(a), must include the following items:

- Applicable Hazardous Waste Codes and Manifest Numbers

- Statement that the waste is not prohibited from land disposal, or statement that the waste is subject to the LDRs with a listing of constituents of concern (for F001–F005, F039) and underlying hazardous constituents (for characteristic wastes) that must be addressed by treatment

- Applicable treatment category (*e.g.*, wastewater/nonwastewater per 40 268.2[d] and [f]), treatment subcategory within a waste code as identified in Table 268.40)

- Available waste analysis data

- Date an exempted waste will again become prohibited

- Contaminants intended to be treated under alternative standards for debris at 40 CFR 268.45

- For contaminated soil subject to LDRs, the constituents subject to treatment as described in 40 CFR 268.49(d) and statement indicating its status with respect to exhibiting a characteristic or containing listed hazardous waste, and status with respect to meeting LDR treatment standards

- Where applicable, certification that the waste meets the applicable treatment standards

- Certification that a lab pack contains no chemicals excluded under 40 CFR 268, Appendix IV, and that it will be sent to a combustion facility in compliance with alternative treatment standards for lab packs at 40 CFR 268.42(c).

See additional discussion of LDR requirements later in this chapter.

Generator Recordkeeping and Reporting

Generators must keep prescribed records and submit prescribed reports pertinent to certain hazardous waste activities for which they are responsible. The following list of the main recordkeeping and reporting requirements are detailed in 40 CFR 262, Subpart D.

Reports and Notifications. *Biennial Report.* All LQGs (SQGs are exempt) must submit an account of their waste management activities, including types and amounts of hazardous waste generated, shipped, and otherwise managed, during the previous calendar year by March 1 of each even-numbered year using EPA Form 8700–13A or an approved equivalent (40 CFR 262.41). This is often included in annual reporting required by the Emergency Planning and Community Right-to-Know Act (EPCRA). Therefore, generators must have an effective system in place to inventory and track waste in order to ensure compliance. States may have more stringent reporting requirements; for example, South Carolina requires quarterly inventory and management activity reports.

Exception Reports. Exception Reports must be submitted to EPA if the generator fails to receive a signed manifest from the designated consignee within the prescribed time from the date accepted by the initial transporter (45 days for LQGs, 60 days for SQGs)(40 CFR 262.42).

LDR Notifications. LQGs and SQGs must submit required LDR notifications with waste shipments (40 CFR 268.7[a]).

Records. Unless stated otherwise, the following records must be kept for a minimum of three years or until any related enforcement actions are resolved, whichever is longer. A longer retention period may be advisable in view of potential liability issues.

Waste Determination Records. LQGs and SQGs must retain records of test results, waste analyses, or other determinations made in accordance with 40 CFR 262.11. Although not explicitly required, CESQGs should also maintain such records to support their status (40 CFR 262.40[c]).

Biennial Reports and Exception Reports. Required of both LQGs and SQGs (40 CFR 262.40[b]).

Original Manifest. LQGs and SQGs must retain until receipt of signed manifest from designated facility (40 CFR 262.40[a]).

Signed Manifest from Designated Facility. Required of both LQGs and SQGs (40 CFR 262.40[a]).

LDR Notifications. including certifications, waste analysis data, and related documentation must be retained by both LQGs and SQGs (40 CFR 268.7[a]).

Contingency Plan. LQGs accumulating waste in 90-day accumulation areas must maintain a contingency plan until closure (40 CFR 265; Subpart D).

Training Records. LQGs accumulating waste in 90-day accumulation areas must maintain training records for former employees for three years from their last date of employment and for current employees until closure (40 CFR 265.16).

Refusals of Emergency Response Arrangements. Refusals by response organizations to enter into emergency response arrangements under 40 CFR 265, Subpart C must be documented in the facility's operating record (40 CFR 265.37).

Requirements for Importers and Exporters of Hazardous Waste

Any person who imports hazardous waste must also comply with the generator standards in 40 CFR 262; Subparts E, F, and H. The importer basically assumes the role of the generator for purposes of manifesting the waste, providing the name, address, and EPA ID Number, signed certification, *etc.* (40 CFR 262.60).

The basic requirements for exporters of hazardous waste at 40 CFR 262, Subpart D include the following:

- Written notification to EPA describing the intended activity

- Issuance of an EPA Acknowledgement of Consent to the exporter, a copy of which must accompany the shipment

- Written consent of the receiving country

- Special manifest requirements

- Conformance of the shipment to the terms of the receiving country's written consent

- The transporter must comply with the applicable requirements of 40 CFR 263

The special requirements detailed in 40 CFR 262, Subpart H apply to the export or import of hazardous waste subject to the manifest requirements of 40 CFR 262 or the universal waste management standards of 40 CFR 273 to or from designated member countries of the Organization for Economic Cooperation and Development (currently the United States, Japan, Australia, New Zealand, and most European countries) for purposes of recovery (*i.e.,* resource recovery, reclamation, recycling, reuse, *etc.*).

Special Rules for Pesticide Waste Disposed of by Farmers

Farmers disposing of waste pesticides from their own use which qualify as hazardous wastes are not subject to any RCRA regulations for those wastes, provided each emptied pesticide container is triple-rinsed per 40 CFR 261.7 requirements and the residues are disposed on their own farm in a manner consistent with disposal instructions on the pesticide label (40 CFR 262, Subpart G).

Standards for Transporters

General Scope and Applicability

The RCRA requirements for hazardous waste transport (40 CFR 263) basically defer to DOT's hazardous materials transportation regulations at 49 CFR 171–179, and consist of relatively few additional requirements. A review of the definition of *on-site* at 40 CFR 260.10 and rules for manifesting shipments at 40 CFR 262.20(e) and (f) are instructive in determining the applicability of these RCRA transporter regulations. In summary:

- The RCRA transporter standards apply to persons engaged in offsite transport of hazardous wastes, regardless of the mode of transport, unless the shipment does not require a hazardous waste manifest.

- 40 CFR 263 does not apply to shipments within the generator or TSDF site, *i.e.*, on-site shipments. This basically means any transport across contiguous site property, even if dissected by roads or other public or private rights-of-way, provided that the transport is across rather than along such rights-of-way. Transport between noncontiguous generator or TSDF properties within a right-of-way between them that is under their control and from which the public is excluded is also considered on-site.

- Transport of hazardous waste along public or private rights-of-way within or along the border of a generator's contiguous site is considered off-site transport. However, a manifest is not required for such transport, and the only RCRA transporter regulations that apply are 40 CFR 263.30 and 263.31, which establish transporter responsibilities in the event of a hazardous waste discharge.

Summary of Transporter Requirements

A summary of the key RCRA hazardous waste transporter requirements is presented below.

EPA Identification Number. As with SQGs and LQGs, transporters of hazardous waste must notify EPA of their activities and obtain an EPA ID Number (40 CFR 263.10).

Temporary Storage in Transfer Facilities. Transporters may store manifested waste they are transporting in containers that meet DOT packaging standards (49 CFR 173, 178, 179) (per 40 CFR 262.30) at transfer facilities for up to 10 days without a RCRA permit (40 CFR 263.12).

Manifest System and Associated Recordkeeping.

Transporters

- May not accept hazardous waste without a manifest from a generator or exporter.

- Must sign and date the manifest upon receipt of the waste and provide a copy to the generator.

- Must comply with specific manifesting and tracking requirements related to exports, rail, and waterborne shipments as detailed in 40 CFR 263.20.

- Must deliver all of the waste accepted from the generator or intermediate transporter to the designated facility (or if precluded from doing so due to an emergency, the alternate desig-nated facility) listed on the manifest, to the next transporter, or (for exports) to the designated destination facility outside the United States. (If the transporter is unable to deliver the waste as specified, he/she must contact the generator for further directions and modify the manifest as directed by the generator.)

- Are generally required to keep copies of manifests and other shipping papers for a minimum of 3 years.

(40 CFR 263.20–22)

Hazardous Waste Discharges. Response to releases of hazardous waste in transit is the transporter's responsibility. Response actions include appropriate immediate action to protect human health and the environment, and cleanup of the discharge, including notification and containment.

Notification and Containment. The transporter must take appropriate immediate action to protect human health and the environment. Such actions must include notification of appropriate authorities and emergency responders (*e.g.*, Fire Department, Hazmat Team) and may include other appropriate action to contain the discharge. The National Response Center ([800] 424–8802) must be notified if the discharge results in death, hospitalization, property damage > $50,000, potential releases or exposure to radiation/radioactive materials or etiologic agents, life-threatening conditions, and releases of reportable quantities of a CERCLA hazardous substance (40 CFR 302.6[a]).

Cleanup. The transporter must clean up the discharge or take such action as may be required or approved by Federal, State, or local officials so that there is no longer a threat to human health or the environment (40 CFR 263.30, 31).

Requirements for Treatment, Storage, and Disposal Facilities

RCRA regulations applicable to TSDFs are contained in 40 CFR 264/265. The 40 CFR 264 *Permitted Facility Standards* are applicable to TSDFs that have received approval of their Part B permit application and have received a RCRA permit. The 40 CFR 265 *Interim Status Standards* are mostly identical to, but in some respects more lenient than, the Part 264 standards. These latter standards apply to TSDF facilities or activities that are brought under RCRA Subtitle C by changes in the RCRA regulations, and govern continued operations, closure, and or postclosure care activities pertinent to that facility, as applicable, unless or until a RCRA permit is issued. Finally, as noted previously (see Table 8), certain 40 CFR 265 standards must be met by generators accumulating waste on-site in accordance with 40 CFR 262.34. A summary of the 40 CFR 264/265 regulations, with particular note of those requirements pertinent to generators, follows.

General Provisions

Applicability. Subpart A of 40 CFR 264/265 explains the applicability of these regulations and RCRA's imminent hazard action provisions.

With some important exceptions, the 40 CFR 264/265 standards apply to facilities that treat, store, or dispose of hazardous waste (*i.e.*, TSDFs) through their operating life, closure, and, if applicable, postclosure care period. Each of these terms has special meanings within RCRA that are essential to know for full interpretation (40 CFR 260.10). Simplistically, however:

- *Ttreatment*—Virtually any action that changes a hazardous waste to neutralize it, recover material or energy from it, or make it less hazardous, safer, or easier to manage.

- *Storage*—Holding waste temporarily, pending future management.

- *Disposal*—Virtually any action that introduces waste or waste constituents into the environment (*i.e.*, on/into air, water or land). Includes placement in waste piles outside of a RCRA-compliant containment building.

- *Closure/postclosure*—Prescribed activities that place TSDFs in an environmentally safe

condition with minimum maintenance requirements. If hazardous waste or constituents above allowable risk levels remain after closure (*e.g.*, waste or waste residuals, contaminated soil, ground water) then prescribed postclosure care is required (40 CFR 264/265, Subpart G).

Several hazardous waste management activities are regulated primarily under other statutes and are not generally subject to 40 CFR 264/265 standards, including: ocean disposal regulated under the Marine Protection, Research, and Sanctuaries Act; underground injection of hazardous waste regulated under the Safe Drinking Water Act underground injection control (UIC) regulations; publicly owned treatment works (POTWs) and other wastewater treatment facilities as defined in the Clean Water Act (40 CFR 260.10).

Several other treatment, storage, and disposal activities are not subject to 40 CFR 264/265 standards (40 CFR 264/265.1). Of potential interest to generators are

- Treatment, storage, or disposal activities of generators accumulating waste onsite, except for the 40 CFR 265 requirements invoked by 40 CFR 262.34

- Operation of elementary treatment units; *i.e.*, containers, tank systems, transport vessels, *etc.* used for neutralizing wastes that are hazardous only because they are corrosive (*i.e.*, D002 characteristic waste, listed wastes with listing basis of *C* only)

- Addition of absorbent material to hazardous waste (or *vice versa*) in a container, provided it is done when waste is first placed in the container and containers are in good condition and compatible with these materials

Imminent Hazard Action. In Subpart A of 40 CFR 264/265, EPA reminds us of Section 7003 of the RCRA statute which states that, regardless of any other RCRA provision, EPA can bring enforcement action for any past or present management of solid or hazardous waste that poses imminent and substantial endangerment to human health or the environment.

General Facility Standards

The General Facility Standards in 40 CFR 264/265, Subpart B (applicable to virtually all TSDFs) consist of the basic requirements listed below. Of these, the training requirements of 40 CFR 265.16 are also applicable to LQGs who accumulate waste in 90-day accumulation areas (Table 8).

EPA ID Number. TSDFs must apply for and obtain from EPA an EPA ID Number in a similar manner to that previously discussed for SQGs and LQGs. EPA notification procedures are discussed in 45 FR 12746 (40 CFR 264/265.11).

Notifications. TSDFs must provide the following written notifications:

- Notice to generators intending to send hazardous waste to the TSDF, informing them that the facility has the appropriate permit(s) for, and will accept, that waste

- Notice to EPA of plans to receive waste from a foreign source at least four weeks in advance of expected arrival of the initial shipment of that waste

- Notice to EPA of change in ownership

(40 CFR 264/265.12)

General Waste Analysis. Before treating, storing, or disposing of a hazardous waste, a TSDF must obtain a detailed chemical and physical analysis of a representative sample of that waste in accordance with its written waste analysis plan. The analysis must contain all information necessary to perform the intended waste management activities in accordance with applicable RCRA requirements. The TSDF may request the generator to supply waste determination and characterization data for this purpose. In addition to physical and chemical data, generator-supplied information would typically include a description of the process generating the waste as necessary to confirm any listed waste codes that may apply and, for containerized liquid wastes sent to landfills, evidence that the sorbents used are not biodegradable (40 CFR 264/265.13).

Security. In general, a TSDF must prevent unknowing entry and minimize the possibility of unauthorized entry to active areas of the facility by personnel and livestock through use of 24-hour surveillance or barriers (*e.g.*, fence) and controlled entry, with warning signs in English and the predominant local language stating *Danger– Unauthorized Personnel Keep Out* or equivalent wording at all entrances and approaches (40 CFR 264/265.14).

General Inspection Requirements. All TSDFs must inspect their facilities for conditions that are causing or may lead to release of hazardous waste constituents to the environment or threats to human health in accordance with a written inspection checklist. This includes such items as operating and structural equipment; safety, security, and emergency equipment; and monitoring equipment; areas prone to spills; *etc.* The inspections must be conducted often enough to preclude significant problems and minimally at frequencies specified elsewhere in 40 CFR 264/265 for specific types of facilities. Repairs or other deficiencies must be promptly effected or corrected. Inspections, deficiencies, repairs, and other corrective actions must be documented and records must be maintained (40 CFR 264/265.15).

Personnel Training. Facility (*i.e.*, TSDF and LQG 90-day accumulation area) personnel must complete a program of classroom or on-the-job instruction that includes the following elements (40 CFR 264/265.16):

- The program must be a written program. Job titles, job descriptions, required qualifications, and training to be provided for each job position must be documented.

- The objectives and scope of the training program must be to enable personnel to perform the duties of their job positions in a way that ensures compliance with RCRA requirements, including their responsibilities under the facility contingency plan (see discussion related to 40 CFR 264/265, Subpart D, below). At a minimum, the training program must be designed to ensure that personnel are able to effectively respond to emergencies.

- Training must be conducted by a person trained in hazardous waste management.

- Personnel must be trained within 6-months of assuming a job position, and may not work unsupervised until they are trained.

- Annual refresher training is required.

- Records documenting that required training was provided to facility personnel are required. Therefore, in addition to documentation noted in the first bullet above, be sure to document personnel attending and dates of training (*e.g.*, by signature and date on attendance rosters) and dates persons begin and end work in a job position. The records must be kept for 3 years from the date an employee last worked at a facility. An individual's training records may accompany them if the employee is transferred within the same company. Inspectors can be expected to require access to any required record within a reasonable time, so it's a good idea to keep copies of such records at the facility in such cases.

The regulations are not specific about which facility personnel are subject to this required RCRA training program, so practical judgement is required. Clearly, some element of training is appropriate for personnel directly involved in hazardous waste management operations or other substantive compliance-related activities at a TSDF or 90-day accumulation area. The need for formal RCRA training for personnel whose duties are not directly related to compliance can be less clear. Similarly, differences of opinion are possible with regard to the 6-month and one-year requirements. If in doubt about such things, a check with your regulator may be advisable.

General Requirements for Ignitable, Reactive, and Incompatible Wastes. Ignitable, reactive, and incompatible wastes pose special hazards. The following two precautions must be taken by TSDFs, and are undoubtedly advisable for generators at hazardous waste accumulation areas:

1) Ignitable and reactive waste must be separated and protected from sources of ignition or reaction, including open flames, smoking, cutting and welding, sources of sparking (static, electrical, mechanical), hot surfaces, frictional heat, spontaneous ignition, radiant heat, *etc.* Such wastes generally should be kept remote from electrical devices (*e.g.,* outlets, thermostats, motors, instruments), and non-sparking tools should be used.

2) *No Smoking* signs must be conspicuously placed wherever there is a hazard from ignitable or reactive waste.

The following requirement is invoked by many of the specific management standards in 40 CFR 264/265, including standards for containers and tank systems (Subparts I and J, respectively), and in such cases are applicable to TSDFs and generators accumulating waste onsite in accordance with 40 CFR 262.34:

3) Where ignitable or reactive waste is treated, stored, or disposed; or where mixing of incompatible waste is mixed with other materials, precautions must be taken to prevent hazardous reactions, including those which generate extreme heat or pressure; fire or explosions; toxic or flammable gas, vapor, mist, or dusts; *etc.*

For permitted TSDFs only, the effectiveness of all of the above precautions must be substantiated with supporting documentation; for example, from the published technical literature, results of trial tests or waste analysis (40 CFR 264/265.17).

Location Standards. Permitted TSDFs may not place noncontainerized or bulk liquid hazardous waste in a salt dome or salt bed, underground mine, or cave; interim status TSDFs may not place any hazardous wastes in such places. Permitted TSDFs are also prohibited from locating within the 100-year floodplain (unless special conditions are met) or within 200 feet of an active geologic fault. Some states have imposed additional standards, some of which are very strict (40 CFR 264/265.18).

Construction Quality Assurance Program. Land disposal units (*i.e.,* surface impoundments, waste piles, and landfills) subject to the design and operating requirements of 40 CFR 264 or 265 must develop and implement a construction quality assurance (CQA) program developed under the direction of a registered professional engineer to ensure the effectiveness of containment and related provisions of the facility (40 CFR 264/265.19).

Preparedness and Prevention

40 CFR 264/265, Subpart C sets forth basic general requirements, applicable to all TSDFs, LQGs, and SQGs, for preventing fire, explosion, or unplanned release of hazardous waste or hazardous waste constituents, and for preparing to respond to such events if they occur. A summary of these requirements follows.

Design and Operation. Facilities must be designed, constructed, maintained, and operated to minimize the possibility of fires, explosions, or unplanned releases.

Required Equipment. Facilities must be equipped with the following specified equipment, unless EPA is convinced that it is not needed in a particular situation.

- *Internal communications or alarm system* that can provide immediate emergency instructions to facility personnel (must be available to all involved personnel when hazardous waste is being moved or otherwise handled)

- *External communications device* that is immediately available and capable of summoning emergency assistance from local police/ fire departments, local/State emergency response teams (if only one employee is at an operating facility, this device must be immediately accessible to them)

- *Fire extinguishers* and other appropriate fire control equipment

- *Spill control and decontamination equipment* appropriate to the waste being managed

- *Fire water* at adequate volume and pressure for fire-fighting equipment (*e.g.*, hoses, sprinklers)

Note that communications systems must be immediately accessible, so portable devices (*e.g.*, radio or cell phone) may be needed at remotely located accumulation areas or remote parts of TSDFs. Also, location of all such equipment should be reasonably nearby, but not so close as to be rendered inaccessible or inoperable by a fire, explosion, or release!

Maintenance and Testing. Communications, fire control, spill control, and decontamination equipment must be maintained and tested as necessary to ensure it is operational in an emergency.

Aisle Space. Adequate aisle space must be maintained at a facility to allow the unobstructed movement of personnel, fire protection, spill control, and decontamination equipment to any area of facility operations unless it is demonstrated to EPA that aisle space is not needed for these purposes.

Arrangements with Local Authorities. To the extent appropriate to the waste and potential need for their services, an attempt must be made to make prior arrangements with emergency response organizations, including

- Familiarization of fire and police departments, spill response teams, *etc.* with pertinent aspects of the facility (*e.g.*, types and amounts of waste, hazards posed, worker locations, ingress/egress routes)

- Establishment of primary response authority where more than one organization may respond

- Agreements with State emergency response teams, contractors, and equipment suppliers

- Familiarization of local hospitals with hazardous wastes managed at the facilities and the types of injuries that could result from a facility emergency

Refusals by response organizations to enter into such arrangements must be documented in the facility's operating record. Arrangements that are made must be documented in the facility's contingency plan.

Contingency Plans and Emergency Procedures

40 CFR 264/265, Subpart D requires that owners/ operators of all TSDFs and LQGs operating 90-day accumulation areas under 40 CFR 262.34 develop and implement a contingency plan, assign an Emergency Coordinator, and implement specific procedures to minimize hazards to human health or the environment from fires, explosions, or unplanned release of hazardous waste or hazardous constituents from their facilities to air, surface water, or soil.

Contingency Plan. A contingency plan designed to minimize hazards, as described above, must be in place for all TSDFs and 90-day accumulation areas operated by LQGs. In practice, the plan may be integrated with spill prevention, control, and countermeasure (SPCC) plans or other facility emergency plans. The plan must be implemented immediately whenever there is a fire, explosion, or release of hazardous waste/constituents at the facility which could threaten human health or the environment (40 CFR 264/265.51). A copy of the contingency plan and all revisions to the plan must be maintained at the facility and provided to all emergency responders that may be called upon to provide emergency services.

The following five items are required in a RCRA contingency plan (40 CFR 264/265.52):

1) *Emergency Response Actions.* Response actions facility personnel must take to minimize threat to human health and the environment, including actions to implement emergency procedures specifically required in 40 CFR 264/265.56 (summarized below).

2) *Emergency Responder Arrangements.* As developed in accordance with Subpart C requirements, discussed above.

3) *Emergency Coordinators.* Names, addresses, telephone numbers (work and home) for primary and alternates in the order they will assume responsibility.

4) *Emergency Equipment.* A list of emergency equipment maintained at the facility in compliance with Subpart C requirements, with location, brief physical description, and outline of capabilities for each item on the list.

5) *Evacuation Plan.* Where evacuation may be necessary, a plan must be provided which describes signals to be used to initiate the evacuation, as well as primary and (where needed) alternate evacuation routes.

The contingency plan must be reviewed and revised, if necessary, if any of the following five conditions occurs (40 CFR 264/265.54):

1) The permit is revised (TSDF only).

2) The plan fails in an emergency.

3) The facility changes in a way that increases potential hazard or ability to respond.

4) Emergency Coordinators change, or

5) Emergency equipment changes.

Emergency Coordinator and Emergency Procedures. An employee designated as Emergency Coordinator must be onsite or on call (*i.e.*, able to reach the facility in a short time in response to emergencies) at all times (40 CFR 264/265.55, 265.56). The Emergency Coordinator must be

- Responsible for coordinating all emergency response measures.

- Thoroughly familiar with the contingency plan, the facility, and its operations, including hazardous waste management operations.

- Authorized to commit resources necessary to implement the contingency plan.

In the event of an emergency, the actions specified in 40 CFR 264/265.56 and specifically described in the contingency plan must be carried out by the Emergency Coordinator or owner/operator, as indicated in Figure 2.

Simple, easy-to-follow contingency plans and emergency procedures that anticipate and provide for credible accident scenarios are absolutely essential to safe facility operation.

Manifest System, Recordkeeping, and Reporting

Requirements applicable to generators for manifesting of waste shipments, recordkeeping, and reporting were discussed previously. TSDFs have many of the same requirements as generators. Some additional requirements uniquely applicable to TSDFs are discussed below (40 CFR 264/265, Subpart E).

Manifest Discrepancies. TSDFs must note significant manifest discrepancies (*i.e.*, differences >10% in weight for bulk shipments or any variation in piece count) and attempt to reconcile them with the generator or transporter(s), as appropriate. The TSDF must report to EPA discrepancies not reconciled within 15 days of receiving the waste shipment.

Unmanifested Waste Reports. TSDFs must report to EPA waste shipments received without required shipping papers (*e.g.*, manifest).

Operating Record. TSDFs are required to keep a written operating record at the facility that documents substantive facility operations and activities, including types and amounts of waste managed, management effected, location of waste inventories, waste analysis and determination records, inspection and monitoring records, notices, certifications, and other information as described in Subpart E.

Ground Water Protection and Monitoring

Protection of ground water is EPA's primary goal in its strategy for hazardous waste disposal. This strategy seeks to use liquids management as a primary means to minimize the potential for hazardous waste or hazardous waste constituents to migrate from so-called *land disposal* units (*i.e.*, surface impoundments, landfills and land treatment units) to ground water. Elements of this liquids management approach include limitations on disposal of liquids, requiring the use of double liners and leachate collection systems, stormwater run-on and run-off controls, and impervious covering of wastes left in place at closure. The ground water monitoring programs specified in 40 CFR 264/265, Subpart F are designed to detect ground water contamination from land disposal units if the liquids management strategy is not fully implemented or otherwise fails. It is possible to obtain a waiver from the Subpart F ground water monitoring requirements in specified cases where it can be demonstrated to EPA that there is little or no potential for contamination of potential drinking water supplies or surface waters (40 CFR 264/265.90).

General Program Elements. Requirements for development and installation of groundwater monitoring systems are similar for interim status and permitted facilities. However, some basic program differences exist, and requirements in the regulations for interim status facilities (*i.e.*, 40

Emergency Coordinator Actions

1) **Notify all facility personnel** (Designee may perform if Emergency Coordinator is on call)
   - Internal alarm/communication
   - Initiate evacuation plan, if appropriate (off-site evacuation addressed below)

2) **Notify appropriate emergency responders** (Designee may perform if Emergency Coordinator is on call)
   - Fire department
   - Police/ambulance
   - Spill response/HazMat team
   - Other

3) **Determine nature and extent of releases** (for fires, explosions, releases) using
   - Observations
   - Records (*e.g.*, inventories, manifests)
   - Monitoring, sampling and analysis

4) **Assess potential hazards to human health and the environment**
   - Direct hazards (*e.g.*, from gases, vapors, direct releases to surface water)
   - Indirect hazards (*e.g.*, run-off of contaminated fire water to surface water)

5) **Initiate notifications for off-site releases**
   - If local evacuation may be required, first notify local authorities and assist in evacuation decision.
   - Notify Regional On-Scene Coordinator (if assigned) or the National Response Center [(800) 424–8802].

6) **Keep fire, explosions and releases in check** during emergency by taking all reasonable measures to
   - Stop processes and operations
   - Collect and contain released waste
   - Remove containers of waste

7) **Monitor for incipient hazards in shutdown processes and operations**
   - Leaks, pressure buildup, gas generation, and ruptures

8) **Provide for treatment, storage, and disposal of contamination post-emergency**
   - Waste, contaminated soil, surface water, *etc.*

9) **Prevent management of wastes at the facility that may be incompatible with released materials** until cleanup is completed.

10) **Ensure repair, replacement of all emergency equipment** required in the contingency plan prior to resumption of operations.

Facility Owner/Operator Post-Emergency Actions

1) **Issue restart notification** to EPA and appropriate State and local authorities, indicating that cleanup is complete and emergency equipment is in place prior to resuming operations at affected areas of the facility.

2) **Document all incidents for which contingency plan is implemented** in the operating record, including time, date, and details.

Figure 2. Environmental Coordinator Emergency Response and Post-Emergency Actions

CFR 265) are in some instances more specific because corresponding requirements for permitted facilities are specified in the approved permits. Key program elements are discussed below.

Monitoring Well Design and Location. Both interim status and permitted facilities must have properly constructed monitoring wells in sufficient number and at locations that reasonably ensure detection of ground water contamination resulting from escape of hazardous waste or constituents from the land-based units of interest. The interim status standards at 40 CFR 265.91 specify that monitoring wells must be cased and appropriately screened, annular space must be sealed, and identification plates and locking caps or security devices must be provided.

Wells must be placed both upgradient to enable evaluation of background ground water quality and downgradient to enable detection of any releases from the unit. At interim status facilities, a minimum of one upgradient and three downgradient wells is required, and the downgradient wells generally must be located at the edge of the waste management area unless (for existing units only) precluded by interfering structures and EPA is convinced an alternate location(s) would be effective. At permitted facilities, the design, number, and location of wells is specified in the permit, and downgradient wells are located at the **point of compliance** (POC), (the vertical surface located at the hydraulically downgradient edge of the area where waste will be placed, and extending down to the uppermost aquifer) (40 CFR 264.95).

Background Water Quality. Background water quality must be established at interim status facilities by monitoring the upgradient (background) wells quarterly for one year (40 CFR 265.92). Parameters to be monitored, specified in 40 CFR 265.92, consist of approximately 20 Interim Primary Drinking Water Standards parameters listed in 40 CFR 265, Appendix III to determine suitability as drinking water, six specified parameters to establish general quality of the ground water, and four parameters serving as indicators of ground water contamination. Background water quality for permitted facilities may be established on the basis of extensive data supplied in the Part B permit application and monitoring initially conducted under the permit.

Detection Monitoring. The initial phase of ground water monitoring, generally referred to as *detection monitoring*, has the objective of detecting statistically significant increases in contaminants at the downgradient (detection) wells as compared to upgradient (background) wells. At interim status facilities, approximately 30 parameters including the four contamination indicator parameters are monitored at least semiannually. The detection monitoring program at permitted facilities is specified in the permit, is developed to specifically address existing ground water quality and hazardous waste managed at the unit, and is likely to be considerably more extensive than at interim status units. Detection monitoring continues during the active life of a unit and during the postclosure care period, unless and until statistically significant contamination attributable to releases from the monitored unit is detected (*i.e.*, significant increase in contaminant indicators [or pH decrease] at detection wells as compared to background wells).

Assessment and Compliance Monitoring. The detection of statistically significant ground water contamination as a result of releases from the unit triggers a requirement for an assessment monitoring program at interim status facilities (40 CFR 265.93) or a compliance monitoring program at permitted facilities (40 CFR 264.99). The assessment monitoring program is based on a plan specific to the facility and designed to determine the nature, extent, and movement characteristics of the contamination. Assessment sampling on a quarterly basis is required until the facility undergoes final closure or until the program is replaced by monitoring conducted under a permit and/or corrective action requirements.

The compliance monitoring program at permitted facilities is designed to determine whether the ground water protection standard (GWPS) for the facility has been exceeded at the established POC. The GWPS, established in the permit for a facility when contamination is detected, specifies concentrations at the POC of selected hazardous constituents that are allowable during the compliance period. These are constituents that have been detected in the uppermost aquifer in detection wells and which are reasonably expected to be in or derived from waste in the unit. The corresponding concentration limits for these GWPS constituents are set at background levels or, if established under the Safe Drinking Water Act (SDWA), maximum contaminant levels (MCLs) for drinking water, unless EPA is convinced through a demonstration that an alternate concentration level (ACL) is appropriate (40 CFR 264.92). The compliance period begins when a compliance monitoring program is initiated and extends until closure is complete for the unit or, if a corrective action program is ongoing at a facility, until it is demonstrated that the GWPS has not been exceeded for three consecutive years (40 CFR 264.96). Constituent levels detected at POC wells must be statistically compared to GWPS levels at least semiannually, using prescribed statistical methods (40 CFR 264.97[h] and 264.99[f]).

Corrective Action Program. RCRA Sections 3004(u), 3004(v), and 3008(h), (added by the HSWA) gave EPA the authority to require TSDFs that have or are seeking a RCRA permit to clean up releases from any solid waste management unit (SWMU) at their facilities, regardless of when such release occurred. The basic provisions of these powerful RCRA statutory provisions are codified in Subpart F at 40 CFR 264.100–101 and in 264, Subpart S, most of which is merely proposed as of this writing. For purposes of this chapter, it is sufficient to note that assessment monitoring program results for an interim status TSDF provide the basis for a regulatory determination to require corrective action at that facility. Similarly, corrective action is required when the GWPS is exceeded at a permitted facility, and the owner/operator must attempt to remove the hazardous constituents or treat them in place (40 CFR 264.100[b], [c]). A ground water monitoring program is required as part of the corrective action program to assess its effectiveness (40 CFR 264.100[d]).

Closure and Postclosure

Applicability. Upon completion of operations, permitted and interim status hazardous waste management units (HWMUs) at TSDFs must be taken out of service, or closed, in accordance with requirements 40 CFR 264/265, Subpart G and the *closure* performance standard at 40 CFR 264/265.111. LQG 90-day hazardous waste accumulation areas are also subject to the basic closure performance standard established in 40

CFR 265.111, and requirements to properly decontaminate or dispose of contaminated equipment, structures, and soil as described in 40 CFR 265.14 (40 CFR 262.34[a]). Two main types of closure are commonly recognized, as described below.

Clean Closure. Closure in which all hazardous waste and liners are removed, and contaminated equipment, structures, and soils associated with a unit are removed or decontaminated of hazardous waste and constituents to the extent necessary to protect human health and the environment. This type of closure typically is applicable to nonland-based units (*e.g.*, container, tank, and containment building storage and treatment units, incinerators). Cases in which some contaminants remain in place but are below protective levels are termed ***risk-based clean closure***. Allowable residual contamination is determined in the context of closure plan approval by the regulator.

Closure with Waste in Place. **Dirty closure** applies to all closures that are not clean closures. This type of closure may involve partial removal or decontamination, waste stabilization measures and, usually, installation of low permeability cover or *cap* with run-on/run-off controls to minimize infiltration by precipitation. This type of closure is typical for land disposal units (*e.g.*, landfills) and nonland-based units from which all waste or contamination above allowable risk levels cannot be removed.

The difference between these two forms of closure is far-reaching, in that closures with waste in place incur the potentially costly long-term requirements for postclosure care at 40 CFR 264/265.116–120, while clean closures do not.

Closure Performance Standard. The performance standard for closure of HWMUs states that the owner/operator must close the facility in a manner that

- Minimizes the need for further maintenance

- Controls, minimizes, or eliminates, to the extent necessary to protect human health and the environment, postclosure escape of hazardous waste, hazardous constituents, leachate, contaminated run-off, or hazardous waste decomposition products to the ground, or surface water, or to the atmosphere

- Complies with Subpart G and closure requirements specified in the Subpart specifically applicable to the unit (*e.g.*, Subpart I or J closure requirements applicable to use and management of containers and tank systems, respectively)

(40 CFR 264/265.111)

In carrying out the closure, all contaminated equipment, structures, and soil must be properly disposed of or decontaminated, and in accordance with specific instructions in the Subpart applicable to the unit (40 CFR 264/265.114). A written certification signed by the owner/operator and a registered professional engineer attesting, that the unit has been closed in accordance with its closure plan, must be sent to EPA within 60 days after closure.

Closure Plan. TSDFs must have in place a written closure plan that includes the following key elements:

- Description of the facility and hydrogeologic conditions

- Estimates of the maximum number, size, and capacity of operational units, and maximum hazardous waste inventory

- Detailed description of closure methods, including disposition of hazardous waste and contaminants, decontamination, ground water monitoring, and unit-specific closure operations such as capping

- Closure schedules

- Expected date of closure (for interim status facilities without EPA-approved plans)

Closure plans for new facilities must be submitted as part of their Part B permit applications. In general, closure plans for interim status facilities must be maintained at the facility and either submitted with their Part B permit application or, if the facility is to close under interim status, within 180 days of closure for land disposal units or within 45 days for other units (40 CFR 265.112).

Postclosure Care. A survey plat indicating the location and dimensions of waste left in place for *dirty closures* must be filed with local authorities (*e.g.*, county government) and EPA upon closure, at which time postclosure care begins (40 CFR 264/265.116). Postclosure care is conducted in accordance with a written postclosure plan that is submitted to EPA with the Part B permit application for new facilities and at least 180 days prior to closure for interim status facilities. Postclosure care minimally consists of ground water monitoring and reporting, as well as monitoring and maintenance of containment systems (*e.g.*, cap and run-on/run-off control systems) and associated reporting. The postclosure care period continues for 30 years after closure, but may be shortened or lengthened by EPA as deemed necessary to protect human health and the environment (40 CFR 264/265.117–118).

Financial Requirements

RCRA does not allow interim status or permitted TSDFs to operate unless there is assurance that adequate financial resources will be available to pay for closure, postclosure care, and potential liability for incidents or accidents associated with their facilities. Therefore, EPA has established (at 40 CFR 264/265, Subpart H) detailed financial assurance requirements applicable to all TSDFs except those owned by the Federal or State governments. There are few differences in financial requirements between interim status and permitted facilities. Financial assurances for postclosure apply only to facilities required to undergo postclosure (*i.e.*, disposal units, waste piles, surface impoundments, tank systems and containment systems that would be *dirty closed*). The main provisions of these regulations are summarized below.

Closure and Postclosure Cost Estimates. Written cost estimates for a third party to close and, if appropriate, perform postclosure care in accordance with the facility's closure or postclosure plan must be prepared and maintained at the facility. The estimates must be updated annually and whenever a modification to the closure or postclosure plan is approved.

Closure and Postclosure Financial Assurance. Financial assurance, commensurate with the closure and postclosure cost estimates for the facility, must be maintained by the owner/operator. This assurance generally may be in the form of one or more of the following instruments:

- Trust fund

- Surety bond

- Letter of credit

- Insurance

- Financial test and corporate guarantee

Liability. TSDFs must demonstrate financial responsibility for bodily injury and property damage to third parties for accidental occurrences as listed below. The types of financial instruments noted above for closure and postclosure are also acceptable for this purpose:

Sudden Accidental Occurrences–$1 million per occurrence, $2 million annual aggregate (applicable to all TSDFs)

Nonsudden Accidental Occurrences–$3 million per occurrence, $6 million annual aggregate (applicable to hazardous waste landfills, surface impoundments, and land treatment facilities only)

The above coverage amounts are exclusive of legal defense costs.

Specific Standards for Management in Containers, Tank Systems, and Containment Buildings

Hazardous waste is most commonly managed in containers, tank systems, and containment buildings. Of particular importance to the objective of this chapter, all generators (except wood-preserving operations authorized to use drip pads [see Subpart W]) are restricted to these three types of units for managing their hazardous waste under the 90-day (LQG) or 180/270-day (SQG) storage provisions of 40 CFR 262.34. The following sections highlight the main requirements specific to these management units (40 CFR 264/265; Subparts I, J, and DD), with particular note of Part 265 requirements applicable to generators.

Use and Management of Containers

The container management standards of 40 CFR 264/265, Subpart I consist of relatively few straightforward design and operating requirements. With few exceptions as noted, the 40 CFR 265 requirements apply to interim status container storage TSDFs and to 90-day and 180/270-day waste accumulation areas operated by LQGs and SQGs (see Table 8). These requirements are summarized below.

Container Condition. Containers must be maintained in good condition. Practical but conservative judgement is required here. For example, minor surficial scratches or rust on a metal container are generally acceptable. However, pitting, corroded seams, bulging, or other indications of actual or potential compromised integrity should be remedied as soon as possible. Hazardous waste must be transferred from containers that are not in good condition or are leaking.

Safety is the pre-eminent concern in any action to transfer waste or move waste in potentially compromised containers. Container integrity and other hazards should be evaluated prior to opening or moving containers. For example, severely corroded drums may be unsuitable for transporting by conventional means (*e.g.*, forklift with drum handling attachment) or vapor pressure in bulging drums containing spent solvents may need to be reduced (*e.g.*, by packing in ice) prior to venting. Careful overpacking of the containers may be an appropriate option, particularly where minimal handling is desired in view of safety concerns; however, consideration of future management plans for the waste is also important (40 CFR 264/265.171).

Container/Waste Compatibility. Containers must be compatible with the wastes they contain. For example, metal containers should not be used for corrosive acids, and some plastics and solvents are incompatible (40 CFR 264/265.172).

Container Management. Containers must always be closed except when adding and removing waste. The term *closed* is subject to interpretation. However, containers should be firmly closed in a way that would prevent escape of waste even if the container is upset. All screw-type lids and bungs should be kept tight, drum locking rings should be firmly secured, *etc.* (also see Subpart CC, air emission requirements). In addition, containers must not be handled in a way that may compromise them, as may occur from improper hoisting/rigging or forklift transfer techniques (40 CFR 264/265.173).

Container Marking and Labeling. Subpart I does not include requirements for marking and labeling containers. However, such requirements are nonetheless imposed by 40 CFR 262.34 on generators accumulating waste onsite without permit or interim status. (see Table 8). Specific requirements for container marking and labeling also may be imposed on container storage TSDFs by states (*e.g.*, South Carolina).

Secondary Containment. Permitted container storage facilities which store hazardous wastes that contain free liquids (as determined by the paint filter test, Method 9095 in SW–846) or certain acutely toxic F-listed wastes must be provided with a secondary containment system designed and operated in accordance with specified requirements, including:

- Sufficiently impervious and sloped or otherwise designed to collect waste and precipitation and prevent its pooling around containers until it can be removed

- Capacity to contain the larger of 10% of total volume of containers or volume of the largest container (as computed from containers containing free liquid only)

- Run-on prevention or additional capacity to accommodate any run-on

- Prompt removal of leaked waste and precipitation

Permitted container storage facilities which do not store free liquids must merely be sloped or otherwise designed to keep precipitation from prolonged contact with the waste containers. The Federal regulations do not impose these secondary containment requirements on interim status units or generators accumulating waste on-site in accordance with 40 CFR 262.34(a). However, some states do (40 CFR 264.175).

Ignitable and Reactive Waste. Containers of ignitable and reactive waste must be located at least 15 meters (50 feet) from the facility property line. This provision is applicable to LQGs also, but not to SQGs accumulating waste on-site in accordance with 40 CFR 262.34 (see Table 8) (40 CFR 264.165.176).

Incompatible Waste. Incompatible wastes or waste and material may not be mixed together, and waste may not be added to a container that previously held a waste or material that is incompatible with it unless special precautions specified in 40 CFR 264/265.17(b) are complied with. Further, containers of hazardous wastes at a TSDF or LQG or SQG accumulation area must be isolated from nearby incompatible wastes or materials by means of dikes, berms, walls, or other appropriate devices. Examples of incompatibles are provided in Appendix V to 40 CFR 265 (40 CFR 264/265.177).

Inspections. Container storage facilities must be inspected at least weekly, and must address leaks and container deterioration in addition to items generally subject to inspection as required by General Facility Standards (40 CFR 264/265.15), discussed above (40 CFR 264/265.174).

Air Emission Standards. The air emission standards of 40 CFR 264/265, Subpart CC apply to container management of certain organic hazardous waste as described separately below (see Table 9)(40 CFR 264/265.178).

Tank Systems

The management standards of Subpart J apply to TSDFs that store or treat hazardous waste in tanks or tank systems (see definitions in 40 CFR 260.10). With some minor exceptions, the 40 CFR 265, Subpart J standards also apply to LQGs and SQGs operating 90-day or 180/270-day accumulation tanks, but requirements for the latter are limited to relaxed tank standards at 40 CFR 265.201 (see Table 8). Tank systems that are part of totally enclosed treatment units, wastewater treatment systems regulated under the Clean Water Act (CWA), or elementary neutralization units (*i.e.*, units used to neutralize waste that is hazardous only as a result of its corrosivity) are exempt from Subpart J requirements. The main provisions of Subpart J (40 CFR 264/265, Subpart J) are summarized below.

Containment and Detection of Releases. New tanks cannot be placed in service without meeting the secondary containment provisions of 40 CFR 264.193, and existing tanks generally must meet essentially equivalent requirements within two years of becoming regulated or before 15 years of age, whichever comes later (40 CFR 264/265.193). Required elements of the secondary containment are listed below.

General Requirement. The system must detect and collect waste and accumulated liquids and prevent them from entering the environment.

Minimum Features. The system must be designed to be compatible with the waste, structurally sound and durable, capable of detecting leaks within 24 hours, and enable removal of accumulated wastes and liquids promptly (within 24 hours, if possible).

Design Options. The system must include one or more of the following: external liner, vault, double-walled tank, or equivalent device approved by EPA.

Specific Design Option Requirements. Specifics for external liners and vaults include prevention of run-on/infiltration or provision of capacity to accommodate 25-year, 24-hour storm event, and capacity to accommodate 100% of the largest tank volume.

Ancillary Equipment. Ancillary piping must be provided with full secondary containment (*e.g.*, trench, jacketing, double-wall piping) except for specific reliable components (*e.g.*, piping, welded fittings) that are subject to visual inspection on a daily basis.

A variance from the requirement for secondary containment is available if EPA can be convinced that human health or the environment would be protected (40 CFR 264/265.193[g], [h]).

Integrity Assessment for Existing Tank Systems. Existing tank systems without secondary containment as described above must be assessed by a qualified registered professional engineer to determine if the tank is leaking or unfit for use. If the assessment indicates that the tank is fit and not leaking, it may be used for the interim time allowable under 40 CFR 264/265.193 until it is retrofitted with secondary containment, provided it passes annual leak tests or inspections (40 CFR

264/265.193[i]); if not, the tank system must be closed (40 CFR 264/265.191).

Design and Installation Requirements for New Tank Systems or Components. New tank systems must be designed and installed in accordance with rigorous standards for containment, structural integrity, corrosion protection, *etc.* as set forth in this section. Certification (*e.g.*, by a qualified registered professional engineer) that the design and installation meets standards is also required (40 CFR 264/265.192).

General Operating Requirements. The addition of hazardous waste and reagents to the tank system that could cause leaks or other failures is prohibited. Similarly, appropriate spill and overfill protective devices and practices, including maintenance of adequate freeboard in uncovered tanks, must be employed. Assurance that safe conditions will be maintained is provided by the requirement for general waste analysis (40 CFR 264/265.13) discussed previously under the General Facilities Standards (Subpart B), and by additional waste analysis and trial treatment or storage tests (*e.g.*, bench scale or pilot scale) or documentation of tests (40 CFR 264/265.200) on similar wastes (40 CFR 264/265.194).

Requirements for Ignitable, Reactive, and Incompatible Wastes. The following special requirements apply to these wastes:

- Except in emergencies, ignitable and reactive wastes must be decharacterized before introduction to the tank system, or managed in the system so that ignition or reaction is precluded. Setbacks from public approaches and property lines prescribed by National Fire Protection Association (NFPA) *Flammable and Combustible Liquids Code* must be observed

- Incompatible wastes or waste and material may not be mixed together in a tank, and waste may not be added to a tank that previously held a material or waste that is incompatible with it unless special precautions specified in 40 CFR 264/265.17(b) are followed

(40 CFR 264/265.198–200)

Inspections. Inspections must be conducted once each operating day (*i.e.*, every day when hazardous waste is in the tank system), and must address spill/overfill control equipment, all above-ground portions of the system (for corrosion, leaks), monitoring and leak detection equipment data, materials of construction, secondary containment, and immediate surroundings to ensure continued integrity and detect any releases, in addition to items generally subject to inspection as required by General Facility Standards (40 CFR 264/265.15) discussed above. If the tank system includes a cathodic protection system, it must be checked within six months of installation and annually thereafter. All sources of impressed current for cathodic protection must be inspected or tested, as appropriate, bimonthly (40 CFR 264/265.195).

Response to Leaks and Spills; Disposition of Unfit Tanks. A tank system or secondary containment system from which there has been a leak or spill and is unfit for use must be removed from service immediately. Measures must be taken to stop and contain the leak or spill and assess its cause (*e.g.*, by inspection). Any release to the environment, except for those less than one pound that are immediately cleaned up, must be reported to EPA within 24 hours followed by a written report within 30 days. If the leak to the environment was a result of compromise or lack of secondary containment, such containment must be repaired or provided (unless, as for new systems, it can be readily inspected, *etc.*) before returning the system to service. Unfit tanks must be closed in accordance with 40 CFR 264/265.197 (40 CFR 264/265.196).

Closure and Postclosure Care. Tank systems can normally be closed in accordance with the closure standard of 40 CFR 264/265.111 (*i.e.*, achieve clean closure) as described above. Those which cannot must be closed as landfills and are subject to postclosure care (Subpart G), ground water monitoring (Subpart F) and postclosure financial assurance requirements (Subpart H). Tank systems without secondary containment must have in place a contingent plan for *dirty* closure and postclosure as well as a *clean closure* plan (40 CFR 264/265.197).

Air Emission Standards. The air emission standards of 40 CFR 264/265, Subparts AA, BB, CC apply to management of certain organic hazardous wastes in tank systems as described separately below (see Table 9)(40 CFR 264.201, 265.202).

Alternate Tank Standards for SQGs. The specific tank system requirements applicable to SQGs managing waste onsite in 180/270-day accumulation units are a relaxed version of Subpart J requirements described above and do not include requirements for secondary containment (40 CFR 265.201). Requirements that do apply include the following:

- General operating requirements (as described above, except that additional testing is not required).

- Requirements for ignitable, reactive, and incompatible wastes (as described above).

- Inspections (as described above, except that less critical system components and surroundings can be inspected weekly instead of daily).

- Compliance with closure performance standard (40 CFR 264/265.111) is not required, but all hazardous waste must nonetheless be removed and properly disposed of at closure.

Containment Buildings

The standards of 40 CFR 264/265, Subpart DD provide additional hazardous waste management flexibility to TSDFs and LQGs in the form of ***containment buildings***. Units that meet containment building standards in this subpart are authorized for a variety of operations that cannot be readily performed with containerized waste or bulk waste in tanks. These include storage and manipulation of otherwise unconfined bulk waste (*i.e.*, waste piles) and/or performance of complex treatment processes without the need to otherwise confine the waste to RCRA compliant tanks or containers. The containment building standards are comprised of relatively straight-forward requirements related to design, operation, and closure.

Design Standards. Containment buildings must be designed with specific features, as described below.

Enclosure. The building must be a complete enclosed structure with floor, walls, and roof.

Structural Soundness and Durability. The building must be structurally sound and durable, with respect to stresses imposed by activities in the unit and environmental factors (lightweight doors are allowable if they do not contact waste and are effective barriers to fugitive dust emissions).

Primary Containment. Buildings to be used for management of only solid phase waste and reagents must be provided with a primary barrier appropriate to physical and chemical characteristics of the waste and durable with respect to stresses imposed by activities in the unit.

Secondary Containment. Buildings for management of hazardous waste containing free liquids or treated with free liquids must be provided with primary and secondary containment barriers impervious to the waste with liquid collection/removal system to remove liquids from the top of the primary barrier and a leak detection system to detect leaks of the primary barrier, which are durable with respect to stresses imposed by activities in the unit. A waiver from this requirement is potentially available if EPA is convinced that containment can be ensured without secondary containment.

Certification by a qualified registered professional engineer that the unit meets design requirements must be obtained and maintained in the operating record of the facility (40 CFR 264/265.1101).

Operating Standards. Containment buildings must be operated in accordance with specific requirements, including those listed below.

Incompatible Waste and Reagents. Incompatible hazardous wastes and treatment reagents that may compromise the containment building must not be introduced to the unit.

Containment Practices. Controls and practices must ensure containment of hazardous waste within the unit, including maintenance of primary barrier, controlling level of waste within the containment walls, prevent tracking of waste from the unit, and control of fugitive dust emissions. In containment buildings that include both areas with and without secondary containment, special written procedures are required to ensure that liquids or wet materials are not introduced to areas without secondary containment.

Prompt Repairs and Reporting. Conditions that have caused or may cause a release of hazardous waste must be repaired promptly. Releases must be documented, affected areas must be immediately removed from service and leakage must be removed, a written plan for repairs must be prepared and executed, and EPA must be notified of the release. Repairs must be certified by a qualified registered professional engineer and EPA must be notified of repair completion.

Inspections and Monitoring. Monitoring and leak detection equipment and associated data, the containment building, and the surrounding area must be inspected at least once every seven days.

Records. The operating record, including inspection and repair records, must be maintained for three years.

(40 CFR 264/265.1102)

Closure and Postclosure Care. Containment buildings can normally be closed in accordance with the closure standard of 40 CFR 264/265.111 (*i.e.*, achieve clean closure) as described above. Those which cannot must be closed as landfills and are subject to postclosure care (Subpart G), ground water monitoring (Subpart F) and postclosure financial assurance requirements (Subpart H).

Land-Based Units: Surface Impoundments, Waste Piles, Land Treatment Units, and Landfills

Surface impoundments, waste piles, land treatment units, and landfills, so-called **land-based units**, pose relatively high risk of contaminating ground water. Such units thus have in common special provisions for ground water protection including requirements for ground water monitoring (40 CFR 264/265, Subpart F) and financial assurance requirements for postclosure care (40 CFR 264/265, Subpart H). Additional special requirements for these units are set forth in 40 CFR 264/265; Subparts K, L, M, and N, respectively. Impoundments, waste piles, and landfills have very similar requirements, the key elements of which are summarized below.

Containment Provisions. With minor exceptions, new units and expansions of existing units must be designed to be consistent with RCRA minimum technological requirements (*i.e.*, double liners, leachate collection system (not applicable to impoundments), and leak detection/collection systems. Related containment measures are also required, including dike integrity and overtopping controls for impoundments and run-on/run-off controls for waste piles and landfills. Containment requirements for existing units are less stringent, but are accompanied by more prescriptive management and operating requirements (which are established in permits for new facilities). Interim status impoundments must meet liner requirements for new facilities within 48 months of becoming regulated.

Action Leakage Rates. A maximum allowable flow rate for removal of leakage through the primary liner by the leak detection system is established as a trigger for initiation of response actions.

Response Action. Facilities are required to develop and obtain regulator approval of a response action plan that describes actions to be taken if the action leakage rate is exceeded. Typically required elements of the plan include notification to EPA within seven days, written preliminary assessment and short-term action plan to EPA within 14 days, report to EPA of results of remedial actions taken and planned within 30 days and at 30-day intervals thereafter.

Inspections and Monitoring. Inspection of the containment system (*i.e.*, liner, *etc.*) is required during its installation for new systems. During operation, inspection of containment provisions is typically required weekly and after storms. Leakage removal volumes must typically be recorded weekly during operation and at a lower frequency dependent on leakage rate during postclosure.

Requirements for Special Wastes. Ignitables and reactives are either prohibited in these units or are allowable under narrowly defined conditions. Similarly, introduction of incompatible wastes and materials combinations are prohibited without observing the special precautions of 40 CFR 264.17(b). The acutely toxic dioxin-containing wastes F020, through F023, F026 and F027 are also prohibited without special management controls. Disposal of small containers in over-packed drums (lab packs) may be placed in landfills only if special requirements of 40 CFR 264/265.316 related to waste incompatibility, packaging, use of sorbents for free liquids, and related items are met. As discussed later, if these wastes are LDR prohibited wastes, they must meet applicable treatment standards.

Closure and Postclosure Care. These units must either close by removal (*i.e.*, clean close), or close by eliminating free liquids and stabilizing waste and waste residues in place where required and by installing a low permeability final cover meeting the technical requirements of 40 CFR 264/265.310 (*i.e.*, a RCRA-style cap) which includes appropriate run-on/run-off controls and erosion control measures. Postclosure care is required for units that are not clean-closed.

Other Requirements. The air emission standards of 40 CFR 264/265, Subparts BB and CC, apply to surface impoundments.

In contrast to the land-based units described above, operation of land treatment facilities involves intentional placement of hazardous waste onto the soil surface or incorporating it into the soil for purposes of treatment typically by immobilization or destruction (*e.g.*, by microbial action). Therefore, the protective requirements for land treatment units, provided in 40 CFR 264/265, Subpart M are substantially different than for the land-based units described above. These requirements are focused on ensuring that wastes are effectively treated in the unit and that hazardous constituents do not escape beyond the unit (*e.g.*, by infiltration beyond the soil column established as the treatment zone, erosion by wind or water, or incorporation into food chain crops).

Incinerators and Thermal Treatment Units

Thermal treatment of hazardous waste may be accomplished in hazardous waste incinerators, boilers and industrial furnaces (BIFs), and other thermal treatment units. EPA regulations at 40 CFR 264/265, Subpart O apply primarily to hazardous waste incinerators (as defined in 40 CFR 260.10). However, these requirements also apply to BIFs which burn hazardous waste for purposes of thermal destruction, and BIFs that burn hazardous waste for energy recovery and choose to be regulated under Subpart O rather than 40 CFR 266 as is normally the case (see discussion of 40 CFR 266, below). The main Subpart O requirements are described below.

Waste Analysis. Beyond the waste analysis normally required by 40 CFR 265.13 for interim status units, any new waste must be analyzed to ensure efficient operation of the incinerator and determine emissions that would result, and must minimally include heating value, halogen and sulfur content, and concentration of lead and mercury. Waste analysis requirements for permitted facilities are established in the permitting process. Sufficient waste analysis must be carried out throughout normal operation to verify that waste feed to the incinerator is within permitted limits (40 CFR 264/265.341).

Performance Standards. Permitted incinerators must be designed, constructed, and maintained to achieve the following destruction and removal efficiency (DRE) when operated in accordance with requirements established in its permit:

- 99.99% DRE for principal organic hazardous constituents (POHCs) specified in the permit application for each waste feed

- 99.9999% DRE for the acutely toxic F020–F023, F026, and F027 wastes (if incinerated at the facility). Interim status units must obtain special certification for these wastes

The ability to achieve the standards is demonstrated in a formal trial burn.

(40 CFR 264.343)

Operating Requirements. Hazardous waste must not be fed to either a permitted or interim status incinerator during startup or shutdown unless the incinerator is within normal operating conditions. Permitted incinerators must operate in accordance with conditions established in the permit application and observe specific requirements for waste feed, fugitive emissions control, waste feed cutoff, *etc.* as described in this section (40 CFR 264/265.345).

Monitoring and Inspections. At interim status units, combustion and emission control instruments must be monitored at least every 15 minutes and the incinerator and associated equipment must be inspected at least daily. The following monitoring and inspection activities are required for permitted facilities:

- Continuous monitoring of combustion temperature, waste feed rate, combustion gas velocity

indicator, carbon monoxide downstream of combustion zone and prior to release

- Upon request of EPA, sampling and analysis of waste and exhaust emissions

- At least daily inspection of the incinerator and associated equipment

- At least weekly testing of emergency waste feed cutoff system and associated alarms

(40 CFR 264/265.347).

Exemptions to some of the above Subpart O requirements may be obtained if the wastes burned at the facility can be demonstrated to not pose a hazard intended to be addressed by the requirements (*e.g.*, absence of 40 CFR 261, Appendix VIII hazardous constituents in the waste).

The requirements for thermal treatment units other than those which qualify as incinerators or BIFs, per definitions in 40 CFR 260.10, are set forth at 40 CFR 265, Subpart P. These requirements are very similar to those described above for interim status incinerators. One notable exception is the provision of standards for open burning or detonation of waste explosives. The standard is simple, consisting of a prohibition of open burning of hazardous waste except for open burning and detonation of waste explosives, minimum distance setbacks from open burning or detonation sites to adjacent property, and a requirement to carry out the operation in a way that does not threaten human health or the environment.

Other Units

EPA has thus far established specific standards for four additional categories of units not previously discussed, none of which is common or applicable to a broad spectrum of facilities. Except for Subpart W, which applies only to the wood-preserving industry, these standards are not applicable to generators of hazardous waste. The standards are described in 40 CFR 264/265; Subparts Q, W, X, and EE.

Chemical, Physical, and Biological Treatment. 40 CFR 265, Subpart Q applies to TSDFs that treat hazardous waste by chemical, physical, or biological means in units other than tanks, surface impoundments, and land treatment facilities regulated under Subparts J, K, and M, respectively. The standards are similar to those established for interim status treatment in tanks or incinerators, and include waste analysis and trial test requirements, operating requirements (restrictions on introduction of certain wastes, waste feed control), and inspection requirements.

Drip Pads. Drip pads are units established at wood-preserving operations to convey drips from the wood treatment process, precipitation, and/or surface water run-off to an associated collection system. The standards include specific design and operating requirements to effectively prevent leakage of waste to underlying soil and convey the

Table 9. RCRA Air Emissions Standards (40 CFR 265; Subparts AA, BB, CC) Summary Requirements for Generators[1]

Regulatory Element	Subpart AA Process Vents	Subpart BB Equipment Leaks	Subpart CC Containers	Subpart CC Tanks
Applicable Units	Distillation, fractionation, thin film evaporation, solvent extraction, air/steam stripping units managing Hazardous Waste with organic concentration $\geq$ 10 ppm wt that are • Subject to RCRA permitting, or • Exempt from permitting under 40 CFR 262.34(a) but located at a facility subject to RCRA permitting (including recycling units), or • LQG 90–day accumulation units managed under 40 CFR 262.34	Equipment that contains or contacts hazardous waste with organics $\geq$ 10 wt % that is • Subject to RCRA permitting, or • Exempt from permitting under 40 CFR 262.34(a) but located at facility subject to RCRA permitting (including recycling units), or • LQG 90–day accumulation units managed under 40 CFR 262.34 Each piece of equipment subject to Subpart BB must be so marked	Containers with volume > 0.1 m^3 (> 26.4 gallons) that received hazardous waste after 12/06/96 with volatile organic (VO) concentration $\geq$ 500 ppm wt at the generator's point of generation and • Subject to RCRA permitting, or • LQG 90–day accumulation units managed under 40 CFR 262.34(a)	Operating tanks that received Hazardous Waste after 12/06/96 with VO $\geq$ 500 ppm wt at the generator's point of generation and are • Subject to RCRA permitting, or • LQG 90–day accumulation units managed under 40 CFR 262.34(a)
Exceptions	Units already regulated under the Clean Air Act (40 CFR 60, 61, 63)	Units already regulated under CAA (40 CFR 60, 61, 63). Equipment in vacuum service or contacting hazardous waste with organics >10 wt % for <300 hours/year are partly exempted	• Units already regulated under CAA (40 CFR 60, 61, 63) • Remediation or mixed waste management units • Certain wastes in which VO levels have been reduced by specified amounts (e.g., $\geq$ 95% or to at least < 100 ppm wt) by treatment after point of origination and before entry to unit	
Major Controls	• Reduce emissions to <3 lbs/hr and 3.1 tons/year or by $\geq$ 95% (wt) • Closed vent systems (if used) must operate with no detectable emissions or operate at below atmospheric pressure • Control devices (if used) must comply with specific standards; e.g., >95% efficiency for vapor recovery, combustion systems; no visible emissions for flares	Specific standards for equipment categories, including pumps, valves, compressors, pressure relief devices, sample connection systems, open-ended valves and lines, closed vent systems/control devices, etc. Standards also may vary by type of service (e.g., gas or vapor service, light (i.e., high vapor pressure) material service[2]	• Containers >0.1 m^3 (26.4 gal) and <0.46 m^3 (122 gal), or >0.46 m^3 and not in light material service[2]: Level 1 Controls • Containers >0.46 m^3 in light material service[2]: Level 2 Controls • Containers >0.1 m^3 treating hazardous waste by stabilization: Level 3 Controls • Level 1–meet DOT regulations(49 CFR 173, 178–180); cover/ closure device or vapor barrier • Level 2 –meet DOT regulations(49 CFR 173, 178–180); emissions non-detectable or pass vapor tightness test • Level 3 –vent through closed vent system/control device (directly or from enclosure)	• Tanks w/ maximum organic vapor pressure (MVOP) < specified limits not used for stabilization: Level 1 or 2 Controls • All Other Tanks: Level 2 Controls • Level 1–fixed roof with continuous barrier • Level 2–variable fixed and/or floating roof designs with stringent controls • Hard-piped hazardous waste transfers
Inspection/ Monitoring; Repairs	• Inspections (typically annual); monitoring with portable instrument) as specified to determine leaks, etc. • Repair typically must be tried within 5 days and complete within 15 days	• Visual inspections, monitoring (e.g. with portable instrument) as specified by equipment/service category to detect leaks, etc. • Repair typically must be tried within 5 days and complete within 15 days	• Levels 1 & 2–initial and annual visual inspection; no monitoring • Level 3 –Same as Subpart AA vents • Repairs typically must be attempted within 24 hrs and complete within 5 days	• Visual inspections, initial and typically annually, as specified by control level and tank design • Monitoring of Level 2 vents • Repairs typically must be tried within 5 days and complete within 45 days (except for tanks under negative pressure)
Recordkeeping	Operations, waste characterization, inspection/monitoring methods, results; other compliance documentation as specified	Equipment identification/characterization; waste characterization; inspection/ monitoring methods, results; other compliance documentation as specified	Inspection/monitoring plan, records; waste characterization data for Level 1 containers not in light material service to substantiate waste is not light material[2]; design, maintenance documentation; etc.	Inspection/monitoring plan, records; waste characterization data (e.g., MVOP determination); design, maintenance documentation; etc.

[1] Subpart CC air emission standards for impoundments are omitted from this table since they are not applicable to generators managing hazardous waste in 90-day accumulation areas per 40 CFR 262.34(a)

[2] *In light material service*, is defined for specific applications at 40 CFR 264.1031 and 265.1081; basically denotes liquids organics in which the vapor pressure of one or more organic constituents is >0.3 kPa at 20°C and the total concentration of such constituents is $\geq$ 20% by weight.

waste to the collection system. Also included are requirements to inspect liners and cover systems during construction and the installed facility during operation weekly and after storms (40 CFR 264/265, Subpart W).

Miscellaneous Units EPA has established very general standards at Subpart X to accommodate the permitting of hazardous waste management units not appropriately addressed by other unit-specific standards of 40 CFR 264. The standard provides that miscellaneous units must be located, designed, constructed, operated, and closed in a manner that will ensure protection of human health and the environment. The standard further directs that permit terms and conditions shall include those requirements of other 40 CFR 264 Subparts that are appropriate for the miscellaneous unit. The absence of specific standards and well-established precedents makes it especially important to establish effective communications with the regulator when attempting to permit a unit under Subpart X. Examples of units that have been permitted under Subpart X include facilities for open burning/detonation of munitions and the United States Department of Energy (DOE) Waste Isolation Pilot Plant (WIPP), a geologic repository in New Mexico for transuranic radioactive wastes, some of which are also hazardous (40 CFR 264, Subpart X).

Hazardous Waste Munitions and Explosives Storage. The standards established for hazardous waste munitions and explosives include basic requirements for protective storage only (*i.e.*, the standards do not address treatment or disposal). Included are design and operating standards for minimizing potential for detonation or other means of release, requirements for monitoring, and standard provisions for closure and postclosure care. The standards apply only to military munitions (40 CFR 264/265, Subpart EE).

Air Emission Standards

In response to a HSWA mandate at RCRA Section 3004(n), EPA established regulations in 40 CFR 264 and 265 for monitoring and control of organic air emissions from various categories of TSDFs, as follows:

- Subparts AA address emissions from certain HWMUs that treat hazardous waste ex-

hibiting organic concentrations ≥ 10 ppm by weight.

- Subparts BB address leaks from equipment managing hazardous waste with organics $\geq 10\%$ by weight.

- Subparts CC address emissions from containers, tanks, and surface impoundments containing hazardous waste with volatile organic (VO) concentrations > 500 ppm by weight.

The provisions of 40 CFR 264 and 265 do not differ substantially. However, 40 CFR 265, Subparts AA, BB, and CC (except for impoundments), apply not only to interim status TSDFs but also to LQGs managing hazardous waste in 90-day accumulation areas in accordance with 40 CFR 262.34(a). Satellite accumulation containers and SQG 180/270-day accumulation units are exempt from the air emission standards.

Table 9 summarizes the main requirements of the air emission requirements pertinent to generators. As shown, storage of hazardous waste in containers < 0.1 m³ (26.4 gallons) is not subject to the standards. Virtually the only standard for storage in containers ≥ 0.1 m³ (26.4 gallons) and ≤ 0.46 m³ (122 gallons), including 55-gallon drums, is conformance to DOT packaging requirements and provision of a cover/closure device or vapor barrier, which in any case is advisable if not required for offsite shipment of the waste. More stringent requirements apply to containers in which waste is treated by stabilization and containers > 0.46 m³ (122 gallons) storing *light liquid* organics, *i.e.*, hazardous waste in which the vapor pressure of one or more organic constituents is > 0.3 kPa at 20°C and the total concentration of such constituents is $\geq 20\%$ (wt). The air emission standards of Subpart BB for tanks, Subpart AA process vents, and Subpart CC equipment leaks can be much more complex, requiring substantial allocation of resources to ensure compliance.

Specific Hazardous Wastes and Management Facilities

With the exception of recently promulgated rules at Subpart M for management of waste military munitions, current 40 CFR 266 requirements address a variety of resource recovery facilities and activities. As discussed previously under generator standards, management of certain hazardous wastes that are recycled (*i.e.*, recyclable materials) in accordance with the 40 CFR 266 management standards exempts them from other more stringent RCRA regulations. The 40 CFR 266 management standards are summarized below.

Recyclable Materials Used in a Manner Constituting Disposal. Subpart C applies to recyclable materials that are applied to or placed on the land (*e.g.*, for use as roadbed, fertilizer, *etc*). The standards basically indicate that, with few exceptions, such activities are subject to full Subtitle C regulation unless they are contained in products produced for use by the general public, have undergone a chemical reaction that makes them inseparable from the product, and meet the LDR treatment standards (or applicable prohibition levels if no treatment standards are yet established) for the hazardous waste they contain (40 CFR 266 Subpart C).

Used Oil Burned for Energy Recovery. 40 CFR 266, Subpart E regulations apply only to that used oil which is burned for energy recovery in boilers or industrial furnaces not regulated under 40 CFR 264/265, Subpart O. (Note that used oil not burned for energy recovery is likely subject to other RCRA requirements!) Among the main provisions of Subpart E are the following:

- Used oil burned for energy recovery which is a listed hazardous waste or which is mixed with hazardous waste (other than CESQG hazard-ous waste) is considered hazardous waste fuel and must be managed under boiler and industrial furnace (BIF) regulations at 40 CFR 266, Subpart H (40 CFR 266.40[c], [d]).

- Used oil containing more than 1000 ppm of total halogens is presumed to be a hazardous waste and also must be managed as a hazardous waste fuel under 40 CFR 266; Subpart H; unless this presumption is successfully rebutted (40 CFR 266.40[c]).

- Used oil that is not a hazardous waste fuel per the above criteria, including used oil that is characteristically hazardous (but not listed) waste is a used oil fuel, of which two types are recognized: specification and off-specification. *Specification used oil* fuel meets the following criteria: arsenic (≤ 5 ppm), cadmium (≤ 2 ppm), chromium (≤ 10 ppm), lead (≤ 100 ppm), flash point ($\geq 100°F$), and total halogens (≤ 4000 ppm). *Off-specification used oil* fuel does not meet these criteria (40 CFR 266.40[e]).

- Specification used oil fuel per the above criteria may be burned in a boiler or industrial furnace without further regulation except that analyses or other documentation substantiating its status as specification used oil fuel is obtained and records are maintained for three years (40 CFR 266.40[e]).

- Off-specification used oil fuel per the above criteria is subject to the full requirements of Subpart E. These standards includes specific prohibitions regarding persons authorized to market or burn the fuel and types of facilities in which it can be burned (40 CFR 266.41), and management standards for marketers and burners (40 CFR 266.43 and 266.44). Relatively few requirements apply to generators that burn used oil fuel they generate for space heating.

Recyclable Materials Utilized for Precious Metal Recovery. Subpart F applies to recyclable materials that are reclaimed to recover economically significant amounts of gold, silver, platinum, palladium, iridium, osmium, rhodium, ruthenium, or any combination thereof. Recovery of silver from spent photographic solutions is an example. Applicable requirements include notification to EPA of the activity, observance of manifesting and import/export requirements, and recordkeeping to demonstrate that the material is not being speculatively accumulated. Material that is speculatively accumulated is subject to full Subtitle C regulation (40 CFR 266, Subpart F).

Spent Lead-Acid Batteries Being Reclaimed. Subpart G pertains to spent lead-acid batteries being reclaimed (*i.e.*, that are recyclable materials). The standard basically specifies that persons who generate, transport, collect, or regenerate such batteries, or who store but do not reclaim them (other than those to be regenerated), are not subject to RCRA rules. However, the standard requires that facilities which store such batteries before reclaiming them (other than those to be regenerated) notify EPA of their activity and adhere to most applicable provisions of 40 CFR 264 with regard to that storage (40 CFR 266, Subpart G).

Hazardous Waste Burned in Boilers and Industrial Furnaces. Subpart H applies to hazardous waste burned or processed in boilers or industrial furnaces (BIFs) for the purpose of energy or materials recovery, with certain specified exceptions (*e.g.*, characteristically hazardous used oil fuel regulated by 40 CFR 266, Subpart E). BIFs, defined in 40 CFR 260.10, include a wide array of fuel-burning devices (*e.g.*, cement kilns) which may be economically fueled in part at least with hazardous waste generated on-site or obtained from off-site sources. Subpart H contains lengthy and highly prescriptive permit and interim status standards that are not greatly different from those ultimately applied to hazardous waste incinerators regulated under 40 CFR 264/265, Subpart O (40 CFR 266, Subpart H).

Military Munitions. Subpart M provides criteria for determining when military munitions become a solid waste and provides management standards for solid waste military munitions that qualify as hazardous waste. The standards are narrowly focused in response to peculiar problems faced by the military with regard to the management of military munitions and are of no practical importance to most generators (40 CFR 266, Subpart M).

Land Disposal Restrictions

The land disposal restrictions (LDRs) prohibit land disposal of hazardous waste unless the waste meets treatment standards established by EPA. Mandated by Congress with passage of HSWA, the LDRs are intended primarily to protect groundwater, and with related HSWA provisions (minimum technology requirements, corrective action) provide a powerful economic incentive for waste minimization. EPA has codified the LDR implementing regulations for disposal of hazardous waste in two places in its regulations based on the mode of land disposal. Disposal of hazardous waste by deep well injection into formations that are not drinking water sources (Class I wells) is addressed by Safe Drinking Water Act (SDWA) regulations at 40 CFR 148. All other forms of land disposal are addressed in RCRA regulations at 40 CFR 268. Only the latter regulations are addressed in any detail here.

The term *land disposal* is very broadly defined in 40 CFR 268.2 to include virtually any temporary or permanent placement in or on the land outside of so-called *corrective action management units* (CAMUs) established for temporary storage of remediation waste as described in 40 CFR 264, Subpart S. Placement of hazardous waste in a RCRA land-based unit (*i.e.*, surface impoundment, waste pile, landfill, land treatment facility) is also considered land disposal. Storage in RCRA tanks, containers, or other nonland-based units is not considered to be land disposal.

Unless exempted, a waste is deemed to be an **LDR restricted waste** upon promulgation of a treatment standard for that waste. A restricted waste is prohibited from land disposal (*i.e.*, is a **prohibited waste**) after the effective date of the treatment standard if it does not meet the applicable treatment standard and no other variances are applicable. The LDR effective dates for various wastes are listed in Appendices VII and VIII of 40 CFR 268.

The LDR treatment standards applicable to a waste *attach* at the point of generation of the waste. Therefore, even if a waste is rendered nonhazardous after it is generated, it is possible that the waste does not meet the treatment standard. In such cases the waste remains prohibited. Consequently, it is important to determine LDR requirements that apply to the waste concurrent with the waste determination process, ideally prior to waste generation. The initial step in this process is to establish whether the waste is subject to any exemptions, extensions, or variances from the standards.

Exemptions, Extensions, and Variances

The hazardous wastes described in this section are important to generators because the wastes are not prohibited from land disposal (40 CFR 268.1).

Newly Identified or Listed Wastes Without LDR Standards. EPA is obligated, but not always successful, in establishing treatment standards for newly identified and listed wastes within six months of their identification or listing. In the meantime, this hazardous waste may be land-disposed in an authorized RCRA Subtitle C facility (40 CFR 268.13).

Restricted Waste Prior to Effective Date of Prohibition. The effective date of the prohibition on land disposal may be set well after the standard is established, and the waste may be land-disposed in an authorized RCRA Subtitle C facility in the interim. For example, EPA may unilaterally delay the effective date for a waste on a national basis for up to two years if insufficient capacity exists to treat the waste. EPA may also grant case-by-case extensions of an effective date for up to two years if a generator or TSDF can justify the extension per 40 CFR 268.5 (40 CFR 268.5, 268.30–37).

Wastes Granted a No Migration Variance. EPA allows for an exemption from the land disposal prohibition upon demonstration that there will be no migration of hazardous constituents from the disposal unit for as long as the wastes remain hazardous (10,000 years typically is required as an assumption for associated modeling studies). However, such variances have been granted in only a few instances of deep well injection (40 CFR 268.6).

Decharacterized Wastes Disposed in SDWA Class I Injection Wells. Wastes which are hazardous only because they exhibit a characteristic and from which the characteristic has been removed may be disposed in an injection well meeting the definition of 40 CFR 146.6(a); *i.e.*, a SDWA Class I injection well.

Decharacterized Wastes Managed in CWA or CWA-Equivalent Systems. Wastes which are hazardous only because they exhibit a characteristic (except D003 reactive cyanides) and which are not subject to an LDR treatment standard that is a specified technology other than deactivation (DEACT) may be land-disposed if the characteristic is first removed by at least one of the following:

- Management in a treatment system with an NPDES permitted discharge

- Treatment to meet CWA Section 307 pretreatment requirements

- Management in a CWA-equivalent zero-discharge system

This is an important exemption for generators of characteristic waste who operate their own wastewater treatment facilities or who discharge to a publicly owned treatment works (POTW). However, it is important to know that the systems in which such decharacterization occurs must not qualify as land-based units. For example, decharacterization in an impoundment that is not a RCRA permitted or interim status unit is prohibited except under the special conditions noted at the end of this section (40 CFR 268.1[c][4]).

***De Minimis* Losses of Characteristic Waste.** *De minimis* losses of wastes that are hazardous as a result of characteristic only that are discharged to CWA regulated wastewater facilities are not subject to LDR. ***De minimis* losses**, specifically defined in this section, includes minor leaks and spills, sample purgings, relief device discharges, rinsate from empty containers or containers that are rendered empty by that rinsing, and minor laboratory waste contributions to the wastewater facility (*e.g.*, <1% or 1 ppm at the facility headworks). Therefore, this can be an especially important exemption for generators who operate production and process facilities and facilities with laboratories (40 CFR 268.1[e][4]).

CESQG Hazardous Waste, Waste Pesticide Disposed by Farmers in Accordance with 40 CFR 262.70, and Universal Waste Managed under 40 CFR 273. As previously discussed, these wastes are not subject to any LDR requirements.

An otherwise prohibited waste may be *land-disposed* to a treatment impoundment or series of treatment impoundments under limited conditions (40 CFR 268.4). The impoundments must meet minimum technological requirements, treatment residues must be analyzed in accordance with an EPA-approved Waste Analysis Plan (WAP) to demonstrate that LDR standards are achieved, and residues not meeting standards must be removed annually.

Treatment Standards

The LDR treatment standards developed by EPA are technology based. The standards presented in 40 CFR 268.40 may be applied to any waste, but have proven to be impractical for some waste forms. EPA has developed alternative standards that may be used for qualifying waste forms, including hazardous debris, hazardous contaminated soil, and lab packs. Highlights of LDR standards development and main provisions of the standards are summarized below.

LDR Standards Development. RCRA requires that EPA promulgate treatment standards which substantially diminish the toxicity or mobility of the hazardous waste such that short- and long-term threats to human health and the environment are minimized (Section 3004 [m]). EPA has attempted to comply with this mandate by developing standards based on performance of the best demonstrated available technology (BDAT) for treating various hazardous wastes. Examples of BDAT include:

- 99.99% DRE incineration for most organics

- Biological destruction of dilute organics in wastewater treatment facilities (WWTFs)

- Chemical precipitation of toxic metal ions in wastewater

- High temperature metal recovery (HTMR) for metallic nonwastewaters

- Slag vitrification for arsenic nonwastewaters

- Roasting or retorting for mercury nonwastewaters

- Alkaline dechlorination for cyanides

When standards are published in the *Federal Register*, EPA also indicates how to obtain the associated *Background Document* which documents the basis as BDAT, including treatment technologies considered, types of waste tested, methods, and test data.

The LDR treatment standards are expressed in one of three forms.

Total Concentration of Hazardous Constituents in the Waste. Used mostly for liquid phase wastes.

Concentration of Hazardous Constituent in TCLP Extract. Used mostly for hazardous metals in solid phase waste.

Specific Technologies. Used mostly for wastes which are difficult to treat or analyze to determine treatment effectiveness, or to promote recovery of metal in high-metal-content waste (*e.g.*, roasting/retorting for high mercury content waste, HTMR for K061 emission control dust sludge from the primary production of steel in electric furnaces).

Concentration standards are generally preferred by EPA because of the treatment flexibility they afford. However, specified technologies are often the only practical way to achieve the *minimize threat mandate*.

Primary LDR Treatment Standards. The LDR Treatment standards presented in 40 CFR 268.40 are applicable to any prohibited hazardous waste. Table 268.40 lists these primary treatment standards. In this table, the treatment standards are assigned to wastes on the basis of hazardous waste code and treatment/regulatory subcategory (if any); and for each waste thus defined, wastewater and nonwastewater forms.

Hazardous Waste Code. For example, D, F, K, U, and P codes.

Treatment/Regulatory Subcategory. Subcategories may be established for a hazardous waste code on the basis of its origin (*e.g.*, D007 chromium–radioactive high-level wastes from reprocessing of fuel rods), subsequent management (*e.g.*, D002 corrosive–managed in a CWA or CWA-equivalent system), concentration of regulated hazardous constituent (*e.g.*, D009 mercury $\leq$ 260 mg/kg), or waste form (*e.g.*, D008 lead-lead acid batteries).

Wastewater and Nonwastewater. Wastewater and *nonwastewater* forms are recognized for each of the categories/subcategories defined above. **Wastewaters** are defined as wastes with <1 wt% total organic carbon (TOC) and <1 wt% total suspended solids. All other wastes are considered to be **nonwastewaters**.

For each hazardous waste thus defined in Table 268.40, the treatment standard is expressed as either application of a specified technology (described in 40 CFR 268.42; Table 1), or as maximum allowable concentration(s) of regulated hazardous constituent(s) in the waste or in TCLP extract. In addition, for all characteristic wastes with standards expressed as concentrations (*i.e.*, the D004–D011 metals and many of the D012–D043 organics), the treatment standards also specify that underlying hazardous constituents (UHCs) in the waste must be at or below *universal treatment standard* (UTS) concentrations. Applicable UHCs and corresponding UTS concentrations are listed in 40 CFR 268.48, "Table UTS". In practice, EPA allows one to address only those UHCs *reasonably expected to be present in the waste*, not the entire universe of UHCs in the table.

Table 10 provides summary examples of specified technologies for treating some common wastes, and the standard five-letter technology code assigned to the technology.

Alternate Standards for Hazardous Debris. Impracticalities in treating and demonstrating compliance with the 40 CFR 268.40 standards for relatively large, solid phase hazardous wastes prompted EPA to establish alternative standards (40 CFR 268.45). To qualify for the alternative standards, this waste must meet the 40 CFR 268.2 definition of **debris**, summarized as follows:

Solid material > 60 mm (2.5 in) particle size that is intended for disposal and is a manufactured object, plant or animal matter, or geologic material, except for

- Materials for which a specified technology treatment standard is already established (*e.g.*, D008 lead-acid batteries in Table 268.40)

- Process residuals (*e.g.*, smelter slag, waste treatment residues)

- Intact containers of hazardous waste that retain at least 75 % of their original volume

Mixtures of untreated debris and other material are considered debris if they are comprised of debris by visual inspection. Debris that is hazardous waste is called *hazardous debris*.

The alternative treatment standards for debris that is hazardous (*hazardous debris*) is provided in 40 CFR 268.45, Table 1. They consist exclusively of the treatment technology types described below.

Extraction Technologies.

- Physical (*e.g.*, abrasive blasting, grinding, spalling, high-pressure spraying)

- Chemical (*e.g.*, water washing, liquid and vapor phase solvent extraction)

- Thermal (high-temperature metals recovery, thermal desorption)

Destruction Technologies.

- Biological (*e.g.*, biodegradation)

- Chemical (chemical oxidation, chemical reduction)

- Thermal (incineration in Subpart O, incinerator, and 40 CFR 266, Subpart H; BIF)

Immobilization Technologies.

- Macroencapsulation (*e.g.*, surface coating or jacket of inert materials)

Table 10. Examples of Common LDR Treatment Technologies Specified in 40 CFR 268.40

Technology Name	Technology Code	Description	Waste Treated
Biodegradation	BIODG	Chemical breakdown of organics, nonmetallic inorganics by microbes	Many dilute organics; phosphorus, nitrogen, or sulfur-containing inorganics.
Chemical Oxidation	CHOXD	Oxidation with hypochlorite, chlorine, peroxide, ozone, or other oxidizer	Many organics, alternate specified technology for many organic wastewaters and some non-wastewaters
Combustion	CMBST	Treatment in 40 CFR 264/265, Subpart O; incinerators and 40 CFR 266, Subpart H; BIFs	Many dilute and concentrated organics; specified technology for many nonwastewater organics
Deactivation	DEACT	Removal of the hazardous characteristics, usually by chemical means (*e.g.*, neutralization, open burning, detonation)	ICR wastes; specified technology for D002, D003 except cyanides
Stabilization	STABL	Immobilization of metals or inorganics in Portland cement, lime/pozzolans, or similar matrices.	RCRA metals, inorganics (*e.g.*, metal containing sludge from industrial WWTF; incinerator ash

- Microencapsulation (*e.g.*, stabilization with lime/pozzolans, Portland cement)
- Sealing (*e.g.*, epoxy, silicone coating)

Associated performance and/or design and operating standards, and restrictions on contaminant or debris types are also specified in 40 CFR 268.45, Table 1. For example, physical extraction must be applied to achieve a *clean surface* for nonporous debris, but for porous debris at least 0.6 cm (0.5 in) of surface must also be removed.

Hazardous debris must be treated using one or more of the above technologies to address the following, as applicable to the waste:

- Ignitability, corrosivity, and reactivity (for debris that qualifies as D001, D002, or D003)
- Constituents for which the waste exhibits the toxicity characteristic (for debris that is characteristically hazardous)
- Constituents for which treatment standards are established in Table 268.40 (for debris contaminated with listed waste)
- Cyanide (for debris that is reactive for cyanide)

Mixtures of debris types or contaminant types must be treated to achieve treatment standards applicable to all types represented in the waste. If an immobilization technology is used in a series of treatments (*i.e.*, **treatment train**), it must be applied last. Hazardous debris that is also a waste polychlorinated biphenyl (PCB) must be treated to either the above standards or TSCA standards at 40 CFR 761, whichever are more stringent. Residues from the treatment of hazardous debris may also require treatment (see 40 CFR 268.45[d]).

Alternate Standards for Contaminated Soil. The LDRs have proven to be a substantial economic disincentive to remediation of contaminated sites where the required cleanup involves excavation of contaminated soil that exhibits one or more hazardous waste characteristics or contains a listed hazardous waste. When such soil exits the remediation unit, associated CAMU or temporary unit (TUs) (40 CFR 264, Subpart S), it becomes waste and, if prohibited, must meet LDR treatment standards before it can be replaced, disposed of in a landfill, or otherwise land-disposed. EPA has developed the alternate standards in con-

sideration of the technical difficulties of treating this waste form, and the fact that remediation processes are regulated in a way that controls risks posed by these soils (40 CFR 268.49).

Applicability. Hazardous soils that exhibit one or more characteristic or which did so when it was generated (*e.g.*, excavated and managed outside of an allowed corrective action unit) must meet LDR treatment standards. Hazardous soils that are deemed by EPA to contain a listed waste upon generation, and where that listed waste is prohibited, must also meet LDR standards.

Constituents Subject to Treatment. With minor exceptions, the alternate treatment standards must address all constituents listed in 40 CFR 268.48, "Table UTS," that are reasonably expected to be present in the soil and present in concentrations greater than 10 times the UTS.

Standards for All Soils. For all soils, constituents subject to treatment may be reduced to below UTS levels and satisfy the standard. Alternatively, these constituents must be reduced by at least 90% as measured by total constituent concentration for nonmetals and metals for which a metal removal technology is applied, and TCLP for all other metals; treatment to below 10 times UTS is not required.

Additional Standards for ICR Soils. ICR wastes must also be decharacterized prior to land disposal.

Additional Standards for Soils with Constituents that Are Not Analyzable. Soils containing constituents subject to treatment that cannot be analyzed must also be treated using appropriate specified technologies listed in Table 268.42.

Treatment Residues. Residues from the treatment of hazardous debris may also require treatment (see 40 CFR 268.49[e]).

Alternate Standards for Lab Packs. EPA has provided alternate LDR treatment standard for **lab packs**, a form of packaging often used for small containers of waste chemicals described earlier in this chapter in the discussion of packaging requirements for generators (40 CFR 268.42[c]). The alternative standard is treatment in an incinerator in accordance with 40 CFR 264/265, Subpart O standards as described earlier, and is available under the following conditions:

- The lab packs must be packaged in accordance with 40 CFR 264/265.316 as previously described for disposal (e.g., with absorbant in accordance with DOT standards, segregated incompatibles, deactivated reactives, etc.).

- The lab packs must not contain any wastes listed in Appendix IV of 40 CFR 268, which consists of D009 mercury, chromium pigment K-listed wastes, arsenic containing P-listed wastes, and a few others.

- Incinerator residues from lab packs containing characteristic metal wastes are treated to meet applicable standards in 40 CFR 268.40.

Alternative Standards by Petition. EPA has made provisions for establishing an alternative LDR treatment standard for a waste in the form of treatability variances and equivalency demonstrations (40 CFR 268.42, 268.44). Under treatability variance procedures, generators or treaters can obtain approval of an alternative LDR standard (alternative concentration limit or application of alternative technology) by successfully demonstrating that their waste is fundamentally different than that evaluated by EPA in establishing the existing standard. Any person may seek an alternative standard through an equivalency demonstration, which seeks to convince EPA that the proposed alternative offers protection equivalent to the existing standard.

Other LDR Provisions

The following provisions are also important to full understanding of LDR requirements.

Dilution Prohibition and Storage Prohibition. EPA has imposed the prohibitions on dilution and storage to discourage efforts to circumvent the LDR treatment standards.

Dilution. Dilution is prohibited as a substitute for proper treatment. However, as reflected in the previous discussion, dilution as part of legitimate treatment in CWA regulated systems, except for cyanides, is generally permissible (40 CFR 268.3).

Storage. Storage of LDR restricted wastes is only permissible to accumulate sufficient quantities to facilitate proper recovery, treatment, or disposal (40 CFR 268.50).

Disposal of LDR-Compliant Waste. The following hazardous wastes are rendered nonhazardous by LDR treatment and may be disposed of in a nonhazardous RCRA Subtitle D landfill:

- Waste that is hazardous by characteristic only, after LDR-compliant treatment has removed the characteristic(s)

- Listed hazardous debris after LDR-compliant treatment using extraction or destruction technologies has been applied

All other LDR treated wastes remain hazardous wastes and must be subsequently managed in RCRA Subtitle C (hazardous waste) TSDFs.

Waste Analysis, Recordkeeping, and Notifications. Waste analysis, recordkeeping, and notification requirements for TSDFs are comparable to those previously described for generators (40 CFR 268.7). The following points are pertinent:

- Both generators and TSDFs applying LDR treatment must analyze the wastes in accordance with a written WAP. Testing or process knowledge may be used by generators to determine if a treatment standard is met. Testing must be done in accordance with SW–846 methods. The appropriate sampling methods (composite or grab) are specified in the regulations.

- LDR notification/certification must accompany at least the initial shipment of untreated and treated hazardous wastes to TSDFs. A one-time notification/certification to EPA is also required for treated waste that is shipped to a Subtitle D facility.

- Copies of waste analysis information and LDR notifications must be retained for at least three years.

For detailed requirements, see 40 CFR 268.7.

Hazardous Waste Permits

SQGs and LQGs managing hazardous waste on-site in accordance with 40 CFR 262.34, and CESQGs, are not required to obtain RCRA permits. However, such generators often send their hazardous waste to TSDFs that are subject to RCRA permitting requirements. In addition, generators can be required to obtain a RCRA permit if the nonhazardous wastes they manage become subject to regulation as hazardous waste. It is therefore useful for generators to know the basics of the RCRA hazardous waste permit program. Main provisions of this program, set forth in 40 CFR 270, are summarized in this section. For information on the administrative process by which permit applications, modifications, *etc.* are processed and permits are issued, see 40 CFR 124.

Applicability

Unless specifically exempted, facilities that treat, store, or dispose of hazardous waste must be permitted for their entire active life, including closure (40 CFR 270.1[b]). Land-based hazardous waste facilities (*e.g.*, landfills) that have not been *clean-closed* which received hazardous waste after July 26, 1982 or that were certified closed after January 26, 1983 must also be permitted for the postclosure care period. The detailed requirements for such facilities are set forth in their RCRA permits. However, certain qualifying hazardous waste management facilities that are permitted under other laws (injection wells permitted under SDWA, POTWs with NPDES permits, ocean disposal authorized under the Marine Protection, Research, and Sanctuaries Act) are deemed to have a RCRA *permit by rule* (no RCRA permit application required), and requirements are set forth in those other permits (40 CFR 270.60).

The following persons or activities are specifically exempted from the requirement to obtain a RCRA permit:

- CESQGs, LQGs, and SQGs managing waste onsite in accordance with 40 CFR 262.34.

- Farmers who dispose of hazardous waste pesticides from their own use as specified (40 CFR 262.70).

- Treatment, storage, or disposal of hazardous waste exempted from regulation under 40 CFR 261.4 or 261.5.

- Totally enclosed treatment units and elementary treatment units as defined in 40 CFR 260.10.

- Transporters operating in compliance with RCRA. (Some states permit transporters also.)

- Persons adding absorbant to waste in a container or waste to an absorbant in a container under specified conditions (40 CFR 270.1[c][2][vii]).

- Treatment or containment activities taken in immediate response to actual or imminent and substantial threat of hazardous waste release (40 CFR 270.1[c][3]).

It is the facility operator's duty to obtain the RCRA permit; however, specifically authorized representatives of both the owner and operator must sign and certify the permit applications (40 CFR 270.10[a]).

Interim Status Authorization (Existing Facilities)

Existing TSDFs may become subject to RCRA hazardous waste permit requirements as a result of a regulatory change (*e.g.*, new listing or other change to 40 CFR 261 that changes the status of waste they manage from nonhazardous to hazardous). As explained above in the introduction to 40 CFR 264/265 standards, these TSDFs may continue to operate under interim status until they receive a RCRA permit, provided they do so in compliance with the 40 CFR 265 standards. Existing facilities that cannot qualify for interim status must close. Existing TSDFs wishing to continue operations must take the following permitting actions.

Notification. No later than 90 days after becoming regulated, provide notification of the hazardous waste activity to EPA under RCRA Section 3010, as described above for generators and TSDFs.

RCRA Part A Permit Application. No later than six months after the date of publication of regulations that cause the change to regulated status or 30 days after the effective date of those regulations, whichever first occurs, submit Part A of the RCRA Permit Application to EPA (40 CFR 270.10[d]).

RCRA Part B Permit Application. Submit Part B of the RCRA permit application (described below) to EPA by the due date established by EPA. EPA is required to provide at least six months notice.

Permitting New Facilities

New TSDFs cannot begin construction until a RCRA permit has been obtained. To initiate the permitting process, the owner/operator must submit both Part A and Part B of the RCRA permit application to EPA. The regulations (40 CFR 270.1[b]) indicate that the application must be submitted at least 180 days before construction is expected to begin. However, the actual length of time required to obtain the permit can be much longer, even several years. Actual time required depends on many factors, including available regulator resources, complexity of the facility or activity, quality of the submittal, and public acceptability. Therefore, development of a realistic schedule for the facility requires coordination with the permitting agency far in advance.

Part A Permit Application

The Part A Permit Application consists of Forms 1 and 3 of the Consolidated Permit Application (EPA Form 8700–23). Information to be supplied in the Part A application is relatively simple and straightforward, and includes the items listed below (40 CFR 270.1[b], 270.13).

Facility, Owner/Operator Identification Information. Includes EPA ID Number, name, location, address, point of contact.

General Facility Information. Provides the standard industrial classification (SIC) codes, existing environmental permits, and business description.

Hazardous Waste Management Facility Information. Includes hazardous waste descriptions; description of hazardous waste processes (using standard process codes) and design capacities; photographs and scaled drawing(s) showing the location of all past, present, and planned treatment, storage, and disposal facilities; topographic vicinity map showing the facility and salient environmental features, including potential receptors (*e.g.*, surface waters, wells).

Certification. Both the owner and operator must sign and certify to the completeness and accuracy of the application.

Part B Permit Application

The Part B Permit Application consists of detailed information about the hazardous waste, proposed hazardous waste management facilities and activities, and pertinent natural and cultural features of the facility environs in a manner that demonstrates how the proposed facility and activity will conform to the applicable management standards for the facility (*i.e.*, applicable 40 CFR 264 standards). Part B information required for virtually all types of facilities is detailed in 40 CFR 270.14, 270.15–270.29 and describes information required for specific types of facilities. The information is not provided on any special form. However, Part B applications are often prepared to conform to the following format established by a *Regulatory Completeness Checklist* used by EPA and at least some State regulators to review the applications:

- Section A: "Part A" Permit Application

- Section B: Facility Description

- Section C: Waste Characteristics

- Section D: Process Information

- Section E: Groundwater Monitoring

- Section F: Procedures to Prevent Hazards

- Section G: Contingency Plan

- Section H: Personnel Training,

- Section I: Closure Plan and (for land-based units), Postclosure Plan, Financial Requirements

- Section J: Other Federal Laws

- Section K: Certification

- Section L: Information Requirements for Solid Waste Management Units

- Section M: Closure Equivalency Demonstration

The "Part B" permit application typically consists of one or more lengthy volumes of information, and can be very costly to prepare.

RCRA Permit Conditions

The RCRA permit requirements include generally applicable conditions as well as conditions specific to the permitted facility (40 CFR 270, Subpart C). Generally applicable conditions include the following list (40 CFR 270.30).

Duty to Comply. The permittee is required to comply with all conditions in the permit. This typically includes conformance with conditions in the approved permit application and applicable RCRA regulations.

Duty to Reapply. RCRA permits issued by EPA are effective for up to 10 years; durations of permits issued by states vary. TSDFs that wish to continue the permitted activity beyond the expiration date must reapply at least 180 days prior to permit expiration (40 CFR 270.50, 270.51).

Need to Halt or Reduce Activity. The need to halt or reduce the permitted activity to maintain compliance is disallowed as a defense in an enforcement action.

Minimize Releases and Impact. In the event of noncompliance, permittees are required to minimize releases and take reasonable measures to prevent significant adverse impact to human health or the environment.

Proper Operation and Maintenance. A permittee is required to operate and maintain the facility in a manner that ensures compliance with permit conditions.

Duty to Provide Information. The permittee must provide compliance-related information to EPA upon request within a reasonable time. What is reasonable depends on the circumstances, but it may be useful to establish an acceptable time with your regulator, particularly if records are difficult to access.

Inspection and Entry. The permittee must allow EPA or an authorized representative to enter the facility premises or record storage areas at reasonable times for inspection, sampling, or monitoring. Such access may be requested with little or no notice. Therefore, typically it is advisable for facilities with substantial access requirements for security or other purposes to avoid potential inconvenience and embarrassment (at least) or noncompliance (at worst) by develop-

ing mutually agreeable access protocols with the regulator well in advance.

Conditions specific to the facility may include applicable technical requirements from 40 CFR 264 and 266, compliance schedules (*e.g.*, for corrective action as described under Subparts F and S), ground water protection standard (GWPS) limits for ground water monitoring at land-based units, *etc.*

Changes to RCRA Permits

RCRA permits may be terminated by EPA upon satisfactory closure of a facility or during the term of a permit for cause (40 CFR 270.43). The major provisions for changing a permit are described in (40 CFR 270, Subpart D).

Revocation and Reissuance. EPA may revoked and reissue permits for specifically identified causes, particularly if cause exists to terminate a permit or the permittee has notified EPA that they propose to transfer the permit to another person. When a permit is revoked and reissued, the entire permit is reopened and subject to revision and the permit is reissued for a new term (40 CFR 270.41).

Modification. If EPA chooses, permits may be changed by modification for the same reasons cited above for revocation and reissuance. EPA may also modify a permit for the following reasons: substantial changes in the permitted facility or activity, previously unavailable information important to permit limits becomes available, new statutory requirements or regulations become applicable, compliance schedule modifications are justified, or (for land disposal facility permits only) change is needed to ensure RCRA compliance. When a permit is modified, only the conditions subject to modification are reopened (40 CFR 270.41, 270.42).

Permit Modification at the Request of the Permittee. Permittees seeking permit modifications must do so by formal written request. Three classes of modification are established as follows: Class 1 Modifications, applicable to minor changes that do not substantially alter permit conditions; Class 2 Modifications, applicable to changes necessary to respond to common variations in waste, technological advances, or compliance with new regulations without substantially changing permitted design specifications or management practices; and Class 3 Modifications, applicable to changes that substantially alter the facility or its operation. Most Class 1 modifications may be put into effect without response from EPA; however, some changes must await a written response. The requirements for Class 2 and Class 3 modifications are increasingly more stringent, and include public notice, public comment period, and other substantial administrative provisions (40 CFR 270.42).

Special Forms of Permits

EPA issues special forms of permits to accommodate variations in the type of unit being permitted or the circumstances of permit issuance (40 CFR 270, Subpart F).

Permits by Rule. As previously discussed, these apply to injection wells permitted under SDWA (40 CFR 144, 145), POTWs with an NPDES permit, and ocean disposal authorized under the Marine Protection, Research, and Sanctuaries Act (40 CFR 220), provided they comply with RCRA regulations specified in this section (40 CFR 270.60).

Emergency Permits. A temporary permit vehicle for addressing a condition posing imminent and substantial endangerment of human health or the environment (*e.g.*, neutralization of reactive peroxide crystals formed on ether or picric acid containers, in the likely event they are intended for discard, and thus hazardous waste, upon discovery) (40 CFR 270.61).

Hazardous Waste Incinerator Permits. Incinerators require a separate form of permit because the permit process is phased to separately allow construction (for new facilities), trial burn, and operation (40 CFR 270.62).

Land Treatment Demonstrations Using Field Test or Laboratory Analyses. A separate form of permit may be issued as a treatment or disposal permit for these facilities based on data from field tests or laboratory analyses, or the permit may be issued in two phases (*i.e.*, field testing/laboratory analyses followed by facility construction and operation) (40 CFR 270.63).

Research, Development, and Demonstration Permits. A temporary permit may be issued to allow the development and demonstration of innovative and experimental hazardous waste treatment technology for up to one year, renewable for three additional one-year periods (40 CFR 270.65).

Boilers and Industrial Furnaces. BIF permits require a separate form of permit because the permit process is phased to separately allow construction (for new facilities), trial burn, and operation (40 CFR 270.66).

Bibliography

General Reference

McCoy and Associates. *RCRA Regulations and Keyword Index*. New York, NY: Elsevier Science, Inc. (Annual publication that provides an overview of RCRA programs, text of current regulations, detailed discussions of requirements, and a keyword index).

McCoy and Associates. *RCRA Land Disposal Restrictions: A Guide to Compliance*. New York, NY: Elsevier Science, Inc. (Annual publication that provides an overview of the Land Disposal Restrictions [LDRs], text of current LDR regulations, detailed discussions of LDR requirements, and a glossary and keyword index).

RCRA Statute, Regulations

<http://www.epa.gov/fedrgstr/> (EPA's searchable *Federal Register* compilation from October 1994 – present)

<http://www.epa.gov/docs/epacfr40/chapt-I.info/subch-I/> (EPA Solid and Hazardous Waste Regulations, 40 CFR, Chapter I, Subchapter I [annual codification])

<http://www.access.gpo.gov/su_docs/aces/aaces002.html> (Government Printing Office site that provides access database to statutes)

<http://www.access.gpo.gov/su_docs/fedreg/frcont99.html> (Government Printing Office site

that provides *Federal Register* Table of Contents daily with links to notices)

<http://tis-nt.eh.doe.gov/oepa> (Department of Energy, Office of Environmental Policy and Assistance [OEPA] homepage, includes *Federal Registers* 1995–Present and links to other *Federal Register* compilations; new regulations; OEPA *Weekly Federal Register Digest*; summary of environmental laws, including RCRA. Also contains links to RCRA regulations; guidance documents; downloadable computer software for waste identification, generator category determination; *etc.*)

<http://envirotext.eh.doe.gov/> (Department of Energy, Office of Environment, Safety and Health, Envirotext searchable database containing all of the Federal environmental, safety, and health requirements located under one homepage. Included are full-text copies of the *United States Code*, most of the *Code of Federal Regulations*, the *Federal Register* and Presidential Executive Orders from 1945 to the present)

RCRA Guidance

<http://tis-nt.eh.doe.gov/oepa/guidance/rcra.htm> (Department of Energy, Office of Environmental Policy and Assistance [OEPA]'s substantial compilation of RCRA environmental guidance documents)

<http://www.eh.doe.gov/> (Department of Energy, Office of Environment, Safety, and Health, Technical Information Services [TIS] homepage, a comprehensive collection of environmental, safety, and health [ES&H] information services; provides access to wide array of ES&H information)

<http://www.epa.gov/epaoswer/general/catalog/catalog1.pdf>(EPA Office of Solid Waste, Catalog of Hazardous and Solid Waste Publications. 11th ed. EPA530–B–98–1001, 1998)

<http://www.epa.gov/epaoswer/osw/index.htm> (EPA Office of Solid Waste homepage)

<http://www.epa.gov/epaoswer/osw/infoserv.htm> (EPA Office of Solid Waste, Information Services, provides information on various

sources of OSW information, including publications, links to other sources)

<http://www.epa.gov/epaoswer/osw/publicat.htm> (EPA Office of Solid Waste publications on numerous RCRA topics)

<http://www.epa.gov/rcraonline/> (RCRA Online, an electronic database of selected letters, memoranda, and questions and answers written by the EPA's Office of Solid Waste since 1980)

<http://www.epa.gov/epaoswer/hotline/index.htm> (RCRA, Superfund, and EPCRA Hotline, publicly accessible service with web site that provides numerous links to RCRA sources of information and access by telephone to information specialists who can assist with requests for factual RCRA information and take orders for documents [Hotline Telephone Numbers: (800) 424–9346 or Washington, DC Area Local (703) 412–9810, national toll-free for the hearing impaired (TDD): (800) 553–7672])

Hazardous Waste Treatment, Storage, and Disposal Facilities

<http://www.epa.gov/epaoswer/hazwaste/data/tsd.htm>(EPA Office of Solid Waste RCRIS Database, OSW's list of RCRA Hazardous Waste Treatment, Storage and Disposal Facilities, including Facility ID number, Facility name, Facility address, contact name and telephone number [as reported by the facility] and a flag indicating TSD processes at the facility)

Waste Minimization

Anonymous. "Waste Minimization—What, Why, and How." In *The Hazardous Waste Consultant*. New York, NY: Elsevier Science Inc., September/October 1995.

<http://www.epa.gov/envirosense/>(EPA, Enviro-ene homepage, provides pollution prevention information, including pollution prevention/cleaner production solutions, innovative technology and policy options, and environmental research publications)

<http://www.epa.gov/minimize/> (EPA Office of Solid Waste, Waste Minimization National Plan Homepage)

Solid and Hazardous Waste Determination and Characterization

Environmental Protection Agency. *Test Methods for the Evaluation of Solid Waste—Physical/Chemical Methods*, EPA Publication SW–846. Washington, DC: GPO, 1998.

<http://www.epa.gov/sw-846/sw846.htm> (EPA methods for hazardous waste sampling, analytical methods, statistical analysis [online version of SW–846])

<http://www.epa.gov/sw-846/> (EPA Office of Solid Waste Methods Team Homepage)

Hazardous Waste Compatibility and Related

Hatayama, H. K., *et al. A Method for Determining the Compatibility of Hazardous Wastes*. EPA–600/2–80–076. (National Technical Information Service, PB80–221005). Cincinnati, OH: Environmental Protection Agency, Office of Research and Development, April 1980.

National Institute for Occupational Safety and Health. *Pocket Guide to Chemical Hazards*. PB 97–177–604. Atlanta, GA: NIOSH, 1997.

<http://tis.eh.doe.gov/web/chem_safety/doe_reg.html> (Department of Energy, Office of Environment, Safety, and Health chemical safety documents, requirements, and guidelines)

<http://tis.eh.doe.gov/web/chem_safety/> (Department of Energy, Office of Environment, Safety, and Health Homepage)

<http://www.cdc.gov/niosh/npg/npg.html>(National Institute for Occupational Safety and Health, online version of *Pocket Guide to Chemical Hazards*, 1997 with links to International Chemical Safety Cards)

Gregory C. DeCamp has been a practicing environmental scientist for over 25 years. Mr. DeCamp is a Charter Member of the Magnolia Chapter of ACHMM in South Carolina. He has managed or conducted numerous environmental projects for industry and government, including siting and routing studies, ecological baseline studies and monitoring, impact analyses, facility design and operations, consultation, compliance program development, auditing, as well as facility closure and remediation plan development. Since 1988, Mr. DeCamp's professional activities have been focused on US Department of Energy facilties and operations, with particular emphasis on hazardous and mixed waste management as well as RCRA compliance issues. Mr. DeCamp currently is an Environmental, Safety, and Health manager for Burns and Roe Enterprises, Inc. in Aiken, South Carolina. Mr. DeCamp would like to thank Alan Eckmyre for posing the challenge of writing this chapter, as well as Alison Dean and long-time colleague Lisa Matis for careful technical review and helpful contributions. A special thank-you goes to Karen, Alison, and Andy, for the time and encouragement.

CHAPTER 36

Pollution Prevention

Patricia A. Kandziora, CHMM
K. Leigh Leonard, CHMM

Introduction

With increased regulatory scrutiny following the cascade of regulations in the 1970s and 1980s, and greater public visibility of their environmental impacts, companies are taking a fresh look at their environmental management programs. Many find that undertaking a serious pollution prevention effort is a successful business response because it reduces their regulatory burden, increases public confidence, and nearly always cuts costs.

Unlike most areas of environmental management, pollution prevention is driven more by business interests than by the existing laws and regulations. Pollution prevention projects with powerful and positive impacts on the bottom line have become the hallmark of the progressive environmental manager. These successes demonstrate that environmental managers can add real business value and go beyond forestalling liabilities and costs through compliance and preparedness. At the same time, since pollution prevention is a relatively new frontier as an environmental management strategy, there is a lot of latitude for individual contribution at all organizational levels.

In the 1990s a new paradigm, referred to as *sustainable development*, is emerging that

529

encompasses, and by some measures, surpasses pollution prevention. Although sustainable development is in its infancy, it warrants mention as an important trend that strongly influences the selection of pollution prevention strategies for institutions that have adopted sustainable development as a goal.

This chapter provides a thorough treatment of pollution prevention, including definitions of key terms used by the Environmental Protection Agency (EPA) and other agencies. After reading the chapter, readers will have a working knowledge of the regulatory history of pollution prevention and they will be able to identify the business impacts of pollution prevention projects. They will also understand how pollution prevention fits into the emerging sustainability paradigm.

Defining the Terms

Over time, EPA has drawn distinctions between the terms *waste minimization* and *pollution prevention*. These nuances can be important to the environmental managers who need to communicate their efforts to EPA or its State counterpart. The term *sustainable development* is more youthful and has yet to be strictly defined by EPA. However sustainable development, as it is currently understood, possesses characteristics that set it apart from pollution prevention and waste minimization. The purpose of this section is to explain the nuances in definitions of these terms as a foundation for the rest of the chapter.

Waste minimization, as currently defined by EPA, pertains to wastes regulated under the Resource Conservation and Recovery Act (RCRA), particularly hazardous wastes, and includes *source reduction* and *environmentally sound recycling* (EPA, 1994).

Source reduction is defined in Section 6605(5)(A) of the Pollution Prevention Act. It is any practice which

- Reduces the amount of any hazardous substance, pollutant, or contaminant entering any waste stream or otherwise released into the environment (including fugitive emissions) prior to recycling, treatment, or disposal

- Reduces the hazards to public health and the environment associated with their release

Environmentally sound recycling in the context of the Resource Conservation and Recovery Act means activities defined under the hazardous waste regulations as *recycling* (40 CFR 261.1[c][4], [5], and [7]). This includes materials that are used, reused, or reclaimed. A material is ***used*** or ***reused*** if it is employed as an ingredient in an industrial process to make a product, or if it is employed in a particular function or application as an effective substitute for a commercial product. A material is ***reclaimed*** if it is processed to recover a usable product, or if it is regenerated.

EPA also defines in the solid and hazardous waste regulations the term ***by-product,*** which means a material that is not one of the primary products of a production process and is not solely or separately produced by the production process (40 CFR 262.1[c][3]).

Pollution prevention is defined in the Pollution Prevention Act and subsequent EPA publications (EPA, 1994). Unlike waste minimization, the context of pollution prevention is not limited to solid and hazardous waste. Rather, it encompasses releases to all media: air emissions, wastewater and storm water discharges and spills, and releases to soil or ground water, as well as solid and hazardous waste generation. Pollution prevention means *source reduction* as defined above. In addition, it includes other practices that reduce or eliminate the creation of pollutants through

- Increased efficiency in the use of raw materials, energy, water, or any other resources

- Protection of natural resources by conservation

So, strictly speaking, there are practices (like recycling) that are considered to be waste minimization, but are not pollution prevention. In addition there are pollution prevention measures that are not considered to be waste minimization because they impact on media other than solid and hazardous waste. Figure 1 provides a graphical representation of the relationship between pollution prevention and waste minimization.

Sustainable development has been defined by the United Nations World Commission on Environment and Development as meeting the

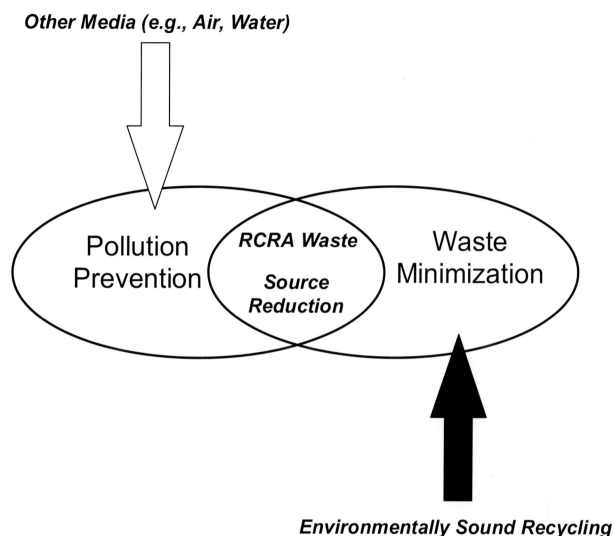

**Figure 1. Relationship Between Pollution
Prevention and Waste Minimization**

needs of the world's current population without making it impossible for the world's future citizens to meet their needs. In addition to consideration of residuals (air emissions, waste water, solid and hazardous waste) sustainable development focuses on energy and material inputs and outputs associated with a process and the life-cycle environmental costs of the product or service produced.

Obviously, the realm of pollution prevention and sustainability overlap. Pollution prevention can certainly be a driving element of a sustainability initiative. However, by comparison, the scope of pollution prevention is narrower, as it concerns

itself with the waste (and other process residuals, such as wastewater and air emissions) without asking about the existence of the process or product itself. The scope of change contemplated from a sustainable development perspective goes to the heart of the strategic goals and mission of the business or organization, which pollution prevention may or may not do. Also, the timeframe for pollution prevention is shorter. (Generally, pollution prevention project proposals with payback periods exceeding two years will be passed over.) Sustainable development forces organizations to look at how their present practices will play out twenty or fifty years from now, and may include moral considerations that go beyond

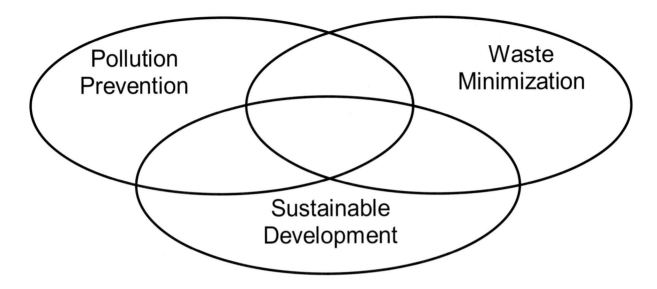

**Figure 2. Relationship Between Pollution Prevention,
Waste Minimization, and Sustainability**

process analysis. Figure 2 graphically relates pollution prevention, waste minimization, and sustainability.

This chapter is primarily about pollution prevention. As such, the authors will generally use the term *pollution prevention* or the abbreviation *P2* for the remainder of the chapter, unless referring to examples that accentuate waste minimization or sustainable development.

Waste Minimization and Pollution Prevention at the Federal Level

As chronicled in this book, environmental laws initially focused on *end of pipe* pollution control, cleanup of contaminated sites, and putting in place national and local spill response plans. Beginning with the Hazardous and Solid Waste Amendments in 1984, a new trend started to emerge. EPA and state regulatory agencies started to shift their focus toward reducing the amount of waste generated and toxics emitted.

Regulations and policies concerning waste minimization and pollution prevention have evolved dramatically since the mid-1980s. EPA's

initial approach was to incorporate pollution prevention into the existing hazardous waste regulatory requirements. In contrast, the present approach is multimedia and more policy-driven than regulatory in nature. Policies emphasize technical assistance, grants, partnerships and voluntary efforts. Regulations continue to be fine-tuned, however, in ways that support EPA's agency-wide emphasis on P2 and the concept has gradually become more integrated with EPA's mainstream media programs. P2 is negotiated into permits and settlements, and incorporated into annual environmental reports required of industry. However, EPA enforcement policies do not yet provide a means for inspectors to credit a company's P2 achievements during the compliance inspection process.

Table 1 shows the major milestones at the Federal level that have stimulated pollution prevention activities. Each of these milestones and its effects on pollution prevention are discussed in detail below.

Resource Conservation and Recovery Act

Waste minimization was incorporated into the Resource Conservation and Recovery Act as part of the 1984 Hazardous and Solid Waste Amendments (HSWA). This resulted in three minor regulatory provisions that stimulated the waste minimization efforts of hazardous waste generators and licensed treatment, storage and disposal (TSD) facilities.

§3002(a)(6). Codified as 40 CFR 262.41(a) and 40 CFR 262.87(a), requires generators to describe, as part of their biennial report, their efforts to reduce the volume or toxicity of hazardous waste they generate, and to describe the changes actually achieved during the present reporting year as compared with previous reporting years.

§3002(b). Codified as 40 CFR 262, Appendix to Part 262, "Uniform Hazardous Waste Manifest and Instructions" (EPA Forms 8700-22 and 8700-22A and their instructions), requires generators (except Conditionally Exempt Small Quantity generators) to certify on each hazardous waste manifest that they have a program in place to reduce the volume or quantity and toxicity of their waste to the degree determined by the generator to be economically practicable.

§3005(h). Codified as 40 CFR 264.74(b)(9), requires the owner/operator of any Treatment,

Storage, and Disposal facility to sign the same certification at least annually, and to file the certification in the facility operating record.

To provide more substance to these reporting and certification requirements, in May 1993, EPA issued its *Guidance to Hazardous Waste Generators on the Elements of a Waste Minimization Program* (58 FR 31114). It lays out a model framework for waste minimization programs including these elements:

- Top management support

- Characterization of waste generation and waste management costs

- Periodic waste minimization assessments

- Appropriate cost allocation

- Encouragement of technology transfer

- Program implementation and evaluation

The purpose of the guidance is to describe EPA's idea of what it means to have a hazardous waste minimization program "in place," as stated by the certification. EPA published the guidance as Interim Final guidance and, to date, has not finalized it.

Table 1. Milestones in Pollution Prevention

Year	Title of Act	Effect on P2
1984	Resource Conservation and Recovery Act	• Requires generator and Treatment, Storage, and Disposal (TSD) facilities to certify their waste minimization program • Requires biennial reporting on waste minimization
1986	Emergency Planning and Community Right-to-Know Act (Parts 311 – 313 of the Superfund Amendments and Reauthorization Act)	• Requires public disclosure of chemical inventories • Requires public disclosure of chemical releases
1990	Pollution Prevention Act	• Set national policy for P2 • Expanded Toxic Release Inventory (TRI) report to include P2 progress and goals • Made grant money available for P2 initiatives

In November 1994, EPA issued *The Waste Minimization National Plan* (EPA, 1994) which established three goals:

1) To reduce, as a nation, the presence of the most persistent, bioaccumulative, and toxic constituents by 25 percent by the year 2000 and by 50 percent by the year 2005

2) To avoid transferring these constituents across environmental media

3) To ensure that these constituents are reduced at their source whenever possible, or, when not possible, that they are recycled in an environmentally sound manner

While waste minimization remains the purview of RCRA, initiatives described in the Waste Minimization National Plan reflect increasing integration between EPA's pollution prevention and waste minimization programs.

Emergency Planning and Community Right-to-Know Act

In keeping with a general movement toward public disclosure, Congress enacted, in 1986, the Emergency Planning and Community Right-to-Know Act (EPCRA), Sections 311–313 of the Superfund Amendments and Reauthorization Act. This program requires certain companies to disclose, through mandatory reporting, the amounts and kinds of toxic materials they store and use (called the Tier II report) and the amount and kinds of toxic chemicals they release into the environment (called the Toxic Release Inventory or TRI report). EPA and its State counterparts have facilitated public access to these annual reports by compiling the data and making summaries available on the World Wide Web and by request. Meeting compliance with EPCRA requirements served as a wake-up call to many plant managers as they became aware of the toxic chemicals their companies use and process, and grew more sensitive to public concerns about their plant's environmental impacts. Even prior to the Pollution Prevention Act, these reporting requirements provided incentive for facilities to make reductions in the storage, use, and release of chemicals regulated under EPCRA.

Pollution Prevention Act

In 1990, Congress passed the Pollution Prevention Act. It set forth national policy regarding the preferred hierarchy of waste management options:

> The Congress hereby declares it to be the national policy of the United States that pollution should be prevented or reduced at the source whenever feasible; pollution that cannot be prevented should be recycled in an environmentally safe manner, whenever feasible; pollution that cannot be prevented or recycled should be treated in an environmentally safe manner whenever feasible; and disposal or other release into the environment should be employed only as a last resort and should be conducted in an environmentally safe manner. (PL 101–58–Nov 5, 1990, Omnibus Budget Reconciliation Act of 1990 [Pollution Prevention Act of 1990], §6602[b].)

Figure 3 provides a graphical representation of the P2 hierarchy set forth in the Pollution Prevention Act.

Four of the policy provisions of the Pollution Prevention Act are as follows: (1) it required EPA to establish an office to oversee and coordinate the agency's pollution prevention activities; (2) it provided for grants to states for technical assistance programs on source reduction; (3) it required EPA to institute a Source Reduction Clearinghouse to facilitate information and technology transfer; and (4) it required EPA to report to Congress on its progress on these initiatives biennially.

The only significant regulatory authority enacted by the Pollution Prevention Act was an expansion of requirements for Toxic Release Inventory (TRI) reports required under EPCRA. Under the expanded provisions, for each toxic chemical covered by TRI requirements, facility owner/operators are required to file a source reduction and recycling report. The report must include, for the current reporting year,

• The quantity of the chemical entering any wastestream (or otherwise released into the environment), the percentage change from the

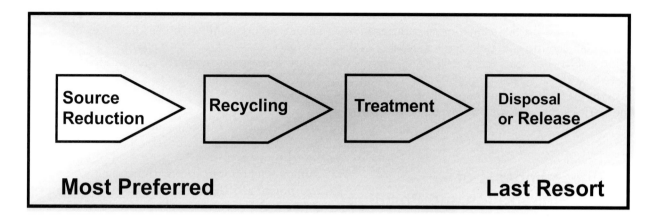

Figure 3. Pollution Prevention Hierarchy as Expressed by Congress in the
Pollution Prevention Act of 1990

previous year, and the amount the owner/operator expects to report for the next two calendar years

- The amount of the chemical which is recycled, the percentage change from the previous year, and the amount the owner/operator expects to report for the next two calendar years

- Source reduction practices used with respect to that chemical during such year at the facility, and techniques the owner/operator used to identify source reduction opportunities

- A ratio of production in the reporting year to production in the previous year

- The amount of the chemical released into the environment due to a catastrophic event, remedial action, or other one-time event not associated with production

- The amount of the chemical treated and the percentage change from the previous year

The Pollution Prevention Act also resulted in specific EPA programs and initiatives. The Office of Prevention, Pesticides, and Toxic Substances was created to oversee EPA's fulfillment of its responsibilities under the Pollution Prevention Act. For example, it is responsible for EPA's Source Reduction Review Project (SRRP), initiated in 1992, which is charged with assessing proposed regulations for Pollution Prevention opportunities. Its results include pollution prevention considerations being incorporated into two new

pesticide regulations (EPA, 1997). EPA has also injected pollution prevention into its compliance activities, most directly through its Supplemental Environmental Projects (SEPs). These incorporate pollution prevention projects into enforcement settlements in exchange for reductions in fines or penalties for alleged violations.

In addition to coordinating the pollution prevention efforts of EPA's various branches, the Office of Prevention, Pesticides and Toxic Substances is directly responsible for EPA's Design for Environment (DfE) partnerships and its Pesticide Environmental Stewardship Program. The DfE program provides technical assistance to businesses to help them incorporate environmental considerations into product design, process change, and technical and management systems. The Pesticide Environmental Stewardship program is a cooperative effort involving EPA, the US Department of Agriculture, and the Food and Drug Administration. The program strives for reduced pesticide use and risk. Its goals include the adoption of environmentally safer integrated pest management methods for more than half of agricultural acreage in the United States by the year 2000 (EPA, 1997).

In the regulatory realm, EPA has been expanding the scope of the Toxic Release Inventory (TRI). In recent years, EPA has increased the number of industry sectors and toxic chemicals subject to the requirements. Also, in 1994, EPA nearly doubled the number of chemicals covered.

A related EPA program is the 33/50 program, which challenged 1,300 companies, operating more than 6,000 facilities, to reduce releases of 17 high-priority TRI chemicals (EPA, 1997). The goal was to reduce these releases by 33% by 1992 and by 50% by 1995. Due to cooperation of affected industries, this goal was achieved a year early.

Business Impacts of Waste Minimization and Pollution Prevention

Historically, waste minimization and pollution prevention projects have been detached from their corporation's or organization's larger purpose. P2 and waste minimization projects are often stand-alone efforts like paper and can recycling or a shift from a more to a less toxic raw material. The context of P2 and waste minimization has most often been that of compliance (enforcement mandate), liability (reduce exposure), or immediate cost reduction. Indeed, the widely cited American Chemical Society's publication, *Less is Better*, prepared in 1985 by EPA's Task Force on RCRA heralds the value of waste minimization as a source of multiple benefits. The premise is that by keeping a small, practical chemical inventory in the laboratory, direct procurement-disposal savings can be as much as 50 percent, as shown by Figure 4.

The laboratory scale scenario demonstrates that, in this case, any waste generation outstrips the superficial savings of low dollar ordering. This is often true in an industrial setting as well.

Volume purchases usually require a primary cash outlay that may necessitate a loan at interest—which could prohibit another investment. Other secondary benefits of a utilitarian inventory include: reduced inventory management overhead (Occupational Safety and Health Act [OSH Act] requirements for grounding systems and fire cabinets, equipping staff with personal protective equipment [PPE]); reduced potential for injury, human suffering, and lost production time; potential decrease in worker's compensation premium modifier; simpler in-house and community emergency response preparation.

A good community image may result from promoting these initiatives. In his presentation to the Chemical Heritage Foundation in September 1998, Eastman Chemical's Chief Executive Officer, Earnest Deavenport, Jr. observes that the United State's chemical industry is doing well with a 50% reduction in releases of toxic chemicals to air, land, and water since 1988; and a 50% reduction in similar materials transferred off-site for disposal or treatment over the same period (Storcks, 1998).

However, he points out, perception of the chemical industry remains a problem; that chemical companies must address societal issues important to communities, extend stewardship beyond merely counseling customers, incorporate a proactive philosophy that ensures that the industry is making health, safety, and environmental issues a priority. Only then will the industry be able "to change public opinion and

Disposal costs per package are less with smaller units (than the larger "economical size")—usually only 15-20% of the larger.		
Type of Cost	**Smaller Unit (500 ml)**	**Larger Unit (2500 ml)**
Purchase Price	$15.50	$52.00
Unit Purchase Price	$.03/ml	$.02/ml
When 1000 ml of product are used, the cost will be . . .		
Unit Purchase Cost	$.03/ml	$.052/ml
Package Disposal Cost	$0	$22.67
Total Cost	$31.00	$74.67
(Purchase and disposal)	(Purchase only)	
Unit Total Cost	$.031/ml	$.075/ml

ACS, 1985.

Figure 4. Comparison of Disposal Cost per Unit Size

enable the chemical industry to 'survive the witch trials' [of modern-time misplaced phobias and misunderstood actions of the chemical industry]." As people become more conscientious of the ethics and records of those with whom they do business, it is imperative for companies to examine ways to operate in a sustainable manner; one way to do that is through a commitment to P2.

Top management support is essential for P2 to thrive. To support a P2 program, top management must first understand how it fits the corporate mission. Customarily, capital investment in environmental programs has a limited appeal. Often viewed as a threshold expenditure, environmental compliance is simply one of the costs of doing business and nothing more. Unfortunately, this perception carries through for P2 when P2 is not marketed well internally; but it need not be true. Who has the responsibility to bring P2 opportunities into the boardroom; to present them as a way of doing business—from design to market? Traditionally the environmental manager's role has been to integrate environmental compliance with every affected aspect of the organization and to keep top management informed.

P2 is most effective when initiated at the line level. The effective environmental manager listens, identifies opportunities, provides support, and stays out of the way! The effective environmental manager counts the success of the P2 project, then communicates it to upper management in a way that encourages continued support.

One example of such a success story involves a Wisconsin company that explored the feasibility of developing a carpet recycling operation can be found at the following website:

<http://www.uwsa.edu/capbud/abstract/ gonring.htm>

The company learned that each year 67 million yards of carpet are sold in Wisconsin; 45 million yards are landfilled, annually. Disposal liability through potential Superfund actions, landfill tipping fees per ton of waste disposed, and useless consumption of valuable landfill space were immediate motivators to examine alternatives to throwing away the used carpet. The economics of material pickup, sorting, cutting, conduxing (mechanical compressing to reconstruct fibers), cleaning and labor, concluded that it is indeed

feasible to operate a used carpet take-back operation. The company is developing a 10,000-square-foot facility for recycling. Now, this innovative company can house a recycling operation near its manufacture site and has a ready raw material source. The company has the opportunity to advertise their ability to pass along any associated savings to their consumers—their market edge!

Environmental laws may change, but they are here to stay. Why not creatively leverage off of compliance—generate less waste, produce better yield, avoid environmental liability and safety hazards, and pass the savings on to the customer? There are untapped market opportunities within business-as-usual compliance. The challenge to the environmental manager is to find, mine, and capitalize on them.

Anyone with doubt as to the business impacts of pollution prevention need only turn to any number of waste management companies' audited corporate annual reports. These reports have shown diminishing revenue for hazardous waste landfills in the 1990s, and an overcapacity for incinerators. Their employees have credited waste minimization by generators as one of the reasons for these conditions. The waste source is drying up. Also, with the land ban regulations in place, specialty treatment business opportunities have evolved (*e.g.*, mercury retort, lead waste smelting, secondary fuels operations for cement kilns), further curtailing the amount of wastes available for traditional disposal/destruction technologies.

One of these "specialty" areas, as mentioned, is metal recovery. No one knows it, but Y2K, the "Millenium Bug," is currently at our doorstep. At 11:59 p.m. December 31, 1999, computers will roll from '99 to the year '00, rather than 2000. Some computers won't "know" what century they are in. Imbedded microchips will halt time-dependent functions in computers and equipment everywhere, and incalculable chaos will ensue—or so some are saying. *The Year 2000: Social Chaos or Social Transformation?* (Petersen, *et al.*, 1998) describes the millennium bug as: "the first-time nonnegotiable deadline" (to which modern functionality is inextricably attached) and "the greatest opportunity to simplify and redesign major systems." If Y2K precipitates a colossal migration of Central Processing Units (CPUs) from our desktops to our dumpsters, the winners may well be those small, labor-intensive recyclers who are in business today. These are the companies that are gleaning minute concentrations of precious metals from discarded circuit boards and pulling apart monitors to smelt lead from cathode ray tubes. Because, in some states, CPUs are deemed hazardous waste, recycling is the preferred option for them. Recycling companies with an eye on their own sustainability and competitiveness are reconsidering the way they will do business in the wake of Y2K; and innovators are capitalizing on the opportunity to intercept an ample waste stream. The computer manufacturing industry is turning to modular assembly where only outmoded components need replacement, reducing need for disposal or reclamation. The manufacturers then advertise module replacement as a convenience to customers, but it is also a P2

initiative of great magnitude, nationwide. It is another way that companies can enhance their competitive edge and sustain themselves in a unique niche.

Feasibility requires that each P2 initiative be analyzed for a return on investment, as a key early step. The P2 model outlined for return on investment and profiled in Table 2 suggests parameters that cover all P2 project needs. The payback can be both explicit and indirect. The parameters create a balance sheet for the P2 investment opportunity and a more tangible depiction of the initiative.

Once these figures are weighed comprehensively, the P2 initiative can be presented to management

Table 2. Balance Sheet for P2 Investment

Investment	Debit $_?_	Credit $_?_
Process equipment (capital improvement, deconstruction)		
Construction material (utility infrastructure, HVAC)		
Research and development (to integrate the project with existing processes)		
O&M (labor, analyticals)		
Site preparation		
Input material: changes in source material		
Start-up		
Training: relative to the new P2 process		
Engineering		
Utility connections		
Process equipment		
Capital: new construction to accommodate the P2 process		
By-product recovery		
Inventory management costs: emergency response		
PPE		
Medical (surveillance)		
Long-term liability		
Environmental risk program overhead: auditing, insurance		
Waste management, packaging, transportation, treatment/disposal long-term liability for cradle-to-grave management of the waste, personnel exposure monitoring, emergency response costs		
Totals		
Net Cost		

as a business priority just like marketing, design, and manufacturing initiatives.

In summary, P2 can be an integral part of the corporation's focus. To approach pollution prevention and waste minimization in this manner infuses environmental management deeper into the mission of the corporation by honing market opportunities. Pursuing P2 opportunities with this type of motivation positions the environmental function as a top management consideration, and reconstitutes the environmental manager's role in the context of the corporate mission.

Bibliography

American Chemical Society. *Less is Better.* Washington, DC: ACS, 1985.

Environmental Protection Agency. *Pollution Prevention 1991: Progress on Reducing Industrial Pollutants.* NTIS PB93–157725 (formerly EPA 21p–3003). Springfield, VA: National Technical Information Service, 1991.

Environmental Protection Agency. *Pollution Prevention 1997. A National Progress Report.* EPA/742/S–97/00. Washington DC: EPA, 1997.

Environmental Protection Agency. *Preventing Pollution through Regulations. The Source Reduction Review Project: An Assessment.* EPA–742–R–96–001. Washington, DC: EPA, 1996.

Environmental Protection Agency. *The Waste Minimization National Plan.* EPA 530–D–94–001. Washington DC: EPA, 1994.

Habicht, Henry F., II. "Memorandum of May 28, 1992". EPA/742/F–92/001. Washington, DC: EPA, 1992.

"Hazardous Waste Identification and Listing; Carbamate Production. Final Rule. (40 CFR 261, 271, and 302)." *Federal Register* 60 (9 February 1995): 7824.

"Identification and Listing of Hazardous Waste." *Code of Federal Regulations.* Title 40, Pt. 261.1, 1980.

"Pesticide Chemicals Category; Formulating, Packaging and Repackaging Effluent Limitations Guidelines (40 CFR 455)." *Federal Register* 61 (6 November 1996): 57518.

Petersen, John. L., Margaret Wheatley, and Myron Kellner-Rogers. "The Year 2000: Social Chaos or Social Transformation?" *The Futurist.* October 1998.

"Standards Applicable to Generators of Hazardous Waste." *Code of Federal Regulations.* Title 40, Pt. 262.1, 1980.

William Storcks. "Deavenport: Industry Will Survive 'Witch Trials.' " *Chemical and Engineering News.* Vol. (76), No. (26), 5 Oct 1998.

Patricia A. Kandziora is the charter Manager of the University of Wisconsin System's Environmental/Occupational Health and Safety Department. The department has supported the academic and research mission of the University's 28 campuses and administrative locations since 1987, providing services in environmental liability, Industrial Hygiene, and occupational safety. Ms. Kandziora began her environmental career in 1976 at the bench of a large municipal laboratory, and spent seven years in the field with the state's hazardous waste inspection agency. Ms. Kandziora specializes in chemical safety, liability, and management systems. She has been the select chapter editor for the National Safety Council's Fundamentals of Industrial Hygiene, *1994; co-author of the state's TSD licensing and auditing protocols; and guest columnist on third-party liability in local and trade publications.*

K. Leigh Leonard is the Associate Environmental/Occupational Health and Safety Manager for the University of Wisconsin System Administration. She coordinates pollution prevention pilot projects and provides policy development as well as technical oversight for the University's hazardous waste management program. Other professional specialties include management of high-hazard waste, chemical safety, safety in the arts, and biosafety. Ms. Leonard has worked in environmental health and safety for over 12 years. She also co-edited the book Pollution Prevention and Waste Minimization in Laboratories.

The authors would like to acknowledge Timothy Anderson, Rayovac Corporate Manager in Madison, Wisconsin, for his review of and contributions to this chapter.

CHAPTER 37

RCRA Corrective Action

Louis Martino, CHMM
David Green, MS

Introduction

The Resource Conservation and Recovery Act (RCRA), as amended, directs the United States Environmental Protection Agency (EPA) and those States authorized to implement RCRA to require corrective action to address releases of hazardous wastes or hazardous waste constituents from solid waste management units (SWMUs). A *solid waste management unit* is any discernible unit in which solid wastes have been placed at any time, regardless of whether or not the unit was intended for the management of solid or hazardous wastes. Such units include any area at a facility where solid wastes have been routinely and systematically released.

RCRA Section 3004(u) requires treatment, storage, and disposal facility (TSDF) permits issued under the provisions of RCRA Section 3005(c) after November 8, 1984 to include the requirements for corrective action. RCRA Section 3004(u) also requires the permits to contain schedules of compliance when corrective action activities cannot be completed before the issuance of a final operating permit to a facility. Further, RCRA Section 3004(v) authorizes EPA to require owners and operators of permitted or interim status facilities to remediate releases that have migrated beyond the boundary of the facility. RCRA Section 3008(h) gives EPA the authority to issue administrative orders requiring corrective action to facilities operating under the interim status provisions of RCRA Section 3010.

Regulatory Provisions

RCRA requirements that address releases or potential releases of hazardous wastes or hazardous waste constituents have existed since RCRA's inception in 1976. These stipulations are found in the requirements for corrective action that address releases to ground water from regulated units. They also exist in the RCRA closure and post-closure care requirements. The RCRA Corrective Action program, as it exists today, came into being with the Hazardous and Solid Waste Amendments of 1984. On July 15, 1985, EPA issued regulations codifying the statutory provisions of RCRA Section 3004 (u) and (v) (50 FR 28702). EPA promulgated a rule to regulate the units that can be used to manage hazardous waste either temporarily (Temporary Unit [TU]), or in permanent disposal sites (Corrective Action Management Unit [CAMU]) (55 FR 8658).

EPA also promulgated a rule for RCRA hazardous waste generated, stored or disposed of during cleanup activities (63 FR 65873). These regulations, found in 40 CFR 260, 261, 264, 265, 268, 270 and 271, the almost 10-year old Proposed Subpart S Rule (55 FR 30798), and the now 3-year old Advance Notice of Proposed Rulemaking (ANPR) (61 FR 19432), are the basis of the present EPA corrective action program under RCRA. Although the regulations contemplated in the Proposed Subpart S Rule and the ANPR were not promulgated by EPA, the basic corrective action framework described has served as guidance and, in some cases, has become the basis of many State corrective action programs.

Corrective action requirements are typically triggered by: (1) applying for an RCRA permit, (2) discovery of a release of hazardous waste or hazardous waste constituents from a SWMU at a permitted or interim status TSDF, (3) discovery of additional SWMUs or releases at a facility already conducting corrective action, or (4) discovery of an imminent and substantial endangerment of human health and the environment at any type of RCRA facility (*i.e.*, generator, transporter, or TSDF). In any case, when a release of hazardous waste or hazardous waste constituents is discovered, corrective action typically is required, either through modification of the facility's permit or through a RCRA administrative order issued under RCRA Sections 3008(h), 3013 or 7003.

The Subpart S proposed rule describes a 4-phase approach to investigating and responding to releases of hazardous wastes and hazardous waste constituents at RCRA facilities. These phases are

1) RCRA Facility Assessment (RFA)

2) RCRA Facility Investigation (RFI)

3) Corrective Measures Study (CMS)

4) Corrective Measures Implementation (CMI)

Interim measures may be required to address immediate threats while a long-term comprehensive corrective action strategy is being developed.

The RCRA Facility Assessment

The RFA is the first phase of the conventional RCRA Corrective Action process. Usually EPA or the State regulatory agency will conduct the RFA; however, in some rare cases, the regulatory agency may direct the facility to conduct the RFA. The RFA consists of a review of existing information about a facility, a visit to the facility, and if warranted, limited sampling of the environmental media. The intent of the RFA is to identify SWMUs and to determine if there is an actual or potential release of hazardous wastes or hazardous waste constituents from these SWMUs (USEPA, August 1986). RFAs have been completed at nearly all TSDFs.

Interim Measures

If there is an actual or potential release of hazardous waste or hazardous waste constituents, EPA or the authorized state may require the facility to conduct interim measures. Interim measures may be conducted during any phase of the Corrective Action process, but are most often required following the RFA or the initial phases of the RFI. ***Interim measures*** are actions used to mitigate any immediate threats while a comprehensive corrective action strategy is developed and implemented. Interim measures range from simple actions, such as fencing an area to prevent access, to complex ground water pump-and-treat operations to prevent further contaminant migration.

The RCRA Facility Investigation

The RFI is the second phase of the conventional RCRA Corrective Action process. The RFI is a detailed investigation, conducted by the facility, to determine the nature, extent, direction, and migration rate of the release. RFIs are investigations that are focused on characterizing a release from a specific SWMU; however, RFIs for large areas or multiple SWMUs are not uncommon.

The Corrective Measures Study

The CMS is the third phase of the conventional RCRA Corrective Action process. If the RFI identifies a release that may pose a threat to human health or the environment, the facility will be required to conduct a CMS. A CMS involves the identification and evaluation of alternatives for remediating the release, but may also include an evaluation confirming the need for corrective measures. Included in a CMS are a preliminary and detailed screening of alternatives (including a *no action* alternative), bench and pilot scale testing of one or more alternatives (if warranted), and the tentative selection of the alternative to be implemented at the SWMU.

Corrective Measures Implementation

Implementation of the corrective measure selected to address the release is the final phase of the conventional RCRA Corrective Action process. This phase includes all aspects of the design, construction, operation, monitoring, and demonstration of completion of the corrective measure. After the facility has successfully demonstrated its completion of the corrective measure, the facility's permit or the RCRA administrative order will be modified by EPA or the authorized state, bringing to an end the facility's obligation to conduct corrective action.

Recent Developments

The term *recent* is a relative term in the corrective action program, given that the Proposed Subpart S Rule is almost a decade old. The corrective action program as first proposed by EPA in 1990 and as adopted by many authorized states generally follows each of these four phases. However, EPA and the states have embarked on approaches to accomplish corrective action goals that conserve the resources of both the regulator and the facility. These approaches include the implementation of EPA's 1991 Stabilization Initiative. The ***Stabilization Initiative*** is a management philosophy in which the overall goal is to control or abate threats, or to prevent or minimize the further migration of contaminants while long-term remedies are pursued. Another approach includes a new emphasis on the early identification of technical impracticability and presumptive remedies. Even casual observers can see that the structure of the RCRA corrective action program has evolved considerably since the 1984 amendments and the July 1990 proposed Subpart S rule, and that it continues to evolve today.

Although the Subpart S Rule was proposed more than 9 years ago, rulemaking for the corrective action rule remains incomplete. The Subpart S Proposed Rule, guidance documents, agency directives, Proposed Subpart S Rule–influenced state corrective action programs, as well as site-specific order and permit requirements have been used to implement the corrective action program. EPA published the Advance Notice of Proposed Rulemaking (ANPR) (61 FR 19432) in 1996. The ANPR has several purposes: It introduces EPA's strategy for improvement of the program, requests information to improve the program, provides a status report on the program, and highlights areas of flexibility within the program. A brief summary of the key points articulated in the ANPR has been provided, because EPA has at least some expectation that the ANPR will be used as guidance for the Corrective Action program.

Highlights of the Corrective Action Advance Notice of Proposed Rulemaking

The following section is a summary of the major changes described in the ANPR.

Use of State Cleanup Programs

Independent State cleanup programs are usually modeled after the Federal Superfund under the Comprehensive Environmental Response, Compensation and Liability Act (CERCLA) program. Currently over half of the states have Superfund-like cleanup authorities. In many cases, EPA believes these authorities are substantively equivalent in scope and effect to the Federal CERCLA program and to the RCRA Corrective Action program. EPA is, therefore, considering the use of State Superfund–like cleanup programs to compel or conduct cleanup at facilities subject to RCRA Corrective Action.

Consideration of Land Use

EPA has been criticized for frequently assuming that the future use of RCRA sites will be residential. Residential use is typically the least restrictive land use, and as such carries with it the greatest potential for exposures. It consequently demands the most conservative exposure assessments. EPA is considering changes in the regulatory program that would help promote consideration of future land use in the corrective action context). In the document, *EPA Guidance on Land Use in the CERCLA Remedy Selection Process* (OSWER Directive No. 9355.7–04, May 25, 1995), EPA stated that

> Discussions with local use authorities and other locally affected parties to make assumptions about future land use are also appropriate in the RCRA context. EPA recognizes that RCRA facilities typically are industrial properties that are actively managed, rather than the abandoned sites that are often addressed under CERCLA. Therefore, consideration of nonresidential uses is especially likely to be appropriate for RCRA facility cleanups.

Point of Compliance

EPA requested comments on the process for establishing the ***point of compliance*** (POC) for corrective action. POCs are the location or locations at which media cleanup standards are achieved, and are generally set on a site and media-specific basis. For air releases, the POC is the location of the person most exposed, or other specified point(s) of exposure close to the source of the release. For surface water, the POC is at the point at which releases could enter the surface water body (for sediments affected by releases to surface water, a sediment POC is also established). For soils, the POC is generally set to ensure protection of human and environmental receptors against direct exposure and to take into account protection of other media from cross-media transfer (*e.g.*, *via* leaching, run-off, or airborne emissions) of contaminants. For groundwater, the POC is the area of contaminated groundwater or, when waste is left in place, at and beyond the boundary of the waste management area encompassing the original source(s) of groundwater contamination. This last approach is often referred to as the *throughout the plume/unit boundary* POC.

Self-Implementing Corrective Action

Due to limited resources, many of the lower-risk facilities which may require some form of corrective action have been passed over by the corrective action program. This has raised concerns about the pace of corrective action cleanups. In order to address these concerns, EPA is considering shifting more of the responsibility for conducting corrective action to the regulated community and is examining a number of approaches to self-implementation. The EPA's intent is to reduce the inefficiencies created by the current command-and-control review and approval process, to increase the number of facilities conducting corrective actions, and to speed up the cleanups.

Standardized Lists of Action Levels and Media Cleanup Standards

While EPA provided a table of action levels for constituents in the 1990 Subpart S proposed rule, it does not maintain a list of nationally accepted action levels or media cleanup standards. Some individuals have opined that clearly defining these numbers will promote reduced transaction costs, national consistency, and voluntary actions. The EPA is concerned lest the use of standardized lists of action levels or media cleanup standards oversimplify risk management decisions and reduce the ability to address site-specific circumstances. For example, EPA indicates that standardized lists of constituent concentrations would, of necessity, be developed using standard land use and exposure assumptions. These assumptions would not be appropriate to all facilities, potentially allowing the media cleanup or action levels to be misapplied. EPA is nevertheless considering whether it should develop and maintain lists of action levels, media cleanup standards, or both.

Performance Standards

EPA is considering establishing performance standards for the corrective action program in lieu of a detailed review and approval of work plans and other documents. EPA believes that focusing the corrective action program on clear, measurable performance standards rather than a prescriptive corrective action process could significantly increase the pace and quality of corrective action cleanups.

Voluntary Corrective Action

The 1990 Subpart S proposal discussed voluntary corrective actions that could be conducted by site owner/operators who were not seeking RCRA permits, and voluntary actions that could be initiated by facility owner/operators who had not yet received permits. Site and facility owner/operators, on the other hand, are concerned that any action they may take voluntarily would be second-guessed by the regulator. EPA is considering ways of encouraging voluntary corrective action and of providing incentives to facility owner/operators who are willing to voluntarily initiate cleanup.

Consistency with the CERCLA Program

Federal facilities have long complained of the duplication of effort required by the various cleanup programs. EPA believes the coordination of cleanup activities at facilities with overlapping jurisdictions is good, and is considering ways to improve coordination. In addition, the EPA is weighing whether the use of the same terms for remedial activities, such as investigations or remedy selection, would improve coordination at sites with overlapping jurisdiction.

Less Focus on Solid Waste Management Units

EPA is concerned that the proposed definition of SWMU deters facilities from addressing contamination on a site-wide basis. In addition, EPA indicates that more than 60 corrective action permits have been appealed on the basis of the improper application of the definition of a SWMU (61 FR 19432). Numerous other permits and orders involve prolonged discussions of the SWMU definition and its site-specific application. The primary areas of discussion and appeals have been in response to whether releases were routine and systematic and whether units had any solid or hazardous waste management activities.

EPA is looking at modifying the definition of SWMU in order to reduce transaction costs and to promote facility-wide approaches to remediation.

Public Participation and Environmental Justice

EPA intends for the final corrective action regulations to be consistent with the Agency's efforts to improve permitting and public participation while providing sufficient flexibility to meet site-specific goals. EPA is trying to improve public participation, especially the participation of communities which have not been effectively involved in the corrective action process to date.

When Permits Can Be Terminated

The 1990 Subpart S proposed rule contained a provision requiring owners and operators to obtain RCRA permits for the entire period needed to comply with the requirements of Subpart S (proposed 40 CFR 270.1[c]). EPA intended this to apply even where the hazardous waste management activities that originally triggered the permit requirement no longer applied. This provision sought to ensure that corrective action would be carried to its conclusion. EPA is considering whether extended permitting is the best approach to ensuring that corrective action is carried out over the long term, or whether other alternatives should be considered. For example, one approach might be to terminate the permit when active hazard waste management ceased, but to continue the cleanup obligation through some other vehicle, possibly an enforcement order.

Area-Wide Contamination Issues

In some cases facilities are located in areas of widely dispersed contamination. For example, facilities may be affected by releases from off-site sources that are not subject to RCRA corrective action (*e.g.*, sources at an adjacent facility not seeking a RCRA permit). In some of these circumstances, contamination from off-site or otherwise unrelated sources would quickly re-contaminate the facility after cleanup efforts were completed. So, while cleanup would be desirable in the long term, it might not make sense in the short term. EPA is considering changes in the application of corrective action requirements for areas of widely dispersed contamination and when the RCRA facility is otherwise impacted by releases from off-site sources.

Definition of Facility for Corrective Action

EPA's definition of a facility for purposes of corrective action has been problematic in some situations. In some circumstances the concept of contiguity, as EPA currently interprets that term, could bring large tracts of land that are only incidentally involved with hazardous waste management under corrective action jurisdiction. In many cases, these large tracts of land could be

addressed using another cleanup authority (*e.g.*, CERCLA or State cleanup programs). EPA has found that imposing this interpretation of contiguity in situations involving large areas of Federally owned land (*e.g.*, Department of Energy [DOE] facilities, Department of the Interior [DOI] lands) can lead to results that arguably were not what Congress anticipated when it enacted the corrective action requirements, forcing the Agency to address contamination that would otherwise be low priority. As a result, EPA is considering changes to its definition of facility.

Future EPA Regulatory Actions and Guidance

EPA indicates in the ANPR that it is planning to promulgate the Subpart S rule in pieces. The first rulemaking component will consider comments received on the 1990 proposal and information and data submitted in response to the ANPR. The second rulemaking component will include the program elements that are to be proposed in 1997 and finalized as a resources permit.

At the time the ANPR was published, the regulated community was concerned about EPA's plans to finalize the Subpart S rule in components. Representatives of the regulated community believed that establishing rules, especially large complicated rules such as Subpart S, in a piecemeal fashion may lead to confusion and difficulty in implementation. These concerns are relative today. The original 1990 Subpart S proposal is over 9 years old, and there have been a number of very fundamental changes in the RCRA program since then. In light of these concerns, some have suggested that EPA consider redrafting or withdrawing the Subpart S Proposed Rule in its entirety.

In the ANPR, EPA also suggests the possibility of relying on mechanisms other than rulemaking (such as guidance documents and directives) to drive the corrective action program. The reality is that nonrulemaking mechanisms have been relied upon to date.

Most recently, EPA promulgated a rule likely to have a significant impact on both the national and State corrective action programs. The Hazardous Remediation Waste Management Requirements

(HWIR–Media), Final (FR 63 65873) establishes new requirements for Resource Conservation and Recovery Act (RCRA) hazardous wastes from remediation activities that are treated, stored, or disposed of during cleanup actions. These new requirements make three key changes: (1) they ease permitting requirements for treating, storing and disposing of remediation wastes; (2) they provide that obtaining these permits will not subject the owner and/or operator to facility-wide corrective action; and, (3) they create a new kind of unit termed a *staging pile* that allows more flexibility in storing remediation waste during cleanup.

The Future of RCRA Corrective Action

As is obvious from the above discussion, RCRA corrective action has always been an evolving program, and will continue to be an evolving program for the foreseeable future. The regulated community is urged to stay tuned to both host-state and EPA information sources in order to remain current.

Techniques for Accelerating RCRA Corrective Action

There are a number of techniques that can be used to accelerate RCRA corrective action. Several useful techniques are described in the remainder of this chapter. We will discuss the development of programmatic project documents as a technique for minimizing the development of documentation. We will also weigh the importance of presumptive remedies and technical impracticability determinations, since these techniques can accelerate corrective action. Finally, we described how the *a priori* development of land use determinations can help quicken the pace of corrective action.

Develop Programmatic Project Documents

For expansive facilities with a large number of SWMUs, site managers should develop a core of

programmatic documents common to all SWMUs in site-wide Program Management Plans. Project documents likely to be applicable to all SWMUs include

- Data Collection Quality Assurance Plan
- Data Management Plan
- Site-Wide Site Conceptual Model
- Public Involvement Plan
- Remediation Waste Management Plan
- Health and Safety Plan

Actions at specific SWMUs or SWMU-clusters that are not reflected in the programmatic documents can be proposed/detailed in SWMU or SWMU-cluster specific documents. For example, a programmatic site conceptual model, a quality assurance project plan, and/or a health and safety plan may need to be amended to reflect the unique aspects of the contamination scenario at a given SWMU cluster.

Develop Programmatic Points of Compliance

The location at which media cleanup standards (MCS) must be attained (*i.e.*, point of compliance [POC]) has significant implications for the scope, magnitude, and cost of corrective action. The facility should seek to establish the most favorable POCs that will protect human health and the environment and that are consistent with the reasonably anticipated future use of the site.

For example, remediation of ground water to a Maximum Contaminant Level (MCL) in the immediate vicinity of the source of the contamination will require more energy and resources than would be the case if the POC were established at a point some distance down-gradient from the source. Natural remediation processes (*e.g.*, absorption, adsorption, chemical degradation, bioremediation) often act significantly to reduce concentrations of contaminants. If the facility were prevented from capitalizing on the benefits of these natural processes, then a more complex and energy-intensive system would have to be implemented to offset the loss of the passive effects.

Likewise, it would not be reasonable to establish a POC down-gradient from one SWMU, when that POC would be influenced by other contamination sources including other SWMUs. In this case the POC for both units should be located down-gradient from both units because the ground water, corrective measure will probably have to address contamination from multiple SWMUs.

Establish Programmatic Action Levels and Media Cleanup Standards That Reflect Anticipated Land Use

The 1990 Proposed Subpart S Rule, described action levels as media-specific, health- or environment-based concentrations of contaminants that are derived from the contaminants' toxicity data. Action levels also use standard assumptions about exposure, assumptions that are presumed to indicate a potential threat to human health or the environment.

EPA's attempts to produce both a flexible program and a nationally consistent program have been especially complicated when setting media-specific action and cleanup levels. While some argue that lists of clearly defined action and cleanup levels will increase the pace of cleanups, many others argue the opposite point, asserting that standardized lists of action or cleanup levels are based on conservative residential exposure scenarios, are often misapplied, and result in overly stringent cleanup requirements.

As described in the 1990 proposal, action levels are a *de facto* determination of the need for further study. Conversely, concentrations of contaminants that are less than the action level are presumed to not be a threat to human health or the environment and would receive a determination of no further action (DNFA). Action levels have not been used to their greatest potential because critics of the approach suggest that a site-specific risk assessment is always necessary to establish protective standards.

Action levels should be developed on a facility-specific basis or be taken from a standardized list. (Some states' regulatory agencies and EPA Regions have developed such lists.) If a standardized list is not available from the appropriate regulatory

agency, the facility should consider developing action levels based on the methodology outlined in the 1990 Subpart S proposed rule.

Other sources for action levels include EPA's draft *Superfund Soil Screening Guidance* and the EPA document *Revised Interim Soil Lead Guidance for CERCLA Sites and RCRA Corrective Action Facilities* (OSWER Directive No. 9355.4-12, July 14, 1994). These documents establish a streamlined approach for determining protective levels for lead in soil, and recommend screening levels for residential land use (400 ppm). They also describe how to develop site-specific media cleanup standards for residential land use.

Media cleanup standards (MCS) state the concentrations of hazardous waste or hazardous waste constituents that must be achieved in groundwater, surface water, soils, and air in order to comply with the standards for corrective measures. MCSs are the established concentrations set for each medium that ensures the protection of human health and the environment.

MCSs define how clean "clean" is, and are established and incorporated into the facility permit during the remedy selection process. They may sometimes be developed by the regulatory authority. An MCS will usually be set for each compound or other hazardous constituent found in excess of an action level. The facility should consider developing its own programmatic MCS values for use during the evaluation process.

These programmatic MCSs should reflect the exposure scenario associated with the reasonably anticipated land use. For example, a facility may develop multiple lists of MCSs to reflect multiple land use/exposure scenarios including: residential, industrial, recreational, and agricultural land uses.

Establish Criteria for Determinations of No Further Action

One outcome of establishing a land use scenario, POCs for each media, and action levels and cleanup standards for target media, is that the facility should also be able to develop and document the criteria for a determination of no further action (DNFA). A Determination requires extensive discussion and negotiation with the regulatory agencies and the other stakeholders, but some typical DNFA criteria indicate that

- There is no actual or potential release of hazardous waste or hazardous waste constituents

- The release does not pose any threat to human health or the environment

- The release, even though it exceeds action levels, is found to be at levels below action levels, or significantly below background levels

- Following a self-implemented action, the limit no longer has constituents above the action levels of an MCS

- The release is to an aquifer that is not, and will not be, used as a source of drinking water

Presumptive Remedies

Presumptive remedies are corrective measures that are based on the positive results of previous actions. Similar remedies are assumed to be both protective and effective when applied under similar circumstances. Presumptive remedies are used to streamline site investigation, speed up the selection of cleanup actions, ensure consistency in remedy selection, and reduce the cost and time of cleanup at similar types of sites.

EPA has developed guidance for applying presumptive remedies for various types of sites, including soil contaminated by volatile organic compounds (VOCs), wood treatment facilities, municipal landfills, releases of polychlorinated biphenyls (PCBs), grain storage units, coal gasification units, and contaminated ground water. EPA also has a publication that discusses the general application of presumptive remedies entitled *Guidance on Presumptive Remedies Policy and Procedures* (OSWER Directive No. 9355.0–47 FS, September 1993a).

Although developed and applied in the CERCLA program, presumptive remedies that are used as a means of accelerating corrective action should offer the same benefits. In the corrective action process, the application of a presumptive remedy can speed the Corrective Measures Study (CMS) more than any other phase. This is because a presumptive remedy eliminates the technology identification and screening step from the CMS.

Instead, the CMS is limited to consideration of the no action alternative and the presumptive remedy technologies that are available. In the 1996 ANPR on corrective action, EPA stated that

> The CMS does not necessarily have to address all potential remedies for every corrective action facility. . . . In cases where EPA has identified a presumptive remedy . . . the purpose of the CMS will be to confirm that the presumptive remedy is appropriate to facility specific conditions (61 FR 19432, 19447).

This is possible because EPA has already conducted the analysis of potentially available technologies and has determined that certain technologies are routinely and appropriately screened out on the basis of effectiveness, implementability, excessive cost, or the alternative is usually not selected for implementation.

Substantiating Technical Impracticability

Although EPA prefers for contamination sources to be treated and/or controlled, the Agency recognizes that some contamination scenarios may be impossible to clean up because of technical barriers. Experience in the Superfund and RCRA Corrective Action programs shows one excellent example of this instance: the contamination of ground water by nonaqueous phase liquids (NAPL). The difficulty of remediating NAPL-contaminated ground water led EPA to issue a guidance document entitled *Memorandum on Ground Water Remediation at Superfund Sites* (May 27, 1992, OSWER Directive No. 9283.1–06). In the document, EPA recognizes that

> Both residual and free phase NAPLs dissolve slowly, supplying potentially significant concentrations of contaminants to ground water over very long time periods. Therefore, the presence of NAPLs will have a significant influence on the time frame required or likelihood of achieving cleanup standards, and should be evaluated when selecting appropriate remedial actions.

The difficulty of restoring NAPL-contaminated ground water eventually led EPA to issue another guidance memorandum entitled *Guidance on*

Evaluating the Technical Impracticability of Ground Water Restoration (September 1993b, OSWER Directive No. 9234.2–25). The purpose of this memorandum was to establish EPA's policy for evaluating the technical impracticability of restoring ground water; it applied equally to the RCRA Corrective Action and the Superfund programs.

Technical impracticability applies to situations other than those where remediation is not feasible from an engineering perspective. As was stated in the preamble to the 1990 Subpart S proposed rule:

> In other situations a determination [of technical impracticability] may be made when remediation may be technically possible, but the scale of operations required might be of such a magnitude and complexity that the alternative would be impracticable. . . The concept of technical impracticability may in some cases also apply to situations in which use of available remedial technologies would create unacceptable risks to workers or surrounding populations, or where cleanup would create unacceptable cross-media impacts (55 FR 30798).

This statement indicates that a TI determination may be obtained in instances other than contamination of groundwater by NAPLs. There are other cases where environmental restoration is not feasible.

Integrate Land Use Planning Into the Environmental Restoration Process

One of the most important factors driving decisions about environmental restoration is the issue of subsequent land use. Risk is a function of toxicity and exposure, and one key determinant of exposure is land use. As a result, land use assumptions can have a significant impact on decisionmaking throughout the environmental restoration process. Land use considerations can impact the scope of the environmental restoration process, the development of sampling and analysis and risk assessment work plans, and the development of remedial alternatives. Also discussed is how restricted land use considerations can, in some cases, increase the importance of ecological risk in the environmental restoration process.

The Relationship Between Land Use and the Development of Investigation Work Plans

A primary objective of Corrective Action Investigation Work Plans is to quantify the risks based on current land use and to quantify the predicted risks based on anticipated land use. In many cases, future land use at DOE/DOD facilities is uncertain; a number of land use scenarios can reasonably be anticipated. This range of reasonably likely future land uses must be considered when developing remedial action objectives. Thus, a suite of anticipated land use scenarios could necessitate expanding the scope of Investigation Work Plans to allow for the development of a commensurate suite of remedial action objectives.

Current land use and assumptions regarding future land use can help focus the Sampling and Analysis Plans (SAPs) associated with the RFI Phase. The Environmental Protection Agency recommends approaches which focus on collecting the data needed to support decisionmaking.

Under EPA's Superfund Accelerated Cleanup Model (SACM), Data Quality Objectives (DQOs) are used to collect the appropriate data to support a site decision. DQOs apply to the entire characterization process including sampling locations, sampling techniques, sample handling, and analytical methods. By linking DQOs with reasonably anticipated future land use, site investigators can help focus and accelerate data collection efforts in the characterization phase.

In general, as land use assumptions become less restrictive, DQOs become more rigorous. For example, if current or future land use is unrestricted, DQOs in the Characterization Phase could necessitate the use of Level V or IV Analytical Methods (Level V and IV Analytical Methods have rigorous QA/QC controls and low method detection limits, and can be used to obtain highly documented data), and standardized sample collection techniques in order to determine whether or not the affected ground water is safe, or will be safe, for human consumption.

Restricted land use assumptions and the associated limitations on exposure could allow site investigators to integrate more lenient DQOs into sampling and analysis plans, with higher method detection limits than could be achieved with Level

II or Level III field screening technologies (*e.g.* X-ray fluorescence, field gas chromatography/mass spectrometry, immuno-assay test kits). Some of the benefits of using such innovative characterization techniques include: rapid sample collection and analysis that allows for on-site decisionmaking, better identification of actual or potential risks to human health and ecological resources, and more rapid assessment of the need for interim measures.

Thus for DOE/DOD sites or portions of DOE/DOD sites frozen in restricted land use patterns, the Characterization Phase and the selection/implementation of interim measures can be accelerated by using the rapid characterization technologies in the SAP. However, if unrestricted land use such as residential land use is anticipated, more rigorous sample collection and analytical methods would likely be required in the SAP in order to provide data of sufficient quality for the conduct of human health risk assessments.

The Impact of Land Use Assumptions on the Development of Corrective Measures

Corrective measure alternatives developed for the RCRA Corrective Action Program are based upon reasonably anticipated land use assumptions. The corrective action program intends for corrective measures to result in site conditions which are consistent with reasonably anticipated future land use.

As specified in EPA's Directive on Land Use in the CERCLA Remedy Selection Process, site investigators rely on future land use assumptions to focus on the development of practicable and cost-effective remedial alternatives. Consequently, the more limited the suite of anticipated future land uses, the more limited the suite of corrective measures and remedial alternatives will be, and the quicker the CMS can be completed.

Conclusions

The core of the RCRA corrective action program is a traditional, stepwise approach involving an RFA, RFI, CMS, and CMI as dictated by either an enforcement mechanism or a permit mechanism.

In general, the corrective action program is driven by proposed regulations, guidance documents, agency directives, and site-specific order and permit requirements rather than promulgated regulations.

The status of corrective action rulemaking remains uncertain. EPA is weighing two approaches: (1) go forward with rulemaking, and (2) reform RCRA cleanups through guidance rather than rules. The developmental nature of the corrective action program creates opportunities to accelerate corrective action. In the absence of a structured regulatory program (and with regulatory authority buy-ins), owner/operators can craft their own site-specific corrective action programs and take advantage of the flexibility in EPA's regulatory reform proposals.

Bibliography

"Codification Rule." *Federal Register* 50 (15 July 1985): 28702.

"Corrective Action Management Units and Temporary Units; Corrective Action Provisions: Final Rule." *Federal Register* 58 (16 February 1993): 8658.

"Corrective Action for Releases From Solid Waste Management Units at Hazardous Waste Management Units; Advance Notice of Proposed Rulemaking." *Federal Register* 61 (1 May 1996): 19432.

"Corrective Action for Solid Waste Management Units (SWMUS) at Hazardous Waste Management Facilities, Proposed Subpart S Rule." *Federal Register* 55 (27 July 1990): 30798.

Environmental Protection Agency. *EPA Guidance on Land Use in the CERCLA Remedy Selection Process.* OSWER Directive No. 9355.7–04. Washington, DC: EPA, May 1995.

Environmental Protection Agency. *Guidance on Evaluating the Technical Impracticability of Ground Water Restoration.* OSWER Directive No. 9234.2–25. Washington, DC: EPA, September 1993b.

Environmental Protection Agency. *Guidance on Implementation of the Superfund Accelerated Cleanup Model (SACM) Under CERCLA and the NCP.* OSWER Directive 9203.1–03. Washington DC: EPA, July 1992.

Environmental Protection Agency. *Guidance on Presumptive Remedies—Policy and Procedures.* OSWER Directive No 9355.04–47 FS. Washington, DC: EPA, September 1993a.

Environmental Protection Agency. *Implementation of RCRA Facility Assessments.* OWSER Directive No. 9502.00–04. Washington DC: EPA, August 1986.

Environmental Protection Agency. *Interim Soil Lead Guidance for CERCLA Sites and RCRA Corrective Action Facilities.* OSWER Directive No. 9355.4–12. Washington, DC: EPA, July 1994.

Environmental Protection Agency. *RCRA Corrective Action Interim Measures Guidance (Interim Final).* OSWER Directive 9902.4. Washington, DC: EPA, June 1988.

Environmental Protection Agency. *RCRA Facility Investigation Guidance.* OSWER Directive 9502.00–6c, OSW: 530/SW–89–031.Washington DC: EPA, May 1989.

Environmental Protection Agency. *The Nation's Hazardous Waste Management Program at a Crossroads—The RCRA Implementation Study.* EPA/530/–SW–90–069. Washington DC: EPA, July 1990.

Environmental Protection Agency. *Use of the ANPR as Guidance.* OSWER Directive Memorandum by Elliott P. Laws. Washington, DC: EPA, January 1997.

"Weekly Report." *Inside EPA.* Vol (18) No. (18): p. 4, May 1997.

Louis Martino is an Environmental Systems Engineer at Argonne National Laboratory. Mr. Martino has expertise in the following areas: project management, site investigation techniques, field laboratory analytical methods, pollution prevention, property transfer, and compliance auditing, as well as regulatory policy analysis. He has over 20 years of experience in investigating uncontrolled waste disposal sites and consulting to the hazardous waste management industry. Mr. Martino has published papers on in situ analysis of soil at an open burning/open detonation disposal facility (9th Annual Conference on Contaminated Soils, University of Massachusetts at Amherst), managing compliance and risk on military testing and training ranges (Federal Facilities Environmental Journal), as well as assessment and remediation of doxin-contaminated soil (Environmental Claims Journal).

David Green is a Senior Engineer with Radian International. Mr. Green has over 12 years experience in the hazardous materials management field. His primary area of expertise includes management of waste munitions and energetics, unexploded ordnance (UXO) remediation, and military range management. For the last six years his primary focus of effort has been on improving the policies, procedures, and regulations applicable to the management of waste military munitions and the conduct of response actions to UXO. Specific efforts include contributions to the development of the EPA Munitions Rule and the proposed DOD Range Rule, and to a variety of related DOD policy efforts, development of strategic and implementation plans, and participation in various public outreach and partnering efforts. Mr. Green has been an author or co-author of several documents and articles on RCRA Corrective Action and CERCLA response actions for DOE, as well as one of the developers and instructors for DOE's RCRA Corrective Action Workshop and the US Army Training and Doctrine Command RCRA Corrective Action Workshop. He has served as compliance audit team member addressing waste management and remediation for multimedia compliance audits at Federal and private facilities. He has recently co-authored US Laws and Regulations: The Handbook of Groundwater Engineering *and an article on the military munitions rule in the* Federal Facilities Environmental Journal.

Treatment Technologies in Hazardous Waste Management

Gazi A. George, PhD, CHMM

Introduction

Proper treatment of hazardous wastes has become an art and/or a concern to all generators and Treatment, Storage, and Disposal facilities (TSDs). This is because of the enactment of the Resource Conservation and Recovery Act (RCRA) in 1976, and the subsequent passage of the "Superfund" Cleanup law in 1980, as well as a complex series of amendments to cover new hazardous wastes, methods of characterization, treatment standards, and specific treatment technologies.

Hazardous Waste Treatment facilities must address all of the following factors:

1) Health and safety of operators and neighbors

2) Process economics

3) Rate of success

4) Flexibility to include the treatment of numerous parameters, such as toxic metals and organics

553

covered by the Land Disposal Restrictions (LDRs)

5) Public exposure from air emissions, water discharges and land disposal as a result of facility operations

6) Impact of future regulations on the available process(es); can they be expanded?

7) Market support in providing volumes of hazardous wastes sufficient to justify continuity of business and associated research and development

Readers must familiarize themselves with the RCRA hazardous waste regulations presented in 40 CFR 260–272, in order to determine whether or not a waste is hazardous. This is achieved by referring to 40 CFR 261, which is confusing to "newcomers" due to terminology and foggy regulatory language. An overview of the RCRA requirements can be found in the RCRA chapter of this book.

Hazardous Waste

A waste will be subject to the hazardous waste regulations if it meets any of the following conditions.

1) **Characteristic Waste.** Waste exhibiting any of the four characteristics of a hazardous waste: ignitibility, corrosivity, reactivity, or toxicity.

   The various categories of characteristic hazardous wastes are detailed below.

   Characteristic of ignitability (D001)—A waste that exhibits any of the following properties:

   - It is a liquid other than aqueous solution containing less than 24% alcohol by volume and has a flash point less than 140°F (60°C), as determined by a Pensky Martens closed cup tester (American Society for Testing and Materials [ASTM] Method D–93–79 or D–93–80) or Setaflash closed cup tester using method ASTM D–3278–78.

   - It is not a liquid and is capable under standard temperature and pressure of causing a fire through friction, absorption of moisture, or spontaneous chemical changes; and when ignited burns so vigorously and persistently, that it creates a hazard.

   - It is an ignitable compressed gas as defined in 49 CFR 173.300 and as determined by the test methods described in the above regulations or equivalent test methods approved under 40 CFR 260.20 and 260.21.

   - It is an oxidizer as defined in 49 CFR 173.151.

   Characteristic of corrosivity (D002)—A waste exhibits the characteristic of corrosivity if a representative sample of the waste has either of the following properties:

   - It is aqueous and has a pH of less than or equal to 2, or greater than or equal to 12.5, as determined by a pH meter, using the Environmental Protection Agency (EPA) publication SW–846.

   - It is a liquid and corrodes steel at a rate greater than 6.35 mm (0.25 inch) per year at a temperature of 55°C.

   Characteristic of reactivity (D003)—A waste exhibits the characteristic of reactivity if a representative sample of the waste has any of the following properties:

   - It normally is unstable and readily undergoes violent changes without detonating.

   - It reacts violently with water.

   - It forms a potentially explosive mixture with water.

   - It generates toxic gases, vapors, or fumes when mixed with water.

   - It is a cyanide, or a sulfide-bearing waste which, when exposed to pH conditions between 2–12.5, can generate toxic gases, vapors, or fumes.

   - It is capable of detonation or causing an explosive reaction if subjected to heat or initiators.

   - It is readily capable of explosive decomposition or reaction at standard temperature and pressure.

   - It is a forbidden explosive defined in 49 CFR 173.51, 49 CFR 173.53, and 49 CFR 173.88.

   Waste codes D004–D043—Characteristic codes and their relevant characteristic level

Table 1. TCLP Concentration Limits for Characteristic Wastes

Metals	mg/L	Base Neutral Extract	mg/L
D004 Arsenic	5.0	D027 1,4–Dichlorobenzene	7.5
D005 Barium	100.0	D030 2,4–Dinitrololuene	0.13
D006 Cadmium	1.0	D032 Hexachlorobenzene	0.13
D007 Chromium	5.0	D033 Hexachlorobutadiene	0.5
D008 Lead	5.0	D034 Hexachloroethane	3.0
D009 Mercury	0.2	D036 Nitrobenzene	2.0
D010 Selenium	1.0	D038 Pyridine	5.0
D011 Silver	5.0	D023 m–Cresol	200.0
		D024 o–Cresol	200.0

Z HE Organics	mg/L	D025 p–Cresol	200.0
D018 Benzene	0.5	D026 Total Cresol	200.0
D019 Carbon tetrachloride	0.5	D037 Pentachlorophenol	100.0
D021 Chlorobenzene	100.0	D041 2,4,5–Trichlorophenol	400.0
D022 Chloroform	6.0	D042 2,4,6–Trichlorophenol	2.0
D028 1,2–Dichloroethane	0.5		

Pesticides	mg/L
D020 Chlordane	0.03
D012 Endrin	0.02
D031 Heptachlor and its epoxide	0.008
D013 Lindane	0.4
D014 Methoxychlor	10.0
D015 Toxaphene	0.5

Z HE Organics continued:
D029 1,1–Dichloroethylene 0.7; D035 Methyl ethyl ketone 200.0; D039 Tetrachloroethylene 0.7; D040 Trichloroethylene 0.5; D043 Vinyl Chloride 0.2

Herbicides	mg/L
D016 2,4–D	10.0
D017 2,4,5–TP (Silvex)	1.0

(concentrations) as determined by the TCLP (Toxicity Characteristic Leaching Procedure) for concentration-based constituents in waste, as shown in Table 1.

2) **Listed Hazardous Waste.** Wastes specifically listed in Subpart D of the regulations:

- Non Specific Source (F-Listed)
- Specific Source (K-Listed)
- Acute Hazardous Waste (P-Listed)
- Toxic Hazardous Waste (U-Listed)

3) **Mixtures.** The waste is a mixture of a listed hazardous waste and a non nonhazardous waste.

4) **Declared to be Hazardous.** The waste has been declared to be hazardous by the generator.

It is important that generators of hazardous wastes investigate, experiment with, and implement waste minimization techniques through recycling, such as reclaiming of usable products or regeneration; source reduction, such as product substitutions, or source control; and finally, treatment using approved technologies.

Treatment Technologies

Treatment of hazardous wastes to eliminate their toxicity and/or their hazardous characteristics can be achieved using a wide spectrum of technologies which revolve around two main pathways—namely chemical and mechanical (physical). **Chemical treatment** includes all forms of chemical reactions such as reduction, oxidation, thermal oxidation, precipitation, neutralization, electrochemical, photolytic, biological degradation, *etc*. **Mechanical or physical treatment** includes those procedures that modify the physical properties of waste materials; for example: filtration, phase separation, filter pressing of suspended materials, centrifugation, agitation, adsorption, *etc*.

Filtration of Suspended Materials

Filtration can be achieved using a porous medium subjected to a pressure gradient such as gravity or pumping. This technology is normally used as a final polishing step following treatment processes such as neutralization, chemical precipitation, biological treatment of organic wastes, and oily wastewater separation by emulsion breaking and dissolved air flotation.

Solid separation can be conducted using simple screening techniques such as in-line strainers, filters and screens. More efficient filtration can be achieved using deep bed filtration, such as specialized cartridge filters and gradient granular beds for slurries that need to be dewatered (typically 1–25% by weight solids). Centrifugal filters, vacuum filters, belt filters as well as plate and frame filter presses generate a filter *cake* reaching or exceeding 50% solids by weight.

Since filtration is a physical process, it has numerous limitations. These limitations include but are not limited to

- Inability to separate dissolved solids unless coupled with another process.

- Except for membrane filtration, this process is not chemically selective for various waste constituents.

- High solids, viscous liquids, such as tars, cannot be filtered.

In all filtration processes, the filtrate is tested for compliance with the relevant discharge permits and the solids can either be deposited in a licensed landfill or sent for further treatment, such as chemical fixation with lime or cement.

Simple Acid–Base Neutralization

This treatment involves mainly wastes carrying characteristic hazardous waste code D002 or similar listed codes such as K062, U123, U134, *etc*.

This treatment method utilizes the simple reaction between an acid and a base. An **acid** is any substance that dissociates in solution to produce a proton (H+) and a **base** is any substance that accepts a proton. Strong acids or bases are those completely dissociating in solution such as nitric acid or potassium hydroxide, respectively; whereas weak acids and bases only partially dissociate in solution, such as acetic acid and ammonium hydroxide, respectively.

$$\underset{\text{Nitric Acid}}{HNO_3} + \underset{\substack{\text{Sodium} \\ \text{Hydroxide}}}{NaOH} \rightarrow \underset{\text{Salt}}{NaNO_3} + \underset{\text{Water}}{H_2O} + heat \qquad (1)$$

Most neutralization reactions involve the adjustment of the acidic or alkaline waste stream with the relevant suitable reagent to obtain a final pH of 6–9 to meet the water discharge requirements established under the Clean Water Act (CWA).

To select an efficient neutralization system, the waste constituents must be properly identified. For example, while hydrochloric acid can be directly neutralized with sodium hydroxide, hydrofluoric acid must be neutralized with lime to yield calcium fluoride, which is insoluble. Furthermore, some acids must be pretreated. For example, chromic acid must be reduced from chromium (+6) to chromium (+3) at an acidic pH using sodium metabisulfite prior to introducing lime slurry. The lime slurry precipitates the chromium (+3) hydroxide as a greenish sludge which can be directed to the process clarifier.

Neutralization of acids or bases which contain high solids such as tank bottoms can be routinely performed using chemical fixation/stabilization with lime to achieve an acceptable residue for landfilling. However, initial laboratory scale determinations must be conducted to avoid excessive heat generation, which can be done through dilution and/or slow additions.

Chemical Precipitation

This is a widely used process utilizing the solubility products of certain compounds of the hazardous constituent (solute) to obtain an insoluble form, either via chemical reaction or changes in the composition of the solvents. The precipitated portion (solids) can then be separated by settling and/or filtration.

Metals and certain anions are commonly removed using such reactions. However not all metals are directly amenable to precipitation. Examples include chromium (+6), salts, arsenic salts, and certain amphoteric ions.

Industries generating liquid wastes suitable for this process include metal plating, metal polishing, inorganic pigments, mining, steel and nonferrous metals, circuit board and wood treatment using chromium, copper, and arsenic (CCA).

Chemical precipitation can be achieved using the reactions described below.

Hydroxide Precipitation. When the solubility product of a metal hydroxide is suitable for precipitation, the use of sodium hydroxide (caustic) or calcium hydroxide (lime slurry) results in a metal hydroxide residue:

$$M^{+2} + 2(OH^-) \rightarrow M(OH)_2 \downarrow \qquad (2)$$

The precipitation of chromium, however, requires that all hexavalent chrome-containing ions be reduced to the trivalent state, since hexavalent chromium cannot be removed directly by hydroxide introduction. Using sulfurous acid as a reducing agent:

$$3H_2SO_3 + H_2Cr_2O_7 \rightarrow Cr_2(SO_4)_3 + 4H_2O \qquad (3)$$

$$Cr_2(SO_4)_3 + 3Ca(OH)_2 \rightarrow 2Cr(OH)_3 \downarrow + 3CaSO_4 \qquad (4)$$
Green Precipitate

Chelated wastes, such as those containing ammonia, ethylenediaminetetraacetic acid (EDTA), citric acids, amines, and sulfamates, require pH adjustment into higher alkaline ranges to effectively precipitate the chelated metals. The use of flocculants and coagulants, such as aluminum sulfate, ferric chloride, ferrous sulfate, dithiocarbamate, cationic and anionic polymers, is essential in chelated metals treatment. Also, the choice between caustic versus lime is a function of the types of chelators present.

Sulfide Precipitation. The use of sulfides to precipitate metals is superior to the use of hydroxides since the metal sulfide precipitate has a much lower solubility than the hydroxide of the same metal. For example, lead hydroxide has a solubility of slightly less than 0.1 g/l compared to about 10^{-8} g/l for lead sulfide. Furthermore, the sulfide ligand is a reducing agent in addition to being a precipitating agent, which can be advantageous in treating chromium wastes as follows:

$$Cr_2O_7^{-2} + 2FeS + 7H_2O \rightarrow \qquad (5)$$
$$2Fe(OH)_3 + 2Cr(OH)_3 \downarrow + 2S + 2OH^-$$

Soluble mercury salts are best precipitated using sulfide addition to achieve solubilities of less than 10^{-45} g/l.

Again, as with hydroxide precipitation, certain metals do not respond to the introduction of sulfides. These metals include arsenic, which has to be co-precipitated with iron. When using sulfide as a precipitating agent, extreme care must be exercised to avoid hydrogen sulfide generation. This can be achieved by ensuring that the operating pH is greater than 8. In addition sulfide must be closely monitored in the effluents, since most precipitation reactions use excess reagents to ensure reaction completion.

Other Precipitation Processes. Certain metals such as cadmium (D006) and lead (D008) respond to carbonate precipitation comparably to hydroxide precipitation, with the advantage of a lower pH and a denser, easier-to-filter sludge at a pH of 8.

Soda ash is used as the source of carbonate in this treatment:

$$Cd(NO_3)_2 + Na_2CO_3 \rightarrow CdCO_3 \downarrow + 2NaNO_3 \qquad (6)$$

Phosphate precipitation is effective in selectively removing trivalent cations such as chromium, aluminum, and iron from wastes containing divalent or monovalent cations.

Sulfate precipitation is effective in the removal of barium (D005) as barium sulfate:

$$Ba^{+2} + SO_4^{-2} \rightarrow BaSO_4 \downarrow \qquad (7)$$
White Precipitate

Selenides can be removed by reaction with iron to form iron selenide, an insoluble compound, whereas other selenium salts can be treated with sulfur dioxide to precipitate elemental selenium.

Chemical Oxidation and Reduction (Redox Reactions)

This technology is utilized as a pretreatment step for numerous waste codes in liquid, sludge, or solid phases prior to other treatment trains such as chemical precipitation and/or chemical fixation.

Oxidation is the process of losing electrons whereas *reduction* is the process of gaining electrons. Therefore in each oxidation reaction there is a material that is oxidized and another that is consequently reduced.

The above processes are best suited for liquids. However, sludges and solids can be similarly treated via wetting, increasing contact time, and the introduction of surface tension modifiers such as surfactants.

Oxidation. The following examples of oxidative reactions demonstrate the effectiveness of this process in treating both inorganic and/or organic species to either render the waste nonhazardous or to create waste that passes the Land Disposal Restriction (LDR) requirements by meeting the relevant treatment standards.

Treatment of Cyanides. The major sources of cyanides can be attributed to metal finishing and the expanding circuit board industries. Cyanides are present as sodium cyanides, potassium cyanides, zinc cyanides, copper cyanides, silver cyanides, and gold cyanides.

Dilute waste streams have up to 1000 ppm reactive cyanides, while concentrated streams containing plating bath materials reach in excess of 200,000 ppm of both simple and/or complex cyanides. Other contributors to these wastes are the photographic industry, heat treating, mining and coking operations.

The most common chemical process utilized in the destruction of cyanides is alkaline chlorination or simply the use of household bleach (sodium hypochlorite):

$$2NaCN + 5NaOCl + H_2O \rightarrow \qquad (8)$$
$$N_2 + 2NaHCO_3 + 5NaCl$$

This reaction is carried out at a pH > 10 to avoid the liberation of the dangerous gas cyanogen chloride.

Another oxidation process utilizes hydrogen peroxide to convert the cyanide into cyanate as in the following equation:

$$NaCN + H_2O_2 \rightarrow NaCNO + H_2O \qquad (9)$$

Potassium permanganate has been used in cyanide treatment; however, due to the cost of permanganate and the incomplete destruction of the cyanide ligand, it has been rarely used in industrial applications:

$$NaCN + 2KMnO_4 + 2KOH \rightarrow \qquad (10)$$
$$2K_2MnO_4 + NaCNO + H_2O$$

Cyanide oxidation is instantaneous in simple cyanides such as sodium or potassium cyanides. However in complex cyanides, such as $Cu(CN)_2$, and $Ni(CN)_2$, an initial step is required in order to dissociate the complex and remove the complexing metal; converting the cyanide into a simple cyanide that is easily amenable to alkaline chlorination:

$$K[Cu(CN)_2] \quad \rightarrow \quad CuCN \ + \ KCNO \qquad (11)$$
Bleach   Insoluble   Cyanate
pH>12

$$CuCN + Na_2S \xrightarrow{[OH]^-} CuS\downarrow \ + \ NaCN + NaOH \qquad (12)$$
Sodium   Remove Sludge Simple
Sulfide   by Filtration   Cyanide

$$2NaCN + 5NaOCl + H_2O \rightarrow \qquad (13)$$
$$N_2\uparrow + 2CO_2\uparrow + 2NaOH + 5NaCl$$

Equation (13) is derived from the simple alkaline chlorination sequence:

$$CN^- + H^+ + OCl^- \rightarrow CNCl + OH^- \qquad (13a)$$

$$CNCl + 2OH^- \rightarrow CNO^- + Cl^- + H_2O \qquad (13b)$$

$$2CNO^- + 3OCl^- + H_2O \rightarrow \qquad (13c)$$
$$2CO_2\uparrow + N_2\uparrow + 3Cl^- + 2(OH^-)$$

The above reaction can be simplified if bleach is added under alkaline conditions with heat and pressure (reactor) to produce nitrogen and carbon dioxide gases.

Organic Compounds. Treatment of low concentrations of organics (<1% by weight to volume) can be achieved using commercially available oxidizers such as permanganates, bleaches, peroxides, ozone, chlorine gas, *etc.*

The following generalized reactions are examples of oxidation of certain organic species:

Phenols.

$$C_6H_5OH + 14H_2O_2 \xrightarrow{Fe^{+2}} 6CO_2 + 17H_2O \qquad (14)$$
Phenol   Hydrogen
            Peroxide

Alcohols.

$$CH_3CH_2OH \rightarrow CH_3COOH \qquad (15)$$
Ethanol   Acetic Acid
(Primary Alcohol)

$$R_2CH(OH) \rightarrow RCOR \qquad (16)$$
Secondary Alcohol

Aldehydes.

$$RCHO \rightarrow RCOOH \qquad (17)$$
        Organic Acid

$$HCHO \rightarrow HCOOH \qquad (18)$$
Formaldehyde   Formic Acid

Sulfur-bearing organics.

$$2RSH \rightarrow RSSR + H_2O \qquad (19)$$

$$RSSR' \rightarrow RSO_3H + R'SO_3H \qquad (20)$$

Organic Halides: R-X (Non Oxidative).

$$RCH_2XR \xrightarrow[HOH]{} CH_2OH + HX \qquad (21)$$
      Hydrolysis   Alcohol   Acid
$$X = F, Cl, Br, I$$

These reactions must be catalyzed by either hydroxyl or hydrogen ions, and are also a function of solubility in water.

Metals Precipitation. Certain metals require an oxidation step (higher valence) to be precipitated more efficiently and to achieve the least solubility.

For example:

$$Hg^{+1} \rightarrow Hg^{+2} + e^- \qquad (22)$$

$$Hg^{+2} + S^{-2} \rightarrow HgS\downarrow \qquad (23)$$
(solubility product close to 10^{-45} g/l)

Other metals, such as arsenic (+3), require oxidation to arsenic (+5) to avoid any possibilities of the formation of the deadly gas arsine, AsH_3, which occurs in a reducing medium in the presence of hydrogen.

Chemical Reduction. Chemical reduction has been used to treat both hazardous and nonhazardous waste. The most common reaction is that of chromium (+6) reduction to chromium (+3) prior to precipitation as the hydroxide

$$4CrO_3 + 6NaHSO_3 + 3H_2SO_4 \rightarrow \qquad (24)$$
           $2Cr_2(SO_4)_3 + 3Na_2SO_4 + 6H_2O$
    Sodium Bisulfite

Certain wastes such as halogenated organic compounds in concentrations less than 1% by weight can be reduced using iron:

$$Fe + 2H_2O + 2RX \rightarrow \qquad (25)$$
$$2ROH + Fe^{+2} + 2X^- + H_2\uparrow$$
$$X = F, Cl, Br, I$$

Chemical reduction has been recently used to desensitize (deactivate) certain shock-sensitive nitro-compounds such as DNT (dinitrotoluene) at concentrations less than 5000 ppm using patented reducing mixtures. This reduction step converts the nitro groups into the respective amine groups hence rendering these residues harmless.

Wet Oxidation

The introduction of a rich source of oxygen such as air (about 20% oxygen) to a liquid waste containing organics or oxidizable species at temperatures ranging from 150°C to 350°C and a pressure ranging from 300–3000 psi results in oxidation. This process can be either batched or continuous flow, and is widely used compared to other technologies due to the relatively low associated cost.

This technology was also modified to be conducted in deep shafts (wells) instead of tanks. Because this process is a *plug-flow* process, where oxygen and liquid move together, it is a great deal more efficient than conventional wet air oxidation. Heat, pressure, reaction, cooling, and depressurizing all occur in one vertical subsurface shaft, thus ensuring safety.

Biological Treatment

Most municipal and domestic wastewaters, including some industrial effluents, are treated using biological processes to detoxify these wastes prior to final discharge. It is anticipated that by the early 21st century, most industrial wastewaters will be biologically treated due to changing regulations under the Clean Water Act.

Despite the fact that numerous forms of biological treatment exist, two main treatment trains are more frequently used by municipalities and Publicly Owned Treatment Works (POTWs):

- *Aerobic* (use of air or oxygen)

- *Anaerobic* (absence of air or oxygen)

The above two types can be used in series, parallel, or individually.

Aerobic systems.

Activated Sludge. The basic design categories of activated sludge systems include conventional, complete mix, and step aeration. The addition of powdered activated carbon (PACT) to a biological process has been used with success in treating hazardous wastes such as solvents.

Organic wastes and recycled sludge are introduced to a reactor where aerobic bacteria are maintained in suspension. The residence time in the bioreactor (hydraulic residence time) is obtained by dividing the reactor volume by the volumetric flow rate. Sludge then leaves the reactor for settling tanks, where it is separated from the waste stream. Most of the sludge is filter-pressed, the remainder being recycled to the reactor.

The recycled fraction volume is determined by the desired *food* to mass ratio (F/M) measured as BOD (biochemical oxygen demand) divided by the mixed liquor suspended solids (MLSS): kg BOD / kg MLSS.

Aerated Ponds and Lagoons. This method of treatment is similar to an extended aeration activated sludge process where a basin is used instead of a reactor. Air injection is performed through diffuse aerators and the biomass is kept in suspension. Modern lagoons recycle the biomass. Aerated lagoons are used for low- to medium-strength organic wastes and efficiency is linked to adequate mixing, annual air temperature over 5°C, and a good inlet/outlet design.

Aerobic Digestion. Two versions exist, varying only in the oxygen supply: conventional, and pure oxygen. These systems can handle many types of wastes. Some examples are waste-activated or trickling filter sludge, mixtures of waste-activated or trickling filter sludge and primary sludge, or waste sludge from activated sludge treatment plants designed without primary settling.

Trickling filters.

Rotating Biological Contactors.

Bioaugmentation.

Anaerobic Digestion. This treatment is one of the oldest processes used for sludge stabilization. It involves the oxidation of organic and inorganic matter with anaerobic and facultative organisms in the absence of molecular oxygen carried out in sealed containers. The by-product of such activity is methane gas (55%), CO_2, N_2, and H_2S produced at the rate of 1 liter per gram of volatile acids

consumed. Optimum conditions include a pH >
6.5 and a temperature of 85–90°F.

Thermal Treatment Processes

Incineration. *Incineration* can be defined as a unit
of operation that utilizes thermal decomposition via
oxidation to reduce carbonaceous matter. Principal
products of such process are carbon dioxide, water,
ash, and heat energy. Various by-products are also
generated including sulfur, nitrogen compounds,
halogens and their compounds.

These units are integrated systems comprised of the
incinerator itself, a waste delivery system, gas
scrubbing equipment, effluent management system,
ash discharge system, and energy recovery
operations.

In the field of hazardous waste incineration there
are more liquid injection incinerators than all other
types combined. Wastes are aspirated and burned
directly in the flame or combustion zone. The
injection nozzle location is determined primarily by
the heating value (Btu) of acceptable wastes.

Incinerators operate at temperatures ranging from
1000°C to 1700°C (1832°F–3092°F), with a waste
residence time in the *hot zone* from milliseconds to
about 3 seconds.

In any incinerator design, the chemical, physical,
and thermodynamic properties of the components
must be considered. This includes storage tanks,
pumps, mixers, valves, nozzles, refractory, heat
recovery, quench systems, and all related pollution
control equipment.

Since wastes in most cases are heterogeneous, the
characteristics must be carefully defined before
finalizing the design. The chemical and physical
parameters to be considered include

- Corrosivity
- Ignitability
- Reactivity
- Toxicity
- Btu value (heating value)
- Moisture and volatile matter
- Viscosity and specific gravity
- Polymerization
- Solid content
- Ash content
- Explosiveness
- Nitrogen, halides, sulfur, and metals, *etc.*

Many factors must be considered in selecting an
economical yet efficient system for waste disposal
and volume reduction through combustion; these
factors are

- Segregation
- Transportation
- Solid and liquid preparation
- Storage and equalization
- Handling and feeding
- Residue handling and disposal
- Environmental impact
- Regulatory requirements

There are many incinerator configurations used to
burn nonhazardous and hazardous wastes in liquid,
sludge, and solid forms; these include:

- Multiple hearth furnaces
- Fluidized bed incinerators
- Liquid waste incinerators
- Waste gas flares
- Direct flame incinerators
- Catalytic combustion incinerators
- Rotary kilns
- Cement kilns
- Molten salt incinerators
- Multiple-chamber incinerators
- Oceanic (ship-mounted) incinerators
- Infectious waste incinerators
- Industrial boilers
- Pyrolysis units

A description of the most commonly applicable
incinerators in the hazardous waste industry will
be briefly discussed. Selection criteria and operating
parameters are shown in Tables 2 and 3.

Multiple Hearth Incinerators. Incineration usually
requires a minimum of six hearths while pyrolysis
requires a greater number. Wastes enter the furnace

Table 2. Waste/Incinerator Selection Criteria

Waste Type	Rotary Kiln	Multiple Hearth	Fluidized Bed	Liquid Incinerators	Catalytic Combustor	Multiple Chamber	Wet Air Oxidation	Molten Salt
Aqueous organic sludge		X	X				X	
Halogenated aromatics	X		X	X If pumpable				X
Organic liquids				X			X	
Granular	X	X	X					
Irregular–bulky	X					X		
Low melting point	X		X	X If pumpable				
Organics with fusible ash	X	X						X
Organics–vapor laden					X			
Toxic high organics				X	X		X	

by dropping through a port at the top of the furnace. A spiral shaft located centrally rotates the waste counterclockwise across the various hearths through the furnace, thus dropping the waste from hearth to hearth. Burners and combustion air ports are located in the walls of the furnace.

Air is cleaned by passing the hot gases from the furnace through a precooler, which sprays fine water spray droplets. The cool gas is then passed through a Venturi throat into which additional water is sprayed. At this time, gases are subcooled (120°F) and stripped of water prior to discharge.

Multiple-hearth furnaces operating in pyrolysis modes can handle feed material with a heat release potential greater than 25,000 Btu/lb of water and still maintain a temperature of about 1500°F.

Pyrolysis. Incineration requires high air input whereas **pyrolysis** is theoretically a zero air, indirect heat process (air starved incineration).

Waste organics are distilled or vaporized to form combustion gases, which can be used as fuel for an external combustion chamber. Pyrolysis is used when wastes have a high Btu (thermal) content greater than 3500 Btu/lb.

Fluidized Bed. This is a simple device consisting of a refractory lined vessel containing inert granular material. Gases are blown through this inert material at a rate to cause the bed to expand and act as a fluid. Gases are injected through nozzles, which restrict downflow. Hot gases leave the

Table 3. Operating Parameters for Various Thermal Units

Process	Temperature Range (°F)	Residence Time
Pyrolysis	900 –500	12 – 15 minutes
Multiple-Hearth	1400 –1800	0.25 to 1.5 hours
Fluidized Bed	1400 –1800	A few seconds
Liquid Incinerators	1200 –3000	0.1 – 2 seconds
Direct Flame	1000 –1500	0.3 – 0.5 seconds
Catalytic Combustor	600 –1500	1 second
*Rotary Kiln (liquid)	1500 – 3000	2 seconds
*Rotary Kiln (solid)	1500 –3000	hours
Molten Salt	1500 – 1800	0.75 seconds
Multiple Chamber	1500 – 1800	secs. (gases), mins. (solids)
Wet Air Oxidation	300 – 550 (at 1500 psig)	10 – 30 minutes

* Cement kiln data are similar to rotary kiln data.

fluidized bed and enter heat-recovery or gas cleaning devices similar to the multiple-hearth incinerators.

Wastes enter the bed through nozzles located either above or within the bed. Excess air requirements for normal incineration are limited to about 40% above the stoichiometric combustion requirements of the waste.

These units experience problems caused by low ash fusion temperatures, which can be avoided by keeping temperatures below the ash fusion level or by adding chemicals that raise the fusion temperature.

Liquid Incinerators. These are the most flexible and economical (labor-free) units. The feed waste acts as a liquid at a viscosity below 10,000 Saybolt Second Units (SSU). Waste is atomized through a suspension burner, which is normally a rotary cup or a pressure atomizer. The burner nozzle is mounted at one end of the refractory lined chamber. Gases exit to the scrubbing equipment. These units are normally equipped with storage and blending tanks to ensure homogeneous flows.

Cement Kilns. A wide variety of hazardous wastes can be treated in cement kilns. Properly managed, they can accomplish what rotary kiln incinerators can without an increase in emissions or a negative environmental impact. Many of the candidate waste streams provide an alternative source of fuel resulting in resource recovery rather than simple waste destruction. Cement kilns can accommodate a variety of materials both as fuels and feed in the cementation process, however there are limitations on the use of hazardous waste materials, which can be divided into two categories:

1) Materials presenting process constraints such as high sodium and potassium

2) Materials unacceptable due to public opinion, such as polychlorinated biphenyl (PCB) and dioxin waste

Rotary Kilns. The rotary kiln is a cylindrical horizontal, refractory-lined shell that is mounted at an incline (about 5°). Rotation of the kiln causes mixing of the waste with combustion air. The ratio of length to diameter of the kiln ranges between 2:1 to 10:1 and it rotates at a speed ranging from 1 to 5 ppm. The combustion temperature ranges from 1500°F–3000°F, and generally samples have residence times ranging from seconds to hours depending on the waste introduced.

These units are effective when the size and nature of the waste precludes the use of other types of incinerators. A few examples of likely materials are debris, bottles, boxes, and pharmaceutical containers. Kilns can operate either in batch or continuous feeding modes.

Exhaust gases from the kiln pass through an afterburner chamber to ensure complete destruction of organic compounds, and then through a precooler, Venturi scrubber, and packed column prior to atmospheric release.

Molten Salt Incinerators. Molten salts baths comprised of 90% sodium carbonate and 10% sodium sulfate operate at temperatures ranging from 1500°F–1800°F. Sometimes lithium and potassium carbonates are used to improve encapsulation of heavy metals. Spent salts can be either recycled or landfilled.

Industrial Boilers. In **fire tube boilers**, heat is transferred from hot combustion products flowing inside tubes to the water surrounding them. Combustion takes place in a cylindrical furnace within the shell. Fire tubes run the length of the shell above and around the furnace. Gases from combustion travel forward through the tubes to the front of the boiler.

Water tube boilers circulate water throughout the combustion chamber through thousands of feet of steel tubing. Heat is transferred from the path of the flue gas into the adjacent water tubes.

Oceanic and Ship-Mounted Incinerators. Ship-mounted incinerators are used to dispose of organic or organometallic wastes that cannot be burned in conventional incinerators. A downside to these units is the stringent EPA scrutiny of their effluents.

Chemical Fixation and Solidification

Treatment of hazardous and nonhazardous wastes using chemical fixation/solidification (CFS) involves the use of raw materials such as fly ash, silica fume, slag, natural pozzolans from concrete, cement kiln dust, and lime flue dusts.

The terms **fixation** and **stabilization,** abundantly used in this industry, refer to the reduction or elimination of the toxicity or the hazard potential of a certain waste stream by lowering the solubility and leachability of the toxic or hazardous components. Another term most commonly used is **solidification**, which references the formation of monoliths with considerable structural strength. The difference between solidification and fixation is that the former does not necessarily involve a chemical reaction between waste and raw material, but only encapsulatation of the waste particles to limit the surface area exposed to leaching.

Micro- and macroencapsulation are used to define processes designed to minimize leachability. **Microencapsulation** yields a semi-homogeneous product resulting from the treatment of finely divided powders, sludges, or viscous liquids, whereas **macroencapsulation** is the encapsulation of solid waste with an impervious layer to eliminate the exposure to leaching. In general, chemical fixation/stabilization/solidification act together to bind the active components of waste in numerous fashions as described below.

Sorption. This treatment involves the addition of a sponge-like solid (such as sodium silicate, gypsum, activated carbon, *etc.*) to absorb any free liquid that might be present in the waste. This process acts only in a physical fashion to merely acquire the liquid fraction onto the surfaces of the solid. The solid materials do not interact chemically with the waste nor do they reduce the leachability of the hazardous components.

pH Adjustment. In this process, the wetted calcium oxide portion of the reagent contributes to neutralization of the hydronium ion, raising the pH to over 10 and hence precipitating the vulnerable heavy metals as their respective hydroxides. The availability of carbonates, phosphates, and sulfates also enhances precipitation of metals such as cadmium, nickel, and barium, respectively.

Pozzolan Interaction. Fine, noncrystalline silica from fly ash or cement kiln dust (CKD), and calcium from lime products give rise to a weak cementation process, thus providing a primary containment to isolate waste from the outside environment.

In this process, the available soluble silicates react with the polyvalent metal ions resulting in highly insoluble nonstoichiometric metal silicates in an amorphous matrix.

A pozzolanic reaction has certain limitations that are a function of particular variables. These are:

1) At least 30% lime is required to give sufficient product strength.

2) Water and wetting for dry wastes is a must to initiate reactions.

3) Oil and grease can negatively impact the quality of the residue.

4) Borates, chromates/dichromates, and sugars interfere with the bonding in the silicate matrix.

5) Pretreatment is required in most cases to meet the relevant LDRs such as D007 wastes and most organic species.

6) When strong lime is used (high calcium oxide), high pH results in two adverse phenomena: the formation of gas and/or leachate and the possible dissolution of some amphoteric metals such as lead, zinc and arsenic.

It is well established that cement-based processes are more effective than lime-based processes in displaying chemical resistance and integral strength.

Chemical Fixation of Organic Compounds. For a short period of time, the EPA introduced treatment standards for organic constituents (D012–D043) measured by TCLP, which allowed numerous treatment facilities to be inventive in finding methods to "hide" the organics. Such methods were: sorption on the surface, dilution, adsorption onto powdered or granular carbon added to the fixation medium, and/or volatilization by introducing "hot" lime into the formulation.

Due to the fact that the above treatment patterns were not effective in reducing the long-term leaching of these organics, EPA published the new LDRs (Land Disposal Restrictions) as measured by total concentration—thus requiring a destructive step prior to solidification.

Pretreatment technologies include: hydrolysis, biodegradation, chemical oxidation, chemical reduction, salt formation, low-temperature decomposition, dehalogenation, and catalytic destruction.

In general, the most common organic wastes encountered in chemical fixation/solidification facilities are:

1) Aqueous wastes containing less than 1000 ppm of organic constituents that are considered hazardous

2) Aqueous wastes containing less than 1000 ppm of nonhazardous organic constituents

3) Solvent- and oil-based organic wastes that are considered RCRA wastes

4) Aqueous wastes containing large concentrations of emulsified organics (up to 25%) that are considered hazardous under RCRA

5) Same as (4) but nonhazardous

6) Solid wastes contaminated with <1% of hazardous organics to cover all the above categories of waste mixtures

7) Mixed inorganic wastes with low organic constituents above the LDR levels

Chemical Fixation of Inorganic Compounds. This discussion will be limited to metals and nonmetals (and their compounds), which are hazardous wastes as defined under RCRA in 40 CFR 261.

Arsenic. Due to the fact that arsenic is classified as a nonmetal, (*i.e.*, it combines with metals to form arsenides, arsenites, and arsenates) it is difficult to precipitate in regular hydroxide precipitation reactions. It has been found that the most effective pathway for treatment of arsenic is the co-precipitation with iron as the arsenate prior to fixation:

$$4As(ONa)_3 + 5FeSO_4 + 4Ca(OH)_2 + 5H_2SO_4 \rightarrow \quad (26)$$
$$FeAs_2O_4 + 2FeAsO_4 + 6Na_2SO_4 + 2Fe(OH)_3 +$$
$$4CaSO_4 + 6H_2O$$

Barium and Compounds. Barium is a widely used element in processes such as heat treating, brick manufacturing, chemical diagnostics, pigments, plastics, fire-extinguishing foams, and pyrotechnics. Barium can be easily treated using chemical fixation in the presence of sulfates and carbonates and silicates to yield a solubility of about 1–5 mg/l, well below the TCLP regulatory level of 100 mg/l.

Cadmium and Its Compounds. Cadmium associates itself with lead, zinc, and copper. It is widely used in electroplating, batteries, pigments, catalysts, phosphors, photography, electronics, and plastic stabilizers.

Lime fixation media are used to take the operating pH up to about 10, where solubility is at its minimum for cadmium hydroxide (0.1 mg/l). To improve

precipitation, iron salts can render the leachable cadmium concentrations nondetectable. Carbonates present within the fixation medium will also improve cadmium fixation especially in low calcium oxide media.

Chromium and Its Compounds. Chromium compounds are widely used in plating/metal finishing, pigments, leather tanning, wood preservation (CCA), drilling muds, catalysts, coatings, *etc.*

The ideal chemical fixation medium is sulfide-rich pozzolan, which will cause chemical reduction of Cr (+6) to Cr (+3), and consequent precipitation as $Cr(OH)_3$, locked in a tight matrix. Alternatively, ferrous sulfate can be added as a reducing agent.

Lead and Its Compounds. Lead is widely used in numerous industries as the metal-lead oxides, halides, tellurides, nitrates, acetates, carbonates, sulfates, peroxides, silicates, *etc.* Lead is also used as an organometallic product, such as tetraethyl lead. Since lead is amphoteric, pH must be closely monitored to avoid solubilization. The addition of phosphates to divalent lead compounds renders them highly insoluble (0.1 mg/l). The presence of active silicates with phosphate precipitates leads to non-detectable concentrations.

Mercury and Its Compounds. Compounds of mercury can only be treated at total concentrations less than 250 ppm due to RCRA regulatory limitations. Despite their high cumulative toxicity, mercury compounds are widely used in agriculture, catalysis, dental work, electrical equipment, chlorine production, magnets, paints, pigments, paper, and pharmaceuticals.

Mercury salts have been successfully treated at low and medium concentrations using chemical fixation technology coupled with a solubility modifier, such as sulfides, thiourea, and thiocarbamates. These reduce the soluble mercury to levels reaching 10^{-48} g/l. In many instances, activated carbon is introduced to adsorb mercury, specifically when organo-mercury compounds are present.

Selenium and Its Compounds. Selenium coexists in sulfur- and copper-bearing deposits and is widely used in ceramics, glass, pigments, rubber, lubricants, electronics, photocopying machines, and pharmaceuticals. Selenium treatment is based on chemical fixation with high-silicate, iron-rich matrices, which can be improved by the introduction of sulfide ions.

Silver and Its Compounds. Most silver-bearing wastes are associated with those bearing high concentrations of lead, gold, copper, and zinc.

Silver is used widely in jewelry, photographic processes, electronics, catalysis, solders, mirrors, and dental products.

Most silver wastes sent to TSDs are of low unrecoverable concentrations. Treatment can be relatively easy (except in cases of silver cyanides). Silver precipitates as silver chloride when high chlorides are present with a solubility product of 1.7×10^{-10}. Silver complexes are common in the photographic industry. Silver ammonia complexes are more difficult to treat and require an oxidation step such as alkaline chlorination to achieve precipitation. Chemical fixation agents showing high efficiency toward silver are those having high silicates with calcium chloride. These agents enhance both speed of cementation and, due to the common-ion effect, the precipitation of silver as AgCl.

Bibliography

Brunner, Calvin R. *Handbook of Incineration Systems.* Reston, VA: Incinerator Consultants, Inc., 1991.

Conner, Jesse R. *Chemical Fixation and Solidification of Hazardous Wastes.* New York, NY: Van Nostrand Reinhold, 1990.

Cheremisinoff, Paul. *Waste Incineration Pocket Handbook.* Northbrook, IL: Pudvan Publishing, 1991.

Environmental Protection Agency. *Test Methods for Evaluating Solid Waste, Physical/Chemical Methods, SW–846,* 4th ed. Washington, DC: EPA, 1996.

Freeman, H. M. *Standard Handbook of Hazardous Waste Treatment and Disposal,* 2nd ed. New York, NY: McGraw-Hill, 1997.

Malhotra, V. M., Ed. *Fly Ash, Silica Fume, Slag and Natural Pozzolans in Concrete.* Vols. 1 and 2. Proceedings, Third International Conference. Trondheim, Norway: American Concrete Institute, 1989.

Manahan, Stanley E. *Environmental Chemistry,* 5th ed. Chelsea, MI: Lewis Publishers, 1991.

Tropy, Michael F. *Anaerobic Treatment of Industrial Wastewaters.* Pollution Technology Review #154. Park Ridge, NJ: Noyes Data Corp., 1988.

Wilk, L., S. Palmer, and M. Brelon. *Corrosive Containing Waste Treatment Technology.* Pollution Technology Review #159. Park Ridge, NJ: Noyes Data Corp., 1988.

Gazi A. George *is a Senior Vice President for US Liquids of Houston, Texas, and is responsible for five hazardous waste treatment facilities: USL City Environmental, Inc., Detroit; USL City Environmental Services of Florida, Tampa; ReClaim Environmental, Shreveport; and ROMIC Environmental of Arizona and California. Dr. George obtained his PhD in Chemistry in 1976 from the University of Lancaster, England, and spent five years in nuclear waste management in Austria, Belgium, France, Germany, Holland, England, and the Middle East. In 1981, he was appointed as the Technical Services Manager and Industrial Hygienist for DEVROJohnson & Johnson International facilities in Scotland. From 1985-88 he was the laboratory and technical manager for American NUKEM at their Detroit CyanoKem facility. From 1988-82, Dr. George worked as the Corporate Director then Vice President of Envotech, Michigan (currently EQ). He then joined City Management Corporation to develop the City Environmental business at their Detroit (CEI) and Florida (CESF) facilities.*

Mixed Waste Management

Thomas Hillmer, CHMM

Chapter Review

This chapter covers in detail the regulations that are applicable to the commercial nuclear industry regarding mixed waste. It reviews the regulatory history of mixed waste and what should be included in a mixed waste management program. It discusses the dilemma that faces facilities that store mixed waste. Every nuclear power facility that relies on the Environmental Protection Agency's notice of nonenforcement may find itself in violation of storage restrictions for mixed waste, because there are no approved disposal facilities that are licensed to take all the various types and forms of mixed waste that are being generated. The only choices for facilities that have generated mixed waste in a form that cannot be disposed of is to either obtain a Part B Permit (40 CFR 264) for storage of mixed waste or store the waste in violation of the law.

Introduction

Mixed waste is defined as any waste that contains a hazardous waste component and also is radioactive. Both the Environmental Protection Agency (EPA) and the Nuclear Regulatory Commission (NRC) would regulate this mixed waste. The NRC generally regulates commercial and non-Department of Energy (DOE) Federal facilities. DOE currently is self-regulating, and its orders apply to DOE sites and contractors. Using their authority, NRC and DOE regulate mixed waste with regard to radiation safety. Using its authority, EPA regulates mixed waste with regard to hazardous waste safety. The NRC is authorized by the Atomic Energy Act (AEA) to issue licenses to commercial users of radioactive materials. The Resource Conservation and Recovery Act (RCRA) gives EPA the authority to control hazardous waste from *cradle-to-grave*. Once a waste is determined to be a mixed waste, the waste handlers must comply with both AEA and RCRA statutes and regulations. However, the provisions in Section 1006(a) of RCRA allow the AEA to take precedence in the event provisions of requirements of the two acts are found to be inconsistent.

The issue of mixed waste disposal is becoming critical to the nuclear power generating industry. When the United States Congress envisioned a country served by nuclear power facilities in the mid-1960s, they failed to envision that regulatory interpretations would create disposal problems for mixtures of nuclear and hazardous waste in the decade of the 1990s. It is an issue that is both technical and political but one that the regulatory agencies are reluctant to provide direction on because of the potential political ramifications.

Yet, every nuclear power facility that relies on the EPA's notice of nonenforcement now finds itself in violation of storage restrictions for mixed waste, because there are no approved disposal facilities that are licensed to take all the various types and forms of mixed waste that have been generated (EPA, 1997). Both the NRC and EPA are fully aware of this dilemma but continue to delay decisionmaking that would enable the generator to meet a single set of regulations.

Currently, as noted above, no single facility in the United States can accept, process, and dispose of all the various types or classifications of mixed waste that are generated. In recognition of this fact, the EPA issued a statement of nonenforcement policy, which carries no legal weight (56 FR 43730, *et seq.,* 63 FR 59989–92, and 63 FR 59991–2). On November 6, 1998, the Environmental Protection

Agency (EPA) extended for three years, until October 31, 2001, its policy on enforcement of RCRA Section 3004(j) (Storage Prohibition at Facilities Generating Mixed Radioactive/Hazardous Waste or *mixed waste policy*). EPA will continue to consider as *relatively low priority* enforcement actions against individuals who store mixed waste when: (1) there is no "available treatment or disposal capacity" for the mixed wastes and (2) the wastes are stored according to "prudent waste management practices" to minimize risk to public health and the environment. This management in the policy includes

- Inventory and compliance assessment of storage areas, including recordkeeping, regular inspections, and voluntary corrective action in the event of noted deficiencies.

- Identification of mixed wastes, including identification, source, generation rate and volumes, and process information of the hazardous components of the waste.

- Waste minimization plans, including active measures to avoid the generation of mixed wastes.

- Good-faith efforts, including demonstration of ongoing good-faith efforts to locate and utilize mixed waste treatment technology and disposal capacity.

The extended policy is limited in scope to only those categories of mixed waste for which there is no treatment or disposal capacity. EPA states in the extended policy that it believes that

> (1) currently treatment is available for most low-level mixed wastes, but treatment continues to be unavailable for a few wastes, such as mixed wastes containing dioxins, PCBs, and lead-based paint solids, and wastes with very high levels of radioactivity; and (2) where treatment technology is available, there is excess capacity at the commercial mixed waste treatment facilities.

EPA defines ***available treatment technology and disposal capacity*** to mean any facility that is

> commercially available to treat or dispose of a particular waste and has either (1) a RCRA permit or interim status, (2) a research, development, and demonstration permit under 40 CFR 270.65; or (3) a land treatment permit under 40 CFR 270.63.

Few management options exist for the mixed waste generators who are at risk of being in noncompliance with the storage prohibition regulations. One of the problems that has been generated by the EPA position is that waste which once could be disposed of as low-level radioactive waste now must also meet EPA land ban requirements. These requirements stipulate that waste be processed in a licensed treatment facility with a process that has been approved for the specific waste in question. Nuclear facilities have been designed with processing capabilities necessary to meet licensing requirements for radioactive waste and not EPA land ban requirements. A significant effort in both time and money will be required to develop a treatment method, obtain the needed NRC license, and retrofit that process and equipment into existing nuclear facilities. Another problem with meeting EPA regulations is the time necessary to obtain permits, which can vary depending on whether the generator is regulated directly by the EPA or by an authorized state. Also, will the state dictate additional restrictions that the EPA might not require?

There are many other questions and concerns created by the EPA interpretation, such as:

- Which other environmental regulations that are not now applicable become applicable after a facility becomes RCRA-permitted?

- The issue of closure of a permitted storage facility, and how that affects an operating nuclear facility and how the closure affects the NRC decommissioning of a facility.

All these issues, along with the severe and continuing shortage of off-site mixed waste treatment, storage, or disposal facilities available for management of mixed waste, ensures that generators of even the smallest quantities of mixed waste will be forced to accumulate those wastes on-site for indefinite periods. Facilities should not depend on this policy past October 31, 2001 because it has not been shown in court that it will block citizen suits under the environmental regulations such as Resource Conservation and Recovery Act or the Superfund Amendments and Reauthorization Act Title II. This in turn will force continued expenditures of resources, by both generators and regulators, in pursuit of hazardous waste permits that add no

measures of environmental protection over that already provided by existing NRC regulations.

Applicable Regulations

The regulations that affect mixed waste stem from many laws. The most important are Atomic Energy Act (AEA, 1954), the Resource Conservation and Recovery Act (RCRA, 1976), the Clean Water Act (CWA, 1977), the Hazardous and Solid Waste Amendments (HSWA, 1984), the Superfund Amendments and Reauthorization Act (SARA, 1986), and the Pollution Prevention Act (PPA, 1990). From these and other acts were derived Federal regulations such as the *Code of Federal Regulations*, Energy, Title 10 (10 CFR); the *Code of Federal Regulations*, Protection of the Environment, Title 40 (40 CFR); and the *Code of Federal Regulations*, Transportation, Title 49 (49 CFR).

Table 1 lists applicable NRC, EPA, Occupational Safety and Health Administration, (OSHA), and Department of Transportation (DOT) regulatory references that must be reviewed in order to develop a complete mixed waste management plan.

The Federal government may grant authority to the states to regulate in its place. When the NRC gives that authority to a state, the state is called an ***agreement state***. When the EPA grants the state authority, the state is called an ***authorized state***. A list of agreement states is located in 57 FR 54932.

Regulatory History

Regulation of waste that contains both a radioactive component and also a RCRA hazardous component came into effect on July 3, 1986 (Laswell and Doohan, 1993). On that date, the EPA, in a *Federal Register* notice, stated that the EPA would regulate any waste that contained hazardous waste and was also radioactive. This was regardless of whether some other regulatory body already regulated the waste. In 1987 the NRC and EPA issued a letter giving additional guidance on the definition and identification of mixed waste (EPA, 1987).

Table 1.
Summary of Regulatory Requirements for the Management of Mixed Waste

NRC Regulations Regarding the Management of Mixed Waste	
10 CFR 20	Standards for protection against radiation
10 CFR 61	Licensing requirements for the land disposal of radioactive waste
10 CFR 71	Packaging and transportation of radioactive material
EPA Regulations Regarding the Management and Storage of Mixed Waste	
40 CFR 260	Hazardous waste management system: General
40 CFR 261	Identification and listing of hazardous waste
40 CFR 262	Standards applicable to generators of hazardous waste
40 CFR 264	Standards for owners and operators of hazardous waste treatment, storage, and disposal facilities
40 CFR 265	Interim status standards for owners and operators of hazardous waste treatment, storage, and disposal facilities
40 CFR 268	Land disposal restrictions
40 CFR 270	EPA administered permit programs: The Hazardous Waste Permit Program
40 CFR 272	Approved state hazardous waste management programs
DOT Regulations Regarding Mixed Waste Transportation	
49 CFR 171	General information, regulations, and definitions
49 CFR 172	Hazardous materials tables and hazardous materials communication regulations
OSHA Regulations Regarding Management of Mixed Waste	
29 CFR 1910.1200	Hazardous waste operations, emergency response
29 CFR 1910.120	Hazard communication and the training of workers

Prior to the EPA determining that they had to regulate the hazardous waste portion of any nuclear waste generated, all radioactive waste was processed, usually on-site, and disposed of as radioactive waste. After the EPA made its determination it became necessary for any facility that processed or stored its radioactive waste, which also contained any hazardous waste, to obtain a treatment, storage, or disposal (TSD) permit. This also meant that any radioactive disposal site would also become a TSD facility if they accepted mixed waste. When the EPA made their determination, no facility was permitted by both the NRC and EPA to process or dispose of mixed waste. With no place licensed to accept all types or forms of mixed waste, the generator had no options but to begin to store mixed waste on its own site.

A literature review was undertaken to determine if a detailed comparison between the NRC regulations and the EPA regulations, which cover mixed waste, existed and was part of the public record. The review found that the EPA had received a detailed comparison as part of submitted comments on the proposed Hazardous Waste Management Rule (Hillmer, 1996).

Historically, the volume of waste affected by the EPA decision is very small. During 1990 the NRC and EPA conducted a survey with the objective of compiling a national profile on the volumes, characteristics, and treatability of commercially generated low-level mixed waste (NRC, 1992). The

Table 2.
Generation of Mixed Waste by Facility Type

Facility Type	Amount of Mixed Waste Generated (in Cubic Feet/Year)
Academic	20,421
Industrial	19,056
Governmental	18,324
Nuclear Utilities	13,275
Medical	10,151

profile was divided into five major facility categories: academic, industrial, medical, NRC-Agreement State-licensed government facilities, and nuclear utilities. A breakdown of the annual generation of mixed waste by reporting facility grouping is shown in Table 2.

The study found that for 1990, an estimated 81,227 cubic feet of mixed waste was generated by reporting facilities across the United States. Using weighting

factors, a statistically valid estimate of the national profile yielded a generation of approximately 140,000 cubic feet. Of this amount, approximately 71% was organic solvents such as chlorofluorocarbons (CFCs), corrosive organics. Waste oil made up 18%, toxic metals made up 3%, and "Other" waste made up the remaining 8%. This study indicates that almost half the waste being generated is not being properly classified as mixed waste and that a significant amount of mixed waste is not being stored as mixed waste but being disposed of as nonmixed waste. The total volume of classified mixed waste in storage, at that time, was estimated at 75,000 cubic feet. These values did not include Federal Government facilities controlled by the Department of Defense (DoD) and Department of Energy (DOE).

Today, the generation of mixed waste has decreased in the commercial electric utility sector to nearly zero. This result is primarily due to pollution prevention programs which have been very successful in finding nonhazardous chemicals to replace hazardous chemicals previously used. To put the waste values in perspective, a comparison with the generation and storage of mixed waste at one DOE site is included. In November of 1995, the Idaho National Engineering Laboratory (INEL) estimated that it will generate an average of 20,000 cubic feet per year over the next 5 years. This value represents a significant reduction in the past generation rate. This new waste generation will be added to the 2,700,000 cubic feet already in storage at INEL. This represents about 12% of the total of all DOE mixed waste now in storage (INEL, 1995). The DOE has projected that it will generate and store, at 37 DOE sites in 22 states, an estimated 8,000,000 cubic feet of low-level mixed waste (LLMW) over the next 20 years.

DOE waste contains an additional subset of mixed waste. ***Mixed transuranic waste (MTRU)*** is waste that has a hazardous component and radioactive elements heavier than uranium. The radioactivity in the MTRU must be greater than 100 nCi/g and co-mingled with RCRA hazardous constituents. The principal hazard from MTRU is from alpha radiation through inhalation or ingestion. MTRU is primarily generated from nuclear weapons fabrication, plutonium-bearing reactor fuel fabrication, and spent-fuel reprocessing. Approximately 55% of DOE's transuranic (TRU) waste is MTRU. MTRU currently is being treated and stored at six DOE sites.

DOE has developed Site Treatment Plans to handle its mixed wastes under the review of EPA or its authorized states. These are being implemented by orders issued by EPA or the State regulatory authority. DOE is also developing a Waste Management Programmatic Environmental Impact Statement for managing treatment, storage, and disposal of radioactive and hazardous waste. This plan will provide environmental input for DOE's proposed action of identifying future configurations for selecting waste management facilities.

In 1992 there appeared to be no short-term solution to the mixed waste problem, and today the situation has not changed. As Laswell and Doohan pointed out in 1992, the best strategy at that time was to eliminate the production of radioactive waste that also contained hazardous waste. This is still the best policy, but it is not always possible. In some cases, where cleaning or maintenance is performed using a specified reagent, there are only hazardous chemicals available or suitable for use. If this function takes place in the presence of radioactivity or radioactive contamination, the result is the generation of a mixed waste.

Dual Regulation

Is dual regulation necessary? That question has been asked since the inception of the EPA position that they must also regulate mixed waste. The NRC conducted an analysis of low-level radioactive waste and EPA hazardous regulations in view of the generation of mixed waste and found that in general there was no conflict in the regulations (NRC, 1985). This fact needs to be clarified, since dual regulation has, in fact, meant that waste that could once have been shipped and disposed of at a low-level waste site now has to remain on site, in storage, since the waste may not meet the specific burial requirements of the single commercial mixed waste burial facility. The nuclear industry attempted to show that dual regulation provided no more protection to the public and the environment. They claimed that adequate regulations existed under the NRC regulatory framework by comparing the EPA regulations for storage of hazardous waste in tanks to the regulations governing the storage of low-level radioactive liquid (Envirosphere, 1988). This study was presented to the NRC and EPA, but the EPA maintained that even though they believed the regulations were equivalent they were legally required to regulate the hazardous portion of low-level radioactive waste. This position has been challenged as being only a political position rather than a legal fact.

In 1993, another study by two prominent lawyers also reviewed the regulations and again supported the conclusion that either the EPA or NRC regulations adequately protected the human health and environment. The researchers also stated that no discernible benefit to human health or the environment resulted from dual regulation. They concluded that until regulators recognized that the mixed waste crisis stemmed from jurisdictional attitudes (whether the NRC or EPA had the right to regulate this waste) and not a real physical property (whether there was a higher risk to the public from the radioactive or hazardous component of the waste), the regulatory system could only become more unworkable (Thompson and Goo, 1993).

In lieu of a regulatory solution, the generator of mixed waste must comply with dual regulations in the best way possible. In cases where mixed waste cannot be processed or disposed, requirements for storage must be met. In states that have taken the responsibility to regulate mixed waste, the first step is to apply for a permit.

Mixed Waste Management

The *Mixed Waste Management Guidelines* (EPRI, 1993), developed by utility and industry experts, offers a complete overview of the development and management of mixed waste programs and provides guidance for developing or assessing the adequacy of mixed waste management programs. The major management issues covered included generator status, waste characterization, permit requirements, mixed waste transportation requirements, land disposal restrictions, and mixed waste source reduction. An earlier study by Nuclear Management and Resources Council, Inc. (NUMARC) detailed similar information and management recommendations (NUMARC, 1990). This information, if followed, will allow the generator to be in compliance with EPA regulations. The EPA has written a document that outlines how to conduct inspections of mixed waste facilities and also is a useful tool to ensure compliance by utilizing the inspection document to do self-auditing (EPA, 1991). In 1994 Edison Electric Institute (EEI), while working with the Electric Power Research Institute (EPRI), issued a compliance manual for the nuclear industry (EEI, 1994). Most recently, the EPA and NRC issued additional guidance on storage of mixed waste (60 FR 40204).

One of the key elements of a comprehensive mixed waste management program is a program to properly characterize waste streams that have the potential to be a mixed waste. Proper characterization of a potential mixed waste stream includes a determination of whether the waste is radioactive, and a solid waste and/or hazardous waste. In addition, the characterization process should consider how the waste would ultimately be managed so that appropriate information is available and documented relative to proper handling, treatment, storage, and/or disposal of the waste.

Currently, there are only a few facilities that are authorized to receive certain categories of mixed waste for treatment or storage. EPA also identifies those facilities that it believes are capable of providing mixed waste treatment, storage, and disposal on the EPA mixed waste homepage:

<http://www.epa.gov/radiation/mixed-waste>

The facilities identified include Envirocare of Utah, Inc. in Utah; Diversified Scientific Services, Inc. (DSSI) in Kingston, Tennessee; Molten Metal Technology (MMT), Massachusetts (although MMT's Tennessee facility reportedly is no longer processing mixed waste); NSSI/Source and Services, Inc. in Houston, Texas; and Perma-Fix Environmental Services, Florida. Other facilities that may offer the same services are: Ramp Industries (RAMP) in Denver, Colorado; and Allied Technology Group, Inc. (ATG) with headquarters in Fremont, California. It is very important that an adequate quality (due diligence) review be performed before using any treatment, storage, or disposal facility, in order to ensure that your waste will be handled in accordance with all applicable regulations.

DSSI has a license to process liquid mixed waste in RCRA categories F001 through F005, as well as D001 solvents and some RCRA D, U, and P listed waste. Envirocare is licensed and permitted to dispose of mixed waste. They can dispose of characteristic waste codes D001 through D043, listed waste F001 through F012, F019, F024, F028, F039, K011, K013, K050 through K052, K061, K069, and most P- and U- listed waste. NSSI is licensed and permitted to store and treat some mixed waste. RAMP can treat a variety of materials, along with ATG who are in the RCRA Part B Permitting process (40 CFR 264).

Mixed Waste Identification

In, *Nuclear Utility Mixed Waste Stream Characterization Study* (EPRI, 1994), EPRI outlines the characterization process, summarizes industry experience relative to key characterization issues, and provides industry data (including more than 400 actual sample results) on the majority of plant processes identified as having the potential to generate mixed waste.

The regulatory guidance for identification (57 FR 11798, *et seq.*), for sampling (EPA, 1986), and for testing (57 FR 10508) have been written and were used in identifying and evaluating the proper management of mixed waste. Additionally, the EPA and NRC supplied guidance in a jointly issued letter of October 4, 1989, which contained an attachment entitled "Definition and Identification of Commercial Low-Level Radioactive and Hazardous Waste."

Mixed Waste Determination

The key to the entire need for a permit is whether or not a mixed waste has been generated and then must be stored. In most nuclear facilities, the determination as to whether a waste is a mixed waste is made after the material in question has been identified as a waste material. Most facilities have a chemical control program or a pollution prevention program, which is the primary method of controlling the use of chemicals. These programs utilize Material Safety Data Sheets to determine how to handle any waste generated using the chemical. One of these program functions is to ensure that hazardous material, which could become a hazardous waste, does not enter a radiological controlled area

(RCA) and is the main method of ensuring that mixed waste is not generated.

When hazardous materials are used in an RCA and have become a waste, the first determination that is made is whether the material has become radioactively contaminated. The methodology developed in this chapter along with the permit checklist comprise the program necessary to determine if you have generated mixed waste and how to write a Mixed Waste Storage Permit Application. The following questions are included as a guide to determine if an unknown material is a mixed waste. The questions are only guidelines to be used to initially characterize and classify the waste. A detailed review of the use of the material, a review of any chemical analysis, and a review of the State and Federal regulations should be made to ensure proper final characterization.

There are three waste types that the waste must be in order for it to be a mixed waste. The first question that must be answered is: Is the waste radioactive? The second question is: Is the waste a solid waste or is it exempted? The last question is: Does the waste meet the requirements of a hazardous waste? The legal definitions of these types of waste are the key to determining if your waste is *mixed waste*.

Proper characterization of potential mixed waste streams is the first step in proper management of mixed waste. After you have characterized the waste stream you will know if your waste must be processed to meet land ban restrictions, shipped off-site for processing or disposal within the required time frame, or stored in a permitted mixed waste facility.

Special Management Concerns

Mixed waste has other special management concerns that also must be considered. Because of dual regulation mixed waste must be stored, so as to be in compliance with NRC and EPA guidance on storage of the waste. EPRI wrote a management guideline on the proper management and compliance with NRC storage regulations (EPRI, 1992). The EPA regulations on storage are contained in 40 CFR 264 and 265. The radiation as well as the chemical hazard associated with mixed waste must also be

considered. Basic radiation protection concerns and management practices associated with sampling and storage must be incorporated in operating procedures (Gollnick, 1988). These procedures will need to be incorporated into a contingency plan. How the facility meets the EPA requirements for Pollution Prevention (PL 101–508) and the NRC policy on volume reduction must be included in station procedures (46 FR 51100–1).

When mixed waste is to be processed or shipped offsite, compliance with EPA, NRC, and DOT regulations must be met (49 CFR 171 and 172, 10 CFR 71 and 57 FR 14500, *et seq.*). Training requirements for personnel working with mixed waste are specified by the EPA, NRC, and OSHA (29 CFR 1910.1200). The training must also include specific training on DOT regulations associated with the shipment of hazardous waste (EPRI, 1993).

State Regulations

Each state has its own set of regulations that are based on the Federal requirements. You must review your State regulations to ensure that you are managing your mixed waste, since a state may add additional requirements to those hazardous waste aspects of your waste. It is important to note that the enforcement policy is not applicable in states that are authorized for the RCRA base program but are not authorized for mixed wastes. In these states, mixed waste is not subject to the federal RCRA program and therefore not subject to the Section 3004(j) storage prohibition. For specific information on a state's authorization status, see the EPA's State Authorization Homepage:

<www.epa.gov/epaoswer/hazwaste/state/>

Regulatory Outlook

As pointed out above, there is no single facility in the United States that can accept, process, and dispose of mixed waste. Because of that fact, the EPA issued a statement of nonenforcement policy. Since that policy carries no legal weight in EPA-authorized states, the generators of mixed waste in authorized states are at risk of being found in noncompliance by the state or of having a citizen suit brought against those generators. Still, without the enforcement policy, the industry would be at a greater risk and that risk would extend to all states. Only a few facilities in the United States can process, store, or dispose of some kinds of mixed waste. Therefore the need to be permitted is paramount, from a liability standpoint, for any mixed waste generator. However, obtaining a permit is an answer that most generators do not want to utilize.

The history of mixed waste is one of continual issuance of guidance documents to assist in the management of mixed waste. These guidance documents offer no real solution but are aimed at reducing the regulatory confusion. The only real assistance that has been given to the generators is the EPA Mixed Waste Enforcement Policy. That policy was scheduled to terminate on October 31, 2001. On March 1st, 1999, the EPA issued an Advance Notice of Proposed Rulemaking entitled, "Approach to Reinventing Regulations on Storing Mixed Low-Level Radioactive Waste," which deals with the possible options the EPA has developed for alternative management and storage of mixed waste. This new rulemaking could give some valuable relief from some of the aspects of dual regulation.

There are a few possible options that may be available in the future. One option is a proposal by the NRC to eliminate certain waste from their jurisdiction and defer them to the authority of RCRA (EPA, 1995a). This is not an option that the nuclear industry supports. The Department of Transportation in its regulations for the transport of hazardous materials has addressed the risk associated with mixed waste. It has concluded that a more restrictive hazard classification should be placed on radioactive shipments than just that of hazardous waste shipments (49 CFR 173.2). These classifications are based on risk to the public due to the consequences of a transportation accident. Because another Federal agency has determined that radioactive material is more of a danger to the public than hazardous material, it gives credence to the belief that the NRC is the appropriate organization to regulate mixed waste.

The EPA has also proposed a new regulation called the Hazardous Waste Management System: Identification and Listing of Hazardous Waste: Hazardous Waste Identification Rule (HWIR).

According to the *EPA News Watch* (EPA, 1995b), this rule offers a cost-effective means of regulating low-risk hazardous waste. The rule would set exit levels for listed hazardous waste. Exit levels are concentration levels that have been determined to be low enough so as to pose little risk of adverse effect to the environment or the public. These levels are based on risk assessments for the waste. One benefit to the mixed waste generator would be that most of the mixed waste has been deemed mixed waste based on the mixture rule. This rule requires any waste (either radioactive or nonradioactive) that has come into contact with a listed waste to be classified as a listed waste even though the waste has low concentrations of listed substances.

Another management option contained in the proposed rule is the contingent management option for the DOE. The EPA is proposing the establishment of conditional exit levels for DOE mixed waste managed in AEA-regulated disposal facilities based on cancer risk to the public for that site.

The EPA has stated that it intends to sign a Notice of Proposed Rule Making by the end of October 1999 that will seek comment on an exemption from the hazardous waste disposal regulation and other regulatory relief as appropriate for commercial mixed waste. It will also send a written recommendation to the EPA Regions and States saying that Part B permits should generally not be called which are solely related to the storage of commercial mixed waste (EPA, 1997).

Summary

The future actions of the EPA will relieve some of the concern over the management of mixed waste. However until specific regulatory action takes place, no real relief for the generators of mixed waste is expected. The reality that dual regulation is not necessary has been documented, and the EPA is aware of the protection afforded mixed waste under NRC regulations (Hillmer, 1996). The complexity and number of regulations involved in the management of mixed waste has been increasing since dual regulation of mixed waste began; thus it is imperative that mixed waste be managed properly and that all precautions be made to adequately identify and correctly characterize mixed waste. Historically, various organizations have attempted to assist in the management of mixed waste by publishing guidance documents. These documents should be reviewed when implementing a mixed waste management program.

Bibliography

"Approach to Reinventing Regulations on Storing Mixed Low-Level Radioactive Waste." *Federal Register* 64 (1 March 1999): 10064.

Edison Electric Institute. *Mixed Waste Storage and Treatment Regulatory/Compliance Manual.* EPRI TR–104223. Washington, DC: EPRI, 1994.

Electric Power Research Institute. *Guidelines for Interim Storage of Low-Level Waste.* Final Report, TR–101669, Research Project 3800. Palo Alto, CA: EPRI, 1992.

Electric Power Research Institute. *Mixed Waste Management Guidelines.* EPRI TR–103344. Palo Alto, CA: EPRI, 1993.

Electric Power Research Institute. *New DOT Training Requirements for Hazardous Material Employees and Implementation by Nuclear Utilities.* TR–102662, Project 2691–13. Palo Alto, CA: EPRI, 1993.

Electric Power Research Institute. *Nuclear Utility Mixed Waste Stream Characterization Study.* EPRI TR–104400. Washington DC: EPRI, 1994.

Environmental Protection Agency. "EPA Newswatch." *EH&S Developments Newsletter.* CPI Electronic Publishing. pp 5-6, 30 November 1995.

Environmental Protection Agency. "EPA, NRC Eyeing Option to Delete Mixed Waste Rules." *Environmental Policy Alert*, p. 2, 13 September 1995.

Environmental Protection Agency. *Letter from EPA Deputy Administrator, Mr. Fred Hansen, to Mr. Doug Green, legal consul for the firm Piper & Marbury.* April 7, 1997.

Environmental Protection Agency. *Test Methods for Evaluating Solid Waste, Physical/Chemical Methods.* SW–846. Washington DC: GPO, 1986.

Environmental Protection Agency, Office of Waste Programs Enforcement. *Conducting RCRA Inspections at Mixed Waste Facilities.* OSWER 9938.9. Washington DC: GPO, 1991.

Environmental Protection Agency and Nuclear Regulatory Commission. *Guidance on the Definition and Identification of Commercial Mixed Low-Level Radioactive and Hazardous Waste and Answers to Anticipated Questions.* Washington DC: EPA, 1/8/87.

Envirosphere Company. *Comparative Assessment of the Environmental Protection Agency's Regulations for Hazardous Waste Tank Systems (40 CFR 265, Subpart J) and Comparable Nuclear Regulatory Commission Requirements.* Washington, DC: Edison Electric Institute, 1988.

"Extension of the Policy on Enforcement of RCRA Section 3004(J) Storage Prohibition at Facilities Generating Mixed Radioactive/Hazardous Waste." Enforcement Policy." *Federal Register* 63 (6 November 1998):59989–92.

Gollnick, D. *Basic Radiation Protection Technology.* ISBN 0-916339–04–1. Altadena, CA: Pacific Radiation Corporation, 1988.

"Guidance on the Storage of Mixed Radioactive and Hazardous Waste." *Federal Register* 60 (7 August 1995) : 40204, *et seq.*

"Hazardous Waste Management System; Identification and Listing of Hazardous Waste; Toxicity Characteristic Revisions; Final Rule." *Federal Register* 55, (29 March 1990): 11798, *et seq.*

Hillmer, T. P. *A Case Study on the Development of a RCRA Part B Permit for Storage of Mixed Waste at a Nuclear Facility.* Unpublished master degree practicum. Phoenix, AZ: Arizona State University, 1996.

Idaho National Engineering Laboratory. *Site Treatment Plan Summary.* Idaho Falls, ID: Department of Energy, Idaho Operations Office, 1995.

Laswell, D. and M. Doohan. "Mixed Waste Dilemma: A Mixed Regulatory Bag?" *Environmental Protection*, pp. 37–42, March 1993.

"Low-Level Waste Shipment Manifest Information and Reporting." *Federal Register* 57 (21 April 1992): 14500, *et seq.*

"Mixed Waste Enforcement Policy." *Federal Register* 56 (29 August 1991): 43730, *et seq.*

"NRC Volume Reduction Policy (Generic Letter No. 81-39)." *Federal Register*. 46 (16 October 1981): 51100–1.

Nuclear Management and Resources Council, Inc. *The Management of Mixed Low-Level Radioactive Waste in the Nuclear Power Industry*. NUMARC/NESP–006. Washington, DC: NUMARC, 1990.

Nuclear Regulatory Commission. *An Analysis of Low-Level Waste: Review of Hazardous Waste Regulations and Identification of Radioactive Mixed Waste*. NUREG–CR–4406. Washington, DC: GPO, 1985.

Nuclear Regulatory Commission. *National Profile on Commercially Generated Low-Level Radioactive Mixed Waste*. NUREG/CR–5938 ORNL–6731. Washington, DC: GPO, 1992.

"Pollution Prevention Act of 1990." (PL 101–508), *United States Statutes at Large*. 104 Stat. 1388.

"Proposed Guidance Document on the Testing of Mixed Radioactive and Hazardous Waste." *Federal Register* 57 (26 March 1992): 10508.

"State Authorization for Mixed Waste Programs." *Federal Register* 57 (23 November 1992): 54932, *et seq.*

Thompson, A. and M. Goo. "Mixed Waste: A Way to Solve the Quandary." *Environmental Law Reporter*, pp. 10705–10719, December, 1993.

Thomas Hillmer *is an Environmental Consultant for Arizona Public Service Company at the Palo Verde Nuclear Generating Station. He has received a Master of Technology degree in Industrial Technology with emphasis in hazardous materials management from Arizona State University, a MFA in Management from Western International University, Phoenix, Arizona (where he graduated with distinction), and a BS degree in Earth Science with major concentration in Environmental Studies from the University of Wisconsin-Parkside, Kenosha, Wisconsin. Mr. Hillmer has worked in the nuclear industry for more than 25 years, managing radioactive and hazardous waste. He has served on various technical advisory committees of the Electric Power Research Institute, Edison Electric Institute's Utility Nuclear Waste Management Group, and the Nuclear Management Resource Council, dealing with pending waste disposal regulations. Mr. Hillmer has published and presented nearly two dozen papers on environmental management, waste processing, volume reduction, decontamination, storage, and disposal of nuclear and mixed waste.*

Part IX

Chemical Perspectives

CHAPTER **40**

Introduction to the Chemistry of Hazardous Materials

Richard E. Hagen, PhD, CHMM

Introduction

Goals of this Chapter

This chapter is entitled an "Introduction to the Chemistry of Hazardous Materials" because a complete presentation of this topic would fill a small library. The identification of salient points in this field, and their presentation, in a manner concise enough for a study guide, is always difficult. Therefore the reader is urged to have a textbook on fundamentals of chemistry available. See the Bibliography at the end of this chapter for a very useful chemistry textbook by Burns and several other useful references for hazardous materials managers.

The primary goal of this chapter is to provide an overview and brief refresher of chemistry for the hazardous materials manager. More specifically, this chapter targets the needs of hazardous materials managers with responsibilities for

- A facility
- Hazardous materials manufacture, distribution, or handling
- Chemical emergency response
- Environmental field or remedial operations
- Hazardous materials consulting, environmental science, and related environmental responsibilities

This chapter is further designed to lay the chemistry foundation for this desk reference. The chapter will provide an explanation of typical physical, chemical, and biological properties and their significance. Since hazardous materials ultimately are chemicals or composed of chemical mixtures, it is critically important that the hazardous materials manager have a firm grasp of chemical principles. Failure to understand these principles will put the hazardous materials manager at a serious disadvantage and often in a *cookbook* mode of simply following *recipes* prescribed by others.

Hazardous Chemicals Definitions and Lists

Before we get into the chemistry of hazardous materials, we need to look at the bewildering array of terms and definitions associated with the word *hazardous* In one sense or another, all chemicals are hazardous. Sugar or wheat flour, in the *wrong forms,* can be hazardous. For example, ground sugar dust in a food plant can cause respiratory problems for line workers. Flour dust in a flourmill is a potent explosion hazard even though the same dust, when collected, can be used to bake a loaf of bread.

A Key Reference on Chemical Hazards

While this chapter outlines an incredible array of facts and chemical principles, the hazardous materials manager is forewarned that wherever possible, he or she should not rely totally on memory when taking action in a hazardous materials incident or release. The National Institute of Occupational Safety and Health (NIOSH) has compiled an excellent handbook on the many chemical and hazard properties discussed below. The reader is urged to obtain and use this reference entitled *Pocket Guide to Chemical Hazards.* This guide presents critical information and data in abbreviated form for 677 or more chemicals and substance groupings found in the work environment. Headings used in the *Pocket Guide* include but are not limited to

- Chemical name/Chemical Abstracts Service (CAS) Registry number
- Synonyms/trade names
- Exposure limits/immediate dangers
- Physical descriptions
- Chemical and physical properties
- Incompatibilities
- Personal protective equipment (PPE) and respirators
- Health hazards

Regulatory Definitions of Hazardous

Chemicals have exhibited their hazardous properties long before regulatory definitions were promulgated. But because hazardous chemicals and their mixtures are highly regulated in much of the world, regulatory definitions must necessarily become controlling in many situations. Definitions applicable to the United States will be used generally throughout this text.

Occupational Health and Safety Administration Definitions of Hazardous Chemicals

The United States Occupational Health and Safety Administration (OSHA) has defined hazard categories in recognition of the problem of hazard identification. Other agencies in the United States have also promulgated regulations that have the effect of defining or identifying hazardous chemicals. The fine points of these various regulations should be recognized and understood by the hazardous materials manager.

Hazardous Chemical. The Occupational Health and Safety Administration, whose purview is worker safety, has perhaps the most comprehensive definition of hazardous chemicals. In 29 CFR 1910.1200 we find hazard definitions.

Physical hazard—Physical hazard means a chemical for which there is scientifically valid evidence that it is a combustible liquid, a compressed gas, an organic peroxide, an oxidizer, or is otherwise explosive, flammable, pyrophoric, unstable (reactive) or water reactive.

Health hazard—Health hazard means a chemical for which there is statistically significant evidence based on at least one study conducted in accordance with established scientific principles that acute or chronic health effects may occur in exposed employees. The term health hazard includes chemicals which are carcinogens, toxic or highly toxic agents, reproductive toxins, irritants, corrosives, sensitizers, hepatotoxins, nephrotoxins, neurotoxins, agents which act on the hematopoietic system, and agents which damage the lungs, skin, eyes, or mucous membranes.

Environmental Protection Agency Definitions and Lists of Hazardous Chemicals

The Environmental Protection Agency (EPA) has promulgated regulations which set forth much of the basic vocabulary used by persons engaged in regulatory aspects of hazardous materials management. In the United States, key laws directed EPA to write regulations which list and/or define hazardous chemicals, including the

- Resource Conservation and Recovery Act (RCRA)

- Comprehensive Environmental Response, Compensation and Liability Act of 1980 (CERCLA or Superfund)

- Superfund Amendments and Reauthorization Act of 1987–Title III (also known as SARA Title III)

The *hazard categories* that have been developed by EPA under the authority of these congressional mandates include

Extremely Hazardous Substances. *Extremely hazardous substances* (EHS) were identified as a published list by EPA, pursuant to Section 302 of SARA Title III. In the United States, when a facility inventories an EHS in quantities equal to or greater than a trigger quantity, the facility is subject to emergency planning and notification requirements. These trigger quantities are termed *threshold planning quantities* pursuant to this so-called *Community Right-to-Know* regula-tion. Chemicals appearing on the EPA EHS list are subject to EPA spill reporting requirements under Section 304 of SARA Title III.

Hazardous Chemicals. Sections 311 and 312 of SARA Title III list pure chemical substances (using chemical nomenclature) which must be reported annually by facilities when inventory quantities equal or exceed 10,000 lbs. It is noteworthy that when the Material Safety Data Sheet (MSDS) prepared according to OSHA regulations lists a hazard for a chemical, the chemical also is deemed hazardous by EPA in this regulation.

EPA's Tier II Community Right-to-Know report required under the SARA Title III, Sections 311 and 312 regulations (40 CFR 370.2), identifies the following hazard categories:

- *Immediate (acute) health hazard, including highly toxic, toxic, irritant, sensitizer, corrosive* (as defined under 29 CFR 1910.1200 OSHA regulations). Has a short-term adverse effect to a target organ, and the effect is of short duration.

- *Delayed (chronic) health hazard, including carcinogens* (as defined under 29 CFR 1910. 1200 OSHA regulations). Has a long-term adverse effect to a chronic target organ, and the effect is of long duration.

- *Fire hazard including flammable, combustible liquid, pyrophoric, and oxidizer* (as defined under 29 CFR 1910.1200 OSHA regulations).

- *Reactive, including unstable reactive, organic peroxide, and water-reactive* (as defined under 29 CFR 1910.1200 OSHA regulations).

Toxic Chemicals. EPA, under authority of Section 313 of SARA Title II Community Right-to-Know legislation, has prepared a list of toxic chemicals. Generally, if a facility *uses* 10,000 lb or manufactures 25,000 lb of a chemical on the Section 313 list, the facility is subject to annual reporting of emissions and releases.

Hazardous Substances. *Hazardous substances* are listed in EPA regulation 40 CFR 302.4 under CERCLA in a table that specifies reportable quantities (RQs) for each substance. When the RQ is exceeded in a chemical release incident, a facility owner is required to report the release to the National Response Center.

The environmental laws in the United States have clearly forced EPA to use various terms and lists to describe and categorize chemicals and substances deemed to be hazardous. Many chemicals fall into more than one list. Consequently, EPA and private publishers issue a consolidated chemical *List of Lists* which specify the categories a given chemical may have. These lists generally contain the following information:

- Chemical name

- CAS Number

- Section 302 (EHS) threshold planning quantity (TPQ)

- EHS RQ (Reportable Quantity)

- CERCLA RQ

- Section 313 Toxic Release Inventory status (yes or no)

- Hazardous waste code (under RCRA) where applicable

Hazardous Waste. *Hazardous waste* is defined pursuant to RCRA in 40 CFR 261. EPA has published a series of lists identifying wastes that are RCRA hazardous. For all other potentially hazardous chemicals and substances, EPA specifies four hazard properties for the characterization of hazardous wastes:

1) Ignitability

2) Corrosivity

3) Reactivity

4) Toxicity

Wastes not found on an EPA hazardous waste list but which exhibit one or more of these characteristics when tested under EPA protocols are termed *characteristic hazardous wastes*.

Other EPA Hazardous Chemical Lists. It should be noted that EPA has written other regulations that list hazardous substances and pollutants. Such regulations stem from the

- Clean Air Act (CAA), under which EPA has developed a list that currently contains 188 hazardous air pollutants

- Clean Water Act (CWA), under which Congress required EPA to develop a list of *toxic* pollutants, now known as a list of *priority pollutants* consisting of some 126 specific chemical substances.

- Toxic Substances Control Act (TSCA), whereby EPA regulates certain chemicals posing unreasonable risks such as asbestos, lead in paint, polychorinated biphenyls (PCBs), dioxin, and other chemicals

Department of Transportation Definition of Hazardous Material

On the premise that practical hazards in transportation of materials may be different from other hazardous material handling and storage situations, the United States Department of Transportation (DOT) and international transportation authorities have a somewhat different set of definitions and categories for those materials deemed hazardous.

Hazardous material—As used by the Department of Transportation (DOT) in 49 CFR 171.101, which provides hazardous material descriptions and shipping names. The nine broad DOT hazard categories are

1) Explosives

2) Gases

3) Flammable Liquids

4) Flammable Solids

5) Spontaneous Combustibles, and Dangerous When Wet

6) Oxidizers and Organic Peroxides

7) Poisons and Infectious Substances

8) Corrosives

9) All other hazardous materials

Key Chemical Principles and the Periodic Table

Definition of Chemistry

Chemistry is the science of matter, energy, and their reactions. Matter is anything that occupies space and has mass. Some 112 types of matter, described as chemical *elements,* have been identified; 92 of those elements are found naturally on the earth, while the remaining 20 elements are manmade. Chemists have found that in many instances, certain groups or families of elements have similar properties.

The *Periodic Table* provides very useful information to a hazardous materials manager by organizing the elements into *Groups* and *Periods* of similar chemical and physical properties. The reader is urged to reference the Periodic Table (see Figure 1) throughout the remainder of this chapter as a convenient *memory hook.*

Atoms, Elements, Molecules, and Compounds.
Elements are composed of extremely small, normally indivisible particles called *atoms*. Atoms of a particular element such as sodium have the same average mass, and other properties are the same. Atoms of different elements generally have different average masses (or *atomic weights*) and different properties.

A *molecule* is a group of atoms that are chemicaly bonded together. A chemically combined substance that is composed of more than one element is a *compound*. For example, sodium and potassium are atoms, but sodium chloride (table salt) or potassium hydroxide are molecules which, as compounds, exhibit characteristic physical properties.

Atoms of two or more elements may combine in more than one ratio to form compounds with very different hazard properties. In many instances, this is important from a hazard assessment standpoint. For example potassium, manganese, and oxygen can combine to form potassium permanganate ($KMnO_4$) or potassium manganate ($KMnO_3$); the former is a very strong oxidizing agent and the latter is a substantially weaker oxidizer. Oxidation and reduction reactions are discussed in more detail later on in this chapter.

Key differences exist between elements that are pure or uncombined and those same elements that have combined with other elements to form molecules or compounds. Metallic sodium, as an example, is highly reactive and hazardous when immersed in water. Sodium chloride, a common compound known as *table salt*, is relatively harmless and nonreactive in water. *Knowing the exact chemical form of an element or a compound is absolutely critical in assessing its hazards, establishing methods to render it less hazardous, and defining conditions for its safe disposal!*

The Periodic Table of Elements. The Periodic Table has been used by chemists for over a century (in one form or another) to organize and simplify their understanding of the elements according to their chemical properties. Hazard properties align closely with chemical properties in many of the situations encountered by hazardous materials managers.

The key goal of this discussion about the Periodic Table is to begin to simplify the overwhelming numbers of elements and chemical compounds that the hazardous materials manager encounters. Common reactions involving salts, acids, and other compounds, when discussed below, should be related to the Periodic Table whenever possible. Compounds from a given family will have many similar properties such as solubility, electrical conductance, pH, *etc.*

In order to understand the Periodic Table better, we will first look at a few basics of subatomic structure. Atoms are composed of a *nucleus* plus *electrons*. The nucleus resides in the center of an atom and contains all the positive charges (protons) of the atom as well as the atom's weight. Most atomic nuclei contain *neutron*s. Neutrons contribute mass or weight to the atom but do not have an electrical charge. Electrons, which surround the nucleus, are negatively charged and contribute an insignificant amount to atomic weight.

The number of protons in the nucleus defines the element's *atomic number*. The total number of protons plus neutrons constitutes the atomic weight. The number of protons in a nucleus of a given element is always the same. But the number of neutrons for that same element may vary. When this happens, atoms that have the same number of protons but different numbers of neutrons are

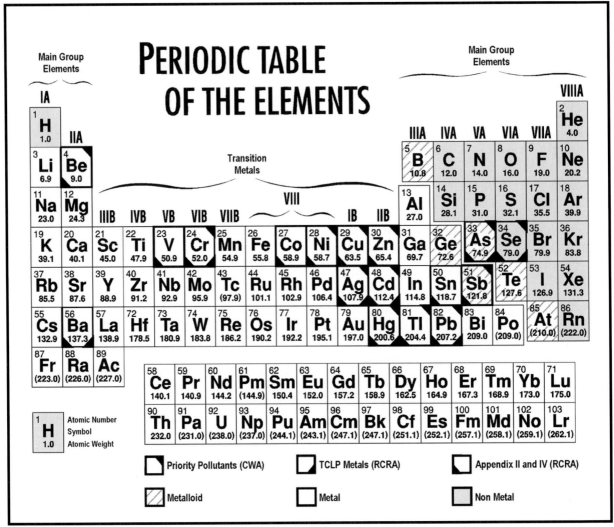

Reproduced with the permission of Columbia Analytical Services, Inc. Kelso, Washington (www.caslab.com).

Figure 1. The Periodic Table

called **isotopes**.   For example, elemental carbon contains six protons and usually six neutrons. In this isotopic form the atomic weight of the element is twelve. But in nature, a certain proportion of carbon atoms contain eight neutrons such that this isotopic form of carbon has a mass of fourteen (six protons plus eight neutrons). This makes for a very important hazard distinction, since carbon-14 is *radioactive* while carbon-12 is not. In nature, the ratios of the two isotopes are such that the average atomic weight of all forms of carbon is 12.0115 as shown in the Periodic Table. Note that certain other isotopes are radioactive, such as uranium-235, lithium-9, *etc.* All elements with atomic numbers greater than bismuth (83) are radioactive.

The **average atomic weight** (as found in nature) is shown in the Periodic Table along with the atomic number. From a chemical reactivity standpoint, isotopic forms of an element generally undergo the same chemical reactions and form the same compounds.

Radioactivity can be defined as the spontaneous emission of certain types of radiation (alpha, beta, and gamma) by unstable atomic nuclei. Radiation chemistry and radiation health effects are discussed in more detail in another chapter of this book. Radioactivity may cause detrimental, or even fatal, health effects. A number of hazardous substances are radioactive and can cause major environmental problems. Conversely, radioactive carbon-14 has been used beneficially for *carbon dating* of archaeological finds. Radioactive materials are used in a number of other ways such as medical research, cancer treatment, electric power generation, *etc.*

Atoms are electrically neutral. They achieve this electrical balance through equal numbers of protons (+ or positive charges) and electrons (– or negative charges). The way the electrons organize themselves in regions of space or **orbitals** around the nucleus determines the chemistry, reactivity, and often the hazards of an element. In general, when an element's outer valence orbital contains the maximum allowable number of electrons for that region, the element or compound is stable and will resist chemical reactions. However, when an outer orbital is incomplete with regard to electron-holding capacity, the element is potentially reactive.

During chemical reactions, electrons are added by the other reacting elements until the valence orbital is filled to capacity, normally eight electrons. The outer orbitals of *inert gases* contain eight electrons in their natural state. Thus they do not ordinarily undergo chemical reactions.

An **ion** is a charged particle that is produced when an atom or group of atoms gains or loses one or more electrons. Metal ions, nearly all of which have fewer than four valence electrons, tend to lose their valence electrons to form positive ions called **cations**. Nonmetal atoms are those that tend to gain electrons to form negative **anions**. For example, the *halogens* (shown in the Periodic Table Group VIIA [fluorine and below] have seven valence electrons. They readily gain an electron to fill an outer valence orbital completely with eight electrons (an *octet* or stable set of electrons).

The Periodic Table helps us organize our chemical knowledge, since the physical and chemical properties of the elements are *periodic* (or repeating) functions of their atomic numbers (their number of protons). The Periodic Table uses the official abbreviations of the elements and arranges the elements left to right by increasing atomic number. For example, H (hydrogen) has an atomic number of one while C (carbon) has an atomic number of six. Atomic numbers are shown above the symbol and the atomic weight (which is not used to determine periodicity) is shown below the element's symbol.

Some Key Facts Regarding the Periodic Table and Families of Elements.

- All the elements to the left of and below the heavy line on the periodic table are *metals*. Aluminum is a metal; silicon is a nonmetal.

- All the elements above and to the right of the line are *nonmetals*. The elements to the right of the heavy line are nonmetals that in one form or another also form compounds with metals.

- *Metalloids* are found in the Periodic Table in the intermediate region between metals and nonmetals. Their properties are generally intermediate in character. For example, metals are good conductors of electricity, nonmetals are nonconductors. Metalloids, a class that includes silicon (Si), germanium (Ge), arsenic (As), and boron (B), are semiconductors.

- The vertical groups labeled IA, IIA, *etc.* are *families* or *groups* of elements with many similar properties.

- *Group IA* elements, known as the *alkali metals*, have a valence of +1 and include lithium (Li), sodium (Na), potassium (K), rubidium (Rb), cesium (Cs) and francium (Fr). These are silver-gray metals that are very hazardous in the elemental form. They are soft such that they can be cut with a knife. Group IA metals are quick to react with water, oxygen, and other chemicals. They are never found free (uncombined) in nature. Other typical properties of alkali metal compounds include high water solubility and presence in salt water and salt deposits. Free alkali metals, being extremely reactive with oxygen, are usually stored under mineral oil or kerosene to minimize explosion hazards and maintain the purity of the metals.

Although hydrogen is shown at the top of Group IA in the Periodic Table, hydrogen is not considered to be part of the 1A family. In fact, hydrogen is a nonmetal with many other incredibly unique properties to be discussed later in this chapter under such topics as corrosives, acids/bases, and organic chemistry.

- *Group IIA, alkaline earth metals*, have a valence of +2 and include beryllium (Be), magnesium (Mg), calcium (Ca), strontium (Sr), barium (Ba), and radium (Ra). Their melting points are higher than Group IA metals. They have low densities even though their densities are a bit higher than Group IA metals. As metals, the alkaline earth metals are also reactive. However, they are less reactive than the alkali metals.

- The first element in *Group IIIA* is boron (B), a metalloid with predominantly nonmetal properties. The remainder of the elements in this group—aluminum (Al), gallium (Ga), *etc.*—form ions with a +3 charge. The density and metallic character increase as the atomic number increases within this group. Aluminum is adjacent to two metalloids in the Periodic Table and is predominantly metallic in character.

   Aluminum, along with several other metals, exhibits a property important to hazardous materials managers, namely that of being *amphoteric*. In their oxide forms aluminum (Al), chromium (Cr), and zinc (Zn) will react with *either* strong acids *or* strong bases, thus exhibiting both metal and nonmetal properties.

- The *Group IVA Carbon Family* embraces carbon (C), silicon (Si), germanium (Ge), tin (Sn), and lead (Pb). Carbon is totally nonmetallic in character. But as the atomic number increases in this family, the elements become more metallic. The last two elements, tin and lead, are typical metals. The outer orbitals of this family have four electrons.

   Carbon is the fundamental element in the huge field of *organic chemistry* which is discussed later in this chapter. But it also is part of a very important inorganic chemical when it is in the form of a *carbonate*. Potassium carbonate (K_2CO_3), for example, consists of a carbon joined to three oxygen atoms. Carbonates are found widely in nature combined with metals.

- In *Group VA*, as with Group IVA, definite nonmetals are found at the top of the table—nitrogen (N) and phosphorus (P)—followed by metalloids as atomic numbers increase—arsenic (As) and antimony (Sb)—and the heavy metal bismuth (Bi). The outer orbitals of this family have five electrons.

- *Group VIA* is the *Oxygen Family.* As one would by now expect from the group name, the outer shells of this family contain six electrons. The group comprises an unusual cast of characters including oxygen (O), sulfur (S), selenium (Se), tellurium (Te), and polonium (Po). The properties of these elements vary as atomic number increases, from the definitely nonmetallic for oxygen to the somewhat metallic for polonium.

- *Group VIIA* is called the *halogens.* This family is somewhat more predictable than the elements in the preceding two groups. Fluorine (F), chlorine (Cl), bromine (Br), iodine (I), and astatine (At) are all *salt formers* (which is what halogen means in Greek). For example, chlorine (Cl) reacts with lithium (Li) to form lithium chloride, a water-soluble salt.

   In the elemental form, all halogens are **diatomic,** meaning they have two atoms per molecule. Three other common elements are also diatomic: oxygen, hydrogen, and nitrogen. Like the alkali metals, halogens are too reactive to be found free in nature. *Fluorine is the most nonmetallic of all the elements* and is thus highly reactive in the elemental form. Both wood and rubber ignite spontaneously in fluorine gas.

   Fluorine and chlorine exist as gases at room temperature. Chlorine is a greenish-yellow gas that reacts with nearly all elements. Bromine is the only halogen to exist in the liquid form at room temperature. Iodine and astatine are solids at room temperature.

- *Group VIIIA* are the *Noble Gases.* The electron shells of the noble or *inert gases* contain eight electrons in their natural state. As a result, they tend to be unreactive. These elements include helium (He), neon (Ne), argon (Ar), krypton (Kr), xenon (Xe) and radon (Ra). Unlike the halogens, these gases are all *monatomic*—one atom per molecule.

   Although Group VIIIA metals are termed *noble* due to their nonreactivity, they can be very hazardous owing to physical factors when stored under pressure in cylinders. Additionally, gases, noble or otherwise, can present an asphyxiation hazard due to displacement of oxygen in a confined-space environment.

- The *transition metals or "B" Series* are located in the central region of elements in the Periodic Table. In general the properties of these elements are quite similar. The "B" elements (IB–VIIB) are all metals and exhibit metallic physical and chemical properties such as, malleability, ductility, *etc.* The transition metals are more brittle, harder, and have higher melting and boiling points than do other metals. A number of metals familiar to the hazardous materials manager reside in this region—chromium (Cr), nickel (Ni), cobalt

(Co), silver (Ag), and mercury (Hg). Note that mercury is the only liquid metal at room temperature, and as such has a high vapor pressure. Chemically, the transition metals differ from those discussed in earlier groups in that they can lose a variable number of electrons to form positive ions with different charges. The hazardous materials manager must recognize that these different valence states can have different hazard and chemical properties. Some common examples of different valence states include iron +2 and iron +3—the ferrous and ferric forms of iron—and copper +1 and copper +2—the cuprous and cupric forms.

Final Comments and Recap on Atomic Theory. The reason each of these families has similar chemical properties within themselves is that the outer electron orbital configurations are identical. It should be noted that the number at the top of each column is equal to the number of electrons in the outer orbital. Thus calcium (Ca, Group IIA) has 2 electrons in the outer orbital while phosphorus (P, Group VA) has 5 electrons in the outer orbital. There are family effects within the transition metals, but such effects are weaker.

A very important rule alluded to earlier explains the family effect. The *Octet Rule* states that all atoms strive to reach an ultimate state of stability. In order to reach this state, all atoms must have eight electrons in the outer orbital. Once this state occurs, the atom will not react with anything else. Please note, however, that hydrogen and helium require only two electrons to be stable.

The Noble Gases (Group VIIIA), which include neon (Ne), radon (Rn), *etc.*, have eight electrons in their outer shell. Helium, while shown in Group VIIIA, is similar to hydrogen, in that it is stable with the outer orbital containing only two electrons. Thus chemists have identified the *Duet Rule,* which states that hydrogen and helium reach their ultimate state of stability when two electrons fill the outer orbital.

Ionic and Covalent Bonding of Compounds. An ionic compound is one that is made up of positive and negative ions. Ionic compounds exist by virtue of an ionic bond which is the attraction between ions having opposite charges. In hazardous materials chemistry, this usually translates to some degree of water solubility.

Covalent bonds exist when two atoms in a molecule share a pair of electrons. Thus methane, an organic compound, contains covalent bonds, while sodium chloride, NaCl, an inorganic compound, is bonded ionically.

Organic vs. Inorganic Compounds. Any compound that is covalent and contains carbon is said to be an *organic compound*. An inorganic compound is any compound that is *not* classified as organic. These two fundamental classes of compounds are discussed in more detail in sections that follow.

Physical, Chemical, and Biochemical Properties

All chemicals, including hazardous ones, are described in terms of their physical, chemical, and biological properties. In order to use this information fully, it is necessary to understand the meaning and importance of the various individual properties, and also to have some grasp of the significance of the various numerical values within the context of chemicals at large. These properties can then be used along with other information to predict the likely behavior of hazardous chemicals, and to recognize and avoid potentially dangerous situations. The first step is to define and comment on several of the more critical properties that are useful in the handling of hazardous materials (Table 1).

Physical State at 20°C. The *physical state* is the nature of the chemical (solid, liquid, or gas) at a defined temperature. (*i.e.*, 20°C or room temperature). Changing the temperature may alter the physical state, depending on the magnitude and direction of the change relative to the melting and boiling points of the chemical. For example, water changes from ice to water when the temperature goes above 32°F (0°C).

Boiling Point. The temperature at which a liquid changes to gas under standard atmospheric pressure (760 mm mercury) is the *boiling point* (bp). The bp of water is 100°C at sea level, while the bp's of ethyl alcohol and *n*-hexane are 78.4°C and 68.7°C, respectively. Lowering the atmospheric pressure (*e.g.*, by applying a vacuum) will lower the BP; conversely, higher pressures result in elevated boiling points. Generally the lower the boiling point of a flammable liquid, the lower the ignition or flashpoint of the liquid. (See section which follows on the chemistry of flammables.)

Melting Point. The temperature at which a solid changes to a liquid is the *melting point* (mp). The melting point is not particularly sensitive to atmospheric pressure, but it is responsive to dissolved salts that depress the melting point. Thus, in winter, it is common to use one of several available salt compounds to keep water from freezing on sidewalks.

Vapor Pressure. *Vapor pressure* (vp) is a measure of the relative volatility of chemicals. It is pressure exerted by the vapor in equilibrium with its liquid at a given temperature. Flammable liquids with high vapor pressures generally represent a greater fire hazard than those with lower vapor pressures. For a given liquid, the vapor pressure increases with increasing temperature. Consequently, drummed materials with

high vapor pressures should not be stored in direct sunlight, as overheating of the materials and resultant increases in vapor pressures could result in bulging drums with failed or weakened seams.

Vapor pressure values can be used to predict the relative rate of evaporation of dissolved solvents from water. At 20°C, water, ethanol, and benzene have vapor pressures of 17.5 mm, 43.9 mm, and 74 mm of mercury, respectively. Evaporation of benzene can be expected to be about 4 times faster than water, with ethanol evaporating at an intermediate rate.

Vapor Density. The *vapor density* (vd) is the mass per unit volume of a given vapor/gas relative to that of air. Thus, acetaldehyde with a vapor density of 1.5 is heavier than air and will accumulate in low spots, while acetylene with a vapor density of 0.9 is lighter than air and will rise and disperse. Heavy vapors present a particular hazard because of the way they accumulate; if toxic, they may poison workers; if

Table 1. List of Commonly Measured Physical and Chemical Properties

Property	Abbreviation
Physical State @ 20°C	—
Boiling Point	bp
Melting Point	mp
Vapor Pressure	vp
Vapor Density	vd
Density	ρ
Specific Gravity	SG
Solubility (in water and other solvents)	K_{sp}
Flashpoint	fp
Auto-ignition Temperature (point)	—
Flammable or Explosive Limits	EL
Heat Content	Btu
Octanol/Water Partition Coefficient	K_{ow}
Biochemical Oxygen Demand after 5 days	BOD_5
Theoretical Oxygen Demand	ThOD
Threshold Limit Value	TLV
pH	—
pK_a	—
Molecular Weight	MW
Chemical Formula	—
Fire point	—
Color	—
Odor	—

nontoxic, they may displace air and cause suffocation by oxygen deficiency; if flammable, once presented with an ignition source, they represent a fire or explosion hazard. Gases heavier than air include carbon dioxide, chlorine, hydrogen sulfide, and sulfur dioxide. *Beware; most common gases encountered by hazardous materials managers are heavier than air and thus dangerous!* Key exceptions to this are hydrogen (H_2), helium (He), acetylene (C_2H_2), methane (CH_4), and ammonia (NH_3); these are hazardous gases that are lighter than air.

Density. The density is the mass per unit volume of any substance, including liquids. The density of a liquid determines whether a spilled material that is insoluble in or immiscible with water will sink or float on water. Knowledge of this behavior is essential in checking whether to use water to suppress a fire involving the material.

Specific Gravity. The ratio of the density of a liquid as compared with that of water is the ***specific gravity*** (SG). Insoluble materials will sink or float in water depending on the SG. Materials heavier than water have SGs > 1, and materials lighter than water have SGs < 1. Thus lead, mercury, and carbon tetrachloride, with SGs of 11.3, 13.6, and 1.6, respectively, will sink in water, whereas gasoline with a SG of 0.66 to 0.69 will float on water.

Solubility. The amount of a given substance (the solute) that dissolves in a unit volume of a liquid (the solvent) is the ***solubility***. This property is of importance in the handling and recovery of spilled hazardous materials. Water-insoluble chemicals are much easier to recover from water than spills of water-soluble chemicals. Acetone, which is soluble (miscible) in water in all proportions, is not readily recoverable from water. In contrast, benzene, which is lighter than water and insoluble as well, can be readily trapped with a skimmer. For organic compounds, solubility tends to decrease with increasing molecular weight and chlorine content. Many inorganic compounds are quite soluble in water, *e.g.*, acids, bases, many salts, *etc.*

Solubility Prediction. The solubility product constant, commonly referred to as the *solubility product*, provides a convenient method of predicting the solubility of a material in water at equilibrium. Copper hydroxide, for example, dissolves according to the following equilibrium:

$$Cu(OH)_2(s) \rightarrow Cu^{2+} + 2OH^- \qquad (1)$$

The resultant solubility product is represented in the following manner:

$$[Cu^{2+}]\,[OH^-] = K_{sp} \qquad (2)$$

Where:

$[Cu^{2+}]$ = molar concentration of copper ions

$[OH^-]$ = molar concentration of hydroxide ions

K_{sp} = the solubility product constant

The solubility product constant (K_{sp}) is commonly used in determining suitable precipitation reactions for removal of ionic species from solution. In the same example, the pH needed to remove copper to any specified concentration can be determined by substituting the molar concentration into the following equation—

$$[OH^-] = \frac{K_{sp}}{[Cu^{2+}]} \qquad (3)$$

and then applying the derived values in turn to these other equations:

$$[OH^-][H^+] = 10^{-14} \qquad (4)$$

$$pH = -\log [H^+] \qquad (5)$$

The use of the K_{sp} for precipitation information is often complicated by a number of interfering factors, including complex metallic ions, high ionic strength solutions, and high solids contents. The solubility principle is applicable solely to ionic compounds, *i.e.*, primarily inorganic compounds.

Flashpoint. The ***flashpoint*** (fp) is the lowest temperature of a liquid at which it gives off enough vapor to form an ignitable mixture with air near the surface of the liquid. Two tests are used to determine flashpoint: Open Cup and Closed Cup. Generally, the Open Cup method results in flashpoints 5° to 10° higher than the Closed Cup method. A flashpoint <140°F (Closed Cup) is the criterion used by EPA to decide whether a waste is hazardous by virtue of its *ignitability*. With the exception of some special cases, DOT regulates materials with flashpoints of <141°F as *flammable* and between 141°F and 200°F as *combustible*.

Fire Point. The *fire point* is the temperature at which a liquid gives off enough vapor to continue to burn when ignited.

Auto-Ignition Temperature. The temperature at which ignition occurs without an ignition source and the material continues to burn without further heat input is called the *auto-ignition temperature*.

Flammable or Explosive Limits. The *flammable or explosive limits* are the upper and lower vapor concentrations at which a mixture will burn or explode. The lower explosive limit of *p*-xylene is 1.1 percent by volume in air, whereas the upper explosive limit is 7.0 percent in air. A mixture of *p*-xylene vapor and air having a concentration of < 1.1 percent in air is too lean in *p*-xylene vapor to burn. By subtraction (7.0 – 1.1 = 5.9), *p*-xylene is said to have a flammable range of 5.9.

Heat Content. The *heat content* is the heat released by complete combustion of a unit-weight of material. Methane has a heat content of about 21,500 Btu/lb., while benzene contains about 17,250 Btu/lb.

Octanol/Water Partition Coefficient. The equilibrium ratio of the concentrations of material partitioned between octanol and water is called the *octanol/water partition coefficient* or (K_{ow}). This coefficient is considered to be an index of the potential of a chemical to be bioaccumulated. Higher values of K_{ow} are associated with greater bioaccumulation potential.

Biochemical Oxygen Demand at Five Days (BOD$_5$). The quantity of oxygen required by microbes for the oxidative breakdown of a given waste material during a 5-day test period is called the *biochemical oxygen demand at five days* (BOD$_5$). The BOD$_5$ is usually taken as an index of the ultimate oxygen demand (*i.e.*, oxygen required when sufficient time is allowed to achieve maximum microbial decomposition). BOD$_5$ is used to predict the impact of a spill or release of material on the oxygen content of a body of water.

Theoretical Oxygen Demand. The theoretical oxygen demand (Th$_{OD}$) is the cumulative amount of oxygen needed to completely oxidize a given material. The Th$_{OD}$ is the upper limit for BOD$_5$ values, although it seldom is achieved. A comparison of the BOD$_5$ and Th$_{OD}$ values for a given chemical provides an indication of the biodegradability of that chemical.

Threshold Limit Value. The *threshold limit value* (TLV) is the exposure level under which most people can work for eight hours a day, day after day, with no harmful effects. A table of the values and accompanying precautions for most common industrial materials is published annually by the American Conference of Governmental Industrial Hygienists.

pH and pK$_a$. The *pH* of a solution is the negative logarithm of the hydrogen ion concentration. Remember that a hydrogen ion (H^+) is a proton, and that a hydronium ion is a proton associated with a water molecule.

The *pK$_a$* is the negative logarithm of the equilibrium constant for acids or bases. Strong acids, such as sulfuric and hydrochloric acids, have low pK$_a$s (*i.e.*, <1.1). Bases such as potassium hydroxide and sodium hydroxide have pK$_a$ values closer to 14.0 Weak acids and weak bases have pK$_a$ values that fall between these two extremes.

Additional Key Chemical Concepts of Environmental Significance

The processes of dissolution/precipitation (for inorganic chemicals), dissolution/phase separation (for organics), adsorption, and volatilization control the distribution of a spilled material in the environment. Conversely, knowledgeable manipulation of these same processes can be advantageous in either cleaning up or mitigating the effects of spilled material. For example, ground water contaminated with volatile organics of limited aqueous solubility can be decontaminated by air stripping these compounds. The air-stripped compounds can then be concentrated by adsorption on activated carbon for subsequent disposal.

With this in mind, the hazardous materials manager needs some additional chemical tools in his or her arsenal. A few of these tools are discussed below.

Concentrations. Chemists, especially environmental chemists, seldom work with pure solutions of materials dispersed in environmental media. Knowledge of units of concentration is required. Units of concentration in common usage include

- *Parts per million* (ppm). Identical with milligrams per liter (mg/l for aqueous solutions) or mg/kg (milligrams per kilogram for solids).

- *Parts per billion* (ppb). Identical with micrograms per liter (μg/l for liquids) or micrograms per kilogram (μg/kg for solids).

- *Moles per liter* or *molar solutions* (a weight of substance equivalent to the gram-molecular or gram-atomic weight in a liter of solution).

- *Equivalents per liter* (commonly used for acids and bases, a one-equivalent-per-liter solution is stated to be a *one normal solution*).

- *Percent by weight* or *volume.*

- Ppm, micrograms per m^3, and percent by volume are often used for vapors and gases, mists, and particulates in air.

Adsorption. Adsorption is an important physical-chemical phenomenon used in the treatment of hazardous wastes or in predicting the behavior of hazardous materials in natural systems. Adsorption is the concentration or accumulation of substances at a surface or interface between media. Adsorption of organic hazardous materials onto soils or sediments is an important factor affecting their mobility in the environment. Adsorption may be predicted by use of a number of equations most commonly relating the concentration of a chemical at the surface or interface to the concentration in air or in solution, at equilibrium. These equations may be solved graphically using laboratory data to plot *isotherms*. The most common application of the adsorption property is the removal of organic compounds from water by activated carbon.

Volatilization. *Volatilization* is the tendency of a material to transfer from a liquid phase (either pure or dissolved, as in aqueous systems) to a gaseous phase (commonly mixed with air). Volatilization, or *evaporation* as it is more commonly called, is controlled by a number of factors, the most important of which are

- The vapor pressure of the material and temperature (vapor pressure increases with temperature)

- The air/material interfacial surface area

- The action of active mass transfer agents such as wind

Hazardous Waste Categories. From the earlier discussion of regulatory definitions of *hazardous*, please recall that RCRA hazardous wastes not otherwise listed fall into one or more of the four following categories:

1) Reactivity

2) Ignitability

3) Corrosivity

4) Toxic characteristic leaching procedure (TCLP) toxicity

Commercial chemical products when they become wastes, and wastes from specific processes, may be *listed* as hazardous wastes because they are known to present hazards related to toxicity. When wastes do not appear on an EPA hazardous waste list (and the generator does not have process information to use in determining a waste's hazard category), EPA requires the generator to analyze the waste using official laboratory protocols. These protocols determine if a waste is *RCRA hazardous* according to limits set for each of the above four categories. In the discussion to follow, various chemical groups will be examined primarily in the context of three of these categories, namely reactivity, ignitability, and corrosivity.

(Note: TCLP toxicity determinations rely on a laboratory protocol specified by EPA that determines the extractability of certain identified *toxic* chemicals in a weakly acid medium. See the chapter on industrial toxicology for background on how a chemical is determined to be *toxic* to humans and other live organisms.)

Acids, Bases, and Corrosivity

Acids, bases, and corrosive chemicals are encountered with very high frequency in the field of

594   Chemical Perspectives

hazardous materials management. As a hazard class, corrosives are among the most common materials. Acids and bases are primary contributors to corrosion and play a number of other roles in hazardous materials management and pollution control.

While the RCRA definition of corrosivity, used in hazardous waste characterization, is limited to pH ranges of <2.0 and >12.5, the *chemical* property of corrosivity can occur at intermediate pH levels. For example, liquids with pHs in the ranges of 2–5 and 8–12 can be chemically corrosive. Salts of strong acids and weak bases such as ammonium sulfate and ammonium chloride hydrolyze in water to form acids which have corrosive activity. More generally, materials in water that exhibit intermediate pH ranges should be evaluated carefully from the standpoint of handling, treatment, personal safety, and environmental impact.

Acids

The acidic nature of a given solution is often represented by its pH, where pH has been defined as the negative log of the hydrogen ion (H^+) or the concentration of the hydronium ion (H_3O^+):

$$pH = -\log [H^+] \tag{6}$$

or

$$pH = -\log[H_3O^+] \tag{7}$$

In actuality, hydrogen ions do not exist to any large extent in solution. Typically, the H^+ has been represented to exist as the *hydronium ion,* H_3O^+, but it probably exists in coordination with four or more water molecules. Although H_3O^+ may better represent the actual state of the proton in solution, it is usually represented as simply H^+ in discussions of acidity. A solution with pH < 7 is acidic, a solution with pH = 7 is neutral, and a solution with pH > 7 is basic. For example, the pH of lemon juice is ≈ 2, while the pH of lye is ≈ 14. Table 2 shows the relative strengths of acids in water.

Neutralization of Bases by Acids, Three Examples.

Example (1)

$$H^+ + OH^- \rightarrow H_2O \tag{8}$$

Hydrogen ion (proton) neutralizes hydroxyl ion (a base) to form water. This is the fundamental neutralization reaction.

Example (2)

$$HCl + NaOH \rightarrow H_2O + NaCl \tag{9}$$

Hydrochloric acid (containing the proton) neutralizes sodium hydroxide (containing the basic OH^- group). This differs from the previous equation only in that the chloride and sodium ions are shown.

Example (3)

$$CaCO_3 + 2HCl \rightarrow$$
$$CaCl_2 + H_2O + CO_2\uparrow \tag{10}$$

This illustration shows that calcium carbonate in water is basic in nature. But following neutralization, neutral calcium chloride salt and water are formed with the evolution of carbon dioxide as a gas.

Bases. A *base* is any material that produces hydroxide ions (OH^-) when it is dissolved in water. The words *alkaline, basic,* and *caustic* are often used interchangeably. Common bases include

Table 2. Relative Strengths of Acids in Water

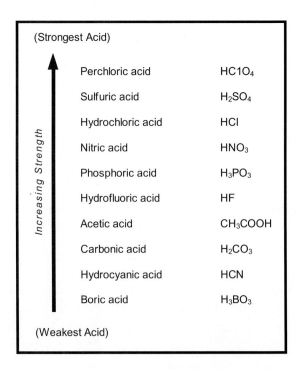

(Strongest Acid)	
Perchloric acid	$HClO_4$
Sulfuric acid	H_2SO_4
Hydrochloric acid	HCl
Nitric acid	HNO_3
Phosphoric acid	H_3PO_3
Hydrofluoric acid	HF
Acetic acid	CH_3COOH
Carbonic acid	H_2CO_3
Hydrocyanic acid	HCN
Boric acid	H_3BO_3
(Weakest Acid)	

Increasing Strength

sodium hydroxide (lye, NaOH), potassium hydroxide (potash lye, KOH), and calcium hydroxide (slaked lime, CaOH$_2$).

The concepts of strong versus weak bases and concentrated versus dilute bases are exactly analogous to those for acids. Strong bases such as sodium hydroxide dissociate completely, while weak bases such as the organic amines (R–NH$_2$, where R is a carbon compound and –NH$_2$ is an amine) dissociate only partially. As with acids, bases can be either inorganic or organic. Typical reactions of bases include neutralization of acids, reaction with metals, and reaction with salts. The reactions of bases with acids were illustrated in the previous section.

Reaction of a Base with Metals.

$$2Al^\circ + 6NaOH \rightarrow 2Na_3AlO_3 + 3H_2\uparrow \qquad (11)$$

The reaction creates heat and may explode due to formation of hydrogen gas.

Reaction of a Base with Salts.

$$Pb(NO_3)_2 + 2NaOH \rightarrow$$
$$Pb(OH)_2\downarrow + 2NaNO_3 \qquad (12)$$

In this reaction lead hydroxide is insoluble and thus reduces the hydroxyl ion concentration in the solution.

Some of the properties and good practices for common acids and bases are presented in Table 3.

Concentrated and *dilute* refer to the concentrations of a chemical in solution. Mixing a concentrated acid with enough water will produce a dilute acid. For example, concentrated HCl is approximately 12 normal (N) in HCl, while a solution of HCl used in a titration may be only 0.5 N. The latter is a dilute acid solution. (Normality is a term used to express the *concentration* of solutions of acids and bases without needing to know the chemical equation for the chemical equation of neutralization. For a more complete discussion see Ebbing, p. 354.)

Strong and *weak* **acids** or **bases** are classified by how completely they ionize in solution. For example, hydrochloric acid (HCl) is classified as a strong acid because it is completely ionized to H$^+$

and Cl$^-$ ions. Acetic acid is classified as a weak acid because it does not totally ionize in solution. Strong acids include perchloric, hydrochloric, sulfuric, nitric, and hydriodic acids. Examples of weak acids include boric, hydrocyanic, carbonic, and acetic acids.

Strong bases include sodium hydroxide, potassium hydroxide, other alkali and alkaline earth hydroxides, and more. Weak bases include ammonium hydroxide, sodium carbonate, certain organic amines, and more.

Be aware that some salts may have corrosive properties in water. As a benchmark, the salt of a strong acid and strong base yields a neutral solution because the ions do not react with water. But the salt of a strong acid and weak base yields an acidic solution because the cation acts as a weak acid. For example, ammonium chloride in water is acidic because the ammonium ion immobilizes the hydroxyl (OH$^-$) portion of the dissociated water molecule (H$^+$ + OH$^-$ = H$_2$O).

Conversely, the salt of a weak acid and strong base yields a basic solution in water because the anion acts as a weak base. As an example, sodium acetate in water is basic because the acetate anion immobilizes hydrogen ions from water.

Thus, the terminology *strong acid* versus *weak acid* may bear little relationship to the nature or extent of potential hazard of a corrosive in water, while the terms *concentrated* versus *dilute* most often do.

The amount of acid or base required for neutralization of a spill of acid or base can be determined mathematically if exact concentrations are known. Often this must be determined empirically by running a titration curve. If calculations are made, it is important to remember that the pK$_a$ for an acid represents the midpoint of the titration curve. For di- and triprotic acids (acids with 2 or 3 protons available), each pK$_a$ represents the midpoint of the titration curve for that proton (an analogous situation holds for pKb's and bases). At the endpoint of the titration curve the addition of a very small amount of acid or base can result in a large change in pH. In neutralization of spills, knowledge of the pK$_a$ (or pK$_b$) and close monitoring of the system pH can prevent additional damage to the environment that might result if the equivalence point is overshot.

Table 3. Properties and Good Practices for Some Common Acids and Bases

---

Acids–Sulfuric, Nitric, Hydrochloric, Acetic

- These acids are highly soluble in water. Concentrated solutions are highly corrosive and will attack materials and tissue.

- If spilled on skin, flushing with lots of water will dilute the acids and reduce tissue damage.

- Sulfuric and nitric acids are strong oxidizers and should not be stored or mixed with any organic material.

- Sulfuric, nitric, and hydrochloric acids will attack metals upon contact and generate hydrogen gas, which is explosive.

- Acetic acid (glacial), an organic acid, is extremely flammable. Its vapors form explosive mixtures in the air. It is dangerous when stored with any oxidizing material, such as nitric and sulfuric acids, peroxides, sodium hypochlorite, *etc.*

- Breathing the concentrated vapors of any of these acids can be extremely harmful. Wear appropriate personal protective equipment.

- When mixing with water, always add acids to water, never water to acids.

Bases (Caustic)–Sodium Hydroxide, Ammonium Hydroxide, Calcium Hydroxide (Slaked Lime), Calcium Oxide (Quick Lime)

- These bases are highly soluble in water.

- Concentrated solutions are highly corrosive. They are worse than most acids because they penetrate the skin and essentially make "soap" out of body fat. That is, caustics turn tissue fatty acids into salts. (In emulsion chemistry, the reaction is termed *saponification*.)

- If spilled on skin, flush immediately with lots of water.

- When mixed with water, they generate a significant amount of heat—especially sodium hydroxide and calcium oxide.

- Do not store or mix concentrated acids and bases, as this gives off much heat—dilute, then mix.

- Do not store or mix ammonium hydroxide with other strong bases. It can release ammonia gas, which is extremely toxic.

- Do not store or mix ammonium hydroxide with chlorine or other oxidizing compounds (*i.e.*, sodium hypochlorite). It can release chlorine gas, which is extremely toxic.

Corrosion and Corrosivity. *Rust* is the most common form of corrosion. Rust is formed when metallic iron oxidizes in the presence air and moisture. However, corrosive agents other than air and water exist. Corrosive agents can be placed in four main groups: oxygen and oxidants, acidic materials, salts, and alkalis (bases).

EPA, under the RCRA hazardous waste regulations, defines corrosivity in terms of pH (*i.e.*, liquid wastes with pH < 2.0 and > 12.5). DOT defines corrosivity in terms of a material's ability to corrode steel (SAE 1020) at a rate of > 6.35 mm (0.250 in.) per year at a temperature of 55°C (131°F), (as determined by the National Association of Corrosion Engineers [NACE] test method TM–01–69). OSHA defines a corrosive as any material that causes visible destruction or irreversible alterations in living tissue at the site of contact. Under both OSHA and DOT, acids and bases are grouped under the *corrosive* definition. For more on other corrosive materials, see the section on substances that produce acidic solutions.

Chemistry of Water-Reactive Materials

The characteristics of a solid waste that would categorize it as a reactive hazardous waste as defined under RCRA include

- It reacts violently with water.

- It forms potentially explosive mixtures with water.

- When mixed with water, it generates toxic gases, vapors, or fumes in a quantity sufficient to present a danger to human health or the environment.

Because water is the most common fire suppressant, the characteristic of reactivity is especially relevant since the application of water to eliminate or prevent the spread of fires involving reactives may be counterproductive rather than helpful. Several categories of chemicals will be discussed from this standpoint.

Substances that Produce Hydrogen Gas: Metals. Several metals in a pure state react with water and air. The extent of reactivity depends upon the physical state of the metal. The highly reactive metals such as lithium, sodium, and potassium are pyrophoric (*i.e.*, they ignite spontaneously in air without an ignition source). In contrast, the less reactive metals such as magnesium, zirconium, titanium, aluminum, and zinc, are highly pyrophoric only as dusts.

Lithium, sodium, and potassium (alkali metals in Group IA on the Periodic Table) react rapidly with water to release hydrogen (H_2) gas. For example:

$$2Na + 2H_2O \rightarrow 2Na^+ + 2OH^- + H_2\uparrow \qquad (13)$$

Sufficient heat is generated during the reaction to ignite the hydrogen gas so that it can react explosively with the oxygen in air.

Metals like magnesium, aluminum, titanium, and zirconium in pure form also react with water to release H_2, but heat must be supplied to initiate the reaction. The generalized representation is:

metal + water + heat $\rightarrow$
      metal oxide or hydroxide + $H_2\uparrow$ (14)

Substances that Produce Hydrogen Gas: Hydrides. True hydrides (*i.e.*, those in which the hydrogen is in its anionic or most reduced form) are salt-like compounds in which the hydrogen is combined with alkali metals, either alone as simple hydrides or in association with other elements as complex hydrides. Hydrides react with water to release hydrogen:

Simple hydrides.

$$2LiH_2 + 2H_2O \rightarrow 3H_2\uparrow + 2LiOH \qquad (15)$$

Complex hydrides.

$$LiAlH_4 + 4H_2O \rightarrow$$
$$Al(OH)_3 + LiOH + 4H_2\uparrow \qquad (16)$$

Peroxides. Compounds containing the O^{2-} ion are hazardous primarily as oxidizing agents but also as water reactives. An example is the liberation of oxygen from the mixture of sodium peroxide and water:

$$2Na_2O_2 + 2H_2O \rightarrow 4NaOH + O_2\uparrow \qquad (17)$$

Substances that Produce Alkaline Aqueous Solutions and Explosive Gases. This group of nonmetals (see Periodic Table Groups IIIA, IVA and VA) is exemplified by nitrides, carbides, and phosphides. Nitrides will react with water to generate ammonia (NH_3), which can be released to the atmosphere depending on how alkaline the solution becomes.

$$Mg_3N_2 + 6H_2O \rightarrow 3Mg(OH)_2 + 2NH_3\uparrow \qquad (18)$$

Carbides, which are binary compounds containing anionic carbon, occur as covalent and as salt-like compounds. The salt-like carbides are water-reactive and, upon hydrolysis, yield flammable hydrocarbons. Typical hydrolysis reactions include

$$CaC_2 + 2H_2O \rightarrow Ca(OH)_2 + C_2H_2\uparrow \qquad (19)$$
(acetylene)

$$Al_4C_3 + 12H_2O \rightarrow 4Al(OH)_3 + 3CH_4\uparrow \qquad (20)$$
(methane)

Other similar carbide-like compounds include Be_2C and Mg_2C_3. Notably, each reaction is sufficiently exothermic to ignite the specific gas formed upon hydrolysis.

Phosphides are binary compounds containing anionic phosphorous (P^{-3}). Heavy metal, alkali, and alkaline earth metal phosphides exist, but few of them are commercially important. Phosphides may be encountered, however, in laboratory or chemical research situations. Phosphides hydrolyze to the flammable and toxic gas phosphine (PH_3). The hydrolysis reaction of aluminum phosphide is

$$AlP + 3H_2O \rightarrow PH_3\uparrow + Al(OH) \qquad (21)$$

Substances that Produce Acidic Aqueous Solutions

Inorganic Chlorides/Halides. Metallic salts are formed from the reaction of a weak base with the strong acid HCl. Salts such as these dissolve in water to produce a markedly acidic solution (see earlier section on acids, bases, and corrosivity). This is exemplified by aluminum chloride, which is corrosive owing to acidity. Acidity, in this instance, results from hydrolysis that produces aluminum oxide and chloride ions. Anhydrous $AlCl_3$ hydrolyzes violently when contacted by water.

Several nonmetallic chlorides also react with water with varying degrees of violence to produce hydrochloric acid. Although these compounds are themselves nonflammable, the heat generated by hydrolysis is sufficient to ignite adjacent flammable materials. These nonmetallic chlorides include antimony pentachloride ($SbCl_5$), boron trichloride (BCl_3), phosphorus oxychloride ($POCl_3$), phosphorus pentachloride (PCl_5), phosphorus trichloride (PCl_3), silicon tetrachloride ($SiCl_4$), thionyl chloride ($SOCl_2$), sulfuryl chloride (SO_2Cl_2) and titanium tetrachloride ($TiCl_4$). Because of their acid-producing tendencies, many of these chlorides are considered to be corrosive.

Organic Chlorides/Halides. Several organic compounds also are hydrolyzed (or react with water) to produce corrosive materials. Notable inclusions among these compounds are acetic anhydride ($[CH_3CO_2]_2O$) and acetyl chloride (CH_3COCl), both of which produce acetic acid upon reaction with water. Both acetic anhydride and acetyl chloride are corrosive. In addition, mixtures of the vapors of acetic anhydride and acetic acid are flammable in air, and acetyl chloride itself is flammable.

Oxidation/Reduction Phenomena

Oxidation/reduction (*redox*) reactions, the bane of high school chemistry students, can be similarly unkind to hazardous materials managers. The explosive potential of oxidation/reduction reactions has resulted time and time again in chemical disasters. Perhaps the largest of these was the explosion of the *S.S. Grandcamp* at Texas City, Texas, in 1947, where thermal decomposition (redox reactions of ammonium nitrate and subsequent oxidation reactions of the decomposition products) led to the deaths of over 600 people and over $33 million (1947 dollars) damage. The addition or loss of electrons in redox reactions involves an accompanying transfer of energy, often a violently **exothermic** (heat-releasing) transfer. The substance that gives up electrons (and is therefore oxidized) is the reducing agent. The substance that gains electrons (and is therefore reduced) is the oxidizing agent.

Oxidizing agents generally are recognizable by their structures or names. They tend to have oxygen in their structures and often release oxygen as a result of thermal decomposition. Oxidizing agents often have *per-* prefixes (perchlorate, peroxides, and permanganate) and often end in *ate* (chromate, nitrate, chlorate).

Strong oxidizers have more potential incompatibilities than perhaps any other chemical group (with the possible exception of water-reactive substances). It is safe to assume that oxidizers should not be stored or mixed with any other material except under carefully controlled conditions. Common oxidizing agents listed in decreasing order of oxidizing strength include

- Fluorine

- Chlorine

- Ozone

- Sulfuric acid (concentrated)

- Hydrogen peroxide

- Oxygen

- Hypochlorous acid

- Metallic iodates

- Metal chlorates

- Bromine

- Lead dioxide

- Ferric salts

- Metallic permanganates

- Iodine

- Metallic dichromates

- Sulfur

- Nitric acid (concentrated)

- Stannic salts

Reducing agents present similar problems. They react with a broad spectrum of chemical classes, and the reactions can be exothermic and violent. Reducing agents are, by definition, highly oxidizable, and may react with air or moisture in the air. Common reducing agents include

- Hydrogen

- Sulfides

- Metals (Li, Na, K, Ca, Sr, Ba)

- Sulfites

- Hydrazine

- Iodides

- Metal acetylides

- Nitrides

- Complex hydrides

- Nitrites

- Metal hydrides

- Phosphites

- Metal hypoborates

- Metallic azides

- Metal hypophosphites

Nomenclature and General Properties of Organic Chemicals

Most compounds in which carbon is the key element are classified as *organic*. Organic chemicals encountered in daily life include fingernail polish remover (acetone), vodka (50% ethanol in water), engine coolant (ethylene glycol) and many more. Common industrial examples of organic compounds include degreasing solvents, lubricants, and heating and motor fuels.

This section will highlight some of the more common characteristics of organic chemicals as they relate to hazards. Various relevant classes of organics will be presented in terms of chemical behavior and physical properties. In order to facilitate the discussion to follow, a few basic definitions will be presented first.

Key Organic Chemical Classes and Definitions

Hydrocarbons—Chemical compounds consisting primarily of carbon and hydrogen.

Aliphatic—A class of organic compounds with the carbon backbone arranged in branched or straight chains (*e.g.*, propane or octane).

Aromatic—Organic molecular structures having the benzene ring (C_6H_6) as the basic unit (*e.g.*, toluene, xylene, polynuclear aromatics [PNAs]).

Note: When burning, aromatic compounds often give off large quantities of black smoke, owing to incomplete combustion.

Saturated—The condition of an organic compound in which each constituent carbon is covalently linked to four different atoms. This is generally a stable configuration (*e.g.*, propane [$CH_3CH_2CH_3$]). CH_3—CH_3 is a chemical depiction of a *saturated alkane*, namely ethane. However, if a hydrogen is removed from each carbon through a chemical *reduction* process, ethene (also known as ethylene) is formed with a double bond between the two carbons. Ethylene ($CH_2=CH_2$) is termed *unsaturated* because it has the capacity to bond with additional hydrogens.

Unsaturated—An organic compound containing double or triple bonds between carbons (*e.g.*, ethylene [$CH_2=CH_2$]). Multiple bonds tend to be sites of reactivity.

Isomers—Different structural arrangements with the same chemical formulas (*e.g.*, *n*-butane and *t*-butane):

n-butane $CH_3—CH_2—CH_2—CH_3$

```
                          CH3
                          |
t-butane          CH3—C—CH3
(tertiary-butane)         |
                          CH3
```

Functional group—An atom or group of atoms, other than hydrogen, bonded to the chain or ring of carbon atoms (*e.g.*, the –OH group of alcohols, the –COOH group of carboxylic acids, the –O– group of ethers). Functional groups determine the behavior of molecules. Consequently, the unique hazards of an organic compound are often determined by its functional group(s).

General Properties of Organic Compounds. Most organic compounds tend to melt and boil at lower temperatures than most inorganic substances. Because many organic compounds volatilize easily at room temperature and possess relatively low specific heats and ignition temperatures, they tend to burn easily. Many organic compounds are flammable. Moreover, organic vapors often have high heats of combustion. Therefore upon ignition, vapors facilitate the ignition of surrounding chemicals, thus compounding the severity of the hazard.

Most organic compounds are less stable than inorganic compounds. However, the presence of one or more halogen atom (F, Cl, Br, I) in the molecular structure of an organic compound increases its stability and inertness to combustion. Thus, partially halogenated hydrocarbons burn with less ease than their nonhalogenated analogues. Fully halogenated derivatives, such as carbon tetrachloride (CCl_4), chlorofluorcarbon refrigerants, and certain polychlorinated biphenyls (PCBs), are almost noncombustible.

Most organic compounds are water-insoluble. Notable exceptions are certain lower-molecular-weight alcohols, aldehydes, ketones, and carboxylic

acids, all known to be *polar* molecules. This characteristic is of importance in fire fighting, because the specific gravity of the compound will then be a major determinant of the suitability of water for the suppression of fires involving the chemical. Conversely, many organic compounds are soluble in one or more organic solvents such as acetone, ethanol, trichloroethylene, gasoline, *etc.*

Except for alkanes and organic acids, organic compounds tend to react easily with oxidizing agents such as hydrogen peroxide or potassium dichromate. Moreover, a mixture of an oxidizing agent and organic matter is usually susceptible to spontaneous ignition. Conversely, except for flammability and oxidation, organic compounds tend to react slowly with other chemicals.

Organic Nomenclature. This section will familiarize the reader with the naming system for some of the more common and simple organic groups, and present the salient characteristics of these groups. The basic system of aliphatic organic nomenclature is shown in Table 4.

The prefix for the name of an organic compound is based on the number of carbons involved and remains the same for each type of compound described. The suffix is determined by the type of compound and is independent of the number of carbons in the molecule. Thus methane, methanol, methanal (formaldehyde), and methanoic (formic) acid represent an alkane, an alcohol, an aldehyde, and a carboxylic acid, respectively; each with one carbon per molecule. The differences are nothing more than increasing oxidation states of the basic alkane hydrocarbon. In contrast, methanol, ethanol, and propanol are all alcohols, but with one, two, and three carbons per molecule, respectively.

The boiling points provided in Table 4 illustrate systematic trends in physical properties of this *homologous series* of alkanes. As the number of carbons (and the molecular weight) per alkane molecule increases, boiling and melting points increase.

The same trends hold in general, within a homologous group substituted to form alcohols, aldehydes, ketones, or carboxylic acids. For example octanol is less volatile and has a higher melting point than a lower-molecular-weight

Table 4. Nomenclature and Physical Properties of Straight-Chain Alkanes

Name	Number of Carbons	Formula	Melting Point (°C)	Boiling Point (°C)
Methane	1	CH_4	−183	−162
Ethane	2	CH_3CH_3	−172	−89
Propane	3	$CH_3CH_2CH_3$	−187	−42
Butane	4	$CH_3(CH_2)_2CH_3$	−138	0
Pentane	5	$CH_3(CH_2)_3CH_3$	−130	36
Octane	8	$CH_3(CH_2)_6CH_3$	−57	126
Decane	10	$CH_3(CH_2)_8CH_3$	−30	174

(Adapted from Ebbing, 1987)

alcohol such as butanol. Systematic trends can also be observed for other properties, such as water solubility. It should be noted that the boiling points provided in Table 4 are for the straight-chain isomers of the molecules. When the values for branched chain molecules are included, the comparisons become complicated.

Alkenes and alkynes are similar in structure to the alkanes except that the alkenes contain a carbon-to-carbon double bond (C=C) and the alkynes contain a carbon-to-carbon triple bond (C≡C). These two types of bonds represent intermediate oxidation states of the basic alkane hydrocarbon (similar to aldehydes, ketones, alcohols, and organic acids). The name prefixes are exactly the same as for the alkanes with the same number of carbons, but the endings are *ene* for compounds with double bonds and their derivatives and *yne* for compounds with triple bonds and their derivatives. Ethene (ethylene) and propene (propylene) are alkenes. Ethyne (acetylene) is an alkyne.

Aromatics are molecules based on single or multiple benzene rings. Some of the more common aromatics include benzene, toluene, xylene, and phenol. As previously mentioned, benzene is a six-carbon ring with the formula C_6H_6. The ring is generally depicted with alternating double and single bonds (now known as a *pi*-bond or *electron cloud*) and is quite stable. The substitution of a methyl group (–CH_3) for one of the hydrogens gives methylbenzene, known more commonly as toluene. The substitution of another methyl group gives dimethylbenzene or xylene. Substitution of a hydroxyl (–OH) for a hydrogen on the benzene ring yields hydroxybenzene, more commonly known as phenol.

Aromatics can also be named more specifically based on a system of assigning names or numbers to various positions on the benzene ring. By using the numbering system for the carbons on single or multiple benzene rings in combination with the names of the relevant substituents, any aromatic compound can be assigned a unique name.

Properties of Individual Functional Groups

Alkanes (C_nH_{2n+2}) are saturated hydrocarbons. The lower-molecular-weight alkanes (ethane through butane) are gases at standard temperature and pressure. The remaining alkanes are water-insoluble liquids that are lighter than water and thus form films or oil slicks on the surface of water. Hence, water is not used to suppress fires involving materials, such as gasoline, that include substantial proportions of liquid alkanes. Alkanes are relatively unreactive with most acids, bases, and mild oxidizing agents. However, with addition of sufficient heat, alkanes will react and burn in air or oxygen when ignited. In fact, low-molecular-weight alkanes (liquid petroleum gas [LPG], butane, and gasoline) are commonly used as fuels. Consequently, the biggest hazard from alkanes is flammability.

Organic carboxylic acids (RCOOH) usually are weak acids that can be very corrosive to skin. (Note: "R" is used to depict a generic organic chain to which the substituent group is bonded.) However, the substitution of Cl atoms on the carbon next to the carboxylic carbon produces a stronger acid. Thus, trichloroacetic acid is a stronger acid than acetic acid.)

Organic sulfonic acids (RSO_2H) generally are stronger acids than organic carboxylic acids.

Organic bases (such as amines, RNH_2) are weak bases but can be corrosive to skin or other tissue.

Alcohols (ROH) are not very reactive. The lower-molecular-weight alcohols (methanol, ethanol, and propanol) are completely miscible with water, but the heavier alcohols tend to be less soluble. Most common alcohols are flammable. Aromatic alcohols like phenol (flashpoint = 79°C) are not as flammable and are fairly water-soluble (solubility = 9 g/l).

Alkenes (C_nH_{2n}) are also known as *olefins*. Alkene compounds are unsaturated hydrocarbons with a single carbon-to-carbon double bond per molecule. The alkenes are very similar to the alkanes in boiling point, specific gravity, and other physical characteristics. Like alkanes, alkenes are essentially nonpolar. Alkenes are insoluble in water but quite soluble in nonpolar solvents like benzene. Because alkenes are mostly insoluble liquids that are lighter than water and flammable as well, water is not used to suppress fires involving these materials. Because of the double bond, alkenes are more reactive than alkanes.

$$\overset{\displaystyle O}{\overset{\displaystyle \|}{}}$$

Esters (RCOR) are not very reactive. Esters are formed when a carboxylic acid is reacted (*neutralized*) by an alcohol. Only the lowest-molecular-weight esters have appreciable solubility in water (ethyl acetate, solubility = 8%). Methyl and ethyl esters are more volatile than the corresponding unesterified acids. Most common esters are flammable. Esters are often easily recognizable due to their sweet to pungent odors.

Ethers (R–O–R) are low on the scale of chemical reactivity but they can be very toxic. Aliphatic ethers generally are volatile, flammable liquids with low boiling points and low flashpoints. Well-known hazardous ethers include diethyl ether, dimethyl ether, and tetrahydrofuran. Beyond flammability, ethers present an additional hazard because they react with atmospheric oxygen in the presence of light to form explosive organic peroxides (See next paragraph.). This is why old vessels/bottles of ethers should not be handled unless precautions for explosions are taken; *i.e.*, approach old ethers as you would a bomb and using *bomb squad* precautions!

Organic peroxides (R–O–O–R) are very hazardous. Most of the compounds are so sensitive to friction, heat, and shock that they cannot be handled without dilution. As a result, organic peroxides present a serious fire and explosion hazard. Commonly encountered organic peroxides include benzyl peroxide, peracetic acid, and methyl ethyl ketone peroxide.

Aldehydes and ketones ($R-\overset{\displaystyle O}{\overset{\displaystyle \|}{C}}H$ and $R-\overset{\displaystyle O}{\overset{\displaystyle \|}{C}}-R$, respectively) share many chemical properties because they possess the carbonyl (C=O) group as a common structural feature. Aldehydes and ketones have lower boiling points and higher vapor pressures than their alcohol counterparts. Aldehydes and ketones through C_4 are soluble in water and have pronounced odors. Ketones are relatively inert, while aldehydes are easily oxidized to their counterpart organic acids.

The Chemistry of Flammables

The elements required for combustion are few—a substrate, oxygen, and a source of ignition. The substrate, or flammable material, occurs in many classes of compounds but most often is organic. Generally, compounds within a given class exhibit increasing heat contents with increasing molecular weights (MW) (see Table 5).

Other properties specific to the substrate that are important in determining flammable hazards are the autoignition temperature, boiling point, vapor pressure, and vapor density.

The *autoignition temperature* (the temperature at which a material will spontaneously ignite) is important in fire prevention (*e.g.*, knowing what fire protection is needed to keep temperatures below the ignition point). This information can also be important in spill or material handling situations. For example, gasoline has been known to spontaneously ignite when spilled onto an overheated engine or manifold.

The *boiling point* and *vapor pressure* of a material are important not only because vapors are more easily ignited than liquids, but also because vapors are more readily transportable than liquids (they may disperse, or when heavier than air, flow to a source of ignition). Vapors with densities greater

Table 5. Heat Content/Increasing Weight Relationships

Compound	Molecular Weight (MW)	Heat of Combustion at 20° C, 1 atm (kg-calories/g MW)
Methane	16	210.8
Ethane	30	368.4
Propane	44	526.3
Methanol	32	170.9
Ethanol	46	327.6
Propanol	60	480.7

than one do not tend to disperse but rather to settle into sumps, basements, depressions in the ground, or other low areas, thus representing active explosion hazards.

Oxygen, the second requirement for combustion, generally is not limiting. Oxygen in the air is sufficient to support combustion of most materials within certain limits. These limitations are compound-specific and are called *the explosive limits in air* (see previous section on explosive limits). The upper and lower explosive limits (UEL and LEL) of several common materials are given in Table 6.

The *source of ignition* may be physical (such as a spark, electrical arc, small flame, cigarette, welding operation, or hot piece of equipment), or it may be chemical, such as an exothermic reaction. In any case, when working with or storing flammables, controlling the source of ignition is often the easiest and safest way to avoid fires or explosions.

Flammability, the tendency of a material to burn, can only be subjectively defined. Many materials

that we normally do not consider flammable will burn, given high enough temperatures. Nor can flammability be gauged by the heat content of materials. Fuel oil has a higher heat content than many materials that are considered to be more flammable because of their lower flashpoints. In fact, flashpoint has become the standard for gauging flammability.

The most common systems for designating flammability are found in definitions of the

- National Fire Protection Association's (NFPA) system

- Department of Transportation (DOT)

- Environmental Protection Agency's (EPA) Resource Conservation and Recovery Act (RCRA) definition of ignitable wastes

All of the above systems are based on a material's flashpoint. The NFPA diamond, which comprises the backbone of the NFPA Hazard Signal System, uses a four-quadrant diamond to display the hazards of a material. The top quadrant (red quadrant) contains flammability information in the form of numbers ranging from zero to four. Materials designated as zero will not burn. Materials designated as four rapidly or completely vaporize at atmospheric pressure and ambient temperature, and will burn readily (flashpoint < 73°F, boiling point < 100°F). The NFPA defines a flammable liquid as one having a flashpoint of 200°F or lower, and divides these liquids into five categories:

1) Class IA: liquids with flashpoints below 73°F and boiling points below 100°F. An example of a Class IA flammable liquid is *n*-pentane (NFPA Diamond: 4).

Table 6. Explosive Limits of Hazardous Materials

Compound	Lower Explosive Limit (%)	Upper Explosive Limit (%)	Flashpoint (°F)	Vapor Density
Acetone	2.15	13	–4	2.0
Acetylene	2.50	100	Gas	0.9
Ammonia (anhydrous)	16	25	Gas	0.6
Benzene	1.30	7.1	12	7.8
Carbon monoxide	12.4	74	Gas	1.0
Gasoline	1.4	7.6	–45	3-4
Hexane	1.1	7.5	–7	3.0
Toluene	1.2	7.1	40	3.1
Vinyl chloride	3.6	33	Gas	2.2
p–xylene	1.0	6.0	90	3.7

2) Class IB: liquids with flashpoints below 73°F and boiling points at or above 100°F. Examples of Class IB flammable liquids are benzene, gasoline and acetone (NFPA Diamond: 3).

3) Class IC: liquids with flashpoints at or above 73°F and below 100°F. Example of Class IC flammable liquids are turpentine and *n*-butyl acetate (NFPA Diamond: 2).

4) Class II: liquids with flashpoints at or above 100°F but below 140°F. Examples of Class II flammable liquids are kerosene and camphor oil (NFPA Diamond: 2).

5) Class III: liquids with flashpoints at or above 140°F but below 200°F. Examples of Class III liquids are creosote oils, phenol, and naphthalene. Liquids in this category generally are termed combustible rather than flammable (NFPA Diamond: 2).

The DOT system generally designates those materials at ambient temperature with a flashpoint of 140°F or less as flammable, those between 140°F and 200°F as combustible, and those with a flashpoint of greater than 200°F as nonflammable. There are some exceptions to these designations. Refer to 49 CFR 173.120 for more information. EPA designates those wastes with a flashpoint of less than 140°F as ignitable hazardous wastes. For more discussion on flammability, please see the references in the Bibliography.

It is quite apparent that the various terms describing *flammability* can be confusing. Please refer to Figure 2, which shows in graphic format how the various terms discussed above interrelate based on flashpoints.

These designations serve as useful guides in storage, transport, and spill response. However, they do have limitations. Since these designations are somewhat arbitrary, it is useful to understand the basic concepts of flammability.

Petroleum Chemistry

Crude oil is the frame of reference of petroleum chemistry. Crude oil is a complex mixture of organic chemicals containing mostly paraffins (high-molecular-weight waxy compounds), napthalenes (polynuclear benzene–based com-pounds),

and other aromatics with molecular weights ranging from the lightest molecular weights (benzene, MW = 150) to molecular weights of over five hundred.

Originally, crude oil was separated into products according to boiling points through simple fractional distillation. Now, crude oil is refined to more useful products through the process of catalytic cracking and reblending. The most common petroleum products in commercial use are gasoline, kerosene, and diesel fuel.

Gasoline is a complex mixture of branched chain aliphatics (C_5–C_{12}), and low-molecular-weight aromatics. Until the mid-1980s tetraethyl lead (TEL) was added to gasoline to improve its performance. Aromatic additives replaced TEL because of its toxicity. Additives designed to clean the injectors such as methytertiarybutylether (MTBE) have further enhanced gasoline's performance in fuel-injected engines. MTBE has received a lot of attention in some states for its impact due to its aroma and taste in contaminated groundwater.

Kerosene is similar in composition to gasoline except that the primary aliphatics have higher molecular weights (C_{12}–C_{16}). Its primary uses are for jet fuel and as a heating fuel.

Diesel fuel is very similar to kerosene except that it contains even higher-molecular-weight aliphatics (C_{15}–C_{18}). Its primary use is as a fuel for motor vehicles.

A number of other categories of products are derived from petroleum:

- The fuel gases: methane, ethane, propane, and butane (C_1–C_4)

- The petroleum ethers (C_5–C_7)

- The lubricating oils and greases (C_{16}–C_{18})

- From the residuum (the solid residue left after processing): paraffin wax (C_{20} – C_{30}), asphalts, petrolatum (vasoline), petroleum coke; and when blended with distillate, #4, #5, and #6 fuel oils

The regulation of underground storage tanks (UST) has made petroleum chemistry more important to a wide spectrum of environmental

Figure 2. Definitions of Flammable and Combustible Liquids[1]

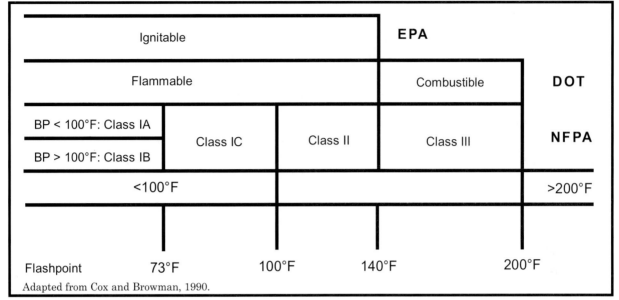

Adapted from Cox and Browman, 1990.

[1] DOT definitions for liquids at ambient temperature. Refer to 49 CFR 173.120 for more information.

professionals. Analysis of petroleum in soils and in groundwater by gas chromatography into diesel-range organics (C_{15}–C_{18}) and gasoline-range organics (C_5–C_{12}) is essential in UST work.

Once a petroleum fire has started, control of the fire can be accomplished in several ways: through water systems (by reducing the temperature), carbon dioxide or foam systems (by limiting oxygen), or through the removal of the substrate (by shutting off valves or other controls).

Toxic Chemicals

Toxicity is broadly defined and encompases a large spectrum of chemical classes, Therefore it would be beyond the scope of this study guide to do more than discuss the general characteristics of a few important classes of toxic chemicals.

Toxic Metals

The most common toxic metals in industrial use are cadmium, chromium, lead, silver, and mercury. Arsenic, selenium, (both metalloids), and barium are less commonly used.

Cadmium, a metal commonly found in alloys and myriad other industrial uses, is fairly mobile in the environment and is responsible for many maladies including renal failure and a degenerative bone disease called *itai itai*.

Chromium, most often found in plating wastes, is also environmentally mobile and is most toxic in the Cr^{+6} valence state.

Historically, lead has been used as a component of an antiknock compound in gasoline and, along with chromium (as lead chromate), in paint and pigments. Because of these and other historical uses, lead is ubiquitous. It is fairly mobile and is particularly soluble in acid environments.

Silver is used widely in the electronics industry. Intake of silver compounds can result in permanent discoloration of the skin and may result in damage to kidneys, lungs, mucous membranes, and other organs.

Mercury enjoys its seeming environmental ubiquity due to its use as a fungicide and as an electrode in the chlorine production process. Elemental mercury is relatively immobile in soils, but is readily transformed to more mobile organometallic compounds through microbial action. Mercury is the responsible agent for the infamous Minimata as well as the Mad Hatter's syndrome that is characterized by degeneration of the central nervous system.

Arsenic and selenium are both commonly used to decolorize glass or to impart a desirable color. Arsenic occurs in a number of important forms, many of which have been used as contact herbicides. Important forms of arsenic include arsenic trioxide and pentoxide. Arsenic also reacts to from arsenic acids, arsenites, and arsenates, and various organic arsenic compounds. Selenium often occurs as selenous acid. Both arsenic and selenium are fairly mobile in soils.

In general, toxic metals can be readily removed from aqueous solution through precipitation reactions, either as the sulfide or (more commonly) as the hydroxide. Various processes are available to stabilize metals in contaminated soil, but all the processes are expensive. It should be pointed out that metals cannot be destroyed, only concentrated, contained, or reduced in toxicity.

Cyanides

Cyanides (CN^-) are dangerously toxic materials that can cause instantaneous death. They are encountered in a number of industrial situations such as in sludges and baths from plating operations. Cyanide is extremely soluble, and many cyanide compounds, when mixed with acid, release deadly hydrogen cyanide gas. Cyanide is sometimes formed during the combustion of various nitrile, cyanohydrin, and methacrylate compounds. Cyanides commonly are treated by chlorine oxidation to the less toxic cyanate (CNO^-) form, then acid-hydrolyzed to CO_2 and N_2. Obviously, care should be taken that the cyanide oxidation is complete prior to acid hydrolysis of the cyanate.

Hydrogen Sulfide

Hydrogen sulfide is a commonly occurring decomposition product of organic matter. It is relatively water-soluble at higher pHs where it dissociates into H^+ and S^{2-} ions. As the pH is decreased below 7, gaseous H_2S begins to predominate and is released. Since its vapor density is greater than 1.0, H_2S gas tends to settle in low places and creates a toxic hazard. H_2S is readily

oxidized by a number of means to less toxic SO_3^{2-} or SO_4^{2-} forms.

Pesticides

Pesticides include the broad categories of insecticides, fungicides, rodenticides, and herbicides.

Insecticides. Insecticides, in common use, fall into three categories. The *chloroinsecticides* have chlorine in their structure. They are less soluble than the other insecticide forms and much less biodegradable (*i.e.*, more persistent). While they are less acutely toxic, several have been identified as potential carcinogens. *Carbamates* are less persistent and less toxic than chloroinsecticides, but some are also suspected carcinogens. *Organophosphate insecticides* generally are more acutely toxic than the other categories but they are not persistent.

Herbicides. Many formerly common herbicides now have been banned or restricted in their use; *e.g.*, 2,4-D and 2,4,5-T. However, the number and diversity of herbicides far exceeds that of insecticides. There are both organic and inorganic herbicides. Examples of inorganic herbicides are $CuSO_4$ and $NaClO_4$. There are at least 22 chemical families of organic herbicides. Even a cursory treatment of the chemistry of these materials would be extensive. Herbicides of limited toxicity (Treflan, Atrazine) as well as extremely toxic ones (Paraquat, Dinoseb) are in use. They range from water-soluble to -insoluble. The detailed chemistry of each should be determined prior to handling.

Chemical Incompatibilities

Chemical incompatibility can manifest itself in many ways; however, we will confine our discussions to those combinations resulting in fires, explosions, extreme heat, evolution of gas (both toxic and nontoxic), and polymerization.

Because of the number of chemicals and subsequent multiple number of potential reactions, it is impractical and (perhaps impossible) to list all potential reactions. Several systems exist for determining the reactions between classes of chemicals. The incompatibilties of the most broadly distributed chemicals are provided in two publications listed in the Bibliography (Bretherick and Hatayaya).

The volume by Bretherick is divided into two sections. The first lists general classes of compounds and gives reactivity information regarding interactions of these classes with other classes and with specific chemicals. The second and much larger section lists specific compounds and references specific adverse reactions as they have been observed or reported in the chemical literature. The work by Hatayaya provides a matrix format compatibility chart listing 40 classes of chemicals. While both of these volumes are extremely helpful, they are not and do not claim to be definitive works on material compatibility. They are, however, useful guides for identifying potential reactions.

Because all of the potential reactions for individual chemicals are not catalogued and because there are few pure solutions of waste materials, laboratory compatibility testing is recommended for most materials. An appropriate protocol for compatibility testing would involve the following steps:

- Obtain all available information about the material. If it is a surplus or off-specification product, obtain an analysis or a Material Safety Data Sheet. If it is a waste, check for previous analyses, and if none exists, arrange for one to be done. Even if a previous analysis exists for this stream, consider running a few screening-type field analyses for confirmation of important properties such as pH, redox potential or other oxidizer test, cyanide, sulfide, and flashpoint.

- Once the identity of the material is known, one of the cited references can be consulted to determine potential reactions. At this point, incompatibility may be obvious. If not, then laboratory testing for compatibility is required.

Compatibility testing is almost by nature an experiment with the unknown. As such, safety must be the watchword. Procedures for compatibility testing should take into account the most severe adverse reaction possible, not just that expected. Such testing should always be performed under a vent hood while wearing, as a minimum, face shield, rubber apron, and gloves. Generally, compatibility testing entails mixing a small volume of one substance with another and observing for heat, gas generation, or polymerization.

Polymerization need not be violent to cause problems. Anyone who has ever had to chisel out or replace a tank of solidified material can attest to this. Often it is advisable to heat the mixture to expected storage or process temperature and then observe for further heat, gas, or polymerization.

Observation of a reaction does not necessarily preclude mixing. Moderate heat or gas generation may not present a problem. However, a number of safety precautions should be taken before mixing the material if any heat or gas generation occurs. If heat is generated, the amount should be determined and a heat balance calculated so that effects of heating on the storage tank and tank base can be calculated. Expansion of the material with heating should also be considered, so as to avoid overfilling the receiving tank.

Generation of gas requires a gas analysis before mixing. If the gas is toxic or if discharge of the resultant gas violates an air-quality constraint, the materials should not be mixed. If the gas is nontoxic, care should still be taken to ensure that the gas generation rate does not exceed the design venting capacity of the tank. Remember that most tanks are designed to withstand a water gauge internal pressure of only about 8 inches (a typical person can provide pressure of 24 inches water gauge by blowing). Also, even if the gas is nontoxic, it may still displace air and, for inside tanks especially, create an asphyxiation hazard.

The Chemistry of Gases

Gases by their very nature are mostly empty space. Thirty-two grams of liquid oxygen occupy a space of 0.026 liters. At Standard Temperature and Pressure (STP) of 0°, 1 atmosphere, this same weight of oxygen occupies 22.4 liters; over 861 times as much volume. If the oxygen is pure, the molecules represent less than 0.1% or 1/1000 of the volume of the gas.

Avogadro's Law tells us that the volume of one mole of any gas at STP will be 22.4 liters. This allows us to calculate the volume of gas produced by a specific reaction. For example, if we accidentally release 130 grams of potassium cyanide into a liter of hydrochloric acid, we can estimate the volume of gas generated. The reaction is illustrated by the equation presented below:

$$KCN + HCl \rightarrow HCN(g) + KCl \qquad (22)$$

The molecular weight of KCN is 65.112, so we have approximately 2 moles of KCN which will react with the HCl (HCl is clearly in excess, since concentrated HCl contains 12 moles per liter) to produce about 2 moles of HCN or about 44.8 liters.

Suppose the reaction does not take place at STP but in mid-summer in Nepal, where the temperature is 70°F and the atmospheric pressure is 0.92. In this case we could use the ideal gas equation. The ideal gas equation incorporates Boyle's Law (volume is inversely proportional to pressure), Avogadro's Law (volume is dependent upon the number of molecules), and the Law of Charles and Gay-Lussac (volume is proportional to temperature) into a single expression. The Ideal Gas Law is represented as:

$$PV = nRT \qquad (23)$$

Where:

P   =   pressure (atm)
V   =   volume (liters)
n   =   the number of moles of gas
T   =   temperature (Kelvin)
R   =   0.0820 (the gas constant which is dependent upon the units used)

The calculation of our example would be as follows:

$$(0.92)(V) = (2)(0.082)(294.1)$$

$$V = 52.43 \text{ liters}$$

Final Comments

The concepts presented in this overview of chemistry come into play in one form or another in the practice of hazardous materials manage-ment. Those who wish to excel in this field are encouraged to *understand and memorize* as many of the terms, concepts, and principles as time and interest permit. They constitute a significant portion of the everyday vocabulary and considerations of this profession. Based on the writer's personal experience, the time invested will be well spent.

Bibliography

Bretherick, L., P.G. Urben, Eds., and M. J Pitt. *Bretherick's Handbook of Reactive Chemical Hazards: An Indexed Guide to Published Data,* 5th ed. Boston, MA: Butterworth-Heinemann, 1995.

Budavari, S., M. J. O'Neil, Eds., and A. Smith, Contributor. *The Merck Index: An Encyclopedia of Chemicals, Drugs, and Biologicals (Annual)* 12th ed. Whitehouse Station, NJ: Chapman & Hall, 1996.

Burke, R. *Hazardous Materials Chemistry for Emergency Responders.* Boca Raton, FL: Lewis Publishers, Inc., 1997.

Burns, Ralph A. *Fundamentals of Chemistry,* 2nd ed. Englewood Cliffs, NJ: Prentice Hall, 1995.

Carson, P. A., Ed., and C. J. Mumford. *Hazardous Chemicals Handbook.* Boston, MA: Butterworth-Heinemann, 1994.

Cooper, A. R., Sr., Ed. *Cooper's Toxic Exposures Desk Reference.* Boca Raton, FL: Lewis Publishers, Inc., 1996.

Dean, J. A., Ed. *Lange's Handbook of Chemistry,* 14th ed. New York, NY: McGraw-Hill, 1992.

Department of Transportation, Transport Canada. *North American Emergency Response Guidebook.* Nenaah, WI: J. J. Keller & Associates, 1998.

Cox, Doye B. and M. G. Browman. "Chemistry of Hazardous Materials." in *Handbook on Hazardous Materials Management.* H. T. Carson and D. B. Cox, Eds. Rockville, MD: Institute of Hazardous Materials Management, 1990.

Ebbing, Darrell D. *General Chemistry,* 6th ed. Boston, MA: Houghton Mifflin Company, 1987.

Hatayaya, H. K., *et al. A Method for Determining the Compatibility of Hazardous Wastes.* EPA – 600/2–80–076. Washington, DC: EPA.

International Conference of Building Officials. *Uniform Fire Code,* 10th ed. Whittier, CA: International Conference of Building Officials, 1997.

Kent, J. A., Ed. *Riegel's Handbook of Industrial Chemistry,* 9th ed. New York, NY: Van Nostrand Reinhold, 1992.

Lewis, R. J., Ed. *Hawley's Condensed Chemical Dictionary,* 13th ed. New York, NY: Van Nostrand Reinhold, 1997.

Lewis, R. J. *Hazardous Chemicals Desk Reference,* 4th ed. New York, NY: Van Nostrand Reinhold, 1997.

Lide, D. R., Ed. *CRC Handbook of Chemistry and Physics: A Ready-Reference Book of Chemical and Physical Data,* 78th ed. Boca Raton, FL: CRC, 1997.

Manahan, S. E. *Environmental Chemistry,* 6th ed. Boca Raton, FL: Lewis Publishers, 1994.

Meyer, E. *Chemistry of Hazardous Materials,* 2nd ed. Upper Saddle River, NJ: Prentice-Hall, 1989.

National Institute for Occupational Safety and Health. *NIOSH Pocket Guide To Chemical Hazards.* NIOSH Publication No. 94–116. Washington, DC: NIOSH.

Perry, R. H., Ed., *et al. Perry's Chemical Engineers' Handbook,* 7th ed. New York, NY: McGraw Hill, 1997.

Pohanish, Richard P. and S. A. Greene. *Hazardous Materials Handbook.* New York, NY: Van Nostrand Reinhold, 1996.

Sittig, M. *Handbook of Toxic and Hazardous Chemicals and Carcinogens,* 3rd ed. Park Ridge, NJ: Noyes Publications, 1991.

Richard E. Hagen currently serves as Vice President and Program Manager, Koester Environmental Services (KES) in Fort Collins, Colorado. Prior to joining KES, he served as Director of Environmental Management, Bristol-Myers Squibb (BMS) Nutritionals Group, and as Chariman of the BMS World-Wide Corporate Environmental Committee. His areas of expertise and interest include industrial hazardous materials management, environmental management systems, remediation quality assurance and control, and environmental program management. Dr. Hagen has been active in environmental technical affairs for over 20 years. Dr. Hagen would like to acknowledge Doye B. Cox, PE, CHMM, Koester Environmental Services, and Michael G. Browman PhD, Tennessee Valley Authority, for preparing portions of this chapter used in the Handbook on Hazardous Materials Management. *Much of the original chapter has been used, with editing to fit the present format. Although direct reference is not made to specific sources, many ideas and concepts for this chapter have been adapted from sources in the Bibilography. The author wishes to thank Mr. Cox and Ms. Adriane P. Borgias for their helpful comments in the development of this review of an exciting and rewarding field.*

CHAPTER 41

Multimedia Sampling

W. Scott Butterfield, MS, CHMM

Introduction

Site investigations frequently require the collection and analysis of multimedia samples (*e.g.*, soil, sediments, air, aqueous, waste, *etc.*) in order to

- Identify contaminants

- Identify the source(s) of contaminants

- Determine the extent of contamination

- Prepare and implement Health and Safety plans

- Establish threat to public health and environment

- Evaluate remedial options

- Confirm site cleanup

One or more of the above reasons for sampling is the Data Use Objective(s) (DUO) of the sampling program. DUOs are important drivers in the planning process, since they provide a basis for determining the sampling procedures required, analytical methodologies to be used by the laboratory, and quality assurance (QA) procedures necessary to provide analytical data needed to meet the DUO(s). DUOs will be discussed in more detail later in this chapter.

Sampling and subsequent analysis can result in a significant expenditure of time and money during the assessment of a particular site. Decisions based on the interpretation of sampling results can have significant financial impacts on the parties involved in the planning and implementation of site remediation programs. Thus it is extremely important to prepare for and conduct a cost-effective sampling program that provides a valid representation of site conditions. This is true whether there is concern about the contents of a few drums or the extent of contamination on a multiacre property. The United States Environmental Protection Agency (EPA) has developed the Data Quality Objectives (DQO) process to help their staff in determining the type and quantity of data needed to support EPA decisions. The reader is encouraged to review the references listed at the end of this chapter to gain a better understanding of this planning process. The blueprint for accomplishing the established goal of the DQO process is the **sampling plan**. A properly prepared and executed sampling plan will result in the generation of data of known quality. This means all components associated with the generation of the data are thoroughly documented and the documentation is verifiable and defensible.

This chapter provides an introduction to the process of planning and executing multimedia sampling programs. Basic terminology is defined and a suggested format for sampling plans is described. Emphasis is placed on the importance of quality assurance/quality control throughout the process. Also provided are significant references, which contain detailed information on quality assurance/ quality control, sampling plan preparation, and sample collection techniques. Upon completion of this chapter the reader should be able to understand the planning process, be able to identify appropriate sampling procedures, and prepare a basic sampling plan. In order to gain proficiency in multimedia sampling, the reader must develop his/her skills under the supervision of experienced multimedia samplers.

Multimedia Samples

The assessment of sites to determine the presence of hazardous materials may involve the collection and analysis of samples from various media on and adjacent to the site. These media can vary from liquid and solid materials contained in on-site containers (*e.g.*, drums, tanks, *etc.*), to contaminated site soils and sludges, to ground and surface waters (on and off-site), to air gases and particulates. The various media can be broken down into the following general categories:

- Surface water

- Ground water

- Soil/sediment

- Air/soil gas

- Wastes

These media may be contaminated with hazardous materials to various degrees. In general, when planning to collect samples from and around hazardous waste sites, **samples** are categorized as one of two types: **hazardous** or **environmental**. Hazardous samples are sometimes further categorized based on the level of contamination, such as high and moderate hazard samples. The determination of sample type is based on existing analytical data, background information (*e.g.*, past activities conducted at the site, container labels, *etc.*) and/or policy. Policy determinations reflect an organization's desire to ensure the health and safety of all involved with sampling programs and to avoid the liability of improper shipment of samples. Some organizations have a simple policy of "if it is on-site the sample is hazardous" and "if it is off-site the sample is environmental". In any event the relative concentrations and/or hazards must be determined prior to sample collection in order to determine

- Proper health and safety precautions

- Required sampling procedures

- Required sample shipping procedures

- Appropriate analytical procedures

The collection of high-hazard samples will normally take longer and cost more than environmental samples because of the higher level of personal protection required (*i.e.*, Level C or B versus Level D) and the more stringent sample shipping requirements. Analytical procedures also differ for high-hazard samples and subsequently cost more.

Types of Samples

There are basically two types of multimedia sample collection techniques utilized during hazardous waste site investigations: grab and composite sampling. A *grab sample* is a discrete aliquot taken from one specific sampling location at a specific point in time. The analytical results from a grab sample should provide a good representation of the concentrations of the contaminants present at the point the sample was taken. An example is a surface soil sample collected from a specific point or *station* on a site. A *composite sample* is a nondiscrete sample composed of two or more equal aliquots collected at various points or times. The analytical results from a composite sample provide average concentrations of the contaminants present. The collection and mixing (homogenizing) of two aliquots of surface material from a pile of *waste* material would result in the generation of a composite sample of the pile.

Grab sampling is preferred in many situations because it minimizes the time and expense that would be required to collect the multiple aliquots required for composite samples. It also reduces the potential health risks of combining unknown and potentially reactive materials. Grab samples are appropriate when the objective is to document the condition of a particular location such as an area of stained soil or the contents of individual containers. Since the sample is taken from a particular point or container, that point or container can be relocated for further investigation if required. Grab sampling may be appropriate to characterize stratified materials that may occur in drums, tanks, lagoons, or wells. Under these circumstances, a grab sample from each layer or phase may be collected.

Composite sampling is utilized when sampling areas or containers of similar material (*e.g.*, pile of homogeneous waste material, drums of similar material, *etc,*) or when a large area is being surveyed to determine the extent or presence of contamination. An example where composite sampling would be appropriate is the evaluation of a multiacre site where contaminants are suspected of being disposed but there are no obvious signs of disposal areas (*e.g.*, stained soil, distressed vegetation, *etc.*). For such a situation a grid could be superimposed over the site and composite samples could be taken from all or a portion of the grids. If contaminants are detected in any of the grids, a more rigorous but localized sampling program can be conducted within the

contaminated grids using grab sampling techniques to better define the extent and magnitude of contamination. It is important to remember that the act of combining materials during composite sampling causes a dilution effect on the components of the aliquots combined. Thus the resultant concentration of contaminants generally is lower than the maximum of any of the aliquots. This dilution effect must be considered when determining the required detection limits of the analytical procedures selected for sampling programs, to avoid false negative analytical results.

Four types of composite samples are generally considered: areal, vertical, flow-proportional, and time. Areal and vertical composite samples are routinely collected during the investigation of hazardous waste sites. *Areal composites* are composed of equal aliquots (grab samples) collected from a defined area such as the surface of a waste pile or an established grid. *Vertical composites* are composed of equal aliquots collected along a defined vertical interval such as a borehole or test pit (for example, the combination of five aliquots taken every foot during the installation of a 5-foot bore hole). *Flow-proportional composite sampling* is generally associated with stream or wastewater discharge samples and is thus not normally utilized in the investigation of hazardous waste sites. *Time composite sampling* is also normally not utilized during hazardous waste site investigations, because the intent of such investigations is to determine as quickly as possible what is present on site at the *present* time. For health and safety reasons, a form of time compositing is utilized to determine air quality on and around sites. Here ambient air is drawn over a filter or adsorbent material at a set flow rate for a period of time. The result is the generation of an average concentration of contaminant(s) per volume of ambient air.

It is very important to remember that composite sampling generates average concentrations within the composited interval. Thus, one cannot draw conclusions concerning the maximum and minimum concentrations of the contaminant in question.

Sampling Strategies

In addition to the types of samples and sample collection techniques, the sampling strategies must also be considered. There are three basic strategies to consider during the development of sampling programs: judgmental, random, and systematic.

Judgmental sampling is subjective. Sampling points are located where contaminants are thought to exist, such as in areas of stained soil or in containers. Judgmental sampling locations may also be identified through historical site information (including photographs), which identifies areas of concern based on past site operations/conditions (*e.g.*, waste discharge points, storage areas, lagoons). ***Judgmental sampling*** generally provides information on worst-case conditions which is useful for identifying contaminants of concern at a site but not the extent of contamination which will require additional sampling utilizing other approaches.

Random sampling is the opposite of judgmental sampling, where locations are arbitrarily assigned using a random selection procedure such as random number tables. Random sampling would be used when statistical evaluations of the data are required (*e.g.*, probability statements). Random sampling, alone, generally is not appropriate for evaluating hazardous waste sites, because a key assumption of homogeneity cannot be made due to the heterogeneity of hazardous waste sites. Random sampling in combination with other strategies such as grid sampling is frequently used on large, relatively homogeneous areas of extensive sites to help determine the extent of contamination (see below).

Systematic sampling involves the establishment of a reproducible scheme such as a grid or transect to which sampling points can be referenced. Samples are then taken at the points grid lines intersect (nodes) or defined distances along the transect line. The number of samples taken will depend on the area covered by the grid and the cell size established (the smaller the cells, the larger the number of nodes generated) or by the spacing of the points along the transects. As discussed above, random sampling can be combined with a systematic approach when the areas within the grids are relatively homogeneous. This approach allows probability statements to be developed about the extent of contamination within the individual grids.

Detailed information on the above topics can be found in the publications listed at the end of this chapter. The majority of these publications can be accessed via the Internet at the sites listed at the end of this chapter.

Multimedia Sampling Plan

Sampling and subsequent analysis can result in a significant expenditure of time and money during the assessment of a particular site. Decisions based on the interpretation of sampling results can have significant financial impacts on the parties involved in the planning and implementation of site remediation programs. Thus, it is extremely important to prepare for and conduct a cost-effective sampling program that provides a valid representation of site conditions. This is true whether there is a concern about the contents of a few drums or the extent of contamination on a multiacre property. The blueprint for attaining this goal is the ***sampling plan***. A properly prepared and executed sampling plan should provide data of known quality, which means all components associated with the generation of the data are thoroughly documented and the documentation is verifiable and defensible.

The EPA and various state environmental departments have prepared many good guidance documents on the preparation of sampling plans. An excellent series on representative sampling is available from the EPA Environmental Response Team, which is available from their Web page (listed in the Internet Resources section) at no charge. Other references are provided at the end of this chapter. In general sampling plans should cover the following topics:

- Introduction/background
- Objectives
- Data use
- Quality assurance
- Sampling approach and methods
- Organization and responsibilities
- Quality assurance requirements
- Deliverables

A discussion of each of the six topics is presented below.

Introduction/Background

The development of an effective sampling plan requires a thorough evaluation of current as well as past site conditions. A thorough file search should be conducted with appropriate/potential regulatory agencies (Federal, State, and local) to determine if information exists on past environmental issues (*e.g.*, spills, fires, citations, permits, *etc.*). It is very important to determine land use histories of sites to gain a better understanding of potential hazards (*e.g.*, contaminants of concern, buried lagoons, underground pipe runs and tanks, existing and past building uses and locations, *etc.*). Historical aerial photographs can be very useful in this regard. In addition to identifying types of contaminants, the volumes and concentrations of these materials should be quantified. This information will assist planners in identifying appropriate sampling and analytical methodologies and detection limits.

Not only is it important to determine the location of potential sources of contaminants, the potential migration pathways should also be identified. Topographic maps such as United States Geological Survey (USGS) topographic maps and aerial photographs provide good information for identifying these pathways. A site reconnaissance should also be conducted prior to the finalization of the sampling plan in order to confirm site conditions and to help refine the evaluation of potential migration pathways. An off-site reconnaissance is appropriate when the site is relatively small, potential sources of contaminants are visible, access has not yet been attained, and/or current site information is required to prepare for an on-site reconnaissance. An on-site reconnaissance can provide a great deal more information about site conditions, but it also requires proper planning, since the investigators will be working in areas potentially contaminated with hazardous materials. This is where the gathered background information becomes extremely valuable.

In order to conduct an on-site reconnaissance, a health and safety plan (HASP), which identifies potential on-site hazards and specifies levels of personal protection, must be prepared and implemented. During the on-site reconnaissance, portable monitoring and analytical equipment can be utilized to better define the areas of contamination and help identify contaminants of concern. A site map should be developed during the reconnaissance that denotes the locations of on-site structures, areas of potential contamination, containers and tanks, *etc.* On-site drainage patterns and potential targets near the site should also be identified and indicated on the site map.

Objectives

The objectives of the sampling plan set the stage for the sampling strategy, the selection of sampling procedures and analytical methods, and Quality Assurance/Quality Control (QA/QC) requirements. There are two subjects that must be addressed in this section: Data Use Objectives and the Quality Assurance Objectives. ***Data Use Objectives*** (DUO) address specific goals of the sampling program. The goal may be as simple as determining the contents of a single drum in order to arrange for proper disposal, or as complex as determining the impact of a site with multiple sources of unknown contaminants on numerous human and ecological resource targets. The Data Use Objectives should be stated as clearly as possible, including what the data is needed for, the questions that must be answered, and what decisions have to be made based on the data generated. Examples of these follow.

The Data Use Objective(s) of this sampling program is (are)

- To determine the Resource Conservation and Recovery Act (RCRA) characteristics of materials contained in all on-site drums in order to classify these materials as RCRA hazardous or nonhazardous materials

- To determine the concentration of volatile organic compounds in residential tap water samples in order to compare these concentrations to State (specify the State agency) drinking water standards and EPA Maximum Contaminant Levels (MCLs)

- To verify the attainment of the site-specific soil cleanup goal for lead (specify the level), prior to backfilling the excavation with clean fill material

If appropriate, the statement should include what the analytical results will be compared to, such as State or Federal regulatory requirements, permits, guidelines, standards, *etc.* These benchmarks will provide a basis for determining Quality Assurance (QA) requirements such as analytical methods and detection limits.

Quality assurance objectives are statements about the desired reliability of the data to be generated. They are defined by determining how precise, accurate, representative, complete, and comparable the data must be to satisfy the data use objective. Detailed discussions on the above quality assurance terms can be found in the references listed at the end of this chapter. Various agencies such as the EPA have established quality assurance *Levels* which address predefined quality assurance objectives. For example the EPA Office of Solid Waste and Emergency Response (OSWER) utilizes a three-level quality assurance objective system (QA1, QA2, and QA3) in the development of their sampling plans. In brief, the objectives of these levels are as follows:

> *QA1*—Screening objective used to afford a quick, preliminary assessment of site contamination. Data collected for this objective provide neither definitive identification of pollutants nor definitive quantification of concentration levels.
>
> *QA2*—Verification objective used to verify analytical results. Here a small percentage of the sample results are verified to provide a certain level of confidence for a portion of the results. An inference is then made on the quality of the remainder of the data.
>
> *QA3*—Definitive objective used to assess the accuracy of the concentration level as well as the identity of the analytes of interest. This objective is appropriate when a high level of qualitative and quantitative accuracy is required for all sampling results.

The above QA Levels in turn specify the types and numbers of QA samples (*e.g.*, blanks, duplicates, *etc.*) required to satisfy the selected QA Level.

Sampling Approach and Methods

This section of the sampling plan should address the sampling strategy, the sampling procedures (including decontamination procedures), sample management, sample containers, sample shipment and documentation. It should be of sufficient detail to permit an experienced sampling team to carry out the sampling program. The sampling strategy should discuss the types and numbers of samples to be collected, their locations and reasons for their collection (*e.g.*, background or contaminant source sample, *etc.*). A sample numbering/identification system should also be developed to ensure sample identity. When numerous samples are required, a tabular presentation of this information should be provided. A site map should also be developed to show the locations of the sampling points or stations. Included in the approach should be the sequence of sample collection. For example, stream sampling for water and sediments should be initiated at the farthest downstream collection point and then conducted in a sequential, upstream direction to the farthest upstream sampling point to prevent the influence of disturbed sediments on the downstream sampling points. Consideration should also be given to collecting background and low concentration samples prior to the collection of high concentration samples to minimize the potential for cross-contamination.

The sampling procedures for all the samples required should be specified. This can be accomplished by either detailing the procedures in this section or by referencing standard operating procedures (SOPs) and attaching the detailed SOP to the Sampling Plan. Procedures should be specified for all aspects of the sampling program including procedures for sampling equipment decontamination, sample collection, sample preparation (*e.g.*, homogenization, compositing, *etc.*), and the collection of QA samples. The handling procedures for waste materials generated during the sampling program (*e.g.*, decontamination fluids, used disposable sampling equipment, personal protective equipment, *etc.*) should also be described.

An important consideration is the composition of the sampling equipment to be utilized during the sampling program. Because the levels of concern for contaminants can be very low, the materials utilized in the construction of sampling equipment must not impact the final results through the release of minute amounts of the contaminants of concern. As a rule of thumb, sampling equipment should be constructed of stainless steel, Teflon, or glass. The sampling SOPs generally will specify the materials that are acceptable and appropriate decontamination procedures.

Once the samples are collected, the effort turns to maintaining the integrity of the samples; that is, ensuring that the material collected continues to be representative of the point or area sampled. This is the role of the sample management program. It begins with the selection of the appropriate sample containers. Specifications for sample containers include size/volume, material (glass, plastic, *etc.*), lid/cap construction/material, *etc.* The sample containers must be cleaned according to procedures that have been established by either the agency (EPA, state, *etc.*,) that requires the results or the analytical laboratory utilized for the analyses. Pre-cleaned sample containers are readily available from commercial suppliers or the analytical laboratories utilized for the analyses.

To prevent the reduction in analyte concentration within the sample due to decomposition, vaporization, *etc.*, samples generally are preserved shortly after collection. Preservation techniques include temperature reduction (down to 4° C) and pH adjustment. Maintaining samples at 4°C during transport or shipping can be difficult at times, particularly during the summer or when working in warmer climates (*e.g.*, Puerto Rico). To help ensure the proper temperature is maintained, the samples should be chilled on ice prior to packing them into transport/shipping containers. When ready for transport or shipping, the samples should be packed with fresh ice. The fresh ice should be double-bagged in plastic zipper-lock bags and placed on top and around the sample containers. Packaging material, usually vermiculite, should then be placed in the shipping container (usually a cooler) to insulate the samples and the ice.

Once a sample has been preserved, there is a specified holding time that must not be exceeded prior to sample analysis. Holding times vary according to analytes of concern and the agency desiring the analytical results. Attention must also be paid to the acceptable method of holding time calculation, which can be either from the time of sample collection or the time of sample receipt by the laboratory.

A good way to keep track of the above is to develop a table which lists sample number, analytical parameter, matrix, container types and numbers, preservation requirements, and holding times. Quality Control samples (blanks, duplicates, *etc.*) can also be included in the table.

Sample transport or shipping procedures should be included as part of the sample management plan. Once the samples are collected and preserved (if required) the sampling team must ensure the samples reach the analytical laboratory in good condition and in a timely manner. Sample bottles must be intact, samples held at the proper temperature (if required), and analyzed within a sufficient time frame to ensure compliance with prescribed holding times. When analytical laboratories are relatively close to the site of sample collection, the samples may be transported to the lab by the sampling team. The sampling team must be aware that the transport or shipment of samples

containing hazardous or potentially hazardous materials over roadways subject to United States Department of Transportation (DOT) regulations, requires compliance with current Federal regulations, as specified in the *Code of Federal Regulations* (CFR) (specifically, Title 49, Parts 171 through 178 [49 CFR 171–178]). To be in compliance the proper shipping name, identification number, packaging, labeling, packing, and paper-work requirements must be determined. In addition, personnel responsible for shipping hazardous materials must meet DOT training requirements. If the shipment of potentially hazardous materials involves overnight shipment by air, International Air Transport Association (IATA) regulations must be applied. Following current IATA regulations ensures compliance with DOT requirements. It is also a good practice to include with the sample shipping papers a notice to laboratory personnel as to the relative hazard of the samples, so they can take proper precautionary measures.

Finally, the documentation procedures for the sampling program should be specified. These procedures can be as simple as the maintenance of a handwritten log book in which sampling activities are described (who, what, where, when). Photography and videotaping can also be useful in addition to written records. In addition to recording all sampling activities, any variations to SOPs specified in the sampling plan should be documented.

Organization and Responsibilities

This section is fairly straightforward and can be satisfied with a table that shows who is responsible for appropriate sampling program activities, such as Project Manager, Sample Management, Health and Safety, *etc.* Depending on the magnitude of the sampling program, sampling personnel may assume a multitude of roles. It is very important that the sampling personnel know their roles prior to implementation of the program, so they can prepare properly.

Quality Assurance/Quality Control

This section should address procedures necessary to ensure the desired quality of the data generated by field sampling activities. Included should be a discussion on field monitoring equipment calibration and preventative maintenance procedures. These procedures normally are based on the manufacturer's instructions. Also included are discussions on the number, types, and method of collection of Quality Assurance/Quality Control (QA/QC) samples required for the sampling program. QA/QC sample requirements are determined by a number of factors, including the data use and quality assurance objectives, total number of samples and sampling procedures (dedicated versus nondedicated sampling equipment). These samples may include blanks (trip, equipment rinsate), duplicates, splits, *etc*. It is a good practice to list all of the QA/QC samples in a table, to ensure that the sampling team is aware of the requirements.

The analytical procedures that the laboratory will follow must be specified, along with a discussion of the appropriate analytical data validation procedures. These procedures are also directly related to the data use and quality objectives of the sampling program. For example, analytical method detection limits must be below any decision criteria concentrations.

Deliverables

Finally, reports, or deliverables, required from the sampling program should be identified. This could be a basic data summary or an in-depth report that discusses all aspects of the sampling program. It is a good practice to prepare an interim or trip report which details the field activity conducted and describes any problems encountered and any variations/modifications to SOPs specified in the sampling plan.

Internet Resources

There are two significant, no-cost, sources of information on sample planning, sampling methods, and quality assurance on the Internet. Both are provided by the EPA. The first is the Office of Research and Development's National Center for Environmental Research and Quality Assurance, (NCERQA) located at

<http://es.epa.gov/ncerqa/qa/index.html>

Many of the quality assurance documents listed below can be accessed through this site. Also available are several computer-based training

courses on quality assurance planning and field sampling that can be downloaded. The second site is maintained by the Environmental Response Team (ERT) and is located at

<http://204.46.140.12>

EPA representative sampling guidance documents, along with compendiums of sampling procedures, are available from this site. Also available from the ERT are individual SOPs on sampling methods, and a computer program for generating sampling plans.

Bibliography

American Society for Testing and Materials, *ASTM Standards on Environmental Sampling,* 2nd ed. W. Conshohocken, PA: ASTM, 1997.

Environmental Protection Agency. *Compendium of ERT Surface Water and Sediment Sampling Procedures.* EPA/540/P–91/005. Washington, DC: EPA Office of Solid Waste and Emergency Response, 1991.

Environmental Protection Agency. *Compendium of ERT Waste Sampling Procedures.* EPA/540/P–91/008. Washington, DC: EPA Office of Solid Waste and Emergency Response, 1991.

Environmental Protection Agency. *A Compendium of Superfund Field Operations Methods.* EPA/540/P–87/001 (OSWER Directive 9355.0–14). Washington, DC: EPA, Office of Emergency and Remedial Response, 1987.

Environmental Protection Agency. *EPA Guidance for Quality Assurance Project Plans, EPA QA/G–5.* EPA/600/R–98/018. Washington, DC: EPA, Office of Research and Development, 1998.

Environmental Protection Agency. *EPA Requirements for Quality Assurance Project Plans for Environmental Data Operations, EPA QA/R–5.* Washington, DC: EPA, Quality Assurance Division, 1997.

Environmental Protection Agency. *Guidance for the Data Quality Objectives Process, EPA QA/G–4.* EPA/600/R–96/055. Washington, DC: Office of Research and Development, 1994.

Environmental Protection Agency. *Handbook for Sampling and Sample Preservation of Water and Wastewater.* EPA–600/4–82–029. Cincinnati, OH: EPA, Environmental Monitoring and Support Laboratory, 1982.

Environmental Protection Agency. *Management of Investigation-Derived Wastes During Site Inspections.* OERR Directive 9345.3–02. Washington, DC: EPA Office of Emergency and Remedial Response, 1991.

Environmental Protection Agency. *Quality Assurance/Quality Control Guidance for Removal Activities, Sampling QA/QC Plan and Data Validation Procedures.* EPA/540/G–90/004. Washington, DC: EPA, Office of Emergency and Remedial Response, 1990.

Environmental Protection Agency. *Removal Program Representative Sampling Guidance, Volume 1: Soil.* OSWER Directive 9360.4–10. Washington, DC: EPA, Office of Solid Waste and Emergency Response, 1991.

Environmental Protection Agency. *Superfund Program Representative Sampling Guidance, Volume 4: Waste.* OSWER Directive 9360.4-14. Washington, D.C.: EPA Office of Solid Waste and Emergency Response, 1995.

Environmental Protection Agency. *Superfund Program Representative Sampling Guidance, Volume 5: Water and Sediment, Part I–Surface Water and Sediment.* OSWER Directive 9360.4–16. Washington, DC: EPA Office of Solid Waste and Emergency Response, 1995.

Environmental Protection Agency. *Sampler's Guide to the Contract Laboratory Program.* EPA/540/R–96/032. Washington, DC: EPA Office of Solid Waste and Emergency Response, 1996.

Environmental Protection Agency. *Superfund Program Representative Sampling Guidance, Volume 5: Water and Sediment, Part II–Ground Water.* OSWER Directive 9360.4–16. Washington, DC: EPA Office of Solid Waste and Emergency Response, 1995.

Environmental Protection Agency. *Test Methods for Evaluating Solid Waste, Volume II: Field Manual, Physical/Chemical Methods,* SW–846, Washington, DC: EPA Office of Solid Waste and Emergency Response, 1986.

"Hazardous Materials Regulations." *Code of Federal Regulations.* Title 49. Pt. 172–178, 1996.

International Air Transport Association. *Dangerous Goods Regulations,* 37th ed. Montreal–Geneva: IATA, 1996.

New Jersey Department of Environmental Protection. *Field Sampling Procedures Manual.* Trenton, NJ: NJDEP, 1992.

W. Scott Butterfield, CHMM, is a Senior Program Manager with Roy F. Weston, Inc. Mr. Butterfield has over 18 years experience in the hazardous waste field conducting and managing emergency response activities, removal operations, and remedial projects, as well as R & D programs. Mr. Butterfield currently is the Site Assessment Team Leader on the EPA Region II Superfund Technical Assessment and Response Team (START) contract. Mr. Butterfield has authored papers on biological sampling at abandoned hazardous waste sites (Proceedings of the 3rd National Conference on Management of Uncontrolled Hazardous Waste Sites, 1982) *and effects of exposure to municipal and chemical landfill leachate on aquatic organisms* (Proceedings of the International Symposium on Industrial and Hazardous Solid Wastes, 1983).

Acronyms

A

AAS	Atomic Absorption Spectrometry
ACBM	Asbestos-Containing Building Material
ACGIH	American Conference of Governmental Industrial Hygienists
ACL	Alternate Concentration Level
ACM	Asbestos-Containing Material
ADI	Lifetime Acceptable Daily Intake
AEA	Atomic Energy Act
AHERA	Asbestos Hazard Emergency Response Act
AHM	Acutely Hazardous Material
AIChE	American Institute of Chemical Engineers
AIDS	Acquired Immune Deficiency Syndrome
AIHA	American Industrial Hygiene Association
aka	also known as
AL	Action Level
ALARA	As Low As Reasonably Achievable
ANPR	Advance Notice of Proposed Rulemaking
ANSI	American National Standards Institute
API	American Petroleum Institute
APR	Air-Purifying Respirator
ARS	Alternative Release Scenario

ASHARA	Asbestos School Hazard Abatement Reauthorization Act
ASME	American Society of Mechanical Engineers
AST	Aboveground Storage Tank
ASTM	American Society for Testing and Materials
atm	Atmosphere

B

BACM	Best Available Control Measures
BACT	Best Available Control Technology
BART	Best Available Retrofit Technology
BAT	Best Available Technology
BATEA	Best Available Technology Economically Achievable
BCPC	Biological and Chemical Protective Clothing
BCT	Best Control Technology
BCT	Best Conventional Technology
BDAT	Best Demonstrated Available Technology
BEI	Biological Exposure Indices
BEMR	Baseline Environmental Monitoring Report
BIF	Boilers and Industrial Furnaces
BLEVE	Boiling Liquid Expanding Vapor Explosion
BMR	Baseline Monitoring Report
BOD	Biochemical or Biological Oxygen Demand
BOD_5	Biochemical Oxygen Demand at Five Days
bp	Boiling Point
BUN	Blood-Urea-Nitrogen

C

C	Corrosive (RCRA)
CAA	Clean Air Act
CAAA	Clean Air Act Amendments
CAER	Chemical Accident Emergency Response
CAM	Continuous Air Monitor
CAMU	Corrective Action Management Unit
CAR	Corrective Action Report
CAS	Chemical Abstracts Service
CASAC	Clean Air Scientific Advisory Committee

CBI	Confidential Business Information
cc	cubic centimeters
CCA	Chromium, Copper, and Arsenic
CDC	Centers for Disease Control and Prevention
CDI	Chronic Daily Intake
CEMS	Continuous Emissions Monitoring System
CEPPO	Chemical Emergency Preparedness and Prevention Office
CEQ	Council on Environmental Quality
CERCLA	Comprehensive Environmental Response, Compensation, and Liability Act
CERCLIS	Comprehensive Environmental Response, Compensation, and Liability Information System
CESQG	Conditionally Exempt Small Quantity Generator
CFC	Chlorofluorocarbon
CFR	Code of Federal Regulations
CGA	Compressed Gas Association
CHEMTREC	Chemical Transportation Emergency Center
CHMM	Certified Hazardous Materials Manager
CHP	Chemical Hygiene Plan
CIH	Certified Industrial Hygienist
CISD	Critical Incident Stress Debriefing
CKD	Cement Kiln Dust
CLP	Contract Laboratory Program
CMA	Chemical Manufacturers Association
CMI	Corrective Measures Implementation
CMS	Corrective Measures Study
CNS	Central Nervous System
CoC	Certificates of Completion
COI	Chemicals of Interest
COMAR	Code of Maryland Regulations
CORRACT	Corrective Action Report System
CPC	Chemical Protective Clothing
cpm	counts per minute
CPR	Cardiopulmonary Resuscitation
CPSC	Consumer Product Safety Commission
CPU	Central Processing Unit
CQA	Construction Quality Assurance

CSF	Confidential Statement of Formula
CSHO	Compliance Safety and Health Officer
CWA	Clean Water Act

D

DCI	Data Call-In
DDESB	Department of Defense Explosives Safety Board
DDT	Dichlorodiphenyltrichloroethane
DEACT	Deactivation
DfE	Design for Environment
DFO	Disaster Field Office
DHHS	Department of Health and Human Services
DNA	Deoxyribonucleic Acid
DNAPL	Dense Non-Aqueous Phase Liquid
DNFA	Determination of No Further Action
DNT	Dinitrotoluene
DoD	Department of Defense
DOE	Department of Energy
DOI	Department of the Interior
DOJ	Department of Justice
DOT	Department of Transportation
dpm	disintegrations per minute
DQO	Data Quality Objective
DRE	Destruction/Removal Efficiency
DUO	Data Use Objective

E

E	Toxicity (RCRA)
EA	Environmental Assessment
EC	European Community
ECOS	Environmental Council of the States
ED_{50}	Median Effective Dose
EDTA	Ethylene Diamine Tetraacetic Acid
EEGL	Emergency Exposure Guidance Limits
EEI	Edison Electric Institute
EHS	Environmental Health and Safety
EHS	Extremely Hazardous Substance(s)

EIHR	Earthquake Induced Hazardous Materials Releases
EIS	Environmental Impact Statement
EISOPQAM	Environmental Investigation Standard Operating Procedures and Quality Assurance Manual
EL	Excursion Limit
EL	Exposure Level
EMAS	Eco-Management and Audit Scheme
EMP	Environmental Management Program
EMS	Emergency Medical Services
EMS	Environmental Management System
ENRD	Environment and Natural Resources Division
EOC	Emergency Operation Center
EPA	Environmental Protection Agency
EPCRA	Emergency Planning and Community Right-To-Know Act
EPRI	Electric Power Research Institute
ERAP	Emergency Response Action Plan
ERPG	Emergency Response Planning Guidelines
ERT	Environmental Response Team
ESF	Emergency Support Function
ESLI	End-of-Service-Life Indicator

F

FAQ	Frequently Asked Questions
FCO	Federal Coordinating Officer
FDA	Food and Drug Administration
FEMA	Federal Emergency Management Agency
FFCA	Federal Facility Compliance Agreement
FFCAct	Federal Facility Compliance Act of 1992
FFDCA	Federal Food, Drug, and Cosmetic Act
FIFRA	Federal Insecticide, Fungicide, and Rodenticide Act
FMEA	Failure Mode and Effects Analysis
FOIA	Freedom of Information Act
FONSI	Finding of No Significant Impact
FOTW	Federally Owned Treatment Works
FQPA	Food Quality Protection Act
FR	Federal Register
FRP	Facility Response Plan

FRP	Federal Response Plan
FTA	Fault Tree Analysis

G

GACT	Generally Achievable Control Technology
GI	Gastrointestinal
G-M	Geiger-Mueller
GWPS	Ground Water Protection Standard

H

H	Acutely Hazardous (RCRA)
HA	Hazard Assessment
HAP	Hazardous Air Pollutant
HASP	Health and Safety Plan
HAZMAT	Hazardous Material
HAZOP	Hazard and Operability Study
HAZWOPER	Hazardous Waste Operations and Emergency Response
HBV	Hepatitis B Virus
HCS	Hazard Communication System
HCV	Hepatitis C Virus
HDD	Halogenated Dibenzodioxin
HDF	Halogenated Dibenzofuran
HHW	Household Hazardous Waste
HIV	Human Immunodeficiency Virus
HLW	High-Level Waste (Radioactive)
HMIS	Hazardous Materials Information System
HMR	Hazardous Materials Regulations
HMTA	Hazardous Materials Transportation Act
HMTUSA	Hazardous Materials Transportation Uniform Safety Act
HP Tech	Health Physics Technician
HSWA	Hazardous and Solid Waste Amendments
HTMR	High Temperature Metal Recovery
HUD	Department of Housing and Urban Development
HVAC	Heating Ventilation and Air Condition
HWIR	Hazardous Waste Identification Rule
HWMU	Hazardous Waste Management Unit

I

I	Ignitability (RCRA)
I/M	Inspection and Maintenance
IARC	International Agency for Research on Cancer
IATA	International Air Transport Association
ICP	Inductively Coupled Plasma
ICP	Integrated Contingency Plan
ICP-AES	Inductively Coupled Plasma-Atomic Emission Spectroscopy
ICR	Ignitable, Corrosive, Reactive (RCRA)
ICS	Incident Command System
IDLH	Immediately Dangerous to Life and Health
IHR	Induced Hazardous Materials Release
INEL	Idaho National Engineering Laboratory (See INEEL)
INEEL	Idaho National Engineering and Environmental Laboratory (Formerly INEL)
IRIS	Integrated Risk Information System
IUPAC	International Union of Pure and Applied Chemistry

L

LAER	Lowest Achievable Emissions Rate
LBPPPA	Lead-Based Paint Poisoning Prevention Act
LC_{50}	Median Lethal Concentration 50 percent
LD_{50}	Median Lethal Dose 50 percent
LDR	Land Disposal Restriction
LEL	Lower Explosive Limit
LEPC	Local Emergency Planning Committee
LGAC	Local Government Advisory Committee
LLMW	Low-Level Mixed Waste
LLRW	Low-Level Radioactive Waste
LLW	Low-Level Waste (LLRW)
LOC	Level of Concern
LP	Liquid Petroleum
LPG	Liquid Petroleum Gas
LQG	Large Quantity Generator

M

MACT	Maximum Achievable Control Technology
MAI	Maximum Allowable Increase
MAP	Model Accreditation Plan
MCL	Maximum Contaminant Level
MCS	Media Cleanup Standards
MHLW	Mixed High-Level Waste
MLSS	Mixed Liquor Suspended Solids
MMR	Military Munitions Rule
MMS	Minerals Management Service
MOU	Memorandum of Understanding
mp	Melting Point
MPRSA	Marine Protection, Research, and Sanctuaries Act
MSDS	Material Safety Data Sheet
MTBE	Methytertiarybutylether
MTD	Maximum Tolerated Dose
MTRU	Mixed Transuranic Waste
MW	Mixed Waste (Radioactive and RCRA-Hazardous)
MW	Molecular Weight

N

N.O.S.	Not Otherwise Specified
NA	North America
NAAQS	National Ambient Air Quality Standards
NACE	National Association of Corrosion Engineers
NAICS	North American Industry Classification Code
NAIN	National Antimicrobial Information Network
NAMS	National Air Monitoring Stations
NAPL	Non-Aqueous Phase Liquid
NAS	National Academy of Sciences
NCEPI	National Center for Environmental Publications and Information
NCERQA	National Center for Environmental Research and Quality Assurance
NCP	National Contingency Plan
NEC	National Electrical Code
NEPA	National Environmental Policy Act
NEPPS	National Environmental Performance Partnership System

NESHAPS	The National Emissions Standards for Hazardous Air Pollutants
NFA	No Further Action
NFPA	National Fire Protection Association
NIOSH	National Institute for Occupational Safety and Health
NIST	National Institute of Standards and Technology
NMOG	Non-Methane Organic Gas
NOA	Notice of Availability
NOAA	National Oceanic and Atmospheric Administration
NOAEL	No Observed Adverse Effect Level
NOEL	No Observed Effect Level
NOI	Notice of Intent
NORA	National Organization Research Agenda
NPDES	National Pollutant Discharge Elimination System
NPFC	National Pollution Funds Center
NPL	National Priorities List
NRC	Nuclear Regulatory Commission
NRC	National Response Center
NRR	Noise Reduction Rating
NRT	National Response Team
NSPS	New Source Performance Standards
NSR	New Source Review
NTP	National Toxicology Program
NUMARC	Nuclear Management and Resources Council, Inc.

O

OCA	Off-Site Consequence Analysis
OCIR	Office of Congressional and Intergovernmental Relations
OCSLA	Outer Continental Shelf Lands Act Amendments of 1978
OEL	Occupational Exposure Limit
OPA	Oil Pollution Act
OPIM	Other Potentially Infectious Materials
OPPTS	Office of Prevention, Pesticides, and Toxic Substances
ORM	Other Regulated Material
OSC	On-Scene Coordinator
OSH	Occupational Safety and Health
OSH Act	Occupational Safety and Health Act
OSHA	Occupational Safety and Health Administration

OSLR	Office of State and Local Relations
OSWER	Office of Solid Waste and Emergency Response
OTAG	Ozone Transport Assessment Group
OTI	OSHA Training Institute

P

P&ID	Piping and Instrumentation Diagram
P2	Pollution Prevention
P-450	Cytochrome P-450 Mono-Oxygenase
PACM	Presumed Asbestos-Containing Material
PACT	Powdered Activated Carbon
PAMS	Photochemical Assessment Monitoring Stations
PAPR	Powered Air-Purifying Respirator
PCB	Polychlorinated Biphenyl
PCM	Phase Contrast Microscopy
PEIS	Programmatic Environmental Impact Statement
PEL	Permissible Exposure Limit
PFD	Process Flow Diagram
PG	Packing Group
PHA	Process Hazard Analysis
PL	Public Law
PLM	Polarized Light Microscopy
PM	Particulate Matter
PM_{10}	Particulate Matter less than 10 Microns in diameter
$PM_{2.5}$	Particulate Matter less than 2.5 Microns in diameter
PMN	Premanufacture Notification
PNA	Polynuclear Aromatics
PNS	Peripheral Nervous System
POC	Point of Compliance
POHC	Principal Organic Hazardous Constituent
POTW	Publicly Owned Treatment Works
PPA	Performance Partnership Agreement
PPA	Pollution Prevention Act
PPA	Prospective-Purchaser Agreement
ppb	parts per billion
PPE	Personal Protective Equipment
ppm	parts per million

PRP	Potentially Responsible Party
PSD	Prevention of Significant Deterioration
psi	Pounds per Square Inch
psia	Pounds per Square Inch Absolute
PSM	Process Safety Management
PSN	Proper Shipping Name
PTC	Permit to Construct
PVC	Polyvinyl Chloride

Q

QA	Quality Assurance
QA/QC	Quality Assurance/Quality Control
QMS	Quality Management System

R

R	Reactivity (RCRA)
RACM	Reasonably Available Control Measures
RACM	Regulated Asbestos-Containing Material
RACT	Reasonably Achievable Emissions Control Technology
RBCA	Risk-Based Corrective Action
RCA	Radiological Controlled Area
RCRA	Resource Conservation and Recovery Act
RCT	Radiation Control Technician
RED	Re-registration Eligibility Document
REI	Restricted Entry Interval
RFA	RCRA Facility Assessment
RfD	Reference Dose
RFI	RCRA Facility Investigation
RFP	Reasonable Further Progress
RI/FS	Remedial Investigation/Feasibility Study
RM Plan	Risk Management Plan
RMP	Risk Management Program
RMP Rule	Risk Management Program Rule
RMW	Radioactive Mixed Waste
ROD	Record of Decision
RQ	Reportable Quantity
RRT	Regional Response Team

RSPA	Research and Special Programs Administration
RTECS	Registry of Toxic Effects of Chemical Substances
RWP	Radiation Work Permit

S

SACM	Superfund Accelerated Cleanup Model
SAP	Sampling and Analysis Plan
SAR	Supplied-Air Respirator
SARA	Superfund Amendments and Reauthorization Act
SBREFA	Small Business Regulatory Enforcement Fairness Act
SCAC	Small Community Advisory Committee
SCBA	Self-Contained Breathing Apparatus
SDWA	Safe Drinking Water Act
SEM	Strategic Environmental Management
SEP	Supplemental Environmental Project
SERC	State Emergency Response Commission
SG	Specific Gravity
SI	International System of Units
SIC	Standard Industrial Classification
SIP	State Implementation Plan
SLAMS	State and Local Air Monitoring Stations
SNAP	Significant New Alternatives Policy Program
SNUR	Significant New Use Rule
SOP	Standard Operating Procedure
SPCC	Spill Prevention, Containment, and Countermeasure (CWA)
SPCC	Spill Prevention, Control, and Countermeasure
SPEGL	Short-Term Public Emergency Guidance Levels
SPMS	Special Purpose Monitoring Stations
SQG	Small Quantity Generator
SRD	Self-Reading Dosimeter
SRRP	Source Reduction Review Project
SSU	Saybolt Second Unit
STP	Site Treatment Plan
STP	Standard Temperature and Pressure
STTF	Small Town (Environmental Planning) Task Force
SWDA	Solid Waste Disposal Act
SWMU	Solid Waste Management Unit

T

T	Toxic (RCRA)
TAPAA	Trans-Alaska Pipeline Authorization Act of 1973
TC	Technical Committee
TCDD	Tetrachlorodibenzo-p-Dioxin
TC_{LO}	Toxic Concentration Low
TCLP	Toxicity Characteristic Leaching Procedure
TDI	Toluene Diisocyanate
TD_{LO}	Toxic Dose Low
TEL	Tetraethyl Lead
TEM	Transmission Electron Microscopy
ThOD	Theoretical Oxygen Demand
TI	Technical Impracticability
TIP	Tribal Implementation Plan
TLD	Thermoluminescent Dosimeter
TLV	Threshold Limit Value
TMDL	Total Maximum Daily Load
TPQ	Threshold Planning Quantity
TQ	Threshold Quantity
TRI	Toxic Release Inventory
TRU	Transuranic Waste
TSCA	Toxic Substances Control Act
TSD	Treatment, Storage, and Disposal
TSDF	Treatment, Storage, and Disposal Facility
TU	Temporary Unit
TWA	Time-Weighted Average

U

UEL	Upper Explosive Limit
UHC	Underlying Hazardous Constituent
UIC	Underground Injection Control
UMTRCA	Uranium Mill Tailings Radiation Control Act
UN	United Nations
URL	Universal Resource Locator
US	United States
USC	United States Code

USCG	Coast Guard
USDA	United States Department of Agriculture
USGS	United States Geological Survey
USPS	United States Postal Service
USSG	United States Sentencing Guidelines
UST	Underground Storage Tank
UTS	Universal Treatment Standard

V

VD	Vapor Density
VOC	Volatile Organic Compound
VP	Vapor Pressure
VPP	Voluntary Protection Program
VSD	Virtually Safe Dose

W

WAP	Waste Analysis Plan
WCS	Worst-Case Release Scenario
WEEL	Workplace Environmental Exposure Limit
WIPP	Waste Isolation Pilot Plant
WPS	Worker Protection Standard
WSR	Waste Shipment Record
WWTF	Wastewater Treatment Facility

X

| XRF | X-Ray Fluorescence Analysis |

Glossary

A

Abandoned—Under RCRA, a material that has been disposed of, burned or incinerated; or accumulated, stored, or treated before, or in lieu of, being disposed of, burned, or incinerated.

Absorbed dose—The amount of energy deposited in any material by ionizing radiation. The unit of absorbed dose, the rad, is a measure of energy absorbed per gram of material. The unit used in countries other than the US is the gray. One gray equals 100 rad.

Absorption—The process by which chemicals cross cell membranes and enter the bloodstream. Major sites of absorption are the skin, lungs, gastrointestinal (GI) tract, and parenteral (*e.g.*, intravenous, subcutaneous).

Accidental release—Under the Risk Management Program rules, an unanticipated emission of a regulated substance or other extremely hazardous substances (EHS) into the atmosphere from a stationary source.

Accumulated speculatively—Under RCRA, a material (except commercial chemical products) that is supposedly ***recycled*** but for which no recycling mechanism is in place and/or less than 75% of the material is recycled in a calendar year.

Acid—In chemistry, any substance that dissociates in solution to produce a proton (H+).

Acid deposition or ***acid rain***—The acidic fallout of rain, gases, or dust caused by the emissions of SO_2 and NO_x originating from the combustion of fossil fuels.

Acquired immune deficiency syndrome (AIDS)—An ultimately fatal illness caused by the Human Immunodeficiency Virus (HIV). When the HIV attacks the immune system, the symptoms of AIDS emerge.

635

Acts—Actions taken by Congress, such as the Clean Air Act Amendments of 1990, the Superfund Amendments and Reauthorization Act of 1986, or the Toxic Substances Control Act of 1976, and considered to be law.

Acutely toxic—In toxicology, the characteristic of a substance that produces an immediate reaction in the body, characterized by rapid onset and short duration of symptoms. Damage to the body from acute reactions may be reversible or irreversible.

Addition—As interpreted under the Clean Water Act by EPA and the courts, any introduction of a pollutant into a body of water with the exception of pollutants that are present in the discharge only because they were present in the intake water, if the discharge is released back into the same body of water; and the discharge of water from dams.

Additive effect—In toxicology, the combined effect of exposure to two chemicals that have both the same mechanism of action and target organ. The effect is equal to the sum of the effects of exposure to each chemical when given alone (*e.g.*, 3 + 5 = 8).

Administrative enforcement—In law, all enforcement actions taken by the agency.

Administrator—In the environmental laws and regulations, a term that generally refers to the Administrator of the United States Environmental Protection Agency.

Aerobic—The use of air or oxygen in a process, such as the treatment of certain types of hazardous wastes.

Affected employees—Under OSHA's lockout/tagout rules, employees who operate or use the machinery, processes, or systems that are being maintained or serviced and contain energy.

Agreement state—A state which has been granted the authority by the Nuclear Regulatory Commission to regulate the management and disposal of radioactive waste.

Air-purifying respirator (APR)—A respirator that removes contaminants from the ambient air.

ALARA—The guiding principle behind radiation protection is that radiation exposures should be kept *As Low As Reasonably Achievable* (ALARA), economic and social factors being taken into account. This commonsense approach means that radiation doses for both workers and the public typically are kept lower than their regulatory limits.

Alcohols—In chemistry, alkanes with a substituted hydroxyl group.

Aldehydes and ketones—In chemistry, organic compounds that possess the carbonyl (C=O) group as a common structural feature.

Aliphatics—In chemistry, a class of organic compounds with the carbon backbone arranged in branched or straight chains (*e.g.*, propane or octane).

Alkaline—In chemistry, the property of material that produces hydroxide ions (OH⁻) when it is dissolved in water. The terms *alkaline, basic,* and *caustic* are used interchangeably.

Alkanes—In chemistry, saturated hydrocarbons such as butane, octane, *etc.*

Alkenes—In chemistry, organic compoundes also known as *olefins* that are unsaturated hydrocarbons with a single carbon-to-carbon double bond per molecule.

Allergic (hypersensitive, sensitization) reactions—In toxicology, the production of antibodies in response to the allergen. This includes the production of antigens, (reaction Types I, II, and III) or mediation via specialized immune cells (Type IV) in response to a substance.

Allowable emissions—Under the CAA, the levels of pollutants that a source may emit as specifically identified in the source's air permit.

Allowance—Under the CAA, the authorization for a unit regulated under Title IV, Acid Deposition Control, to emit one ton of SO_2 during a calendar year

Alpha (α) *particle*—One form of ionizing radiation. Alpha (α) particles are composed of two protons and two neutrons. Alpha particles do not travel very far from their radioactive source. They cannot pass through a piece of paper, clothes or even the layer of dead cells that normally protects the skin. Because alpha particles cannot penetrate human skin they are not considered an *external exposure hazard* (this means that if the alpha particles stay outside the human body they cannot harm it). However, alpha particle sources located within the body may pose an *internal* health hazard if they are present in great enough quantities. The risk from indoor radon is due to inhaled alpha particle sources that irradiate lung tissue.

Alternative release scenario (ARS) analysis—Under the Risk Management Program rules, an analysis of potential releases that are more likely to occur than the *worst-case release scenario* (WCS) and that will reach an end point off-site. The ARS is required of an operator of a stationary source for all flammable and toxic substances in covered processes.

Anaerobic—The absence of air or oxygen in a process, such as the treatment of certain types of hazardous wastes.

Anions—In chemistry, negative *ions.*

Antagonistic effect—In toxicology, when an exposure to two chemicals together interferes each with the other's toxic actions (*e.g.*, 4 + 6 = 8), or one chemical interferes with the toxic action of the other chemical, such as in antidotal therapy (*e.g.*, 0 + 4 = 2).

Aquifer—Any geologic formation capable of holding water in sufficient quantities to produce *free water.* Aquifers can be unconsolidated soil or rock formations.

Area sources—Under the CAA, *stationary* or nonroad sources that are too small and/or too numerous to be included in a stationary source inventory. Examples of area sources include: water heaters, gas furnaces, fireplaces, and wood stoves.

Areal composite samples—Environmental samples that are composed of equal aliquots (grab samples) collected from a defined area such as the surface of a waste pile or an established grid.

Aromatic—In chemistry, organic molecular structures having single of multiple benzene rings (C_6H_6) as the basic unit (*e.g.*, toluene, xylene, polynuclear aromatics [PNAs]).

Articles—Under the Occupational Safety and Health Act, manufactured items which are formed to specific shapes during manufacture. They have end-use functions dependent upon their shapes during end use. They do not release or otherwise result in exposure to a hazardous chemical under normal conditions of use.

Asbestos—A noncombustible chemical-resistant mineral that includes chrysotile, amosite, crocidolite, tremolite asbestos, anthophyllite asbestos, actinolite asbestos, and any of these minerals that have been chemically treated and/or altered.

Asbestos-containing building material (ACBM)—Asbestos-containing material found in or on interior structural members or other parts of a building.

Asbestos-containing material (ACM)—Any material or product containing greater than 1 percent asbestos.

Asbestosis—Also known as *white lung*, a scarring of lung tissue which forms from an accumulation of asbestos fibers in the lung. Asbestosis is the most common asbestos-related disease. The symptoms of asbestosis are similar to those of emphysema.

Atmosphere-supplying respirator—A respirator that provides air from a source other than the surrounding atmosphere.

Atomic number—In chemistry, the number of protons in the nucleus.

Atoms—In chemistry, the smallest and generally indivisible unit of matter. Atoms of a particular element such as sodium have the same average mass, and other properties are the same. Atoms of different elements generally have different average masses (or *atomic weights*) and different properties.

Attainment areas—Under the CAA, geographic areas in which the air quality is cleaner than the NAAQS (*i.e.,* the concentrations of the *criteria pollutants* are below the standard).

Audit—In an audit process, the comparison of actual conditions to expected conditions, and a determination as to whether the actual conditions are in conformance or not in conformance with the expected conditions.

Audit objective(s)—In an audit process, the reason an audit is being conducted. Usually the reason is to demonstrate conformance to stated criteria.

Audit scope—In an audit process, the nature of the audit—a company, a site, or unit within a site or company.

Auditee—In an audit process, the entity being audited.

Auditor—In an audit process, the person or team actually collecting evidence and determining findings.

Authorized employees—Under OSHA's lockout/tagout rules, employees who actually perform service or maintenance on systems that contain energy.

Authorized state—Under RCRA, a state that has been granted the authority by the Environmental Protection Agency to regulate the management and disposal of hazardous waste.

Auto-ignition temperature—The temperature at which ignition occurs without an ignition source and the material continues to burn without further heat input.

Available treatment technology and disposal capacity—As defined by the EPA, any facility that is "commercially available to treat or dispose of a particular waste and has either (1) a RCRA permit or interim status; (2) a research, development, and demonstration permit under 40 CFR 270.65; or (3) a land treatment permit under 40 CFR 270.63."

Average atomic weight—In chemistry, the atomic weight (as generally found in nature) shown in the Periodic Table.

B

Backsliding—As defined by the Clean Water Act, a term that refers to the lowering of a discharge effluent limit or a pollution control requirement.

Base—In chemistry, any material that produces hydroxide ions (OH^-) when it is dissolved in water. A base can also be thought of as any substance that accepts a proton.

Baseline conditions—In an Environmental Management System (EMS), the state of the existing EMS relative to the requirements of the a system standard like ISO 14000.

Basic—In chemistry, the property of material that produces hydroxide ions (OH^-) when it is dissolved in water. The terms ***alkaline, basic,*** and ***caustic*** are used interchangeably.

Best available control measures (BACM)—Under the CAA, emissions control technologies (typically used for fugitive dust, residential wood burning, and prescribed burning) that may be used for the management of particulate matter emissions.

Best available control technology (BACT)—Under the CAA, an emission limitation based on the maximum achievable degree of reduction for a pollutant, taking into account energy, environmental, and economic impacts, and other costs.

Best available retrofit technology (BART)—Under the CAA, an emissions control technology that may be implemented to control regional haze. BART takes into account the availability of the control technology, cost of compliance, current pollution control equipment used by the source, and the expected improvement in visibility.

Best practices—Under OSHA, the best possible engineering and administrative means (over and above the specific requirements of law) that is provided by the employer to control hazards in the workplace.

Beta (β) particle—One form of ionizing radiation. Beta (β) particles are similar to electrons except that they come from the atomic nucleus and are not bound to any atom. Beta particles cannot travel very far from their radioactive source. For example, they can travel only about one half an inch in human tissue, and they may travel a few yards in air. They are not capable of penetrating something as thin as a pad of paper.

Bioactivation—In toxicology, the biotransformation of a relatively nontoxic chemical into one or more toxic metabolites.

Biochemical oxygen demand at five days (BOD₅)—The quantity of oxygen required by microbes for the oxidative breakdown of a given waste material during a 5-day test period. The BOD_5 usually is taken as an index of the ultimate oxygen demand (*i.e.*, oxygen required when sufficient time is allowed to achieve maximum microbial decomposition).

Biological effects of complex chemical or other mixtures—In toxicology, one approach to *in vitro* toxicological studies that examines the synergistic, additive, and antagonistic effects of exposure to a number of chemicals, as well as the correlation between chemical and physical agents, (*i.e.*, noise, temperature, humidity, vibration, and stress), and between chemical and environmental interactions, (circadian rhythms, seasonal changes), *etc.*

Biological indicators—Under OSHA's HAZWOPER rules, dead animals or vegetation that may indicate the presence of hazards that could be Immediately Dangerous to Life and Health.

Biological plausibility—One of the factors used in epidemiological studies to determine the confidence level of the data. Biological plausibility means that the investigated agent is known to cause the health effect.

Biotransformation—In toxicology, the process by which living organisms can chemically change a substance (also known as metabolism).

Bloodborne pathogens—Pathogenic microorganisms that are present in human blood and can cause disease in humans.

Body of standing law—Laws that are amended must then be revised and republished to reflect all permanent amendments. Also called *statutes*.

Boiling point (bp)—The temperature at which a liquid changes to gas under standard atmospheric pressure (760 mm mercury). Under the *Process Safety Management rules*, this refers specifically to the boiling point of a liquid at a pressure of 14.7 pounds per square inch absolute (psia).

Brownfields—Abandoned, idled, or underutilized industrial and commercial facilities where expansion or redevelopment is complicated by real or perceived environmental contamination (GAO, *Superfund: Barriers to Brownfield Redevelopment*, 1996).

Burned for energy recovery or used to produce a fuel product—Under RCRA, a material (except commercial chemical products that are already fuels) that is used as a fuel and considered *recycled,* and therefore is *discarded.*

By-product—Under RCRA, a material that is not one of the primary products of a production process and is not solely or separately produced by the production process.

C

Cancer—The unrestrained proliferation of immature or abnormal cells that leads to tumor formation and, eventually, to the inhibition of the normal function of the organ or tissue.

Catastrophic injuries—Under OSHA, any single incident resulting in the hospitalization of three or more employees.

Catastrophic release—Under OSHA's Process Safety Management rules, a major uncontrolled emission, fire, or explosion involving one or more highly hazardous chemicals that presents a serious danger to employees in the workplace.

Categorical exclusion—Under NEPA, the lowest level of action a Federal agency may take on a detailed environmental analysis if it meets certain criteria, which have been previously determined as having no significant environmental impact.

Cations—In chemistry, positive *ions*.

Caustic—In chemistry, the property material that produces hydroxide ions (OH^-) when it is dissolved in water. The terms *alkaline, basic,* and *caustic* are used interchangeably.

Ceiling—The highest concentration of air contaminants to which any employee may be intentionally exposed.

Central nervous system (CNS) depression (narcosis)—A reversible effect that may be caused by exposure to a substance which is characterized by drowsiness, headache, nausea, slurred speech, difficulty in concentrating, dizziness, loss of coordination, and possibly coma and death in extreme situations.

Central nervous system (CNS) toxicity—The damage a substance may cause to the brain and spinal cord.

Characteristic hazardous wastes—Under RCRA, wastes not found on an EPA hazardous waste list but which exhibit one or more hazardous characteristics (ignitability, corrosivity, reactivity, or toxicity) when tested under EPA protocols.

Characteristic of corrosivity (D002)—Under RCRA, one of the characteristics of hazardous waste in which the waste exhibits either of the following properties: (1) it is aqueous and has a pH of less than or equal to 2 or greater than or equal to 12.5, as determined by a pH meter, using the Environmental Protection Agency (EPA) publication SW-846; (2) it is a liquid and corrodes steel at a rate greater than 6.35 mm (0.25 inch) per year at a temperature of 55°C.

Characteristic of ignitability (D001)—Under RCRA, one of the characteristics of hazardous waste in which the waste exhibits any of the following properties: (1) it is a liquid other than aqueous solution containing less than 24% alcohol by volume and has a flash point less than 140°F (60°C), as determined by a Pensky Marten closed cup tester (American Society for Testing and Materials [ASTM] method D–93–79 or D–93–80) or Setaflash closed cup tester using method ASTM–D–3278–78; (2) it is a liquid and is capable under standard temperature and pressure of causing a fire through friction, absorption of moisture, or spontaneous chemical changes; and when ignited burns so vigorously and persistently that it creates a hazard; (3) it is an ignitable compressed gas as defined in 49 CFR 173.300 and as determined by the test methods described in the above regulations or equivalent test methods approved under 40 CFR 260.20 and 260.21; (4) it is an oxidizer as defined in 49 CFR 173.151.

Characteristic of reactivity (D003)—Under RCRA, one of the characteristics of hazardous waste in which the waste has any of the following properties: (1) it normally is unstable and readily undergoes violent changes without detonating; (2) it reacts violently with water; (3) it forms a potentially explosive mixture with water; (4) it generates toxic gases, vapors, or fumes when mixed with water; (5) it is a cyanide- or a sulfide- bearing waste which, when exposed to pH conditions between 2–12.5, can generate toxic gases, vapors, or fumes; (6) it is capable of detonation or causing an explosive reaction if subjected to heat or initiators; (7) it is readily capable of explosive decomposition or reaction at standard temperature and pressure; (8) it is a forbidden explosive defined in 49 CFR 173.51, 49 CFR 173.53, and 49 CFR 173.88.

Chemical hygiene plan (CHP)—A comprehensive plan that is a major component of the OSHA Laboratory Standard and includes the development and implementation of a chemical hygiene plan capable of (1) protecting employees from health hazards associated with hazardous chemicals in that laboratory, and (2) keeping exposures below specified OSHA permissible exposure limits.

Chemical substance—Under TSCA, any organic or inorganic substance of a particular molecular identity, including any combination of substances occurring in whole or in part as a result of a chemical reaction or occurring in nature. It does not include mixtures; pesticides as defined by FIFRA; tobacco; *source, special nuclear* or *by-product materials* as defined in the 1954 Atomic Energy Act; any article the sale of which is subject to the tax imposed by the 1986 Internal Revenue Code; and any food, food additive, drug, cosmetic or device as defined by the Federal Food, Drug, and Cosmetic Act.

Chemical treatment—A method of treating hazardous waste which includes all forms of chemical reactions such as reduction, oxidation, thermal oxidation, precipitation, neutralization, electrochemical, photolytic, biological degradation, *etc.*

Chemistry—The science of matter, energy, and their reactions.

Chronic toxin—In toxicology, substance that exhibits no symptoms or mild symptoms to the body at the time of exposure, but may build after a series of exposures. Chronic reactions may be triggered by a build-up of the material (*e.g.,* lead) in body tissues, or by the immune system becoming sensitized to specific toxins, such as organic solvents.

Class I—Under the OSHA asbestos rules, activities involving the removal of Thermal System Insulation (TSI) and surfacing asbestos-containing material and presumed asbestos-containing material.

Class I areas—Under the CAA, geographic areas that allow for very little deterioration of air quality and include all international parks; national wilderness areas and memorial parks larger than 5000 acres; and national parks exceeding 6000 acres.

Class II—Under the OSHA asbestos rules, activities involving the removal of asbestos-containing material that is not thermal system insulation or surfacing material (miscellaneous materials). Examples are removal of floor or ceiling tiles, siding, roofing, and transite panels.

Class II areas—Under the CAA, all geographic areas other than Class I areas or areas that have been redesignated from ***Class II*** to ***Class III***. Class II areas allow for moderate deterioration of air quality.

Class III—Under the OSHA asbestos rules, repair and maintenance operations where asbestos-containing material (ACM) including thermal system insulation and surfacing materials is likely to be disturbed. It includes repair and maintenance activities involving intentional disturbance of ACM or presumed asbestos-containing material (PACM). It is limited to ACM/PACM.

Class III areas—Under the CAA, the least restrictive class of geographic areas that result from redesignation of ***Class II areas***. To date, no areas in the United States have been redesignated to Class III.

Class IV—Under the OSHA asbestos rules, maintenance and custodial work, including cleanup, during which employees contact asbestos-containing material or presumed asbestos-containing material.

Clean alternative fuel—Fuel that meets certain low emissions standards and is useable by a certified clean-fuel vehicle. Examples include ethanol, gasohol mixtures >85% by alcohol by volume, reformulated gasoline, diesel, natural gas, liquified petroleum gas, hydrogen, and electricity.

Client—In an audit process, the party commissioning the audit.

Closure—Under RCRA, the prescribed activities that take place at treatment, storage, and disposal facilities (TSDFs) in order to leave the facilities in an environmentally safe condition with minimum maintenance requirements after operations have ceased.

Code of Federal Regulations (CFR)—The body of standing regulation (48 Titles), accurate as of the annual publication date of each title.

Cold substances—Substances that are considered hazardous to the touch because they are 32°F (0°C) or colder.

Commerce clause of the Constitution—A clause in the Constitution that authorizes Congress to regulate interstate and foreign commerce.

Common law—The various rules of law that have evolved over centuries from judicial decisions that relied on usages and customs of the people. Also call *case law*.

Competent person—Under the OSHA asbestos standards, a person who directs the establishment and supervision of regulated work areas established where airborne concentrations of asbestos fibers may be expected to equal or exceed the PEL. The competent person (1) establishes negative pressure enclosures where necessary, (2) supervises exposure monitoring, (3) designates appropriate personal protective equipment, (4) ensures training of workers with respect to proper use of said equipment, (5) ensures the establishment and use of hygiene facilities, and (6) ensures that proper engineering controls are used throughout the project.

Composite sample—In environmental sampling, a nondiscrete sample composed of two or more equal aliquots collected at various points or times. The analytical results from a composite sample provide average concentrations of the contaminants present. The collection and mixing (homogenizing) of two aliquots of surface material from a pile of "waste" material would result in the generation of a composite sample of the pile.

Concordance—One of the factors used in epidemiological studies to determine the confidence level of the data. Concordance occurs when the same end point or type of pathological change (*e.g.*, type of tumor) arises in more than one species, strain, or sex, and/or in both animals and humans.

Cone of depression—In hydrogeology, the cone-shaped depression of the water table that results from removal of ground water from a pumping well.

Confined spaces—As defined by OSHA, spaces which are: (1) large enough for an employee to enter, (2) have a limited or restricted means of entry or exit, and (3) have not been designed for continuous occupancy.

Connected piping—As defined by the Underground Storage Tank rules, all buried piping, including valves, elbows, joints, and flexible connectors, attached to tank systems through which regulated substances flow.

Consistency—One of the factors used in epidemiological studies to determine the confidence level of the data. Consistency occurs when two or more studies yield comparable findings of excess risk, preferably using different test models.

Contained-in policy—Under RCRA, the policy that a mixture of hazardous waste and material other than solid waste, such as soil, ground water, and debris, must be managed as if the entire mixture were hazardous waste unless, or until the hazardous waste is determined to no longer be present.

Containers—Under RCRA rules, portable devices of various sizes up to and including transport vehicles.

Containment buildings—Under RCRA, units that meet certain design, operation, and closure standards and are constructed for a variety of operations that cannot be readily performed with containerized waste or bulk waste in tanks (such as storage and manipulation of otherwise unconfined waste piles and/or performance of complex treatment processes without the need to otherwise confine the waste to RCRA-compliant tanks or containers).

Contract-based remedies—Actions that are based on an agreement between two parties.

Corrosive chemicals—In toxicology, substances that disintegrate body tissues. Corrosive chemicals affect in particular the water and fatty tissues of the body, and are capable of causing rapid and deep destruction of tissue.

Covalent—In chemistry, bonds that exist when two atoms in a molecule share a pair of electrons.

Covered process—Under the Process Safety Management and Risk Management Program rules, any activity involving the use, storage, manufacturing, handling, or on-site movement of a regulated substance, or combination of these activities with a highly hazardous chemical in excess of a TQ amount.

Cradle-to-grave—Under RCRA, a term used to describe the generator's responsibility for tracking hazardous waste from generation through final disposition.

Criteria pollutants—Under the CAA, six pollutants (ozone, carbon monoxide, particulate matter [PM_{10} and $PM_{2.5}$], lead, nitrogen dioxide, and sulfur oxides [SO_2]) that generally are associated with urban areas and have associated National Ambient Air Quality Standards.

Cryogenics—Extremely cold substances that are considered hazardous to the touch because they are −148°F (−100°C) or colder.

D

Data use objectives (DUO)—In environmental sampling, objectives that address specific goals of the sampling program including what the data is needed for, the questions that must be answered, and what decisions have to be made based on the data generated. The goal may be as simple as determining the contents of a single drum in order to arrange for proper disposal, or as complex as determining the impact of a site with multiple sources of unknown contaminants on numerous human and ecological resource targets.

De minimis *losses*—Under RCRA, generally denotes minor leaks and spills, sample purgings, relief device discharges, rinsate from empty containers or containers that are rendered empty by that rinsing, and minor laboratory waste contributions normal to a well-designed and -operated facility.

Debris—Under RCRA LDR regulations, solid material that is greater than 60 mm (2.5 in) particle size that is intended for disposal and is a manufactured object, plant or animal matter, or geologic material, except for: (1) materials for which a specified technology treatment standard is already established, (2) process residuals, and (3) intact containers of hazardous waste that retain at least 75% of their original volume.

DECIDE process—A decision-making process under 29 CFR 1910.120(q) used to train Hazardous Materials Technicians for hazardous materials incidents.

Deed restrictions—One method of managing risk on brownfields properties. Deed restrictions provide public notification of use constraints and grant a variety of options for enforcing compliance with them.

Deferred UST—According to the Underground Storage Tank rules, an UST for which EPA has deferred the Subpart B (design, construction, installation, and notification), Subpart C (general operations requirements), Subpart D (release detection), Subpart E (release reporting, investigation, and confirmation), and Subpart G (closure) regulations. Until EPA decides how to regulate these USTs fully, the only regulations that apply are Subpart A (interim prohibition) and Subpart F (release response and corrective action). Examples of deferred tanks include underground, field-constructed, bulk storage tanks, and UST systems that contain radioactive wastes.

Delaney clause—Part of the Federal Food Drug and Cosmetic Act, added to the law in 1958. The clause basically stated that carcinogens could not purposely be added to food. The Delaney clause prohibited the setting of tolerances for carcinogens on processed foods, even though a tolerance may already have existed for the raw commodity from which the processed food was derived.

Delayed or chronic toxicity—In toxicology, a reaction by the body that manifests itself after a period of latency (may be many years) following exposure.

Delegation—The authorization of a state by an agency (EPA) to establish permitting and enforcement programs which would be operated by the state in lieu of the Federal program.

Delisting petition—Under RCRA, a process used by generators to exclude a waste at a particular facility from the hazardous wastes listed by EPA in 40 CFR 261, Subpart D.

Derived-from rule—Under RCRA, a rule that states that any solid waste generated from the treatment, storage, or disposal of a listed hazardous waste (such as sludge from a wastewater treatment plant and incinerator ash) is also a listed waste and bears the same listed waste codes regardless of whether or not the hazardous waste was listed on the basis of characteristic(s) only. Any solid waste generated from the treatment, storage, or disposal of a waste that is hazardous by characteristic only, is hazardous only if it exhibits one or more hazardous characteristic.

Dermal and eye corrosion—In toxicology, an irreversible destruction of tissue with pain, ulcerations, and scarring.

Dermal irritation—In toxicology, a localized, nonimmune, inflammatory response of the skin that is characterized by reversible redness, swelling, and pain at the site of contact.

Dermal sensitization—In toxicology, a skin reaction triggered by an immune response to allergens. This reaction is characterized by redness, swelling, crusting/scaling and vesicle formation.

Developmental toxicity—In toxicology, the damage a substance may cause to the developing organism (due to the exposure of either parent).

Diatomic—In chemistry, having two atoms per molecule.

Dip tank—Under OSHA, any tank or other container which contains a flammable or combustible liquid and in which articles or materials are immersed for any process.

Directions for use—Under FIFRA, statements on a pesticide label that describe how much of the product to use, for what applications the pesticide can be used (*e.g.*, which crops, pests, locations, *etc.*), and how to apply it.

Dirty closure—Under RCRA, applies to all closures that are not clean closures and may involve partial removal or decontamination, waste stabilization measures and, usually, installation of low permeability cover or *cap* with run-on/run-off controls to minimize infiltration by precipitation. This type of closure is typical for land disposal units (*e.g.*, landfills) and nonland-based units from which all waste or contamination above allowable risk levels cannot be removed. Also known as *closure with waste in place.*

Discarded—Under RCRA, materials that have been ***abandoned***, ***recycled***, are ***inherently waste-like*** or a ***military munition specifically identified as a solid waste,*** as specifically defined in 40 CFR 261.2.

Discharge—Under the CWA, any addition of any pollutant to navigable waters from any point source.

Disposal—Under RCRA, virtually any action that introduces waste or waste constituents into the environment (*i.e.*, on/into air, water or land) including placement in waste piles outside of a RCRA-compliant containment building.

Distribution—In toxicology, the movement of chemicals throughout the body after they have been absorbed.

Diversity jurisdiction—In law, the ability of a defendant to remove a case from a State court to the Federal court due to the party's diverse citizenship.

Docket—The official administrative record that is established for a particular regulation in order to document that all administrative procedures, as well as any procedures specified in the enabling legislation, are followed. All comments that are submitted in response to a proposed regulation are kept with this docket.

Doctrine of Respondeat Superior—In law, gives employers responsibility for the acts of their employees in the course of their employment.

Dosage—In toxicology, the most critical factor in determining whether a toxic response will take place. Dosage is defined as the unit of chemical per unit of biological system (*e.g.*, mg/kg body weight, mg/body surface area, ml/kg body weight, *etc.*)

Dose—In radiation protection, the effect that radiation has on any material. Radiation dose is simply the quantity of radiation energy deposited in a material. There are several additional descriptive terms used in radiation protection which provide precise information on how the dose was deposited, the method used to calculate the dose, and how the radiation energy deposited in tissue will affect humans. See ***absorbed dose*** and ***equivalent dose***.

Dose-response—In toxicology, the quantitative (measurable) relationship between the dose of a chemical or agent (*e.g.*, mg chemical/kg body weight) and an effect (response) caused by the chemical or agent. The dose-response relationship is the most fundamental concept in toxicology.

Due diligence—In property assessments, all appropriate inquiry that is made into the previous ownership and use of property that is consistent with good commercial or customary practice at the time of land acquisition.

Duration of exposure—In toxicology, the length of time a person or animal is exposed. The duration can be acute (*e.g.*, usually a single dose), subchronic (*e.g.*, days to years), and chronic (*e.g.*, years to a lifetime).

E

Electrons—In chemistry, the particles that surround the nucleus, are negatively charged, and contribute an insignificant amount to atomic weight.

Elements—In chemistry, extremely small, normally indivisible particles called ***atoms***.

Emergency—As defined by FEMA, any event which threatens to, or actually does, inflict damage to property or people.

Emergency response action plan (ERAP)—Under the Oil Pollution Act of 1990, the part of the Facility Response Plan that contains all information that is needed to combat a spill; organized for easy reference such as: (1) information on the individual who certified the plan, (2) emergency notification phone list, (3) spill response notification form, (4) list of response equipment and its location, (5) schedule of response equipment testing and deployment drills, (6) list of response team members, (7) evacuation plan, (8) description of the response resources for small, medium, and worst-case discharges, and (9) the facility diagram.

Emission standard or limitation—Under the CAA, State or Federal requirements relating to the quantity or rate of emissions and including: (1) a schedule or timetable of compliance, and emission limitation, standard of performance, or emissions standard; (2) a control or prohibition of a motor vehicle fuel or fuel additive; (3) a permit condition or requirement relating to attainment and nonattainment areas; any state implementation plan requirement relating to transportation control measures, air quality maintenance plans, vehicle inspection and maintenance programs, vapor recovery, visibility, and ozone protection; and (5) a standard, limitation, or schedule issued under the Title V program.

Emissions offsets—Under the CAA, surplus emissions reductions that are required in order for a ***new source*** to obtain a permit in a ***nonattainment area.*** The emissions offset are intended to compensate for the sources' impact on air quality and further the progress toward achieving clean air.

End use product—Under FIFRA, a pesticide product whose labeling includes directions for use of the product (as distributed or sold, or after combination by the user with other substances) for controlling pests or defoliating, desiccating, or regulating the growth of plants, and does not state that the product may be used to manufacture or formulate other pesticide products.

Environment—Under NEPA, the natural and physical environment (air, water, geography, and geology) as well as the relationship of people with that environment including health and safety, socioeconomics (jobs, housing, schools, transportation), cultural resources, noise, and aesthetics.

Environmental aspects—In ISO 14000, any element of the organization's activities, products, and services that can interact with the environment, that it can control and over which it can be expected to have a significant impact. The impact is the change that occurs in the environment, positive or negative, current as well as in the future.

Environmental assessment (EA)—Under NEPA, a concise public document that is prepared in order to determine whether an *environmental impact statement* (EIS) is necessary or whether a *finding of no significant impact* (FONSI) can be made.

Environmental impact statement (EIS)—Under NEPA, a detailed statement required as part of Federal agency decision-making. The statement describes (1) the environmental impact of the proposed project, (2) any adverse environmental effects which cannot be avoided, (3) alternatives to the proposed action, (4) the relationship between local short-term uses of man's environment and the maintenance and enhancement of long-term productivity and (4) any irreversible and irretrievable commitments of resources that would be involved in the proposed action should it be implemented.

Environmental management system (EMS)—The organizational structure, planning activities, responsibilities, practices, procedures, processes, and resources for developing, implementing, maintaining, reviewing, and correcting/improving an organization's approach to environmental management.

Environmental receptor—Under the Risk Management Program rules, natural areas such as national or state parks, forests, or monuments; officially designated wildlife sanctuaries, preserves, refuges, or areas; and Federal wilderness areas which could be exposed to an accidental release.

Environmental sample—In environmental sampling, samples that present a relatively low risk to health or the environment because of the potential level and type of contamination. Policy determinations reflect an organization's desire to ensure the health and safety of all involved with sampling programs and to avoid the liability of improper shipment of samples. Some organizations have a simple policy of "if it is on-site, the sample is hazardous" and "if it is off-site, the sample is environmental".

Environmentally sound recycling—In the context of RCRA, activities defined under the hazardous waste regulations as *recycling* (40 CFR 261.1[c][4], [5], and [7]). This includes materials that are used, reused, or reclaimed. A material is reclaimed if it is processed to recover a usable product, or if it is regenerated.

Equivalent dose (dose equivalent)—In radiation safety, the concept of equivalent dose involves the impact that different types of radiation have on humans. Not all types of radiation produce the same effect in humans. The equivalent dose takes into account the type of radiation and the absorbed dose. For example when considering beta, x-ray, and gamma ray radiation, the equivalent dose (expressed in rems) is equal to the absorbed dose (expressed in rads). For alpha radiation, the equivalent dose is assumed to be twenty times the absorbed dose.

Esters—In chemistry, compounds that are formed when a carboxylic acid is reacted (*neutralized*) by an alcohol. Esters often are easily recognizable due to their sweet to pungent odors.

Ethers—In chemistry, compounds that generally are volatile, flammable liquids with low boiling points and low flashpoints.

Evaporation—In chemistry, the tendency of a material to transfer from a liquid phase (either pure or dissolved, as in aqueous systems) to a gaseous phase (commonly mixed with air). Also known as *volatilization.*

Excretion—In toxicology, the elimination of chemicals from the body.

Executive branch—The part of the United States government where the regulatory agencies are located. The Executive Branch is headed by the President, who names the administrators of the regulatory agencies.

Executive order—An order issued by the President that can be used to delegate to agencies authority that the President may have as a direct result of the office or as a result of a Congressional delegation of authority.

Existing source—Under the CAA, any source other than a new source.

Existing USTs—As defined by the Underground Storage Tank rules, tanks for which installation commenced on or before December 22, 1988.

Exothermic—In chemistry, the characteristic of a chemical reaction in which heat is released.

Experimental animal-human comparisons—One approach to *in vitro* toxicological studies that examines existing human data and compares exposure levels and effects to studies performed on animals. This experimental methodology attempts to establish correlations between the effects observed in animals and those observed in humans.

Exposure—In *toxicology*, the amount of an agent available for absorption into the body. In *radiation safety*, the measure of the amount of ionization produced by x-rays or gamma rays as they travel through air. The unit of radiation exposure is the roentgen.

Extremely hazardous substance (EHS)—Substances identified in a published list by EPA, pursuant to Section 302 of SARA Title III. In the United States, when a facility inventories an EHS in quantities equal to or greater than a trigger quantity, the facility is subject to emergency planning and notification requirements. Under the *Risk Management Program* rules, any substance that, if released to the atmosphere, could cause death or serious injury because of its acute toxic effect, or as a result of an explosion or fire, or which causes substantial property damage by blast, fire, toxicity, reactivity, flammability, volatility, or corrosivity.

Eye irritation—In toxicology, a localized, nonimmune, inflammatory response of the eye that is characterized by reversible redness, swelling, pain, and tearing of the eyes.

F

Facility—Under the *Oil Pollution Act* of 1990, any structure, group of structures, equipment, or device (other than a vessel) that is used for exploring, drilling, producing, storing, handling, transferring, processing, or transporting oil. The term also includes any motor vehicle, rolling stock, or pipeline used for these purposes. Facilities are further subdivided into onshore and offshore facilities. The boundaries of a facility may depend on several site-specific factors, such as the ownership or operation of buildings, structures, and equipment on the same site and the types of activity at the site. Under the *Process Safety Management* rules, the buildings, containers or equipment which contain a process.

Facility response plan (FRP)—Under the Oil Pollution Act of 1990, a plan that addresses how a facility will respond to a worst-case oil discharge and a substantial threat of such a discharge.

Facility siting—Under the Process Safety Management rules, the physical location of the covered processes on the plant's property.

Federally owned treatment works (FOTW)—Water treatment facilities owned by the Federal government that operate under a permit issued under Section 402 of the Federal Water Pollution Control Act (Clean Water Act).

Female reproductive toxicity—In toxicology, the damage a substance may cause to the female reproductive system.

Finding—In environmental auditing, the determination, when evidence is compared to the applicable criteria, of whether the audited entity does or does not conform to a standard.

Finding of no significant impact (FONSI)—Under NEPA, one of the outcomes of the ***Environmental Assessment*** that indicates no significant environmental impacts for a proposed project.

Fire point—The temperature at which a liquid gives off enough vapor to continue to burn when ignited.

Fire tube boilers—A method of treatment for hazardous waste using industrial boilers. Heat is transferred from hot combustion products flowing inside tubes to the water surrounding them. Combustion takes place in a cylindrical furnace within the shell. Fire tubes run the length of the shell above and around the furnace.

Fixation—A method of treatment for hazardous waste which reduces or eliminates the toxicity or the hazard potential of a certain waste stream by lowering the solubility and leachability of the toxic or hazardous components. Also called ***stabilization.***

Flammable liquid substances—Liquid substances that are easily ignitable. Various Federal agencies and industry organizations have different technical definitions. Which standard (and which definition) applies to a specific situation will depend on which agency has jurisdiction (*i.e.,* Occupational Safety and Health Administration [OSHA], National Fire Protection Association [NFPA], Environmental Protection Agency [EPA], Department of Transportation [DOT], *etc.*).

Flammable or explosive limits—The upper and lower vapor concentrations at which a mixture will burn or explode.

Flashpoint—The lowest temperature of a liquid at which it gives off enough vapor to form, with air, an ignitable mixture near the surface of the liquid. Two tests are used to determine flashpoint: Open Cup and Closed Cup.

Flow-proportional composite sampling—In environmental sampling, a type of sample that generally is associated with stream or wastewater discharge samples and thus is not normally utilized in the investigation of hazardous waste sites.

Foreseeable emergency—Under OSHA regulations, any potential occurrence (*e.g.*, equipment failure, rupture of containers, *etc.*) that could result in an uncontrolled release of a hazardous chemical into the workplace.

Free water—In hydrogeology, ground water not incorporated into the soil as moisture.

Frequency of exposure—In toxicology, the number of exposures per time period. The frequency influences the amount of substance available for absorption into the body.

Friable asbestos-containing material—Material which, when dry, may be crumbled, pulverized, or reduced to powder by hand pressure and includes any damaged nonfriable material.

Full emergency (major event alert)—An emergency that impacts the public and necessitates full mobilization of facility and community resources, such as a major plant explosion.

Functional group—In chemistry, an atom or group of atoms, other than hydrogen, bonded to the chain or ring of carbon atoms (*e.g.*, the –OH group of alcohols, the –COOH group of carboxylic acids, the –O– group of ethers). Functional groups determine the behavior of molecules. Consequently, the unique hazards of an organic compound often are determined by its functional group(s).

G

Gaining streams—In hydrogeology, streams that act as discharge points for ground water.

Gamma (γ) rays—Gamma (γ) rays are an example of electromagnetic radiation, as is visible light. Gamma rays originate from the nucleus of an atom. They are capable of traveling long distances through air and most other materials. Gamma rays require more *shielding* material, such as lead or steel, to reduce their numbers than is required for alpha and beta particles.

Gap analysis—In ISO 14000, a process used to develop a clear understanding of the gap between current environmental management activities and documentation and ISO 14001 requirements. This is accomplished through a section-by-section review of current environmental management activities against the ISO 14001 criteria.

General duty clause—Section 5 (a) 1 of the OSH Act which declares that each employer has a duty to furnish to his employees a workplace free from recognized hazards that are likely to cause serious physical harm or death. The implication of the clause is that employers must use common sense in providing a safe and healthful workplace. Owners and operators of stationary sources at which regulated or any other EHSs are present in a process, regardless of the amount of the substance, are subject to the *RMP general duty clause*.

Generally achievable control technology (GACT)—Under the CAA, an alternative emissions control standard that may be applied to **area sources** of Hazardous Air Pollutants in order to control the most significant HAPs emissions.

Grab sample—In environmental sampling, a discrete aliquot taken from one specific sampling location at a specific point in time. The analytical results from a grab sample should provide a good representation of the concentrations of the contaminants present at the point the sample was taken.

Greenfields—Uncontaminated properties outside of urban areas that have not previously been sites of industrial or commercial uses (as contrasted to **brownfields**).

H

Half-life—The time required for a population of atoms of a given radionuclide to decrease, by radioactive decay, to exactly one-half of its original number is called the radionuclide's half-life. No operation, either chemical or physical, can change the decay rate of a radioactive substance. Half-lives range from much less than a microsecond to more than a billion years. The longer the half-life, the more stable the nuclide. After one half-life, half the original atoms will remain; after two half-lives, one fourth (or ½ of ½) will remain; after three half-lives, one eighth of the original number (½ of ½ of ½) will remain; and so on.

Hazard—In *toxicology,* an inherent property of the chemical that would exist no matter what quantity was present. Under the *Process Safety Management rules,* an inherent characteristic of a material, system, process, or plant that must be controlled in order to avoid undesirable consequences. Typical hazards associated with processes involving highly hazardous chemicals include: combustible/

flammable, explosive, toxic, simple and chemical asphyxiant, corrosive, chemical reactant, thermal, potential energy, kinetic energy, electrical energy, and pressure source hazards.

Hazard assessment (HA)—Under the *Risk Management Program* rules, the evaluation of the off-site consequences of accidental releases to public health and the environment. Under *OSHA safety standards,* a survey of the job area to identify sources of hazards, and consideration of the following categories of hazards: impact, penetration, compression, chemical, heat, harmful dust, light radiation, falling, *etc.*; the sources or potential problem areas for these hazards; the potential for injuries.

Hazard identification—The first step in risk assessment. Hazard Identification is used to to determine the hazardous chemical's relative toxicity, concentration, extent of contamination, and toxicological end point(s) of concern.

Hazard or toxicity assessment (dose-response assessment)—A methodology used to determine whether dose-response information is available for the chemical of interest. A hazard assessment relies on toxicological and epidemiological data to ascertain specific threshold exposure levels. It examines and attempts to quantify the severity of a potential hazard. This assessment requires information to establish risks for both carcinogens and noncarcinogenic agents.

Hazard quotient—For noncarcinogens, a quantitative appraisal of the levels of exposure that may pose risks because they exceed the ambient levels found in an unexposed population. The hazard quotient is calculated using the Maximum Allowable Daily Intake, the Chronic Daily Intake, and the Reference Dose.

Hazardous chemicals—Under the OSH Act, all chemicals listed by OSHA with a permissible exposure limit (PEL) or by the American Conference of Governmental Industrial Hygienists (ACGIH) with a Threshold Limit Value (TLV) or those listed in the National Toxicology Program (NTP) Annual Report on Carcinogens (latest edition) or have been found to be a potential carcinogen in the International Agency for Research on Cancer (IARC) Monographs (latest editions), or by OSHA. Hazardous waste; tobacco or tobacco products; wood or wood products; food, drugs, or cosmetics that are intended for personal use; or ***articles*** are not considered to be hazardous chemicals.

Hazardous material—Under DOT rules, a substance or material which has been determined by the Secretary of Transportation to be capable of posing an unreasonable risk to health, safety, and property when transported in commerce, and which has been so designated. The term includes hazardous substances, hazardous wastes, marine pollutants, and elevated-temperature materials, materials designated as hazardous under the provisions of 49 CFR 172.101, and materials that meet the defining criteria for hazard classes and divisions in 49 CFR 173. The hazard categories include: (1) explosives, (2) gases, (3) flammable liquids, (4) flammable solids, (5) spontaneous combustibles, and dangerous when wet, (6) oxidizers and organic peroxides, (7) poisons and infectious substances, (8) corrosives and (9) all other hazardous materials.

Hazardous samples—In environmental sampling, samples that present a relatively high risk to health or the environment because of the potential level and type of contamination. Hazardous samples can be classified as high and moderate hazard samples. The determination of sample type is based on existing analytical data, background information (*e.g.*, past activities conducted at the site, container labels, *etc.*) and/or policy. Some organizations have a simple policy of "if it is on-site, the sample is hazardous" and "if it is off-site, the sample is environmental".

Hazardous substances—Under the Comprehensive Environmental Response, Compensation, and Liability Act, elements, compounds, mixtures, solutions, and substances which when released into the environment may present substantial danger to public health and welfare or the environment. The term includes substances listed in EPA's regulation 40 CFR 302.4 under CERCLA, specifying reportable quantities (RQs) for each substance.

Hazardous substance USTs—One of the two broad categories of underground storage tanks that contain either: (1) hazardous substances as defined in Section 101(14) of the Comprehensive Environmental Response, Compensation, and Liability Act (CERCLA), 40 CFR 302.4, but not including any of the substances regulated as a hazardous waste under RCRA Subtitle C, or (2) any mixture of such CERCLA-listed substances and petroleum that is not a petroleum UST system.

Hazardous waste—Under RCRA, and defined in 40 CFR 261, a solid waste that, because of quantity, concentration, or physical, chemical, or infectious characteristics: (a) causes or significantly increases mortality or serious irreversible or incapacitating reversible illness, or (b) poses a substantial present or potential hazard to human health or the environment when improperly managed.

Hazardous waste manifest—Under RCRA, a chain of custody for a hazardous waste that tracks the waste in transport.

Hazmat employee—Under the DOT hazardous materials regulations, an employee who handles hazardous materials and directly affects hazardous material transportation safety.

Health hazard—Under the Occupational Safety and Health Act, a hazard associated with a chemical for which there is statistically significant evidence, based on at least one study conducted in accordance with established scientific principles, that acute or chronic health effects may occur in exposed workers. Health hazards are categorized as: toxic or highly toxic, reproductive toxins, carcinogens, irritants, corrosives, sensitizers, hepatotoxins (liver), nephrotoxins (kidney), neurotoxins (nerve), hematopoietic (blood) or other agents that damage the lungs, eyes, skin, and mucous membranes.

Heat content—In chemistry, the heat released by complete combustion of a unit-weight of material.

Hematotoxicity—In toxicology, the damage a substance may cause to the formed elements of the blood (*e.g.*, red blood cells, white blood cells, platelets, *etc.*) and/or the bone marrow (area of blood cell formation),

Hepatitis B virus (HBV)—A disease caused by a highly infectious virus that attacks the liver. HBV infection can lead to severe illness, liver damage, and in some cases death, and is the most common cause of liver cancer worldwide.

Hepatitis C virus (HCV)—A viral illness that affects the liver that is spread by blood-to-blood contact, and is therefore a bloodborne pathogen. Some of the most common ways of spreading the virus are: (1) transfusion of blood products, (2) intravenous drug use, (3) tattooing and body piercing, (4) sharing needles. Menstrual blood can contain the virus.

Hepatotoxicity—In toxicology, the damage a substance may cause to the liver. Hepatotoxicity can be caused by cell death (necrosis), accumulation of lipids, damage to the bile ducts, inflammation (hepatitis), and the presence of fibrotic tissue in the liver.

Highly hazardous chemical—Under the Process Safety Management rules, a substance possessing toxic, reactive, flammable, or explosive properties and listed in the PSM Rule.

Hot substances—Substances that are considered thermally hazardous because they are 120°F (49°C) or hotter to the touch.

Hot work—Under the Process Safety Management rules, work involving electric or gas welding, cutting, brazing, or similar flame, or spark-producing operations.

Hydraulic conductivity—In an aquifer, the length (distance traveled)/ time (potential velocity) value.

Hydrocarbons—In chemistry, chemical compounds consisting primarily of carbon and hydrogen.

Hydrologic cycle—In hydrogeology, the constant movement of water by precipitation, overland runoff, and evaporation.

I

Identification number—Under the DOT hazardous materials regulations, a four-digit number assigned to the commodity, or group of similar commodities, by the United Nations or the United States.

Ignitable, corrosive, reactive (ICR) wastes—Under RCRA, wastes that exhibit any one or more of the characteristics, ignitability, corrosivity, and reactivity; chemical properties that can be immediately dangerous if improperly managed and thus merit special protective measures in the regulations.

Immediate or acute toxicity—In toxicology, a reaction by the body that occurs rapidly after a single exposure to a toxic substance.

Imminent danger—Under OSHA, any condition where there is the reasonable certainty that an immediate danger exists that may cause death or serious physical harm before the danger can be eliminated through normal enforcement procedures.

In-process wastes—Under RCRA, hazardous wastes generated in product or raw material tanks, transport vehicles or vessels, pipelines, manufacturing process units (except surface impoundments), until removed or until 90 days after the process ceases.

In vitro *methods*—In toxicology, laboratory scale toxicological studies performed in test tubes. These tests are performed on tissue or cells in order to observe changes in the structure or growth of cells.

Incident command system (ICS)—A system used for the management of an emergency response and based upon basic business management practices. The main functional areas of planning, directing, organizing, coordinating, communicating, delegating, and evaluating are under the overall direction of the Incident Commander.

Incineration—A process used for the treatment of hazardous waste that utilizes thermal decomposition via oxidation to reduce carbonaceous matter.

Independent effect—In toxicology, toxicity of substances that are asserted independently of each other

Indirect sources—Under the CAA, sources that do not produce pollution but have the ability to attract vehicles, or other mobile sources of pollution. Examples include: buildings, parking lots, garages, roads, and highways.

Inherently waste-like—Under RCRA, materials that ordinarily are managed as waste or that pose a substantial hazard when recycled, including specific dioxin- and furan-containing wastes.

Initial training—Under OSHA's Process Safety Management rules, training that is required before an employee begins participating in operations that use highly hazardous chemicals.

Injunction—A court order directing the defendant to take some action or to refrain from an action.

Inspections—Under the Risk Management rules, agency visits to a facility in order to check on the accuracy of the RM Plan data and on the implementation of the Risk Management rules.

Interim measures—Under RCRA, actions used to mitigate any immediate threats to the environment from a waste management unit while a comprehensive corrective action strategy is developed and implemented. Interim measures range from simple actions, such as fencing an area to prevent access, to complex groundwater pump-and-treat operations to prevent further contaminant migration.

Into the environment—Under CERCLA, the status of a release of a hazardous substance when it is not completely contained within a building or structure, even if it remains on the plant or facility grounds.

Ion—In chemistry, a charged particle that is produced when an atom or group of atoms gains or loses one or more electrons.

Ionized compounds—Chemical materials (acids, bases, and salts) that are present in a charged state.

Ionizing radiation—Radiation that has enough energy to cause atoms to lose electrons and become ions. Alpha and beta particles, as well as gamma and x-rays, are all examples of ionizing radiation. Ultraviolet, infrared, and visible light are examples of non-ionizing radiation. Radiation can cause the body's molecules to break apart and form electrically charged particles called ***ions*** that bond readily with other body chemicals and form toxicants such as hydrogen peroxide.

Irreversible effects—In toxicology, adverse effects to the body that do not reverse after the exposure ceases.

Isomers—In chemistry, different structural arrangements with the same chemical formulas (*e.g.*, *n*-butane and *t*-butane).

Isopleths—In hydrogeology, contour lines drawn between points of equal ground water elevations.

Isotopes—In chemistry, atoms of a given element that have the same number of protons but different numbers of neutrons.

J

Judgmental sampling—A method of environmental sampling that provides information on worst-case conditions and is useful for identifying contaminants of concern at a site but not the extent of contamination that will require additional sampling utilizing other approaches.

Judicial branch—The part of the United States government where the court system is located and where environmental laws are interpreted and enforced.

Judicial enforcement—Enforcement that occurs once the agency takes a case to the courthouse for lawsuit or criminal prosecution.

K

K_{ow}—See Octanol/Water Partition Coefficient.

L

Lab packs—Under RCRA and DOT standards, a means of preparing numerous small containers of hazardous waste (*e.g.*, discarded chemicals) for transportation and disposal.

Label—Under FIFRA, all labels and other written, printed, or graphic matter that is attached to the pesticide device or any of its containers or wrappers and that accompanies the pesticide or device at any time. A reference may be made on the label or in literature accompanying the pesticide or device.

Laboratory—As defined in the OSHA Laboratory Standard, a facility where the laboratory use of hazardous chemicals occurs and where relatively small quantities of hazardous chemicals are used on a nonproduction basis.

Laboratory scale—As defined in the OSHA Laboratory Standard, includes work with substances in which the containers used for reactions, transfers, and other handling of substances are designed to be easily and safely manipulated by one person and excludes commercial quantities of materials.

Laboratory use of hazardous chemicals—The handling or use of chemicals in the OSHA Laboratory Standard in which: (1) chemical manipulations are carried out on a *laboratory scale*, (2) multiple chemical procedures or chemicals are used, (3) the procedures involved are not part of a production process, nor in any way simulate a production process, and (4) protective laboratory practices and equipment are available and in common use to minimize the potential for employee exposure to hazardous chemicals.

Land disposal units—Under RCRA, surface impoundments, landfills, and land treatment units.

Land disposal—Broadly defined under RCRA to include virtually any temporary or permanent placement of hazardous waste in or on the land (such as in surface impoundments, waste piles, landfills, or land treatment facilities).

Land-based units—Under RCRA, units such as surface impoundments, waste piles, land treatment units, and landfills, that are used for land treatment, storage, or disposal of hazardous waste and pose relatively high risk of contaminating ground water.

LDR-restricted waste—Under RCRA, a waste that has a promulgated a treatment standard.

Level A—Under the OSHA regulations, the level of protection that is required when there is the greatest potential for exposure to skin, respiratory and eye hazards. Level A includes respiratory protection with positive pressure, full face-piece self-contained breathing apparatus (SCBA), or positive pressure supplied-air respirator with escape SCBA; totally encapsulated chemical- and vapor-protective suits; inner and outer chemical-resistant gloves; and disposable protective suits, gloves, and boots.

Level B—Under the OSHA regulations, the level of protection that is required when the highest level of respiratory protection but a lesser level of skin protection is needed. Level B includes respiratory protection with positive pressure, full face-piece self-contained breathing apparatus (SCBA) or positive pressure supplied air respirator with escape SCBA; inner and outer chemical-resistant gloves; faceshield; hooded chemical-resistant clothing, coveralls, and outer chemical-resistant boots.

Level C—Under the OSHA regulations, the level of protection that is required when the concentration and type of airborne substances are known and the criteria for using air-purifying respirators are met. Typical Level C equipment includes full-face air-purifying respirators, inner and outer chemical-resistant gloves, hard hat, escape mask and disposable chemical-resistant outer boots.

Level D—Under the OSHA regulations the minimum level of protection. Level D may be sufficient when no contaminants are present or work operations preclude splashes, immersion, or the other potential for unexpected inhalation or contact with hazardous chemicals. Appropriate Level D protective equipment may include gloves, coveralls, safety glasses, faceshield, and chemical-resistant, steel-toe boots or shoes.

Lifetime acceptable daily intake (ADI)—A "safe" lifetime dose of a substance which has been interpolated from dose-response curves, taking into account safety and other modifying factors.

Limited emergencies (standby notification alert)—Emergencies that have no public danger and require a limited staff for the Command Post, which may be a temporary setup in an office or near the scene, such as a break in a water pipe in an office area requiring immediate maintenance attention.

Listed hazardous wastes—Under RCRA, wastes that are specifically listed in 40 CFR 261, Subpart D as hazardous on the basis that they may exhibit hazardous characteristic(s) such as ignitability, corrosivity, reactivity, and toxicity or otherwise may be toxic or acutely hazardous.

Local toxicity—In toxicology, the reaction of a body to a substance that occurs at the site of chemical contact. The chemical need not be absorbed to cause this reaction.

Long-term recovery—In disaster response, the activities that continue until the entire disaster area is completely redeveloped, either as it was in the past or for entirely new purposes that are less disaster-prone.

Losing streams—In hydrogeology, streams that can actually help recharge the ground water, frequently found in carbonate areas.

Lowest achievable emissions rate (LAER)—Under the CAA, the most stringent achievable emissions limitation for a particular source. Unlike ***BACT,*** LAER does not take into account economic impacts and other costs.

M

Macro-encapsulation—A method of treatment for hazardous waste in which the solid waste is encapsulated within an impervious layer in order to eliminate the exposure to leaching.

Major fiber release—Under the EPA asbestos regulations, an event in which "greater than three square or linear feet of asbestos-containing building material becomes dislodged from its substrate."

Major source—Under the CAA, a ***stationary source*** that emits or has the potential to emit a significant quantity (as defined in the CAA) of one or more air pollutants.

Male reproductive toxicity—In toxicology, the damage a substance may cause to the male reproductive system.

Management—As defined by FEMA, the coordination of an organized effort to attain specific goals or objectives.

Management representative—In ISO 14000, the person who champions the effort to implement the EMS by taking the lead in coordinating the implementation of the EMS, overseeing development of EMS documentation, and obtaining employee buy-in into the program.

Manufacture—Under the Toxic Substances Control Act, *manufacture* means to "import, produce, or manufacture" a toxic substance.

Manufacturing use product—Under FIFRA, any pesticide product that is not an ***end use product***.

Material safety data sheet (MSDS)—Under OSHA regulations, a document that provides the user with important safety, health, and environmental information about a chemical product.

Maximum achievable control technology (MACT)—Under the CAA, a control technology based on the average emissions limitation achieved by the best 12% of existing sources in a particular category. If there are fewer than 30 sources in the category, then the emissions limitation is based on the best performing 5 sources.

Maximum allowable increase (MAI)—Under the CAA, the upper limit for the level of increase for permitted emissions from a *stationary source*. In the regulations, MAI is referred to as an *ambient air increment*, and the permitted emissions levels result in *consumption* of the increment.

Mechanical treatment—In the treatment of hazardous waste, those procedures that modify the physical properties of waste materials; for example: filtration, phase separation, filter pressing of suspended materials, centrifugation, agitation, adsorption, *etc.* Also known as *physical treatment.*

Media cleanup standards (MCS)—Under RCRA, the concentrations of hazardous waste or hazardous waste constituents that must be achieved in groundwater, surface water, soils, and air in order to comply with the standards for corrective measures. MCSs are the established concentrations set for each medium that ensures the protection of human health and the environment.

Median effective dose (ED$_{50}$)—The single dose of a substance that can be expected to cause a particular effect (other than lethality) to occur in 50% of the exposed population.

Median lethal concentration (LC$_{50}$)—The single dose of a chemical, quantified as an exposure concentration (*e.g.*, concentration in air such as ppm or mg/m^3) that can be expected to cause death in 50% of the exposed population. The LC$_{50}$ is similar to the LD$_{50}$ except that the chemical is quantified as an exposure concentration rather than a dose (*e.g.*, mg/kg-body weight).

Median lethal dose (LD$_{50}$)—The single dose of a chemical that can be expected to cause death in 50% of the exposed population.

Melting point (mp)—The temperature at which a solid changes to a liquid.

Mesothelioma—A malignancy of the lining of the chest or abdominal cavity. Pleural mesothelioma is a malignant growth of the pleura, the exterior lining of the lungs. Peritoneal mesothelioma is a malignancy of the peritoneum of the abdominal cavity. Either form of the disease spreads quickly and is always fatal.

Micro-encapsulation—A method of treatment for hazardous waste in which a semi-homogeneous product results from the treatment of finely divided powders, sludges, or viscous liquids.

Military munition specifically identified as a solid waste—Under RCRA, a specific category of hazardous waste specified in 40 CFR 266, Subpart M.

Minor emergencies (minor event alert)—Emergencies that are handled on a regular day-to-day basis by routine procedure, such as an employee with a laceration requiring first aid.

Miscellaneous materials—Under the EPA asbestos regulations, mostly nonfriable asbestos products and materials such as floor tile, ceiling tile, roofing felt, concrete pipe, outdoor siding, and fabrics.

Mitigation—According to *FEMA*, those actions and activities taken to reduce or eliminate the chance of occurrence or the effects of a disaster. Under *NEPA*, actions taken to reduce otherwise significant environmental effects, which includes (1) avoiding the impact altogether by not taking a certain action or parts of an action, (2) minimizing impact by limiting the degree or magnitude of the action and its implementation, (3) rectifying the impact by repairing, rehabilitating, or restoring the affected environment, (4) reducing or eliminating the impact over time by preservation and maintenance operations during the life of the action, (5) compensating for the impact by replacing or providing substitute resources or environments.

Mitigation system—Under the Risk Management Program rules, specific activities, technologies, or equipment designed or deployed to capture or control substances upon loss of containment to minimize exposure to the public or the environment. Passive mitigation means equipment, devices, or technologies that function without human, mechanical, or energy input; common examples include dikes, berms, and flame arresters. Active mitigation systems are equipment, devices, or technologies that require human, mechanical, or energy input to function. Examples include relief and isolation valves, sprinkler systems, and fire brigades.

Mixed transuranic waste (MTU)—Waste that has a hazardous component and radioactive elements heavier than uranium.

Mixed waste (MW)—Under RCRA, a hazardous waste that also contains certain radioactive constituents subject to the Atomic Energy Act or constituents (PCBs) subject to TSCA. Under the *Federal Facilities Compliance Act* of 1992, waste that is a mixture of hazardous wastes (as defined under RCRA), and source, special nuclear, or by-product (*i.e.*, radioactive) material subject to the Atomic Energy Act (AEA) of 1954.

Mixture—Under the Toxic Substances Control Act, any combination of two or more chemical substances if the combination does not occur in nature and is not, in whole or in part, the result of a chemical reaction. There are two exceptions to this definition: (1) The combination is *not* a mixture if it occurs as the result of a chemical reaction in which there are no new chemical substances; and (2) the combination is *not* a mixture if it could have been manufactured for commercial purposes without a chemical reaction at the time the combination's chemical substances were combined.

Mixture rule—Under RCRA, a rule that applies to listed hazardous waste mixed with solid waste, and specifies that the waste codes of the hazardous waste apply to the mixture unless the hazardous waste was listed on the basis of hazardous characteristic(s) only and the mixture exhibits none of the characteristics.

Mobile sources—Under the CAA, sources that are not *stationary sources*, such as vehicles, airplanes, and other forms of transportation.

Model accreditation plan (MAP)—Under the EPA asbestos regulations, a plan that established five areas of accreditation as the appropriate training requirements for individuals who would: (1) perform inspections, (2) prepare management plans for buildings, (3) supervise abatement projects, (4) serve as workers on abatement projects, or (5) design abatement projects.

Modified source—Under the CAA, a *stationary source* that has been physically changed or is operated in a manner that increases the emissions of existing or new pollutants.

Molecular and experimental embryology—In toxicology, one approach to *in vitro* studies which is used to detect abnormal changes at the molecular and embryonic levels.

Molecules—In chemistry, a group of atoms that are chemically bonded together; *e.g.,* O_2, BF_3, CH_4, *etc.*

Monte Carlo analysis—Used for evaluating the risk analyses for site conditions that involve many complex variables. In Monte Carlo analysis, the probability curves for various uncertainty parameters are inserted into the calculations of risk.

Multistate ozone nonattainment areas—Under the CAA, single ozone nonattainment areas that cover more than one state and are required to coordinate their implementation plans with other states.

Mutagens—In toxicology, substances that alter the genetic code in the gametes (sperm and/or eggs) of the body so that the appearance or function of the affected part of the body is changed and the change is passed along to following generations as a permanent alteration of the genetic structure.

N

National air monitoring stations (NAMS)—Under the CAA, 1080 key ambient air monitoring sites, located in areas of high population within the State and Local Air Monitoring Stations network, that are used to measure maximum pollutant concentrations.

Navigable waters—Under the Clean Water Act, all "waters of the United States," including waters used in interstate commerce, waters subject to tides, interstate waters, and intrastate lakes, rivers, streams, wetlands, sloughs, prairie potholes, wet meadows, playa lakes, or natural ponds. Waters used by interstate travelers and migratory birds are also included. Ground water is not included unless there is a hydrological connection between the ground water and surface water or unless ground water is specifically included under a state program.

Negligence—Under tort law, a breach of duty or duty owed by the defendant to the plaintiff which causes damage to the plaintiff.

Nephrotoxicity—In toxicology, the damage a substance may cause to various portions of nephrons (the functional units of the kidney that produce urine including the glomerulus and the proximal tubule).

Neutron—The part of the nucleus of an atom that contributes mass or weight to the atom that is neutral in charge (neither a positive nor a negative charge). Neutrons are about the same size as protons.

New chemical substance—Under the Toxic Substances Control Act, any chemical substance that is not included in the chemical substance list compiled and published under Title 15, Section 2607 (b) of the United States Code.

New source—Under the CAA, a *stationary source* that intends to but has not yet started construction.

New source performance standards (NSPS)—Under the CAA, source-specific emission control limitations and requirements intended to promote the best technological system of continuous emissions reduction.

New source review (NSR)—Under the CAA, a scoping process for a proposed project that includes a review of emissions limitations, and public involvement similar to the PSD process.

New tank system—As defined in the Underground Storage Tank rules, tanks for which installation commenced after December 22, 1988. All new tank systems must meet the requirements of Subparts B, C, and D before they can commence operation.

No action alternative—Under NEPA, the action that would happen to the environment if the Federal agency's proposed action was not implemented.

No migration—The concept that waste constituents will not migrate from the facility; *i.e.*, there is no migration pathway.

No observed adverse effect level (NOAEL)—In toxicology, the highest dose at which no adverse effects have been detected.

Nonattainment areas—Under the CAA, geographic areas in which the air quality is poorer than the NAAQS (*i.e.,* the concentrations of the *criteria pollutants* are above the standard).

Non-ionized compound—In chemistry, a material that is not electrically charged.

Nonpoint sources—In the Clean Water Act, sources of pollution that are diffuse, not emanating from a specific location or pipe.

Nontransportation-related facilities—Under the Oil Pollution Act of 1990, all fixed facilities, including support equipment (but excluding certain pipelines) railroad tank cars en route, transport trucks en route, and equipment associated with the transfer of bulk oil to or from water transportation vessels. The term also includes mobile or portable facilities such as drilling or workover rigs, production facilities, and portable fueling facilities while in a fixed, operating mode.

Nonwastewaters—Under RCRA LDR regulations, wastes which have >1 wt% total organic carbon (TOC) and >1 wt% total suspended solids.

Not otherwise specified (NOS)—Under the DOT hazardous materials regulations, a material that meets the definition of a DOT hazard class but is not listed by a specific name in the hazardous materials table 172.101.

Notice of intent (NOI)—Under NEPA, a notice published in the *Federal Register* that describes the proposed action, possible alternatives the agency is considering, background information on issues and potential impacts, and the proposed scoping process including whether, when, and where any scoping meeting will be held. The NOI also provides a contact point within the agency for further information.

Nuclear radiation—Energy in the form of waves or particles (also known as *radiation*). Radiation is a part of our natural world. People have always been exposed to radiation that originates from within the Earth (*terrestrial* sources) and from outer space (*cosmogenic* or *galactic* sources). Radiation also comes from sources such as radioactive material or from equipment such as x-ray machines, or accelerators.

Nucleus—The center of an atom that contains all the positive charges (protons) and neutrons of the atom, as well as all of the atom's weight.

O

Objectives—In an Environmental Management System (like ISO 14000) the goals the organization sets for itself based on its environmental policy and significant aspects.

Octanol/water partition coefficient (K_{ow})—The equilibrium ratio of the concentrations of material partitioned between octanol and water. This coefficient is considered to be an index of the potential of a chemical to be bioaccumulated. Higher values of K_{ow} are associated with greater bioaccumulation potential.

Off-site consequence analysis (OCA)—Under the Risk Management Program rules, an analysis of at least one worst-case release scenario and one alternative release scenario that addresses the impacts to off-site populations and the environment in areas beyond the property boundary of the stationary source or areas within the property boundary to which the public has routine and unrestricted access, both during or outside business hours.

Off-specification used oil—Under RCRA, fuel (used oil that is not classified under the regulations as a hazardous waste fuel) that does not meet the criteria for *specification used oil*.

Oil—Under the Clean Water Act, oil of any kind or in any form, including, but not limited to, petroleum, sludge, oil refuse, and oil mixed with wastes other than dredged spoil.

Operator—As defined in the Underground Storage Tank rules, any person in control of or having responsibility for the daily operation of the UST system.

Orbitals—In chemistry, the way the electrons organize themselves in the regions of space around an atom.

Organic bases—In chemistry, weak bases, such as amines, that can be corrosive to skin or other tissue.

Organic carboxylic acids—In chemistry, generally weak acids that can be very corrosive to skin.

Organic peroxides—In chemistry, very hazardous and shock-sensitive compounds that present a serious fire and explosion hazard. Commonly encountered organic peroxides include benzyl peroxide, peracetic acid, and methyl ethyl ketone peroxide.

Organic sulfonic acids—In chemistry, generally stronger acids than organic carboxylic acids.

Other potentially infectious materials (OPIM)—Under OSHA's Bloodborne Pathogen standard, bodily fluids which may contain pathogenic organisms; includes (1) human body fluids such as semen, vaginal secretions, cerebrospinal fluid, synovial fluid, pleural fluid, pericardial fluid, peritoneal fluid, amniotic fluid, (2) saliva in dental procedures, (3) any body fluid visibly contaminated with blood, (4) all body fluids in situations where differentiation between body fluids is difficult. This would include waste mixtures such as vomit, toilet overflow, sewage, and leachate associated with refuse containers.

Overpacks—Under the DOT hazardous materials regulations, a type of packaging used for materials that are consolidated in accordance with the provisions of 49 CFR 172.404.

Owners/operators—Under the Risk Management Program rules, any person who owns, operates, leases, controls, or supervises a stationary source and its processes. For RM Plan registration purposes, the owner or operator normally is the highest-ranking company executive on-site.

Oxidation—In a chemical reaction, the process of losing electrons

P

P2—Under the Pollution Prevention Act and subsequent EPA publications, applies to releases of contaminants to all media (air emissions, wastewater and storm water discharges and spills, releases to soil or groundwater, as well as solid and hazardous waste generation) and means *source reduction* as well as other practices that reduce or eliminate the creation of pollutants through 1) increased efficiency in the use of raw materials, energy, water or any other resources and 2) protection of natural resources by conservation. Also known as *pollution prevention*.

Performance standard—A standard, usually promulgated by a regulatory agency, that requires the employers to *effectively* implement a program (as opposed to following specific rules).

Peripheral nervous system (PNS) toxicity—In toxicology, the damage a substance may cause to the sensory and motor nerves of the extremities; is often reversible.

Permit by rule—Under RCRA, qualifying hazardous waste management facilities that are permitted under other laws (injection wells permitted under SDWA, POTWs with NPDES permits, ocean disposal authorized under the Marine Protection, Research, and Sanctuaries Act) where requirements are set forth in those other permits and thus are also deemed to have a RCRA permit.

Permit-required confined space—As defined by OSHA, a *confined space* that: (1) contains or has the potential to contain a hazardous atmosphere, (2) contains a material that has the potential to engulf an entrant, (3) has inwardly converging walls or a floor which slopes downward and tapers to a smaller cross-section so that an entrant could be trapped or asphyxiated inside, (4) contains any other recognized serious safety or health hazard.

Permits—Permissions issued by Federal agencies or by State agencies that have been delegated authority under Federal law that allow a permittee to conduct an activity that has an impact on the environment.

Permit shield—Under Title V, Operating Permits, of the CAA, a determination made by the regulating agency that a source is in compliance with all aspects of the Clean Air Act. The permit shield does not extend to noncompliance with permit conditions, applicable requirements that were not covered in the permit application, or new requirements promulgated after issuance of the Title V permit.

Personal protective equipment (PPE)—Items of clothing and equipment which are used by themselves or in combination with other protective clothing and equipment to isolate the individual wearer from a particular hazard or a number of hazards or to protect the environment from the individual.

Pest—Under FIFRA, an organism deleterious to man or the environment. This includes any vertebrate animal other than man; any invertebrate animal, including but not limited to any insect, other arthropod, nematode, or mollusk such as a slug and snail, but excluding any internal parasite on living man or other animals; any plant growing where not wanted, including any mosses, alga, liverwort, or other plant of any higher order, and any plant part such as a root; or any fungus, bacterium, virus, or other microorganisms, except those on or in living man or other living animals and those on or in processed food or processed animal feed, beverages, drugs and cosmetics (as defined under FFDCA Sections 201[g][1] and 201[i]).

Pesticide—Under FIFRA, any substance or mixture of substances intended for preventing, destroying, repelling, or mitigating any pest, or intended for use as a plant regulator, defoliant, or desiccant, other than any article that: is a new animal drug under the FFDCA Section 201(w); is an animal drug that has been determined by regulation of the Secretary of Health and Human Services not to be a new animal drug; or an animal feed under the FFDCA Section 201(x) that bears or contains any substances described above.

Pesticide product—Under FIFRA, a pesticide in a particular form (including composition, packaging, and labeling) in which the pesticide is, or is intended to be, distributed or sold. The term includes any physical apparatus used to deliver or apply the pesticide if distributed or sold with the pesticide.

Petroleum USTs—One of the two broad categories of underground storage tanks that contain petroleum or a mixture of petroleum with *de minimis* quantities of other regulated substances. These systems include tanks containing motor fuel, jet fuel, distillate fuel oil, residual fuel oil, lubricants, petroleum solvents, or used oil.

pH—In chemistry, the negative logarithm of the hydrogen ion (H^+) concentration and a measure of a chemical's acidity or basicity.

Photochemical assessment monitoring stations (PAMS)—Under the CAA, ambient air monitoring stations that are required in any ozone *nonattainment area* that has been designated as serious, severe, or extreme.

Physical hazard—As defined by OSHA, a hazard associated with a chemical for which there is scientifically valid evidence that it is a combustible liquid, a compressed gas, an organic peroxide, an oxidizer, or is otherwise explosive, flammable, pyrophoric (may spontaneously ignite in air at 130°F or less), unstable (reactive), or water-reactive.

Physical state—The nature of the substance (solid, liquid, gas, fume, mist dust, vapor, *etc.*) at a defined temperature (*i.e.*, 20°C or room temperature).

Physical treatment—In the treatment of hazardous waste, those procedures that modify the physical properties of waste materials—for example: filtration, phase separation, filter pressing of suspended materials, centrifugation, agitation, adsorption, *etc.* Also known as *mechanical treatment.*

pK$_a$—In chemistry, the negative logarithm of the equilibrium constant for acids or bases. Strong acids; such as sulfuric and hydrochloric acids have low pK$_a$s (*i.e.*, < 1.1). Bases such as potassium hydroxide and sodium hydroxide have pK$_a$s closer to 14.0   Weak acids and weak bases have pK$_a$s that fall between these two extremes.

Plan, Do, Check, Improve—The fundamental philosophy of Total Quality Management.

Pleural effusion—A collection of fluid around the lung and the most common effect of inhalation of asbestos dust.

Pleural plaque—A thickening of tissue under the parietal pleura, which can become calcified.

Pleural thickening—A thickening of the visceral (lung) and/or parietal (chest wall) pleura. The thickening can vary from 0.5 to 2 cm in thickness and results in increased difficulty in breathing.

Pneumatic and hydraulic high-pressure hazards—Hazards resulting from pressurized gases and liquids.

Point of compliance (POC)—The point at which the permitted concentrations of hazardous waste are enforced at a RCRA-permitted hazardous waste treatment, storage, and disposal facility (TSDF). The point of compliance is the vertical surface located at the hydraulically downgradient edge of the area where waste will be placed, and extending down to the uppermost aquifer. Under *RCRA corrective action*, the location or locations at which media cleanup standards are achieved, generally set on a site- and media-specific basis.

Point source—Under the Clean Water Act, any discernible, confined, and discrete conveyance, including but not limited to, any pipe, ditch, channel, tunnel, conduit, well, discrete fissure, container, rolling stock, concentrated animal feeding operation, or vessel or other floating craft from which pollutants are or may be discharged. Point sources can include vehicles, natural conveyances, and intermittent sources such as stormwater outfalls.

Pollutant—Under the Clean Water Act, any material which can contaminate water including: dredge spoil, solid waste, incinerator residue, sewage, garbage, sewage sludge, munitions, chemical wastes, biological materials, radioactive materials, heat, wrecked or discarded equipment, rock, sand, cellar dirt, and industrial, municipal, and agricultural waste discharged into water.

Pollution prevention—Under the Pollution Prevention Act and subsequent EPA publications, applies to releases of contaminants to all media (air emissions, wastewater and storm water discharges and spills, releases to soil or groundwater, as well as solid and hazardous waste generation) and means *source reduction* as well as other practices that reduce or eliminate the creation of pollutants through 1) increased efficiency in the use of raw materials, energy, water or any other resources and 2) protection of natural resources by conservation. Also known as *P2.*

Porosity—In hydrogeology, the ratio of empty spaces (voids) in between the particles of soil, sand, gravel, or fractured rock to the total volume, which tells us the maximum amount of water a rock or soil can hold if it is saturated.

Potential disaster (response alert)—An emergency that is one step beyond a *limited emergency* and has a potential public impact. Under these conditions, additional staff is required at the Command Post and the Incident Command System in its more formal structure is implemented. A hazardous materials release being handled by First Responder Operations trained employees would be a possible example.

Potentiating effect—In toxicology, when exposure to one substance, having very low or no significant toxicity, enhances the toxicity of another (*e.g.*, 0 + 5 = 15); the result is a more severe injury than that which the toxic substance would have produced by itself.

Potentiometric maps—In hydrogeology, visual images of the top of water tables that are generated by mapping the elevation of the water at numerous locations.

Potentiometric surface—In an aquifer, the boundary between the unsaturated and saturated zones where the pressure exerted by the surface of the water table is equal to the atmospheric pressure at that depth.

Preamble—The part of a proposed regulation that describes the rationale for the regulation, how it complies with and/or fulfills the requirements of the underlying statutes. This section also summarizes the input received from interested parties to date and how this feedback has impacted development of the regulation.

Preparedness—In disaster response, the planning how to respond in case an emergency or disaster occurs, and working to increase the resources that are available to respond effectively.

Presumed asbestos-containing material (PACM)—Thermal insulation and surfacing material found in buildings constructed no later than 1980.

Prevention—In safety, a nonoccurrence of a harmful event that ultimately provides additional long-term benefits of employee dedication, productivity, efficiency, and safety.

Private nuisance—Under tort law, an intentional act by the defendant which substantially and unreasonably interferes with the plaintiff's use and enjoyment of his land.

Process—Under the Risk Management Program rules, any activity involving a regulated substance, including any use, storage, manufacturing, handling, or onsite movement of such substances, or combination of these activities.

Process hazard analysis (PHA)—Under the Risk Management Program and OSHA Process Safety Management rules, an analysis that identifies, evaluates, and controls the hazards involved with the design, operation, and maintenance of processes containing regulated substances.

Program 1, 2, 3 eligibility—Three program levels established under the Risk Management Program rules. All stationary sources must assign each covered process a program level, based on the size and complexity of a source's processes.

Prohibited waste—Under RCRA, a waste that is an *LDR-restricted waste*, or actually prohibited from land disposal. Generally, a restricted waste is a prohibited waste upon the effective date of the treatment standard if the waste does not meet the applicable treatment standard and no other variances are applicable.

Proper shipping name (PSN)—Under the DOT hazardous materials regulations, an assigned name to a hazardous material that determines whether the material is legal to move, how to package, label, mark, and placard the shipment, whether special provisions apply, *etc.*

Proton—The part of the nucleus of an atom that has a single positive charge. While protons and neutrons are about 2,000 times heavier than electrons, they are still very small particles. A grain of sand weighs about a hundred million trillion (100,000,000,000,000,000,000) times more than a proton or a neutron.

Public nuisance—Under tort law, a use of one's property to intentionally cause or permit a condition to exist that injures or endangers the public health, welfare, or safety.

Public receptor—Under the Risk Management Program rules, receptors that include off-site residences, institutions (schools, hospitals), industrial, commercial, and office buildings, parks, or recreational areas occupied by the public at any time without restriction where members of the public could be exposed to toxic concentrations, radiant heat, or overpressure, as a result of an accidental release of a regulated substance. Roads are not included as public receptors but should be considered when coordinating with emergency planning organizations.

Public scoping—Under NEPA, a process in which the agency involves the public and other State and Federal agencies in determining the range of alternatives and actions to be discussed in an *environmental assessment* or *environmental impact statement,* to identify alternatives or other significant environmental issues that may have been overlooked by the agency in their development of the proposed action.

Public vessel—A vessel that (1) is owned, or demise-chartered, and operated by the United States government or a government of a foreign country; and (2) is not engaged in commercial service (33 CFR 151.1006).

Pulmonary toxicity—In toxicology, the damage a substance may cause to the lungs. Pulmonary toxicity can be caused by obstruction of the airways (*e.g.*, by swelling or constriction), damage to the area of gas exchange (alveolar area) in the lungs, and various types of immune reactions.

Punitive damages—Damages imposed in addition to actual damages that compensate the plaintiff for his injury.

Pyrolysis—A process used for the treatment of hazardous waste that is theoretically a zero-air indirect-heat process (air-starved incineration).

Q

QA1—In environmental sampling, a screening objective used to obtain a quick, preliminary assessment of site contamination. Data collected for this objective provide neither definitive identification of pollutants nor definitive quantification of concentration levels.

QA2—In environmental sampling, a verification objective used to verify analytical results. Here a small percentage of the sample results are verified to provide a certain level of confidence for a portion of the results. An inference is then made on the quality of the remainder of the data.

QA3—In environmental sampling, a definitive objective used to assess the accuracy of the concentration level as well as the identity of the analytes of interest. This objective is appropriate when a high level of qualitative and quantitative accuracy is required for all sampling results.

Quality assurance (QA) objectives—In environmental sampling, statements about the desired reliability of the data to be generated. The statements are defined by determining how precise, accurate, representative, complete and comparable the data must be to satisfy the data use objective.

R

Radiation—The emission and propagation of energy through space. Radiation can be classified as either ionizing radiation (alpha particles, beta particles, gamma rays, x-rays, neutrons, high-speed electrons, high-speed protons, and other particles capable of producing ions) or non-ionizing radiation (sound, radio, microwaves, or visible, infrared, and ultraviolet light).

Radioactive decay—The process where an energetically unstable atom transforms itself to a more energetically favorable, or stable, state. The unstable atom can emit ionizing radiation in order to become more stable. This atom is said to be *radioactive*, and the process of change is called ***radioactive decay***.

Radioactive mixed waste (RMW)—Waste that contains both a hazardous component regulated under RCRA and radioactive material regulated under the Atomic Energy Act. The applicability of Subtitle I to RMW depends on whether those USTs are regulated under RCRA Subtitle C.

Radioactivity—The spontaneous emission of certain types of radiation (alpha, beta, and gamma) by unstable atomic nuclei.

Random sampling—Environmental sampling in which locations are arbitrarily assigned using a random selection procedure such as random number tables. Random sampling would be used when statistical evaluations of the data are required (*e.g.*, probability statements). Random sampling is the opposite of judgmental sampling.

Reactive chemicals—Substances that produce a violent reaction when exposed to or mixed with another substance, sometimes even water or air.

Reasonably achievable emissions control technology (RACT)—Under the CAA, a control technology that is reasonably available and which may be incorporated into a SIP for devices, systems process modifications, or other apparatus or techniques. RACT takes into account the necessity of imposing the controls as well as their social, environmental, and economic impacts.

Reasonably available control measures (RACM)—Under the CAA, emissions control technologies (typically used for fugitive dust, residential wood burning, and prescribed burning) that may be used for the management of particulate matter emissions.

Reclamation—Under RCRA, the processing of a material to regenerate that material or to recover a usable product (except sludges and byproducts that exhibit a characteristic of hazardous waste but that are not listed as a hazardous waste, and commercial chemical products).

Recognized environmental conditions—As defined by CERCLA, the presence or likely presence of any hazardous substances or petroleum products on a property under conditions that indicate a release, a past release, or a material threat of any hazardous substances or petroleum products into structures on the property or into the ground, groundwater, or surface water of the property.

Record of decision (ROD)—Under NEPA, the notification of the public of the agency's decision on a proposed action and the reasons for that decision.

Recovery activities—In disaster response, the activities that occur until all community systems return to normal or nearly normal conditions. This includes both ***short-term*** and ***long-term*** recovery actions.

Recyclable materials—Under RCRA, hazardous wastes that are legitimately recycled and generally are subject to less than full Subtitle C regulation in order to encourage recycling.

Recycled—Under RCRA, a material that has been ***used in a "manner constituting disposal***, burned for energy recovery, or used to produce a fuel product, reclaimed, accumulated speculatively or accumulated, stored, or treated before recycling."

Reduction—In a chemical reaction, the process of gaining electrons.

Reference dose (RfD)—In toxicology, a dosage value calculated from the NOAEL using uncertainty and modifying factors, given the differences in animal and human responses and the varying sensitivities of humans.

Refresher training—Under OSHA's Process Safety Management rules, training that is required a minimum of every three years by employees and contractors who work on or around the process equipment containing highly hazardous chemicals.

Regulated substance—The list of regulated substances under the EPA List Rule (40 CFR 68.130) that includes 77 volatile toxic substances and 63 flammable substances.

Regulations—Specific procedures promulgated by agencies for the administration and enforcement of environmental laws. The terms ***rules*** and ***regulations*** are synonymous, and both terms are used interchangeably in the law.

Regulatory compliance review—A type of audit that evaluates an organization's compliance with environmental regulations.

Release—As defined in the Comprehensive Environmental Response, Compensation, and Liability Act, any spilling, leaking, pumping, pouring, emitting, emptying, discharging, injecting, escaping, leaching, dumping, or disposing of hazardous substances to the environment.

Remedial action—Under the Comprehensive Environmental Response, Compensation, and Liability Act, remedies that are protective of human health and the environment, that maintain protection over time, and minimize untreated wastes.

Removal action—Under the Comprehensive Environmental Response, Compensation, and Liability Act, an immediate action taken over the short term to address a release or threatened release of a hazardous substance.

Reopeners—Legal requirements setting forth the bases on which additional cleanup of the brownfield property can be required in addition to what was originally agreed upon with the State agency. These reopeners vary from state to state, but can include: (1) fraudulent or misrepresented information has been provided to the agency, (2) an imminent threat to human health or the environment occurs on the property, (3) a previously unknown condition or new information comes to the agency's attention, and (4) the risk on which the agreed-upon cleanup was based significantly changes so that additional cleanup to prevent endangerment to human health and the environment is required.

Replacement in kind—Under the Process Safety Management rules, a replacement which satisfies the design specification.

Reportable quantity (RQ)—The amount of a hazardous substance which, when released to the environment, must be reported to governmental authorities under the CWA, CERCLA, SARA Title III, or RCRA.

Reregistration—A process started after FIFRA went through a major revision in 1988; active ingredients of pesticides that had been initially registered before 1984 are put through a process of complete reevaluation. This evaluation covers labels, formulae, and especially supporting data. A determination is made whether or not an active ingredient should be eligible for reregistration, and what additional supporting data, if any, is needed to complete the cost-benefit picture.

Respiratory sensitization—In toxicology, a pulmonary (lung) reaction triggered by an immune response to allergens. Pulmonary reactions are characterized by coughing, labored breathing, tightness in the chest, and shortness of breath.

Response—As defined by FEMA, the effective and efficient application of assets and activities to resolve the immediate impacts of an event.

Responsible Corporate Officer Doctrine—In law, a doctrine that allows the criminal conviction of a corporate officer based purely on the fact that he was in a position of responsibility and authority within the corporation which would have allowed him to prevent the violation.

Responsible party—Under the Oil Pollution Act of 1990, the owner or operator of a facility from which oil is discharged who is liable for the costs associated with the cleanup of the spill and any damages resulting from an oil spill.

Restricted-use pesticide—Under FIFRA, a pesticide whose toxicity (typically, acute toxicity) is high enough that the EPA restricts its use.

Reuse—Under RCRA, the recycling of a material in a particular function or application as an effective substitute for a commercial product. Reuse is also a form of *use*.

Reversible effects—In toxicology, adverse effects to the body that wear off (or reverse), given sufficient time after the exposure ceases.

Right-to-Know—Under OSHA's Hazard Communication standard, the fundamental philosophy that employees have a right to know about the chemical hazards present in their workplace.

Risk—In toxicology, the likelihood (probability) that an adverse effect (injury or harm) will occur in a given situation. OSHA defines risk as a function of both hazard and the amount of exposure (see 59 FR 6126). Risk can be expressed as a value that ranges from zero (no injury or harm will occur) to one (harm or injury will definitely occur).

Risk analysis—As defined by the *Society for Risk Analysis*, the detailed examination including risk assessment, risk evaluation, and risk management alternatives, performed to understand the nature of unwanted, negative consequences to human life, health, property, or the environment; an analytical process of quantification of the probabilities and expected consequences for identified risks. In *toxicology,* a quantitative appraisal of the levels of exposure that may pose risks because they exceed the ambient levels found in an unexposed population. In *emergency response decision-making*, a tool that is used to evaluate the hazard, the exposure vulnerability, and the risk in the proposed action or inaction.

Risk-based clean closure—Under RCRA, closure in which some contaminants remain in place but are at protective levels. Allowable residual contamination is determined by the regulator in the context of closure plan approval.

Risk-based corrective action (RBCA)—A technically defensible risk assessment strategy used for decision making in a variety of remediation scenarios.

Rotary kiln—A cylindrical, horizontal, refractory-lined shell used in the treatment of hazardous waste.

Routes of exposure—The manner in which a substance enters a body. Routes of exposure include the gastrointestinal tract (ingestion), lungs (inhalation), and skin (dermal, topical, or percutaneous).

Rules—Specific procedures promulgated by agencies for the administration and enforcement of environmental laws. The terms ***rules*** and ***regulations*** are synonymous, and both terms are used interchangeably in the law.

Rural transport areas—Under the CAA, rural (ozone) nonattainment areas associated with a larger metropolitan area.

S

Safe condition—Under OSHA, the requirement that compressed gas containers be maintained in accordance with the requirements of the DOT Hazardous Materials Regulations and the Compressed Gas Association standards.

Safety—the probability that adverse effects (harm) will *not occur* under specified conditions (the inverse of risk).

Sampling plan—In environmental sampling, the blueprint for accomplishing the established goal of the Data Quality Objectives that results in the generation of data of known quality. The sampling plan ensures that all components associated with the generation of the data are thoroughly documented and the documentation is verifiable and defensible.

Satellite accumulation—The provision under RCRA that allow generators to accumulate small quantities of hazardous waste at the worksite.

Saturated—The condition of an organic compound in which each constituent carbon is covalently linked to four different atoms.

Saturated zone—In soil, an area beneath the unsaturated zone where soil particles are completely surrounded by water and no air is present and which makes up an aquifer.

Sensitizers—In toxicology, substances that change the body's proteins so that the body fails to recognize them as its own and reacts by stimulating the immune system to produce antibodies. Sensitizers can cause an allergic reaction to occur as the body's antibodies try to destroy the altered proteins.

Short titles—Those portions of the law that are enacted by Congress at a point in time. These laws may be original acts that create an entirely new regulatory area, or (more often) acts that amend existing laws.

Short-term recovery—In disaster response, the activities aimed at returning vital life-support systems to minimum operating standards.

Solid waste—As defined by RCRA, any garbage, refuse, sludge, and other discarded material, including solids, semisolids, liquids, and contained gases. Note that solid waste is not restricted to solid phase material. The statutory definition includes specific exceptions for several wastes, chiefly on the basis that they are regulated under other statutes.

Solid waste management unit (SWMU)—Under RCRA, any discernible unit in which solid wastes have been placed at any time, regardless of whether or not the unit was intended for the management of solid or hazardous wastes. Such units include any area at a facility where solid wastes have been routinely and systematically released.

Solidification—A method of treatment for hazardous waste in which monoliths with considerable structural strength are formed.

Solubility—The amount of a given substance (the solute) that dissolves in a unit volume of a liquid (the solvent). In biological fluid/material such as blood and lipids (fat), the solubility may affect the absorption of a substance into the body, its movement (distribution) within the body and its potential for storage in the body (*e.g.* in fat or bone).

Source reduction—As defined in Section 6605(5)(A) of the Pollution Prevention Act, any practice which 1) reduces the amount of any hazardous substance, pollutant, or contaminant entering any waste stream or otherwise released into the environment (including fugitive emissions) prior to recycling, treatment, or disposal and 2) reduces the hazards to public health and the environment associated with their release.

Special purpose monitoring stations (SPMS)—Under the CAA, monitoring stations installed by State and local air pollution control agencies in order to support their air programs.

Specific gravity (SG)—The ratio of the density of a liquid as compared with that of water.

Specific retention—In hydrogeology, the volume of water that is retained as a coating on soil or rock particles after gravity draining.

Specific yield—The volume of water that will drain by gravity from a specific volume of rock or soil.

Specification used oil—Under RCRA, fuel (used oil that is not classified under the regulations as a hazardous waste fuel) that meets the following criteria: arsenic ($\leq$5 ppm), cadmium ($\leq$2 ppm), chromium ($\leq$10 ppm), lead ($\leq$ 100 ppm), flash point ($\geq$100°F), and total halogens ($\leq$4000 ppm).

Stabilization—A method of treatment for hazardous waste that reduces or eliminates the toxicity or the hazard potential of a certain waste stream by lowering the solubility and leachability of the toxic or hazardous components. Also call *fixation.*

Stabilization initiative—A RCRA corrective action management philosophy in which the overall goal is to control or abate threats, or to prevent or minimize the further migration of contaminants while long-term remedies are pursued.

State and local air monitoring stations (SLAMS)—Under the CAA, a network of 4000 ambient air monitoring stations generally set up in urban areas to meet the needs of State and local air pollution control agencies.

State implementation plan (SIP)—Under the CAA, a plan, prepared by a state, that contains a detailed description regarding how the state will fulfill its responsibilities toward achieving and maintaining clean air. The plan is essentially a collection of rules that the state uses to control air pollution.

Stationary sources—Sources of air pollution which don't move, such as a building, structure, facility, or other installation. Under the *Risk Management Program* rules, all buildings, structures, equipment, or substance-emitting stationary activities which belong to the same industrial group, located on one or more contiguous properties, under the control of the same person(s), from which accidental release may occur.

Statutes—Laws that are amended must then be revised and republished to reflect all permanent amendments. Also called the *body of standing law*.

Statutory law—Federal statutes enacted by the legislative bodies.

Storage—Under RCRA, holding waste temporarily, pending future management.

Strategic environmental management (SEM)—The adaptation and integration of traditional business and engineering theory and practice toward the maximization of profit, to the responsible management of environmental issues with the goal of minimizing costs and potential liability.

Strict liability—Abnormally dangerous activities which may cause a defendant to be liable for harm to the person, land, or personal property of another resulting from that activity, although the defendant has exercised the utmost care to prevent the harm.

Structure/activity relationships—In toxicology, one approach to *in vitro* toxicological studies which involves the assessment of the toxicity of a chemical based upon its molecular structure. A ***hazard or toxicity assessment*** is also referred to as a ***dose-response assessment*** because it seeks to determine whether dose response information is available for the chemical of interest.

Substantial authority personnel—High-level personnel or individuals within a corporation who exercise substantial supervisory authority (such as a plant manager) and any other individuals, who, although not part of an organization's management, nevertheless exercise substantial discretion when acting within the scope of their authority (such as an individual with authority to set, negotiate, or approve price levels or contracts).

Substantial harm facility—Under the Oil Pollution Act of 1990, a facility that meets the following: (1) the facility transfers oil over water to or from vessels and has a total oil storage capacity of at least 42,000 gallons, or (2) the total oil storage capacity at the facility is at least 1 million gallons and at least one of the following criteria is met: (a) the facility's secondary containment for each AST area will not hold the volume of the largest single AST plus sufficient freeboard for precipitation, (b) a discharge could injure fish, wildlife, or sensitive environments, (c) a discharge would shut down operations at a public drinking water intake, (d) the facility has had a reportable spill of at least 10,000 gallons within the last five years.

Surfacing material—Asbestos-containing material that has been sprayed or troweled on surfaces (walls, ceilings, and structural members) for acoustical decorative or fireproofing purposes. This includes plaster and fireproofing insulation.

Sustainable development—As defined by the United Nations World Commission on Environment and Development, meeting the needs of the world's current population without making it impossible for the world's future citizens to meet their needs. In addition to consideration of residuals (air emissions, waste water, solid and hazardous waste) sustainable development focuses on energy and material inputs and outputs associated with a process and the life-cycle environmental costs of the product or service produced.

Synergistic effect—In toxicology, when exposure to two chemicals has a combined effect much greater than the sum of the effects of each substance when given alone (*e.g.*, 3 + 5 = 30); each substance magnifies the toxicity of the other.

Synthetic minor source—Under the CAA, a source having the capacity to operate as a ***major source*** but voluntarily operating below an enforceable limitation in order to keep its emissions below the significant quantities for major sources, as defined in the Act.

System—A combined or organized whole from many parts.

Systematic sampling—Environmental sampling that involves the establishment of a reproducible scheme such as a grid or transect to which sampling points can be referenced. Samples are then taken at the points where grid lines intersect (nodes) or at defined distances along the transect line.

Systemic toxicity—In toxicology, the reaction of a body to a substance that occurs at a site or sites distant from the site of chemical absorption.

T

Tanks—Under RCRA, stationary devices made primarily of nonearthen materials.

Tank vessels—Vessels that are constructed, adapted to carry, or that carry oil or hazardous materials in bulk as cargo or cargo residue and that are US documented vessels, operate in United States waters, or transfer oil or hazardous material in a place subject to the jurisdiction of the United States.

Target organs—In toxicology, the organs which will sustain major adverse effects as a result of exposure to a certain chemical.

Targets—In an environmental management system, like ISO 14000, the detailed performance requirements arising from identified ***objectives***.

Teratogens—In toxicology, substances that produce genetic changes or tumors in developing fetuses so that there is some change in the appearance or function of the body but the genetic code of the gametes is not changed and so the changes are not passed on to following generations.

Theoretical oxygen demand (Th_{OD})—The cumulative amount of oxygen needed to completely oxidize a given material. The Th_{OD} is the upper limit for BOD_5 values, although it is seldom achieved. A comparison of the BOD_5 and Th_{OD} values for a given chemical provides an indication of the biodegradability of that chemical.

Thermal hazards—Hazards associated with substances that are cold or hot. These divisions are *not* related to heat or cold stress from exposure to the environment.

Thermal system insulation—Under the OSHA asbestos regulations, insulation used to inhibit heat transfer or prevent condensation on pipes, boilers, tanks, ducts, and other components of hot and cold water systems and heating ventilation and air condition (HVAC) systems. This includes pipe lagging and pipe wrap; block, batt, and blanket insulation; cement and "mud," and other products such as gaskets and ropes.

Threshold dose—The lowest dose of a chemical at which a specified measurable effect is observed and below which it is not observed (*e.g.*, no observed effect level [NOEL]).

Threshold limit value (TLV)—The exposure level under which most people can work for eight hours a day, day after day, with no harmful effects. A table of the values and accompanying precautions for most common industrial materials is published annually by the American Conference of Governmental Industrial Hygienists.

Threshold planning quantities (TPQ)—Under the *Community Right-to-Know* regulation, trigger quantities for chemicals appearing on EPA's Extremely Hazardous Substance (EHS) list. When a facility inventories an EHS in quantities equal to or greater than a trigger quantity, the facility is subject to emergency planning and notification requirements.

Threshold quantity (TQ)—Under the Risk Management Program rules, the maximum amount of a substance used in a single process that triggers the requirement for the substance to be reported and included in the emergency response plans for chemical spills and releases.

Time composite sampling—Environmental sampling that is used to determine the concentrations of contaminants over time. Time composite sampling is normally not utilized during hazardous waste site investigations because the intent of such investigations is to determine as quickly as possible what is present on site at the "present" time. For health and safety reasons, a form of time compositing is utilized to determine air quality on and around sites.

Time-weighted average (TWA)—The measured concentration of the chemical multiplied by the duration of the exposure.

Tort—A legal wrong committed upon the person or property of another, independent of contract.

Toxic concentration low (TC$_{LO}$)—The lowest concentration of a substance in air to which humans or animal have been exposed for any given period of time that has produced any toxic effect in humans or has produced teratogenic or reproductive effects in animals.

Toxic dose low (TD$_{LO}$)—The lowest concentration of a substance introduced by any route, other than inhalation, over any given period of time and reported to produce any toxic effect in humans or to produce teratogenic or reproductive effects in animals

Toxicity—In *toxicology*, the ability of a chemical, physical, or biological agent to cause acute or chronic damage to biological material. Under *RCRA*, a hazardous waste characteristic based on the toxic properties of 8 metals and 32 organic compounds.

Toxicokinetics—In toxicology, the movement of substances (chemicals) within the body. This movement is usually divided into four inter-related processes termed absorption, distribution, biotransformation (metabolism), and excretion.

Toxicology—The study of the adverse effects of chemical, physical, and biological agents on living organisms.

Trade secret—As defined by OSHA and included in the *Hazard Communication* and *Process Safety Management* rules, any confidential formula, pattern, process, device, information, or compilation of information that is used in a particular company, that gives the company an opportunity to gain an advantage over competitors who do not know or use it.

Transmissivity—In an aquifer, the measure of an aquifer's capacity to deliver water. Transmissivity is the product of the hydraulic conductivity and the aquifer thickness.

Treatment—Under RCRA, virtually any action that changes a hazardous waste to neutralize it, recover material or energy from it, or make it less hazardous, safer, or easier to manage.

Treatment train—A series of treatments used to treat hazardous waste effluents.

Trespass—Under tort law, a physical invasion of another's rights, whether in his person, personal property, or land.

Tribal implementation plan (TIP)—Under the CAA, a plan, prepared by a tribe, that contains a detailed description regarding how the tribe will fulfill its responsibilities toward achieving and maintaining clean air. The plan is essentially a collection of rules that the tribe uses to control air pollution.

Trigger task—Under the OSHA lead standard, any task, such as sanding, scraping, manual demolition, heat gun removal of paint, spray painting, abrasive blasting, torch burning, welding, cutting, and tool cleaning that results in an occupational exposure to lead above the permissible exposure limit of 50 μg/m^3.

Tumorogens—In toxicology, substances that cause *benign* (noncancerous) or *malignant* (cancerous) growths, usually seen as a swelling or enlargement, in the body by altering the genetic code in body tissues.

U

Unclassified area—Under the CAA, a geographic area (primarily rural) that has insufficient available air monitoring data for making an *attainment* or *nonattainment* determination.

Underground storage tank (UST)—As defined in RCRA, Subtitle I, any tank or combination of tanks (including connected underground pipes) that is used to contain an accumulation of regulated substances, the volume of which (including the volume of connected underground pipes) is 10% or more beneath the surface of the ground.

United States Code (USC)—The official cumulative set of all Federal statutes with all amendments up to the date of publication. The USC is organized into *Titles* that place statutes of similar subject matter in the same volume.

United States Sentencing Guidelines (USSG)—Guidelines developed by the United States Sentencing Commission intended to increase the uniformity of sentencing, remove the uncertainty of sentences caused by the parole system, and ensure proportionality in sentencing criminal conduct of different severity.

Universal precautions—Under OSHA's Bloodborne Pathogens standard, a means of infection control that requires employees to assume that all human blood and body fluids contain potentially infectious bloodborne pathogens, and to handle those articles and substances as if they were pathogenic.

Universal wastes—Under RCRA, certain hazardous wastes (in particular batteries, pesticide stocks that have been recalled or which are being collected and managed under a waste pesticide collection program, and certain mercury-containing thermostats) that are generated in large quantity by a variety of generators, that may, if adopted by the authorized state, be managed in accordance with EPA's Standards for Management of Universal Waste.

Unsaturated—In chemistry, the property in which an organic compound containing double or triple bonds between carbons (*e.g.*, ethylene [CH_2=CH_2]). Multiple bonds tend to be sites of reactivity.

Unsaturated zone—In hydrogeology, an area directly beneath the soil surface in which the soil particles are surrounded in varying degrees by air and water. This area is also called the *vadose zone.*

Upper respiratory tract irritation—In toxicology, a reversible inflammatory reaction that occurs in the nose and throat. This reaction is characterized by sneezing, nasal discharge, coughing, hoarseness, the production of phlegm, and nasal inflammation (rhinitis).

Use—Under RCRA, the recycling of a material in a particular function or application as an effective substitute for a commercial product. *Reuse* is also a form of use.

Used in a "manner constituting disposal"—Under RCRA, a material that has been placed on the land directly or as an ingredient in a product (except commercial chemical products ordinarily applied to the land).

V

Vadose zone—In hydrgeology, an area directly beneath the soil surface in which the soil particles are surrounded in varying degrees by air and water. This area is also called the *unsaturated zone.*

Vapor density (VD)—The mass per unit volume of a given vapor/gas relative to that of air.

Vapor pressure (VP)—A measure of the relative volatility of chemicals. It is pressure exerted by the vapor in equilibrium with its liquid at a given temperature.

Vertical composite samples—Environmental samples that are composed of equal aliquots collected along a defined vertical interval such as a borehole or test pit (for example, the combination of five aliquots taken every foot during the installation of a 5-foot bore hole).

Vessel—Every description of watercraft or other artificial contrivance used, or capable of being used, as a means of transportation on water.

Virtually safe dose (VSD)—That dose in which the risk is very low (10^{-6} or one excess tumor, *etc.* per one million persons). This is the value accepted as the "safe dose" for humans by regulatory policy.

Volatilization—The tendency of a material to transfer from a liquid phase (either pure or dissolved, as in aqueous systems) to a gaseous phase (commonly mixed with air). Also known as *evaporation.*

Voluntary disclosure—One of the actions which can be taken by a defendant and used by the Department of Justice as a mitigating factor in deciding whether to bring a criminal prosecution for a violation of an environmental statute.

W

Waste codes D004 through D043—Under RCRA, the characteristic codes for hazardous waste and their relevant characteristic levels (concentrations) as determined by TCLP (Toxicity Characteristic Leaching Procedure) for concentration-based constituents.

Waste minimization—As currently defined by EPA, includes *source reduction* and *environmentally sound recycling* of wastes regulated under the Resource Conservation and Recovery Act (RCRA), particularly hazardous wastes.

Wastewaters—Under RCRA LDR regulations, wastes with <1 wt% total organic carbon (TOC) and <1 wt% total suspended solids. All other wastes are considered to be *nonwastewaters*.

Water table aquifers—Unconfined aquifers in which the water table forms the upper boundary. Water table aquifers are usually in unconsolidated material and where the water table rises and falls seasonally. Water tables tend to rise in winter when more rainfall infiltrates, and tend to fall in summer due to less rainfall infiltration.

Water tube boilers—A method of treatment for hazardous waste using industrial boilers. Water is circulated throughout the combustion chamber in thousands of feet of steel tubing. Heat is transferred from the path of the flue gas into the adjacent water tubes.

Wetlands—Under the Clean Water Act, areas that are inundated or saturated by surface or groundwater at a frequency and duration sufficient to support a prevalence of vegetation typically adapted for life in saturated soil conditions. In order for a site to be classified as a wetland, the appropriate vegetation, soils, and hydrology must be present.

Worst-case release scenario (WRS) analysis—Under the Risk Management Program rules, an analysis of the potential releases that reach the greatest distance beyond the stationary source's boundary.

X

X-rays—an example of electromagnetic radiation that arises as electrons are deflected from their original paths (inner orbital electrons change their orbital levels around the atomic nucleus). X-rays, like gamma rays, are capable of traveling long distances through air and most other materials. Like gamma rays, x-rays require more shielding to reduce their intensity than do beta or alpha particles. x- and gamma rays differ primarily in their origin: x-rays originate in the electronic shell, gamma rays originate in the nucleus.

Conversion Table

To Convert	into	Multiply by	To Convert	into	Multiply by
A					
Acres	Sq ft	43,560.0	Btu	Hp-hrs	3.931×10^{-4}
	Sq meters	4,047		Joules	1,054.8
	Sq miles	1.562×10^{-3}		Kilogram-calories	0.2520
	Sq yards	4,840		Kilogram-meters	107.5
Acre-feet	Cu feet	43,560.0		Kilowatt-hrs	2.928×10^{-4}
	Gallons	3.259×10^{5}			
Ares	Acres	0.02471	Btu/hr	Foot-pounds/sec	0.2162
	Sq meters	100.0		Gram-calories/sec	0.0700
Atmospheres	Cm of Hg (@ 0°C)	76.0		Horsepower	3.929×10^{-4}
	Ft of water (@ 4°C)	33.90		Kilowatts	0.01757
	In of Hg(@ 0°C)	29.92		Watts	17.57
	Kg/sq cm	1.0333	Bushels	Cu feet	1.2445
	Kg/sq meter	10,332		Cu inches	2, 150.4
	Pounds/sq inch	14.70		Cu meters	0.03524
	Tons (short)/sq ft	1.058		Liters	35.24
B				Pecks	4.0
Barrels (oil)	Gallons (oil)	42.0		Pints (US dry)	64.0
Bars	Atmospheres	0.9869		Quarts (US dry)	32.0
	Dynes/sq cm	10^{6}			
	Kg/sq meter	1.020×10^{4}	**C**		
	Pounds/sq ft	2,089.0			
	Pounds/sq inch	14.50	Centares	Sq meters	1.0
	Ergs	1.0550×10^{10}	Centigrade	Fahrenheit	(C° x 9/5) + 32
	Foot-pounds	778.3	Centigrams	Grams	0.01
	Gram-calories	252.0			

677

To Convert	into	Multiply by	To Convert	into	Multiply by
Centiliters	Liters	0.01	Cubic inches	Gallons (US liq)	4.329×10^{-5}
Centimeters	Feet	3.281×10^{-2}		Liters	0.01639
	Inches	0.3937		Pints (US liq)	0.03463
	Kilometers	10^{-5}		Quarts (US liq)	0.01732
	Meters	0.01	Cubic meters	Bushels (dry)	28.38
	Miles	6.214×10^{-6}		Cu centimeters	10^6
	Millimeters	10.0		Cu feet	35.31
	Mils	393.7		Cu inches	61,023.0
	Yards	1.094×10^{-2}		Cu yards	1.308
Centimeter-dynes	Cm-grams	1.020×10^{-3}		Gallons (US liq)	264.2
	Meter-kg	1.020×10^{-8}		Liters	1000.0
	Pound-ft	7.376×10^{-8}		Pints (US liq)	2,113.0
Centimeter-grams	Cm-dynes	980.7		Quarts (US liq)	1,057
	Meter-kg	10^{-5}	Cu yards	Cu cm	7.646×10^5
	Pound-ft	7.233×10^{-5}		Cu feet	27
Cm of Hg (@ 0°C)	Atmospheres	0.01316		Cu inches	46,656.0
	Ft of water (@ 4°C)	0.4461		Cu meters	0.7646
	Kg/sq meter	136.0		Gallons (US liq)	202.0
	Pounds/sq ft	27.85		Liters	764.6
	Pounds/sq in	0.1934		Pints (US liq)	1,615.9
Centimeters/sec	Feet/min	1.1969		Quarts (US liq)	807.9
	Feet/sec	0.03281			
	Kilometers/hr	0.036	**D**		
	Knots (international)	0.1943			
	Meters/min	0.6	Days	Hours	24.0
	Miles/hr	0.02237		Minute	1,440.0
	Miles/min	3.728×10^{-4}		Second	86,400.0
Cubic centimeters	Cu feet	3.531×10^{-5}	Decigrams	Grams	0.1
	Cu inches	0.06102	Deciliters	Liters	0.1
	Cu meters	10^{-6}	Decimeters	Centimeters	0.1
	Cu yards	1.308×10^{-6}	Degrees (Angles)	Minutes	60.0
	Gallons (US liq)	2.642×10^{-4}		Quadrants	0.01111
	Liters	0.001		Radians	0.01745
	Pints (US liq)	2.113×10^{-3}		Seconds	3,600.0
	Quarts (US liq)	1.057×10^{-3}	Dekagrams	Grams	10.0
Cubic feet	Bushels (dry)	0.8036	Dekaliters	Liters	10.0
	Cu centimeters	28,320.0	Dekameters	Meters	10.0
	Cu inches	1,728.0	Drams	Grams	1.7718
	Cu meters	0.02832		Grains	27.3437
	Cu yards	0.03704		Ounces (avdp)	0.0625
	Gallon (US liq)	7.48052			
	Liters	28.32	**E**		
	Pints (US liq)	59.84			
	Quarts (US liq)	29.92	Ergs	Btu	9.480×10^{-11}
Cubic inches	Cu centimeters	16.39		Foot-pounds	7.367×10^{-8}
	Cu feet	5.787×10^{-4}		Gram-calories	0.2389×10^{-7}
	Cu meters	1.639×10^{-5}		Gram-cm	1.020×10^{-3}
				Hp-hrs	3.7250×10^{-14}

To Convert	into	Multiply by	To Convert	into	Multiply by
Ergs	Joules	10^{-7}	**G**		
	Kg-calories	2.389×10^{-11}			
	Kg-meters	1.020×10^{-8}	Gallons (US liq)	Cu centimeters	3,785.0
	Kw-hours	0.2778×10^{-10}		Cu feet	0.1337
				Cu inches	231.0
F				Cu meters	3.785×10^{-3}
				Cu yards	4.951×10^{-3}
Fahrenheit	Centigrade	(F°-32) x 5/9		Liters	3.785
Fathoms	Feet	6.0		Pints (US liq)	8.0
Feet	Centimeters	30.48		Quarts (US liq)	4.0
	Kilometers	3.048×10^{-4}	Gallons (Imperial)	Gallons (US liq)	1.20095
	Meters	0.3048	Gallons (US liq)	Gallons (Imperial)	0.83267
	Miles (nautical)	1.645×10^{-4}	Gallons of water	Pounds of water	8.3453
	Miles (statute)	1.894×10^{-4}	Gallons (US liq)/min	Cu feet/sec	2.228×10^{-3}
	Millimeters	304.8		Liters/sec	0.06308
	Mils	1.2×10^4		Cu feet/hr	8.0280
Ft of water (@ 4°C)	Atmospheres	0.02950	Gills (US)	Liters	0.1183
	In of Hg(@ 0°C)	0.8826		Pints (US liq)	0.25
	Kg/sq cm	0.03048		Grams	0.06480
	Kg/sq meter	304.8		Ounces (avdp)	2.0833×10^{-3}
	Pounds/sq foot	62.43		Pennyweight (troy)	0.04167
	Pounds/sq inch	0.4335	Grains/gal (US)	Parts/million	17.118
Feet/minute	Cm/sec	0.5080		Pounds/million gal	142.86
	Feet/sec	0.01667	Grains/gal (Imperial)	Parts/million	14.286
	Kilometers/sec	0.01829	Grams	Grains	15.43
	Meters/min	0.3048		Kilograms	0.001
	Miles/hr	0.01136		Milligrams	1,000.0
	Miles/min	0.01136		Ounces (avdp)	0.03527
Feet per 100 feet	Per cent grade	1.0		Ounces (troy)	0.03215
Foot-pounds	Btu	1.286×10^{-3}		Poundals	0.07093
	Ergs	1.356×10^7	Grams	Pounds (avdp)	2.205×10^{-3}
	Gram-calories	0.3238	Grams/cm	Pounds/inch	5.600×10^{-3}
	Hp-hours	5.050×10^{-7}	Grams/cu cm	Pounds/cu ft	62.43
	Kg-calories	3.24×10^{-4}		Pounds/cu inch	0.03613
	Kg-meters	0.1383		Pounds/circ mil-ft	3.405×10^{-7}
	Kw-hours	3.766×10^{-7}	Grams/liter	Grains/gal	58.417
Foot-pounds/min	Btu/min	1.286×10^{-3}		Lb/1,000 gal	8.345
	Foot-pounds/sec	0.01667		Parts/million	1,000.0
	Horsepower	3.030×10^{-5}	Grams/sq cm	Pounds/sq ft	2.0481
	Kg-calories/min	3.24×10^{-4}	Gram-calories	Btu	3.9683×10^{-3}
	Kilowatts	2.260×10^{-5}		Ergs	4.1868×10^7
Foot-pounds/sec	Btu/hr	4.6263		Foot-pounds	3.0880
	Btu/min	0.07717		Hp-hours	1.5596×10^{-6}
	Horsepower	1.818×10^{-3}		Kw-hours	1.1630×10^{-6}
	Kg-calories/min	0.01945		Watt-hours	1.1630×10^{-3}
	Kilowatts	1.356×10^{-3}	Gram-calories/sec	Btu/hour	14.286
Furlongs	Rods	40.0	Gram-centimeters	Btu (IST)	9.297×10^{-8}
	Feet	660.0		Ergs	980.7

To Convert	into	Multiply by	To Convert	into	Multiply by
Gram-centimeters	Kg-calorie	2.343×10^{-8}	In of Water (@ 4°C)	In of Hg(@ 0°C)	0.07355
	Kg-meters	10^{-5}		Kg/sq cm	2.540×10^{-3}
				Ounces/sq inch	0.5781
H				Pounds/sq ft	5.204
				Pounds/sq inch	0.03613
Hectares	Acres	2.471			
	Sq feet	1.076×10^5	**J**		
Hectograms	Grams	100.0			
Hectoliters	Liters	100.0	Joules	Btu	9.480×10^{-4}
Hectometers	Meters	100.0		Ergs	10^7
Hectowatts	Watts	100.0		Foot-pounds	0.7376
Horsepower	Btu/min	42.44		Kg-calories	2.389×10^{-4}
	Foot-pounds/min	33,000		Kg-meters	0.1020
	Foot-pounds/sec	550.0		Watt-hours	2.778×10^{-4}
Horsepower (metric) (542.5 ft-lb/sec)	Horsepower (550 ft-lb/sec)	0.9863	**K**		
Horsepower (550 ft-lb/sec)	Horsepower(metric) (542.5 ft-lb/sec)	1.014	Kilograms	Dynes	980,665
Horsepower	Kg-calories/min	10.68		Grams	1,000.0
	Kilowatts	0.7457		Poundals	70.93
	Watts	745.7		Pounds	2.205
Horsepower (Boiler)	Kilowatts	9.803		Tons (long)	9.842×10^{-4}
Horsepower-hours	Btu	2,547		Tons (short)	1.102×10^{-3}
	Ergs	2.6845×10^{13}	Kilograms/cu meter	Grams/cu cm	0.001
	Foot-pounds	1.98×10^6		Lb/cu ft	0.06243
	Gram-calories	641,190		Lb/cu inch	3.613×10^{-5}
	Kg-calories	641.19	Kilograms/meter	Pounds/ft	0.6720
	Kg-meters	2.737×10^5	Kilograms/sq cm	Atmospheres	0.9678
	Kw-hours	0.7457		Ft of water (@39.2°F)	32.81
Hours	Days	4.167×10^{-2}		In of Hg (@ 0°C)	28.96
	Minutes	60		Pounds/sq ft	2,048
	Seconds	3,600		Pounds/sq inch	14.22
	Weeks	5.952×10^{-3}	Kilograms/sq meter	Atmospheres	9.678×10^{-5}
				Bars	98.07×10^{-6}
I				Ft of water (@39.2°F)	3.281×10^{-3}
				In of Hg (@ 0°C)	2.869×10^{-3}
Inches	Centimeters	2.540		Pounds/sq ft	0.2048
	Feet	8.333×10^{-2}		Pounds/sq inch	1.422×10^{-3}
	Meters	2.540×10^{-2}	Kilogram./sq mm	Kg/cu meter	10^6
	Miles	1.578×10^{-5}	Kilogram-calories	Btu	3.968
	Millimeters	25.40		Foot-pounds	3,088
	Mils	1,000.0		Hp-hours	1.560×10^{-3}
	Yards	2.788×10^{-2}		Joules	4,186
In of Hg(@ 0°C)	Atmospheres	0.03453		Kg-meters	426.9
	Ft of water	1.133 (at 39.2°F)		Kilojoules	4.186
	Kg/sq cm	0.03453		Kw-hours	1.163×10^{-3}
	Kg/sq meter	345.3	Kilogram-meters	Btu	9.294×10^{-3}
	Pounds/sq ft	70.73		Ergs	9.804×10^7
	Pounds/sq inch	0.4912		Foot-pounds	7.233
In of water (@ 4°C)	Atmospheres	2.458×10^{-3}			

To Convert	into	Multiply by	To Convert	into	Multiply by
Kilogram-meters	Joules	9.804	Links (surveyor's)	Inches	7.92
	Kg-calories	2.342×10^{-3}	Liters	Bushels (dry)	0.03531
	Kw-hours	2.723×10^{-6}		Cu centimeters	1,000.0
Kiloliters	Liters	1000.0		Cu feet	0.03531
Kilometers	Centimeters	10^5		Cu inches	61.02
	Feet	3,281		Cu meters	0.001
	Inches	3.937×10^4		Cu yards	$1.308\ 10^{-3}$
	Meters	1000.0		Gallon (US liq)	0.2642
	Miles	0.6214		Pints (US liq)	2.113
	Millimeters	10^6		Quarts (US liq)	1.057
	Yards	1,094.0	Liters/min	Cu ft/sec	5.886×10^{-4}
Kilometers/hr	Cm/sec	27.78		Gallons/sec	4.403×10^{-3}
	Feet/min	54.68			
	Feet/sec	0.9113	**M**		
	Knots	0.5396			
	Meters/min	16.67	Meters	Centimeters	100.0
	Miles(statute)/hr	0.6214		Feet	3.281
Kilowatts	Btu/min	56.92		Inches	39.37
	Foot-pounds/min	4.426×10^4		Kilometers	0.001
	Foot-pounds/sec	737.6		Miles (nautical)	5.396×10^{-4}
	Horsepower	1.341		Miles (statute)	6.214×10^{-4}
	Kg-calories/min	14.34		Millimeters	1000.0
	Watts	1000.0		Varas	1.179
Kilowatt-hour	Btu	3,413		Yards	1.094
	Ergs	3.600×10^{13}	Meters/min	Cm/sec	1.667
	Foot-pounds	2.655×10^6		Feet/min	3.281
	Gram-calories	859,850		Feet/sec	0.05468
	Hp-hours	1.341		Kilometers/hr	0.06
	Joules	3.6×10^6		Knots	0.03238
	Kg-calories	859.85		Miles (statute)/hour	0.03728
	Kg-meters	3.671×10^5	Meters/sec	Feet/min	196.8
	Lb H_2O evaporated			Feet/sec	3.281
	from and at 212°F	3.53		Kilometers/hr	3.6
	Lb H_2O raised from			Kilometers/min	0.06
	62° to 212°F	22.75		Miles (statute)/hour	2.237
Knots	Feet/hr	6,080		Miles/min	0.03728
	Kilometers/hr	1.8532	Micrograms	Grams	10^{-6}
	Miles (nautical)/hr	1.0	Microliters	Liters	10^{-6}
	Miles (statute)/hr	1.151	Miles (nautical)	Feet	6,076.103
	Yards/hr	2,027		Kilometers	1.852
	Feet/sec	1.689		Meters	1,852
				Miles (statute)	1.1508
				Yards	2,025.4
L			Miles (statute)	Centimeters	1.609×10^5
				Feet	5, 280
League (statute)	Mile (statute)	3.0		Inches	6.336×10^4
Links (engineer's)	Inches	12.0		Kilometers	1.609
				Meters	1,609

To Convert	into	Multiply by	To Convert	into	Multiply by
Miles (statute)	Miles (nautical)	0.8684		Grams	28.349527
	Yards	1,760		Pounds	0.0625
Miles/hr	Cm/sec	44.70		Ounces (troy)	0.9115
	Feet/min	88.0		Tons (long)	2.790×10^{-5}
	Feet/sec	1.467		Tons (metric)	2.8355×10^{-5}
	Kilometers/hr	1.609	Ounces (US liq)	Cu inches	1.805
	Kilometers/min	0.02682		Liters	0.02957
	Knots	0.8684	Ounces (troy)	Grains	480.0
	Meters/min	26.82		Grams	31.103481
	Miles/min	0.01667		Ounces (avdp)	1.09714
Miles/min	Cm/sec	2,682		Pennyweights (troy)	20.0
	Feet/sec	88.0		Pounds (troy)	0.08333
	Kilometers/min	1.609	Ounces/sq inch	Pounds/sq inch	0.0625
	Miles (nautical)/min	0.8684			
	Miles/hr	60.0	**P**		
Milliers	Kilograms	1000.0			
Milligrams	Grams	0.001	Parts/million	Grains/gal (US liq)	0.0584
Milligrams/Liter	Parts/million	1.0		Grains/gal (Imperial)	0.07016
Milliliters	Liters	0.001		Lbs/million gallons	8.345
Millimeters	Centimeters	0.1	Pennyweights (troy)	Grains	24.0
	Feet	3.281×10^{-3}		Grams	1.55517
	Inches	0.03937		Ounces (troy)	0.05
	Kilometers	10^{-6}		Pounds (troy)	4.1667×10^{-3}
	Meters	0.001	Pints (dry)	Cu inches	33.60
	Miles	6.214×10^{-7}	Pints (US liq)	Cu cm	473.2
	Mils	39.37		Cu feet	0.01671
	Yards	1.094×10^{-3}		Cu inches	28.87
Million gallons/day	Cu feet/sec	1.54723		Cu meters	4.732×10^{-4}
Mils	Centimeters	2.540×10^{-3}		Cu yards	6.189×10^{-4}
	Feet	8.333×10^{-5}		Gallons (US liq)	0.125
	Inches	0.001		Liters	0.4732
	Kilometers	2.540×10^{-8}		Quarts (US liq)	0.5
	Yards	2.788×10^{-5}	Poundals	Grams	14.10
Miner's inches	Cu feet/min	1.5		Kilograms	0.01410
Minutes (angles)	Degrees	0.01667		Pounds	0.03108
	Quadrants	1.852×10^{-4}	Pounds (avdp)	Drams (avdp)	256.0
	Radians	2.909×10^{-4}		Grains	7,000
	Seconds	60.0		Grams	453.5924
Myriagrams	Kilograms	10.0		Kilograms	0.4536
Myriameters	Kilometers	10.0		Ounces (avdp)	16.0
				Ounces (troy)	14.5833
N				Poundals	32.17
				Pounds (troy)	1.21528
Nepers	Decibels	8.686		Tons (short)	0.0005
			Pounds (troy)	Grains	5,760
O				Grams	373.24177
				Ounces (avdp)	13.1657
Ounces (avdp)	Drams (avdp)	16.0		Ounces (troy)	12.0
	Grains	437.5			

To Convert	into	Multiply by	To Convert	into	Multiply by
Pounds (troy)	Pennyweights (troy)	240.0		Minutes	3,438
	Pounds (avdp)	0.822857		Quadrants	0.6366
	Tons (long)	3.6735×10^{-4}		Seconds	2.063×10^5
	Tons (metric)	3.7324×10^{-4}	Rods	Feet	16.5
	Tons (short)	4.1143×10^{-4}			
Pounds of water	Cu feet	0.01602	**S**		
	Cu inches	27.68			
	Gallons	0.1198	Seconds (angles)	Degrees	2.788×10^{-4}
Pounds of water/min	Cu feet/sec	2.670×10^{-4}		Minutes	0.01667
Pound-feet	Cm-dynes	1.356×10^7		Quadrants	3.087×10^{-6}
	Cm-grams	13,825		Radians	4.848×10^{-6}
	Meter-kg	0.1383	Square centimeters	Sq feet	1.076×10^{-3}
Pounds/cu feet	Grams/cu cm	0.01602		Sq inches	0.1550
	Kg/cu cm	16.02		Sq meters	0.0001
	Pounds/cu inch	5.787×10^{-4}		Sq miles	3.861×10^{-11}
Pounds/cu inch	Grams/cu cm	27.68		Sq millimeters	100.0
	Kg/cu meter	2.768×10^4		Sq yards	1.196×10^{-4}
	Pound/cu ft	1,728.0	Square feet	Acres	2.296×10^{-5}
Pounds/foot	Kg/meter	1.488		Sq centimeters	929.0
Pounds/inch	Grams/cm	178.6		Sq inches	144.0
Pounds/sq foot	Atmospheres	4.725×10^{-4}		Sq meters	0.09290
	Feet of water	0.01602		Sq miles	3.587×10^{-8}
	In of Hg (@ 0°C)	0.01414		Sq millimeters	9.290×10^4
	Kgs/sq meter	4.882		Sq yards	0.1111
	Pounds/sq inch	6.944×10^{-3}	Square inches	Sq centimeters	6.452
Pounds/sq inch	Atmospheres	0.06804		Sq feet	6.944×10^{-3}
	Feet of water	2.307		Sq millimeters	645.2
	In of Hg (@ 0°C)	2.036		Sq mils	10^4
	Kgs/sq meter	703.1		Sq yards	7.716×10^{-4}
	Pounds/sq ft	144.0	Square kilometers	Acres	247.1
				Sq centimeters	10^{10}
Q				Sq feet	10.76×10^6
				Sq inches	1.550×10^9
Quadrants (angles)	Degrees	90.0		Sq meters	10^6
	Minutes	5,400.0		Sq miles	0.3861
	Radians	1.571		Sq yards	1.196×10^6
	Seconds	3.24×10^5	Square meters	Acres	2.471×10^{-4}
Quarts (dry)	Cu inches	67.20		Sq centimeters	10^4
Quarts (US liq)	Cu cm	946.4		Sq feet	10.76
	Cu feet	0.03342		Sq inches	1,550
	Cu inches	57.75		Sq miles	3.861×10^{-7}
	Cu meters	9.464×10^{-4}		Sq millimeters	10^6
	Cu yards	1.238×10^{-3}		Sq yards	1.196
	Gallons (US liq)	0.25	Square miles	Acres	640.0
	Liters	0.9463		Sq feet	27.88×10^6
				Sq km	2.590
R				Sq meters	2.590×10^6
Radians	Degrees	57.30		Sq yards	3.098×10^6

To Convert	into	Multiply by
Square Millimeters	Sq cms	0.01
	Sq feet	1.076×10^{-5}
	Sq inches	1.550×10^{-3}
Square Yards	Acres	2.066×10^{-4}
	Sq cms	8,361
	Sq feet	9.0
	Sq inches	1,296
	Sq meters	0.8361
	Sq miles	3.228×10^{-7}
	Sq millimeters	8.361×10^{5}

T

To Convert	into	Multiply by
Temp (°C) + 273	Absolute Temp (°C)	1.0
Temp (°C) + 17.78	Temp (°F)	1.8
Temp (°F) + 460	Absolute Temp (°F)	1.0
Temp (°F) – 32	Temp (°C)	5/9
Tons (long)	Kilograms	1,016
	Pounds	2,240
	Tons (short)	1.120
Tons (metric)	Kilograms	1,000.0
	Pounds	2,205
Tons (short)	Kilograms	907.1848
	Ounces	32,000.0
	Ounces (troy)	29,166.66
	Pounds	2,000.0
	Pounds (troy)	2,430.56
	Tons (long)	0.89287
	Tons (metric)	0.9078
Tons (short)/sq ft	Kg/sq meter	9,765.0
	Pounds/sq in	2,000.0
Tons of H_2O/ 24 hrs	Pounds of H_2O/hr	83.333
	Gallons/min	0.16643
	Cu feet/hr	1.3349

Y

To Convert	into	Multiply by
Yards	Centimeters	91.44
	Feet	3.0
	Inches	36.0
	Kilometers	9.144×10^{-4}
	Meters	0.9144
	Miles (nautical)	4.934×10^{-4}
	Miles (statute)	5.682×10^{-4}
	Millimeters	914.4

Index

B

Backsliding 443
Ban and phase out rule 362
Base 556, 592–594
  organic 602
  reaction with metals 595
  reaction with salts 595
Baseline Environmental Management Report 39
Benign tumorgen 84
Best Available Control Measures 380
Best Available Control Technology 379
Best Available Retrofit Technology 380
Best Available Technology 443
Best Conventional Pollutant Control Technology 443
Best management practices 45
  release reporting 206
Best Practicable Control Technology 443
Best practices 86
Beta particles 339
Biochemical oxygen demand 592
Biological airborne pathogens, respirator selection 127
Biological effects of complex chemicals 74
Biological exposure indices 158
Biological indicators 90
Biological plausibility 73
Biological protective equipment, trends 141
Biological treatment of hazardous waste 560
Biotransformation 147
Blasting agents 88
Bloodborne pathogens 96, 185
  Acquired Immunodeficiency Syndrome 187
  diseases 187–188
  disinfection 190
  engineering controls 97, 189
  exposure control program 97, 186
  exposure to 190
  Hepatitis B virus 96, 187
    Vaccination 188–189
  Hepatitis C virus 187
  housekeeping 189
  Human Immunodeficiency Virus 96, 187
  labeling 190
  other potentially infectious materials 186
  personal protective equipment 189

  recordkeeping, medical and occupational exposure 191
  safe work practices 97, 189
  training 96, 97, 188
  universal precautions 97
Boiling point 156, 589
Brownfields 247
  CERCLA liability 248
  CERCLIS 249
  data quality objectives 255
  definition 247
  insitutional controls, deed restrictions 250
  problems caused by 248
  project team 256
  risk analysis 71
  State actions 249–252
BS 7750 49
BS 8800 49
Burned for energy recovery (RCRA) 479
Business impacts, pollution prevention and waste minimization 536
By-product 530

C

Cancer 152
  mesothelioma 353
Carbon monoxide 374, 384
  NAAQS 384
  nonattainment areas 384
Carcinogens
  ACGIH classification 74
  asbestos 352
  EPA guidelines for risk assessment 77
  malignant carcinogens 84
  slope factors 75
  weight-of-evidence classification 74
Carcinogensis 75
Case law 17
Catastrophic and fatal accidents
  OSHA inspections 105
  process safety management rule 156
Categorical exclusion determinations 230–231
Cations 587
Caustic 594
CDL hazardous materials endorsement 271
Ceiling 95, 152
Central nervous system depression 152
Central nervous system toxicity 152

Certification
  of hazmat employee (DOT) 271
  of risk management plan submittal 410
  of SPCC plan 455
Change in service 290
Change management
  process safety management 170
  risk management program 407
Characterization samples 485, 489
Checklist
  process safety management 162
  risk management program 403
Chemical cartridge respirators 124
Chemical compatibility 178, 607
  Department of Transportation requirements 215
Chemical emergency response, selection of respiratory protection 127
Chemical fixation and solidification 563–565
Chemical hazards 83, 582
  explosive 159
  flammable 159, 603
  reactive 158
Chemical hygiene plan 99, 181–182
Chemical interactions
  toxicology, effects from chemical exposure 149
Chemical precipitation 557
Chemical processes 67
  auditing 67
Chemical procurement 182
Chemical protective clothing 128
  chemical warfare agents 128
  materials of construction 128
  permeation resistance 135
  reusable 128
  selection of 134–135
  splash/particulate 128
  trends 141
Chemical Safety and Hazard Investigation Board 402
Chemical storage 178
Chemical substances (TSCA) 296, 298
  importation of 298
  information rules 299
Chemical Transportation Emergency Center (CHEMTREC) 315
Chemical treatment, of hazardous waste 556, 558
Chemistry
  adsorption 593
  concentration of chemicals in solution 593
  evaporation 593
  flammable liquids 603

No observed effect level 149
Nomenclature of organic compounds 600
Noncarcinogens
   benign tumorgens 84
   dose-response curve 75
   hazard quotient 76
   risk assessment 74
Non-ionized chemicals 147
Nonroad engines 387
Nonattainment areas 374, 382
   carbon monoxide 384
   emissions offsets 378
   new source review 378
   ozone 383, 389
      multistate 384
      rural transport areas 384
   particulate matter 384
Nonfriable asbestos 362
Nonpoint sources 443
Nontransportation-related facilities
   451
Nonwastewaters 519
North American Emergency
   Response Guidebook 275
Not otherwise specified 272
Notice of Intent 233
Nuclear radiation 339
Nuclear Regulatory Commission
   33, 38
Nucleus 585
Nuisance 19

O

Objectives, ISO 14001 51, 55
Occupational carcinogens, risk
   assessment 72
Occupational exposure limits 152
Occupational Safety and Health Act
   85, 103, 107, 152, 179
   access to employee exposure and
      medical records 86, 215
   administrative controls 96
   air contaminants 86, 95, 215
   anhydrous ammonia 88
   asbestos 354–358
   bloodborne pathogens 96
   compressed gases 87, 215
   confined spaces 93
   construction industry standards
      100, 107
      gases, vapors, fumes, dusts,
         and mists 100
      hazard communication 100
      lead 349
      process safety management
         100
      safety training and education
         100
      ventilation 100

control of hazardous energy 94
decontamination 91
design standards 179
dip tanks 87, 215
engineering controls 90, 96, 114
explosives and blasting agents
   88
fall protection 133
Federal agencies 105
general duty clause 85, 104, 108
general industry standards
   86, 107
hazard assessment 114, 118–
   119
hazard communication 98, 209–
   212
   laboratory standard 100, 180
hazardous chemicals 583
hazardous materials 87, 270,
   582
horizontal standards 179
ionizing radiation 86
jurisdiction 105
laboratory standard 99, 180,
   210
   chemical hygiene plan 99, 181
   Material Safety Data Sheets
      100
lead 346
   permissible exposure levels
      347
   standard for general industry
      346
   standard for the construction
      industry 349
levels of protection 135
lockout/tagout 94
maritime standards 108
Material Safety Data Sheet
   format 214
medical services and first aid
   215
performance standards 179
personal protective equipment
   90
   electrical protection 117
   eye and face protection
      92, 118
   foot protection 117–118
   hand protection 118
   hazard assessment 115, 118–
      119
   head protection 117–118
   respiratory protection 92,
      113, 117–118, 215
process safety management
   88, 155, 215, 397
research 104
spray finishing using flammable
   and combustible liquids
   87, 215

ventilation 86, 215
   construction industry 100
   vertical standards 179
   work practices 90
Occupational Safety and Health
   Administration 103
   citations 106–107
   consultation 108
   informational resources 104,
      109
   inspections 104–106
   OSHA Training Institute 104
   risk analysis 72
   State plan states 104
Octanol/water partition coefficient
   592
Off-site consequence analysis
   399, 403, 410, 421, 431
Offshore, oil pollution prevention
   457
Oil 200
   spill history in SPCC plans 455
Oil Pollution Act 12, 449–450
   area response plans 452
   definitions 451
   enforcement 452, 458
   facility 451
   Facility Response Plan 456
   financial responsibility 451
   history of 449
   international provisions 450
   liability 450–451, 457
   local and facility spill response
      453
   major elements of 450
   National Contingency Plan 452
   offshore oil pollution prevention
      457
   Oil Spill Liability Trust Fund
      451, 457
   penalties 458
   prevention 450
      offshore oil pollution 457
      SPCC 453
      vessel oil pollution 457
Oil Spill Liability Trust Fund
   451, 457
One-plan concept 417, 434–435
Operating procedures
   process safety management
      164, 165, 168, 170
   review and update 165
   underground storage tanks 288
Operations and maintenance,
   asbestos 364
Operator 25, 284
   of stationary sources and the
      RMP 398–399
Orbitals 587
Organic carboxylic acids 602
Organic chemicals 599–602